KW-222-339

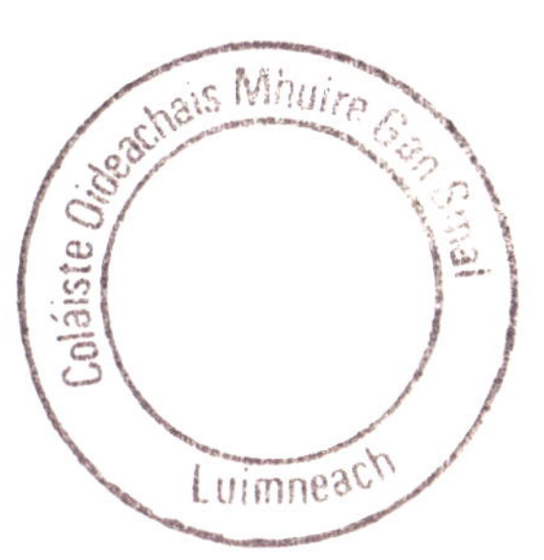
Coláiste Oideachais Mhuire Gan Smál
Luimneach

Algorithms and Combinatorics 15

Springer
Berlin
Heidelberg
New York
Barcelona
Budapest
Hong Kong
London
Milan
Paris
Santa Clara
Singapore
Tokyo

Michel Marie Deza
Monique Laurent

Geometry of Cuts and Metrics

With 94 Figures

Springer

Michel Marie Deza
Monique Laurent
Département de Mathématiques et d'Informatique
Laboratoire d'Informatique de l'Ecole Normale Supérieure
45 rue d'Ulm
F-75230 Paris Cedex 05
France
e-mail: deza@dmi.ens.fr
laurent@dmi.ens.fr

Visiting addresses

Michel Marie Deza
Department of Mathematics
Moscow Pedagogical State University
Krasnoprudnaya, 14
107140 Moscow
Russia

Monique Laurent
CWI
Kruislaan 413
1098 SJ Amsterdam
The Netherlands

Cataloging-in-Publication Data applied for.

Die Deutsche Bibliothek - CIP-Einheitsaufnahme
Deza, Michel Marie:
Geometry of cuts and metrics / Michel Marie Deza ; Monique Laurent. - Berlin ; Heidelberg ; New York ; Barcelona ; Budapest ; Hong Kong ; London ; Milan ; Paris ; Santa Clara ; Singapore ; Tokyo : Springer, 1997
(Algorithms and combinatorics ; 15)
ISBN 3-540-61611-X Gb.

Mathematics Subject Classification (1991):
Primary: 05-02, 05Cxx, 11Hxx, 52Bxx, 52Cxx, 90Cxx
Secondary: 05C12, 05C85, 11H06, 11P21, 15A57, 52B12, 52C07, 90C27

ISSN 0937-5511
ISBN 3-540-61611-X Springer-Verlag Berlin Heidelberg New York

Typesetting: Camera-ready copy produced from the authors' output file
SPIN 10471041 41/3143 - 5 4 3 2 1 0 - Printed on acid-free paper

Preface

Cuts and metrics are well-known and central objects in graph theory, combinatorial optimization and, more generally, discrete mathematics. They also occur in other areas of mathematics and its applications such as distance geometry, the geometry of numbers, combinatorial matrix theory, the theory of designs, quantum mechanics, statistical physics, analysis and probability theory. Indeed, cuts are many faceted objects, and they can be interpreted as graph theoretic objects, as metrics, or as probabilistic pairwise correlations. This accounts for their fecund versatility.

Due to the wealth of results, in writing this book we had to make a selection. We focus on polyhedral and other geometric aspects of cuts and metrics. Our aim is to collect different results, established within diverse mathematical fields, and to present them in a unified framework. We try to show how these various results are tied together via the notions of cuts and metrics and, more specifically, the cut cone and the cut polytope. One of our guidelines for selecting topics was to concentrate on those aspects that are less well-known and are not yet covered in a unified way elsewhere. The book has, moreover, been written with a special attention to interdisciplinarity. For this reason, while some topics are treated in full detail with complete proofs, some other topics are only touched, by mentioning results and pointers to further information and references.

The book is intended as a source and reference work for researchers and graduate students interested in discrete mathematics and its interactions with other areas of mathematics and its applications. In particular, it is of interest for those specializing in algebraic and geometric combinatorics and in combinatorial optimization.

The book is subdivided into five parts, in which we consider the following topics: ℓ_1-metrics and hypercube embeddable metrics, hypermetrics and Delaunay polytopes in lattices, the metric structure of graphs, designs in connection with hypercube embeddings, and further geometric questions linked with cut polyhedra.

We do not cover extensively the topic of optimization. Indeed, research on cuts and metrics in this direction is already well-documented. Some survey papers are available; for instance, by Frank [1990] on multicommodity flows and by Poljak and Tuza [1995] on the max-cut problem. Nevertheless, we do present some of the recent breakthroughs. For instance, we discuss the new semidefinite programming approximative algorithm of Goemans and Williamson, as well as the result of Bourgain on Lipschitz ℓ_1-embeddings with small distortion, and its application to approximating multicommodity flows by Linial, London and Rabinovich. Moreover, we mention en route a number of further results relevant to the max-cut problem and binary matroids.

We made each of the five parts of the book as self-contained as possible; in principle, each of them can be read independently of the other. Moreover, we have chosen a common labeling system for all items such as theorems, examples, figures, etc., in order to simplify the search throughout the text. Some portions of text that contain side information or lengthy proofs are in small print and can be avoided at first reading.

The part of the book treating links with the geometry of numbers, is based on a survey paper coauthored by the authors with V.P. Grishukhin of the Academy of Sciences of Russia in Moscow (cf. Deza, Grishukhin and Laurent [1995]). We are grateful to Slava Grishukhin for kindly permitting us to include the material in this book.

There are several institutions that we wish to thank for their support while we were working on the book. We would particularly like to thank our home institute LIENS at the Department of Mathematics and Computer Science of Ecole Normale Supérieure in Paris, for offering us a helpful environment and stimulating working conditions. The help of Jacques Beigbeder in solving various LaTeX puzzles in the presentation of the book was greatly appreciated. The first author is grateful to the Tokyo Institute of Technology and the Institute of Mathematics of the Academia Sinica in Taipei. The second author is very thankful to CWI (Center for Mathematics and Computer Science) in Amsterdam for the hospitality and fruitful interactions on the occasion of several long-term visits. We wish also to express special thanks to the University of Augsburg and the Konrad-Zuse-Zentrum in Berlin; in particular, to Martin Grötschel for the many stimulating discussions during our visits and for sharing with us his time and enthusiasm on the topic.

We are grateful to several further people for useful discussions and cooperation on the topic of this book. In particular, the first author thanks warmly Jin Akiyama, Peter Cameron, Walter Deuber, Antoine Deza, Komei Fukuda, Marie Grindel, Zeev Jabotinsky, Ko-Wei Lih, Sergey Shpectorov, Navin Singhi, Ivo Rosenberg, Pierre Rosenstiehl, and Maximillian Voloshin. The second author is very grateful to Laci Lovász for insightful advice and for offering her the opportunity to present parts of the material at the Department of Computer Science of Yale University. She also thanks David Avis, Hans-Jürgen Bandelt, Victor Chepoi, Michele Conforti, Caterina De Simone, Cid De Souza, Bert Gerards, François Laburthe, Franz Rendl, Lex Schrijver, András Sebö, and Bruce Shepherd.

We have a special memory to our dear friend and colleague Svata Poljak, who was always enthusiatic, generous and eager to share interest and questions on this topic.

Contents

Chapter 1. Outline of the Book

This chapter gives an overview of the topics treated in this book. The central objects in the book are polytopes and cones related to cuts and metrics. Interesting problems concerning these polyhedra arise in many different areas of mathematics and its applications. Surprisingly, these polyhedra have been considered independently by a number of authors with various mathematical backgrounds and motivations. One of our objectives is to show on the one hand, the richness and diversity of the results in connection with these polyhedra, and on the other hand, how they can be treated in a unified way as various aspects of a common set of objects. Research on cuts and metrics profits greatly from the variety of subjects where the problems arise. Observations made in different areas by independent authors turn out to be equivalent, facts are not isolated, and views from different perspectives provide new interpretations, connections and insights.

This book is subdivided into five parts, each treating seemingly diverse topics. Namely, Parts I to V contain results relevant to the following areas:

1. the theory of metrics; more precisely, isometric embeddings into the Banach ℓ_1-space,

2. the geometry of numbers; more precisely, lattices and Delaunay polytopes,

3. graph theory; more precisely, the hypercube and its isometric subgraphs,

4. design theory; more precisely, the designs arising in connection with the various hypercube embeddings of the equidistant metric, together with complexity aspects of the hypercube embeddability problem,

5. geometry of polyhedra; more precisely, geometric questions on the cut and metric polyhedra (e.g., description of their facets, adjacencies, symmetries, etc.); applications to the solution of some problems such as Borsuk's problem, or completion problems for positive semidefinite matrices and Euclidean distance matrices.

We have made each of the five parts as self-contained as possible. For this reason, some notions and definitions may be repeated in different parts if they are central there. In principle, a reader who is interested, for instance, only in the aspects of geometry of numbers of cuts may consult Part II without any prior reading of

Part I. Chapter 2, however, contains some basic notation on graphs, polyhedra, matrices and algorithms that will be used throughout the book.

In what follows we give a brief overview of the material covered in Parts I to V. This introductory treatment is meant to provide an orientation map through the book for the reader. We already define here several notions, but all of them will be redefined later in the text as they are needed.

1.1 Outline of Part I. Measure Aspects: ℓ_1-Embeddability and Probability

In Part I we study the distance spaces that can be isometrically embedded into the ℓ_1-space $(\mathbb{R}^m, d_{\ell_1})$ for some integer $m \geq 1$. Here, d_{ℓ_1} denotes the ℓ_1-distance defined by

$$d_{\ell_1}(x,y) := \sum_{1 \leq i \leq m} |x_i - y_i| \quad \text{for } x, y \in \mathbb{R}^m.$$

One of the basic results is a characterization in terms of cut semimetrics. Given a subset S of the set $V_n := \{1, \ldots, n\}$, the *cut semimetric* $\delta(S)$ is the distance on V_n where two elements $i \in S$, $j \in V_n \setminus S$ are at distance 1, while two elements $i, j \in S$, or $i, j \in V_n \setminus S$, are at distance 0. Every cut semimetric is obviously isometrically ℓ_1-embeddable. In fact, a distance d is isometrically ℓ_1-embeddable if and only if it can be decomposed as a nonnegative linear combination of cut semimetrics. In other words, if CUT_n denotes the cone generated by the cut semimetrics on V_n, then

$$d \text{ is isometrically } \ell_1\text{-embeddable} \iff d \in \mathrm{CUT}_n.$$

The cone CUT_n is called the *cut cone.* We also consider isometric embeddings into the hypercube. Call a distance d on V_n *hypercube embeddable* if the distance space (V_n, d) can be isometrically embedded into the space $(\{0,1\}^m, d_{\ell_1})$ (for some $m \geq 1$), i.e., if we can find n binary vectors $v_1, \ldots, v_n \in \{0,1\}^m$ such that

$$d(i,j) = d_{\ell_1}(v_i, v_j) \quad \text{for all } i, j \in V_n.$$

In fact, the hypercube embeddable distances on V_n are the members of the cut cone CUT_n that can be written as a nonnegative integer combination of cut semimetrics.

Let $\mathrm{CUT}_n^\square$ denote the *cut polytope*, which is defined as the convex hull of the cut semimetrics $\delta(S)$ for $S \subseteq V_n$. That is, $\mathrm{CUT}_n^\square$ consists of the distances that can be decomposed as convex combinations of cut semimetrics. The cut cone and polytope also admit the following characterization in terms of measure spaces: A distance d belongs to the cut cone CUT_n (resp. the cut polytope $\mathrm{CUT}_n^\square$) if and only if there exist a measure space (resp. a probability space) $(\Omega, \mathcal{A}, \mu)$ and n events $A_1, \ldots, A_n \in \mathcal{A}$ such that

$$d(i,j) = \mu(A_i \triangle A_j) \quad \text{for all } i, j \in V_n.$$

(See Section 4.2 for the above results.)

There is another set of polyhedra that are closely related to cut polyhedra and for which the above interpretation in terms of measure spaces takes a nice form. Given a subset S of V_n, its *correlation vector* $\pi(S)$ is defined by $\pi(S)_{ij} = 1$ if both i, j belong to S and $\pi(S)_{ij} = 0$ otherwise, for $i, j \in V_n$. The cone generated by the correlation vectors $\pi(S)$ for $S \subseteq V_n$ is called the *correlation cone* and is denoted by COR_n. Similarly, $\mathrm{COR}_n^{\square}$ denotes the *correlation polytope*, defined as the convex hull of the correlation vectors. These polyhedra admit the following characterization (see Section 5.3): A vector p belongs to the correlation cone COR_n (resp. the correlation polytope $\mathrm{COR}_n^{\square}$) if and only if there exist a measure space (resp. a probability space) $(\Omega, \mathcal{A}, \mu)$ and n events $A_1, \ldots, A_n \in \mathcal{A}$ such that

$$p_{ij} = \mu(A_i \cap A_j) \quad \text{for all } i, j \in V_n.$$

Hence, the members of the correlation polytope are nothing but the pairwise joint correlations of a set of n events; this explains the name "correlation" polyhedra.

In fact, this result is an analogue of the similar result mentioned above for the cut polyhedra. The point is that the correlation polyhedron COR_n (or $\mathrm{COR}_n^{\square}$) is the image of the cut polyhedron CUT_{n+1} (or $\mathrm{CUT}_{n+1}^{\square}$) under a linear bijective mapping (the covariance mapping; see Section 5.2). This is a simple but interesting correspondence as it permits to translate results between cut polyhedra and correlation polyhedra. One of our objectives in this book will be to bring together and give a unified presentation for results that have been obtained by different authors in these two contexts (cut/correlation).

The correlation polytope provides the right setting for a classical question in probability theory, often referred to as the Boole problem and which can be stated as follows: Given n events $A_1, \ldots, A_n$ in a probability space $(\Omega, \mathcal{A}, \mu)$ find a good estimate of the probability $\mu(A_1 \cup \ldots \cup A_n)$ that at least one of these events occurs using the fact that the pairwise correlations $\mu(A_i \cap A_j)$ are known. Tight lower bounds for $\mu(A_1 \cup \ldots \cup A_n)$ can be derived from the valid inequalities for $\mathrm{COR}_n^{\square}$ (see Section 5.4).

We have now seen that the ℓ_1-embeddable distances on V_n are the members of the cut cone CUT_n. Hence, testing ℓ_1-embeddability amounts to testing membership in the cut cone. This problem turns out to be NP-complete. Moreover, characterizing ℓ_1-embeddability amounts to finding a description of the cone CUT_n by a set of linear inequalities. As CUT_n is a polyhedral cone (since it is generated by the finite set of cut semimetrics), we know that it can be described by a finite list of inequalities. However, finding the full list for arbitrary n is an 'impossible' task if NP $\neq$ co-NP. Nevertheless, large classes of valid inequalities for CUT_n (or $\mathrm{CUT}_n^{\square}$) are known. We give an up-to-date survey of what is known about the linear description of the cut polyhedra in Part V. Among the known inequalities, the most important ones are the hypermetric inequalities and the negative type inequalities, which are introduced in Section 6.1. They are the inequalities of the form:

$$\sum_{1 \leq i < j \leq n} b_i b_j x_{ij} \leq 0$$

where $b_1, \ldots, b_n$ are integers with sum $\sum_{i=1}^n b_i = 1$ (*hypermetric* case) or $\sum_{i=1}^n b_i = 0$ (*negative type* case). Part II will be entirely devoted to hypermetric inequalities and, more specifically, to their link with Delaunay polytopes in lattices. The hypermetric inequalities provide necessary conditions for ℓ_1-embedability. In fact, hypermetricity turns out to be a sufficient condition for ℓ_1-embeddability for several classes of metrics. Several such classes are presented in Chapter 8; they consist of metrics arising from valuated poset lattices, semigroups and normed vector spaces. The negative type inequalities are implied by the hypermetric inequalities. Hence, they provide a weaker necessary condition for ℓ_1-embeddability.

Negative type inequalities are classical inequalities in analysis. They were already used by Schoenberg in the thirties for characterizing the distance spaces that are isometrically ℓ_2-embeddable; namely, Schoenberg proved that a distance d is isometrically ℓ_2-embeddable if and only if the squared distance d^2 satisfies the negative type inequalities. Moreover, the negative type inequalities define a cone which is nothing but the image of the cone of positive semidefinite symmetric matrices (of order $n-1$ if the inequalities are on n points) under a linear bijective mapping (in fact, the same mapping that made the link between cut and correlation polyhedra). These results are presented in Sections 6.2 and 6.3, together with further basic facts on ℓ_2-spaces.

Several additional aspects are treated in Part I, including: operations and functional transforms of distance spaces preserving some metric properties such as ℓ_1-embeddability, hypermetricity, etc. (see Chapters 7 and 9); for given n, the minimum dimension of an ℓ_1-space permitting to embed any ℓ_1-embeddable distance on n points; for given m, the minimum number of points to check in a distance space in order to ensure embeddability in the m-dimensional ℓ_1-space (see Chapter 11).

We consider in Chapter 10 the question of finding Lipschitz embeddings where a small distortion of the distances is allowed. Bourgain [1985] shows that every semimetric on n points can be embedded into some ℓ_1-space with a distortion in $O(\log_2 n)$. We present this result together with an application by Linial, London and Rabinovich [1994] to approximations of multicommodity flows.

1.2 Outline of Part II. Hypermetric Spaces: an Approach via Geometry of Numbers

Part II is entirely devoted to the study of hypermetric inequalities and of their link with some objects of the geometry of numbers, namely, lattices and Delaunay polytopes.

Hypermetric inequalities are the inequalities of the form:

$$\sum_{1 \leq i < j \leq n} b_i b_j x_{ij} \leq 0$$

where $b_1, \ldots, b_n$ are integers with sum $\sum_{i=1}^n b_i = 1$. They define a cone, called the *hypermetric cone* and denoted by HYP_n. Note that triangle inequalities are a special case of hypermetric inequalities (obtained by taking all components of

b equal to 0 except two equal to 1 and one equal to -1). Hence, hypermetricity is a strengthening of the notion of semimetric. As every cut semimetric satisfies the hypermetric inequalities, we have

$$\text{CUT}_n \subseteq \text{HYP}_n.$$

This inclusion holds at equality if $n \leq 6$ and is strict for $n \geq 7$. For instance, the path metric of the graph $K_7 \backslash P_3$ is hypermetric but not ℓ_1-embeddable. Actually, the graphs whose path metric is hypermetric are characterized in Chapter 17.

A typical example of a hypermetric space arises from point lattices. Let L be a point lattice in $\mathbb{R}^k$, that is, a discrete subgroup of $\mathbb{R}^k$. Take a sphere $S \subseteq \mathbb{R}^k$ in one of the interstices of L, i.e., such that no point from L lies in the closed ball with boundary S. Blow up S until it is 'held rigidly' by lattice points. Then, the set of lattice points lying on S endowed with the square of the Euclidean distance forms a distance space which is semimetric and, moreover, hypermetric. The convex hull of the set $S \cap L$ of lattice points lying on S forms a polytope, called a *Delaunay polytope*. Hence, Delaunay polytopes have the interesting property that their set of vertices can be endowed with a metric structure which is hypermetric. In other words, for any Delaunay polytope P with set of vertices $V(P)$, the distance space $(V(P), (d_{\ell_2})^2)$ is a hypermetric space. Even more striking is the fact that, conversely, every hypermetric distance space on n points can be isometrically embedded into the space $(V(P), (d_{\ell_2})^2)$ for some Delaunay polytope P of dimension $k \leq n-1$. These results are presented in Section 14.1.

The hypermetric cone HYP_n is defined by infinitely many inequalities. However, it can be shown that a finite number of them suffices to describe HYP_n. In other words, HYP_n is a polyhedral cone. See Section 14.2 where several proofs are given for this result. One of them relies essentially on the above link between hypermetrics and Delaunay polytopes and on Voronoi's finiteness result for the number of types of lattices in fixed dimension.

The correspondence between hypermetrics and Delaunay polytopes permits the translation of several notions from the hypermetric cone to Delaunay polytopes. For instance, one can define the *rank* of a Delaunay polytope as the dimension of the smallest face of the hypermetric cone containing the corresponding hypermetric distance. One can then define, in particular, extreme Delaunay polytopes which correspond to extreme rays of the hypermetric cone. This notion of rank and the correspondence between Delaunay polytopes and faces of the hypermetric cone are investigated in Chapter 15.

The various types of Delaunay polytopes that may arise in root lattices are described in Section 14.3. The extreme Delaunay polytopes among them are classified; there are three of them, namely, the 1-dimensional simplex, the Schläfli polytope 2_{21} (of dimension 6) and the Gosset polytope 3_{21} (of dimension 7) (see Section 16.2). Further examples of extreme Delaunay polytopes are described in Sections 16.3 and 16.4; they arise from other lattices such as the Leech lattice and the Barnes-Wall lattice. Some connections between extreme Delaunay polytopes and equiangular sets of lines or perfect lattices are also mentioned in Sections 16.1 and 16.5.

Chapter 17 studies *hypermetric graphs* in detail, i.e., the graphs whose path metric is hypermetric. These graphs are characterized as the isometric subgraphs of Cartesian products of three types of graphs, namely, half-cube graphs, cocktail-party graphs and the Gosset graph G_{56}. Moreover, ℓ_1-graphs are those for which no Gosset graph occurs in the Cartesian product. Several refined results are presented; in particular, for suspension graphs and for graphs having some regularity properties. Further characterizations are discussed for bipartite graphs equipped with the truncated distance (taking value 1 on edges and value 2 on non-edges).

We encounter in this context the class consisting of the connected regular graphs whose adjacency matrix has minimum eigenvalue greater than or equal to -2. This class is well studied in the literature. Beside line graphs and cocktail-party graphs, it contains a list of 187 graphs, which is subdivided into three groups. Each of these three groups is characterized by some parameter. Interestingly, this parameter has an interpretation in terms of some associated Delaunay polytope using hypermetricity (see Section 17.2). Hence this is an example of a situation where a new approach: hypermetricity, sheds new light on a classical notion.

1.3 Outline of Part III. Embeddings of Graphs

In Part III we study various metric and embeddability properties of graphs. For a connected graph $G = (V, E)$ we consider the associated *path metric* d_G defined on the node set V of G, where the distance between two nodes $i, j \in V$ is defined as the length of a shortest path connecting i and j in G. Our objective in Part III is to investigate the structure of the graphs whose path metric enjoys some metric properties such as ℓ_1-embeddability, hypercube embeddability, hypermetricity, etc.

The graphs which are isometric subgraphs of hypercubes are well understood. Several characterizations are presented in Chapter 19. One of them states that the isometric subgraphs of the hypercube are precisely the bipartite graphs whose path metric satisfies a restricted class of hypermetric inequalities, namely, the pentagonal inequalities (hypermetric inequalities on five points). Being an isometric subgraph of a hypercube means being an isometric subgraph of a Cartesian product of copies of K_2. In Chapter 20 we consider isometric embeddings into arbitrary Cartesian products. The following is a well-known result in the metric theory of graphs: Every graph can be isometrically embedded in a canonical way into a (smallest) Cartesian product, called the *canonical metric representation* of the graph. For bipartite graphs, this representation permits to obtain a decomposition of the path metric as a linear combination of primitive semimetrics. One of the main tools underlying these various results is an equivalence relation defined on the edge set of the graph. The number of equivalence classes is an invariant of the graph, called its *isometric dimension.* The number of factors in the canonical representation is precisely the isometric dimension. Moreover, for a bipartite graph G the isometric dimension of G is equal to the (linear) dimension of the smallest face of the semimetric cone that contains d_G.

In Chapter 21 we study *ℓ_1-graphs* in detail, i.e., the graphs whose path metric is ℓ_1-embeddable. This constitutes a relaxation of hypercube embeddability. Indeed a graph G is an ℓ_1-graph if and only if its path metric d_G is hypercube embeddable up to scale, i.e., if ηd_G is hypercube embeddable for some integer η. The smallest such η is called the *minimum scale* of the graph. It is shown that the minimum scale of an ℓ_1-graph is equal to 1 or to an even number and that it is less than or equal to $n-2$ (n is the number of nodes of the graph). For ℓ_1-graphs the factors in the canonical representation are of a very special type; indeed, they are either half-cube graphs or cocktail-party graphs. This result is already proved in Section 14.3, using the connection with Delaunay polytopes. Another proof is given in Chapter 21 which is elementary and has several important applications. In particular, it yields a polynomial time algorithm for recognizing ℓ_1-graphs as well as as a characterization for ℓ_1-rigid graphs (the graphs having an essentially unique ℓ_1-embeding).

The ℓ_1-graphs with minimum scale 1 or 2 are precisely those that can be isometrically embedded into some half-cube graph. They can, moreover, be characterized in terms of some forbidden isometric subspaces (see Section 21.4).

1.4 Outline of Part IV. Hypercube Embeddings and Designs

In Part IV we investigate in detail the hypercube embeddability problem. Given a distance d on V_n, one may ask the following questions: Is d hypercube embeddable ? If yes, does d admit a unique hypercube embedding ? If this is the case we say that d is h-rigid. (Here, "unique" means unique up to certain trivial operations.) If d is not h-rigid, then what are the possible hypercube embeddings of d ?

There are some classes of metrics for which the first question has trivially a positive answer. This is the case, for instance, for the *equidistant metric* $2t\mathbb{1}_n$ where $t \geq 1$ is an integer; $2t\mathbb{1}_n$ denotes the metric on V_n taking value $2t$ for every pair of distinct points. Then, only the last two questions about the number of hypercube embeddings are of interest.

On the other hand, there are some classes of metrics for which deciding hypercube embeddability is a hard task. In fact, testing hypercube embeddability for general metrics is an NP-hard problem. Nevertheless, for some classes of metrics, one is able to characterize their hypercube embeddability by a set of conditions which can be tested in polynomial time. Several such classes are presented in Chapter 24. Among them, we examine the classes of metrics taking two distinct values of the form: a, $2a$ ($a \geq 1$ integer), or three distinct values of the form: $a, b, a+b$ ($a, b \geq 1$ integers not both even). For instance, testing hypercube embeddability for the class of distances on n points with values in the set $\{2,4\}$, or $\{1,2,3\}$, or $\{3,5,8\}$ can be done in time polynomial in n. On the other hand, this same problem is NP-complete for the class of distances with values in the set $\{2,3,4,6\}$. We also examine the class of metrics having a "bipartite structure"; by this we mean the metrics on V_n for which there exists a subset S of points such that any two points of S (or of its complement) are at distance 2. One of the main tools used for recognizing hypercube embeddability for the above classes of metrics is that they contain large equidistant submetrics that are h-rigid; this fact allows us to infer information on the structure of the metrics from the local structure of some of their submetrics.

Chapters 22 and 23 deal essentially with the equidistant metric $2t\mathbb{1}_n$. In Chapter 22 we give some conditions on n and t under which the metric $2t\mathbb{1}_n$ is h-rigid. For instance, if $n \geq t^2+t+3$, then $2t\mathbb{1}_n$ is h-rigid. Moreover, for $n = t^2+t+2$ with $t \geq 3$, the metric $2t\mathbb{1}_n$ is h-rigid if and only if there does not exist a projective plane of order t. In Chapter 23 we examine the possible hypercube embeddings of the metric $2t\mathbb{1}_n$ when it is not h-rigid. An easy observation is that the possible hypercube embeddings of $2t\mathbb{1}_n$ correspond to the $(2t,t,n-1)$-designs (a $(2t,t,n-1)$-design being a collection $\mathcal{B}$ of subsets of V_{n-1} such that every point of V_{n-1} belongs to $2t$ members of $\mathcal{B}$ and every two points of V_{n-1} belong to t common members of $\mathcal{B}$). This leads to the question of finding such designs with specified parameters. This topic is treated in detail in Chapter 23. For instance, a well-known result by Ryser asserts that any design corresponding to a hypercube embedding of $2t\mathbb{1}_n$ has at least $n-1$ blocks, with equality if and only if $n = 4t$ and there exists a Hadamard matrix of order $4t$. Hence, two important classes of designs: projective planes and Hadamard designs, play an important role in the study of the variety of embeddings of the equidistant metric. An explicit description of all the possible hypercube embeddings of $2t\mathbb{1}_n$ is given in Section 23.4 for the following restricted parameters: $t \leq 2$ and $(t=3, n=5)$.

In Chapter 25 we group results related to cut lattices. The cut lattice consists of the vectors that can be written as an integer combination of cut semimetrics. Note that belonging to both the cut cone and to the cut lattice is a necessary condition for a distance d to be hypercube embeddable. In Section 25.3 we study the graphs whose family of cuts forms a Hilbert basis; this amounts to studying (in the context of arbitrary graphs) the case when the above necessary condition is also sufficient. In Section 25.1 we give a description of the cut lattice and of several related lattices. Constructions are presented in Section 25.2 for distances that belong to the cut cone and to the cut lattice but that are not hypercube embeddable.

1.5 Outline of Part V. Facets of the Cut Cone and Polytope

In Part V we survey known results about the facial structure and the geometry of the cut cone and of the cut polytope.

A fundamental property is that all the facets of the cut polytope $\mathrm{CUT}_n^\square$ can be obtained from the facets of $\mathrm{CUT}_n^\square$ that contain a given vertex; this is derived by the so-called switching operation. In particular, all the facets of $\mathrm{CUT}_n^\square$ can be derived from the facets of the cut cone CUT_n. Therefore, for the purpose of investigating the facial structure, it suffices to consider the cut cone. As we have already mentioned, finding a complete linear description of the cut polyhedra is probably a hopeless task. Nevertheless, large classes of inequalities are known. Two classes have already been introduced; they are the hypermetric inequalities and the negative type inequalities. The negative type inequalities never define facets of the cut cone as they are implied by the hypermetric inequalities. On the other hand, the hypermetric inequalities contain large subclasses of facets; they are investigated in Chapter 28.

Triangle inequalities, a very special case of hypermetric inequalities, are considered in detail in Chapter 27. Despite their simplicity, the triangle facets already contain a considerable amount of information about the cut polyhedra. For instance, they provide an integer programming formulation for cuts. Moreover, the triangle inequalities provide the complete linear description of the cut cone CUT_n for $n \leq 4$. Their projections suffice to describe the cut polyhedron of an arbitrary graph G if G does not have K_5 as a graph minor (and, hence, if G is planar).

We make in Section 27.4 a detour to cycle polyhedra of binary matroids. Cycle spaces of binary matroids are nothing but set families that are closed under taking symmetric differences. Hence, the family of cuts in a graph is an instance of cycle space. The switching operation applies in the general framework of binary matroids and there are analogues of the triangle inequalities (in fact, of their projections) for the cycle polyhedra. Hence, several questions that are raised for cut polyhedra can be posed in the general setting of binary matroids; for instance, about linear relaxations by the triangle inequalities or about Hilbert bases. We review in Section 27.4 the main results in this area.

Hypermetric and negative type inequalities belong to the larger class of gap inequalities, described in Section 28.4. Although gap inequalities themselves are not well understood, a weakening of them (obtained by loosening their right-hand sides) serves as a basis for obtaining very good approximations for the max-cut problem (see Section 28.4.1).

In Chapter 29 we study the clique-web inequalities, that constitute a generalization of hypermetric inequalities. In Chapter 30 we present several other classes of inequalities: suspended tree inequalities, path-block-cycle inequalities, circulant inequalities, parachute inequalities, etc.. Section 30.6 contains the complete linear description of the cone CUT_n for $n \leq 7$.

Chapter 31 contains several geometric properties of the cut polytope $\mathrm{CUT}_n^\square$ and of its relaxation by the semimetric polytope $\mathrm{MET}_n^\square$ (defined by the triangle inequalities). In Section 31.6 we study adjacency properties of these polytopes. For instance, any two cuts are adjacent on both the cut polytope and the semimetric polytope. Hence, the 1-skeleton graph of the cut polytope is the complete graph. Moreover, $\mathrm{CUT}_n^\square$ has many simplex faces in common with $\mathrm{MET}_n^\square$ of dimension up to $\lfloor \log_2 n \rfloor$. This indicates that $\mathrm{MET}_n^{\square}$ is wrapped quite tightly around $\mathrm{CUT}_n^{\square}$. In Section 31.7, the Euclidean distance from the hyperplane supporting a facet of $\mathrm{CUT}_n^{\square}$ to the barycentrum of $\mathrm{CUT}_n^{\square}$ is considered. It is conjectured that this distance is minimized by triangle facets. The conjecture is verified for all facets defined by an inequality with coefficients in $\{0, 1, -1\}$ and asymptotically for some other cases. Simplex facets are considered in Section 31.8. It turns out that for $n \leq 7$ the great majority of facets of $\mathrm{CUT}_n^{\square}$ are simplices. In fact about 97% of the facets of $\mathrm{CUT}_7^{\square}$ are simplices ! This may well be a general phenomenon for any n.

Further geometric results are presented in Sections 31.1-31.4. Borsuk [1933] asked whether it is possible to partition every set X of points in $\mathbb{R}^d$ into $d+1$ subsets, each having a smaller diameter than X. This question was answered in the negative by Kahn and Kalai [1993] by a construction using cuts, that we present in Section 31.1. The result in Section 31.2 indicates how to obtain valid inequalities for pairwise angles among a set of vectors from the valid inequalities for the cut polytope. This permits in particular to answer an old question of Fejes Tóth [1959] concerning the maximum value for the sum of the pairwise angles among a set of n vectors. Section 31.3 deals with the completion problem for partial positive semidefinite matrices. It turns out that necessary conditions for this completion problem can be obtained from the valid inequalities for the cut polytope, as a reformulation of the result in Section 31.2. Finally, Section 31.4 deals with the completion problem for partial Euclidean matrices; that is, with the study of the projections of the negative type cone. These two completion problems are closely related and have intimate links with the polyhedra under investigation in this book.

Chapter 2. Basic Definitions

In this chapter we introduce some basic definitions about graphs, polyhedra, matrices, and algorithmic complexity. We present here only the very basic notions; further definitions will be introduced later in the text as they are needed. The reader may consult, for instance, the following textbooks for more detailed information: Bondy and Murty [1976] for graphs, Grünbaum [1967], Schrijver [1986], Ziegler [1995] for polyhedra, Lancaster and Tismenetsky [1985] for matrices, and Garey and Johnson [1979] for algorithms and complexity.

2.1 Graphs

A *graph* $G = (V, E)$ consists of a finite set V of *nodes* and a finite set E of *edges*. Every edge $e \in E$ consists of a pair of nodes u and v, called its *endnodes*; we then denote the edge e alternatively as (u, v) or as uv. Two nodes are said to be *adjacent* if they are joined by an edge. Two edges are said to be parallel if they have the same endnodes. Here, we will only consider *simple* graphs, i.e., graphs in which every edge has distinct endnodes and no two edges are parallel. The *degree* of a node $v \in V$ is the number of edges to which v is incident. When every two nodes in G are adjacent, then the graph G is said to be a *complete graph*. It is customary to denote the complete graph on n nodes by K_n; we can suppose that the node set of K_n is the set $V_n := \{1, \ldots, n\}$ and that its edge set is the set $E_n := \{ij \mid i \neq j \in V_n\}$ (where the symbol ij denotes the unordered pair of the integers i, j, i.e., ij and ji are considered identical).

A graph G is said to be *bipartite* if its node set can be partitioned into $V = V_1 \cup V_2$ in such a way that no two nodes in V_1 and no two nodes in V_2 are adjacent. The sets V_1, V_2 are said to form a *bipartition* of G. If G is bipartite with bipartition (V_1, V_2) and if every node in V_1 is adjacent to every node in V_2, then G is called a *complete bipartite graph*. We let K_{n_1,n_2} denote the complete bipartite graph with bipartition (V_1, V_2) where $|V_1| = n_1$ and $|V_2| = n_2$. The complete bipartite graph $K_{1,n}$ $(n \geq 1)$ is sometimes called a *star*.

Given a node subset $S \subseteq V$ in a graph G, let $\delta_G(S)$ denote the set of edges in G having one endnode in S and the other endnode in $V \setminus S$; $\delta_G(S)$ is called the *cut* [1] *determined by* S.

[1] Thus, the symbol $\delta_G(S)$ denotes here an edge set. In fact, the symbol $\delta(S)$ will be mostly used in the book for denoting a 0-1 vector, namely, the cut semimetric determined by S (see Section 3.1). When G is the complete graph K_n, then the incidence vector of the cut $\delta_G(S)$

Let $G = (V, E)$ be a graph. A graph $H = (W, F)$ is said to be a *subgraph* of G if $W \subseteq V$ and $F \subseteq E$. Given a node subset $W \subseteq V$, $G[W]$ denotes the subgraph of G *induced* by W; its node set is W and its edge set consists of the edges of G that are contained in W. The set W is said to induce a *clique* in G if any two nodes in W are adjacent, i.e., if $G[W]$ is a complete graph. A *matching* in G is an edge subset $F \subseteq E$ such that no two edges in F share a common node; a matching F is a *perfect matching* if every node of G belongs to exactly one edge in F.

Given an edge subset $F \subseteq E$ in G, $G \backslash F := (V, E \setminus F)$ is called the graph obtained from G by *deleting* F. When $F = \{e\}$ we also denote $G \backslash \{e\}$ by $G \backslash e$. *Contracting* an edge $e := uv$ in G means identifying the endnodes u and v of e and deleting the parallel edges that may be created while identifying u and v; G/e denotes the graph obtained from G by contracting the edge e. For an edge set $F \subseteq E$, G/F denotes the graph obtained from G by contracting all edges of F (in any order). A graph H is said to be a *minor* of G if it can be obtained from G by a sequence of deletions and/or contractions of edges, and deletions of nodes.

The following graphs will be frequently used in the book:

- The *path* P_n, with node set $V = \{v_1, \ldots, v_n\}$ and whose edges are the pairs $v_i v_{i+1}$ for $i = 1, 2, \ldots, n-1$.
- The *circuit* C_n, with node set $V = \{v_1, \ldots, v_n\}$ and whose edges are the pairs $v_i v_{i+1}$ for $i = 1, 2, \ldots, n-1$ together with the pair $v_1 v_n$.
- The *hypercube graph* $H(n, 2)$, with node set $V = \{0,1\}^n$ and whose edges are the pairs of vectors $x, y \in \{0,1\}^n$ such that $|\{i \in [1, n] \mid x_i \neq y_i\}| = 1$.
- The *half-cube graph* $\frac{1}{2}H(n, 2)$, with node set $V = \{x \in \{0,1\}^n \mid \sum_{i=1}^n x_i$ is even$\}$ and whose edges are the pairs $x, y \in \{0,1\}^n$ such that $|\{i \in [1, n] \mid x_i \neq y_i\}| = 2$.
- The *cocktail-party graph* $K_{n \times 2}$, with node set $V = \{v_1, \ldots, v_n, v_{n+1}, \ldots, v_{2n}\}$ and whose edges are all pairs of nodes in V except the n pairs $v_1 v_{n+1}, \ldots, v_n v_{2n}$; in other words, $K_{n \times 2}$ is the complete graph K_{2n} in which a perfect matching has been deleted.

Two graphs $G = (V, E)$ and $G' = (V', E')$ are said to be *isomorphic* if there exists a bijection $f : V \longrightarrow V'$ such that

$$uv \in E \Longleftrightarrow f(u)f(v) \in E';$$

we write $G \simeq G'$ if G and G' are isomorphic. There are some isomorphisms among the above graphs; for instance,

$$H(2,2) \simeq C_4,\ K_{2\times 2} \simeq C_4,\ \frac{1}{2}H(2,2) \simeq K_2,\ \frac{1}{2}H(3,2) \simeq K_4,\ \frac{1}{2}H(4,2) \simeq K_{3\times 2}.$$

---

coincides with the cut semimetric $\delta(S)$ defined by S. However, the graph notation $\delta_G(S)$ will be used only locally in the book and the reader will then be reminded that $\delta_G(S)$ stands for an edge set. So no confusion should arise between the two symbols $\delta_G(S)$ and $\delta(S)$.

The graphs C_5, $H(3,2)$ and $K_{3\times 2}$ are depicted in Figure 2.1.1. The *Petersen graph* P_{10}, which will also be used on several occasions, is shown in Figure 2.1.2.

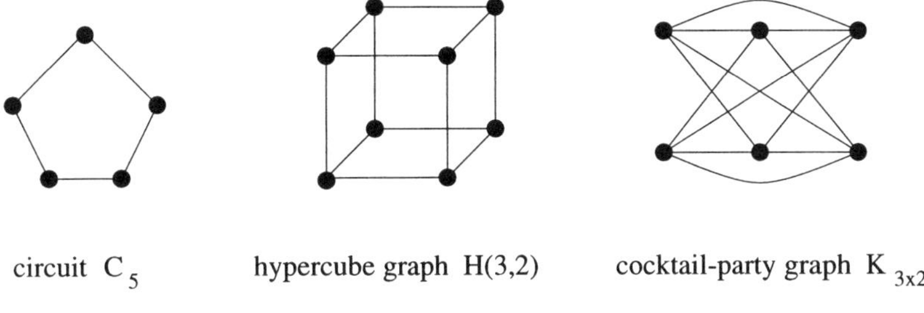

Figure 2.1.1

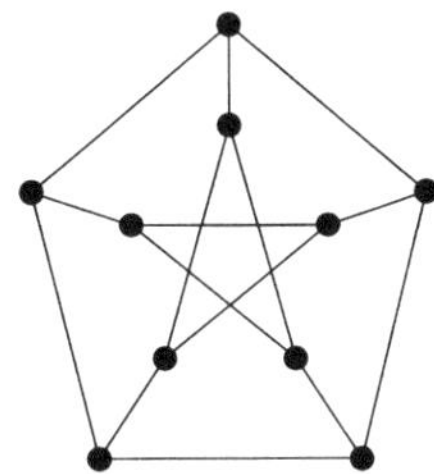

Figure 2.1.2: The Petersen graph P_{10}

A graph G is said to be *connected* if, for every two nodes u, v in G, there exists a path in G joining u and v; a graph which is not connected is said to be *disconnected.* A *forest* is a graph which contains no circuit; a *tree* is a connected forest. A *cycle* or *Eulerian graph* is a graph which can be decomposed as an edge disjoint union of circuits (equivalently, it is a graph in which every node has an even degree).

Let us now consider two operations on graphs: the Cartesian product and the clique sum operation. Let $G_1 = (V_1, E_1)$ and $G_2 = (V_2, E_2)$ be two graphs. Their *Cartesian product* $G_1 \times G_2$ is the graph $G := (V_1 \times V_2, E)$ with node set

$$V_1 \times V_2 = \{(v_1, v_2) \mid v_1 \in V_1 \text{ and } v_2 \in V_2\}$$

and whose edges are the pairs $((u_1, u_2), (v_1, v_2))$ (with $u_1, v_1 \in V_1$ and $u_2, v_2 \in V_2$) such that, either $(u_1, v_1) \in E_1$ and $u_2 = v_2$, or $u_1 = v_1$ and $(u_2, v_2) \in E_2$.

Let $G = (V, E)$ be a graph. Let V_1 and V_2 be subsets of V such that $V = V_1 \cup V_2$ and such that the set $W := V_1 \cap V_2$ induces a clique in G. Suppose moreover that there is no edge joining a node in $V_1 \setminus W$ to a node in $V_2 \setminus W$. Then, G is called the *clique k-sum* of the graphs $G_1 := G[V_1]$ and $G_2 := G[V_2]$, where $k := |W|$. One may say simply that G is the *clique sum* of G_1 and G_2 if one does not wish to specify the size of the common clique.

For a graph G, its *suspension graph* ∇G denotes the graph obtained from G by adding a new node (called the *apex* of ∇G) and making it adjacent to all the nodes in G. Moreover, the *line graph* of G is the graph $L(G)$ whose nodes are the edges of G with two edges adjacent in $L(G)$ if they share a common node.

2.2 Polyhedra

We assume familiarity with basic linear algebra. By $\mathbb{R}$ (resp. $\mathbb{Q}$, $\mathbb{Z}$, $\mathbb{N}$) we mean the set of real (resp. rational, integer, natural) numbers. Given a set E and a subset $S \subseteq E$, the *incidence vector* of S is the vector $\chi^S \in \mathbb{R}^E$ defined by

$$\chi_e^S := \begin{cases} 1 & \text{if } e \in S, \\ 0 & \text{if } e \in E \setminus S. \end{cases}$$

If A and B are subsets of E, then $A \triangle B$ denotes their *symmetric difference* defined by

$$A \triangle B = (A \cup B) \setminus (A \cap B).$$

For a matrix M, we let M^T denote its transpose matrix; similarly, x^T denotes the transpose of a vector x. Hence,

$$x^T y = \sum_{i=1}^{n} x_i y_i \quad \text{for } x, y \in \mathbb{R}^n.$$

We remind that the *dimension* of a set $X \subseteq \mathbb{R}^n$ is defined as the cardinality of a largest affinely independent subset of X minus one; it is denoted as $\dim(X)$. The set $X \subseteq \mathbb{R}^n$ is said to be *full-dimensional* if $\dim(X) = n$. The *rank* of X, denoted as $\operatorname{rank}(X)$, is defined as the cardinality of a largest linearly independent subset of X. Let us introduce some notation for linear hulls. Given a set $X \subseteq \mathbb{R}^n$ and $\mathbb{K} \subseteq \mathbb{R}$, set

$$\mathbb{K}(X) := \{\sum_{x \in X} \lambda_x x \mid \lambda_x \in \mathbb{K} \text{ for all } x \in X\}.$$

(In this definition we suppose that only finitely many λ_x's are nonzero.) When $\mathbb{K} = \mathbb{Z}$, the set $\mathbb{Z}(X)$ is called the *integer hull* of X; when $\mathbb{K} = \mathbb{R}_+$, the set $\mathbb{R}_+(X)$ is called the *conic hull* of X; when $\mathbb{K} = \mathbb{Z}_+$, the set $\mathbb{Z}_+(X)$ is known as the *integer cone* generated by X. We will also consider the *affine integer hull* of X, defined as

$$\mathbb{Z}_{af}(X) := \{\sum_{x \in X} \lambda_x x \mid \lambda_x \in \mathbb{Z} \text{ for all } x \in X \text{ and } \sum_{x \in X} \lambda_x = 1\}$$

and the *convex hull* of X, defined as

$$\operatorname{Conv}(X) := \{\sum_{x \in X} \lambda_x x \mid \lambda_x \geq 0 \text{ for all } x \in X \text{ and } \sum_{x \in X} \lambda_x = 1\}.$$

A set $X \subseteq \mathbb{R}^n$ is said to be *convex* if $\operatorname{Conv}(X) = X$. A *convex body* in $\mathbb{R}^n$ is a convex subset of $\mathbb{R}^n$ which is compact and full-dimensional. The *polar* X° of $X \subseteq \mathbb{R}^n$ is defined as

$$X^\circ := \{x \in \mathbb{R}^n \mid x^T y \leq 1 \; \forall y \in X\}.$$

The set X is said to be *centrally symmetric* if $-x \in X$ for all $x \in X$. We now consider in detail conic and convex hulls.

Cones and Polytopes. We recall here several basic definitions concerning cones and polytopes. Given a subset $X \subseteq \mathbb{R}^n$, the set X is said to be a *cone*[2] if $\mathbb{R}_+(X) = X$. If C is a cone of the form $C = \mathbb{R}_+(X)$, one says that C is *generated* by X and the members of X are called its *generators*. The cone C is said to be *finitely generated* if it admits a finite set of generators, i.e., if $C = \mathbb{R}_+(X)$ for some finite set X. The polar of a cone $C \subseteq \mathbb{R}^n$ can alternatively be described as

$$C^\circ = \{x \in \mathbb{R}^n \mid x^T y \leq 0 \ \forall y \in C.\}$$

Every convex set of the form Conv(X), where X is finite, is called a *polytope.*

Let A be an $m \times n$ matrix and let $b \in \mathbb{R}^m$ be a vector. Then, the set

$$\{x \in \mathbb{R}^n \mid Ax \leq b\}$$

is called a *polyhedron.* Every polyhedron is obviously a convex set and it is a cone when b is the zero vector. Every cone of the form $\{x \in \mathbb{R}^n \mid Ax \leq 0\}$ is called a *polyhedral cone.*

A classical result, generally attributed to Minkowski and Weyl, asserts that polytopes and bounded polyhedra are, in fact, the same notions. (A set $P \in \mathbb{R}^n$ is said to be *bounded* if there exists a constant R such that $\max_{1 \leq i \leq n} |x_i| \leq R$ for all $x = (x_i)_{i=1}^n \in P$.) In other words, every polytope P, which is given as the convex hull of a finite set X, can be expressed as the solution set of a finite system $Ax \leq b$ of linear inequalities; such a system is called a *linear description* of P. The converse statement holds as well. There is a similar result for cones: A cone C is finitely generated if and only if it can be expressed as the solution set of a system $Ax \leq 0$ of linear inequalities; such a system is called a *linear description* of C. In other words, a cone is finitely generated if and only if it is polyhedral.

In practice, finding a linear description of a polytope P or of a cone C (assuming that they are given by their generators) may be a hard task. One of our main objectives in this book will be, in fact, to give information about the linear description of a specific polytope and cone, namely, the cut polytope and the cut cone:

$$\mathrm{CUT}_n^\square := \mathrm{Conv}(\chi^{\delta_{K_n}(S)} \mid S \subseteq V_n), \ \mathrm{CUT}_n := \mathbb{R}_+(\chi^{\delta_{K_n}(S)} \mid S \subseteq V_n).$$

Another well-known result that we will sometimes apply is *Carathéodory's theorem*, which can be stated as follows: Let $X \subseteq \mathbb{R}^n$. If $x \in \mathbb{R}_+(X)$ then $x \in \mathbb{R}_+(X')$, where X' is a linearly independent subset of X. If $x \in \mathrm{Conv}(X)$ then $x \in \mathrm{Conv}(X')$, where X' is an affinely independent subset of X.

The following polytopes will be frequently used in the book:

[2] Hence in this definition a cone is always assumed to be convex.

- The n-dimensional *simplex* α_n; this is the polytope

$$\mathrm{Conv}(0, e_1, \ldots, e_n) = \{x \in \mathbb{R}^n \mid 0 \le x_i \le 1 \ (1 \le i \le n), \sum_{i=1}^{n} x_i \le 1\}.$$

- The n-dimensional *cross-polytope* β_n; this is the polytope

$$\mathrm{Conv}(\pm e_1, \ldots, \pm e_n) = \{x \in \mathbb{R}^n \mid \sum_{i=1}^{n} a_i x_i \le 1 \ \text{ for all } a \in \{\pm 1\}^n\}.$$

- The n-dimensional *hypercube* γ_n; this is the polytope

$$\mathrm{Conv}(\{0,1\}^n) = [0,1]^n.$$

(Here, $e_1, \ldots, e_n$ denote the coordinate vectors in $\mathbb{R}^n$.) Two polytopes P, P' in $\mathbb{R}^n$ are said to be *affinely equivalent* if $P' = f(P)$, where f is an affine bijection of $\mathbb{R}^n$. A *simplex* is any polytope of the form $\mathrm{Conv}(X)$, where the set X is affinely independent. Similarly, a *simplex cone* is a cone of the form $\mathbb{R}_+(X)$, where the set X is linearly independent. We use, in fact, the symbol α_n for denoting any n-dimensional simplex. Similarly, β_n and γ_n denote the above cross-polytope and hypercube, up to affine bijection. Note that α_1, β_1 and γ_1 coincide, and that $\beta_2 = \gamma_2$ (up to affine bijection). Figure 2.2.1 shows α_3 (the tetrahedron), β_3 (the octahedron), and γ_3 (the usual cube).

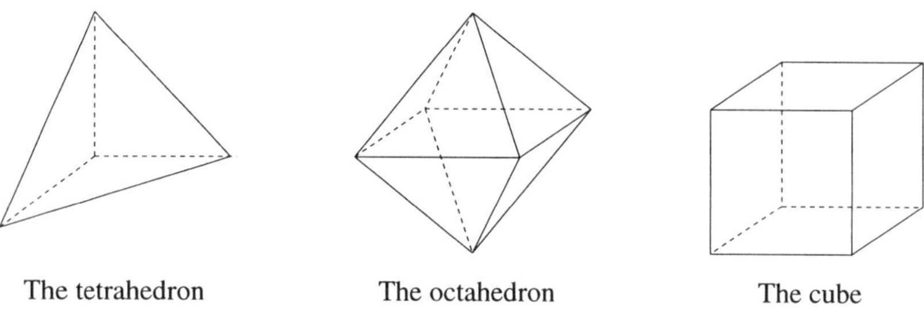

Figure 2.2.1

Faces. Let P be a polytope in $\mathbb{R}^n$. A set $F \subseteq P$ is called a *face* of P if, for every $x \in F$, every decomposition: $x = \alpha y + (1-\alpha)z$ where $0 \le \alpha \le 1$ and $y, z \in P$ implies that $y, z \in F$. The only face of dimension $\dim(P)$ is P itself. Every face of dimension $\dim(P) - 1$ is called a *facet* of P. A face of dimension 0 is of the form $\{x\}$; then, x is said to be a *vertex* of P. A face of dimension 1 is called an *edge* of P. Two vertices x, y of P are said to be *adjacent* on P if the set $\{\alpha x + (1-\alpha)y \mid 0 \le \alpha \le 1\}$ is an edge of P. A graph is attached to every polytope P, called its *1-skeleton graph*, and defined as follows: Its node set is the set of vertices of P with an edge between two vertices if they are adjacent on P. For instance, the 1-skeleton graphs of the polytopes α_n, β_n and γ_n are, respectively, the complete graph K_{n+1}, the cocktail-party graph $K_{n\times 2}$ and the hypercube graph $H(n, 2)$.

A face F of P is called a *simplex face* if F is a simplex, i.e., if the vertices of P lying on F are affinely independent.

Given a vector $v \in \mathbb{R}^n$ and $v_0 \in \mathbb{R}$, the inequality $v^T x \le v_0$ is said to be *valid* for P if $v^T x \le v_0$ holds for all $x \in P$. Then, the set

$$F := \{x \in P \mid v^T x = v_0\}$$

is clearly a face of P; it is called the *face induced by* the inequality $v^T x \le v_0$. In fact, every face of P is induced by some valid inequality.

The above definitions extend to cones in the following way. Let C be a cone in $\mathbb{R}^n$. A *face* of C is any subset $F \subseteq C$ such that, for every $x \in F$, every decomposition: $x = y + z$ where $y, z \in C$ implies that $y, z \in F$. A face of dimension $\dim(C) - 1$ is called a *facet* of C; a face of dimension 1 is called an *extreme ray* of C. Given $v \in \mathbb{R}^n$, the inequality $v^T x \le 0$ is said to be *valid* for C if $v^T x \le 0$ holds for all $x \in C$. Then, the set

$$\{x \in C \mid v^T x = 0\}$$

is called the *face* of C *induced by* the valid inequality $v^T x \le 0$. When C is a polyhedral cone, every face of C arises in this manner. A face F of a cone C is called a *simplex face* if F is a simplex cone.

Linear Programming Duality. A *linear programming problem* consists of maximizing (or minimizing) a linear function over a polyhedron. A typical example of such a problem is:

$$\text{(P)} \qquad \max(c^T x \mid Ax \le b),$$

also written as

$$\begin{array}{ll} \max & c^T x \\ \text{s.t.} & Ax \le b \end{array}$$

where A is an $m \times n$ matrix, $b \in \mathbb{R}^m$ and $c \in \mathbb{R}^n$. The linear function $c^T x$ is often called the *objective function* of the program (P). To the linear program (P) is associated another linear program (D), called its *dual* and defined as

$$\text{(D)} \qquad \min(b^T y \mid y^T A = c^T, y \ge 0).$$

One refers to (P) as to the *primal* program. The following result, known as the *linear programming duality theorem*, establishes a fundamental connection between the linear programs (P) and (D).

Theorem 2.2.2. *Given an $m \times n$ matrix A and vectors $b \in \mathbb{R}^m$, $c \in \mathbb{R}^n$, then*

$$\max(c^T x \mid Ax \le b) = \min(b^T y \mid y^T A = c^T, y \ge 0)$$

provided both sets $\{x \mid Ax \le b\}$ and $\{y \mid y^T A = c^T, y \ge 0\}$ are nonempty. ∎

This theorem admits several other equivalent formulations. For example,

$$\max(c^T x \mid Ax \le b, x \ge 0) = \min(b^T y \mid y^T A \ge c^T, y \ge 0),$$
$$\max(c^T x \mid Ax = b, x \ge 0) = \min(b^T y \mid y^T A \ge c^T).$$

2.3 Algorithms and Complexity

Although complexity is not a central topic in this book, we will encounter a number of problems for which one is interested in their complexity status. We will not define in a precise mathematical way the notions of algorithms and complexity. We only give here some 'naive' definitions, which should be sufficient for our purpose. Precise definitions can be found, e.g., in the textbooks by Garey and Johnson [1979], or Papadimitriou and Steiglitz [1982].

Let (P) be a problem and suppose, for convenience, that (P) is a decision problem (that is, a problem which asks for a 'yes' or 'no' answer). We may take, for instance, for (P) any of the following problems:

(P1) The ℓ_1-embeddability problem.
Instance: A rational valued distance d on a set $V_n = \{1, \ldots, n\}$.
Question: Is d isometrically ℓ_1-embeddable ? That is, do there exist vectors $v_1, \ldots, v_n \in \mathbb{Q}^m$ (for some $m \geq 1$) such that $d(i,j) = d_{\ell_1}(v_i, v_j)$ for all $i, j \in V_n$?

(P2) The hypercube embeddability problem for graphs.
Instance: A graph $G = (V, E)$.
Question: Can G be isometrically embedded into some hypercube ?

An instance of a problem is specified by providing a certain input. (For (P1), the input consists of the $\binom{n}{2}$ numbers $d(i,j)$ while, for (P2), the input consists of a graph G which can, for instance, be described by its adjacency matrix.) The *size* of an instance is the number of bits needed to represent the input data in binary encoding. (For instance, the size of an integer p is size(p) = $\lceil \log_2(|p|+1) \rceil$; the size of a rational number $\frac{p}{q}$ is size(p) + size$(q) + 1$; the size of an $m \times n$ rational matrix A is $mn + \sum_{i,j} \text{size}(a_{ij})$, etc.) Suppose that we have an algorithm for solving (P). Its complexity is measured by counting the total number of elementary steps needed to be performed throughout the execution of the algorithm (elementary steps such as arithmetic operations, comparisons, branching instructions, etc., are supposed to take a unit time). The algorithm is said to have a *polynomial running time* if the total number of elementary steps can be expressed as $p(l)$, where $p(.)$ is a polynomial function and l is the size of the instance.

Problems are classified into several complexity classes. The class P consists of the decision problems which can be solved in polynomial time. The class NP consists of the decision problems for which every 'yes' answer admits a certificate that can be checked in polynomial time (but one does not need to know how to find such a certificate in polynomial time). Similarly, a problem is in co-NP if every 'no' answer admits a certificate that can be checked in polynomial time. Hence, P $\subseteq$ NP $\cap$ co-NP.

For example, the problem (P1) belongs to NP (because it can be shown that, if d is ℓ_1-embeddable, then there exists a set of vectors $v_1, \ldots, v_n \in \mathbb{R}^{\binom{n}{2}}$ providing an ℓ_1-embedding of d and such that the size of the v_i's is polynomially bounded by that of d; see Section 4.4). On the other hand, as we will see in Chapter 19, the problem (P2) belongs to P.

Among the problems in NP, some can be shown to be hardest. A problem is said to be *NP-complete* if it belongs to NP and if a polynomial algorithm for solving it could be used once as a subroutine to obtain a polynomial algorithm for any problem in NP. A problem is *NP-hard* if any polynomially bounded algorithm for solving it would imply a polynomial algorithm for solving an arbitrary NP-problem; note that the problem itself is not required to be in NP and that one permits more than one call to the subroutine. A typical way to show that a problem (P) is NP-hard is to show that some known NP-complete problem polynomially reduces to it. Given two problems (P) and (Q) one says that (Q) *reduces polynomially* to (P) if a polynomial algorithm solving (Q) can be constructed from a polynomial algorithm solving (P). For example, the hypercube embeddability problem for arbitrary distances is NP-hard. Typically, the optimization versions of NP-complete decision problems are also NP-hard. For instance, the max-cut problem:

Given $w \in \mathbb{Q}^{E_n}$, *find a set* $S \subseteq V_n$ *for which* $w^T\delta(S)$ *is maximum*

is NP-hard, because the following problem is NP-complete:

Given $w \in \mathbb{Q}^{E_n}$ *and* $K \in \mathbb{Q}$, *does there exist* $S \subseteq V_n$ *such that* $w^T\delta(S) \geq K$ *?*

(See Section 4.4.)

We remind that, for two functions $f(n)$ and $g(n)$ ($n \in \mathbb{N}$), the notation: $f(n) = O(g(n))$ means that there exists a constant $C > 0$ such that $f(n) \leq Cg(n)$ for all $n \in \mathbb{N}$. Similarly, the notation: $f(n) = \Omega(g(n))$ means that $f(n) \geq Cg(n)$ for all $n \in \mathbb{N}$, for some constant $C > 0$.

2.4 Matrices

We group here some preliminaries about matrices. A square matrix A is said to be *orthogonal* if $A^TA = I$, i.e., if its inverse matrix A^{-1} is equal to its transpose matrix A^T. We let $\mathrm{OA}(n)$ denote the set of orthogonal $n \times n$ matrices. The orthogonal matrices are the isometries of the Euclidean space; that is, the linear transformations of $\mathbb{R}^n$ preserving the Euclidean distance. A well-known basic fact is that any two congruent sets of points can be matched by some orthogonal transformation. We formulate below this fact for further reference. Recall that $\| x \|_2 = \sqrt{x^Tx}$ for $x \in \mathbb{R}^n$.

Lemma 2.4.1. *Let* $u_1, \ldots, u_p \in \mathbb{R}^n$ *and* $v_1, \ldots, v_p \in \mathbb{R}^n$ *be two sets of vectors such that* $\| u_i - u_j \|_2 = \| v_i - v_j \|_2$ *for all* $i, j = 1, \ldots, p$. *Then, there exists* $A \in \mathrm{OA}(n)$ *such that* $Au_i = v_i$ *for* $i = 1, \ldots, p$. *Moreover, such a matrix* A *is unique if the set* $\{u_1, \ldots, u_p\}$ *has affine rank* $n+1$. ∎

Let $A = (a_{ij})_{i,j=1}^n$ be an $n \times n$ symmetric matrix. Then, A is said to be *positive semidefinite* if $x^TAx \geq 0$ holds for all $x \in \mathbb{R}^n$ (or, equivalently, for all

$x \in \mathbb{Z}^n$); then, we write: $A \succeq 0$. Equivalently, A is positive semidefinite if and only if all its eigenvalues are nonnegative, or if $\det(A_I) \geq 0$ for every principal submatrix A_I of A (setting $A_I := (a_{ij})_{i,j \in I}$ for a subset $I \subseteq \{1, \ldots, n\}$ and letting $\det(A_I)$ denote the determinant of A_I).

A symmetric $n \times n$ matrix $A = (a_{ij})$ can be encoded by its upper triangular part (including the diagonal), i.e., by the vector $(a_{ij})_{1 \leq i \leq j \leq n}$. We let PSD_n denote the set of vectors $(a_{ij})_{1 \leq i \leq j \leq n}$ for which the symmetric $n \times n$ matrix (a_{ij}) (setting $a_{ji} = a_{ij}$) is positive semidefinite. The set PSD_n is a cone in $\mathbb{R}^{E_n}$, called the *positive semidefinite cone*.

A *quadratic form* is a function of the form:

$$x \in \mathbb{R}^n \mapsto x^T A x = \sum_{1 \leq i,j \leq n} a_{ij} x_i x_j$$

where $A = (a_{ij})$ is an $n \times n$ symmetric matrix; it is said to be positive semidefinite when the matrix A is positive semidefinite.

The *Gram matrix* $\text{Gram}(v_1, \ldots, v_n)$ of a set of vectors $v_1, \ldots, v_n \in \mathbb{R}^k$ $(k \geq 1)$ is the $n \times n$ matrix whose (i,j)-th entry is $v_i^T v_j$. It is easy to check that the rank of the matrix $\text{Gram}(v_1, \ldots, v_n)$ is equal to the rank of the system $(v_1, \ldots, v_n)$. Every Gram matrix is obviously positive semidefinite. It is well-known that, conversely, every positive semidefinite matrix can be expressed as a Gram matrix. We recall this result below, as it will be often used in our treatment.

Lemma 2.4.2. *Let $A = (a_{ij})_{1 \leq i,j \leq n}$ be a symmetric matrix which is positive semidefinite and let $k \leq n$ be its rank. Then, A is a Gram matrix, i.e., there exist vectors $v_1, \ldots, v_n \in \mathbb{R}^k$ such that $a_{ij} = v_i^T v_j$ for $1 \leq i, j \leq n$. The system $(v_1, \ldots, v_n)$ has rank k. Moreover, if $v_1', \ldots, v_n'$ are other vectors of $\mathbb{R}^k$ such that $a_{ij} = v_i'^T v_j'$ for $1 \leq i, j \leq n$, then $v_i' = U v_i$ $(1 \leq i \leq n)$ for some orthogonal matrix U.*

Proof. By assumption, A has k nonzero eigenvalues which are positive. Hence, there exists an $n \times n$ matrix Q_0 such that $A = Q_0 D Q_0^T$, where D is an $n \times n$ matrix whose entries are all zero except k diagonal entries, say with indices $(1,1), \ldots, (k,k)$, equal to 1. Denote by Q the $n \times k$ submatrix of Q_0 consisting of its first k columns. Then, $A = QQ^T$ holds, i.e., $a_{ij} = v_i^T v_j$ for $1 \leq i, j \leq n$, where $v_1^T, \ldots, v_n^T$ denote the rows of Q. It is easy to see that $(v_1, \ldots, v_n)$ has the same rank k as A. The unicity (up to orthogonal transformation) of $v_1, \ldots, v_n$ follows from Lemma 2.4.1. ∎

Checking whether a given symmetric matrix A is positive semidefinite can be done in polynomial time. The next result is given in (Grötschel, Lovász and Schrijver [1988], page 295). For completeness, we recall the proof here.

Proposition 2.4.3. *Let $A = (a_{ij})$ be a symmetric $n \times n$ matrix with rational entries. There exists an algorithm permitting to check whether A is positive semidefinite and, if not, to construct a vector $x \in \mathbb{Q}^n$ such that $x^T A x < 0$. This* *algorithm runs in time polynomial in n and in the size of A.*

Proof. The algorithm is based on Gaussian elimination and proceeds in the following way. Check first whether $a_{11} \geq 0$. If not, then A is not positive semidefinite and $x^T Ax = a_{11} < 0$ for $x := (1,0,\ldots,0)^T$. Suppose that $a_{11} = 0$. Then, A is not positive semidefinite if $a_{1i} \neq 0$ for some $i \in \{2,\ldots,n\}$. Indeed, $x^T Ax < 0$ if we set $x_i := -1$, $x_j := 0$ for $j \neq 1, i$ and if we choose $x_1 \in \mathbb{Q}$ such that $2x_1a_{1i} - a_{ii} > 0$. If $a_{11} = a_{12} = \ldots = a_{1n} = 0$ then $A \succeq 0$ if and only if its $(n-1) \times (n-1)$ submatrix $A' := (a_{ij})_{i,j=2}^n$ is positive semidefinite. Moreover, if we can find $y \in \mathbb{Q}^{n-1}$ such that $y^T A'y < 0$, then $x^T Ax < 0$ for the vector $x := (0, y) \in \mathbb{Q}^n$.

Suppose now that $a_{11} > 0$. Consider the $(n-1)\times(n-1)$ matrix $A' = (a'_{ij})_{i,j=2}^n$ (obtained by pivoting A with respect to the entry a_{11}) defined by

$$a'_{ij} := a_{ij} - \frac{a_{i1}}{a_{11}}a_{1j} \quad \text{for } i,j = 2,\ldots,n.$$

Then, $A \succeq 0$ if and only if $A' \succeq 0$. Moreover, suppose that we can find a vector $y \in \mathbb{Q}^{n-1}$ such that $y^T A'y < 0$. We indicate how to construct a vector $x \in \mathbb{Q}^n$ such that $x^T Ax < 0$. We set $x := (\alpha, y)$ where α has to be determined. Write $A := \begin{pmatrix} a_{11} & b^T \\ b & B \end{pmatrix}$. Then, $x^T Ax = \alpha^2 a_{11} + 2\alpha y^T b + y^T By$, where

$$y^T By = \sum_{i,j=2}^n y_i y_j a_{ij} = \sum_{i,j=2}^n y_i y_j (a'_{ij} + \frac{a_{i1}}{a_{11}} a_{1j}) = y^T A'y + \frac{1}{a_{11}}(y^T b)^2.$$

Therefore, $x^T Ax = \alpha^2 a_{11} + 2\alpha y^T b + y^T A'y + \frac{1}{a_{11}}(y^T b)^2$. Let D denote the discriminant of this quadratic expression (in the variable α); then, $D = -a_{11} y^T A'y > 0$. Hence, if we choose $\alpha \in \mathbb{Q}$ such that $\frac{-y^T b - \sqrt{D}}{a_{11}} < \alpha < \frac{-y^T b + \sqrt{D}}{a_{11}}$, then $x := (\alpha, y) \in \mathbb{Q}^n$ satisfies $x^T Ax < 0$.

In all cases we have reduced our task to a problem of order $n-1$. Moreover, one can verify that the entries of the smaller matrices to be considered do not grow too large, i.e., that their sizes remain polynomially bounded by the size of the initial matrix A (see Grötschel, Lovász and Schrijver [1988] for details). ∎

Let M be a symmetric $n \times n$ matrix. The *inertia* $\text{In}(M)$ of M is defined as the triple (p,q,s), where p (resp. q, s) denotes the number of positive (resp. negative, zero) eigenvalues of M; hence, $n = p + q + s$. If P is a nonsingular matrix, then it is well-known that the two matrices M and PMP^T have the same inertia; this result is known as *Sylvester's law of inertia.* The following result is useful for computing the inertia of a matrix.

Lemma 2.4.4. *Let M be a symmetric matrix with the following block decomposition:*

$$M = \begin{pmatrix} A & B \\ B^T & C \end{pmatrix},$$

where C is nonsingular. Then,

$$\mathrm{In}(M) = \mathrm{In}(C) + \mathrm{In}(A - BC^{-1}B^T), \ \det M = \det C \cdot \det(A - BC^{-1}B^T).$$

(The matrix $A - BC^{-1}B^T$ is known as the Schur complement of C in M.)

Proof. The following identity holds:

$$\begin{pmatrix} I & BC^{-1} \\ 0^T & I \end{pmatrix} \begin{pmatrix} A - BC^{-1}B^T & 0 \\ 0^T & C \end{pmatrix} \begin{pmatrix} I & 0 \\ C^{-1}B^T & I \end{pmatrix} = \begin{pmatrix} A & B \\ B^T & C \end{pmatrix} = M.$$

By Sylvester's law of inertia, the two matrices M and $\begin{pmatrix} A - BC^{-1}B^T & 0 \\ 0^T & C \end{pmatrix}$ have the same inertia. Therefore, $\mathrm{In}(M) = \mathrm{In}(C) + \mathrm{In}(A - BC^{-1}B^T)$. ∎

We conclude with a lemma that will be needed later.

Lemma 2.4.5. *Let M be a symmetric $n \times n$ matrix and let U be a subspace of $\mathbb{R}^n$ such that $x^T Mx \leq 0$ holds for all $x \in U$. If U has dimension $n - 1$, then M has at most one positive eigenvalue.*

Proof. Suppose, for contradiction, that M has two positive eigenvalues λ_1 and λ_2. Let u_1 and u_2 be eigenvectors for λ_1 and λ_2, respectively, with $u_1^T u_2 = 0$ and $\| u_1 \|_2 = \| u_2 \|_2 = 1$. Let V denote the subspace of $\mathbb{R}^n$ spanned by u_1 and u_2. Then, $x^T Mx > 0$ holds for all $x \in V$, $x \neq 0$; indeed, if $x = a_1 u_1 + a_2 u_2$, then $x^T Mx = a_1^2 \lambda_1 + a_2^2 \lambda_2 > 0$ if $(a_1, a_2) \neq (0, 0)$. As U and V have respective dimensions $n - 1$ and 2, there exists $x \in U \cap V$ with $x \neq 0$. Then, $x^T Mx \leq 0$ since $x \in U$ and $x^T Mx > 0$ since $x \in V$, yielding a contradiction. ∎

Part I

Measure Aspects: ℓ_1-Embeddability and Probability

Introduction

A classical problem in analysis is the isometric embedding problem, which consists of asking whether a given distance space (X, d) can be isometrically embedded into some prescribed important "host" space. The case which has been most extensively studied in the literature is when the host space is the Hilbert space. Work in this area was initiated by some results of Cayley [1841] and later continued, in particular, by Menger [1928, 1931, 1954], Schoenberg [1935, 1937, 1938a, 1938b] and Blumenthal [1953]. One of Schoenberg's well-known results asserts that a distance space (X, d) is isometrically L_2-embeddable if and only if the squared distance d^2 satisfies a list of linear inequalities (the so-called negative type inequalities). These results were later extended in the context of L_p-spaces. Of particular importance for our purpose is a result by Bretagnolle, Dacunha Castelle and Krivine [1966], which states that (X, d) is isometrically L_p-embeddable if and only if the same holds for every *finite* subspace of (X, d).

Our objective here is to study the isometric embedding problem in the case when the host space is the Banach L_1-space. Our approach is mainly combinatorial. The case $p = 1$ turns out to be specific among the various L_p-spaces. Indeed, by the above mentioned finiteness result, we may restrict ourselves to finite distance spaces. Now, one of the basic facts about L_1 is that a finite distance space (X, d) is isometrically L_1-embeddable if and only if d belongs to the so-called cut cone, the cone generated by all cut semimetrics. This cone is finitely generated. Therefore, L_1-embeddability of (X, d) can be characterized by a finite list of linear inequalities.

However, in contrast with the L_2-case, not all inequalities of this list are known, in general. In fact, in view of the complexity status of the related max-cut problem, it is quite unlikely that this list can be completely described. It is one of the objective of this book to survey what is known about this list.

Therefore, the problem of characterizing L_1-embeddability reduces to the purely combinatorial problem of describing the linear structure of the cut cone. Interest in this cone is also motivated by its relevance to several hard combinatorial optimization problems.

Part I is organized as follows. Chapter 3 contains some preliminaries about distance spaces. We present the basic connection existing between the cut cone and L_1-metrics in Chapter 4 and, in an equivalent form in the context of covariances, in Chapter 5. Two of the main known necessary conditions for L_1-embeddability, namely, the hypermetric and negative type conditions, are presented in Chapter 6. Several characterizations for L_2-metrics are discussed there, including Schoenberg's result in terms of the negative type condition and Menger's compactness result. Chapters 7 and 9 describe several operations on L_1-metrics. In Chapter 8 we investigate the metric spaces arising from normed vector spaces, poset lattices, and semigroups; in all cases, a characterization of L_1-embeddability is given in terms of the hypermetric and negative type conditions. Lipschitz embeddings with some distortion are treated in Chapter 10, together with an application to the approximation of multicommodity flows in graphs. Finally, we group in Chapter 11 several additional results related to ℓ_1-embeddings in a space of fixed dimension and to the minimum ℓ_p-dimension and, in Chapter 12, we attempt to illustrate the wide range of use of the L_1-metric.

Chapter 3. Preliminaries on Distances

We introduce here all the notions and definitions that we need in Part I, in particular, about distance spaces, isometric embeddings, measure spaces, and our main host spaces: the Banach ℓ_p- and L_p-spaces for $1 \leq p \leq \infty$. The book will deal, in fact, essentially with the case $p = 1$, also with the Euclidean case $p = 2$ and the semimetric case $p = \infty$. We will consider the general case $p \geq 1$ only episodically. However, we now introduce the definitions for p arbitrary, as this permits us to have a unified setting for the various parameters and, moreover, to emphasize the specificity of the case $p = 1$.

3.1 Distance Spaces and ℓ_p-Spaces

Distances and Semimetrics: Definitions and Examples. Let X be a set. A function $d : X \times X \to \mathbb{R}_+$ is called a *distance* on X if d is *symmetric*, i.e., satisfies $d(i,j) = d(j,i)$ for all $i,j \in X$, and if $d(i,i) = 0$ holds for all $i \in X$. Then, (X,d) is called a *distance space.* If d satisfies, in addition, the following inequalities:

$$(3.1.1) \qquad d(i,j) \leq d(i,k) + d(j,k)$$

for all $i,j,k \in X$, then d is called a *semimetric* on X. Moreover, if $d(i,j) = 0$ holds only for $i = j$, then d is called a *metric* on X. The inequality (3.1.1) is called a *triangle inequality.*

Set $V_n := \{1,\ldots,n\}$ and $E_n := \{ij \mid i,j \in V_n, i \neq j\}$, where the symbol ij denotes the unordered pair of the integers i,j, i.e., ij and ji are considered identical. Let d be a distance on the set V_n. Because of symmetry and since $d(i,i) = 0$ for $i \in V_n$, we can view the distance d as a vector $(d_{ij})_{1 \leq i < j \leq n} \in \mathbb{R}^{E_n}$. Conversely, every vector $d \in \mathbb{R}^{E_n}$ yields a distance d on V_n by symmetry and by taking distance 0 on the diagonal pairs. Hence, a distance on V_n can be viewed alternatively as a (symmetric with zero on the diagonal) function on $V_n \times V_n$ or as a vector in $\mathbb{R}^{E_n}$. We will use freely these two representations for a distance on V_n; the distinction should be clear from the context. Moreover, we will use both symbols $d(i,j)$ and d_{ij} for denoting the distance between two points i and j.

Let d be a distance on the set V_n. Its *distance matrix* is defined as the $n \times n$ symmetric matrix D whose (i,j)-th entry is $d(i,j)$ for all $i,j \in V_n$. Hence, all diagonal entries of D are equal to zero.

The triangle inequalities (3.1.1) (for $i, j, k \in V_n$) define a cone in the space $\mathbb{R}^{E_n}$, called the *semimetric cone* and denoted by MET_n; its elements are precisely the semimetrics on V_n.

Very simple examples of semimetrics can be constructed in the following way. Given an integer $t \geq 1$, let $t\mathbb{1}_n$ denote[1] the *equidistant metric* on V_n, that takes value t on every pair in E_n; $t\mathbb{1}_n$ is obviously a metric on V_n.

Given a subset $S \subseteq V_n$, let $\delta(S)$ denote the distance on V_n that takes value 1 on the pairs (i, j) with $i \in S$, $j \in V_n \setminus S$, and value 0 on the remaining pairs. Clearly, $\delta(S)$ is a semimetric (not a metric if $n \geq 3$); it is called a *cut semimetric*[2].

Given a normed[3] space $(E, \| \,.\, \|)$, a metric $d_{\|.\|}$ can be defined on E by setting

$$d_{\|.\|}(x, y) := \| x - y \|$$

for all $x, y \in E$; the metric $d_{\|.\|}$ is called a *norm metric* or *Minkowski metric.*

For any $p \geq 1$, the vector space $\mathbb{R}^m$ can be endowed with the ℓ_p-norm $\| \,.\, \|_p$ defined by

$$\| x \|_p = \Big(\sum_{1 \leq k \leq m} |x_k|^p \Big)^{\frac{1}{p}}$$

for $x \in \mathbb{R}^m$. Then, the associated norm metric is denoted by d_{ℓ_p} and is called the *ℓ_p-metric.* Thus, $d_{\ell_p}(x, y) = \| x - y \|_p$ for $x, y \in \mathbb{R}^m$. The metric space $(\mathbb{R}^m, d_{\ell_p})$ is abbreviated as ℓ_p^m. Similarly, ℓ_∞^m denotes the metric space $(\mathbb{R}^m, d_{\ell_\infty})$, where d_{ℓ_∞} denotes the norm metric associated with the norm $\| \,.\, \|_\infty$ which is defined by

$$\| x \|_\infty = \max(|x_k| : 1 \leq k \leq m),$$

for $x \in \mathbb{R}^m$.

For $1 \leq p < \infty$, the metric space ℓ_p^∞ consists of the set of infinite sequences $x = (x_i)_{i \geq 0} \in \mathbb{R}^{\mathbb{N}}$ for which the sum $\sum_{i \geq 0} |x_i|^p$ is finite, endowed with the distance $d(x, y) = \left(\sum_{i \geq 0} |x_i - y_i|^p \right)^{\frac{1}{p}}$. In the same way ℓ_∞^∞ is the set of bounded infinite sequences $x \in \mathbb{R}^{\mathbb{N}}$, endowed with the distance

$$d(x, y) = \max(|x_i - y_i| : i \geq 0).$$

Another classical distance on the set $\mathbb{R}^m$ is the *Hamming distance* d_H defined by

$$d_H(x, y) := |\{i \in [1, m] : x_i \neq y_i\}|$$

[1]Thus, in notation $t\mathbb{1}_n$, the letter n refers to the number of points on which the metric is defined and not to the dimensionality of the space containing $t\mathbb{1}_n$.

[2]The cut semimetrics are also called in the literature split metrics, or dissimilarities, or binary metrics (e.g., in Bandelt and Dress [1992], Fichet [1987a], Le Calve [1987]).

[3]We remind that a *norm* on a vector space E is a function $x \in E \mapsto \| x \| \in \mathbb{R}_+$ such that $\| x \| = 0$ if and only if $x = 0$, $\| \lambda x \| = |\lambda| \, \| x \|$ for $\lambda \in \mathbb{R}$, $x \in E$, and $\| x + y \| \leq \| x \| + \| y \|$ for $x, y \in E$.

for all $x, y \in \mathbb{R}^m$. Hence, when computed on binary vectors $x, y \in \{0,1\}^m$, the Hamming distance $d_H(x,y)$ and the ℓ_1-distance $d_{\ell_1}(x,y)$ coincide.

Other examples of metric spaces are the graphic metric spaces, that arise from connected graphs. Let $G = (V, E)$ be a connected graph and let d_G (or $d(G)$) denote the *path metric* of G where, for two nodes $i, j \in V$, $d_G(i,j)$ denotes the shortest length of a path from i to j in G. Then, (V, d_G) is called the *graphic metric space* associated with G. The graphic metric space associated with a hypercube graph $H(m, 2)$ is called a *hypercube metric space.* Observe that the graphic metric space of $H(m,2)$ coincides with the distance space $(\{0,1\}^m, d_{\ell_1})$.

If $G = (V, E)$ is a graph and $w = (w_e)_{e \in E}$ are nonnegative weights assigned to its edges, one can define similarly the path metric[4] $d_{G,w}$ of the weighted graph (G, w). Namely, for two nodes $i, j \in V$, $d_{G,w}(i,j)$ denotes the smallest value of $\sum_{e \in P} w_e$ where P is a path from i to j in G.

Isometric Embeddings. Let (X, d) and (X', d') be two distance spaces. Then, (X, d) is said to be *isometrically embeddable* into (X', d') if there exists a mapping ϕ (the *isometric embedding*) from X to X' such that

$$d(x,y) = d'(\phi(x), \phi(y))$$

for all $x, y \in X$. One also says that (X, d) is an *isometric subspace* of (X', d'). All the embeddings considered here are isometric, so we will often omit to mention the word "isometric".

For two graphs G and H, one writes $G \hookrightarrow H$ and says that G is an *isometric subgraph* of H when $(V(G), d_G)$ is an isometric subspace of $(V(H), d_H)$.

Clearly, if (X, d) is embeddable into (X', d') then the same holds for every subspace (Y, d), where $Y \subseteq X$. It may sometimes be sufficient to check all subspaces (Y, d) of (X, d) on a limited number of points in order to ensure embeddability of the whole space (X, d) into (X', d'). The smallest number of points that are enough to check is called (following Blumenthal [1953]) the order of congruence of (X', d'). More precisely, (X', d') is said to have *order of congruence* p if, for every distance space (X, d),

$$\begin{array}{c} (Y,d) \text{ embeds into } (X',d') \text{ for every } Y \subseteq X \text{ with } |Y| \le p \\ \Downarrow \\ (X,d) \text{ embeds into } (X',d') \end{array}$$

and p is the smallest such integer (possibly infinite).

A distance space (X, d) is said to be *ℓ_p-embeddable* if (X, d) is isometrically embeddable into the space ℓ_p^m for some integer $m \ge 1$. The smallest such integer m is called the *ℓ_p-dimension* of (X, d) and is denoted by $m_{\ell_p}(X, d)$. Then, we define the *minimum ℓ_p-dimension* $m_{\ell_p}(n)$ by

[4]It would be more correct to speak of 'path semimetric' rather than 'path metric', as some distinct points might be at distance zero if some of the edge weights are equal to zero. We will however keep the terminology of 'path metric' for simplicity.

(3.1.2) $$m_{\ell_p}(n) := \max(m_{\ell_p}(X,d) : |X| = n \text{ and } (X,d) \text{ is } \ell_p\text{-embeddable});$$

that is, $m_{\ell_p}(n)$ is the smallest integer m such that every ℓ_p-embeddable distance on n points can be embedded in ℓ_p^m. (It is known that $m_{\ell_p}(n)$ is finite; in fact, $m_{\ell_p}(n) \leq \binom{n}{2}$ for all n and p. Cf. Section 11.2.) The distance space (X,d) is said to be ℓ_p^∞*-embeddable* if it is an isometric subspace of ℓ_p^∞.

A distance space (X,d) is said to be *hypercube embeddable* if it can be isometrically embedded in some hypercube metric space $(\{0,1\}^m, d_{\ell_1})$ for some integer $m \geq 1$. As the hypercube metric space $(\{0,1\}^m, d_{\ell_1})$ is an isometric subspace of ℓ_1^m, every hypercube embeddable distance space is ℓ_1-embeddable. (In fact, if d is rational valued, then the space (X,d) is ℓ_1-embeddable if and only if $(X,\lambda d)$ is hypercube embeddable for some integer λ; see Proposition 4.3.8.)

Basic Observations on ℓ_p-Metrics. Obviously, if a distance d is ℓ_p-embeddable then d is a semimetric and, moreover, αd is ℓ_p-embeddable for any $\alpha > 0$. On the other hand, the sum $d_1 + d_2$ of two ℓ_p-embeddable distances d_1 and d_2 is, in general, not ℓ_p-embeddable. In other words, the set of ℓ_p-embeddable distances on V_n is not a cone in general. However, the two extreme cases $p = 1$ and $p = \infty$ are exceptional. In each of these two cases the set of ℓ_p-embeddable distances on V_n forms a cone that, in addition, is polyhedral.

The case $p = 1$ is directly relevant to the central topic of this book. Indeed, the distances on n points that are ℓ_1-embeddable are precisely the distances that can be decomposed as a nonnegative linear combination of cut semimetrics. Hence, they form a cone, the cut cone CUT_n, which is a polyhedral cone as it is generated by the 2^{n-1} cut semimetrics. (Details are given in Section 4.1.)

Consider now the case $p = \infty$. Clearly every ℓ_∞-embeddable distance is a semimetric. Conversely, every semimetric on V_n is ℓ_∞^{n-1}-embeddable. Indeed, the following n vectors $v_1, \ldots, v_n \in \mathbb{R}^{n-1}$, where

$$v_i := (d(1,i), d(2,i), \ldots, d(n-1,i)) \text{ for } i = 1, \ldots, n,$$

provide an isometric embedding of (V_n, d) into $(\mathbb{R}^{n-1}, d_{\ell_\infty})$. Therefore,

$$\text{MET}_n = \{d \in \mathbb{R}^{E_n} \mid d \text{ is } \ell_\infty\text{-embeddable}\}.$$

Hence, the ℓ_∞-embeddable distances on V_n form a cone, the semimetric cone MET_n. This cone is a polyhedral cone as MET_n is defined by finitely many linear inequalities (namely, the $3\binom{n}{3}$ triangle inequalities). Moreover, the minimum ℓ_∞-dimension satisfies:

(3.1.3) $$m_{\ell_\infty}(n) \leq n - 1.$$

Although the ℓ_p-embeddable distances on V_n do not form a cone if $1 < p < \infty$, the p-th powers of these distances do form a cone. For $1 \leq p < \infty$, set

(3.1.4) $$\mathrm{NOR}_n(p) := \{d \in \mathbb{R}^{E_n} \mid \sqrt[p]{d} \text{ is } \ell_p\text{-embeddable}\},$$

where $\sqrt[p]{d}$ denotes the vector $\left(\sqrt[p]{d_{ij}}\right)_{ij \in E_n}$. Then, the set $\mathrm{NOR}_n(p)$ is a cone which is not polyhedral if $1 < p < \infty$; moreover, every cut semimetric lies on an extreme ray of $\mathrm{NOR}_n(p)$ (see Lemma 11.2.2). Note that the cones $\mathrm{NOR}_n(1)$ and CUT_n coincide. The cone $\mathrm{NOR}_n(2)$ (which corresponds to the Euclidean distance) has also been extensively investigated; results are grouped in Section 6.2. We want to point out that, although $\mathrm{NOR}_n(1) = \mathrm{CUT}_n$ appears to be much nicer than $\mathrm{NOR}_n(2)$ since it is a polyhedral cone with very simple extreme rays, we do not know an inequality description of $\mathrm{NOR}_n(1)$. In fact, the investigation of the facial structure of the cone $\mathrm{NOR}_n(1) = \mathrm{CUT}_n$ will form an important part of this book, taken up especially in Part V.

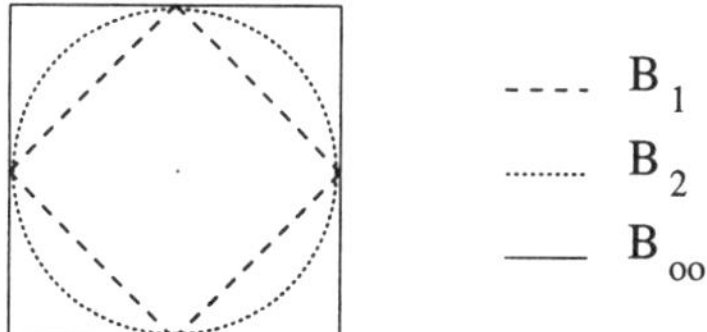

Figure 3.1.5: The unit balls for the ℓ_1, ℓ_2 and ℓ_∞-norms

To conclude let us compare the unit balls of the various ℓ_p-spaces. Let B_p denote the unit ball in ℓ_p^n, defined by

$$B_p := \{x \in \mathbb{R}^n : \| x \|_p \leq 1\}.$$

Then, $B_\infty = [-1,1]^n$ is the n-dimensional hypercube γ_n (with side length 2), B_1 coincides with the n-dimensional cross-polytope β_n, and B_2 is the usual Euclidean unit ball. The following inclusions hold:

$$B_1 \subseteq B_p \subseteq B_\infty \quad \text{for } 1 \leq p \leq \infty,$$

which follow from the well-known Jensen's inequality (see, e.g., Section 2.10 in Hardy, Littlewood and Pólya [1934]):

$$\| x \|_q \leq \| x \|_p \quad \text{for all } x \in \mathbb{R}^n,\ 1 \leq p < q \leq \infty.$$

Figure 3.1.5 shows the three balls B_1, B_2, and B_∞ in dimension $n = 2$. Note that the balls B_1 and B_∞ are in bijection via the mapping $f : \mathbb{R}^2 \longrightarrow \mathbb{R}^2$ defined by

(3.1.6) $$x = (x_1, x_2) \mapsto f(x) := \left(\frac{x_1 - x_2}{2}, \frac{x_1 + x_2}{2}\right)$$

(which rotates the plane by 45 degrees and then shrinks it by a factor $\frac{1}{\sqrt{2}}$). Indeed, one can verify that $\| x \|_\infty = \| f(x) \|_1$ for all $x \in \mathbb{R}^2$. Therefore, the mapping f provides an isometry between the distance spaces $(\mathbb{R}^2, d_{\ell_\infty})$ and $(\mathbb{R}^2, d_{\ell_1})$.

3.2 Measure Spaces and L_p-Spaces

We now define the distance space $L_p(\Omega, \mathcal{A}, \mu)$, which is attached to any measure space $(\Omega, \mathcal{A}, \mu)$. For this, we recall some definitions[5] on measure spaces. Let Ω be a set and let $\mathcal{A}$ be a *σ-algebra* of subsets of Ω, i.e., $\mathcal{A}$ is a collection of subsets of Ω satisfying the following properties:

$$\begin{cases} \Omega \in \mathcal{A}, \\ \text{if } A \in \mathcal{A} \text{ then } \Omega \setminus A \in \mathcal{A}, \\ \text{if } A = \bigcup_{k=1}^{\infty} A_k \text{ with } A_k \in \mathcal{A} \text{ for all } k, \text{ then } A \in \mathcal{A}. \end{cases}$$

A function $\mu : \mathcal{A} \longrightarrow \mathbb{R}_+$ is a *measure* on $\mathcal{A}$ if it is additive, i.e.,

$$\mu(\bigcup_{k\geq 1} A_k) = \sum_{k\geq 1} \mu(A_k)$$

for all pairwise disjoint sets $A_k \in \mathcal{A}$, and satisfies $\mu(\emptyset) = 0$. Note that measures are always assumed here to be nonnegative. A *measure space* is a triple $(\Omega, \mathcal{A}, \mu)$ consisting of a set Ω, a σ-algebra $\mathcal{A}$ of subsets of Ω, and a measure μ on $\mathcal{A}$. A *probability space* is a measure space with total measure $\mu(\Omega) = 1$.

Given a function $f : \Omega \longrightarrow \mathbb{R}$, its L_p*-norm* is defined by

$$\| f \|_p := \left(\int_\Omega |f(\omega)|^p \mu(d\omega) \right)^{\frac{1}{p}}.$$

Then, $L_p(\Omega, \mathcal{A}, \mu)$ denotes the set of functions $f : \Omega \longrightarrow \mathbb{R}$ which satisfy $\| f \|_p < \infty$. The L_p-norm defines a metric structure on $L_p(\Omega, \mathcal{A}, \mu)$, namely, by taking $\| f - g \|_p$ as distance between two functions $f, g \in L_p(\Omega, \mathcal{A}, \mu)$. A distance space (X, d) is said to be L_p*-embeddable* if it is a subspace of $L_p(\Omega, \mathcal{A}, \mu)$ for some measure space $(\Omega, \mathcal{A}, \mu)$.

The most classical example of an L_p-space is the space $L_p(\Omega, \mathcal{A}, \mu)$, where Ω is the open interval $(0, 1)$, $\mathcal{A}$ is the family of Borel subsets of $(0, 1)$, and μ is the Lebesgue measure; it is simply denoted by $L_p(0, 1)$. We now make precise the connections existing between L_p-spaces and ℓ_p-spaces:

(i) If $\Omega = \mathbb{N}$, $\mathcal{A} = 2^\Omega$ is the collection of all subsets of Ω, and μ is the *cardinality measure*, i.e., $\mu(A) = |A|$ if A is a finite subset of Ω and $\mu(A) = \infty$ otherwise, then $L_p(\mathbb{N}, 2^{\mathbb{N}}, |.|)$ coincides with the space ℓ_p^∞.

(ii) If $\Omega = V_m$ is a set of cardinality m, $\mathcal{A} = 2^\Omega$, and μ is the cardinality measure, then $L_p(V_m, 2^{V_m}, |.|)$ coincides with ℓ_p^m.

In other words, ℓ_p^m is an isometric subspace of ℓ_p^∞ which, in turn, is L_p-embeddable.

We consider in detail L_1-embeddable distance spaces in Chapter 4. It turns out that, for a finite distance space, the properties of being ℓ_1-, ℓ_1^∞, or L_1-embeddable are all equivalent to the property of belonging to the cut cone (see Theorem 4.2.6).

[5]For more information concerning measure spaces and integrability of functions, the reader may consult the textbook by Rudin [1966].

Similar results are known for the case $p \geq 1$ (see, e.g., Fichet [1994]). Namely, for a finite distance space (X, d), the properties of being ℓ_p-, ℓ_p^∞-, or L_p-embeddable are all equivalent.

Though we are mainly concerned with finite distance spaces, i.e., with distance spaces (X, d) where X is finite, we also present a number of results involving infinite distance spaces. For instance, we consider in Section 8.3 the normed vector spaces whose norm metric is L_1-embeddable. However, the following fundamental result of Bretagnolle, Dacunha Castelle and Krivine [1966] shows that the study of L_p-embeddable spaces can be reduced to the finite case.

Theorem 3.2.1. *Let $p \geq 1$ and let (X, d) be a distance space. Then, (X, d) is L_p-embeddable if and only if every finite subspace of (X, d) is L_p-embeddable.* ∎

Similarly, the study of ℓ_p^m-embeddable spaces can be reduced to the finite case. The next result follows from the compactness theorem of logic (as observed by Malitz and Malitz [1992] in the case $p = 1$).

Theorem 3.2.2. *Let $p, m \geq 1$ be integers and let (X, d) be a distance space. Then, (X, d) is ℓ_p^m-embeddable if and only if every finite subspace of (X, d) is ℓ_p^m-embeddable.*

Proof. Necessity is obvious. Conversely, suppose that every finite subspace of (X, d) is ℓ_p^m-embeddable. Fixing $x_0 \in X$ we can restrict ourselves to finding an embedding of (X, d) in which x_0 is mapped to the zero vector. Hence, we search for an element $(u_x)_{x \in X}$ of the set $K := \prod_{x \in X} [-d(x_0, x), d(x_0, x)]^m$ such that $\| u_x - u_y \|_p = d(x, y)$ for all $x, y \in X$. For $x, y \in X$, let $K_{x,y}$ denote the subset of K consisting of the elements that satisfy the condition $\| u_x - u_y \|_p = d(x, y)$. By assumption, any intersection of a finite number of $K_{x,y}$'s is nonempty. Therefore, since K is compact (by Tychonoff's theorem, as it is a Cartesian product of compact sets) and since the $K_{x,y}$'s are closed sets, the intersection $\bigcap_{x,y \in X} K_{x,y}$ is nonempty, which shows that (X, d) is ℓ_p^m-embeddable. ∎

Finally, we introduce one more semimetric space. Let $(\Omega, \mathcal{A}, \mu)$ be a measure space. Set

$$\mathcal{A}_\mu := \{A \in \mathcal{A} \mid \mu(A) < \infty\}.$$

One can define a distance d_μ on $\mathcal{A}_\mu$ by setting

$$d_\mu(A, B) = \mu(A \triangle B)$$

for all $A, B \in \mathcal{A}_\mu$. Then, d_μ is a semimetric on $\mathcal{A}_\mu$. We call d_μ a *measure semimetric* and the space $(\mathcal{A}_\mu, d_\mu)$ a *measure semimetric space*. The semimetric d_μ is also called the Fréchet-Nikodym-Aronszajn distance in the literature. We will consider in Section 9.2 the related Steinhaus distance, which is defined by

$$\frac{\mu(A\triangle B)}{\mu(A\cap B)}$$

for $A, B \in \mathcal{A}_\mu$. Note that the measure semimetric space $(\mathcal{A}_\mu, d_\mu)$ is the subspace of $L_1(\Omega, \mathcal{A}, \mu)$ consisting of its 0-1 valued functions. Moreover, if $\Omega = V_m$ is a finite set of cardinality m, $\mathcal{A} = 2^\Omega$, and μ is the cardinality measure, then the space $(\mathcal{A}_\mu, d_\mu)$ coincides with the hypercube metric space $(\{0,1\}^m, d_{\ell_1})$.

We close the section with two remarks grouping a number of results about the isometric embeddings existing among the various L_p-spaces. Figure 3.2.3 summarizes some of the connections existing among ℓ_p-spaces and mentioned in Remark 3.2.4 below.

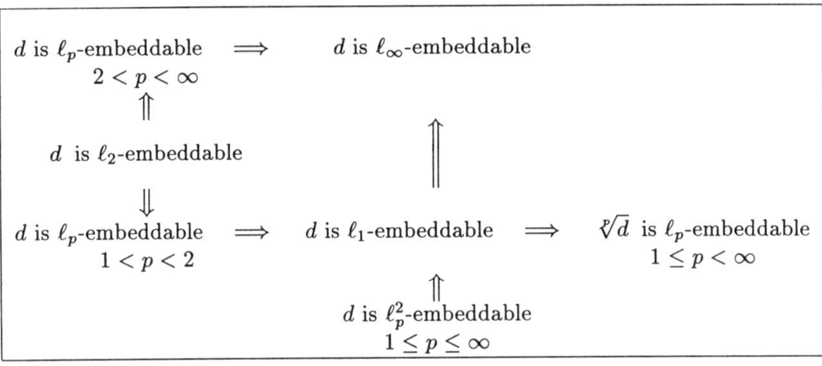

Figure 3.2.3: Links between various ℓ_p-spaces

Remark 3.2.4. Isometric embeddings among the L_p-spaces. There is a vast literature on the topic of isometric embeddings among the various L_p-spaces; see, e.g., Wells and Williams [1975], Dor [1976], Ball [1987] , Lyubich and Vaserstein [1993]. We summarize here some of the main results. The following is shown in Dor [1976]. Let $1 \le p < \infty$, $1 \le r \le \infty$ and $m \in \mathbb{N}$, $m \ge 2$. Then, ℓ_r^m is an isometric subspace of $L_p(0,1)$ if and only if, either (*i*) $p \le r < 2$, or (*ii*) $r = 2$, or (*iii*) $m = 2$ and $p = 1$. Hence, for instance, ℓ_r^3 does not embed isometrically in $L_p(0,1)$ if $r > 2$. It was already shown by Bretagnolle, Dacunha Castelle and Krivine [1966] that $L_p(0,1)$ embeds isometrically in $L_1(0,1)$ for all $1 \le p \le 2$. As a reformulation of the above result, we have the following implications for a distance space (X, d):

- If (X,d) is ℓ_p^2-embeddable for some $1 \le p \le \infty$, then (X,d) is L_1-embeddable.
- If (X,d) is ℓ_p-embeddable for some $1 \le p \le 2$, then (X,d) is L_1-embeddable.
- If (X,d) is ℓ_2-embeddable, then (X,d) is L_p-embeddable for all $1 \le p \le \infty$.

Let $r \neq p$ such that $1 \le r, p < \infty$ and let $m \ge 1$ be an integer. Then, ℓ_r^m embeds isometrically in ℓ_p^n for some integer $n \ge 1$ if and only if $r = 2$ and p is an even integer (Lyubich and Vaserstein [1993]). Given an even integer p, define $N(m,p)$ as the smallest integer $n \ge 1$ for which ℓ_2^m embeds isometrically into ℓ_p^n. It is shown in Lyubich and Vaserstein [1993] that $N(2,p) = \frac{p}{2} + 1$ and that, for any $p \ge 2$ and $m \ge 1$, $\max(N(m-1,p), N(m,p-2)) \le N(m,p) \le \binom{m+p-1}{m-1}$. An exact evaluation of $N(m,p)$ is known for small values of p, m; for instance, $N(3,4) = 6, N(3,6) = 11, N(3,8) = 16, N(7,4) = 28, N(8,6) = 120, N(23,4) = 276, N(23,6) = 2300, N(24,10) = 98280$.

Therefore, given r and $m \in \mathbb{N}$ such that $1 < r \le 2 < m$, we have that ℓ_r^m does not embed isometrically into ℓ_1^n (n positive integer), but ℓ_r^m embeds into $L_1(0,1)$ and, moreover, every finite subspace of ℓ_r^m on s points embeds into $\ell_1^{\binom{s}{2}}$. ∎

Remark 3.2.5. We mention here some observations made by Fichet [1994]. First, it can be easily verified that every semimetric on 3 points embeds isometrically in ℓ_p^2 for any $1 \le p \le \infty$. Moreover, every semimetric on 4 points embeds isometrically in ℓ_1^2. On the other hand, there exist semimetrics on 4 points that are not ℓ_p-embeddable for any $1 < p < \infty$. For such an example consider the distance d on $X := \{1,2,3,4\}$ taking value 2 on the pairs $(1,3)$ and $(2,4)$ and value 1 on all other pairs.

One can verify that d is not ℓ_p-embeddable for $1 < p < \infty$. For, suppose that there exist vectors $u_1, u_2, u_3, u_4 \in \mathbb{R}^N$ providing an ℓ_p-embedding of d. For each coordinate $m \in [1, N]$ consider the distance d^m on X defined by $d_{ij}^m := |u_i(m) - u_j(m)|$ for $i, j \in X$. We now exploit the fact that d satisfies several triangle inequalities at equality. It can be easily observed (using Minkowski's inequality) that every d^m satisfies the same triangle equalities as d. From this follows that, for each coordinate m, either (i) $u_1(m) = u_2(m) \le u_3(m) = u_4(m)$, or (ii) $u_2(m) = u_3(m) \le u_1(m) = u_4(m)$. Denoting by N_1 (resp. N_2) the set of coordinates m for which (i) (resp. (ii)) occurs and setting $a := \sum_{m \in N_1} |u_1(m) - u_3(m)|^p$, $b := \sum_{m \in N_2} |u_1(m) - u_2(m)|^p$, we obtain that $a = b = 1$ and $2^p = a + b$, a contradiction if $p > 1$. ∎

Chapter 4. The Cut Cone and ℓ_1-Metrics

In this chapter, we establish the basic connections existing between the cut cone and metrics. We show how the members of the cut cone can be interpreted in terms of ℓ_1-metrics and measure semimetric spaces; see in particular Theorem 4.2.6, where several equivalent characterizations are stated. We also introduce in Section 4.3 several basic notions which will be used throughout the book, namely, the notions of rigidity, size, scale and realizations. In Section 4.4, we present a number of basic problems about cut polyhedra and indicate their complexity status.

4.1 The Cut Cone and Polytope

We give here the exact definitions for the cut cone and polytope. We start with recalling the notion of cut semimetric. Given a subset S of $V_n = \{1, \ldots, n\}$, let $\delta(S)$ denote the vector of $\mathbb{R}^{E_n}$ defined by

(4.1.1) $$\delta(S)_{ij} = 1 \text{ if } |S \cap \{i,j\}| = 1, \text{ and } \delta(S)_{ij} = 0 \text{ otherwise,}$$

for $1 \leq i < j \leq n$. Obviously, $\delta(S)$ defines a distance on V_n which is a semimetric; for this reason, $\delta(S)$ is called a *cut semimetric*. Note that, in later chapters, we may drop the word "semimetric" and simply speak of "the cut $\delta(S)$" or of "the cut vector $\delta(S)$". We use here the word "semimetric" in order to stress the fact that we are working with distance spaces.

The cone in $\mathbb{R}^{E_n}$, which is generated by the cut semimetrics $\delta(S)$ for $S \subseteq V_n$, is called the *cut cone*[1] and is denoted by CUT_n. The polytope in $\mathbb{R}^{E_n}$, which is defined as the convex hull of the cut semimetrics $\delta(S)$ for $S \subseteq V_n$, is called the *cut polytope* and is denoted by $\mathrm{CUT}_n^\square$. Hence,

(4.1.2) $$\mathrm{CUT}_n = \{ \sum_{S \subseteq V_n} \lambda_S \delta(S) \mid \lambda_S \geq 0 \text{ for all } S \subseteq V_n \},$$

(4.1.3) $$\mathrm{CUT}_n^\square = \{ \sum_{S \subseteq V_n} \lambda_S \delta(S) \mid \sum_{S \subseteq V_n} \lambda_S = 1 \text{ and } \lambda_S \geq 0 \text{ for all } S \subseteq V_n \}.$$

It may sometimes be convenient to consider an arbitrary finite set X instead on

[1] The cone CUT_n is also known in the literature under the name of *Hamming cone*; see, e.g., Avis [1977, 1981], Assouad and Deza [1982], etc.

V_n. One defines in the same way[2] the cut cone $\mathrm{CUT}(X)$ and the cut polytope $\mathrm{CUT}^\square(X)$ on X. Hence,

$$\mathrm{CUT}(V_n) = \mathrm{CUT}_n \text{ and } \mathrm{CUT}^\square(V_n) = \mathrm{CUT}_n^\square.$$

Note that CUT_n and $\mathrm{CUT}_n^\square$ are trivial for $n = 2$, as there is only one nonzero cut semimetric on V_2. From now on, we will suppose that $n \geq 3$.

There is a vast literature on the cut polyhedra; references will be given throughout the text when appropriate. Let us mention a few areas where these polyhedra arise. The cut polytope comes up in connection with the following basic geometric question: What are the linear inequalities that are satisfied by the pairwise angles of a set of vectors ? In fact, the inequalities valid for the cut polytope $\mathrm{CUT}_n^\square$ yield valid relations for the pairwise angles among a set of n unit vectors (see Sections 6.4 and 31.2). As will be explained in the next section, the cut cone arises naturally in the context of ℓ_1-metrics. On the other hand, the cut polytope also plays an important role in combinatorial optimization, as it permits to formulate the max-cut problem. Given weights $w = (w_{ij}) \in \mathbb{R}^{E_n}$ associated with the pairs $ij \in E_n$, the *max-cut problem* consists of finding a cut $\delta(S)$ whose weight $w^T\delta(S)$ is as large as possible. Hence, it is the problem:

$$\text{(4.1.4)} \qquad \begin{array}{ll} \max & w^T\delta(S) \\ \text{s.t.} & S \subseteq V_n, \end{array}$$

which can be reformulated as an optimization problem over the cut polytope:

$$\begin{array}{ll} \max & w^T x \\ \text{s.t.} & x \in \mathrm{CUT}_n^\square. \end{array}$$

This is a hard problem, for which no algorithm is known that runs in a polynomial number of steps (polynomial in n and the size of w); see Section 4.4. The max-cut problem has many applications in various fields. For instance, the problem of determining ground states of spin glasses in statistical physics, or the problem of minimizing the number of vias subject to pin assignment and layer preferences in VLSI circuit design, can both be formulated as instances of the max-cut problem. These two applications are treated in detail in Barahona, Grötschel, Jünger and Reinelt [1988]. We expose the application to spin glasses in Section 4.5. Other applications and connections are mentioned in Section 5.1, within the framework of unconstrained boolean quadratic programming. The reader may consult the survey paper by Poljak and Tuza [1995] for more information about max-cut and the annotated bibliography by Laurent [1997a] for detailed references.

[2] We will use later (in Sections 25.3 and 27.3) the notation $\mathrm{CUT}(G)$ and $\mathrm{CUT}^\square(G)$ for denoting the projections of CUT_n and $\mathrm{CUT}_n^\square$ on the subspace indexed by the edge set of a graph G. It should not be confused with the notation $\mathrm{CUT}(X)$ and $\mathrm{CUT}^\square(X)$ which stands for the cut polyhedra of the complete graph with node set X.

4.2 ℓ_1-Spaces

Every member d of the cut cone CUT_n defines a semimetric on n points, as it satisfies all the triangle inequalities (3.1.1). Hence a natural question is the characterization of the semimetrics that belong to the cut cone. The main result is that a semimetric belongs to the cut cone if and only if it is isometrically ℓ_1-embeddable, a result first found in Assouad [1980b]. Several other equivalent characterizations are stated in Theorem 4.2.6. The results[3] in this section are essentially taken from Assouad [1980b] and from Assouad and Deza [1982]. We start with several intermediate results.

Proposition 4.2.1. *Let $d = (d_{ij})_{1 \le i < j \le n} \in \mathbb{R}^{E_n}$. The following assertions are equivalent.*

(i) $d \in \mathrm{CUT}_n$ *(resp.* $d \in \mathrm{CUT}_n^{\square}$*).*

(ii) *There exist a measure space (resp. a probability space) $(\Omega, \mathcal{A}, \mu)$ and events $A_1, \ldots, A_n \in \mathcal{A}$ such that $d_{ij} = \mu(A_i \triangle A_j)$ for all $1 \le i < j \le n$.*

Proof. (i) $\Longrightarrow$ (ii) Suppose that $d \in \mathrm{CUT}_n$. Then,

$$d = \sum_{S \subseteq \{1,\ldots,n\}} \lambda_S \delta(S),$$

where $\lambda_S \ge 0$ for all S. We define a measure space $(\Omega, \mathcal{A}, \mu)$ in the following way: Let Ω denote the family of subsets of $\{1, \ldots, n\}$, let $\mathcal{A}$ denote the family of subsets of Ω, and let μ denote the measure on $\mathcal{A}$ defined by

$$\mu(A) := \sum_{S \in A} \lambda_S$$

for each $A \in \mathcal{A}$ (i.e., A is a collection of subsets of $\{1, \ldots, n\}$). Set

$$A_i := \{S \in \Omega \mid i \in S\}.$$

Then,

$$\mu(A_i \triangle A_j) = \mu(\{S \in \Omega : |S \cap \{i,j\}| = 1\}) = \sum_{S \in \Omega : |S \cap \{i,j\}|=1} \lambda_S = d_{ij},$$

for all $1 \le i < j \le n$. Moreover, if $d \in \mathrm{CUT}_n^{\square}$, then $\sum_S \lambda_S = 1$, i.e., $\mu(\Omega) = 1$, which shows that $(\Omega, \mathcal{A}, \mu)$ is a probability space.
(ii) $\Longrightarrow$ (i) Conversely, suppose that $d_{ij} = \mu(A_i \triangle A_j)$ for $1 \le i < j \le n$, where $(\Omega, \mathcal{A}, \mu)$ is a measure space and $A_1, \ldots, A_n \in \mathcal{A}$. Set

$$A^S := \bigcap_{i \in S} A_i \cap \bigcap_{i \notin S} (\Omega \setminus A_i)$$

[3] The result from Proposition 4.2.1 was already established by Avis [1977].

for each $S \subseteq \{1, \ldots, n\}$. Then,

$$A_i = \bigcup_{S \mid i \in S} A^S, \; A_i \triangle A_j = \bigcup_{S \mid \; |S \cap \{i,j\}|=1} A_S \; \text{ and } \Omega = \bigcup_S A^S.$$

Therefore,

$$d = \sum_{S \subseteq \{1,\ldots,n\}} \mu(A^S)\delta(S),$$

which shows that d belongs to the cut cone CUT_n. Moreover, if $(\Omega, \mathcal{A}, \mu)$ is a probability space, i.e., if $\mu(\Omega) = 1$, then $\sum_S \mu(A^S) = 1$, implying that d belongs to the cut polytope $\mathrm{CUT}_n^{\square}$. ∎

Proposition 4.2.2. *Let $d \in \mathbb{R}^{E_n}$ and (V_n, d) be the associated distance space. The following assertions are equivalent.*

(i) *$d \in \mathrm{CUT}_n$.*

(ii) *(V_n, d) is ℓ_1-embeddable, i.e., there exist n vectors $u_1, \ldots, u_n \in \mathbb{R}^m$ (for some m) such that $d_{ij} = \| u_i - u_j \|_1$ for all $1 \leq i < j \leq n$.*

Proof. (i) $\Longrightarrow$ (ii) Suppose that $d \in \mathrm{CUT}_n$. Then,

$$d = \sum_{1 \leq k \leq m} \lambda_k \delta(S_k),$$

where $\lambda_1, \ldots, \lambda_m \geq 0$ and $S_1, \ldots, S_m \subseteq V_n$. For $1 \leq i \leq n$, define the vector $u_i \in \mathbb{R}^m$ with components $(u_i)_k = \lambda_k$ if $i \in S_k$ and $(u_i)_k = 0$ otherwise, for $1 \leq k \leq m$. Then, $d_{ij} = \| u_i - u_j \|_1$ for $1 \leq i < j \leq n$. This shows that (V_n, d) is ℓ_1-embeddable.

(ii) $\Longrightarrow$ (i) Suppose that (V_n, d) is ℓ_1-embeddable, i.e., that there exist n vectors $u_1, \ldots, u_n \in \mathbb{R}^m$ (for some $m \geq 1$) such that $d_{ij} = \| u_i - u_j \|_1$, for $1 \leq i < j \leq n$. We show that $d \in \mathrm{CUT}_n$. By additivity of the ℓ_1-norm, it suffices to show the result for the case $m = 1$. Hence, we can suppose that $d_{ij} = |u_i - u_j|$ where $u_1, \ldots, u_n \in \mathbb{R}$. Without loss of generality, we can also suppose that $u_1 \leq u_2 \leq \ldots \leq u_n$. Then, it is easy to check that

$$d = \sum_{1 \leq k \leq n-1} (u_{k+1} - u_k)\delta(\{1, 2, \ldots, k-1, k\}).$$

This shows that $d \in \mathrm{CUT}_n$. ∎

Remark 4.2.3. The proof of Proposition 4.2.2 shows, in fact, the following result: If a distance d on V_n can be decomposed as a nonnegative linear combination of m cut semimetrics, i.e., if $d = \sum_{k=1}^m \lambda_k \delta(S_k)$ where $\lambda_k \geq 0$ for all k, then d is ℓ_1^m-embeddable. ∎

There is a characterization for hypercube embeddable semimetrics, analogous to that of Proposition 4.2.2.

Proposition 4.2.4. *Let $d \in \mathbb{R}^{E_n}$ and (V_n, d) be the associated distance space. The following assertions are equivalent.*

(i) $d = \sum_{S \subseteq V_n} \lambda_S \delta(S)$ *for some nonnegative integers* λ_S.

(ii) (V_n, d) *is hypercube embeddable, i.e., there exist* n *vectors* $u_1, \ldots, u_n \in \{0,1\}^m$ *(for some* m*) such that* $d_{ij} = \| u_i - u_j \|_1$ *for all* $1 \leq i < j \leq n$.

(iii) *There exist a finite set* Ω *and* n *subsets* $A_1, \ldots, A_n$ *of* Ω *such that* $d_{ij} = |A_i \triangle A_j|$ *for all* $1 \leq i < j \leq n$.

(iv) (V_n, d) *is an isometric subspace of* $(\mathbb{Z}^m, d_{\ell_1})$ *for some integer* $m \geq 1$.

Proof. The proof of (i) $\Longleftrightarrow$ (ii) is analogous to that of Proposition 4.2.2. Namely, for (i) $\Longrightarrow$ (ii), assume $d = \sum_{k=1}^m \delta(S_k)$ (allowing repetitions). Consider the binary $n \times m$ matrix M whose columns are the incidence vectors of the sets $S_1, \ldots, S_m$. If $u_1, \ldots, u_n$ denote the rows of M, then $d_{ij} = \| u_i - u_j \|_1$ holds, providing an embedding of (V_n, d) in the hypercube of dimension m. Conversely, for (ii) $\Longrightarrow$ (i), consider the matrix M whose rows are the n given vectors $u_1, \ldots, u_n$. Let $S_1, \ldots, S_m$ be the subsets of $\{1, \ldots, n\}$ whose incidence vectors are the columns of M. Then, $d = \sum_{k=1}^m \delta(S_k)$ holds, giving a decomposition of d as a nonnegative integer combination of cut semimetrics.
The assertion (iii) is a reformulation of (ii), the implication (iii) $\Longrightarrow$ (iv) is obvious, and (iv) $\Longrightarrow$ (i) follows from the proof of the implication (ii) $\Longrightarrow$ (i) in Proposition 4.2.2. ∎

We now make the link with L_1-spaces.

Lemma 4.2.5. *Let* (X, d) *be a distance space. The following assertions are equivalent*

(i) (X, d) *is* L_1*-embeddable.*

(ii) (X, d) *is a subspace of a measure semimetric space* $(\mathcal{A}_\mu, d_\mu)$ *for some measure space* $(\Omega, \mathcal{A}, \mu)$.

Proof. The implication (ii) $\Rightarrow$ (i) is clear, since $(\mathcal{A}_\mu, d_\mu)$ is a subspace of $L_1(\Omega, \mathcal{A}, \mu)$. We check (i) $\Rightarrow$ (ii). It suffices to show that each space $L_1(\Omega, \mathcal{A}, \mu)$ is a subspace of $(\mathcal{B}_\nu, d_\nu)$ for some measure space $(T, \mathcal{B}, \nu)$. For this, set $T := \Omega \times \mathbb{R}$, $\mathcal{B} := \mathcal{A} \times \mathcal{R}$ where $\mathcal{R}$ is the family of Borel subsets of $\mathbb{R}$, and $\nu := \mu \otimes \lambda$ where λ is the Lebesgue measure on $\mathbb{R}$. For $f \in L_1(\Omega, \mathcal{A}, \mu)$, let $E(f) = \{(\omega, s) \in \Omega \times \mathbb{R} \mid s > f(\omega)\}$ denote its epigraph. Then, the mapping $f \longmapsto E(f) \triangle E(0)$ provides an isometric embedding from $L_1(\Omega, \mathcal{A}, \mu)$ to $(\mathcal{B}_\nu, d_\nu)$, since $\| f - g \|_1 = \nu(E(f) \triangle E(g))$ holds. ∎

The next result summarizes the equivalent characterizations that have been obtained for the members of the cut cone CUT_n.

Theorem 4.2.6. *Let* $d \in \mathbb{R}^{E_n}$ *and* (V_n, d) *be the associated distance space. The following assertions are equivalent.*

(i) $d \in \mathrm{CUT}_n$.

(ii) *(V_n, d) is ℓ_1-embeddable.*

(iii) *(V_n, d) is L_1-embeddable.*

(iv) *There exist a measure space $(\Omega, \mathcal{A}, \mu)$ and $A_1, \ldots, A_n \in \mathcal{A}$ such that $d_{ij} = \mu(A_i \triangle A_i)$ for all $1 \leq i < j \leq n$.*

(v) *(V_n, d) is an isometric subspace of ℓ_1^∞.* ∎

At this point let us observe a convexity property of the cut semimetrics entering any decomposition of an ℓ_1-embeddable distance. We need a definition.

Definition 4.2.7. *Let (X, d) be a distance space. A subset $U \subseteq X$ is said to be* d-convex *if $d(x, y) = d(x, z) + d(z, y)$ and $x, y \in U$ imply that $z \in U$.*

Convex sets arise naturally in the context of ℓ_1-metrics. Namely,

Lemma 4.2.8. *Let (X, d) be an ℓ_1-embeddable distance space and suppose that $d = \sum_{A \subseteq X} \lambda_A \delta(A)$ where $\lambda_A \geq 0$ for all $A \subseteq X$. Then, both sets A and $X \setminus A$ are d-convex for every cut semimetric $\delta(A)$ entering the decomposition with a positive coefficient $\lambda_A > 0$.*

Proof. The result follows from the fact that every triangle equality satisfied by d is also satisfied by a cut semimetric $\delta(A)$ entering the decomposition of d. ∎

We conclude this section with a remark on how the equivalence (i) $\Longleftrightarrow$ (ii) from Proposition 4.2.4 can be generalized in the context of Hamming spaces and multicuts. Let $q \geq 2$ be an integer and let $S_1, \ldots, S_q$ be q pairwise disjoint subsets of V_n forming a partition of V_n, i.e., such that $S_1 \cup \ldots \cup S_q = V_n$. Then, the *multicut semimetric* $\delta(S_1, \ldots, S_q)$ is the vector of $\mathbb{R}^{E_n}$ defined by

$$\begin{array}{ll} \delta(S_1, \ldots, S_q)_{ij} = 0 & \text{if } i, j \in S_h \text{ for some } h,\ 1 \leq h \leq q, \\ \delta(S_1, \ldots, S_q)_{ij} = 1 & \text{otherwise} \end{array}$$

for $1 \leq i < j \leq n$. The following can be easily checked.

Proposition 4.2.9. *Let $d \in \mathbb{R}^{E_n}$, let (V_n, d) be the associated distance space and let $q_1, \ldots, q_k \geq 1$ be integers. The following assertions are equivalent.*

(i) *$d = \sum_{h=1}^k \delta(S_1^{(h)}, \ldots, S_{q_h}^{(h)})$, where $(S_1^{(h)}, \ldots, S_{q_h}^{(h)})$ is a partition of V_n, for each $h = 1, \ldots, k$.*

(ii) *(V_n, d) is an isometric subspace of the Hamming space $(\prod_{h=1}^k \{0, 1, \ldots, q_h - 1\}, d_H)$.* ∎

Hence, we obtain again the equivalence (i) $\Longleftrightarrow$ (ii) from Proposition 4.2.4 in the case when $q_1 = \ldots = q_k = 2$. Observe that the Hamming space $(\prod_{h=1}^k \{0, 1, \ldots, q_h - 1\}, d_H)$ coincides with the graphic metric space of the graph

$\prod_{h=1}^{k} K_{q_h}$ (the graph obtained by taking the Cartesian product of the complete graphs $K_{q_1}, \ldots, K_{q_k}$). Note also that

$$\delta(S_1, \ldots, S_q) = \frac{1}{2} \sum_{i=1}^{q} \delta(S_i)$$

for every partition $(S_1, \ldots, S_q)$ of V_n. Therefore, every Hamming space is ℓ_1-embeddable with scale 2 (see the definition below).

As an example, consider the distance $d := (1,1,1,2;1,2,2;2,1;1) \in \mathbb{R}^{E_5}$. Then the distance space (V_5, d) is an isometric subspace of the Hamming space $(\{0,1,2\} \times \{0,1\}, d_H)$. To see it, consider the isometric embedding

$$i \in \{1,2,3,4,5\} \mapsto v_i \in \{0,1,2\} \times \{0,1\}$$

where $v_1 = (0,0)$, $v_2 = (1,0)$, $v_3 = (2,0)$, $v_4 = (0,1)$, and $v_5 = (2,1)$. Equivalently, d can be decomposed as the following sum of multicut semimetrics:

$$d = \delta(\{1,4\},\{2\},\{3,5\}) + \delta(\{1,2,3\},\{4,5\}).$$

4.3 Realizations, Rigidity, Size and Scale

We group here several definitions related to ℓ_1- and hypercube embeddability. Namely, for an ℓ_1-metric, we define the notions of realizations, rigidity, size and scale.

Realizations. We define here the notions of $\mathbb{R}_+$- and $\mathbb{Z}_+$-realizations for an ℓ_1-metric; they will be used especially in Part IV.

Definition 4.3.1. *Let d be a distance on V_n. Any decomposition:*

$$d = \sum_{S \subseteq V_n} \lambda_S \delta(S),$$

where $\lambda_S \geq 0$ (resp. $\lambda_S \in \mathbb{Z}_+$) for all S, is called an $\mathbb{R}_+$-realization *(resp.* $\mathbb{Z}_+$-realization*) of d.*

Hence, for an ℓ_1-embeddable distance d on V_n, we can speak alternatively of an ℓ_1-embedding of d (which consists of a set of n vectors $v_1, \ldots, v_n \in \mathbb{R}^m$ $(m \geq 1)$ such that $d_{ij} = \| v_i - v_j \|_1$ for all $i,j \in V_n$), or of an $\mathbb{R}_+$-realization of d (which is a decomposition of d as a *nonnegative* linear combination of cut semimetrics).

In the same way, if d is hypercube embeddable, we can speak alternatively of a hypercube embedding of d (which consists of a set of n vectors $v_1, \ldots, v_n \in \{0,1\}^m$ $(m \geq 1)$ such that $d_{ij} = \| v_i - v_j \|_1$ for all $i,j \in V_n$), or of a $\mathbb{Z}_+$-realization of d (which is a decomposition of d as a *nonnegative integer* linear combination of cut semimetrics). The binary $n \times m$ matrix M whose rows are the vectors $v_1, \ldots, v_n$ (providing a hypercube embedding of d) is called an *h-realization matrix* (or, simply, *realization matrix*) of d. If $S_1, \ldots, S_m$ denote the subsets of V_n whose incidence vectors are the columns of M, then $d = \sum_{j=1}^m \delta(S_j)$ is a $\mathbb{Z}_+$-realization of d. This simple fact is the basic idea for the proof of the equivalence (i) $\iff$ (ii) in Proposition 4.2.4.

If $v_1, \ldots, v_n$ is an ℓ_1-embedding of d, then the $n \times m$ matrix M whose rows are the vectors $v_1, \ldots, v_n$ is also called a *realization matrix* of d. Note, however, that the associated $\mathbb{R}_+$-realization of d cannot be read immediately from M as in the hypercube case (we have seen in the proof of Proposition 4.2.2 how to construct an $\mathbb{R}_+$-realization).

As an example, consider the distance on V_4 defined by $d := (3,1,3;4,4;2) \in \mathbb{R}^{E_4}$. The vectors $v_1 = (1,1,1,0,0)$, $v_2 = (1,0,0,0,1)$, $v_3 = (0,1,1,0,0)$, $v_4 = (0,1,1,1,1)$ provide a hypercube embedding of d, corresponding to the following $\mathbb{Z}_+$-realization of d:

$$d = \delta(\{1,2\}) + 2\delta(\{2\}) + \delta(\{4\}) + \delta(\{2,4\}),$$

with associated realization matrix $\begin{pmatrix} 1 & 1 & 1 & 0 & 0 \\ 1 & 0 & 0 & 0 & 1 \\ 0 & 1 & 1 & 0 & 0 \\ 0 & 1 & 1 & 1 & 1 \end{pmatrix}$.

Rigidity. Let (V_n, d) be a distance space. Then, (V_n, d) is said to be *ℓ_1-rigid* if d admits a unique $\mathbb{R}_+$-realization. Similarly, (V_n, d) is said to be *h-rigid* if d admits a unique $\mathbb{Z}_+$-realization. The notion of ℓ_1-rigidity is, in fact, directly relevant to the existence of simplex faces for the cut cone, as the next result shows.

Lemma 4.3.2. *Let $C := \mathbb{R}_+(v_1, \ldots, v_p) \subseteq \mathbb{R}^m$ be a cone, where each vector v_i lies on an extreme ray of C. Suppose, moreover, that the cone C is not a simplex cone. Then, each point that lies in the relative interior of C admits at least two (in fact, an infinity of) distinct decompositions as a nonnegative linear combination of the generators $v_1, \ldots, v_p$.*

Proof. We can suppose without loss of generality that C is full-dimensional. Let $x \in C$ lying in the interior of C. By Carathéodory's theorem, we have $x = \sum_{h=1}^k \lambda_h v_{i_h}$ where $\lambda_1, \ldots, \lambda_k > 0$ and $\{v_{i_1}, \ldots, v_{i_k}\}$ is a linearly independent subset of $\{v_1, \ldots, v_p\}$. If $k \leq m-1$, then one can easily construct another decomposition of x. (Indeed, let H be a hyperplane containing $v_{i_1}, \ldots, v_{i_k}, x$. As x is an interior point of C, H cuts C into two nonempty parts and, thus, $x = y + z$ where y belongs to one part and z to the other one.) Suppose now that $k = m$. As C is not a simplex cone, we have $p \geq k+1$, i.e., there exists $v_{k+1} \in \{v_1, \ldots, v_p\} \setminus \{v_{i_1}, \ldots, v_{i_k}\}$. As the set $\{v_{i_1}, \ldots, v_{i_k}, v_{k+1}\}$ is linearly dependent, there exist $\alpha_1, \ldots, \alpha_k \in \mathbb{R}$ such that $\sum_{h=1}^k \alpha_h v_{i_h} + v_{k+1} = 0$. Then, $x = \sum_{h=1}^k (\lambda_h + \lambda \alpha_h) v_{i_h} + \lambda v_{k+1}$ is another decomposition of x, setting $\lambda := 1$ if all α_h's are nonnegative and $\lambda := \min(-\frac{\lambda_h}{\alpha_h} \mid \alpha_h < 0)$ otherwise. ∎

Corollary 4.3.3. *The following assertions are equivalent for a distance space* (V_n, d).

(i) (V_n, d) *is* ℓ_1*-rigid, i.e., there is a unique way of decomposing* d *as a nonnegative linear combination of nonzero cut semimetrics.*

(ii) d *lies on a simplex face of* CUT_n.

Proof. The implication (ii) $\Longrightarrow$ (i) is obvious, while (i) $\Longrightarrow$ (ii) follows from Lemma 4.3.2 applied to the smallest face of CUT_n that contains d. ∎

The notion of h-rigidity can be reformulated in the following way, in terms of h-realization matrices. Let (V_n, d) be a hypercube embeddable distance space. Let M be an h-realization matrix of (V_n, d). The following operations (a), (b), (c) on M yield a new h-realization matrix for (V_n, d):

(a) Permute the columns of M.

(b) Add modulo 2 a binary vector to every row of M.

(c) Add to M, or delete from M, a column which is the zero vector or the all-ones vector.

Two hypercube embeddings of (V_n, d) are said to be *equivalent* if their associated realization matrices can be obtained from one another via the above operations (a),(b),(c). Then, (V_n, d) is h-rigid if and only if it has a unique hypercube embedding, up to equivalence.

ℓ_1- and h-Size. We now turn to the notions of ℓ_1-size and h-size. Suppose d is a distance on V_n. If $d = \sum_{\emptyset \neq S \subset V_n} \lambda_S \delta(S)$ is a decomposition of d as a linear combination of nonzero cut semimetrics, then the quantity $\sum_{\emptyset \neq S \subset V_n} \lambda_S$ is called its *size*. Moreover, if d is ℓ_1-embeddable then the quantity

$$s_{\ell_1}(d) := \min\Big(\sum_{\emptyset \neq S \subset V_n} \lambda_S \mid d = \sum_{\emptyset \neq S \subset V_n} \lambda_S \delta(S) \text{ with } \lambda_S \geq 0 \text{ for all } S\Big) \tag{4.3.4}$$

is called the *minimum* ℓ_1*-size* of d. Similarly, if d is hypercube embeddable, then

$$s_h(d) := \min\Big(\sum_{\emptyset \neq S \subset V_n} \lambda_S \mid d = \sum_{\emptyset \neq S \subset V_n} \lambda_S \delta(S) \text{ with } \lambda_S \in \mathbb{Z}_+ \text{ for all } S\Big) \tag{4.3.5}$$

is called the *minimum* h*-size* of d. This parameter has the following interpretation: The minimum h-size of d is equal to the minimum dimension of a hypercube in which d can be isometrically embedded.

One can easily obtain some bounds on the minimum ℓ_1-size. Suppose that $d = \sum_{\emptyset \neq S \subset V_n} \lambda_S \delta(S)$ with $\lambda_S \geq 0$ for all S. As $n - 1 \leq \sum_{1 \leq i < j \leq n} \delta(S)(i,j) \leq \lfloor \frac{n}{2} \rfloor \lceil \frac{n}{2} \rceil$ for each subset $S \neq \emptyset, V_n$, we obtain

$$\frac{\sum_{1 \leq i < j \leq n} d(i,j)}{\lfloor \frac{n}{2} \rfloor \lceil \frac{n}{2} \rceil} \leq \sum_{\emptyset \neq S \subset V_n} \lambda_S \leq \frac{\sum_{1 \leq i < j \leq n} d(i,j)}{n-1}. \tag{4.3.6}$$

Moreover, equality holds in the lower bound of (4.3.6) if and only if the decomposition of d uses only equicuts (i.e., $\lambda_S \neq 0$ only if $|S| = \lfloor \frac{n}{2} \rfloor, \lceil \frac{n}{2} \rceil$) and equality holds in the upper bound if and only if the decomposition uses only star cuts (i.e., $\lambda_S \neq 0$ only if $|S| = 1$).

Example 4.3.7. Let $2\mathbb{1}_n$ denote the equidistant metric on V_n taking the value 2 on all pairs. Then, $2\mathbb{1}_n$ is ℓ_1-rigid if and only if $n = 3$. Indeed, $2\mathbb{1}_3$ is ℓ_1-rigid because the cut cone CUT_3 is a simplex cone, and $2\mathbb{1}_n$ is not ℓ_1-rigid for $n \geq 4$ as

$$2\mathbb{1}_n = \sum_{1 \leq i \leq n} \delta(\{i\}) = \frac{1}{n-2} \sum_{1 \leq i < j \leq n} \delta(\{i, j\})$$

has two distinct $\mathbb{R}_+$-realizations.

On the other hand, $2\mathbb{1}_n$ is h-rigid if and only if $n \neq 4$, its unique $\mathbb{Z}_+$-realization being given by: $2\mathbb{1}_n = \sum_{i=1}^n \delta(\{i\})$; see Theorem 22.0.6. The metric $2\mathbb{1}_4$ has one additional $\mathbb{Z}_+$-realization, namely, $2\mathbb{1}_4 = \delta(\{1,2\}) + \delta(\{1,3\}) + \delta(\{1,4\})$.

Hence, $s_h(2\mathbb{1}_n) = n$ for $n \neq 4$ and $s_h(2\mathbb{1}_4) = 3$. On the other hand, $s_{\ell_1}(2\mathbb{1}_n) = \frac{4(n-1)}{n}$ if n is even and $s_{\ell_1}(2\mathbb{1}_n) = \frac{4n}{n+1}$ if n is odd. ∎

Using the linear programming duality theorem, the minimum ℓ_1-size of d can also be expressed as

$$s_{\ell_1}(d) = \max(v^T d \mid v \in \mathbb{R}^{E_n},\ v^T \delta(S) \leq 1 \text{ for all } S \subseteq V_n).$$

Clearly, in this maximization problem, it suffices to consider the vectors $v \in \mathbb{R}^{E_n}$ for which the inequality $v^T x \leq 1$ defines a facet of the cut polytope $\mathrm{CUT}_n^\square$.

As an illustration, let us mention an exact formulation for $s_{\ell_1}(d)$ in the case $n \leq 5$. All the facets of CUT_n are known for $n \leq 5$. Namely, they are defined by the triangle inequalities (3.1.1) if $n \leq 4$, and they are defined by the triangle inequalities (3.1.1) together with the pentagonal inequalities (6.1.9) if $n = 5$. This permits to obtain that, for $d \in \mathrm{CUT}_4$,

$$s_{\ell_1}(d) = \max\left(\frac{d(i,j) + d(i,k) + d(j,k)}{2} \mid 1 \leq i < j < k \leq 4\right)$$

and, for $d \in \mathrm{CUT}_5$,

$$s_{\ell_1}(d) = \max \left(\frac{d(i,j)+d(i,k)+d(j,k)}{2} \mid 1 \leq i < j < k \leq 5; \frac{\sum_{1 \leq i < j \leq 5} d(i,j)}{6}; \frac{\sum_{1 \leq i < j \leq 5} d(i,j) - 2\sum_{1 \leq j \leq 5, j \neq i} d(i,j)}{2} \mid 1 \leq i \leq 5\right).$$

Scale. The following result is an immediate consequence of Propositions 4.2.2 and 4.2.4.

Proposition 4.3.8. *Let (V_n, d) be a distance space where d is rational valued. Then, (V_n, d) is ℓ_1-embeddable if and only if $(V_n, \eta d)$ is hypercube embeddable for* *some scalar* η. ∎

Let d be a distance on V_n that is ℓ_1-embeddable and takes rational values. Every integer η for which $(V_n, \eta d)$ is hypercube embeddable is called a *scale* of (V_n, d); then, we also say that d is hypercube embeddable with the scale η. The smallest such integer η is called the *minimum scale* of (V_n, d) and is denoted by $\eta(d)$.

It is easy to see that all integer valued ℓ_1-embeddable distances on V_n admit a common scale.

Lemma 4.3.9. *There exists an integer α such that αd is hypercube embeddable for every ℓ_1-embeddable distance d on V_n that is integer valued.*

Proof. Let X be a set of linearly independent cut semimetrics on V_n, let M_X denote the matrix whose columns are the members of X, and let α_X denote the smallest absolute value of the determinant of a $|X|\times|X|$ nonsingular submatrix of M_X. Then, we define α as the lowest common multiple of the integers α_X (for X arbitrary set of linearly independent cut semimetrics). This integer α satisfies the lemma. Indeed, let d be an integer valued distance on V_n that is ℓ_1-embeddable. By Carathéodory's theorem, d can be decomposed as $d = \sum_{\delta(S)\in X} \lambda_S \delta(S)$, where X is a set of linearly independent cut semimetrics and $\lambda_S > 0$ for all S. Let A be a $|X| \times |X|$ nonsingular submatrix of M_X with $|\det A| = \alpha_X$, let E denote the index set for the rows of A and set $d_E := (d_{ij})_{ij\in E}$. Then, $A\lambda = d_E$, i.e., $\lambda = A^{-1}d_E$. Applying Cramer's rule, we obtain that $(\det A)\lambda$ is integer valued. This shows that αd is hypercube embeddable. ∎

We deduce from the above proof the following (very rough) upper bound

$$\eta(d) \leq \binom{n}{2}!$$

for the minimum scale of an integral ℓ_1-distance d on V_n (as the determinant of a $k \times k$ binary matrix is less than or equal to $k!$ in absolute value).

Let us define η_n as the smallest integer η such that ηd is hypercube embeddable for every ℓ_1-embeddable distance d on V_n that is integer valued and satisfies:

$$d(i,j) + d(i,k) + d(j,k) \in 2\mathbb{Z} \quad \text{for all } i,j,k \in V_n.$$

(This condition ensures that d can be decomposed as an *integer* sum of cut semimetrics, an obvious necessary condition for hypercube embeddability; see Proposition 25.1.1.) Hence, by the above, $\eta_n \leq \prod_{k=1}^{\binom{n}{2}} k!$. In fact, as we will see in Section 25.2, $\eta_n = 1$ for $n \leq 5$, and $\eta_6 = 2$.

4.4 Complexity Questions

We formulate here several basic problems related to cut semimetrics and we indicate their complexity status. A typical way to show that a given problem (P) is NP-hard is to show that some known NP-complete problem (P0) reduces polynomially to it. We use as starting point the following well-known problem:

(P0) The partition problem.
Instance: A nonnegative integer vector $b = (b_1, \ldots, b_n)$.
Question: Can b be partitioned ? That is, does there exist $S \subseteq V_n$ such that $\sum_{i\in S} b_i = \sum_{i\in V_n\setminus S} b_i$?
Complexity: NP-complete (Karp [1972]).

(P1) The max-cut problem (decision version).
Instance: A vector $w \in \mathbb{Z}_+^{E_n}$ and $K \in \mathbb{Z}_+$.
Question: Does there exist $S \subseteq V_n$ such that $w^T\delta(S) \geq K$?
Complexity: NP-complete (Karp [1972]).

Proof. It is clear that (P1) is in NP. To see that (P1) is NP-complete, one can observe (following Karp [1972]) that (P0) polynomially reduces to it. Indeed, given integers $b_1, \ldots, b_n \in \mathbb{N}$, define $w \in \mathbb{Z}^{E_n}$ and $K \in \mathbb{Z}_+$ by $w_{ij} := b_i b_j$ (for $ij \in E_n$) and $K := \frac{1}{4}(\sum_{i=1}^n b_i)^2$. Then, for $S \subseteq V_n$, $w^T\delta(S) = (\sum_{i\in S} b_i)(\sum_{i\in V_n\setminus S} b_i) \geq K$ if and only if $\sum_{i\in S} b_i = \frac{1}{2}\sum_{i=1}^n b_i$, i.e., if b can be partitioned. ∎

(P2) The max-cut problem (optimization version).
Instance: A vector $w \in \mathbb{Z}_+^{E_n}$.
Question: Find $S \subseteq V_n$ for which $w^T\delta(S)$ is maximum.
Complexity: NP-hard.

While NP-hard in general, the max-cut problem may become easy for certain classes of weight functions. It is convenient to represent a weight function by its supporting graph (with edges the pairs with nonzero weights). Note first that the max-cut problem remains NP-hard when restricted to 0, 1-valued weight functions and to each of the following classes of graphs: cubic graphs (Yannakakis [1978]), graphs having a node whose deletion results in a planar graph (Barahona [1982]), chordal graphs, tripartite graphs, complements of bipartite graphs (Bodlaender and Jansen [1994]). On the other hand, the max-cut problem can be solved in polynomial time for several classes of graphs; for instance, for planar graphs (Orlova and Dorfman [1972], Hadlock [1975]), more generally for graphs with no K_5-minor (Barahona [1983]), also for graphs of fixed genus with weights ± 1 (Barahona [1981]). (See also Section 27.3.2.)

Next, we look at the complexity of the separation problem over the cut cone CUT_n, as this will then enable us to derive the complexity of the ℓ_1-embeddability problem.

(P3) The separation problem for the cut cone.
Instance: A rational distance d on V_n.

Question: Does d belong to the cut cone CUT_n ? If not, find $a \in \mathbb{Q}^{E_n}$ such that $a^T d > 0$ and $a^T \delta(S) \leq 0$ for all $S \subseteq V_n$.
Complexity: NP-hard (Karzanov [1985]).

Proof. Set $P_n := \mathrm{CUT}_n \cap \{x \mid \sum_{ij \in E_n} x_{ij} \leq 1\}$. We first show that the problem (P1) can be polynomially reduced to an optimization problem over P_n. For this, let $w \in \mathbb{Z}_+^{E_n}$ and $K \in \mathbb{Z}_+$ be given. For $s \neq t \in V_n$, define a new weight function $w^{st} \in \mathbb{Z}^{E_n}$ by

$$w_{ij}^{st} := \begin{cases} -w_{ij} & \text{if } i,j \in V_n \setminus \{s,t\}, \\ -w_{ij} + M & \text{if } i \in \{s,t\}, j \in V_n \setminus \{s,t\}, \\ -w_{ij} + K - 1 - M(n-2) & \text{if } ij = st, \end{cases}$$

for $ij \in E_n$, where $M := \sum_{ij \in E_n} w_{ij}$. Note that $(w^{st})^T \delta(S) = -w^T \delta(S) + K - 1$ if $\delta(S)_{st} = 1$ and that $(w^{st})^T \delta(S) \geq 0$ if $\delta(S)_{st} = 0$. Hence, if $\delta(S)_{st} = 1$, then $w^T \delta(S) \geq K$ if and only if $(w^{st})^T \delta(S) < 0$. Therefore, in order to solve (P1), it suffices to solve the problem:

$$\begin{array}{ll} \min & (w^{st})^T x \\ \text{s.t.} & x \in P_n \end{array}$$

for each of the $\binom{n}{2}$ pairs $st \in E_n$ and to verify whether the minimum is negative. As (P1) is NP-complete, we obtain that the optimization problem over P_n is NP-hard. Using results from Grötschel, Lovász and Schrijver [1988] (namely, the polynomial equivalence between optimization and separation problems over polyhedra), we deduce that the separation problem for P_n is NP-hard. As the separation problem over P_n is clearly equivalent to the separation problem over CUT_n, we obtain that (P3) is NP-hard. ∎

(P5) The ℓ_1-embeddability problem.
Instance: A rational distance d on V_n.
Question: Does d belong to the cut cone CUT_n (i.e., is d ℓ_1-embeddable) ?
Complexity: NP-complete (Avis and Deza [1991]).

Proof. We first check that (P5) is in NP. Indeed, if $d \in \mathrm{CUT}_n$, then d can be decomposed as $d = \sum_{i=1}^m \alpha_i \delta(S_i)$ for some nonnegative scalars α_i's, where $m \leq \binom{n}{2}$ (by Carathéodory's theorem). Thus, the polyhedron $\{\alpha \in \mathbb{R}_+^m \mid d = \sum_{i=1}^m \alpha_i \delta(S_i)\}$ is nonempty. Therefore, it contains a rational vector α whose size is polynomially bounded by n and the size of d (use Theorem 10.1 in Schrijver [1986]). Hence, this α provides a certificate that can be checked in time polynomial in the size of d. This shows that (P5) is in NP. We can now conclude that (P5) is NP-complete, since the separation problem (P3) is NP-hard. Indeed, consider again the polytope $P_n = \mathrm{CUT}_n \cap \{x \mid \sum_{ij \in E_n} x_{ij} \leq 1\}$ instead of CUT_n. As P_n is a full-dimensional bounded polytope for which we know an interior point (e.g., $\frac{1}{\binom{n}{2}+1} \mathbb{1}_n$), it follows from general results in Grötschel, Lovász and Schrijver [1988] that the separation problem for P_n reduces polynomially to the membership problem for P_n (use, in particular, (4.3.11) and Theorems 6.3.2 and 6.4.9 there). ∎

(P6) The hypercube embeddability problem.
Instance: An integer distance d on V_n.
Question: Is d hypercube embeddable ?
Complexity: NP-hard; NP-complete when d is a distance on V_n having a point at distance 3 from all other points and all other distances belong to $\{0, 2, 4, 6\}$ (Chvátal [1980]; see Theorem 24.1.8).

Many other problems related to cut semimetrics are hard problems. This is the case, for instance, for the problem of computing the minimum ℓ_1-size $s_{\ell_1}(d)$ for an ℓ_1-embeddable distance d, or that of computing the minimum h-size $s_h(d)$ for a hypercube embeddable distance. As an example, consider the equidistant metric $d = 2t\mathbb{1}_{t^2+t+1}$ (on $t^2 + t + 1$ points). Then, computing the minimum h-size of d would yield an answer to the question of existence of finite projective planes. Indeed, it can be shown that $s_h(d) \geq t^2 + t + 1$, with equality if and only if there exists a projective plane of order t (see Section 23.3). In the same vein, computing the minimum scale of $d \in \mathrm{CUT}_n$ is also hard. If d_n denotes the distance on V_n that takes value 1 for every pair except value 2 for a single pair, then

$$\eta(d_n) = 2\min(t \mid 4t \geq s_h(2t\mathbb{1}_n))$$

(see (7.4.5)). Hence, we meet again the question of computing the minimum h-size of the equidistant metric. More details about these examples can be found in Section 7.4.

Finding a complete linear description of the cut cone CUT_n (or of the cut polytope $\mathrm{CUT}_n^{\square}$) is also a hard task. Due to the NP-completeness of the max-cut problem, it follows from a result of Karp and Papadimitriou [1982] that there exists no polynomially concise linear description of CUT_n (or $\mathrm{CUT}_n^{\square}$) unless NP = co-NP.

Hence, many questions about the cut polyhedra turn out to be hard. Nevertheless, we will present in this book a number of results dealing with instances where these questions become tractable. For instance, even though the hypercube embeddability problem is NP-hard for general distances, it can be solved in polynomial time for some classes of metrics; several such classes are described in Chapter 24. A complete linear description of the cut cone CUT_n is not known for general n; yet, large classes of valid inequalities and facets are known (yielding a complete description for $n \leq 7$); see Part V. Moreover, for some classes of metrics, the known classes of inequalities already suffice for characterizing ℓ_1-embeddability; see Remark 6.3.5. For some restricted instances of weight functions, the max-cut problem becomes polynomial-time solvable as noted above. We will also mention several classes of valid inequalities for the cut cone for which the separation problem can be solved in polynomial time (e.g., in Section 29.5); hence, these systems of inequalities define tractable linear relaxations of the cut polytope that could be used for approximating the max-cut problem.

4.5 An Application to Statistical Physics

We describe here an application of the max-cut problem to a problem arising in statistical physics; namely, the problem of determining ground states of spin glasses.

We start with reformulating the max-cut problem (4.1.4) as the following quadratic ± 1-optimization problem:

$$(4.5.1) \qquad \begin{array}{ll} \max & \frac{1}{2}\sum_{1\le i<j\le n} w_{ij}(1-x_i x_j) \\ \text{s.t.} & x \in \{\pm 1\}^n. \end{array}$$

(Indeed, setting $S := \{i \in [1,n] \mid x_i = 1\}$ for $x \in \{\pm 1\}^n$, the weight function does compute the weight $w^T\delta(S)$ of the corresponding cut.) The applied problem that we discuss below will be of the form (4.5.1).

A central topic in physics is the investigation of properties of spin glasses. A number of theories have been developed over the years in order to model spin glasses and try to explain their behaviour. One of the most commonly used models is the Ising model both for its simplicity and its accuracy in representing real world situations. Some aspects in this model yield to optimization problems of the form (4.5.1) as we see below[4].

A spin glass is an alloy of magnetic impurities diluted in a nonmagnetic metal. Alloys that show spin glass behaviour are, for instance, CuMn, the metallic crystal AuFe, or the insulator EuSrS. At very low temperature the spin glass system attains a minimum energy configuration, referred to as a ground state. This state can be found by minimizing the hamiltonian representing the total energy of the system; this problem turns out to be a max-cut problem in a certain graph.

Let us assume that the given spin glass system consists of n magnetic impurities (atoms). Each magnetic atom i has a magnetic orientation (spin) which, in the Ising model, is represented by a variable $s_i \in \{\pm 1\}$ (meaning magnetic north pole 'up' or 'down'). Between any pair i, j of atoms there is an interaction J_{ij}, which depends on the nonmagnetic material and the distance r_{ij} between the atoms. The orientation of the exterior magnetic fiel is represented by $s_0 \in \{\pm 1\}$ and its strength by a constant h. Then the energy of the system is given by the hamiltonian:

$$H := -\sum_{1\le i<j\le n} J_{ij} s_i s_j - h\sum_{i=1}^{n} s_0 s_i.$$

Finding a state of minimum energy is the problem:

$$\begin{array}{ll} \min & -\sum_{1\le i<j\le n} J_{ij} s_i s_j - h\sum_{i=1}^{n} s_0 s_n \\ \text{s.t.} & s \in \{\pm 1\}^{n+1}, \end{array}$$

[4]The reader may consult Barahona, Grötschel, Jünger and Reinelt [1988], or Godsil, Grötschel and Welsh [1995] for more details on this topic and Mezard, Parisi and Virasoro [1987] for an introduction to the general theory of spin glasses.

which can be easily brought in the form (4.5.1). Thus, we find an instance of the max-cut problem.

As the interaction J_{ij} decreases rapidly as the distance r_{ij} increases, it is common to set $J_{ij} := 0$ if the atoms are far apart. Moreover, there are two ways of generating the interactions J_{ij} that have been intensively studied: the Gaussian model where the J_{ij}'s are chosen from a Gaussian distribution, and the $\pm J$-model where interactions take only two values $\pm J$ (according to some distribution). One also assumes that the atoms are regularly located on a 2- or 3-dimensional grid; then interactions between atoms are nonzero only along the edges of the grid. In this model, the problem of determining ground states of the spin glass system is formulated as a max-cut problem on a graph which is a 2-dimensional or 3-dimensional grid plus a universal node (corresponding to the exterior magnetic field) joined to all nodes in the grid.

This instance of the max-cut problem is NP-hard, as mentioned earlier in Section 4.4. In fact, if one neglects the exterior magnetic field, the problem remains NP-hard in the 3-dimensional case (Barahona [1982]), but it becomes polynomial in the 2-dimensional case (since a 2-dimensional grid is a planar graph). Interestingly, the classical technique used for solving max-cut on a planar graph (reduction to a Chinese postman problem in the dual graph) by Orlova and Dorfman [1972] and Hadlock [1975]) was independently discovered in the field of physics by Toulouse [1977].

Toulouse's paper together with the papers by Bieche, Maynard, Rammal and Uhry [1980] and by Barahona, Maynard, Rammal and Uhry [1982] have pioneered the study of spin glasses from an optimization point of view. Since then, lots of efforts have been made for designing algorithms permitting to compute exact ground states of spin glass systems, in the various cases mentioned above. These algorithms are essentially of the type 'branch-and-cut' and use knowledge of the cut polytope (in particular, the cycle inequalities presented in Section 27.3.1). Computational results can be found in Grötschel, Jünger and Reinelt [1987], Barahona, Grötschel, Jünger and Reinelt [1988], Barahona, Jünger and Reinelt [1989], Barahona [1994], De Simone, Diehl, Jünger, Mutzel, Reinelt and Rinaldi [1995, 1996] and in references therein.

Chapter 5. The Correlation Cone and $\{0,1\}$-Covariances

We introduce here another set of polyhedra: the correlation cone and the correlation polytope, that have been considered in the literature in connection with several different problems (relevant, among others, to probability theory, quantum logic, or optimization). The correlation polyhedra turn out to be equivalent - via a linear bijection - to the cut polyhedra. As a consequence, any result about the cut polyhedra has a direct counterpart for the correlation polyhedra and vice versa. These connections are explained in detail in Section 5.2 and 5.3, and an application to the Boole problem in probabilities is described in Section 5.4.

5.1 The Correlation Cone and Polytope

As before, we set $V_n = \{1, \ldots, n\}$ and $E_n = \{ij \mid i,j \in V_n, i \neq j\}$ denotes the set of unordered pairs of elements of V_n. In the following, we often identify V_n with the set of diagonal pairs ii for $i = 1, \ldots, n$. In other words, a vector $p \in \mathbb{R}^{V_n \cup E_n}$ can be supposed to be indexed by the pairs ij for $1 \leq i \leq j \leq n$.

The main objects considered in this section are the correlation cone and polytope, that we now introduce. Let S be a subset of V_n. Let us define the vector $\pi(S) = (\pi(S)_{ij})_{1 \leq i \leq j \leq n} \in \mathbb{R}^{V_n \cup E_n}$ by

$$\pi(S)_{ij} = 1 \text{ if } i,j \in S \text{ and } \pi(S)_{ij} = 0 \text{ otherwise} \tag{5.1.1}$$

for $1 \leq i \leq j \leq n$; $\pi(S)$ is called a *correlation vector.* The cone in $\mathbb{R}^{V_n \cup E_n}$, generated by the correlation vectors $\pi(S)$ for $S \subseteq V_n$, is called the *correlation cone* and is denoted by COR_n. The polytope in $\mathbb{R}^{V_n \cup E_n}$, defined as the convex hull of the correlation vectors $\pi(S)$ for $S \subseteq V_n$, is called the *correlation polytope* and is denoted by $\mathrm{COR}_n^{\square}$. Hence,

$$\mathrm{COR}_n = \{ \sum_{S \subseteq V_n} \lambda_S \pi(S) \mid \lambda_S \geq 0 \text{ for all } S \subseteq V_n \}, \tag{5.1.2}$$

$$\mathrm{COR}_n^{\square} = \{ \sum_{S \subseteq V_n} \lambda_S \pi(S) \mid \sum_{S \subseteq V_n} \lambda_S = 1 \text{ and } \lambda_S \geq 0 \text{ for all } S \subseteq V_n \}. \tag{5.1.3}$$

It is sometimes convenient to consider an arbitrary finite subset X instead of V_n. Then, the correlation cone is denoted by $\mathrm{COR}(X)$ and the correlation polytope by $\mathrm{COR}^{\square}(X)$. Hence,

$$\mathrm{COR}(V_n) = \mathrm{COR}_n \text{ and } \mathrm{COR}^{\square}(V_n) = \mathrm{COR}_n^{\square}.$$

Note that $\pi(S)$ coincides with the upper triangular part (including the diagonal) on the matrix $(\chi^S)(\chi^S)^T$, where $\chi^S \in \{0,1\}^n$ denotes the incidence vector of the set S. Hence, the valid inequalities for the correlation cone COR_n can be interpreted as the symmetric $n \times n$ matrices that are nonnegative on binary arguments; this point of view is taken in Deza [1973a].

The correlation polytope has been considered in the literature in connection with many different problems, arising in various fields. We mention some of them below.

The correlation polytope plays, for instance, an important role in combinatorial optimization. Indeed, it permits to formulate a well-known NP-hard optimization problem, namely, the *unconstrained quadratic 0-1 programming problem:*

$$(5.1.4) \qquad \begin{array}{ll} \max & \sum\limits_{1\le i\le j\le n} c_{ij}x_ix_j \\ \text{s.t.} & x \in \{0,1\}^n \end{array}$$

(where $c_{ij} \in \mathbb{R}$ for all i,j). Clearly, this problem can be reformulated as:

$$\begin{array}{ll} \max & c^T p \\ \text{s.t.} & p \in \mathrm{COR}_n^{\square}. \end{array}$$

There are many papers studying the unconstrained quadratic 0-1 programming problem; we just cite a few of them, e.g., De Simone [1989], Isachenko [1989], Padberg [1989], Boros and Hammer [1991, 1993], Boissin [1994]. There, the polytope $\mathrm{COR}_n^{\square}$ is mostly known under the name of *boolean quadric polytope.*

As will be explained in Section 5.3, the members of $\mathrm{COR}_n^{\square}$ can be interpreted as joint correlations of events in some probability space. This fact explains the name "correlation polytope", which was introduced by Pitowsky [1986]. For $n = 3$, the correlation polytope $\mathrm{COR}_3^{\square}$ is called there the *Bell-Wigner polytope.* In this context, the correlation polytope occurs in connection with the Boole problem, which will be discussed in Section 5.4. This interpretation was independently discovered by several authors, in particular, by McRae and Davidson [1972], Assouad [1979, 1980b], Pitowsky [1986, 1989, 1991], etc. Interestingly, these authors came to it from different mathematical backgrounds, ranging from mathematical physics, quantum logic to analysis.

The correlation polytope also arises in the field of quantum mechanics, in connection with the so-called representability problem for density matrices of order 2. These matrices were introduced as a tool for representing physical properties of a system of particles (see Löwdin [1955]). It turns out that the study of some of their properties (in particular, of their diagonal elements) leads to considering the correlation polytope. See, e.g., Yoseloff and Kuhn [1969], McRae and Davidson [1972]. There is a large literature on this topic; we refer, e.g., to Deza and Laurent [1994b, 1994c] where this connection has been surveyed in detail with an extended bibliography. One more example where the correlation polytope (in fact, its polar) occurs, is in connection with the study of two-body operators (see Erdahl [1987]).

It turns out that the correlation cone (or polytope) is very closely related to the cut cone (or polytope). In fact, it is nothing but its image under a linear bijective mapping. We describe this mapping in Section 5.2. As a consequence, we obtain several characterizations for the members of the correlation cone and polytope, which are counterparts of the characterizations given in the preceding section for the cut polyhedra; see Section 5.3. We present an application to the Boole problem in Section 5.4. The Boole problem can be stated as follows: Given n events $A_1, \ldots, A_n$ in a probability space, find good bounds for the probability $\mu(A_1 \cup \ldots \cup A_n)$ of their union in terms of the joined probabilities $\mu(A_i \cap A_j)$ (or in terms of higher order joined probabilities).

Another consequence of this correspondence between cut and correlation polyhedra is the equivalence of the max-cut problem (4.1.4) and of the unconstrained quadratic 0-1 programming problem (5.1.4). In particular, the latter problem is also NP-hard.

5.2 The Covariance Mapping

A simple but fundamental property is that the cut cone CUT_{n+1} (resp. the cut polytope $\mathrm{CUT}^{\square}_{n+1}$) is in one-to-one correspondence with the correlation cone COR_n (resp. the correlation polytope $\mathrm{COR}^{\square}_n$) via the covariance mapping, defined below.

Consider the mapping

$$\xi : \mathbb{R}^{E_{n+1}} \longrightarrow \mathbb{R}^{V_n \cup E_n}$$

from the space $\mathbb{R}^{E_{n+1}}$ (indexed by the $\binom{n+1}{2}$ pairs of elements of V_{n+1}) to the space $\mathbb{R}^{V_n \cup E_n}$ (indexed by the n diagonal pairs of elements of V_n and the $\binom{n}{2}$ pairs of elements of V_n) defined as follows:

$$p = \xi(d)$$

for $d = (d_{ij})_{1 \leq i < j \leq n+1}$ and $p = (p_{ij})_{1 \leq i \leq j \leq n}$ with

$$\text{(5.2.1)} \qquad \begin{cases} p_{ii} = d_{i,n+1} & \text{for } 1 \leq i \leq n, \\ p_{ij} = \frac{1}{2}(d_{i,n+1} + d_{j,n+1} - d_{ij}) & \text{for } 1 \leq i < j \leq n \end{cases}$$

or, equivalently,

$$\text{(5.2.2)} \qquad \begin{cases} d_{i,n+1} & = p_{ii} \qquad \text{for } 1 \leq i \leq n, \\ d_{ij} & = p_{ii} + p_{jj} - 2p_{ij} \qquad \text{for } 1 \leq i < j \leq n. \end{cases}$$

The mapping ξ is called the *covariance mapping.* Note that the element $n+1$ plays a special role in the definition of ξ; if we want to stress this fact, we denote ξ by ξ_{n+1} and we say that ξ_{n+1} is the covariance mapping *pointed at the position* $n+1$. The mapping ξ is obviously a linear bijection from $\mathbb{R}^{E_{n+1}}$ to $\mathbb{R}^{V_n \cup E_n}$. One can easily check that, for any subset S of V_n,

$$\xi(\delta(S)) = \pi(S).$$

Therefore,

$$\xi(\mathrm{CUT}_{n+1}) = \mathrm{COR}_n \text{ and } \xi(\mathrm{CUT}^{\square}_{n+1}) = \mathrm{COR}^{\square}_n, \tag{5.2.3}$$

i.e., COR_n (resp. $\mathrm{COR}^{\square}_n$) is nothing but the image of CUT_{n+1} (resp. $\mathrm{CUT}^{\square}_{n+1}$) under the covariance mapping ξ.

In the same way, given a finite subset X and an element $x_0 \in X$, the cut cone $\mathrm{CUT}(X)$ and the correlation cone $\mathrm{COR}(X \setminus \{x_0\})$ (resp. the cut polytope $\mathrm{CUT}^{\square}(X)$ and the correlation polytope $\mathrm{COR}^{\square}(X \setminus \{x_0\})$) are in one-to-one linear correspondence via the covariance mapping ξ pointed at the position x_0 (also denoted as ξ_{x_0} if one wants to stress the choice of the point x_0). For the sake of clarity, we rewrite the definition.

Let X be a set (not necessarily finite) and $x_0 \in X$, let d be a distance on X, and let p be a symmetric function on $X \setminus \{x_0\}$. Then, $p = \xi(d) = \xi_{x_0}(d)$ if

$$p(x,y) = \frac{1}{2}(d(x,x_0) + d(y,x_0) - d(x,y)) \text{ for all } x,y \in X \setminus \{x_0\} \tag{5.2.4}$$

or, equivalently,

$$\begin{cases} d(x,x_0) = p(x,x) & \text{for all } x \in X \setminus \{x_0\}, \\ d(x,y) = p(x,x) + p(y,y) - 2p(x,y) & \text{for all } x,y \in X \setminus \{x_0\}. \end{cases} \tag{5.2.5}$$

Therefore, for X finite,

$$\xi_{x_0}(\mathrm{CUT}(X)) = \mathrm{COR}(X \setminus \{x_0\}) \text{ and } \xi_{x_0}(\mathrm{CUT}^{\square}(X)) = \mathrm{COR}^{\square}(X \setminus \{x_0\}).$$

Note that, if one uses relation (5.2.4) for computing $p(x,x_0)$, then one obtains that $p(x,x_0) = 0$ for all $x \in X$. This explains why we consider p as being defined only on the pairs of elements from $X \setminus \{x_0\}$.

The covariance mapping has appeared in many different areas of mathematics. See, for instance, Bandelt and Dress [1992], Fichet [1987a] (where ξ is named *Farris transform* or *linear generalized similarity function*), Critchley [1988], Coornaert and Papadopoulos [1993] (where, for a metric space (X,d) and its image $p = \xi(d)$, the quantity $p(x,y)$ is known as the *Gromov product* of $x,y \in X \setminus \{x_0\}$).

The connection between cut and correlation polyhedra, which is formulated in (5.2.3), was discovered independently by several authors (e.g., by Hammer [1965], Deza [1973a], Barahona, Jünger and Reinelt [1989], De Simone [1989]).

As a consequence of (5.2.3), every inequality valid for the cut polytope $\mathrm{CUT}^{\square}_{n+1}$ can be transformed into an inequality which is valid for the correlation polytope $\mathrm{COR}^{\square}_n$ and vice versa, via the covariance mapping. We formulate this fact in a precise way in Proposition 5.2.7 below.

"cut side"	"correlation side"
$d \in \mathbb{R}^{E_{n+1}}$ $\delta(S)$ (for $S \subseteq V_n$) CUT_{n+1} $\mathrm{CUT}^{\square}_{n+1}$	$p \in \mathbb{R}^{V_n \cup E_n}$ $\pi(S)$ COR_n $\mathrm{COR}^{\square}_n$
Triangle inequalities: $d(i,j) - d(i,n+1) - d(j,n+1) \leq 0$ $d(i,n+1) - d(j,n+1) - d(i,j) \leq 0$ $d(j,n+1) - d(i,n+1) - d(i,j) \leq 0$ $d(i,n+1) + d(j,n+1) + d(i,j) \leq 2$ $d(i,j) - d(i,k) - d(j,k) \leq 0$ $d(i,j) + d(i,k) + d(j,k) \leq 2$	$0 \leq p_{ij}$ $p_{ij} \leq p_{ii}$ $p_{ij} \leq p_{jj}$ $p_{ii} + p_{jj} - p_{ij} \leq 1$ $-p_{kk} - p_{ij} + p_{ik} + p_{jk} \leq 0$ $p_{ii} + p_{jj} + p_{kk} - p_{ij} - p_{ik} - p_{jk} \leq 1$
Hypermetric inequalities: $\sum_{1 \leq i < j \leq n+1} b_i b_j d(i,j) \leq 0$ with $b \in \mathbb{Z}^{n+1}$, $\sum_{1 \leq i \leq n+1} b_i = 1$	$\sum_{1 \leq i,j \leq n} b_i b_j p_{ij} - \sum_{1 \leq i \leq n} b_i p_{ii} \geq 0$ with $b \in \mathbb{Z}^n$ (i.e., $(\sum_{1 \leq i \leq n} b_i p_i)(\sum_{1 \leq i \leq n} b_i p_i - 1) \geq 0$, setting $p_{ii} := p_i, p_{ij} := p_i p_j$)
Negative type inequalities: $\sum_{1 \leq i < j \leq n+1} b_i b_j d(i,j) \leq 0$ with $b \in \mathbb{Z}^{n+1}$, $\sum_{1 \leq i \leq n+1} b_i = 0$	$\sum_{1 \leq i,j \leq n} b_i b_j p_{ij} \geq 0$ with $b \in \mathbb{Z}^n$
$\sum_{1 \leq i < j \leq n+1} b_i b_j d(i,j) \leq k(k+1)$ with $b \in \mathbb{Z}^{n+1}$, $\sum_{1 \leq i \leq n+1} b_i = 2k+1$	$(\sum_{1 \leq i \leq n} b_i p_i - k)(\sum_{1 \leq i \leq n} b_i p_i - k - 1) \geq 0$ with $b \in \mathbb{Z}^n, k \in \mathbb{Z}$

Figure 5.2.6: Corresponding inequalities for cut and correlation polyhedra

We show in Figure 5.2.6 how this correspondence applies to several classes of inequalities, namely, to the triangle inequalities, the hypermetric inequalities, and to the negative type inequalities (these inequalities will be treated in Section 6.1 and, in full detail, in Chapter 28).

Proposition 5.2.7. *Let $a \in \mathbb{R}^{V_n}$, $b \in \mathbb{R}^{E_n}$, $c \in \mathbb{R}^{E_{n+1}}$ be linked by*

$$\begin{cases} c_{i,n+1} & = a_i + \frac{1}{2} \sum_{1 \leq j \leq n,\ j \neq i} b_{ij} \quad \text{for } 1 \leq i \leq n, \\ c_{ij} & = -\frac{1}{2} b_{ij} \qquad\qquad\qquad \text{for } 1 \leq i < j \leq n. \end{cases}$$

Given $\alpha \in \mathbb{R}$, the inequality

$$\sum_{1 \leq i < j \leq n+1} c_{ij} d(i,j) \leq \alpha$$

is valid (resp. facet defining) for the cut polytope $\mathrm{CUT}_{n+1}^{\square}$ if and only if the inequality

$$\sum_{1 \leq i \leq n} a_i p_{ii} + \sum_{1 \leq i < j \leq n} b_{ij} p_{ij} \leq \alpha$$

is valid (resp. facet defining) for the correlation polytope $\mathrm{COR}_n^{\square}$. ∎

5.3 Covariances

We present here several characterizations for the members of the correlation cone and polytope; they are counterparts to the results of Section 4.2, via the covariance mapping. We first introduce the notion of M-covariance. This notion is studied in Assouad [1979, 1980b] for M being a subset of a Hilbert space. We consider here only the cases when $M = \mathbb{R}$ or $M = \{0,1\}$.

Definition 5.3.1. *Let M be a subset of $\mathbb{R}$. A symmetric function $p : X \times X \longrightarrow \mathbb{R}$ is called an M-*covariance *if there exist a measure space $(\Omega, \mathcal{A}, \mu)$ and functions $f_x \in L_2(\Omega, \mathcal{A}, \mu)$ (for $x \in X$) taking values in M, and such that*

$$p(x,y) = \int_\Omega f_x(\omega) f_y(\omega) \mu(d\omega) \text{ for all } x, y \in X.$$

*In particular, p is a $\{0,1\}$-*covariance *if and only if there exist a measure space $(\Omega, \mathcal{A}, \mu)$ and sets $A_x \in \mathcal{A}_\mu$ (for $x \in X$) such that*

$$p(x,y) = \mu(A_x \cap A_y) \text{ for all } x, y \in X.$$

The next lemma shows how $\mathbb{R}$-covariances and $\{0,1\}$-covariances are related to L_2- and L_1-embeddable distance spaces, respectively (via the covariance mapping).

Lemma 5.3.2. *Let X be a set and $x_0 \in X$. Let d be a distance on X and let $p = \xi_{x_0}(d)$ be the corresponding symmetric function on $X \setminus \{x_0\}$. Then,*

(i) $(X, \sqrt{d})$ *is* L_2*-embeddable if and only if* p *is an* $\mathbb{R}$*-covariance on* $X \setminus \{x_0\}$.

(ii) (X, d) *is* L_1*-embeddable if and only if* p *is a* $\{0,1\}$*-covariance on* $X \setminus \{x_0\}$.

Proof. (i) is an immediate verification; (ii) too, using Lemma 4.2.5. ∎

Therefore, for X finite, p is a $\{0,1\}$-covariance on X if and only if p belongs to the correlation cone COR(X). The following finitude result is a consequence of Lemma 5.3.2 and Theorem 3.2.1.

Proposition 5.3.3. *Let* p *be a symmetric function on* X*. Then,* p *is a* $\{0,1\}$*-covariance on* X *if and only if, for each finite subset* Y *of* X*, the restriction of* p *to* Y *is a* $\{0,1\}$*-covariance on* Y. ∎

We now give an interpretation of the members of the correlation cone and polytope in terms of correlations of events in a measure space; it is the analogue of Proposition 4.2.1 (via the covariance mapping). It was rediscovered in Pitowsky [1986].

Proposition 5.3.4. *Let* $p = (p_{ij})_{1 \leq i \leq j \leq n} \in \mathbb{R}^{V_n \cup E_n}$*. The following assertions are equivalent.*

(i) $p \in \mathrm{COR}_n$ *(resp.* $p \in \mathrm{COR}_n^{\square}$*).*

(ii) *There exist a measure space (resp. a probability space)* $(\Omega, \mathcal{A}, \mu)$ *and events* $A_1, \ldots, A_n \in \mathcal{A}$ *such that* $p_{ij} = \mu(A_i \cap A_j)$ *for all* $1 \leq i \leq j \leq n$. ∎

As a consequence of Proposition 5.3.4, every inequality valid for the correlation polytope $\mathrm{COR}_n^{\square}$ can be interpreted as an inequality that is satisfied by the joined probabilities of a set of n events. Consider, for instance, the inequalities (on the "correlation side") corresponding to the first four triangle inequalities in Figure 5.2.6. They express some very simple properties of joined probabilities. The first three express the fact that the probability $\mu(A_i \cap A_j)$ of the intersection of two events A_i, A_j is nonnegative and less than or equal to each of the probabilities $\mu(A_i)$, $\mu(A_j)$. The fourth one simply says that the probability $\mu(A_i \cup A_j)$ of the union of two events is less than or equal to one.

For the members of the correlation cone which can be written as a nonnegative *integer* linear combination of correlation vectors, we can assume that the measure space in Proposition 5.3.4 (ii) is endowed with the cardinality measure. Namely, we have the following result, which is an analogue of Proposition 4.2.4 (i) $\Longleftrightarrow$ (iii) (via the covariance mapping).

Proposition 5.3.5. *Let* $p = (p_{ij})_{1 \leq i \leq j \leq n} \in \mathbb{R}^{V_n \cup E_n}$*. The following assertions are equivalent.*

(i) $p = \sum_{S \subseteq V_n} \lambda_S \pi(S)$ *for some nonnegative integers* λ_S.

(ii) *There exist a finite set* Ω *and* n *subsets* $A_1, \ldots, A_n$ *of* Ω *such that* $p_{ij} = |A_i \cap A_j|$ *for all* $1 \leq i \leq j \leq n$. ∎

"cut side"		"correlation side"
$d \in \text{CUT}_{n+1}$ (resp. $d \in \text{CUT}^{\square}_{n+1}$)	if and only if there exist a measure space (resp. a probability space) $(\Omega, \mathcal{A}, \mu)$ and $A_1, \ldots, A_n \in \mathcal{A}$ of finite measure such that	$p \in \text{COR}_n$ (resp. $p \in \text{COR}^{\square}_n$)
$d_{ij} = \mu(A_i \triangle A_j)$ for $1 \leq i < j \leq n+1$ (setting $A_{n+1} = \emptyset$)		$p_{ij} = \mu(A_i \cap A_j)$ for $1 \leq i \leq j \leq n$
$d = \sum_S \lambda_S \delta(S)$ for some $\lambda_S \in \mathbb{Z}_+$ for all S	if and only if there exist a set Ω and finite subsets $A_1, \ldots, A_n$ of Ω such that	$p = \sum_S \lambda_S \pi(S)$ for some $\lambda_S \in \mathbb{Z}_+$ for all S
$d_{ij} = |A_i \triangle A_j|$ for $1 \leq i < j \leq n+1$ (setting $A_{n+1} = \emptyset$)		$p_{ij} = |A_i \cap A_j|$ for $1 \leq i \leq j \leq n$

Figure 5.3.6: Corresponding interpretations for members of cut and correlation polyhedra

A vector p satisfying the condition (ii) from Proposition 5.3.5 is called an *intersection pattern*. Hence, the intersection patterns of order n and the hypercube embeddable distances on $n+1$ points are equivalent notions (via the covariance mapping). Testing whether a given vector p is an intersection pattern is a hard problem; Chvátal [1980] shows that this problem is NP-complete when restricted to the vectors p with $p_{ii} = 3$ for all $i \in V_n$. On the other hand, the problem becomes polynomial when restricted to the vectors p with $p_{ii} = 2$ for all $i \in V_n$. We refer to Chapter 24 for further results related to this problem.

Figure 5.3.6 shows in parallel the results from Propositions 5.3.4 and 5.3.5 for covariances, and the corresponding results for distances from Propositions 4.2.1 and 4.2.4.

5.4 The Boole Problem

We describe here an application of the interpretation of the correlation polytope given in Proposition 5.3.4 to the following problem, known as the *Boole problem.* Let $(\Omega, \mathcal{A}, \mu)$ be a probability space and let $A_1, \ldots, A_n$ be n events of $\mathcal{A}$. Classical questions, which go back to Boole [1854], are the following:

> *Suppose we are given the values $p_i := \mu(A_i)$ for $1 \leq i \leq n$, what is the best estimation of $\mu(A_1 \cup \ldots \cup A_n)$ in terms of the p_i's ?*
>
> *Suppose we are given the values $p_i := \mu(A_i)$ for $1 \leq i \leq n$ and the values of the joint probabilities $p_{ij} := \mu(A_i \cap A_j)$ for $1 \leq i < j \leq n$. What is the best estimation of $\mu(A_1 \cup \ldots \cup A_n)$ in terms of the p_i's and the p_{ij}'s ?*

It is easy to see that the first question can be answered in the following manner:

$$\max(p_1, \ldots, p_n) \leq \mu(A_1 \cup \ldots \cup A_n) \leq \min(1, \sum_{1 \leq i \leq n} p_i).$$

As we see below, the answer to the second question involves, in fact, the inequalities that define facets of the correlation polytope $\mathrm{COR}_n^\square$ and of another related polytope. Namely, we have the following lower bound:

$$\mu(A_1 \cup \ldots \cup A_n) \geq \max(w^T p \mid w^T x \leq 1 \text{ is facet defining for } \mathrm{COR}_n^\square)$$

(see Proposition 5.4.3 and relation (5.4.4)) and an upper bound is given by the quantity $z_{\max}$ defined in (5.4.5). These estimations for $\mu(A_1 \cup \ldots \cup A_n)$ can be obtained using linear programming techniques. This approach, that we describe below, was considered, in particular, by Kounias and Marin [1976] and Pitowsky [1991].

Let p denote the vector of $\mathbb{R}^{V_n \cup E_n}$ defined by $p_i := \mu(A_i)$ for $1 \leq i \leq n$ and $p_{ij} := \mu(A_i \cap A_j)$ for $1 \leq i < j \leq n$. By Proposition 5.3.4, p belongs to the polytope $\mathrm{COR}_n^\square$. Thus, we can define the following quantities[1] $z_{\min}$ and $z_{\max}$:

$$z_{\min} := \min \Big(\sum_{\emptyset \neq S \subseteq V_n} \lambda_S \mid \sum_{\emptyset \neq S \subseteq V_n} \lambda_S \pi(S) = p, \ \lambda_S \geq 0 \text{ for } \emptyset \neq S \subseteq V_n \Big), \tag{5.4.1}$$

$$z_{\max} := \max \Big(\sum_{\emptyset \neq S \subseteq V_n} \lambda_S \mid \sum_{\emptyset \neq S \subseteq V_n} \lambda_S \pi(S) = p, \ \lambda_S \geq 0 \text{ for } \emptyset \neq S \subseteq V_n \Big). \tag{5.4.2}$$

The quantities $z_{\min}$ and $z_{\max}$ provide bounds for $\mu(A_1 \cup \ldots \cup A_n)$, as the next result shows.

Proposition 5.4.3. $z_{\min} \leq \mu(A_1 \cup \ldots \cup A_n) \leq z_{\max}$.

[1] Note that the parameter $z_{\min}$ is the analogue for the correlation cone of the notion of minimum ℓ_1-size, defined in (4.3.4).

Proof. For $S \subseteq V_n$, set

$$A^S := \bigcap_{i \in S} A_i \cap \bigcap_{i \notin S} (\Omega \setminus A_i).$$

Then,

$$A_i \cap A_j = \bigcup_{S \subseteq V_n | i,j \in S} A^S, \ \Omega = \bigcup_{S \subseteq V_n} A^S, \text{ and } A_1 \cup \ldots \cup A_n = \bigcup_{S \subseteq V_n | S \neq \emptyset} A^S.$$

Therefore,

$$p = \sum_{S \subseteq V_n | S \neq \emptyset} \mu(A^S) \pi(S),$$

with $\mu(A^S) \geq 0$ for all S. Hence, $(\mu(A^S) \mid \emptyset \neq S \subseteq V_n)$ is a feasible solution to the programs (5.4.1) and (5.4.2), with objective value $\mu(A_1 \cup \ldots \cup A_n)$. This shows the result. ∎

Using linear programming duality, we can reformulate $z_{\min}$ and $z_{\max}$. Namely,

(5.4.4) $$z_{\min} = \max(w^T p \mid w^T \pi(S) \leq 1 \text{ for all } S, \ \emptyset \neq S \subseteq V_n)$$

and, as can be easily verified, it suffices to consider in (5.4.4) the vectors w for which the inequality $w^T x \leq 1$ defines a facet of $\mathrm{COR}_n^{\square}$. Similarly,

(5.4.5) $$z_{\max} = \min(w^T p \mid w^T \pi(S) \geq 1 \ \text{ for all } S, \ \emptyset \neq S \subseteq V_n),$$

where it suffices to consider the vectors w for which the inequality $w^T x \geq 1$ defines a facet of the polytope $\mathrm{Conv}(\{\pi(S) \mid \emptyset \neq S \subseteq V_n\})$ (which is distinct from $\mathrm{COR}_n^{\square}$ since it does not contain the origin).

Therefore, every valid inequality for $\mathrm{COR}_n^{\square}$ yields a lower bound for $\mu(A_1 \cup \ldots \cup A_n)$ in terms of the joint probabilities $p_{ij} = \mu(A_i \cap A_j)$. Many such inequalities are known; cf. Part V for a presentation of large classes of such inequalities. As an illustration, we now mention a few examples of such inequalities together with the corresponding lower bounds.

A first observation is that

$$\frac{n \sum_{1 \leq i \leq n} p_i - 2 \sum_{1 \leq i < j \leq n} p_{ij}}{\lfloor \frac{n+1}{2} \rfloor \lceil \frac{n+1}{2} \rceil} \leq \sum_{\emptyset \neq S \subseteq V_n} \lambda_S \leq \frac{n \sum_{1 \leq i \leq n} p_i - 2 \sum_{1 \leq i < j \leq n} p_{ij}}{n}$$

for any decomposition $p = \sum_{\emptyset \neq S \subseteq V_n} \lambda_S \pi(S)$ with $\lambda_S \geq 0$ for all S. (This follows from (4.3.6), via the covariance mapping.) From the definition of $z_{\min}$, $z_{\max}$ and from Proposition 5.4.3, we obtain:

$$(5.4.6)\qquad \begin{array}{rcl} \dfrac{n \sum\limits_{1\le i\le n} p_i - 2 \sum\limits_{1\le i<j\le n} p_{ij}}{\lfloor \frac{n+1}{2}\rfloor \lceil \frac{n+1}{2}\rceil} & \le & \mu(A_1 \cup \ldots \cup A_n) \\ \mu(A_1 \cup \ldots \cup A_n) & \le & \dfrac{n \sum\limits_{1\le i\le n} p_i - 2 \sum\limits_{1\le i<j\le n} p_{ij}}{n}. \end{array}$$

Consider now the inequality:

$$(5.4.7)\qquad 2k \sum_{1\le i\le n} p_i - 2 \sum_{1\le i<j\le n} p_{ij} \le k(k+1).$$

It is valid for the correlation polytope $\mathrm{COR}_n^\square$ if $1 \le k \le n-1$. (Moreover, it is facet defining if $1 \le k \le n-2$ and $n \ge 4$. Indeed, it corresponds (via the covariance mapping) to the inequality:

$$\sum_{1\le i<j\le n} x_{ij} + (2k+1-n) \sum_{1\le i\le n} x_{i,n+1} \le k(k+1)$$

which defines a facet of $\mathrm{CUT}_{n+1}^\square$ if $1 \le k \le n-2$ and $n \ge 4$; see Theorem 28.2.4.) This yields the following lower bound for $\mu(A_1 \cup \ldots \cup A_n)$:

$$(5.4.8)\qquad \frac{2}{k+1} \sum_{1\le i\le n} p_i - \frac{2}{k(k+1)} \sum_{1\le i<j\le n} p_{ij} \le \mu(A_1 \cup \ldots \cup A_n)$$

for each k, $1 \le k \le n-1$. The bound (5.4.8) was found independently by several authors, including Chung [1941], Dawson and Sankoff [1967], Galambos [1977]. Note that (5.4.8) coincides with the lower bound of (5.4.6) in the case $n = 2k$. The case $k = 1$ of (5.4.8) gives the bound

$$\sum_{1\le i\le n} p_i - \sum_{1\le i<j\le n} p_{ij} \le \mu(A_1 \cup \ldots \cup A_n)$$

which is a special case of the Bonferroni bound (5.4.14) mentioned below. More generally, given integers $b_1, \ldots, b_n$ and $k \ge 0$, the inequality:

$$(5.4.9)\qquad \sum_{1\le i\le n} b_i(2k+1-b_i)p_i - 2 \sum_{1\le i<j\le n} b_i b_j p_{ij} \le k(k+1)$$

is valid for $\mathrm{COR}_n^\square$, which yields the bound:

$$\frac{1}{k(k+1)} \left(\sum_{1\le i\le n} p_i b_i (2k+1-b_i) - 2 \sum_{1\le i<j\le n} b_i b_j p_{ij} \right) \le \mu(A_1 \cup \ldots \cup A_n).$$

To see the validity of inequality (5.4.9), note that it can alternatively be written as

$$(5.4.10)\qquad \left(\sum_{1\le i\le n} b_i p_i - k \right) \left(\sum_{1\le i\le n} b_i p_i - k - 1 \right) \ge 0$$

with the convention that, when developing the product, the expression $p_i p_j$ is replaced by the variable p_{ij} (setting $p_{ii} = p_i$).

Remark 5.4.11. The inequality (5.4.9) (or (5.4.10)) (or special cases of it) was considered independently by many authors; among others, by Kelly [1968], Davidson [1969], Yoseloff [1970], McRae and Davidson [1972], Kounias and Marin [1976], Erdahl [1987], Mestechkin [1987], Pitowsky [1991]. The inequality (5.4.10) appears in Figure 5.2.6; it corresponds (via the covariance mapping and after setting $b_{n+1} := 2k+1-\sum_{i=1}^{n} b_i$) to the inequality:

$$\sum_{1\leq i<j\leq n+1} b_i b_j x_{ij} \leq k(k+1), \tag{5.4.12}$$

which is valid for the cut polytope $\mathrm{CUT}^{\square}_{n+1}$. In order to help the reader understand how this inequality relates with further inequalities to be introduced later, let us mention that the class of inequalities of the form (5.4.12) contains the hypermetric inequalities (to be defined in Section 6.1) as special instances. More precisely, (5.4.12) is a hypermetric inequality if $k=0$. Moreover, (5.4.12) is a switching of a hypermetric inequality if the sequence $b_1,\ldots,b_{n+1}$ has gap 1. (The notions of switching and gap will be defined later in Sections 26.3 and 28.4.) ∎

Generalization to Higher Order Correlations. Clearly, much of the above treatment can be generalized to higher order correlations. Namely, let $\mathcal{I}$ be a family of subsets of V_n. Given a subset S of V_n, its *$\mathcal{I}$-correlation vector* $\pi^{\mathcal{I}}(S) \in \mathbb{R}^{\mathcal{I}}$ is defined by $\pi^{\mathcal{I}}(S)_I = 1$ if $I \subseteq S$ and $\pi^{\mathcal{I}}(S)_I = 0$ otherwise, for all $I \in \mathcal{I}$. Then, the cone $\mathrm{COR}_n(\mathcal{I})$ (resp. the polytope $\mathrm{COR}^{\square}_n(\mathcal{I})$) is defined as the conic hull (resp. the convex hull) of all $\mathcal{I}$-correlation vectors $\pi^{\mathcal{I}}(S)$ for $S \subseteq V_n$.

Given an integer $1 \leq m \leq n$, let $\mathcal{I}_{\leq m}$ denote the collection of all subsets of V_n of cardinality less than or equal to m. Hence, $\mathcal{I}_{\leq 2}$ consists of all singletons and pairs of elements of V_n and, therefore, $\mathrm{COR}_n(\mathcal{I}_{\leq 2})$ and $\mathrm{COR}^{\square}_n(\mathcal{I}_{\leq 2})$ coincide, respectively, with COR_n and $\mathrm{COR}^{\square}_n$.

For $\mathcal{I} = 2^{V_n}$, which consists of all subsets of V_n, $\mathrm{COR}^{\square}_n(2^{V_n})$ is a simplex and $\mathrm{COR}_n(2^{V_n})$ is a simplex cone, both of dimension 2^n-1. This implies, in particular, that every correlation polytope $\mathrm{COR}^{\square}_n(\mathcal{I})$ arises as a projection of the simplex $\mathrm{COR}^{\square}_n(2^{V_n})$ (namely, on the subspace $\mathbb{R}^{\mathcal{I}}$).

The result from Proposition 5.3.4 extends easily to the case of arbitrary $\mathcal{I}$-correlations.

Proposition 5.4.13. *Let $\mathcal{I}$ be a nonempty collection of subsets of $\{1,\ldots,n\}$ and let $p=(p_I)_{I\in\mathcal{I}} \in \mathbb{R}^{\mathcal{I}}$. The following assertions are equivalent.*

(i) $p \in \mathrm{COR}_n(\mathcal{I})$ *(resp.* $p \in \mathrm{COR}^{\square}_n(\mathcal{I})$*).*

(ii) *There exist a measure space (resp. a probability space) $(\Omega,\mathcal{A},\mu)$ and events $A_1,\ldots,A_n \in \mathcal{A}$ such that $p_I = \mu(\bigcap_{i\in I} A_i)$ for all $I \in \mathcal{I}$.* ∎

A more general version of the Boole problem consists of finding estimates for the quantity $\mu(A_1 \cup \ldots \cup A_n)$ in terms of the joined correlations $\mu(\bigcap_{i\in I} A_i)$ for $I \in \mathcal{I}$. There is an obvious analogue of Proposition 5.4.3, where the bounds $z_{\min}$ and $z_{\max}$ are now in terms of the polytopes $\mathrm{COR}^{\square}_n(\mathcal{I})$ and $\mathrm{Conv}(\{\pi^{\mathcal{I}}(S) \mid \emptyset \neq S \subseteq V_n\})$ (instead of $\mathrm{COR}^{\square}_n$ and $\mathrm{Conv}(\{\pi(S) \mid \emptyset \neq S \subseteq V_n\})$).

In the case when $\mathcal{I} = \mathcal{I}_{\leq m}$, several bounds for $\mu(A_1 \cup \ldots \cup A_n)$ have been proposed in the literature in terms of the quantities:

$$S_k := \sum_{1 \leq i_1 < \ldots < i_k \leq n} \mu(A_{i_1} \cap \ldots \cap A_{i_k})$$

for $1 \leq k \leq n$. For instance, the following bounds hold:

$$\text{(5.4.14)} \qquad \begin{cases} \mu(A_1 \cup \ldots \cup A_n) \geq \sum_{1 \leq i \leq m} (-1)^{i-1} S_i & \text{for } m \text{ even}, \\ \mu(A_1 \cup \ldots \cup A_n) \leq \sum_{1 \leq i \leq m} (-1)^{i-1} S_i & \text{for } m \text{ odd}, \end{cases}$$

which were first discovered by Bonferroni [1936]. Several improvements of these bounds have been later proposed; see, e.g., Boros and Prekopa [1989], Grable [1993].

Clearly, if all the quantities S_k $(1 \leq k \leq n)$ are known, then the exact value of $\mu(A_1 \cup \ldots \cup A_n)$ is given by the inclusion-exclusion formula:

$$\mu(A_1 \cup \ldots \cup A_n) = \sum_{1 \leq k \leq n} (-1)^{k-1} S_k.$$

The error with which $\mu(A_1 \cup \ldots \cup A_n)$ can be approximated when knowing S_j only for $j \leq k$ (where $k \leq n$ is given) has been studied by Linial and Nisan [1990] and Kahn, Linial and Samorodnitsky [1997]. Let $A_1, \ldots, A_n, B_1, \ldots, B_n$ be events in a probability space $(\Omega, \mathcal{A}, \mu)$ satisfying

$$\mu(\bigcap_{i \in I} A_i) = \mu(\bigcap_{i \in I} B_i)$$

for all $I \subseteq \{1, \ldots, n\}$ with $|I| \leq k$. Then, Linial and Nisan [1990] show that

$$\frac{\mu(A_1 \cup \ldots \cup A_n)}{\mu(B_1 \cup \ldots \cup B_n)} \leq \left(\frac{\lambda^k + 1}{\lambda^k - 1}\right)^2, \quad \text{where } \lambda := \frac{\sqrt{n} + 1}{\sqrt{n} - 1}.$$

In particular, the ratio $\frac{\mu(A_1 \cup \ldots \cup A_n)}{\mu(B_1 \cup \ldots \cup B_n)}$ is bounded by $1 + O(\exp(-\frac{2k}{\sqrt{n}}))$ if $k = \Omega(\sqrt{n})$ and by $O(\frac{n}{k^2})$ if $k = O(\sqrt{n})$. Recently, Kahn, Linial and Samorodnitsky [1997] show that

$$|\mu(A_1 \cup \ldots \cup A_n) - \mu(B_1 \cup \ldots \cup B_n)| = \exp(-\Omega(\frac{k^2}{n \log_2 n})).$$

Moreover, there exist coefficients λ_j $(1 \leq j \leq k)$ which can be found in time polynomial in n and satisfying

$$|\mu(A_1 \cup \ldots \cup A_n) - \sum_{1 \leq j \leq k} \lambda_j S_j| \leq \exp(-\Omega(\frac{k^2}{n \log_2 n})).$$

The problem of evaluating the probability $\mu(A_1 \cup \ldots \cup A_n)$ has many applications; see, e.g., Kahn, Linial and Samorodnitsky [1997]. An example of application is to the problem of enumerating the satisfying assignments of an n-variable DNF expression $F = C_1 \vee \ldots \vee C_m$, where each clause C_j is in conjonctive form. This is a hard problem; much effort has been done for approximating this number (see Luby and Veličković [1991] and references therein). If we let A_j $(j = 1, \ldots, m)$ denote the set of satisfying assignments for the clause C_j, then the number of satisfying assignments for F is $|A_1 \cup \ldots \cup A_m|$. It is shown in Kahn, Linial and Samorodnitsky [1997] that $|A_1 \cup \ldots \cup A_m|$ is uniquely determined once one knows the number of satisfying assignments for $\wedge_{i \in I} C_i$ for every $I \subseteq \{1, \ldots, m\}$ such that $|I| \leq \log_2 n + 1$.

Chapter 6. Conditions for L_1-Embeddability

We present in this chapter two necessary conditions for L_1-embeddability, namely, the hypermetric and the negative type conditions. There are many other known necessary conditions, arising from the known valid inequalities for the cut cone; they will be described in Part V. We focus here on the hypermetric and negative type conditions. These conditions seem indeed to be among the most essential ones. For instance, there are several classes of distance spaces for which these conditions are also sufficient for ensuring L_1-embeddability; see, in particular, Chapters 8, 19, 24, and Remark 6.3.5 for a summary. Hypermetric inequalities will be treated in detail for their own sake in Part II and, as facets of the cut cone, in Chapter 28.

The present chapter is organized as follows. The hypermetric and negative type conditions are introduced in Section 6.1. Several characterizations for ℓ_2-embeddable spaces are presented in Section 6.2. We present, in particular, a characterization in terms of the negative type condition and Menger's result concerning the isometric subspaces of the m-dimensional Euclidean space. Section 6.3 contains a summary of the implications that exist between the properties of being L_1-, L_2-embeddable, of negative type, or hypermetric, for a distance space. We treat in some detail in Section 6.4 an example: the spherical distance space, which consists of the sphere equipped with the usual great circle distance. This example is, in a sense, intermediate between ℓ_1 and ℓ_2. Indeed, every spherical distance space is ℓ_1-embeddable and, on the other hand, the Euclidean distance can be realized asymptotically as a limit of spherical distances. The spherical distance space will be useful in Section 31.3 for the positive semidefinite completion problem.

6.1 Hypermetric and Negative Type Conditions

6.1.1 Hypermetric and Negative Type Inequalities

Let $n \geq 2$ and let $b_1, \ldots, b_n$ be integers. We consider the inequality:

$$\sum_{1 \leq i < j \leq n} b_i b_j d_{ij} \leq 0 \tag{6.1.1}$$

(in the variable d_{ij}). For convenience, we introduce the following notation. Given

$b \in \mathbb{R}^n$, $Q_n(b)$ denotes the vector of $\mathbb{R}^{E_n}$ defined by

$$Q_n(b)_{ij} := b_i b_j \quad \text{for } 1 \leq i < j \leq n.$$

Hence, the inequality (6.1.1) can be rewritten as $Q_n(b)^T d \leq 0$. When the parameter n is clear from the context we also denote $Q_n(b)$ by $Q(b)$. We can suppose that at least two of the b_i's are nonzero; else, $Q_n(b) = 0$ and the inequality (6.1.1) is void.

When $\sum_{i=1}^n b_i = 1$, the inequality (6.1.1) is called a *hypermetric inequality* and, when $\sum_{i=1}^n b_i = 0$, it is called a *negative type inequality*. The inequality (6.1.1) is said to be *pure* if $|b_i| = 0, 1$ for all $i \in V_n$. The inequality (6.1.8) is said to be a *k-gonal inequality* if $\sum_{i=1}^n |b_i| = k$ holds. Note that k and $\sum_{i=1}^n b_i$ have the same parity.

In particular, the 2-gonal inequality is the inequality of negative type (6.1.1), where $b_i = 1$, $b_j = -1$ and $b_h = 0$ for $h \in V_n \setminus \{i, j\}$, for some distinct $i, j \in V_n$; it is nothing but the nonnegativity constraint $d_{ij} \geq 0$. The pure 3-gonal inequality is the hypermetric inequality (6.1.1), where $b_i = b_j = 1$, $b_k = -1$ and $b_h = 0$ for $h \in V_n \setminus \{i, j, k\}$, for some distinct $i, j, k \in V_n$; it coincides with the triangle inequality (3.1.1). For $\epsilon = 0, 1$, the pure $(2k + \epsilon)$-gonal inequality reads :

$$\sum_{1 \leq r < s \leq k+\epsilon} d_{i_r i_s} + \sum_{1 \leq r < s \leq k} d_{j_r j_s} - \sum_{\substack{1 \leq r \leq k+\epsilon \\ 1 \leq s \leq k}} d_{i_r j_s} \leq 0,$$

where $i_1, \ldots, i_k, i_{k+\epsilon}, j_1, \ldots, j_k$ are distinct indices of V_n.

As an example, the 5-gonal inequalities are the inequalities $Q_n(b)^T d \leq 0$, where b is (up to permutation of its components) one of the following vectors:

$$b = (1, 1, 1, -1, -1, 0, \ldots, 0), \ b = (1, 1, 1, -2, 0, \ldots, 0),$$
$$b = (2, 1, -1, -1, 0, \ldots, 0), b = (3, -1, -1, 0, \ldots, 0),$$
$$b = (2, 1, -2, 0, \ldots, 0), \ b = (3, -2, 0, \ldots, 0).$$

Figure 6.1.2 shows the pure 4-gonal and 5-gonal inequalities or, rather, their left hand sides. It should be understood as follows: a plain edge between two nodes i and j indicates a coefficient +1 for the variable d_{ij} and a dotted edge indicates a coefficient -1. The pure 5-gonal inequality is also called the *pentagonal inequality*.

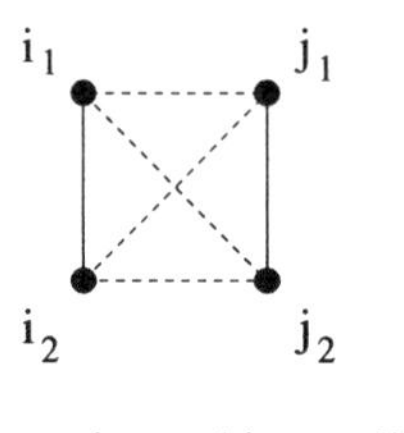

pure 4-gonal inequality

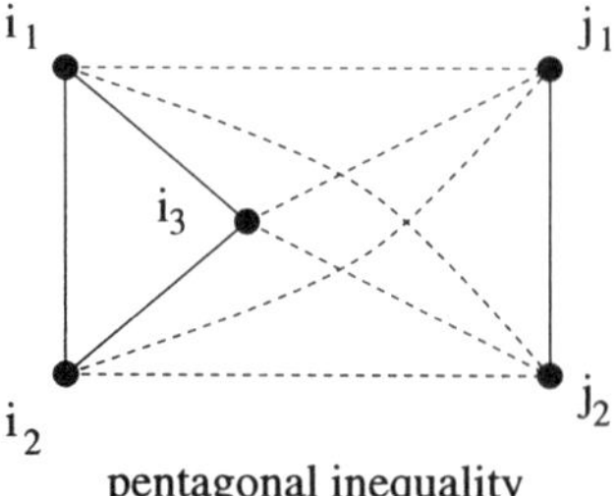

pentagonal inequality

Figure 6.1.2

The negative type inequalities are classical inequalities in analysis; they were used, in particular, by Schoenberg [1937, 1938a, 1938b]. The hypermetric inequalities were considered by several authors, including Deza [1960, 1962], Kelly [1970a], Baranovskii [1971, 1973] and, in the context of correlations or boolean quadratic programming (i.e., under the form indicated in Lemma 6.1.14; see also Figure 5.2.6), by Kelly [1968], Davidson [1969], Yoseloff [1970], McRae and Davidson [1972], Kounias and Marin [1976], Erdahl [1987], Mestechkin [1987], Pitowsky [1991]. (Recall Remark 5.4.11.)

The hypermetric inequalities: $Q_n(b)^T d \leq 0$ for $b \in \mathbb{Z}^n$ with $\sum_{i=1}^n b_i = 1$, define a cone in $\mathbb{R}^{E_n}$, called the *hypermetric cone* and denoted by HYP_n. Similarly, the *negative type cone*[1] NEG_n is the cone in $\mathbb{R}^{E_n}$, which is defined by the negative type inequalities: $Q_n(b)^T d \leq 0$ for $b \in \mathbb{Z}^n$ with $\sum_{i=1}^n b_i = 0$. If we consider an arbitrary finite set X instead of V_n, then we also denote the hypermetric cone by $\mathrm{HYP}(X)$.

In fact, the negative type inequalities are implied by the hypermetric inequalities. In other words,

$$\mathrm{HYP}_n \subseteq \mathrm{NEG}_n \quad \text{for all } n \geq 3.$$

This result can be read immediately from Figure 5.2.6 (by looking at the corresponding inequalities on the "correlation side"). It can also be derived from the following result of Deza [1962] which shows, more precisely, how $(2k+1)$- and $(2k+2)$-gonal inequalities relate.

Proposition 6.1.3. *Let $k \geq 1$ be an integer. The $(2k+2)$-gonal inequalities are implied by the $(2k+1)$-gonal inequalities.*

Proof. Let $b \in \mathbb{Z}^n$ with $\sum_{i=1}^n b_i = 0$ and $\sum_{i=1}^n |b_i| = 2k+2$. We show that the inequality $Q_n(b)^T d \leq 0$ can be expressed as a nonnegative linear combination of $(2k+1)$-gonal inequalities. We can suppose without loss of generality that $b_1, \ldots, b_p > 0 > b_{p+1}, \ldots, b_n$, for some p, $1 \leq p \leq n-1$. For $1 \leq i \leq p$, set

$$c^{(i)} := (-b_1, \ldots, -b_{i-1}, 1 - b_i, -b_{i+1}, \ldots, -b_p, -b_{p+1}, \ldots, -b_n)$$

and, for $p+1 \leq i \leq n$, set

$$c^{(i)} := (b_1, \ldots, b_p, b_{p+1}, \ldots, b_{i-1}, b_i + 1, b_{i+1}, \ldots, b_n).$$

Then, each vector $c^{(i)}$ belongs to $\mathbb{Z}^n$, has sum of entries 1, and sum of absolute values of its entries $2k+1$. Therefore, each inequality $Q_n(c^{(i)})^T d \leq 0$ is a $(2k+1)$-gonal inequality. Observe now that

$$\sum_{1 \leq i \leq n} |b_i| Q_n(c^{(i)}) = 2k Q_n(b).$$

[1] We will consider in Section 31.4 another cone related to NEG_n. Namely, given a graph $G = (V_n, E)$, the cone $\mathrm{NEG}(G)$ is defined as the projection of NEG_n on the edge set on the subspace $\mathbb{R}^E$ indexed by the edge set of G.

This shows that the $(2k+2)$-gonal inequality $Q_n(b)^T d \leq 0$ is implied by the $(2k+1)$-gonal inequalities $Q_n(c^{(i)})^T d \leq 0$ $(1 \leq i \leq n)$. ∎

As an example, the 4-gonal inequality: $Q_4(1,1,-1,-1)^T d \leq 0$ follows by summation of the following 3-gonal inequalities:

$$Q_4(1,1,-1,0)^T d \leq 0,$$
$$Q_4(1,1,0,-1)^T d \leq 0,$$
$$Q_4(-1,0,1,1)^T d \leq 0,$$
$$Q_4(0,-1,1,1)^T d \leq 0.$$

Corollary 6.1.4. *The negative type inequalities are implied by the hypermetric inequalities.* ∎

Remark 6.1.5. Note that the negative type inequalities do *not* imply the triangle inequalities. In other words, a distance may be of negative type without being a semimetric; that is, the negative type cone NEG_n is not contained in the semimetric cone MET_n. To see it, consider for instance the distance d on V_n defined by $d_{1i} = 1$ for $i = 2, \ldots, n$ and $d_{ij} = \frac{2n}{n-1}$ for $2 \leq i < j \leq n$. Then, d violates some triangle inequalities, as $d_{ij} - d_{1j} - d_{1i} = \frac{2}{n-1} > 0$ for any $i \neq j \in \{2, \ldots, n\}$. On the other hand, it is easy to verify that $d \in \mathrm{NEG}_n$ (e.g., because its image $\xi_1(d)$ - under the covariance mapping pointed at position 1 - defines a positive semidefinite matrix). See also Remark 6.1.11, where it is observed that the k-gonal inequalities do not follow from the $(k+2)$-gonal inequalities.

On the other hand, for $d \in \mathrm{NEG}_n$, the condition $d_{1n} = 0$ implies that $d_{1i} = d_{in}$ for all $i = 2, \ldots, n-1$. (Hence, the metric condition is partially satisfied.) Moreover, letting d' denote the distance on the set $V_{n-1} = V_n \setminus \{n\}$ defined as the projection of d (i.e., $d'_{ij} := d_{ij}$ for all $i, j \in V_{n-1}$), then $d \in \mathrm{NEG}_n$ if and only if $d' \in \mathrm{NEG}_{n-1}$. In other words, for testing membership in the negative type cone, we can restrict ourselves to distances taking only positive values. The same holds clearly for the hypermetric cone. ∎

One of the main motivations for introducing hypermetric inequalities lies in the fact that they are valid for the cut cone, i.e.,

$$\text{(6.1.6)} \qquad \mathrm{CUT}_n \subseteq \mathrm{HYP}_n.$$

In other words,

Lemma 6.1.7. *Every distance space that is isometrically ℓ_1-embeddable satisfies all the hypermetric inequalities.*

Proof. It suffices to verify that every cut semimetric satisfies all the hypermetric inequalities. For this, let $S \subseteq V_n$ and $b \in \mathbb{Z}^n$ with $\sum_{i=1}^n b_i = 1$. Then,

$$\sum_{1 \le i < j \le n} b_i b_j \delta(S)_{ij} = \sum_{i \in S, j \notin S} b_i b_j = \left(\sum_{i \in S} b_i\right)\left(\sum_{j \notin S} b_j\right) = \left(\sum_{i \in S} b_i\right)\left(1 - \sum_{i \in S} b_i\right)$$

is nonpositive since $\sum_{i \in S} b_i$ is an integer. ∎

6.1.2 Hypermetric and Negative Type Distance Spaces

Let (X, d) be a distance space. Then, (X, d) is said to be *hypermetric* (resp. of *negative type*) if d satisfies all the hypermetric inequalities (resp. all the negative type inequalities), i.e., if d satisfies

$$\text{(6.1.8)} \qquad \sum_{1 \le i < j \le n} b_i b_j d(x_i, x_j) \le 0$$

for all $b \in \mathbb{Z}^n$ with $\sum_{i=1}^n b_i = 1$ (resp. with $\sum_{i=1}^n b_i = 0$) and for all distinct elements $x_1, \ldots, x_n \in X$ $(n \ge 2)$.

Observe that in the above definition we can drop the condition that the elements $x_1, \ldots, x_n$ be distinct. Indeed, suppose for instance that $x_1 = x_2$. Then, $d(x_1, x_2) = 0$ and $d(x_1, x_i) = d(x_2, x_i)$ for all i. Therefore, the quantity $\sum_{1 \le i < j \le n} b_i b_j d(x_i, x_j)$ can be rewritten as $\sum_{2 \le i < j \le n} b'_i b'_j d(x_i, x_j)$, after setting $b'_2 = b_1 + b_2, b'_3 = b_3, \ldots, b'_n = b_n$.

In other words, (X, d) is hypermetric (resp. of negative type) if and only if d satisfies the inequalities (6.1.8) for all $b \in \{0, -1, 1\}^n$ with $\sum_{i=1}^n b_i = 1$ (resp. $= 0$) and all (not necessarily distinct) elements $x_1, \ldots, x_n \in X$ $(n \ge 2)$.

Given an integer $k \ge 1$ and $\epsilon \in \{0, 1\}$, a distance space (X, d) is said to be $(2k+\epsilon)$*-gonal* if d satisfies the inequalities (6.1.8) for all $b \in \mathbb{Z}^n$ with $\sum_{i=1}^n b_i = \epsilon$ and $\sum_{i=1}^n |b_i| = 2k + \epsilon$ and for all $x_1, \ldots, x_n \in X$ $(n \ge 2)$. Again we obtain the same definition if we require that d satisfies all these inequalities only for b pure, i.e., with entries in $\{0, 1, -1\}$. For instance, (X, d) is 5-gonal if and only if, for all $x_1, x_2, x_3, y_1, y_2 \in X$, the following inequality holds:

$$\text{(6.1.9)} \qquad \sum_{1 \le i < j \le 3} d(x_i, x_j) + d(y_1, y_2) - \sum_{\substack{i=1,2,3 \\ j=1,2}} d(x_i, y_j) \le 0.$$

This is the pentagonal inequality, that we rewrite here for further reference.

Observe that the notion of k-gonal distance spaces is monotone in k, in the sense that $(k+2)$-gonality implies k-gonality. Namely,

Lemma 6.1.10. *Let (X, d) be a distance space.*

(i) *If (X, d) is $(k+2)$-gonal, then (X, d) is k-gonal, for any integer $k \ge 2$.*

(ii) *If (X, d) is $(2k+1)$-gonal, then (X, d) is $(2k+2)$-gonal, for any integer $k \ge 1$.*

Proof. (i) Suppose that (X,d) is $(k+2)$-gonal. Let $b \in \mathbb{Z}^n$ with $\sum_{i=1}^n |b_i| = k$ and $\sum_{i=1}^n b_i = \epsilon$, where $\epsilon = 1$ if k is odd and $\epsilon = 0$ if k is even. Let $x_1, \ldots, x_n \in X$. We show that $\sum_{1\le i<j\le n} b_i d_j d(x_i,x_j) \le 0$. For this, set $b' := (b,1,-1) \in \mathbb{Z}^{n+2}$ and $x_{n+1} = x_{n+2} := x$, where $x \in X$. Then, $\sum_{1\le i<j\le n} b_i d_j d(x_i,x_j) = \sum_{1\le i<j\le n+2} b'_i b'_j d(x_i,x_j)$, which is nonpositive by the assumption that (X,d) is $(k+2)$-gonal. The assertion (ii) follows from Proposition 6.1.3. ∎

Remark 6.1.11. Note that the k-gonal inequalities do *not* follow from the $(k+2)$-gonal inequalities $(k \ge 2)$. (The proof of Lemma 6.1.10 (i) works indeed at the level of distance spaces since we make the assumption that the two points x_{n+1} and x_{n+2} of X coincide.). For instance, the 5-gonal inequalities do *not* imply the triangle inequalities. To see it, consider the distance d on V_5 defined by $d_{ij} = 1$ for all pairs except $d_{12} = \frac{9}{4}$ and $d_{34} = \frac{3}{2}$. Then, d violates some triangle inequality as $d_{12} - d_{13} - d_{23} = \frac{1}{4} > 0$; on the other hand, one can verify that d satisfies all 5-gonal inequalities. ∎

Remark 6.1.12. Equality case in the hypermetric and negative type inequalities. The following question is considered by Kelly [1970a], Assouad [1984], Ball [1990]. What are the distance spaces, within a given class, that satisfy a given hypermetric or negative type inequality at equality ?

For instance, Kelly [1970a] characterizes the finite subspaces of $(\mathbb{R}, d_{\ell_1})$ that satisfy the $(2k+1)$-gonal inequality at equality. Namely, given $x_1, \ldots, x_{k+1}, y_1, \ldots, y_k \in \mathbb{R}$, the equality

$$\sum_{1\le i<j\le k+1} |x_i - x_j| + \sum_{1\le i<j\le k} |y_i - y_j| - \sum_{\substack{1\le i\le k+1\\ 1\le j\le k}} |x_i - y_j| = 0$$

holds if and only if $y_1, \ldots, y_k$ separate $x_1, \ldots, x_{k+1}$, i.e., if there exist a permutation α of $\{1, \ldots, k+1\}$ and a permutation β of $\{1, \ldots, k\}$ such that

$$x_{\alpha(1)} \le y_{\beta(1)} \le x_{\alpha(2)} \le y_{\beta(2)} \le \cdots \le y_{\beta(k)} \le x_{\alpha(k+1)}.$$

This can be easily verified by looking at the explicit decomposition of the distance space $(\{x_1, \ldots, x_{k+1}, y_1, \ldots, y_k\}, d_{\ell_1})$ as a nonnegative sum of cuts (and using the construction from the proof of Proposition 4.2.2 (ii) $\Longrightarrow$ (i)). Generalizations and related results can be found in Kelly [1970a] and Assouad [1984].

Along the same lines, Ball [1990] characterizes the scalars $x_1, \ldots, x_n \in \mathbb{R}$ for which the distance space $(\{x_1, \ldots, x_n\}, d_{\ell_1})$ satisfies the negative type inequality (6.1.1) at equality when $b = (-(n-4), 1, \ldots, 1, -2)$. This result will be used in the proof of Proposition 11.2.4 (i), for deriving a lower bound on the minimum ℓ_1-dimension of a distance space. ∎

6.1.3 Analogues for Covariances

We indicate here how the hypermetric inequalities and the negative type inequalities translate, when transported to the context of correlations (via the covariance mapping). This information has already been mentioned in Figure 5.2.6; Lemma 6.1.14 below validates it. We first introduce a definition.

Definition 6.1.13. *A symmetric function $p : X \times X \longrightarrow \mathbb{R}$ is said to be of* positive type *on X if, for all $n \geq 2$, $x_1, \ldots, x_n \in X$, the matrix $(p(x_i, x_j))_{i,j=1}^n$ is positive semidefinite.*

Lemma 6.1.14. *Let X be a set and $x_0 \in X$. Let d be a distance on X and $p = \xi_{x_0}(d)$ be the corresponding symmetric function on $X \setminus \{x_0\}$.*

(i) *(X, d) is of negative type if and only if p is of positive type on $X \setminus \{x_0\}$.*

(ii) *(X, d) is hypermetric if and only if p satisfies:*

$$\sum_{1 \leq i,j \leq n} b_i b_j p(x_i, x_j) - \sum_{1 \leq i \leq n} b_i p(x_i, x_i) \geq 0,$$

for all $b \in \mathbb{Z}^n$ and all $x_1, \ldots, x_n \in X \setminus \{x_0\}$ $(n \geq 2)$.

Proof. Let $x_1, \ldots, x_n \in X \setminus \{x_0\}$, $b_0, b_1, \ldots, b_n \in \mathbb{Z}$, and set $\epsilon := \sum_{i=0}^n b_i$. The proof is based on the following observation:

$$\begin{aligned} &\sum_{0 \leq i < j \leq n} b_i b_j d(x_i, x_j) \\ &= \sum_{1 \leq i \leq n} b_0 b_i p(x_i, x_i) + \sum_{1 \leq i < j \leq n} b_i b_j (p(x_i, x_i) + p(x_j, x_j) - 2p(x_i, x_j)) \\ &= \epsilon \sum_{1 \leq i \leq n} b_i p(x_i, x_i) - \sum_{1 \leq i,j \leq n} b_i b_j p(x_i, x_j). \end{aligned}$$

∎

As an immediate application, we have:

(6.1.15) $$\xi(\mathrm{NEG}_{n+1}) = \mathrm{PSD}_n.$$

In other words, the negative type cone NEG_{n+1} is in one-to-one linear correspondence with the positive semidefinite cone PSD_n.

6.2 Characterization of L_2-Embeddability

In this section, we study the distance spaces that can be isometrically embedded into some ℓ_2-space. In other words, we consider the distances that can be realized as the pairwise Euclidean distances of some configuration of points in a space $\mathbb{R}^m$ $(m \geq 1)$. Such distance spaces form, in fact, the topic of a long established and active area of research, known as *distance geometry.* Investigations in this area go back to Cayley [1841] who made several observations that were later systematized by Menger [1928], leading, in particular, to the theory of Cayley-Menger determinants. Research in this area was pursued, in particular, by Schoenberg [1935] who discovered a new characterization of ℓ_2-embeddable distances in terms of the negative type inequalities. The monograph by Blumenthal [1953] remains a classic reference on the subject. Interest in the area of distance geometry was stimulated in the recent years by its many applications, e.g., to the theory of multidimensional scaling (cf. the survey paper by de Leeuw and Heiser [1982]) and to the molecular conformation problem (cf. the monograph by Crippen and Havel [1988]).

This section contains several characterizations for ℓ_2-embeddable distance spaces. First, we present Schoenberg's result, which gives a characterization for ℓ_2-embeddability in terms of the negative type inequalities (see Theorem 6.2.2). As an application, checking ℓ_2-embeddability for a finite distance space can be done in polynomial time; this contrasts with the ℓ_1-case where the corresponding ℓ_1-embeddability problem is known to be NP-complete. We then mention an equivalent characterization in terms of Cayley-Menger matrices. Another fundamental result is a result by Menger, concerning the isometric subspaces of the m-dimensional Euclidean space. More precisely, Menger showed that a distance space (X, d) can be isometrically embedded into the m-dimensional Euclidean space $(\mathbb{R}^m, d_{\ell_2})$ if and only if the same property holds for every subspace of (X, d) on $m + 3$ points (see Theorem 6.2.13). Further characterizations for ℓ_2-embeddability can be found in Theorem 6.2.16.

6.2.1 Schoenberg's Result and Cayley-Menger Determinants

In a first step, we make the link between the notions of functions of positive type and of $\mathbb{R}$-covariances. The characterization of L_2-embeddable spaces in terms of the negative type inequalities given in Theorem 6.2.2 below will then follow as an immediate consequence, using Lemmas 5.3.2 and 6.1.14; this result is due to Schoenberg [1935, 1938b]. Figure 6.2.3 summarizes these connections[2].

Lemma 6.2.1. *Let p be a symmetric function on X. Then, p is of positive type on X if and only if p is an $\mathbb{R}$-covariance on X.*

Proof. Suppose first that p is an $\mathbb{R}$-covariance on X. Then,

$$p(x, y) = \int_\Omega f_x(\omega) f_y(\omega) \mu(d\omega)$$

for all $x, y \in X$, where f_x are real valued functions of $L_2(\Omega, \mathcal{A}, \mu)$. Let $b \in \mathbb{Z}^X$ with finite support. Then,

$$\sum_{x,y \in X} b_x b_y p(x, y) = \int_\Omega (\sum_{x \in X} b_x f_x(\omega))^2 \mu(d\omega) \geq 0.$$

This shows that p is of positive type on X. Conversely, suppose that p is of positive type on X. We show that p is an $\mathbb{R}$-covariance on X. In view of the

[2]The equivalence: $\sqrt{d}$ is ℓ_2-embeddable $\Longleftrightarrow$ $p := \xi(d)$ is a positive semidefinite matrix (for a distance d on a finite set) was, in fact, known to several other authors. It is, for instance, explicited in a paper by Young and Householder [1938].

finitude result from Theorem 3.2.1 and Lemma 5.3.2, we can suppose that X is finite. By assumption, the matrix $(p(x,y))_{x,y\in X}$ is positive semidefinite. Hence, by Lemma 2.4.2, it is a Gram matrix. This shows that p is an $\mathbb{R}$-covariance on X. ∎

Theorem 6.2.2. *Let (X,d) be a distance space. Then, (X,d) is of negative type if and only if $(X,\sqrt{d})$ is L_2-embeddable.* ∎

We remind that the cone $\mathrm{NOR}_n(2)$ has been defined in (3.1.4) precisely as the set of distances d on V_n for which $\sqrt{d}$ is ℓ_2-embeddable. Therefore,

$$\mathrm{NEG}_n = \mathrm{NOR}_n(2).$$

The distance matrix associated with a distance space (V_n,d) where $d\in\mathrm{NOR}_n(2)$ (i.e., $\sqrt{d}$ is ℓ_2-embeddable) is also known in the literature as a *Euclidean distance matrix* (see, e.g., Gower [1985], Hayden, Wells, Wei-Min Liu and Tarazaga [1991]). Hence, the set of Euclidean distance matrices identifies with the negative type cone. More information on this cone and on its projections will be given in Section 31.4.

d distance on V_{n+1}		$p := \xi_{n+1}(d)$
$\sqrt{d}$ is ℓ_2-embeddable i.e., $d\in\mathrm{NOR}_n(2)$	$\Longleftrightarrow$	(p_{ij}) is a Gram matrix (setting $p_{ji}=p_{ij}$)
$\Updownarrow$		$\Updownarrow$
$d\in\mathrm{NEG}_{n+1}$	$\Longleftrightarrow$	$(p_{ij})\in\mathrm{PSD}_n$

Figure 6.2.3

As an immediate application, the problem of deciding whether a finite distance space (X,d) is ℓ_2-embeddable can be solved in polynomial time. Indeed, this amounts to checking whether the symmetric matrix (p_{ij}) associated with $p:=\xi(d^2)$ is positive semidefinite (which can be done in polynomial time; recall Proposition 2.4.3). In contrast, checking whether (X,d) is ℓ_1-embeddable is an NP-complete problem (recall problem (P5) in Section 4.4).

We address now the following question: Given an ℓ_2-embeddable distance space (X,d), what is the minimum dimension m of a Euclidean space $\ell_2^m = (\mathbb{R}^m, d_{\ell_2})$ in which (X,d) can be embedded ? This parameter is the ℓ_2-dimension of (X,d), denoted as $m_{\ell_2}(X,d)$. It turns out that this question can be very easily answered, as $m_{\ell_2}(X,d)$ can be expressed as the rank of a related matrix.

We introduce some notation. Let (X, d) be a finite distance space with, say, $X = \{1, \ldots, n\}$. Let $x_0 \in X$ and let $p = \xi_{x_0}(d)$ denote the image of d under the covariance mapping ξ_{x_0} (recall relation (5.2.5)). Then, we let $P_{x_0}(X, d)$ denote the $(n-1) \times (n-1)$ matrix whose (i,j)-th entry is $p(i,j)$ for $i, j \in X \setminus \{x_0\}$. The next result indicates a way to compute the ℓ_2-dimension. For convenience, we formulate it for a distance space whose square root is ℓ_2-embeddable.

Proposition 6.2.4. *Let (X, d) be a finite distance space of negative type, i.e., such that $(X, \sqrt{d})$ is ℓ_2-embeddable. Then, the ℓ_2-dimension $m_{\ell_2}(X, \sqrt{d})$ of $(X, \sqrt{d})$ is given by*

$$m_{\ell_2}(X, \sqrt{d}) = \text{rank } P_{x_0}(X, d).$$

Proof. Apply Lemma 2.4.2. ∎

Corollary 6.2.5. *Let (X, d) be a finite distance space. Then, $(X, \sqrt{d})$ embeds isometrically into ℓ_2^r if and only if the matrix $P_{x_0}(X, d)$ is positive semidefinite and has rank $\leq r$. Moreover, if $m_{\ell_2}(X, \sqrt{d}) = r$, then there exists a subset $Y \subseteq X$ such that $|Y| = r+1$ and $m_{\ell_2}(Y, \sqrt{d}) = r$. (There exists such a set Y containing any given element $x_0 \in X$.)* ∎

Therefore,

$$(6.2.6) \qquad m_{\ell_2}(n) = n - 1,$$

where $m_{\ell_2}(n)$ is the minimum ℓ_2-dimension, defined as the minimum integer m such that every ℓ_2-embeddable distance on n points embeds in ℓ_2^m. Some other formulations for the ℓ_2-dimension of a distance space can be derived using Lemma 6.2.7 and relation (6.2.10) below.

Lemma 6.2.7. *Let (X, d) be a distance space with associated distance matrix D and let $x_0 \in X$. Then,*

$$\text{rank } P_{x_0}(X, d) = \text{rank } (I - \frac{1}{n}J)D(I - \frac{1}{n}J).$$

(Here, J denotes the all-ones matrix and I the identity matrix.)

Proof. One can bring the matrix $(I - \frac{1}{n}J)D(I - \frac{1}{n}J)$ to $P_{x_0}(X, d)$ by performing some row/column manipulations. ∎

Conditions about the matrix $P_{x_0}(X, d)$ can be reformulated in terms of the Cayley-Menger determinants. These determinants are a classical notion in the theory of distance geometry (see Blumenthal [1953]). They are defined in the following manner. Let (X, d) be a finite distance space with, say, $X = \{1, \ldots, n\}$ and let D be its associated distance matrix. Then, the $(n+1) \times (n+1)$ symmetric matrix:

(6.2.8) $$\mathrm{CM}(X,d) := \begin{pmatrix} D & e \\ e^T & 0 \end{pmatrix}$$

(where e denotes the all-ones vector) is called the *Cayley-Menger matrix* of the distance space (X,d) and $\det \mathrm{CM}(X,d)$ is its *Cayley-Menger determinant.* The matrices $\mathrm{CM}(X,d)$ and $P_{x_0}(X,d)$ are, in fact, closely related. To see it, consider the 2×2 submatrix $\begin{pmatrix} 0 & 1 \\ 1 & 0 \end{pmatrix}$ of $\mathrm{CM}(X,d)$, whose first row/column is indexed by the element x_0 of X and whose second row/column corresponds to the new $(n+1)$-th entry in $\mathrm{CM}(X,d)$. One can easily verify that its Schur complement in the matrix $\mathrm{CM}(X,d)$ coincides with the matrix $-2P_{x_0}(X,d)$. Therefore,

(6.2.9) $$\det\ \mathrm{CM}(X,d) = (-1)^{|X|} 2^{|X|-1} \det P_{x_0}(X,d),$$

(6.2.10) $$\mathrm{rank}\ \mathrm{CM}(X,d) = \mathrm{rank}\ P_{x_0}(X,d) + 2.$$

Positive semidefiniteness for $P_{x_0}(X,d)$ can also be expressed in terms of the Cayley-Menger determinants. Indeed, $P_{x_0}(X,d)$ is positive semidefinite if and only if every principal submatrix has a nonnegative determinant. But, a principal submatrix of $P_{x_0}(X,d)$ is of the form $P_{x_0}(Y,d)$ for some $Y \subseteq X$ with $x_0 \in Y$. Hence, we have the following result, which was formulated under this form in Menger [1954].

Proposition 6.2.11. *Let (X,d) be a finite distance space. Then,*

(i) *$(X,\sqrt{d})$ is ℓ_2-embeddable if and only if $(-1)^{|Y|} \det \mathrm{CM}(Y,d) \geq 0$ for all $Y \subseteq X$ (or for all $Y \subseteq X$ containing a given point $x_0 \in X$).*

(ii) *Suppose $|X| = r+1$; $X = \{x_0, x_1, \ldots, x_r\}$. Then, $(X,\sqrt{d})$ is ℓ_2-embeddable with ℓ_2-dimension r if and only if $(-1)^{k+1} \det \mathrm{CM}(\{x_0, x_1, \ldots, x_k\}, d) > 0$ for all $k = 1, \ldots, r$.*

(iii) *Suppose $|X| = r+2$. Then, $(X,\sqrt{d})$ is ℓ_2-embeddable with ℓ_2-dimension r if and only if the elements of X can be ordered as $x_0, x_1, \ldots, x_r, x_{r+1}$ in such a way that $(-1)^{k+1} \det \mathrm{CM}(\{x_0, x_1, \ldots, x_k\}, d) > 0$ for all $k = 1, \ldots, r$ and $\det \mathrm{CM}(X,d) = 0$.*

(iv) *Suppose $|X| = r+3$. Then, $(X,\sqrt{d})$ is ℓ_2-embeddable with ℓ_2-dimension r if and only if the elements of X can be ordered as $x_0, x_1, \ldots, x_r, x_{r+1}, x_{r+2}$ in such a way that $(-1)^{k+1} \det \mathrm{CM}(\{x_0, x_1, \ldots, x_k\}, d) > 0$ for all $k = 1, \ldots, r$, $\det \mathrm{CM}(X \setminus \{x_{r+1}\}, d) = 0$, and $\det \mathrm{CM}(X \setminus \{x_{r+2}\}, d) = 0$.* ∎

Remark 6.2.12. Note that $\det \mathrm{CM}(X,d) = 2d(x_0,x_1) \geq 0$ if $|X| = 2$, $X = \{x_0, x_1\}$. Moreover, if $|X| = 3$, $X = \{x_0, x_1, x_2\}$ and if we set $a := d(x_0,x_1)$, $b := d(x_0,x_2)$ and $c := d(x_1,x_2)$, then one can check that

$$\det \mathrm{CM}(X,d) = (\sqrt{a}+\sqrt{b}+\sqrt{c})(\sqrt{a}-\sqrt{b}-\sqrt{c})(-\sqrt{a}+\sqrt{b}-\sqrt{c})(-\sqrt{a}-\sqrt{b}+\sqrt{c}).$$

Hence, $\det \mathrm{CM}(X,d) \leq 0$ if $\sqrt{d}$ satisfies the triangle inequalities. That is, every semimetric space on 3 points is ℓ_2-embeddable. Moreover, if (X,d) is a distance space on $|X| = 4$ points, then $(X, \sqrt{d})$ is ℓ_2-embeddable if and only if $\sqrt{d}$ is a semimetric and $\det \mathrm{CM}(X, d) \geq 0$. ∎

6.2.2 Menger's Result

For an infinite distance space (X, d), we know from Theorem 3.2.1 that (X, d) is L_2-embeddable if and only if this property holds for every finite subspace of (X, d). Menger [1928] shows[3] that, in order to ensure embeddability in the m-dimensional space ℓ_2^m, it suffices to consider the subspaces of (X, d) on $m + 3$ points.

Theorem 6.2.13. *Given $m \geq 1$, a distance space (X, d) can be isometrically embedded in ℓ_2^m if and only if, for every[4] $Y \subseteq X$ with $|Y| = m + 3$, (Y, d) can be isometrically embedded in ℓ_2^m.*

This result will follow from the following sharper statement.

Theorem 6.2.14. *Given $m \geq 1$, a distance space (X, d) can be isometrically embedded in ℓ_2^m if and only if there exist an integer r $(0 \leq r \leq m)$ and a subset $Y := \{x_0, x_1, \ldots, x_r\}$ of X such that*

(i) *the distance space (Y, d) can be isometrically embedded in ℓ_2^r but not in ℓ_2^{r-1},*

(ii) *for every $x, y \in X$, the distance space $(Y \cup \{x, y\}, d)$ can be isometrically embedded in ℓ_2^r.*

Then, $m_{\ell_2}(X, d) = r$.

Before giving the proof, we introduce some notation. Let (X, d) and (X', d') be two distance spaces and let $x_1, \ldots, x_k \in X$, $x'_1, \ldots, x'_k \in X'$. We write:

$$x_1, \ldots, x_k \sim x'_1, \ldots, x'_k$$

if the corresponding subspaces are isometric, i.e., if $d(x_i, x_j) = d'(x'_i, x'_j)$ for all $i, j = 1, \ldots, k$. When $x_1, \ldots, x_k$ are vectors in $\mathbb{R}^n$ $(n \geq 1)$ then the distance is implicitely supposed to be the Euclidean distance. In other words, the notation: $x_1, \ldots, x_k \sim x'_1, \ldots, x'_k$ means then that $\| x_i - x_j \|_2 = d'(x'_i, x'_j)$ for all i, j.

Proof of Theorems 6.2.13 and 6.2.14. Suppose first that every subspace of (X, d) on $m + 3$ points embeds in ℓ_2^m; we show the existence of r and $x_0, \ldots, x_r \in X$

[3] Details about this topic can also be found in Menger [1931] and in the monographs by Menger [1954] and Blumenthal [1953]. The analogue question in the case of the L_1-space will be addressed in Section 11.1.

[4] In fact, the result remains valid if we only assume that (Y, d) can be isometrically embedded into ℓ_2^m for every $Y \subseteq X$ with $|Y| = m + 3$ and containing a given element $x_0 \in X$. This fact will be used in Section 6.4 for the proof of Theorem 6.4.8.

satisfying the conditions (i),(ii) from Theorem 6.2.14. For this, define r as the smallest integer such that every subspace of (X,d) on $r+3$ points embeds in ℓ_2^r. Then, there exists $Y \subseteq X$ with $|Y| = r+2$ and such that (Y,d) does not embed in ℓ_2^{r-1}. Hence, $m_{\ell_2}(Y,d) = r$. By Corollary 6.2.5, we can find points $x_0, \ldots, x_r \in Y \subseteq X$ for which $m_{\ell_2}(\{x_0, \ldots, x_r\}, d) = r$. Hence, these $r+1$ points satisfy the conditions (i),(ii).
Conversely, let us now suppose that there exist an integer r and some points $x_0, x_1, \ldots, x_r \in X$ satisfying the conditions (i),(ii) from Theorem 6.2.14. We show that (X,d) embeds in ℓ_2^r. By (i), there exist a set of vectors $x'_0, \ldots, x'_r \in \mathbb{R}^r$ (of affine rank $r+1$) such that

$$x_0, x_1, \ldots, x_r \sim x'_0, x'_1, \ldots, x'_r.$$

Given $x \in X$, the distance space $(\{x_0, \ldots, x_r, x\}, d)$ embeds in ℓ_2^r by (ii). Hence, there exist vectors $\overline{x}_0, \ldots, \overline{x}_r, \overline{x} \in \mathbb{R}^r$ such that

$$x_0, x_1, \ldots, x_r, x \sim \overline{x}_0, \overline{x}_1, \ldots, \overline{x}_r, \overline{x}.$$

As $x'_0, \ldots, x'_r \sim \overline{x}_0, \ldots, \overline{x}_r$, we can find (by Lemma 2.4.1) an orthogonal transformation g of $\mathbb{R}^r$ mapping every $\overline{x}_i$ onto x'_i (for $i = 0, 1, \ldots, r$). Setting $x' := g(\overline{x})$, we obtain a vector $x' \in \mathbb{R}^r$ such that

$$x_0, x_1, \ldots, x_r, x \sim x'_0, x'_1, \ldots, x'_r, x'.$$

Such a vector x' is unique (as $x'_0, x'_1, \ldots, x'_r$ have full affine rank $r+1$). Hence, this defines a mapping $x \in X \mapsto x' \in \mathbb{R}^r$. We now show that $d(x,y) = \| x' - y' \|_2$ for all $x, y \in X$. Let $x, y \in X$. By (ii), there exist vectors $x''_0, x''_1, \ldots, x''_r, x'', y'' \in \mathbb{R}^r$ such that

$$x_0, x_1, \ldots, x_r, x, y \sim x''_0, x''_1, \ldots, x''_r, x'', y''.$$

Therefore,

$$x'_0, x'_1, \ldots, x'_r, x' \sim x''_0, x''_1, \ldots, x''_r, x'' \quad \text{and} \quad x'_0, x'_1, \ldots, x'_r, y' \sim x''_0, x''_1, \ldots, x''_r, y''.$$

Using Lemma 2.4.1, we can find an orthogonal transformation f' (resp. f'') of $\mathbb{R}^r$ mapping x'_i to x''_i (for $i = 0, 1, \ldots, r$) and x' to x'' (resp. y' to y''). Therefore, the two mappings f' and f'' coincide (as they coincide on a set of full affine rank). This implies that $x', y' \sim x'', y''$. As $x, y \sim x'', y''$, we deduce that $x, y \sim x', y'$. This concludes the proof. ∎

Let us observe that Theorem 6.2.13 is best possible. Indeed, there exist distance spaces that do not embed in ℓ_2^m while any subspace on $m+2$ points does. In other words, the order of congruence of the distance space $\ell_2^m = (\mathbb{R}^m, d_{\ell_2})$ is equal to $m+3$.

Such an example can be constructed as follows. Let $x_0, x_1, \ldots, x_m \in \mathbb{R}^m$ be the vertices of an equilateral simplex Δ in $\mathbb{R}^m$ with, say, side length a. Denote by x_{m+1} the center of this simplex and let b denote the Euclidean distance from x_{m+1} to any vertex x_i of Δ. Finally, let c denote the Euclidean distance from

x_{m+1} to the hyperplane supporting any facet of Δ. We now define a distance d on the set $X := \{0, 1, \dots, m+1, m+2\}$ by setting

$$\begin{array}{ll} d(i,j) := a & \text{for } i \neq j \in \{0,1,\dots,m\}, \\ d(i,m+1) = d(i,m+2) := b & \text{for } i = 0,1,\dots,m, \\ d(m+1,m+2) := 2c. & \end{array}$$

Now, (X,d) is not ℓ_2^m-embeddable (in fact, not ℓ_2-embeddable). On the other hand, every subspace of (X,d) on $m+2$ points embeds in ℓ_2^m. (Indeed, this is obvious for the subspaces $(X \setminus \{m+2\}, d)$ and $(X \setminus \{m+1\}, d)$. This is also true for the subspace $(X \setminus \{i\}, d)$ where $i = 0, \dots, m$. For this, let x'_i denote the symmetric of x_{m+1} around the hyperplane spanned by $x_0, \dots, x_{i-1}, x_{i+1}, \dots, x_m$; then,

$$0, \dots, i-1, i+1, \dots, m, m+1, m+2 \sim x_0, \dots, x_{i-1}, x_{i+1}, \dots, x_m, x_{m+1}, x'_i.)$$

For instance, for dimension $m = 1$, the distance matrix

$$\begin{pmatrix} 0 & 2 & 1 & 1 \\ 2 & 0 & 1 & 1 \\ 1 & 1 & 0 & 2 \\ 1 & 1 & 2 & 0 \end{pmatrix}$$

provides an example of a non ℓ_2-embeddable metric for which every 3-point subspace embeds on the line ℓ_2^1.

On the other hand, Menger showed that, for a distance space (X,d) on more than $m+3$ points, checking ℓ_2^m-embeddability of its subspaces on $m+2$ points suffices for ensuring ℓ_2^m-embeddability of the whole space (X,d) (cf. Menger [1931] or Blumenthal [1953]).

Theorem 6.2.15. *Let (X,d) be a distance space with $|X| \geq m+4$. Then, (X,d) is isometrically ℓ_2^m-embeddable if and only if (Y,d) is isometrically ℓ_2^m-embeddable for every $Y \subseteq X$ with $|Y| \leq m+2$.* ∎

The proof of this result is based on a careful analysis of the properties of the 'obstructions' to Theorem 6.2.13; that is, of the distance spaces on $m+3$ points which do not embed in ℓ_2^m while all their subspaces on $m+2$ points do.

6.2.3 Further Characterizations

We present here several additional equivalent characterizations for distance spaces of negative type. The equivalence (i) $\Longleftrightarrow$ (ii) in Theorem 6.2.16 below is given in Gower [1982], (i) $\Longleftrightarrow$ (iii) in Hayden and Wells [1988] and (i) $\Longrightarrow$ (iv) is mentioned in Graham and Winkler [1985].

Theorem 6.2.16. *Let (X,d) be a finite distance space with $X = \{1, \dots, n\}$. Let D be the associated $n \times n$ distance matrix and let $\mathrm{CM}(X,d)$ be the Cayley-Menger matrix defined by (6.2.8). Consider the following assertions.*

(i) *(X, d) is of negative type.*

(ii) *The matrix $(I - es^T)(-D)(I - se^T)$ is positive semidefinite for any $s \in \mathbb{R}^n$ with $s^T e = 1$ (e denoting the all-ones vector).*

(iii) *The matrix* $\mathrm{CM}(X, d)$ *has exactly one positive eigenvalue.*

(iv) *The matrix D has exactly one positive eigenvalue (if D is not the zero matrix).*

Then, (i) $\Longleftrightarrow$ (ii) $\Longleftrightarrow$ (iii) $\Longrightarrow$ (iv).

Proof. (i) $\Longleftrightarrow$ (ii) Let $s \in \mathbb{R}^n$ with $s^T e = 1$ and set $K := I - se^T$ and $A := K^T(-D)K$. Then, for $x \in \mathbb{R}^n$, we have that $x^T A x = y^T(-D)y$, setting $y = Kx$. One checks easily that the range of K consists of the vectors $y \in \mathbb{R}^n$ such that $\sum_{i=1}^n y_i = 0$. Therefore, we obtain that A is positive semidefinite if and only if $y^T(-D)y \geq 0$ for all $y \in \mathbb{R}^n$ such that $\sum_{i=1}^n y_i = 0$, i.e., if (X, d) is of negative type.
(i) $\Longleftrightarrow$ (iii) Let $x_0 \in X$. Consider the 2×2 submatrix $C := \begin{pmatrix} 0 & 1 \\ 1 & 0 \end{pmatrix}$ of $\mathrm{CM}(X, d)$ with row/column indices the two elements x_0 and $n+1$. We use the fact, already mentioned earlier, that the Schur complement of C in $\mathrm{CM}(X, d)$ is equal to the matrix $-2P_{x_0}(X, d)$. Hence, applying Lemma 2.4.4 and the fact that C has one positive eigenvalue, we obtain that $P_{x_0}(X, d) \succeq 0$ if and only if the matrix $\mathrm{CM}(X, d)$ has exactly one positive eigenvalue. The result now follows as $P_{x_0}(X, d) \succeq 0$ is equivalent to (X, d) being of negative type.
(i) $\Longrightarrow$ (iv) The matrix D has at least one positive eigenvalue since D has all its diagonal entries equal to 0. If (X, d) is of negative type then, by Lemma 2.4.5, D has at most one positive eigenvalue since $x^T D x \leq 0$ holds for all x in an $(n-1)$-dimensional subspace of $\mathbb{R}^n$. Therefore, D has exactly one positive eigenvalue. ∎

Remark 6.2.17. Let (X, d) be a distance space with $X = \{1, \ldots, n\}$ and let $s \in \mathbb{R}^n$ with $s^T e = 1$. Set $d(i, .) := \sum_{j \in X} s_j d(i, j)$ for $i \in X$ and $d(., .) := \sum_{i,j \in X} s_i s_j d(i, j)$. Then the matrix $A := (I - es^T)(-D)(I - se^T)$ considered in Theorem 6.2.16 (ii) has its entries of the form:

$$a_{ij} := d(i, .) + d(j, .) - d(., .) - d(i, j) \quad \text{for} \quad i, j \in X.$$

In the case when $s := \frac{1}{n}e$, this matrix A was already considered by Torgerson [1952] who showed that $(X, \sqrt{d})$ is ℓ_2-embeddable if and only if A is positive semidefinite.

Observe that, when (X, d) is of negative type, one can choose the vectors u_1, $\ldots$, u_n providing an ℓ_2-embedding of $(X, \sqrt{d})$ in such a way that $\sum_{i \in X} s_i u_i = 0$ and, then, the matrix $\frac{1}{2}A$ coincides with the Gram matrix of $u_1, \ldots, u_n$ (which shows again the equivalence of (i) and (ii) in Theorem 6.2.16). In particular, if x_0 is a given element of X and s is the corresponding coordinate vector, then $\frac{1}{2}A$ coincides with the matrix $P_{x_0}(X, d)$ augmented by a zero row and column in position x_0. ∎

The next result considers the case when the distance matrix D has a constant row sum (necessarily equal to $\frac{e^T De}{e^T e}$); that is, when the all-ones vector e is an eigenvector of D. It can be found in Hayden and Tarazaga [1993]; see also Alexander [1977].

Theorem 6.2.18. *Let (X, d) be a distance space with $X = \{1, \ldots, n\}$ and with nonzero distance matrix D. The following assertions are equivalent.*

(i) *(X, d) is of negative type and D has a constant row sum.*

(ii) *D has exactly one positive eigenvalue and D has a constant row sum.*

(iii) *There exist vectors $u_1, \ldots, u_n$ providing an ℓ_2-embedding of $(X, \sqrt{d})$ that lie on a sphere whose center coincides with the barycentrum of the u_i's.*

Moreover, under these conditions, $m_{\ell_2}(X, \sqrt{d}) = \text{rank } D - 1$ and the radius r of the sphere containing the vectors u_i providing an ℓ_2-embedding of $(X, \sqrt{d})$ is given by: $r^2 = \frac{1}{2n}\sum_{j\in X} d_{ij}$.

Proof. The implication (i) $\Longrightarrow$ (ii) is contained in Theorem 6.2.16.
(ii) $\Longrightarrow$ (iii) By assumption, $De = \lambda e$ where $\lambda := \frac{1}{n}e^T De$ and the matrix $\frac{\lambda}{n}ee^T - D$ is positive semidefinite. Hence, there exist vectors $u_1, \ldots, u_n$ such that $\frac{\lambda}{n} - d_{ij} = v_i^T v_j$ for all $i, j \in X$. Therefore, $d_{ij} = (\| u_i - u_j \|_2)^2$ for $i, j \in X$, after setting $u_i := \frac{1}{\sqrt{2}} v_i$ $(i \in X)$. By construction, the u_i's lie on the sphere centered at the origin with squared radius $\frac{\lambda}{2n}$ and their barycentrum is the origin since

$$(\| \sum_{i\in X} v_i \|_2)^2 = \sum_{i,j\in X} v_i^T v_j = \sum_{i,j\in X} (\frac{\lambda}{n} - d_{ij}) = n\lambda - e^T De = 0.$$

Moreover, $m_{\ell_2}(X, \sqrt{d})$ is equal to the rank of the system $(u_1, \ldots, u_n)$, i.e., to the rank of matrix $\frac{\lambda}{n}ee^T - D$ which can be easily verified to be equal to rank $D - 1$.
(iii) $\Longrightarrow$ (i) Assume that $u_1, \ldots, u_n$ provide an ℓ_2-embedding of $(X, \sqrt{d})$ and that $\sum_{i\in X} u_i = 0$ and $\| u_i \|_2 = r$ for all $i \in X$, for some $r > 0$. Then,

$$\sum_{j\in X} d_{ij} = \sum_{j\in X} ((\| u_i \|_2)^2 + (\| u_j \|_2)^2 - 2u_i^T u_j) = 2nr^2$$

does not depend on $i \in X$. This also shows the desired value for the radius r. ∎

6.3 A Chain of Implications

We summarize in this section the implications existing between the properties of being L_1-, L_2-embeddable, of negative type, and hypermetric.

Theorem 6.3.1. *Let (X, d) be a finite distance space with associated distance matrix D. Consider the following assertions.*

(i) (X, d) *is* L_2*-embeddable.*

(ii) (X, d) *is* L_1*-embeddable.*

(iii) (X, d) *is hypermetric.*

(iv) (X, d) *is of negative type.*

(v) $(X, \sqrt{d})$ *is* L_2*-embeddable.*

(vi) D *has exactly one positive eigenvalue.*

We have the chain of implications: (i) $\Longrightarrow$ (ii) $\Longrightarrow$ (iii) $\Longrightarrow$ (iv) $\Longleftrightarrow$ (v) $\Longrightarrow$ (vi).

Proof. The implication (i) $\Longrightarrow$ (ii) is a classical result in analysis; see Proposition 6.4.12 below for a proof. The implication (ii) $\Longrightarrow$ (iii) follows from Lemma 6.1.7, and (iii) $\Longrightarrow$ (iv) from Corollary 6.1.4. Finally, (iv) $\Longleftrightarrow$ (v) holds by Theorem 6.2.2 and (iv) $\Longrightarrow$ (vi) by Theorem 6.2.16. ∎

Example 6.3.2. Let $K_{2,3}$ denote the complete bipartite graph with node set $\{x_1, x_2, x_3\} \cup \{y_1, y_2\}$ and let $d(K_{2,3})$ denote its path metric. Then, $d(K_{2,3})(x_i, x_j) = d(K_{2,3})(y_1, y_2) = 2$ and $d(K_{2,3})(x_i, y_j) = 1$. Hence, $d(K_{2,3})$ violates the pentagonal inequality (6.1.9). See Figure 6.3.3 where the numbers into parentheses indicate how to choose $b_1, \ldots, b_5$ so as to obtain $Q(b)^T d(K_{2,3}) > 0$. Therefore, $d(K_{2,3})$ is not hypermetric. Hence, $d(K_{2,3})$ is not ℓ_1-embeddable. In fact, $d(K_{2,3})$ is not even of negative type since the distance matrix of $K_{2,3}$ has two positive eigenvalues. ∎

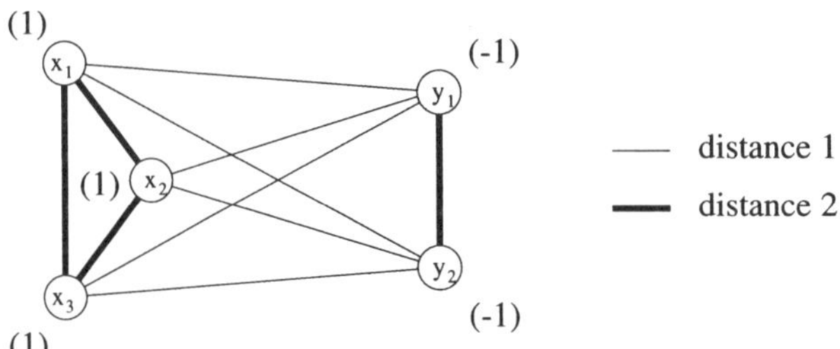

Figure 6.3.3: The path metric of $K_{2,3}$ is not 5-gonal

Remark 6.3.4. Singular ℓ_1-distance matrices. Let (X, d) be a finite distance space. If (X, d) is ℓ_p-embeddable for some $1 \leq p \leq 2$, then (X, d) is ℓ_1-embeddable (recall Remark 3.2.4) and, therefore, its distance matrix has exactly one positive eigenvalue. Let $v_1, \ldots, v_n$ be distinct points of $\mathbb{R}^m$ $(m \geq 2)$. It can be observed from results of Schoenberg [1937, 1938a, 1938b] that, for $1 < p \leq 2$, the ℓ_p-distance matrix $(\| v_i - v_j \|_p)_{i,j=1}^n$ has $n - 1$ negative eigenvalues, i.e., is nonsingular. This is not true for $p = 1$. For instance, the ℓ_1-distance matrix of the vectors $v_1 = (0, 0)$, $v_2 = (1, 0)$,

$v_3 = (0,1)$ and $v_4 = (1,1)$, is singular. Several characterizations have been given for the configurations of distinct points $v_1, \ldots, v_n \in \mathbb{R}^m$ ($m \geq 2$) whose ℓ_1-distance matrix $(\| v_i - v_j \|_1)$ is nonsingular (see Reid and Sun [1993]; see also Lin and Pinkus [1993] for applications to ridge functions interpolation). In particular, let $A = (a_{ij})$ denote the $n \times n$ matrix whose entry a_{ij} is defined as the number of positions where the coordinates of the vectors v_i and v_j coincide (i.e., $m - a_{ij}$ is equal to the Hamming distance between v_i and v_j). The matrix A is positive semidefinite. Moreover, A is positive definite if and only if the matrix $(\| v_i - v_j \|_1)_{i,j=1}^n$ is nonsingular. ∎

The implication (ii) $\Longrightarrow$ (iii) of Theorem 6.3.1 is, in general, strict (as was first observed by Assouad [1977] and Avis [1981]). It is strict[5], in particular, if $7 \leq |X| < \infty$. In other words, the inclusion $\mathrm{CUT}_n \subseteq \mathrm{HYP}_n$ is strict for $n \geq 7$. To see it, it suffices to exhibit an inequality which defines a facet for CUT_n and is not hypermetric. Many such inequalities are described in Part V.

However, there are many examples of classes of distance spaces (X,d) for which the properties of being hypermetric and L_1-embeddable are equivalent. Such examples with X infinite will be presented in Chapter 8. We summarize below what is known about this question.

Remark 6.3.5. We give here a list of distance spaces (X,d) for which L_1-embeddability can be characterized by a set $\mathcal{I}$ of inequalities that are all hypermetric or of negative type.

(i) (V_n, d) with $n \leq 6$; $\mathcal{I}$ consists of the hypermetric inequalities, i.e., $\mathrm{CUT}_n = \mathrm{HYP}_n$ for $n \leq 6$ (see Section 30.6). More precisely, $\mathcal{I}$ consists of the p-gonal inequalities with $p = 3,5$ in the case $n = 5$, and $p = 3,5,7$ in the case $n = 6$.

(ii) A normed space $(\mathbb{R}^m, d_{\|.\|})$; $\mathcal{I}$ consists of the negative type inequalities (see Theorem 8.3.1).

(iii) A normed space $(\mathbb{R}^m, d_{\|.\|})$ whose unit ball is a polytope; $\mathcal{I}$ consists of the 7-gonal inequalities (see Theorem 8.3.2).

(iv) (L, d_v) where $(L, \preceq)$ is a poset lattice, v is a positive valuation on L, and $d_v(x,y) := v(x \vee y) - v(x \wedge y)$ for $x, y \in L$; $\mathcal{I}$ consists of the 5-gonal inequalities or, equivalently, $\mathcal{I}$ consists of the negative type inequalities (see Theorem 8.1.3 and Example 8.2.6).

(v) $(\mathcal{A}, d)$ where $\mathcal{A}$ is a family of subsets of a set Ω which is stable under the symmetric difference, v is a nonnegative function on $\mathcal{A}$ such that $v(\emptyset) = 0$, and $d(A,B) := v(A \triangle B)$ for $A, B \in \mathcal{A}$; $\mathcal{I}$ consists of the inequalities of negative type (see Corollary 8.2.8).

---

[5]This implication remains strict in the case $X = \mathbb{N}$. For this, consider for instance the distance d on $\mathbb{N}$ obtained by taking iterative spherical t-extensions (see Section 7.3) of the path metric of the Schläfli graph G_{27}. Hence, d_{ij} is the shortest length of a path joining i and j in G_{27} if i and j are both nodes of G_{27} and $d_{ij} = t$ otherwise. For $t \geq \frac{4}{3}$, d is hypermetric (by Proposition 14.4.6), but d is not L_1-embeddable (since the path metric of G_{27} lies on an extreme ray of the hypermetric cone on 27 points; see Section 16.2).

(vi) The graphic space $(V, d(G))$ where G is a connected bipartite graph with node set V; $\mathcal{I}$ consists of the 5-gonal inequalities (see Theorem 19.2.1).

(vii) The graphic space $(V, d(G))$ where G is a connected graph on at least 37 nodes and having a node adjacent to all other nodes; $\mathcal{I}$ consists of the negative type inequalities and the 5-gonal inequalities (see Corollary 17.1.10 (i)).

(viii) The graphic space $(V, d(G))$ where G is a connected graph on at least 28 nodes and having a node adjacent to all other nodes; $\mathcal{I}$ consists of the hypermetric inequalities (see Corollary 17.1.10 (ii)). ∎

We conclude with mentioning two examples of application of negative type inequalities to geometric questions, taken from Deza and Maehara [1994]. The first one concerns the following theorem of Rankin [1955] (related to the problem of determining the maximum number of disjoint balls that can be packed in a given ball).

Theorem 6.3.6. *Let $R < \frac{1}{\sqrt{2}}$ and let N_R denote the maximum number N of points $x_1, \ldots, x_N$ that can be placed in a closed (Euclidean) ball of radius R in such a way that $\| x_i - x_j \|_2 \geq 1$ for all $i \neq j = 1, \ldots, N$. Then, $N_R = \lfloor \frac{1}{1-2R^2} \rfloor$.*

Proof. A short proof can be given using negative type inequalities. Suppose $x_1, \ldots, x_N$ lie in the ball of center x_0 and radius R and that $\| x_i - x_j \|_2 \geq 1$ for all $i \neq j = 1, \ldots, N$. Applying the negative type inequality $Q(-n, 1, \ldots, 1)^T x \leq 0$ to the distance d on $\{x_0, x_1, \ldots, x_N\}$ defined by $d(x_i, x_j) := (\| x_i - x_j \|_2)^2$ for all i, j, we obtain the inequality:

$$N \sum_{i=1}^{N} d(x_0, x_i) \geq \sum_{1 \leq i < j \leq N} d(x_i, x_j).$$

As $d(x_0, x_i) \leq R^2$ and $d(x_i, x_j) \geq 1$ for $i \neq j \leq n$ this implies that $N^2 R^2 \geq \binom{N}{2}$; that is, $N \leq N_R := \lfloor \frac{1}{1-2R^2} \rfloor$. Equality is attained by considering for $x_1, \ldots, x_N$ the $N := N_R$ vertices of a regular $(N-1)$-dimensional simplex with side length 1 and $x_0 := \frac{1}{N} \sum_i x_i$. ∎

The next result concerns a generalization of the well-known parallelogram theorem to higher dimensions. This theorem asserts that, given four distinct points x_1, x_2, y_1, y_2, the following inequality holds:

$$\begin{aligned} (\| x_1 - y_1 \|_2)^2 + (\| x_1 - y_2 \|_2)^2 + (\| x_2 - y_1 \|_2)^2 + (\| x_2 - y_2 \|_2)^2 \\ \geq (\| x_1 - x_2 \|_2)^2 + (\| y_1 - y_2 \|_2)^2 \end{aligned}$$

with equality if and only if $x_1 y_1 x_2 y_2$ is a parallelogram.

Given a polytope P, a line segment joining two vertices of P is called a *diagonal* of P if it is not contained in any proper face of P. An *n-parallelotope* is the vector sum of n segments with a common endpoint such that no segment is contained in the affine hull of the others (thus, a parallelogram if $n = 2$).

Theorem 6.3.7. *Let P be a polytope in $\mathbb{R}^n$ that is combinatorially equivalent to an n-dimensional hypercube, let $G(P) := (V_P, E_P)$ denote its 1-skeleton graph, and let D_P denote the set of diagonals of P. Then,*

$$\sum_{xy\in E_P} (\| x-y \|_2)^2 \geq \sum_{xy\in D_P} (\| x-y \|_2)^2,$$

with equality if and only if P is an n-parallelotope.

Proof. The proof is by induction on n. It relies essentially on the following identity:

$$\sum_{i,j=1}^{n} (\| x_i - y_j \|_2)^2 - \sum_{1\leq i<i'\leq n} (\| x_i - x_{i'} \|_2)^2 + (\| y_i - y_{i'} \|_2)^2 = n^2(\| p-q \|_2)^2$$

for any points $x_1, \ldots, x_n, y_1, \ldots, y_n$ and setting $p := \frac{1}{n}\sum_{i=1}^n x_i$, $q := \frac{1}{n}\sum_{i=1}^n y_i$. See Deza and Maehara [1994] for details. ∎

6.4 An Example: The Spherical Distance Space

We describe here a classical example of ℓ_1-embeddable distance space, namely, the spherical distance space. This is the distance space defined on a sphere S, taking as distance between two points x, $y \in S$ the quantity

$$(6.4.1) \qquad r \cdot \arccos\left(\frac{(x-c)^T(x-y)}{r^2}\right),$$

where c denotes the center of S and r its radius. The quantity (6.4.1) is known as the *spherical distance* (or *great circle metric*) between x and y. (This is the geodesic distance on the sphere S between the points x and y, which coincides with the angle between the two vectors from c to x and from c to y.) We may clearly suppose that the sphere S is centered at the origin and, up to scaling of the distances, that it has radius 1. Let S_m denote the m-dimensional unit sphere, i.e.,

$$S_m := \{x \in \mathbb{R}^{m+1} \mid \sum_{i=1}^{m+1} x_i^2 = 1\}.$$

We let $\mathcal{S}_m$ denote the distance space (S_m, d_S), where d_S is the distance defined by (6.4.1), i.e.,

$$(6.4.2) \qquad d_S(x,y) := \arccos(x^T y) \text{ for all } x, y \in S_m.$$

Similarly, we let $\mathcal{S}_{m,r}$ denote the distance space $(S_{m,r}, d_S)$, where $S_{m,r}$ is the m-dimensional sphere centered at the origin with radius r and d_S is defined by (6.4.1). So, $\mathcal{S}_m = \mathcal{S}_{m,1}$. The distance space $\mathcal{S}_{m,r}$ is called a *spherical distance space.* We refer to Blumenthal [1953] for a detailed study of the metric properties of these distance spaces. We mention here some properties that are most relevant to our treatment.

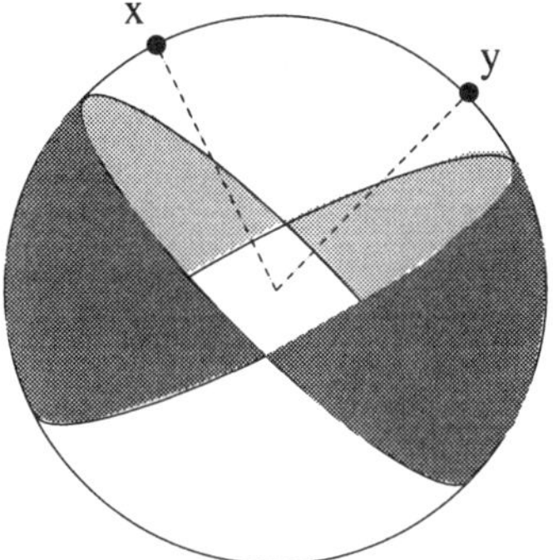

Figure 6.4.3: $H(x)\triangle H(y)$ is the shaded region

A first basic property of spherical distance spaces is that they are ℓ_1-embeddable. For this, let us consider for a point $x \in S_m$ the *hemisphere* $H(x)$ containing x; that is, $H(x)$ consists of the points $y \in S_m$ such that $d_S(x,y) \leq \frac{\pi}{2}$ or, equivalently, $x^T y \geq 0$. We also consider the measure μ on S_m defined by

$$\mu(A) := \frac{\text{vol}(A)}{\text{vol}(S_m)} \quad \text{for } A \subseteq S_m.$$

(Here, vol(A) denotes the m-dimensional volume of A.) Hence, S_m equipped with the measure μ is a probability space. Then,

$$\text{(6.4.4)} \qquad \mu(H(x)\triangle H(y)) = \frac{1}{\pi}\arccos(x^T y) \quad (= \frac{1}{\pi} d_S(x,y)) \quad \text{for } x,y \in S_m.$$

This relation can be easily verified; it was first observed by Kelly [1970b]. For an illustration in the case $m = 2$, see Figure 6.4.3. The result from Theorem 6.4.5[6] [7] below follows as an immediate consequence, using Proposition 4.2.1.

Theorem 6.4.5. *Given $u_1, \ldots, u_n \in S_m$, the vector $(\frac{1}{\pi}\arccos(u_i^T u_j))_{1\leq i<j\leq n}$ belongs to the cut polytope* $\text{CUT}_n^\square$. *Therefore, the spherical distance space $\mathcal{S}_{m,r}$*

[6] In other words, Theorem 6.4.5 indicates how to derive inequalities valid for the pairwise angles among a set of n unit vectors; namely, by considering valid inequalities for the cut polytope $\text{CUT}_n^\square$. This fact will be reminded in Section 31.2.

[7] There is no analogue of Theorem 6.4.5 for the closely related elliptic distance spaces. The *elliptic distance space* $\mathcal{E}\ell_m$ is obtained from the spherical distance space $\mathcal{S}_m$ by identifying antipodal points; thus taking $\min(d_S(x,y), \pi - d_S(x,y))$ as distance between $x, y \in S_m$. The elliptic distance space $\mathcal{E}\ell_m$ is a semimetric space (Blumenthal [1953]) but it is not hypermetric (in fact, not 5-gonal (Kelly [1970b])). One more divergence between spherical and elliptic distance spaces lies in the determination of their congruence orders. As we see in Theorem 6.4.8 the order of congruence of the spherical distance space $\mathcal{S}_m$ can be easily determined; on the other hand the order of congruence of the elliptic distance space $\mathcal{E}\ell_m$ is known only in dimension $m \leq 2$ (see Seidel [1975] and precise references therein).

is isometrically ℓ_1-embeddable for every $r > 0$, $m \geq 1$. ∎

In fact, isometric subspaces of spherical distance spaces and of Euclidean distance spaces are very closely related notions, as the results below indicate.

Let d be a distance on the set $V_n := \{1, \ldots, n\}$. Consider the symmetric matrix[8] $A = (a_{ij})_{i,j=1}^n$ defined by

$$a_{ij} := \cos(d_{ij}) \quad \text{for } i, j = 1, \ldots, n.$$

Hence, the diagonal entries of A are all equal to 1. Moreover, let $\tilde{d}$ denote the distance on the set $V_n \cup \{x_0\}$ (x_0 is a new element not belonging to V_n) defined by

$$\begin{array}{ll} \tilde{d}(x_0, i) = 1 & \text{for } i \in V_n, \\ \tilde{d}(i,j) = 2 - 2\cos(d(i,j)) = 4\sin^2(\frac{d(i,j)}{2}) & \text{for } i, j \in V_n. \end{array}$$

Proposition 6.4.6. *Let $m \geq 1$ be an integer and d, $\tilde{d}$, and A be defined as above. The following assertions are equivalent.*

(i) *The distance space (V_n, d) is an isometric subspace of the m-dimensional spherical distance space $\mathcal{S}_m$.*

(ii) *$d_{ij} \in [0, \pi]$ for all $i, j \in V_n$ and the matrix A is positive semidefinite with* rank $A \leq m + 1$.

(iii) *$d_{ij} \in [0, \pi]$ for all $i, j \in V_n$ and the distance space $(V_n \cup \{x_0\}, \tilde{d})$ can be isometrically embedded into ℓ_2^{m+1}.*

Proof. The equivalence (i) $\Longleftrightarrow$ (ii) is clear. Indeed, (V_n, d) is an isometric subspace of $\mathcal{S}_m$ if and only if there exist $u_1, \ldots, u_n \in S_m$ such that $\cos(d(i,j)) = u_i^T u_j$ for all $i, j \in V_n$; that is, if the matrix A is positive semidefinite with rank $\leq m+1$. The equivalence (ii) $\Longleftrightarrow$ (iii) follows from Corollary 6.2.5 and the fact that A coincides with the matrix $P_{x_0}(V_n \cup \{x_0\}, \tilde{d})$. ∎

Corollary 6.4.7. *A distance space (V_n, d) embeds isometrically in $\mathcal{S}_{m,r}$ (for some $m \geq 1$) if and only if $d_{ij} \in [0, \pi r]$ for all $i, j \in V_n$ and the matrix $A := (\cos(\frac{d_{ij}}{r}))_{i,j=1}^n$ is positive semidefinite. Then, the smallest m such that (V_n, d) embeds into $\mathcal{S}_{m,r}$ is $m =$* rank $A - 1$. ∎

Using Proposition 6.4.6, one can derive a compactness result for spherical distance spaces analogue to the compactness result for Euclidean spaces from Theorem 6.2.13; this result can be found in Blumenthal [1953].

Theorem 6.4.8. *Given $m \geq 1$, a distance space (X, d) can be isometrically embedded into the m-dimensional spherical distance space $\mathcal{S}_m$ if and only if the*

[8] We will study in detail in Section 31.3 the set of positive semidefinite matrices with all diagonal entries equal to 1. We will use, in particular, Theorem 6.4.5.

same holds for every subspace (Y, d) where $Y \subseteq X$ with $|Y| = m+3$. ∎

In particular, the order of congruence of the spherical distance space $\mathcal{S}_m$ is equal to $m+3$. To see it, consider the distance space (X, d), where $|X| \geq m+3$ and $d_{ij} := \arccos(-\frac{1}{m+1})$ for all $i \neq j \in X$. Then, (X, d) does not embed in $\mathcal{S}_m$ (because the matrix $(\cos(d_{ij}))_{i,j \in X}$ is not positive semidefinite). On the other hand, every subspace (Y, d) with $|Y| = m+2$ embeds in $\mathcal{S}_m$ (as $(\cos(d_{ij}))_{i,j \in Y}$ is positive semidefinite with rank $m+1$).

There are further intimate links between the spherical and Euclidean distances. In fact, as Schoenberg [1935] observed, every set of affinely independent vectors equipped with the Euclidean distance can be isometrically embedded into some spherical distance space. And, if the vectors are not independent, their Euclidean distances can be realized asymptotically as limits of spherical distances. This permits to derive the well-known implication: "ℓ_2-embeddable $\Longrightarrow \ell_1$-embeddable".

Proposition 6.4.9. *Let $u_0, u_1, \ldots, u_n \in \mathbb{R}^n$ be affinely independent. Then, the distance space $(\{u_0, u_1, \ldots, u_n\}, d_{\ell_2})$ is an isometric subspace of $\mathcal{S}_{n,r}$ for some r large enough.*

Proof. We can suppose without loss of generality that $u_0 := 0$. By assumption, the matrix $P = (p_{ij} := u_i^T u_j)_{i,j=1}^n$ is positive definite. Let $A(r)$ be the $(n+1) \times (n+1)$ symmetric matrix with entries $a_{ij} := \cos(\frac{d_{ij}}{r})$ for $i, j = 0, 1, \ldots, n$. In view of Corollary 6.4.7, it suffices to show that $A(r) \succ 0$ for r large enough. In what follows, we write $A \sim B$ for two matrices A and B if $A \succ 0 \Longleftrightarrow B \succ 0$. Clearly, $A(r) \sim B(r)$, where the 00-th entry of $B(r)$ is $b_{00} := 1$, its $0i$-th entry is $b_{0i} := a_{0i} - 1$ (for $i = 1, \ldots, n$) and its ij-th entry is $b_{ij} := a_{ij} - a_{0i} - a_{0j} + 1$ (for $i, j = 1, \ldots, n$) (to see it, subtract the row indexed by 0 in $A(r)$ to every other row and, then, the column indexed by 0 to every other column). We now use the fact that $\cos x = 1 - \frac{1}{2}x^2 + o(x^4)$ when $x \longrightarrow 0$. Hence, each entry of $B(r)$ can be expressed as $b_{0i} = -\frac{1}{2r^2}p_{ii} + o(\frac{1}{r^4})$, $b_{ij} = \frac{1}{r^2}p_{ij} + o(\frac{1}{r^4})$. Therefore, after suitably scaling $B(r)$, we obtain that $A(r) \sim B(r) \sim C(r)$, where the 0-th entry of $C(r)$ is $4r^2$, its $0i$-th entry is $p_{ii} + o(\frac{1}{r^2})$, and its ij-th entry is $p_{ij} + o(\frac{1}{r^2})$. One can now easily verify that $C(r) \succ 0$ for any r large enough. (Note here that it suffices to check that the $n+1$ principal subdeterminants of $C(r)$ disposed along the diagonal are positive, which holds for r large enough by the assumption that $P \succ 0$.) ∎

Remark 6.4.10. Proposition 6.4.9 extends to an arbitrary set of vectors in the case when they all lie on a common line. But, it does not extend to an arbitrary set of vectors in dimension ≥ 2. As counterexample, take for $u_1, \ldots, u_{n+1}$ the vertices of an equilateral simplex and for u_0 the barycentrum of the simplex. Then, the distance space $(\{u_0, u_1, \ldots, u_{n+1}\}, d_{\ell_2})$ does not embed in any spherical distance space. ∎

Let us now indicate how the Euclidean distance can be approximated by spherical distances[9]. For this, let $u_1, \ldots, u_n \in \mathbb{R}^m$. We are interested in evaluating their mutual Euclidean distances. So, set $d_{ij} := \| u_i - u_j \|_2$ for $i, j = 1, \ldots, n$. We show how to express d as a limit of distances $d^{(r)}$ (when $r \longrightarrow \infty$) where each $d^{(r)}$ can be isometrically embedded in $\mathcal{S}_{m,r}$. The idea for this is intuitively very simple. Namely, for $r > 0$ consider the sphere $S^{(r)}$ in the space $\mathbb{R}^{m+1}$ with center $c := (0, \ldots, 0, r)$ and radius r. One can visualize $S^{(r)}$ as a sphere lying on top of $\mathbb{R}^m$, being viewed as the hyperplane $x_{m+1} = 0$ in $\mathbb{R}^{m+1}$. Every vector $u \in \mathbb{R}^m$ with $\| u \|_2 \leq r$ can be 'lifted' to a point $u^{(r)} \in S^{(r)}$, by setting

$$u^{(r)} := \left(u, r - \sqrt{r^2 - (\| u \|_2)^2} \right).$$

Let $r \geq \max_{i=1}^n \| u_i \|_2$. For $i = 1, \ldots, n$, let $u_i^{(r)} \in S^{(r)}$ be the 'lifting' of u_i as defined above. Set

$$d^{(r)}(i,j) := r \cdot \arccos \left(\frac{(u_i - c)^T (u_j - c)}{r^2} \right) \quad \text{for } i, j = 1, \ldots, n.$$

So, $d^{(r)}(i,j)$ represents the spherical distance between $u_i^{(r)}$ and $u_j^{(r)}$ on the sphere $S^{(r)}$. Clearly, this length converges to the Euclidean distance $\| u_i - u_j \|_2$ as r tends to infinity, i.e.,

$$\text{(6.4.11)} \qquad \lim_{r \longrightarrow \infty} d^{(r)}(i,j) = \| u_i - u_j \|_2 .$$

In a precise way, one can estimate $d^{(r)}(i,j)$ for large r as follows:

$$\begin{aligned} d^{(r)}(i,j) &= r \cdot \arccos \left(\tfrac{\sqrt{r^2 - (\|u_i\|_2)^2} \sqrt{r^2 - (\|u_j\|_2)^2} + u_i^T u_j}{r^2} \right) \\ &= r \cdot \arccos \left(\sqrt{1 - \tfrac{(\|u_i\|_2)^2}{r^2}} \sqrt{1 - \tfrac{(\|u_j\|_2)^2}{r^2}} + \tfrac{u_i^T u_j}{r^2} \right) \\ &\approx r \cdot \arccos \left(1 - \tfrac{(\|u_i - u_j\|_2)^2}{2r^2} \right) \approx \| u_i - u_j \|_2 . \end{aligned}$$

(Here, we use the fact that $\sqrt{1-x} \approx 1 - \frac{1}{2}x$ and $\arccos(1 - x^2) \approx \sqrt{2}x$ as $x \longrightarrow 0$.) ∎

Proposition 6.4.12. *For a distance space (X, d),*

d is isometrically ℓ_2-embeddable $\Longrightarrow$ d is isometrically ℓ_1-embeddable.

Proof. This follows from the above observations, (6.4.11) and Theorem 6.4.5. ∎

[9]This fact was already observed in Kelly [1975].

6.5 An Example: Kalmanson Distances

We mention here another class of ℓ_1-embeddable semimetrics arising from the so-called Kalmanson distances. These distances present moreover the interesting feature that they yield polynomial-time solvable instances of the traveling salesman problem.

Let d be a distance on the set $V_n = \{1, \ldots, n\}$. We say that d is a *Kalmanson distance* if it satisfies the condition:

$$\max(d_{ij} + d_{rs}, d_{is} + d_{jr}) \leq d_{ir} + d_{js}$$

for all $1 \leq i \leq j \leq r \leq s \leq n$. In this definition, the ordering of the elements is important; so we also say that d is a Kalmanson distance with respect to the ordering $1, \ldots, n$. Moreover, it is convenient to visualize the elements $1, \ldots, n$ as being ordered along a circuit in that circular order. For $1 \leq i < j \leq n$, set

$$\alpha_{ij}(d) := d_{ij} + d_{i+1,j+1} - d_{i,j+1} - d_{i+1,j}$$

(the indices being taken modulo n) and $[i, j] := \{i, i+1, \ldots, j-1, j\}$. Chepoi and Fichet [1996] observed the following fact.

Lemma 6.5.1. *Every distance d on V_n can be decomposed as*

$$2d = \sum_{1 \leq i < j \leq n} \alpha_{ij}(d) \; \delta([i+1, j]).$$

Proof. It is a simple verification. ∎

We introduce one more definition. A distance d on V_n is said to be *circular decomposable* if d can be decomposed as

$$d = \sum_{1 \leq i < j \leq n} \alpha_{ij} \delta([i+1, j])$$

for some nonnegative scalars α_{ij} $(1 \leq i < j \leq n)$; hence, it is ℓ_1-embeddable. As all $\alpha_{ij}(d)$'s are nonnegative if d is a Kalmanson distance, we deduce from Lemma 6.5.1 that every Kalmanson distance is circular decomposable. It can be easily verified that, conversely, every circular decomposable distance is a Kalmanson distance. Therefore, we have the following result[10].

Proposition 6.5.2. *A distance d is a Kalmanson distance if and only if it is circular decomposable (with respect to the same ordering).* ∎

Example 6.5.3. Here are some examples of Kalmanson distances. First, the path metric of a weighted tree is a Kalmanson distance for some ordering of

[10] The equivalence between Kalmanson distances and circular decomposable distances was proved earlier by Christopher, Farach and Trick [1996]; their proof is, however, more complicated than the one presented here, due to Chepoi and Fichet [1996].

the nodes of the tree (Bandelt and Dress [1992]). As another example, consider a set of points $X := \{x_1, \ldots, x_n\}$ in the plane $\mathbb{R}^2$ such that the x_i's lie on the boundary of their convex hull and occur in that circular order along the boundary. Then, the set X equipped with the Euclidean distance[11] d_{ℓ_2} provides a Kalmanson distance with respect to the given ordering (Kalmanson [1975]). To see it, consider four points x_i, x_j, x_r, x_s occurring in that order along the boundary. Then, the segments $[x_i, x_r]$ and $[x_j, x_s]$ intersect in a point y. We obtain that

$$\begin{aligned} d_{\ell_2}(x_i, x_j) + d_{\ell_2}(x_r, x_s) &\leq d_{\ell_2}(x_i, y) + d_{\ell_2}(y, x_j) + d_{\ell_2}(x_r, y) + d_{\ell_2}(y, x_s) \\ &= d_{\ell_2}(x_i, x_r) + d_{\ell_2}(x_j, x_s). \end{aligned}$$

∎

From a computational point of view, Deineko, Rudolf and Woeginger [1995] show that one can test whether a distance on n points is a Kalmanson distance with respect to some ordering of the elements and find such an ordering in time $O(n^2 \log n)$. We conclude with mentioning an application to the traveling salesman problem.

Remark 6.5.4. An application to the traveling salesman problem. The traveling salesman problem[12] can be formulated as follows. Given a distance d on V_n, find a Hamiltonian circuit $(i_1, \ldots, i_n)$ whose weight:

$$\sum_{h=1}^{n-1} d(i_h, i_{h+1}) + d(i_n, i_1)$$

is minimum. This is an NP-hard problem. The problem remains NP-hard for Euclidean distances (that is, if d represents Euclidean distances among a set of points in some space $\mathbb{R}^m$) and, thus, for ℓ_1-distances. However, as was already observed by Kalmanson [1975], the traveling salesman problem can be solved very easily for Kalmanson distances. Indeed, suppose that d is a Kalmanson distance on V_n with respect to the ordering $1, \ldots, n$. Then, the Hamiltonian circuit $(1, \ldots, n)$ has minimum weight. Indeed, from Lemma 6.5.1, we obtain that the weight of the circuit $(1, \ldots, n)$ is equal to $\sum_{1 \leq i < j \leq n} \alpha_{ij}(d)$. On the other hand, the quantity: $\sum_{1 \leq i < j \leq n} \alpha_{ij}(d)$ is a lower bound for the weight of any Hamiltonian circuit (since a circuit and a cut meet in at least two edges). ∎

[11] The same holds if we consider an arbitrary norm metric on $\mathbb{R}^2$ instead of the Euclidean distance or, more generally, a projective metric (to be defined later).

[12] The reader may consult Lawler, Lenstra, Rinnoy Kan and Shmoys [1985] for detailed information on this problem.

Chapter 7. Operations

We describe here several operations which permit to extend a given distance on V_n to a distance on V_{n+1}. Examples of such operations include the gate extension operation, the antipodal extension operation, the spherical extension operation, which are described, respectively, in Sections 7.1, 7.2 and 7.3. We also consider the direct product, the tensor product, and the 1-sum operations in Sections 7.5 and 7.6. We discuss, in particular, conditions under which these operations preserve metric properties such as ℓ_1- or hypercube embeddability, the hypermetric and negative type conditions, or membership in the cut lattice. The *cut lattice* $\mathcal{L}_n$ is defined by

$$\mathcal{L}_n := \{\sum_S \lambda_S \delta(S) \mid \lambda_S \in \mathbb{Z} \text{ for all } S\}.$$

More information on $\mathcal{L}_n$ will be given in Section 25.1. In Section 7.4, we treat in detail the example of the cocktail-party graph. This graph plays, in fact, a central role in the theory of ℓ_1-embeddings. Indeed, it is one of the possible factors (besides the half-cube graph and the Gosset graph) that may enter the canonical metric representation of a hypermetric or ℓ_1-graph (see Theorems 14.3.6 and 14.3.7). The specificity of the cocktail-party graph is also demonstrated by the following result of Cameron, Goethals, Seidel and Shult [1976]: The only connected regular graphs on $n > 28$ nodes whose adjacency matrix has minimum eigenvalue ≥ -2 are line graphs and cocktail-party graphs.

7.1 The Gate Extension Operation

Let d be a distance on $V_n = \{1, \ldots, n\}$ and let $\alpha \in \mathbb{R}_+$. We define a distance d' on $V_{n+1} = V_n \cup \{n+1\}$ by setting

$$\text{(7.1.1)} \qquad \begin{cases} d'(1, n+1) & = \alpha, \\ d'(i, n+1) & = \alpha + d(1, i) \quad \text{for } 2 \leq i \leq n, \\ d'(i, j) & = d(i, j) \quad \text{for } 1 \leq i < j \leq n. \end{cases}$$

The distance d' is called a *gate extension* of d and is denoted by $\text{gat}_\alpha(d)$. This operation will be used especially in the case $\alpha = 0$; then, $\text{gat}_0(d)$ is also called the *gate* 0*-extension* (or, simply, 0-extension) of d. By construction, $\text{gat}_\alpha(d)$ satisfies the following triangle equalities:

$$d'(i, n+1) = d'(1, n+1) + d'(1, i)$$

for all $i = 2, \ldots, n$. From this follows immediately that any $\mathbb{R}_+$-realization of $\text{gat}_\alpha(d)$ is necessarily of the form

$$\text{gat}_\alpha(d) = \sum_{S \subseteq V_n, 1 \notin S} \lambda_S \delta(S) + \alpha \delta(\{n+1\}),$$

where $d = \sum_{S \subseteq V_n, 1 \notin S} \lambda_S \delta(S)$. The next result can be easily checked.

Proposition 7.1.2. *Let d be a distance on V_n and let $\alpha \in \mathbb{R}$.*

(i) $\text{gat}_\alpha(d)$ *is ℓ_1-embeddable (resp. ℓ_1-rigid) if and only if $\alpha \geq 0$ and d is ℓ_1-embeddable (resp. ℓ_1-rigid).*

(ii) $\text{gat}_\alpha(d)$ *is hypercube embeddable (resp. h-rigid) if and only if $\alpha \in \mathbb{Z}_+$ and d is hypercube embeddable (resp. h-rigid).*

(iii) $\text{gat}_\alpha(d) \in \mathcal{L}_{n+1}$ *if and only if $\alpha \in \mathbb{Z}$ and $d \in \mathcal{L}_n$.*

(iv) $\text{gat}_\alpha(d)$ *is hypermetric (resp. of negative type) if and only if $\alpha \geq 0$ and d is hypermetric (resp. of negative type).* ∎

7.2 The Antipodal Extension Operation

Let d be a distance on the set $V_n = \{1, \ldots, n\}$ and let $\alpha \in \mathbb{R}_+$. We define a distance d' on the set $V_{n+1} = V_n \cup \{n+1\}$ by setting

$$\text{(7.2.1)} \qquad \begin{cases} d'(1, n+1) & = \alpha, \\ d'(i, n+1) & = \alpha - d(1,i) \quad \text{for } 2 \leq i \leq n, \\ d'(i,j) & = d(i,j) \quad \text{for } 1 \leq i < j \leq n. \end{cases}$$

The distance d' is called an *antipodal extension* of d and is denoted by $\text{ant}_\alpha(d)$. (Compare with the definition of the gate extension $\text{gat}_\alpha(d)$ from (7.1.1).) Note that $\text{ant}_\alpha(d)$ satisfies the triangle equalities:

$$\text{(7.2.2)} \qquad d'(1, n+1) = d'(1,i) + d'(i, n+1)$$

for all $i = 2, \ldots, n$ (so, the new point "$n+1$" is "antipodal" to the point "1"). If we apply the antipodal extension operation iteratively n times, starting from d, we obtain a distance on $2n$ points, denoted by $\text{Ant}_\alpha(d)$, and called the *full antipodal extension* of d. So, $\text{Ant}_\alpha(d)$ is defined by

$$\text{(7.2.3)} \qquad \begin{cases} \text{Ant}_\alpha(d)(i, n+i) & = \alpha & \text{for } 1 \leq i \leq n, \\ \text{Ant}_\alpha(d)(i, n+j) & = \alpha - d(i,j) & \text{for } 1 \leq i \neq j \leq n, \\ \text{Ant}_\alpha(d)(i,j) & = d(i,j) & \text{for } 1 \leq i \neq j \leq n, \\ \text{Ant}_\alpha(n+i, n+j) & = d(i,j) & \text{for } 1 \leq i \neq j \leq n. \end{cases}$$

These two operations are treated in detail in Deza and Laurent [1992e].

Observe that, if $d = \sum_{S \in \mathcal{S}} \lambda_S \delta(S)$, where $\mathcal{S}$ is a collection of nonempty proper subsets of V_n, then

$$\text{(7.2.4)} \qquad \begin{array}{rl} \mathrm{ant}_\alpha(d) = & \sum_{S \in \mathcal{S} | 1 \not\in S} \lambda_S \delta(S \cup \{n+1\}) + \sum_{S \in \mathcal{S} | 1 \in S} \lambda_S \delta(S) \\ & +(\alpha - \sum_{S \in \mathcal{S}} \lambda_S) \delta(\{n+1\}). \end{array}$$

Conversely, every decomposition of $\mathrm{ant}_\alpha(d)$ as a nonnegative combination of nonzero cut semimetrics is of the form (7.2.4), since $\mathrm{ant}_\alpha(d)$ satisfies the triangle equalities (7.2.2) for $i = 2, \ldots, n$. In particular, if $\mathrm{ant}_\alpha(d)$ is ℓ_1-embeddable, then the size of any of its $\mathbb{R}_+$-realizations is equal to α. Similarly, any $\mathbb{R}_+$-realization of $\mathrm{Ant}_\alpha(d)$ is of the form

$$\text{(7.2.5)} \qquad \mathrm{Ant}_\alpha(d) = \sum_{S \in \mathcal{S}} \lambda_S \delta(S \cup S^*) + (\alpha - \sum_{S \in \mathcal{S}} \lambda_S) \delta(\{n+1, \ldots, 2n\}),$$

where we set $S^* := \{n + i \mid i \in V_n \text{ and } i \not\in S\}$. These observations permit to establish the next result.

Proposition 7.2.6. *Let d be a distance on V_n and $\alpha \in \mathbb{R}$.*

(i) $\mathrm{ant}_\alpha(d)$ *(resp.* $\mathrm{Ant}_\alpha(d)$*) is ℓ_1-embeddable if and only if $\alpha \geq s_{\ell_1}(d)$ and d is ℓ_1-embeddable. Moreover,* $\mathrm{ant}_\alpha(d)$ *(resp.* $\mathrm{Ant}_\alpha(d)$*) is ℓ_1-rigid if and only if d is ℓ_1-rigid.*

(ii) $\mathrm{ant}_\alpha(d)$ *(resp.* $\mathrm{Ant}_\alpha(d)$*) is hypercube embeddable if and only if d is hypercube embeddable, $\alpha \in \mathbb{Z}_+$, and $\alpha \geq s_h(d)$. Moreover,* $\mathrm{ant}_\alpha(d)$ *(resp.* $\mathrm{Ant}_\alpha(d)$*) is h-rigid if and only if d is h-rigid.*

(iii) $\mathrm{ant}_\alpha(d)$ *(resp.* $\mathrm{Ant}_\alpha(d)$*) belongs to $\mathcal{L}_{n+1}$ if and only if $d \in \mathcal{L}_n$ and $\alpha \in \mathbb{Z}$.* ∎

Proposition 7.2.6 is a useful tool; it permits, for instance, to construct examples of semimetrics that are ℓ_1-embeddable and belong to the cut lattice, but are not hypercube embeddable. Indeed, let d be a hypercube embeddable distance and suppose that we can find an integer α such that

$$s_{\ell_1}(d) \leq \alpha < s_h(d).$$

Then, $\mathrm{ant}_\alpha(d)$ is ℓ_1-embeddable and belongs to the cut lattice, but is not hypercube embeddable.

Example 7.2.7. Let $2\mathbb{1}_n$ denote the distance on V_n that takes the value 2 on all pairs. Then, for $n \geq 5$, $s_h(2\mathbb{1}_n) = n$ and $s_{\ell_1}(2\mathbb{1}_n) < 4$ (see Example 4.3.7 and Section 7.4). Therefore, for $n \geq 5$, the metric $\mathrm{ant}_4(2\mathbb{1}_n)$ (which takes value 2 on all pairs except value 4 on one pair) is ℓ_1-embeddable, belongs to the cut lattice, but is not hypercube embeddable. (The metric $\mathrm{Ant}_4(2\mathbb{1}_n)$ has the same properties.) Moreover, the metric $\mathrm{ant}_4(2\mathbb{1}_n)$ is ℓ_1-rigid if and only if $2\mathbb{1}_n$ is ℓ_1-rigid, i.e., $n = 3$. ∎

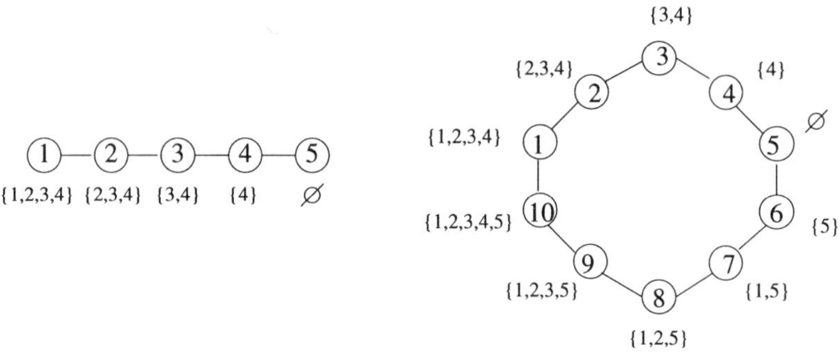

Figure 7.2.8: Hypercube embeddings of the path P_5 and the circuit C_{10}

Example 7.2.9. There are many examples of graphs G whose path metric d_G can be constructed as the antipodal extension of the path metric of another graph. For instance, let $K_{n+1}\backslash e$ denote the complete graph on $n+1$ nodes with one deleted edge. Then,

$$d(K_{n+1}\backslash e) = \text{ant}_2(d(K_n)).$$

This example will be treated in detail in Section 7.4, together with the case of the cocktail-party graph. The path metric d_I of the 1-skeleton graph of the icosahedron (see, e.g., Brouwer, Cohen and Neumaier [1989] for a description of this graph) can be expressed as

$$d_I = \text{Ant}_3\,(d(\nabla C_5))$$

where ∇C_5 is the graph obtained by adding a new node adjacent to all nodes of the 5-circuit C_5. In fact, d_I is ℓ_1-rigid; see Deza and Laurent [1994a]. Also,

$$d(H(n,2)) = \text{Ant}_n\,(d(H(n-1,2))) = \text{Ant}_n\left(2d\left(\frac{1}{2}H(n,2)\right)\right)$$

where $\frac{1}{2}H(n,2)$ is the half-cube graph. Let $P_n = (1,2,\ldots,n)$ and $C_n = (1,2,\ldots,n)$ denote the path and the circuit on n nodes. Then,

$$d(C_{2n}) = \text{Ant}_n(d(P_n)) = \text{Ant}_n(2d(C_n)).$$

One can easily verify (see also Example 19.1.4) that

$$d(P_n) = \sum_{1\le i\le n-1} \delta(\{1,2,\ldots,i-1,i\}),$$

which, by (7.2.5), implies

$$d(C_{2n}) = \sum_{1\le i\le n-1} \delta(\{1,2,\ldots,i-1,i,n+i+1,n+i+2,\ldots,2n-1,2n\}) + \delta(\{1,\ldots,n\}).$$

Figure 7.2.8 shows the path P_5 and the circuit C_{10} together with a labeling of their nodes (by sets) providing the ℓ_1-embeddings shown above. ∎

7.3 The Spherical Extension Operation

Let d be a distance on V_n and let $t \in \mathbb{R}_+$. We define a distance d' on $V_{n+1} = V_n \cup \{n+1\}$ by setting

$$(7.3.1) \qquad \begin{cases} d'(i,n+1) &= t \quad \text{for } 1 \le i \le n, \\ d'(i,j) &= d(i,j) \quad \text{for } 1 \le i < j \le n. \end{cases}$$

The distance d' is called the *spherical t-extension* of d and is denoted by $\mathrm{sph}_t(d)$. This operation is considered in Deza and Grishukhin [1994].

Proposition 7.3.2. *Let d be a distance on V_n and let $t \in \mathbb{R}_+$.*

(i) *$\mathrm{sph}_t(d)$ is a semimetric if and only if $2t \ge \max(d(i,j) \mid 1 \le i < j \le n)$ and d is a semimetric.*

(ii) *$\mathrm{sph}_t(d) \in \mathcal{L}_{n+1}$ if and only if $t \in \mathbb{Z}$ and $d(i,j)$ is an even integer for all $1 \le i < j \le n$.*

(iii) *If d is ℓ_1-embeddable and $2t \ge s_{\ell_1}(d)$, then $\mathrm{sph}_t(d)$ is ℓ_1-embeddable.*

(iv) *If d is hypercube embeddable and $2t \ge s_h(d)$, then $2 \cdot \mathrm{sph}_t(d)$ is hypercube embeddable.*

Proof. (i), (ii) are immediate. For (iii), note that $\mathrm{sph}_t(d) = \frac{1}{2}(\mathrm{ant}_{2t}(d) + \mathrm{gat}_0(d))$. Hence, if d is ℓ_1-embeddable and $2t \ge s_{\ell_1}(d)$, then $\mathrm{ant}_{2t}(d)$ is ℓ_1-embeddable by Proposition 7.2.6 and $\mathrm{gat}_0(d)$ is ℓ_1-embeddable by Proposition 7.1.2. This implies that $\mathrm{sph}_t(d)$ is ℓ_1-embeddable. The proof is identical for (iv). ∎

For an integer $m \ge 2$, define recursively $\mathrm{sph}_t^m(d)$ as $\mathrm{sph}_t(\mathrm{sph}_t^{m-1}(d))$, after setting $\mathrm{sph}_t^1(d) := \mathrm{sph}_t(d)$.

Lemma 7.3.3. *Let d be a distance on V_n and $t \in \mathbb{R}_+$. If d is ℓ_1-embeddable and $2t \ge s_{\ell_1}(d)$, then $\mathrm{sph}_t^m(d)$ is ℓ_1-embeddable for each integer $m \ge 1$. Moreover,*

$$2t - \frac{t}{\lceil \frac{m}{2} \rceil} \le s_{\ell_1}(\mathrm{sph}_t^m(d)) \le 2t - \frac{2t - s_{\ell_1}(d)}{2^m}.$$

Proof. As $\mathrm{sph}_t(d) = \frac{1}{2}(\mathrm{ant}_{2t}(d) + \mathrm{gat}_0(d))$, we can find a $\mathbb{R}_+$-realization of $\mathrm{sph}_t(d)$ of size $\frac{1}{2}(2t + s_{\ell_1}(d))$, i.e.,

$$s_{\ell_1}(\mathrm{sph}_t(d)) \le t + \frac{1}{2} s_{\ell_1}(d).$$

Hence, $s_{\ell_1}(\mathrm{sph}_t(d)) \le 2t$, implying that $\mathrm{sph}_t^2(d)$ is ℓ_1-embeddable. Therefore, $\mathrm{sph}_t^m(d)$ is ℓ_1-embeddable for all $m \ge 1$. The upper bound from Lemma 7.3.3 follows easily by induction. For the lower bound, note that $\mathrm{sph}_t^m(d)$ contains as a subdistance the equidistant metric $t\mathbb{1}_m$ (taking the same value t on m points). Therefore, $s_{\ell_1}(\mathrm{sph}_t^m(d)) \ge s_{\ell_1}(t\mathbb{1}_m)$, where $s_{\ell_1}(t\mathbb{1}_m) = 2t - \frac{t}{\lceil \frac{m}{2} \rceil}$ (see, e.g., Section 7.4). ∎

In particular, if d is ℓ_1-embeddable and $2t \geq s_{\ell_1}(d)$, then

$$\lim_{m \longrightarrow \infty} s_{\ell_1}(\mathrm{sph}_t^m(d)) = 2t.$$

Fichet [1992] defines the ℓ_1-*radius* $r(d)$ of an ℓ_1-embeddable distance d as the smallest scalar $t > 0$ such that $\mathrm{sph}_t(d)$ is ℓ_1-embeddable. Thus,

$$\frac{1}{2} \max_{i,j} d_{ij} \leq r(d) \leq \frac{1}{2} s_{\ell_1}(d).$$

For some examples of metrics (e.g., ultrametrics, path metrics of paths, etc.) equality holds with the lower bound.

We refer to Section 14.4 for the study of conditions under which the spherical extension operation preserves other metric properties, such as hypermetricity or the negative type property.

7.4 An Example: The Cocktail-Party Graph

We present here some examples which apply, in particular, the properties of the antipodal extension operation. Namely, we show how to obtain ℓ_1-embeddings for the path metrics of the graph $K_{n+1} \backslash e$ and of the cocktail-party graph $K_{n \times 2}$ in terms of ℓ_1-embeddings of the complete graph K_n.

Let us consider the equidistant metric $\mathbb{1}_n$, that takes the value 1 for all pairs. Hence, $\mathbb{1}_n$ coincides with the path metric of the complete graph K_n. Let

$$d_n := \mathrm{ant}_2(\mathbb{1}_n)$$

denote the distance on $V_{n+1} = V_n \cup \{n+1\}$ obtained by taking the antipodal extension $\mathrm{ant}_\alpha(\mathbb{1}_n)$ of $\mathbb{1}_n$ with parameter $\alpha = 2$. So, d_n takes the value 1 on all pairs except value 2 on the pair $(1, n+1)$. Hence, d_n coincides with the path metric of the graph $K_{n+1} \backslash e$, obtained by deleting the edge $e := (1, n+1)$ from the complete graph on V_{n+1}. Let

$$d'_n := \mathrm{Ant}_2(\mathbb{1}_n)$$

denote the distance on $V_{2n} = V_n \cup \{n+1, \ldots, 2n\}$ obtained by taking the full antipodal extension $\mathrm{Ant}_\alpha(\mathbb{1}_n)$ of $\mathbb{1}_n$ with $\alpha = 2$ (recall (7.2.3)). So, d'_n takes value 1 on all pairs except value 2 on the n pairs: $(1, n+1), \ldots, (n, 2n)$. Hence, d'_n coincides with the path metric of the cocktail-party graph $K_{n \times 2}$.

We now see how to construct ℓ_1-embeddings of the metrics d_n and d'_n. In order to be able to apply Proposition 7.2.6 and relation (7.2.4), we need to find an ℓ_1-decomposition of $\mathbb{1}_n$ whose size is less than or equal to 2. Clearly, the distance $\mathbb{1}_n$ can always be decomposed as

$$\mathbb{1}_n = \frac{1}{2} \sum_{1 \leq i \leq n} \delta(i), \tag{7.4.1}$$

whose size is equal to $\frac{n}{2}$. For $n = 3, 4$, $\frac{n}{2} \leq 2$; hence, we can apply (7.2.4) and (7.2.5), which yield:

$$d_3 = \frac{1}{2}\left(\delta(\{1\}) + \delta(\{2,4\}) + \delta(\{3,4\}) + \delta(\{4\})\right),$$

$$d'_3 = \frac{1}{2}\left(\delta(\{1,5,6\}) + \delta(\{2,4,6\}) + \delta(\{3,4,5\}) + \delta(\{4,5,6\})\right),$$

$$d_4 = \frac{1}{2}\left(\delta(\{1\}) + \delta(\{2,5\}) + \delta(\{3,5\}) + \delta(\{4,5\})\right),$$

$$d'_4 = \frac{1}{2}\left(\delta(\{1,6,7,8\}) + \delta(\{2,5,7,8\}) + \delta(\{3,5,6,8\}) + \delta(\{4,5,6,7\})\right).$$

If $n \geq 5$, no ℓ_1-embedding of d_n can be constructed from (7.4.1) since its size is greater than 2. Let $\mathcal{E}_n$ denote the collection of all subsets $S \subseteq V_n$ such that $|S| \in \{\lfloor\frac{n}{2}\rfloor, \lceil\frac{n}{2}\rceil\}$ and $1 \notin S$. Setting

$$\alpha_n := \binom{n-2}{\frac{n}{2}-1} \text{ for } n \text{ even, } \alpha_n := 2\binom{n-2}{\frac{n-3}{2}} \text{ for } n \text{ odd,}$$

one can easily check that

$$(7.4.2) \qquad \mathbb{1}_n = \frac{1}{\alpha_n}\sum_{S\in\mathcal{E}_n}\delta(S).$$

As $|\mathcal{E}_n| = \binom{n-1}{\frac{n}{2}}$ for n even and $|\mathcal{E}_n| = \binom{n}{\frac{n+1}{2}}$ for n odd, the decomposition from (7.4.2) has size $\frac{|\mathcal{E}_n|}{\alpha_n}$ which is equal to $\frac{2(n-1)}{n}$ for n even and to $\frac{2n}{n+1}$ for n odd. Observe that (7.4.2) provides the ℓ_1-embedding of $\mathbb{1}_n$ of minimum size, i.e.,

$$s_{\ell_1}(\mathbb{1}_n) = 2 - \frac{1}{\lceil\frac{n}{2}\rceil} = \begin{cases} \frac{2(n-1)}{n} & \text{if } n \text{ is even,} \\ \frac{2n}{n+1} & \text{if } n \text{ is odd.} \end{cases}$$

(Recall (4.3.6).) Applying (7.2.4) and (7.2.5), we obtain

$$d_n = \frac{1}{\alpha_n}\sum_{S\in\mathcal{E}_n}\delta(S\cup\{n+1\}) + (2 - s_{\ell_1}(\mathbb{1}_n))\delta(\{n+1\}),$$

$$d'_n = \frac{1}{\alpha_n}\sum_{S\in\mathcal{E}_n}\delta(S\cup S^*) + (2 - s_{\ell_1}(\mathbb{1}_n))\delta(\{1,2,\ldots,n\}).$$

(We remind that $S^* := \{n+i \mid i \in V_n \text{ and } i \notin S\}$.) This shows that d_n and d'_n are ℓ_1-embeddable. Moreover, by Proposition 7.2.6, d_n, d'_n are ℓ_1-rigid if and only if $\mathbb{1}_n$ is ℓ_1-rigid, i.e., if $n = 3$. (We refer to Part IV for the study of the variety of embeddings of $\mathbb{1}_n$ for $n \geq 4$.) Furthermore, $\alpha_n d_n$ and $\alpha_n d'_n$ are hypercube embeddable (since $\alpha_n(2 - s_{\ell_1}(\mathbb{1}_n)) = 2\alpha_n - |\mathcal{E}_n| \in \mathbb{Z}$) and they embed in the hypercube of dimension $2\alpha_n$. So we have shown:

Proposition 7.4.3. *The path metrics of the graphs $K_{n+1}\backslash e$ and $K_{n\times 2}$ are ℓ_1-embeddable; they are ℓ_1-rigid if and only if $n = 3$. They embed in the hypercube* *of dimension $2\alpha_n$ with the scale α_n.* ∎

For instance, for $n = 5$, the above decompositions for d_5, d'_5 read:

$$\begin{aligned} d_5 = \ & \tfrac{1}{6}\left[\delta(\{2,3,6\}) + \delta(\{2,4,6\}) + \delta(\{2,5,6\}) + \delta(\{3,4,6\}) + \delta(\{3,5,6\})\right] \\ & + \tfrac{1}{6}\left[\delta(\{4,5,6\}) + \delta(\{2,3,4,6\}) + \delta(\{2,4,5,6\}) + \delta(\{2,3,5,6\})\right] \\ & + \tfrac{1}{6}\delta(\{3,4,5,6\}) + \tfrac{1}{3}\delta(6), \end{aligned}$$

$$\begin{aligned} d'_5 = \ & \tfrac{1}{6}\left[\delta(\{2,3,6,9,10\}) + \delta(\{2,4,6,8,10\}) + \delta(\{2,5,6,8,9\}) + \delta(\{3,4,6,7,10\})\right] \\ & + \tfrac{1}{6}\left[\delta(\{3,5,6,7,9\}) + \delta(\{4,5,6,7,8\}) + \delta(\{2,3,4,6,10\}) + \delta(\{2,3,5,6,9\})\right] \\ & + \tfrac{1}{6}\left[\delta(\{2,4,5,6,8\}) + \delta(\{3,4,5,6,7\})\right] + \tfrac{1}{3}\delta(\{1,2,3,4,5\}). \end{aligned}$$

Another ℓ_1-embedding of d_5 is given, for instance, by

$$d_5 = \frac{1}{4}\left(\delta(\{1\}) + \delta(\{6\}) + \sum_{2 \le i < j \le 5} \delta(\{1,i,j\})\right).$$

Therefore, the minimum scale of d_5 is equal to 4. Indeed, the minimum scale $\eta(d_n)$ of d_n is clearly an even integer. Moreover, $\eta(d_5) \neq 2$ since $2d_5$ is not hypercube embeddable (because $2d_5 = \mathrm{ant}_4(2\mathbb{1}_5)$ and $4 < s_h(2\mathbb{1}_5)$, as $s_h(2\mathbb{1}_5) = 5$ by Theorem 22.0.6). In fact, one can check (Deza and Grishukhin [1994]) that $4d_5$ has three distinct (up to permutation) $\mathbb{Z}_+$-realizations. Besides the above one, they are

$$4d_5 = \sum_{2 \le i \le 5} \delta(\{1,i\}) + \delta(\{i,6\}), \text{ and}$$

$$4d_5 = \delta(\{1\}) + \delta(\{1,j\}) + \sum_{i \in \{2,3,4,5\} \setminus \{j\}} \delta(\{i,j,6\}) + \delta(\{i,6\}),$$

for any $j \in \{2,3,4,5\}$.

So we know the minimum scale $\eta(d_n)$ of d_n for $n = 3,4,5$:

$$\eta(d_3) = \eta(d_4) = 2, \ \eta(d_5) = 4.$$

(These facts were already observed by Blake and Gilchrist [1973].) We also have the following (very loose) upper bound: $\eta(d_n) \le \alpha_n$, from Proposition 7.4.3. A better bound: $\eta(d_n) < n$ is provided by the next result from Shpectorov [1993].

Lemma 7.4.4. *For an integer $k \ge 2$, the path metric of the cocktail-party graph $K_{2^k \times 2}$ embeds isometrically into the hypercube of dimension 2^k with the scale 2^{k-1}. Hence, if $2^{k-1} < n \le 2^k$, then $2^{k-1}d(K_{n\times 2})$ is hypercube embeddable. Therefore, $\eta(d_n)$, $\eta(d'_n) < n$.*

Proof. Consider the vector space $GF(2)^k$ over the two-element field $GF(2) = \{0,1\}$. Every hyperplane in $GF(2)^k$ consists of 2^{k-1} points and the symmetric difference of two hyperplanes also contains 2^{k-1} points. We obtain a hypercube embedding of $d(K_{2^k\times 2})$ with the scale 2^{k-1} in the hypercube $\{0,1\}^{2^k}$ by labeling the nodes by the $2^k - 1$ hyperplanes, together with their complements, the full set $GF(2)^k$, and $\emptyset$. ∎

Given an integer $t \in \mathbb{Z}_+$, note that

$$2td_n = 2t\ \mathrm{ant}_2(\mathbb{1}_n) = \mathrm{ant}_{4t}(2t\mathbb{1}_n)$$

is hypercube embeddable if and only if $4t \geq s_h(2t\mathbb{1}_n)$. Hence, the minimum scale of d_n can be expressed as

(7.4.5) $$\eta(d_n) = 2\min(t \in \mathbb{Z}_+ \mid 4t \geq s_h(2t\mathbb{1}_n)).$$

Therefore, in order to determine the minimum scale $\eta(d_n)$, we need to know the minimum h-size $s_h(2t\mathbb{1}_n)$ of the equidistant metric $2t\mathbb{1}_n$, for $t \in \mathbb{Z}_+$. This question is considered in detail in Chapter 23; see, in particular, Section 23.3. The quantity $s_h(2t\mathbb{1}_n)$ is not known in general. Its exact computation is a hard problem. Indeed, for some choices of the parameters n and t, it relies on the question of existence of some classes of designs such as projective planes or Hadamard designs. For instance, $s_h(2t\mathbb{1}_{t^2+t+1}) = t^2+t+1$ if and only if there exists a projective plane of order t, and $s_h(2t\mathbb{1}_{4t}) = 4t-1$ if and only if there exists a Hadamard matrix of order $4t$ (see Proposition 23.3.2). Nevertheless, some results are known. We quote here some of them, more can be found in Part IV.

(i) $s_h(2t\mathbb{1}_n) \geq n-1$ with equality if and only if $n = 4t$ and there exists a Hadamard matrix of order $4t$.

(ii) $s_h(2t\mathbb{1}_n) = \lceil 4t - \frac{2t}{\lceil \frac{n}{2} \rceil} \rceil$ if $n \leq 4t \leq 80$.

(See Theorem 23.3.1 and Corollary 23.3.6.) This implies:

Lemma 7.4.6. *We have: $2\lceil \frac{n}{4} \rceil \leq \eta(d_n) < n$. Moreover, $\eta(d_n) = 2\lceil \frac{n}{4} \rceil$ if $n \leq 80$; and $\eta(d_{4t}) = 2t$ if and only if there exists a Hadamard matrix of order $4t$.* ∎

7.5 The Direct Product and Tensor Product Operations

We present two operations: the direct product and the tensor product, which preserve, respectively, ℓ_1-embeddability and $\{0,1\}$-covariances or, equivalently, the cut cone and the correlation cone.

Definition 7.5.1.

(i) *Let (X_1, d_1) and (X_2, d_2) be two distance spaces. Their* direct product *is the distance space $(X_1 \times X_2, d_1 \oplus d_2)$ where, for $x_1, y_1 \in X_1, x_2, y_2 \in X_2$,*

$$d_1 \oplus d_2((x_1, x_2), (y_1, y_2)) = d_1(x_1, y_1) + d_2(x_2, y_2).$$

(ii) *Let* $p_1 : X_1 \times X_1 \longrightarrow \mathbb{R}$ *and* $p_2 : X_2 \times X_2 \longrightarrow \mathbb{R}$ *be two symmetric functions. Their* tensor product *is the symmetric function*

$$p_1 \otimes p_2 : (X_1 \times X_2) \times (X_1 \times X_2) \longrightarrow \mathbb{R}$$

which is defined, for $x_1, y_1 \in X_1, x_2, y_2 \in X_2$, *by*

$$p_1 \otimes p_2((x_1, x_2), (y_1, y_2)) = p_1(x_1, y_1)p_2(x_2, y_2).$$

For path metrics, the direct product operation corresponds to the Cartesian product of graphs. Namely, if G and H are two connected graphs, then the direct product of their path metrics coincides with the path metric of the Cartesian product of G and H.

Proposition 7.5.2. *Let* d_i *be a distance on the set* X_i, *for* $i = 1, 2$.

(i) $(X_1 \times X_2, d_1 \oplus d_2)$ *is* ℓ_1*-embeddable (resp.* ℓ_1*-rigid, hypercube embeddable, h-rigid) if and only if* (X_1, d_1) *and* (X_2, d_2) *are* ℓ_1*-embeddable (resp.* ℓ_1*-rigid, hypercube embeddable, h-rigid).*

(ii) $(X_1 \times X_2, d_1 \oplus d_2)$ *is hypermetric (resp. of negative type) if and only if* (X_1, d_1) *and* (X_2, d_2) *are hypermetric (resp. of negative type).*

Proof. The proof of (i) is based on the following two observations:

- If $d_1 = \sum_{S \subseteq X_1} \alpha_S \delta(S)$ and $d_2 = \sum_{T \subseteq X_2} \beta_T \delta(T)$, then

$$d_1 \oplus d_2 = \sum_{S \subseteq X_1} \alpha_S \delta(S \times X_2) + \sum_{T \subseteq X_2} \beta_T \delta(X_1 \times T).$$

- Let ρ_i denote the projection from $X_1 \times X_2$ to X_i, for $i = 1, 2$. Suppose that

$$d_1 \oplus d_2 = \sum_{A \in \mathcal{A}} \lambda_A \delta(A)$$

with $\lambda_A > 0$ for $A \in \mathcal{A}$, where $\mathcal{A}$ is a collection of proper subsets of $X_1 \times X_2$. Then, for each $A \in \mathcal{A}$, $A = \rho_1(A) \times \rho_2(A)$ with $\rho_1(A) = X_1$ or $\rho_2(A) = X_2$. This can be seen from the fact that $d_1 \oplus d_2$ satisfies the triangle equalities:

$$d((x_1, x_2), (y_1, y_2)) = d((x_1, x_2), (y_1, x_2)) + d((y_1, x_2), (y_1, y_2)),$$

$$d((x_1, x_2), (y_1, y_2)) = d((x_1, x_2), (x_1, y_2)) + d((x_1, y_2), (y_1, y_2))$$

for all $x_1, y_1 \in X_1, x_2, y_2 \in X_2$. Hence, $d_i = \sum_{A \in \mathcal{A}} \lambda_A \delta(\rho_i(A))$ for $i = 1, 2$.

We prove (ii) in the hypermetric case (the negative type case is similar). If $(X_1 \times X_2, d_1 \oplus d_2)$ is hypermetric, then so is (X_1, d_1), as it is isomorphic to the subspace $(X_1 \times \{x_2\}, d_1 \oplus d_2)$ (where $x_2 \in X_2$). Conversely suppose that both (X_1, d_1) and (X_2, d_2) are hypermetric. Let $b \in \mathbb{Z}^{X_1 \times X_2}$ with $\sum_{(x_1, x_2) \in X_1 \times X_2} b(x_1, x_2) = 1$. Define $a \in \mathbb{Z}^{X_1}$ and $c \in \mathbb{Z}^{X_2}$ by setting

$$a_{x_1} := \sum_{x_2 \in X_2} b(x_1, x_2) \text{ for } x_1 \in X_1, \ c_{x_2} := \sum_{x_1 \in X_1} b(x_1, x_2) \text{ for } x_2 \in X_2.$$

Then, $\sum_{x_1\in X_1} a_{x_1} = \sum_{x_2\in X_2} c_{x_2} = 1$ and

$$\sum_{\substack{x_1,y_1\in X_1\\ x_2,y_2\in X_2}} b(x_1,x_2)b(y_1,y_2)d_1\oplus d_2((x_1,x_2),(y_1,y_2)) =$$

$$\sum_{x_1,y_1\in X_1} a_{x_1}a_{y_1}d_1(x_1,y_1) + \sum_{x_2,y_2\in X_2} c_{x_2}c_{y_2}d_2(x_2,y_2) \le 0.$$

This shows that $(X_1\times X_2, d_1\oplus d_2)$ is hypermetric. ∎

Proposition 7.5.3.

(i) *If* $p_1\in \mathrm{COR}(X_1)$ *and* $p_2\in \mathrm{COR}(X_2)$, *then* $p_1\otimes p_2\in \mathrm{COR}(X_1\times X_2)$.

(ii) *If* $p_1\in \mathrm{COR}(X)$ *and* $p_2\in \mathrm{COR}(X)$, *then* $p_1\circ p_2\in \mathrm{COR}(X)$. (Here, $p_1\circ p_2$ stands for the componentwise product.)

Proof. The proof of (i) is based on the following observation:

$$p_1 = \sum_{S\subseteq X_1}\alpha_S\pi(S) \text{ and } p_2 = \sum_{T\subseteq X_2}\beta_T\pi(T) \Longrightarrow p_1\otimes p_2 = \sum_{S\subseteq X_1, T\subseteq X_2}\alpha_S\beta_T\pi(S\times T).$$

Assertion (ii) follows from the fact that $\pi(S)\circ\pi(T) = \pi(S\cap T)$ for all subsets S,T of X. ∎

7.6 The 1-Sum Operation

Let d_1 be a distance on X_1, d_2 be a distance on X_2 and suppose that $|X_1\cap X_2| = 1$, $X_1\cap X_2 = \{x_0\}$. The *1-sum* of d_1 and d_2 is the distance d on $X_1\cup X_2$ defined by

$$\begin{cases} d(x,y) = d_1(x,y) & \text{if } x,y\in X_1,\\ d(x,y) = d_2(x,y) & \text{if } x,y\in X_2,\\ d(x,y) = d(x,x_0) + d(x_0,y) & \text{if } x\in X_1, y\in X_2.\end{cases}$$

Proposition 7.6.1.

(i) *Let d be the* 1-*sum of* d_1 *and* d_2. *Then, d is ℓ_1-embeddable (resp. ℓ_1-rigid, hypercube embeddable, h-rigid) if and only if d_1 and d_2 are ℓ_1-embeddable (resp. ℓ_1-rigid, hypercube embeddable, h-rigid).*

(ii) *d is hypermetric (resp. of negative type) if and only if d_1 and d_2 are hypermetric (resp. of negative type).*

Proof. The proof of (i) is based on the following two observations:

- If $d_1 = \sum_{S\subseteq X_1\setminus\{x_0\}}\alpha_S\delta(S)$ and $d_2 = \sum_{T\subseteq X_2\setminus\{x_0\}}\beta_T\delta(T)$, then

$$d = \sum_{S\subseteq X_1\setminus\{x_0\}}\alpha_S\delta(S) + \sum_{T\subseteq X_2\setminus\{x_0\}}\beta_T\delta(T).$$

- If $d = \sum_{A \in \mathcal{A}} \lambda_A \delta(A)$ with $\lambda_A > 0$ for $A \in \mathcal{A}$, where $\mathcal{A}$ is a collection of nonempty subsets of $X_1 \cup X_2 \setminus \{x_0\}$, then $A \subseteq X_1$ or $A \subseteq X_2$ for each $A \in \mathcal{A}$. This follows from the fact that d satisfies the triangle equalities: $d(x_1, x_2) = d(x_1, x_0) + d(x_0, x_2)$ for all $x_1 \in X_1, x_2 \in X_2$. Hence, $d_i = \sum_{A \in \mathcal{A} | A \subseteq X_i} \lambda_A \delta(A)$ for $i = 1, 2$.

We prove (ii) in the hypermetric case (the negative type case is similar). The space (X_1, d_1) is a subspace of $(X_1 \cup X_2, d)$ and, hence, it is hypermetric whenever $(X_1 \cup X_2, d)$ is hypermetric. Suppose now that (X_1, d_1) and (X_2, d_2) are hypermetric. Let $b \in \mathbb{Z}^{X_1 \cup X_2}$ with $\sum_{x \in X_1 \cup X_2} b_x = 1$. Define $a \in \mathbb{Z}^{X_1}$, $c \in \mathbb{Z}^{X_2}$ by setting $a_x := b_x$ for $x \in X_1 \setminus \{x_0\}$, $a_{x_0} := \sum_{x \in X_2} b_x$, $c_x := b_x$ for $x \in X_2 \setminus \{x_0\}$, and $c_{x_0} := \sum_{x \in X_1} b_x$. Then,

$$\sum_{x,y \in X_1 \cup X_2} b_x b_y d(x, y) = \sum_{x,y \in X_1} a_x a_y d_1(x, y) + \sum_{x,y \in X_2} c_x c_y d_2(x, y) \leq 0.$$

This shows that $(X_1 \cup X_2, d)$ is hypermetric. ∎

For path metrics of graphs, the 1-sum operation corresponds to the clique 1-sum operation for graphs. Namely, if G_1 and G_2 are two connected graphs and if G denotes their clique 1-sum (obtained by identifying a node in G_1 with a node in G_2 and denoting it as x_0), then the path metric of G coincides with the 1-sum of the path metrics of G_1 and G_2.

Chapter 8. L_1-Metrics from Lattices, Semigroups and Normed Spaces

We present in this chapter several classes of distance spaces for which L_1-embeddability can be fully characterized using only hypermetric or negative type inequalities. These distance spaces arise from poset lattices, semigroups and normed vector spaces. Several of the results mentioned here are given with a sketch of proof only or, at times, with no proof at all. They indeed rely on analytical methods whose exposition is beyond the scope of this book. We do, however, provide references that the interested reader may consult for further information.

8.1 L_1-Metrics from Lattices

In this section, we consider a class of metric spaces arising from (poset) lattices[1]. We start with recalling some definitions. A *poset* (*partially ordered set*) consists of a set L equipped with a binary relation $\preceq$ satisfying the conditions: $x \preceq x$, $x \preceq y$ and $y \preceq x$ imply that $x = y$, $x \preceq y$ and $y \preceq z$ imply that $x \preceq z$, for $x, y, z \in L$. The *meet* $x \wedge y$ (if it exists) of two elements x and y is the (unique) element satisfying $x \wedge y \preceq x, y$ and $z \preceq x \wedge y$ if $z \preceq x, y$; similarly, the join $x \vee y$ (if it exists) is the (unique) element such that $x, y \preceq x \vee y$ and $x \vee y \preceq z$ if $x, y \preceq z$. Then, the poset $(L, \preceq)$ is said to be a *lattice* if every two elements $x, y \in L$ have a join $x \vee y$ and a meet $x \wedge y$.

Given a lattice L, a function $v : L \longrightarrow \mathbb{R}_+$ satisfying

$$v(x \vee y) + v(x \wedge y) = v(x) + v(y) \text{ for all } x, y \in L. \tag{8.1.1}$$

is called a *valuation* on L. The valuation v is said to be *isotone* if $v(x) \leq v(y)$ whenever $x \preceq y$ and *positive* if $v(x) < v(y)$ whenever $x \preceq y, x \neq y$. Set

$$d_v(x, y) := v(x \vee y) - v(x \wedge y) \text{ for all } x, y \in L. \tag{8.1.2}$$

One can easily check that (L, d_v) is a semimetric space if v is an isotone valuation on L and that (L, d_v) is a metric space if v is a positive valuation on L; in the latter case, L is called a *metric lattice* (see Birkhoff [1967]). Clearly, every metric

[1]The word "lattice" is used here in the context of partially ordered sets. Another notion of lattice, referring to discrete subgroups of $\mathbb{R}^n$, will be considered in Part II. A good reference on poset lattices is the textbook by Birkhoff [1967].

lattice is *modular*, i.e., satisfies

$$x \wedge (y \vee z) = (x \wedge y) \vee z$$

for all x, y, z with $z \preceq x$. A lattice is said to be *distributive* if

$$x \wedge (y \vee z) = (x \wedge y) \vee (x \wedge z)$$

for all x, y, z. The following result of Kelly [1970a] gives a characterization of the L_1-embeddable metric lattices. Another characterization in terms of the negative type condition will be given in Example 8.2.6.

Theorem 8.1.3. *Let L be a lattice, let v be a positive valuation on L, and let d_v be the distance on L defined by* (8.1.2). *The following assertions are equivalent.*

(i) *L is a distributive lattice.*

(ii) *(L, d_v) is 5-gonal.*

(iii) *(L, d_v) is hypermetric.*

(iv) *(L, d_v) is L_1-embeddable.*

Proof. It suffices to show the implications (ii) $\Rightarrow$ (i) and (i) $\Rightarrow$ (iv).
(ii) $\Rightarrow$ (i) Using the definition of the valuation v and applying the pentagonal inequality (6.1.9) to the points $x_1 := x \vee y$, $x_2 := x \wedge y$, $x_3 := z$, $y_1 := x$, $y_2 := y$, we obtain the inequality:

$$2(v(x\vee y\vee z)-v(x\wedge y\wedge z)) \leq v(x\vee y)+v(x\vee z)+v(y\vee z)-v(x\wedge y)-v(x\wedge z)-v(y\wedge z).$$

By applying again the pentagonal inequality to the points $x_1 := x$, $x_2 := y$, $x_3 := z$, $y_1 := x \vee y$, $y_2 := x \wedge y$, we obtain the reverse inequality. Therefore, equality holds in the above inequality. In fact, this condition of equality is equivalent to L being distributive (see Birkhoff [1967]).
(i) $\Rightarrow$ (iv) Let L_0 be a finite subset of L; we show that (L_0, d_v) is L_1-embeddable. Let K be the sublattice of L generated by L_0. Suppose K has length n. Then, K is isomorphic to a family $\mathcal{N}$ of subsets of a set X of cardinality $|X| = n$ with the property that $\mathcal{N}$ is closed under taking unions and intersections (see Birkhoff [1967] p. 58). Via this isomorphism, we have a valuation defined on $\mathcal{N}$, again denoted by v. We can assume without loss of generality that $v(\emptyset) = 0$. Then, v can be extended to a valuation v^* on 2^X satisfying $v^*(S) = \sum_{x\in S} v^*(\{x\})$ for $S \subseteq X$. Now, if $x \longmapsto S_x$ is the isomorphism from K to $\mathcal{N}$, then we have the embedding $x \longmapsto S_x$ from (L_0, d_v) to $(2^X, v^*)$ which is isometric. Indeed,

$$\begin{aligned} d_v(x,y) &= v(x \vee y) - v(x \wedge y) = v(S_x \cup S_y) - v(S_x \cap S_y) \\ &= v^*(S_x \cup S_y) - v^*(S_x \cap S_y) = v^*(S_x \triangle S_y). \end{aligned}$$

This shows that every finite subset of (L, d_v) is L_1-embeddable. Therefore, by Theorem 3.2.1, (L, d_v) is L_1-embeddable. ∎

Example 8.1.4. (Assouad, [1979]) Let $(\mathbb{N}^*, \preceq)$ denote the lattice consisting of the set $\mathbb{N}^*$ of positive integers with order relation $x \preceq y$ if x divides y. Then, for $x, y \in \mathbb{N}^*$, $x \wedge y$ is the g.c.d. (greatest common divisor) of x and y and $x \vee y$ is their l.c.m. (lowest common multiple). One checks easily that $(\mathbb{N}^*, \preceq)$ is a distributive lattice. Therefore, $(\mathbb{N}^*, d_v)$ is L_1-embeddable for every positive valuation v on $\mathbb{N}^*$. For instance,

$$x \in \mathbb{N}^* \longmapsto v(x) := \log x$$

is a positive valuation on $\mathbb{N}^*$. Hence, the metric d_v, defined by

$$d_v(x, y) = \log(\frac{\text{l.c.m.}(x, y)}{\text{g.c.d.}(x, y)})$$

for all integers $x, y \geq 1$, is L_1-embeddable. ∎

8.2 L_1-Metrics from Semigroups

We consider now some distance spaces arising in the context of semigroups. We recall that a *commutative semigroup* $(S, +)$ consists of a set S equipped with a composition rule "+" which is commutative and associative. We assume the existence of a neutral element denoted by 0. Let $(S, +)$ be a commutative semigroup and let $v : S \longrightarrow \mathbb{R}_+$ be a mapping such that $v(0) = 0$. We define a symmetric function[2] D_v on S by setting

$$\text{(8.2.1)} \qquad D_v(x, y) := 2v(x + y) - v(2x) - v(2y) \ \text{ for all } x, y \in S.$$

Theorem 8.2.5 below gives some conditions on S and D_v sufficient for ensuring that the distance space (S, D_v) be L_1-embeddable. It is essentially based on a result by Berg, Christensen and Ressel [1976] concerning a certain class of functions of positive type on S. More precisely, given $v : S \longrightarrow \mathbb{R}_+$, let p_v denote the symmetric function on S^2 defined by

$$\text{(8.2.2)} \qquad p_v(x, y) := v(x) + v(y) - v(x + y) \ \text{ for all } x, y \in S.$$

Define $\mathcal{N}$ as the set of functions $v : S \longrightarrow \mathbb{R}_+$ with $v(0) = 0$ and for which p_v is of positive type on S (that is, the matrix $(p_v(x_i, x_j))_{i,j=1}^n$ is positive semidefinite for all $x_1, \ldots, x_n \in S$, $n \geq 2$). The members of $\mathcal{N}$ are completely described in Theorem 8.2.3 below, that we state without proof; this result was proved by Berg, Christensen and Ressel [1976]. A *character* on S is a mapping $\rho : S \longrightarrow [-1, 1]$ satisfying $\rho(x + y) = \rho(x)\rho(y)$ for all $x, y \in S$, and $\rho(0) = 1$. Denote by $\hat{1}$ the unit character, taking value 1 for every element $x \in S$, and let $\hat{S}$ denote the set of characters on S.

[2] At this point, D_v is not necessarily a distance, as it may take negative values. However, D_v is nonnegative if we assume, e.g., that D_v is of negative type.

Theorem 8.2.3. *Let $v \in \mathcal{N}$. Then, there exist* (i) *a function $h : S \longrightarrow \mathbb{R}_+$ satisfying $h(x+y) = h(x) + h(y)$ for all $x, y \in S$, and* (ii) *a nonnegative Radon measure μ on $\hat{S} \setminus \{\hat{1}\}$ satisfying $\int_{\hat{S}-\{\hat{1}\}} (1-\rho(x))\mu(d\rho) < \infty$ for all $x \in S$, such that*

$$v(x) = h(x) + \int_{\hat{S}-\{\hat{1}\}} (1-\rho(x))\mu(d\rho) \quad \text{for all } x \in S. \tag{8.2.4}$$

∎

This allows us to derive conditions for the distance space (S, D_v) to be L_1-embeddable, as was observed by Assouad [1979, 1980b].

Theorem 8.2.5. *Let $(S, +)$ be a commutative semigroup with neutral element 0, let $v : S \longmapsto \mathbb{R}_+$ be a mapping such that $v(0) = 0$, and let D_v be defined by* (8.2.1)*. Assume that one of the following assertions* (i) *or* (ii) *holds.*

(i) *$(S, +)$ is a group.*

(ii) *For each $x \in S$, there exists an integer $n \geq 1$ such that $2nx = x$.*

Then, (S, D_v) is L_1-embeddable if and only if (S, D_v) is of negative type.

Proof. Suppose that (S, D_v) is of negative type. Let $p_v : S^2 \longrightarrow \mathbb{R}$ denote the symmetric function obtained by applying the covariance transformation ξ (pointed at 0) to D_v. Then,

$$p_v(x,y) = \frac{1}{2}(D_v(x,0) + D_v(y,0) - D_v(x,y)) = v(x) + v(y) - v(x+y)$$

for $x, y \in S$. In other words, p_v coincides with the function defined in (8.2.2). By assumption, (S, D_v) is of negative type or, equivalently (by Lemma 6.1.14), p_v is of positive type. That is, the function v belongs to the set $\mathcal{N}$. Applying Theorem 8.2.3, we can suppose that v is of the form (8.2.4). In case (i), $(S, +)$ is a group and, thus, every character on S takes only values ± 1. Setting

$$A_x := \{\rho \in \hat{S} \mid \rho(x) = -1\} \quad \text{for } x \in S,$$

we obtain that $p_v(x,y) = 4\mu(A_x \cap A_y)$ for all $x, y \in S$. In case (ii), every character on S takes only values $0, 1$. Setting

$$A_x := \{\rho \in \hat{S} \mid \rho(x) = 0\} \quad \text{for } x \in S,$$

we obtain that $p_v(x,y) = \mu(A_x \cap A_y)$ for all $x, y \in S$. Therefore, p_v is a $\{0,1\}$-covariance in both cases (i) and (ii). This shows (by Lemma 5.3.2) that (S, D_v) is L_1-embeddable. ∎

Example 8.2.6. Let $(L, \preceq)$ be a lattice and let S be a subset of L which is stable under the join operation $\vee$ of L and contains the least element 0 of L. Then,

$(S, \vee)$ is a commutative semigroup satisfying Theorem 8.2.5 (ii). Therefore, given a mapping $v : S \longrightarrow \mathbb{R}_+$ such that $v(0) = 0$,

$$(S, D_v) \text{ is } L_1\text{-embeddable} \iff (S, D_v) \text{ is of negative type},$$

where

$$D_v(x, y) = 2v(x \vee y) - v(x) - v(y) \text{ for } x, y \in S.$$

We formulate in Corollary 8.2.7 the result obtained in the special case when v is a valuation on L. ∎

Corollary 8.2.7. *Let L be a lattice, let $v : L \longrightarrow \mathbb{R}_+$ be a valuation on L (that is, v satisfies* (8.1.1)*) and let d_v be defined by* (8.1.2)*. Then,*

$$(L, d_v) \textit{ is } L_1\textit{-embeddable} \iff (L, d_v) \textit{ is of negative type.}$$

Proof. Observe that, since v satisfies (8.1.1), the two distances d_v and D_v coincide, where d_v is defined by (8.1.2) and D_v by (8.2.1). Hence, the result follows by applying Theorem 8.2.5 to the semigroup $(L, \vee)$ (condition (ii) applies as $x \vee x = x$ for all $x \in L$). ∎

Another example of application of Theorem 8.2.5 is obtained by taking as semigroup a set family closed under the symmetric difference.

Corollary 8.2.8. *Let $\mathcal{A}$ be a family of subsets of a set Ω which is closed under the symmetric difference (that is, $A \triangle B \in \mathcal{A}$ for all $A, B \in \mathcal{A}$). Let $v : \mathcal{A} \longrightarrow \mathbb{R}_+$ be a mapping such that $v(\emptyset) = 0$, and let $d_{v,\triangle}$ be the distance on $\mathcal{A}$ defined by*

$$d_{v,\triangle}(A, B) := v(A \triangle B) \quad \textit{for all } A, B \in \mathcal{A}. \tag{8.2.9}$$

Then,

$$(\mathcal{A}, d_{v,\triangle}) \textit{ is } L_1\textit{-embeddable} \iff (\mathcal{A}, d_{v,\triangle}) \textit{ is of negative type.}$$

Proof. Observe that $d_{v,\triangle} = \frac{1}{2} D_v$ and apply Theorem 8.2.5 to the commutative group $(\mathcal{A}, \triangle)$. ∎

The distance spaces of the form $(\mathcal{A}, d_{v,\triangle})$ are of particular interest. They are indeed very closely related to the measure semimetric spaces (defined in Section 3.2), which play a central role in the theory of L_1-metrics. We will apply Corollary 8.2.8 in the next chapter for finding metric transforms preserving L_1-embeddability; see Theorem 9.0.3 and its proof.

8.3 L_1-Metrics from Normed Spaces

Let $(E, \| \, . \, \|)$ be a normed space. We consider the associated metric space $(E, d_{\|.\|})$, where $d_{\|.\|}$ is the norm metric defined by

$$d_{\|.\|}(x, y) = \| x - y \|$$

for all $x, y \in E$. In this section, we mention without proof several results characterizing the norms on $E = \mathbb{R}^m$ for which the metric space $(\mathbb{R}^m, d_{\|.\|})$ is L_1-embeddable.

We first recall some definitions. If $\| \, . \, \|$ is a norm on $\mathbb{R}^m$, its *unit ball* B is defined as

$$B := \{x \in \mathbb{R}^m : \| x \| \leq 1\}$$

and B° denotes the polar of B.

A polytope is called a *zonotope* if it is the vector sum of some line segments. (Hence, parallelotopes, that were mentioned in Theorem 6.3.7, are special instances of zonotopes.) A *zonoid* is a convex body that can be approximated by zonotopes with respect to the Blaschke-Hausdorff metric. Zonotopes and zonoids are central objects in convex geometry and they are also relevant to many other fields, in particular, to the topic of L_1-metrics as the results below show. We refer to Bolker [1969] and Schneider and Weil [1983] for detailed exposition on zonoids.

We now present several equivalent characterizations[3] for the L_1-embeddable normed metric spaces $(\mathbb{R}^m, d_{\|.\|})$.

Theorem 8.3.1. *Let $\| \, . \, \|$ be a norm on $\mathbb{R}^m$ and let B be its unit ball. The following assertions are equivalent.*

(i) *$d_{\|.\|}$ is of negative type.*

(ii) *$d_{\|.\|}$ is hypermetric.*

(iii) *$(\mathbb{R}^m, d_{\|.\|})$ is L_1-embeddable.*

(iv) *The polar of B is a zonoid.* ∎

Precise reference for the equivalence (i) $\iff$ (ii) $\iff$ (iv) from Theorem 8.3.1 can be found in Schneider and Weil [1983] and (iii) $\iff$ (iv) is proved in Bolker [1969]. L_1-embeddability of norm metrics can be characterized by much simpler inequalities when the unit ball of the normed space is a polytope; we refer to Assouad [1980a, 1984] and Witsenhausen [1978] for the next result.

Theorem 8.3.2. *Let $\| \, . \, \|$ be a norm on $\mathbb{R}^m$ for which the unit ball B is a polytope. The following assertions are equivalent.*

(i) *$\| \, . \, \|$ satisfies Hlawka's inequality:*

$$\| x \| + \| y \| + \| z \| + \| x+y+z \| \geq \| x+y \| + \| x+z \| + \| y+z \|$$

for all $x, y, z \in \mathbb{R}^m$.

[3] More generally, it is shown in Bretagnolle, Dacunha Castelle and Krivine [1966] that, for $1 \leq p \leq 2$, a norm metric $d_{\|.\|}$ is L_p-embeddable if and only if its p-th power $(d_{\|.\|})^p$ is of negative type.

(ii) $\| \cdot \|$ *satisfies the 7-gonal inequality:*

$$\sum_{1\le i<j\le 4} \| x_i - x_j \| + \sum_{1\le h<k\le 3} \| y_h - y_k \| \le \sum_{\substack{1\le i\le 4\\ 1\le k\le 3}} \| x_i - y_k \|$$

for all $x_1, x_2, x_3, x_4, y_1, y_2, y_3 \in \mathbb{R}^m$.

(iii) $(\mathbb{R}^m, d_{\|\cdot\|})$ *is* L_1*-embeddable.*

(iv) *The polar of* B *is a zonotope.* ∎

In fact, the implication (ii) $\Longrightarrow$ (i) of Theorem 8.3.2 remains valid for general norms. Namely, if an arbitrary norm on $\mathbb{R}^m$ satisfies the 7-gonal inequality, then it also satisfies Hlawka's inequality (Assouad [1984]).

The above results can be partially extended to the more general concept of projective metrics. A continuous metric d on $\mathbb{R}^m$ is called a *projective metric* if it satisfies

$$d(x, z) = d(x, y) + d(y, z)$$

for any collinear points x, y, z lying in that order on a common line. Clearly, every norm metric is projective. The cone of projective metrics is the object considered by Hilbert's fourth problem (cf. Alexander [1988], Ambartzumian [1982]; see also Szabó [1986] for an account of recent progress on this problem). Alexander [1988] gave the following characterization for the L_1-embeddable projective metrics.

Theorem 8.3.3. *Let* d *be a projective metric on* $\mathbb{R}^m$. *The following assertions are equivalent.*

(i) $(\mathbb{R}^m, d)$ *is* L_1*-embeddable.*

(ii) d *is hypermetric.*

(iii) *There exists a positive Borel measure* μ *on the hyperplanesets of* $\mathbb{R}^m$ *satisfying*

$$\begin{cases} \mu([[x]]) = 0 & \text{for all } x \in \mathbb{R}^m, \\ 0 < \mu([[x, y]]) < \infty & \text{for all } x \neq y \in \mathbb{R}^m, \end{cases}$$

and such that d *is given by the following formula (called* Crofton formula*):*

$$2d(x, y) = \mu([[x, y]]) \text{ for } x, y \in \mathbb{R}^m,$$

where $[[x, y]]$ *denotes the set of hyperplanes meeting the segment* $[x, y]$. ∎

In dimension $m = 2$, Theorem 8.3.3 (iii) always holds; that is, every projective metric on $\mathbb{R}^2$ is L_1-embeddable (Alexander [1978]). On the other hand, the norm metric d_{ℓ_∞} arising from the norm $\| x \|_\infty = \max(|x_1|, |x_2|, |x_3|)$ in $\mathbb{R}^3$ is not L_1-embeddable since it is not hypermetric. Indeed, the points $x_1 = (1, 1, 0)$, $x_2 = (1, -1, 0)$, $x_3 = (-1, 1, 0)$, $y_1 = (0, 0, 0)$ and $y_2 = (0, 0, 1)$ violate the pentagonal inequality (6.1.9) (Kelly [1970a]).

Chapter 9. Metric Transforms of L_1-Spaces

Let (X, d) be a distance space and let $F : \mathbb{R}_+ \longrightarrow \mathbb{R}_+$ be a function such that $F(0) = 0$. We define the distance space $(X, F(d))$ by setting

$$F(d)_{ij} = F(d_{ij}) \text{ for all } i, j \in X. \tag{9.0.1}$$

Following Blumenthal [1953], $(X, F(d))$ is called a *metric transform* of (X, d). A general question is to find nontrivial functions F which preserve certain properties, such as metricity, L_1- or L_2-embeddability, of the original distance space.

Metric transforms of L_2-spaces have been intensively studied in the literature, in particular, by Schoenberg [1937, 1938a], von Neumann and Schoenberg [1941] (see also Wells and Williams [1975] where the general case of L_p spaces is considered). One of Schoenberg's results, most relevant to our treatment, concerns the characterization of the continuous functions F for which the metric transform of $L_2(\Omega, \mathcal{A}, \mu)$ is L_2-embeddable; it is formulated in Theorem 9.1.4.

Questions of the same type have been studied for positive semidefinite matrices. Namely, one may ask what are the functions F for which the matrix $(F(a_{ij}))$ is positive semidefinite whenever (a_{ij}) is positive semidefinite. Results in this direction can be found in (Horn and Johnson [1991], Section 6.3). For instance, the exponential function $F(t) = \exp(t)$ and the power function $F(t) = t^k$ $(k \in \mathbb{N})$ preserve positive semidefiniteness. In other words, for any matrix (a_{ij}),

$$(a_{ij}) \succeq 0 \Longrightarrow (\exp(a_{ij})) \succeq 0, \ (a_{ij}^k) \succeq 0.$$

We present in this section several results about metric transforms of ℓ_1-spaces. To start with, let us give some conditions sufficient for ensuring that a function F preserves semimetric spaces.

Lemma 9.0.2. *Let $F : \mathbb{R}_+ \longrightarrow \mathbb{R}_+$ be a monotone nondecreasing concave function, such that $F(0) = 0$. If (X, d) is a semimetric space, then $(X, F(d))$ is also a semimetric space.*

Proof. It suffices to show that, if a, b, c are nonnegative scalars such that $a \leq b+c$, then $F(a) \leq F(b) + F(c)$ holds. We have $F(a) \leq F(b+c)$ as F is nondecreasing. We now verify that $F(b+c) \leq F(b) + F(c)$. Indeed, $F(b) \geq \frac{b}{b+c}F(b+c)$ and $F(c) \geq \frac{c}{b+c}F(b+c)$ as F is concave and $F(0) = 0$. The result follows by summing these two relations. ∎

We are interested in determining some functions F preserving L_1-embeddability. For this purpose, it is convenient to introduce the class $\mathcal{F}$ consisting of the functions $F : \mathbb{R}_+ \longrightarrow \mathbb{R}_+$ such that $F(0) = 0$ and F preserves the property of being of negative type; that is, F satisfies:

$$(X, d) \text{ is of negative type } \Longrightarrow (X, F(d)) \text{ is of negative type}$$

for any distance space (X, d). Using Corollary 8.2.8, we can prove the following result from Assouad [1979, 1980b]:

Theorem 9.0.3. *Let (X, d) be a distance space and let F be a function from the family $\mathcal{F}$. Then,*

$$(X, d) \textit{ is } L_1\textit{-embeddable} \Longrightarrow (X, F(d)) \textit{ is } L_1\textit{-embeddable.}$$

Proof. Let (X, d) be a distance space which is L_1-embeddable. Then, by Proposition 4.2.1, (X, d) is an isometric subspace of a measure semimetric space $(\mathcal{A}_\mu, d_\mu)$ for some measure space $(\Omega, \mathcal{A}, \mu)$. That is, there is a mapping

$$x \in X \mapsto A_x \in \mathcal{A}_\mu$$

such that $d(x, y) = \mu(A_x \triangle A_y)$ for all $x, y \in X$. Let $v : \mathcal{A}_\mu \longrightarrow \mathbb{R}_+$ be defined by

$$v(A) := F(\mu(A)) \quad \text{for } A \in \mathcal{A}_\mu.$$

We apply Corollary 8.2.8 to the pair $(\mathcal{A}_\mu, v)$. So, let $d_{v,\triangle}$ be defined by (8.2.9); that is,

$$d_{v,\triangle}(A, B) = F(\mu(A \triangle B)) \quad \text{for } A, B \in \mathcal{A}_\mu.$$

Then, the distance space $(\mathcal{A}_\mu, d_{v,\triangle})$ is the metric transform of $(\mathcal{A}_\mu, d_\mu)$ under F and $(X, F(d))$ is an isometric subspace of $(\mathcal{A}_\mu, d_{v,\triangle})$.

As $(\mathcal{A}_\mu, d_\mu)$ is of negative type (because it is L_1-embeddable) and as $F \in \mathcal{F}$, we deduce that $(\mathcal{A}_\mu, d_{v,\triangle})$ is of negative type. Therefore, by Corollary 8.2.8, $(\mathcal{A}_\mu, d_{v,\triangle})$ is L_1-embeddable and, thus, its isometric subspace $(X, F(d))$ is L_1-embeddable too. ∎

It is therefore of crucial importance to determine which functions belong to $\mathcal{F}$. Such functions have been completely characterized by Schoenberg [1938a]; we will mention the result in Theorem 9.1.4. For instance, the functions $F(t) = \frac{t}{1+t}$, $\log(1 + t)$, $1 - \exp(-\lambda t)$ (for $\lambda > 0$) (called the *Schoenberg transform*), t^α (for $0 < \alpha \leq 1$) (called the *power transform*), all belong to $\mathcal{F}$. It turns out that the Schoenberg transform plays a central role in the description of the family $\mathcal{F}$.

We consider the Schoenberg transform in detail in Section 9.1. We show that it preserves the negative type property and, thus, L_1-embeddability. Moreover, we describe all the functions preserving the negative type property and, as a consequence, several examples of functions preserving L_1-embeddability. In Section 9.2, we consider the biotope transform, which is yet another way of deforming distances while retaining the property of being L_1-embeddable. In Section 9.3, we consider the following question concerning the power transform: Given an arbitrary semimetric space (X, d), determine the largest exponent α for which the power transform (X, d^α) enjoys some metric properties such that L_1- or L_2-embeddability, the hypermetric property, etc.

9.1 The Schoenberg Transform

We consider here the *Schoenberg transform*:

$$F(t) = 1 - \exp(-\lambda t) \quad \text{for } t \in \mathbb{R}_+,$$

where λ is a positive scalar. The results presented here are based essentially on the work of Schoenberg. In a first step, we show that the Schoenberg transform preserves the negative type property; this fact was proved in Schoenberg [1938b].

Theorem 9.1.1. *Let (X, d) be a distance space. The following assertions are equivalent.*

(i) *(X, d) is of negative type.*

(ii) *The symmetric function $p : X \times X \longrightarrow \mathbb{R}$, defined by*

$$p(x, y) = \exp(-\lambda d(x, y)) \quad \textit{for } x, y \in X,$$

is of positive type for all $\lambda > 0$.

(iii) *$(X, 1 - \exp(-\lambda d))$ is of negative type for all $\lambda > 0$.*

Proof. Note that the properties involved in Theorem 9.1.1 are all of finite type, i.e., they hold if and only if they hold for any finite subset of X. Hence, we can assume that X is finite, say $X = \{1, \dots, n\}$.
(i) $\Longrightarrow$ (ii) Since (X, d) is of negative type then, by Theorem 6.2.2, $(X, \sqrt{d})$ is ℓ_2-embeddable, i.e., there exist $x^{(1)}, \dots, x^{(n)} \in \mathbb{R}^m$ $(m \geq 1)$ such that $d_{jk} = (\| x^{(j)} - x^{(k)} \|_2)^2$ for all $j, k \in X$. Let $b_1, \dots, b_n \in \mathbb{R}$. We show

$$\sum_{1 \leq j,k \leq n} b_j b_k \exp\left(-\lambda(\| x^{(j)} - x^{(k)} \|_2)^2\right) \geq 0.$$

For this, we use the following classical identity:

$$\exp(-x^2) = 2^{-1}\pi^{-\frac{1}{2}} \int_{-\infty}^{\infty} \exp(ixu) \exp(-\frac{u^2}{4}) du.$$

(Here, i denotes the complex square root of unity.) Then,

$$\sum_{j,k\in X} b_j b_k \exp\left(-\lambda(\| x^{(j)} - x^{(k)} \|_2)^2\right) = \sum_{j,k\in X} b_j b_k \prod_{h=1}^{m} \exp(-\lambda(x_h^{(j)} - x_h^{(k)})^2)$$

$$= \sum_{i,j\in X} b_j b_k 2^{-m} \pi^{-\frac{m}{2}} \prod_{h=1}^{m} \int_{-\infty}^{\infty} \exp(i\sqrt{\lambda}(x_h^{(j)} - x_h^{(k)})u_h) \exp(-\frac{u_h^2}{4}) du_h$$

$$= \sum_{j,k\in X} b_j b_k 2^{-m} \pi^{-\frac{m}{2}} \int_{-\infty}^{\infty} \cdots \int_{-\infty}^{\infty} \exp(i\sqrt{\lambda}(x^{(j)} - x^{(k)})^T u) \exp(-\frac{1}{4}\sum_{h=1}^{m} u_h^2) du_1 \ldots du_m$$

$$= 2^{-m}\pi^{-\frac{m}{2}} \int_{-\infty}^{\infty} \cdots \int_{-\infty}^{\infty} \left|\sum_{j\in X} b_j \exp(i\sqrt{\lambda} {x^{(j)}}^T u)\right|^2 \exp(-\frac{1}{4}(\sum_{h=1}^{m} u_h^2)) du_1 \ldots du_m$$

is nonnegative.

(ii) $\Longrightarrow$ (iii) Set

$$d'_{ij} := \exp(-\lambda d_{ii}) + \exp(-\lambda d_{jj}) - 2\exp(-\lambda d_{ij})) = 2(1 - \exp(-\lambda d_{ij}))$$

for $i, j \in X$. That is, d' arises from $p = \exp(-\lambda d)$ by applying the inverse of the covariance mapping (defined in (5.2.5)). Applying Lemma 6.1.14, we obtain that (X, d') is of negative type, i.e., that $(X, 1 - \exp(-\lambda d))$ is of negative type.

(iii) $\Longrightarrow$ (i) Let $b_1, \ldots, b_n \in \mathbb{R}$ with $\sum_{i=1}^n b_i = 0$. Then,

$$\sum_{1\le i<j\le n} b_i b_j (1 - \exp(-\lambda d_{ij})) \le 0,$$

since $1 - \exp(-\lambda d)$ is of negative type. By expanding in series the exponential function, we obtain

$$\sum_{1\le i<j\le n} b_i b_j (1 - \exp(-\lambda d_{ij}))$$

$$= \lambda\left(\sum_{1\le i<j\le n} b_i b_j d_{ij} - \frac{\lambda}{2}\sum_{1\le i<j\le n} b_i b_j d_{ij}^2 + \frac{\lambda^2}{3!}\sum_{1\le i<j\le n} b_i b_j d_{ij}^3 - \ldots\right) \le 0$$

for all $\lambda > 0$. By dividing by λ and, then, taking the limit when $\lambda \to 0$, we deduce that $\sum_{1\le i<j\le n} b_i b_j d_{ij} \le 0$. This shows that (X, d) is of negative type. ∎

Remark 9.1.2. The equivalence (i) $\Longleftrightarrow$ (ii) from Theorem 9.1.1 is a classical result in linear algebra (see, e.g., Theorems 6.3.6 and 6.3.13 in Horn and Johnson [1991]). The proof given above for the implication (i) $\Longrightarrow$ (ii) is the original proof of Schoenberg. Another proof can be given, which uses only the fact that

$$(a_{ij}) \succeq 0 \Longrightarrow (\exp(a_{ij})) \succeq 0 \text{ for any matrix } (a_{ij})$$

It goes as follows. Suppose that the distance space (X, d) is of negative type and let $x_0, x_1, \ldots, x_n \in X$. Set $a_{ij} := d(x_0, x_i) + d(x_0, x_j) - d(x_i, x_j)$ for $i, j = 0, 1, \ldots, n$. Then, the matrix $A := (a_{ij})$ is positive semidefinite (as it coincides with the image of d under the covariance mapping pointed at x_0, up to a factor 2). Therefore, the matrix $B := (\exp(a_{ij}))$ is positive semidefinite. Let D denote the diagonal matrix with ith diagonal entry $\exp(-d(x_0, x_i))$ for $i = 0, 1, \ldots, n$. Then, the matrix DBD is positive semidefinite and its ijth entry is equal to $\exp(-d(x_i, x_j))$ for all $i, j = 0, 1, \ldots, n$. Therefore, the matrix $(\exp(-d(x_i, x_j))_{i,j=0}^n$ is positive semidefinite for all $x_0, x_1, \ldots, x_n \in X$ and $n \geq 1$, as required. ∎

Corollary 9.1.3. *Let (X, d) be a distance space. Then,*

$$(X,d) \text{ is } L_1\text{-embeddable} \iff (X, 1-\exp(-\lambda d)) \text{ is } L_1\text{-embeddable for all } \lambda > 0.$$

Proof. The "only if" part follows from Theorems 9.0.3 and 9.1.1. The proof for the converse implication is analogous to that of Theorem 9.1.1 (iii) $\Longrightarrow$ (i), replacing the negative type inequality by an arbitrary inequality valid for the cut cone CUT(Y) where Y is a finite subset of X. ∎

Remark that Theorem 9.1.1 remains valid if we assume only that (ii) and (iii) hold for a set of positive λ's admitting 0 as accumulation point. The same remark also applies to Corollary 9.1.3.

By Theorem 9.1.1, we know that the function $F(t) = 1 - \exp(-\lambda t)$ (for $\lambda > 0$) belongs to $\mathcal{F}$. Schoenberg[1] [1938a] has described all functions in $\mathcal{F}$ (assuming that all their derivatives exist). Namely,

Theorem 9.1.4. *Let $F : \mathbb{R}_+ \longrightarrow \mathbb{R}_+$ be a function such that $F(0) = 0$ and its n-th derivative $F^{(n)}$ exists on $\mathbb{R}_+ \setminus \{0\}$ for each $n \geq 1$. The following assertions are equivalent.*

(i) *$F \in \mathcal{F}$ (that is, F preserves the negative type property).*

(ii) *F is of the form:*

$$F(t) = \int_0^\infty \frac{1-\exp(-tu)}{u} d\gamma(u) \quad \text{for } t \geq 0, \tag{9.1.5}$$

where γ is a positive measure on $\mathbb{R}_+$ satisfying $\int_1^\infty \frac{d\gamma(u)}{u} < \infty$.

(iii) *$(-1)^{n-1}F^{(n)}(t) \geq 0$ for all $t > 0$ and $n \geq 1$.* ∎

Example 9.1.6. We mention here some examples of functions in the family $\mathcal{F}$.

[1]Schoenberg characterizes, in fact, the metric transforms preserving L_2-embeddability. Clearly, G preserves L_2-embeddability if and only if F preserves the negative type property, where F and G are linked by $F(t) = (G(\sqrt{t}))^2$ for $t \geq 0$. Further results concerning metric transforms in relation with embeddability in various L_2 and L_p spaces have been established, in particular, by Schoenberg [1937, 1938a, 1938b], von Neumann and Schoenberg [1941]. See also the exposition by Wells and Williams [1975].

(i) It can be easily verified that every function of the form (9.1.5) belongs to $\mathcal{F}$. For instance, the power transform: $F(t) = t^\alpha$ belongs to $\mathcal{F}$ for $0 < \alpha \leq 1$. This follows from the integral formula:

$$t^\alpha = e_\alpha^{-1} \int_0^\infty (1 - \exp(-\lambda^2 t))\lambda^{-1-2\alpha} d\lambda \quad \text{for } t \geq 0,$$

where

$$e_\alpha := \int_0^\infty (1 - \exp(-u^2))u^{-1-2\alpha} du$$

(which can be checked by setting: $u = \lambda\sqrt{t}$ in the first integral). Alternatively, this follows from the fact that $(-1)^{n-1}F^{(n)} \geq 0$ for all $n \geq 1$.

(ii) Each of the functions $F(t) = t^\alpha$ $(0 < \alpha \leq 1)$, $\frac{t}{1+t}$, $\log(1 + t)$ belongs to $\mathcal{F}$. Therefore, they all preserve L_1-embeddability. ∎

We conclude with a result of Kelly [1972] on metric transforms of the 1-dimensional ℓ_1-space.

Proposition 9.1.7. *Let $F : \mathbb{R}_+ \longrightarrow \mathbb{R}_+$ be a monotone nondecreasing concave function such that $F(0) = 0$. Then the metric transform of the distance space $\ell_1^1 = (\mathbb{R}, d_{\ell_1})$ under F is hypermetric.* ∎

9.2 The Biotope Transform

We mention here another transformation which preserves L_1-embeddability. It does not belong to the category of metric transforms, as defined by (9.0.1).

Let d be a distance on a set X and let s be a point of X. We define a new distance $d^{(s)}$ on X by setting

$$d^{(s)}(i,j) := \frac{d(i,j)}{d(i,s) + d(j,s) + d(i,j)}$$

for all $i, j \in X$. In particular, if $(\Omega, \mathcal{A}, \mu)$ is a measure space and if (X, d) is the measure semimetric space $(\mathcal{A}_\mu, d_\mu)$, then its transform $d_\mu^{(\emptyset)}$ takes the form

$$d_\mu^{(\emptyset)}(A, B) = \frac{\mu(A \triangle B)}{\mu(A) + \mu(B) + \mu(A \triangle B)} = \frac{\mu(A \triangle B)}{2\mu(A \cup B)}$$

for $A, B \in \mathcal{A}_\mu$. The distance

$$(A, B) \in \mathcal{A}_\mu \times \mathcal{A}_\mu \mapsto \frac{\mu(A \triangle B)}{\mu(A \cup B)}$$

is called the *Steinhaus distance.* The distance

$$(A, B) \mapsto \frac{|A \triangle B|}{|A \cup B|},$$

which is obtained in the special case when μ is the cardinality measure, is also known under the name of *biotope distance.* This terminology comes from the fact that this distance is used in some biological problems for the study of biotopes (see Marczewski and Steinhaus [1958]). As a consequence of the next Proposition 9.2.1, the Steinhaus and biotope distances are L_1-embeddable; (i) is given in Marczewski and Steinhaus [1958] and (ii) in Assouad [1980b].

Proposition 9.2.1.

(i) *If d is a semimetric on X, then $d^{(s)}$ is a semimetric on X.*

(ii) *If (X, d) is L_1-embeddable, then $(X, d^{(s)})$ is L_1-embeddable.*

Proof. (i) follows from (ii) and the fact that a distance space on at most 4 points is L_1-embeddable if and only if it is a semimetric space (see Remark 6.3.5 (i)). (ii) By Lemma 4.2.5, we can suppose that (X, d) is an isometric subspace of some measure semimetric space $(\mathcal{A}_\mu, d_\mu)$, i.e., $d(i,j) = \mu(A_i \triangle A_j)$ where $A_i \in \mathcal{A}_\mu$ for all $i, j \in X$, and we can suppose without loss of generality that $A_s = \emptyset$. Hence, as was already observed,

$$d^{(s)}(i,j) = \frac{\mu(A_i \triangle A_j)}{2\mu(A_i \cup A_j)}$$

for all $i, j \in X$. By Lemma 5.3.2, showing that $(X, d^{(s)})$ is L_1-embeddable amounts to showing that $p := \xi_s(d^{(s)})$ is a $\{0,1\}$-covariance. From (5.2.4), p is defined by

$$p(i,j) = \frac{1}{2}(d^{(s)}(i,s) + d^{(s)}(j,s) - d^{(s)}(i,j))$$

for $i, j \in X \setminus \{s\}$. Hence,

$$p(i,j) = \frac{1}{4} + \frac{1}{4}\frac{\mu(A_i \cap A_j)}{\mu(A_i \cup A_j)}$$

for $i, j \in X \setminus \{s\}$. Therefore, it suffices to show that the symmetric function

$$q : (i,j) \in (X \setminus \{s\})^2 \longmapsto \frac{\mu(A_i \cap A_j)}{\mu(A_i \cup A_j)}$$

is a $\{0,1\}$-covariance. For this, we use the identity

$$\frac{\mu(A \cap B)}{\mu(A \cup B)} = \frac{\mu(A \cap B)}{\mu(\Omega)} \left(\sum_{i \geq 0} \left(\frac{\mu(\bar{A} \cap \bar{B})}{\mu(\Omega)} \right)^i \right)$$

(which follows from the identity $\sum_{i \geq 0}(1-u)^i = \frac{1}{u}$ for all $0 < u \leq 1$). Therefore, q is a $\{0,1\}$-covariance, i.e., belongs to the correlation cone $\mathrm{COR}(X \setminus \{s\})$. This follows from the fact that $\{0,1\}$-covariances are preserved under taking sums, products and limits (for the product operation, recall Proposition 7.5.3). ∎

9.3 The Power Transform

We return here to the study of metric transforms and, more specifically, to that of the *power transform*:

$$F(t) = t^\alpha \quad \text{for } t \geq 0,$$

where $0 < \alpha \leq 1$. We address the following question: What is the largest exponent $\alpha \in [0,1]$ for which the metric transform (X, d^α) of an arbitrary semimetric space (X, d) can be embedded in a certain host space, say, ℓ_1 or ℓ_2 ? As α lies within $[0,1]$, we are at least assured that (X, d^α) is a semimetric space (by Lemma 9.0.2).

For instance, (X, d^α) is ℓ_2-embeddable for every $\alpha \leq 1$, if $|X| = 3$ (recall Remark 6.2.12). Blumenthal [1953] shows that (X, d^α) is ℓ_2-embeddable for every $\alpha \leq \frac{1}{2}$, if $|X| = 4$; moreover, $\frac{1}{2}$ is the largest such exponent. Deza and Maehara [1990] consider the general case of a semimetric space on $n \geq 3$ points. They give some range of values for α (depending on n) for which (X, d^α) enjoys some metric properties such as hypermetricity, ℓ_1 or ℓ_2-embeddability for any semimetric space (X, d) with $|X| = n$. The following function:

$$\gamma(s) := \log_2(1 + \frac{1}{s}) \quad \text{for } s > 0$$

will be useful for formulating the results. Deza and Maehara [1990] show the following result.

Theorem 9.3.1. *Let (X, d) be a semimetric space with $|X| = n$. Then, (X, d^α) is hypermetric for every $0 < \alpha \leq \gamma(n-1)$.*

Corollary 9.3.2. *Let (X, d) be a semimetric space with $|X| = n$. Then, (X, d^α) is ℓ_2-embeddable for every $0 < \alpha \leq \frac{1}{2}\gamma(n-1)$.*

Proof of Corollary 9.3.2. If $\alpha \leq \frac{1}{2}\gamma(n-1)$ then, by Theorem 9.3.1, $(X, d^{2\alpha})$ is hypermetric and, thus, of negative type. Therefore, (X, d^α) is ℓ_2-embeddable. ∎

For the proof of Theorem 9.3.1, we need some preliminary lemmas.

Lemma 9.3.3. *Let $0 < \alpha \leq \gamma(s)$ where $s \geq 1$ and let (X, d) be a semimetric space. Then,*

$$d(x,y) \leq d(y,z) \Longrightarrow d(x,z)^\alpha \leq \frac{1}{s} d(x,y)^\alpha + d(y,z)^\alpha.$$

Proof. As $d(x,z)^\alpha \leq (d(x,y) + d(y,z))^\alpha$, it suffices to show that $(d(x,y) + d(y,z))^\alpha \leq \frac{1}{s} d(x,y)^\alpha + d(y,z)^\alpha$ or, equivalently,

$$\left(1 + \frac{d(y,z)}{d(x,y)}\right)^\alpha \leq \frac{1}{s} + \left(\frac{d(y,z)}{d(x,y)}\right)^\alpha.$$

Let $f(t) := \frac{1}{s} + t^\alpha - (1+t)^\alpha$ for $t \geq 1$. It remains to check that $f(t) \geq 0$ for $t \geq 1$. Indeed, $f(t) \geq f(1)$ (as f is monotone nondecreasing) and $f(1) = \frac{1}{s} + 1 - 2^\alpha \geq 0$ as $\alpha \leq \gamma(s)$. ∎

The next lemmas deal with proving that (X, d^α) is hypermetric; that is, that (X, d^α) satisfies the $(2m+1)$-gonal inequality:

$$\sum_{1 \leq i < j \leq m} d(x_i, x_j)^\alpha + \sum_{1 \leq i < j \leq m+1} d(y_i, y_j)^\alpha \leq \sum_{\substack{1 \leq i \leq m \\ 1 \leq j \leq m+1}} d(x_i, y_j)^\alpha \tag{9.3.4}$$

for every sequence:

$$x_1, \ldots, x_m, y_1, \ldots, y_{m+1} \in X \tag{9.3.5}$$

(with possible repetitions) and every $m \geq 1$. To simplify notation, let us denote $d(x,y)^\alpha$ by the symbol xy, for any $x, y \in X$. We can suppose (after possibly reordering the indices) that the sequence (9.3.5) satisfies:

$$x_k y_k \leq x_i y_j \quad \text{for all } i, j \geq k, \ \ k = 1, \ldots, m. \tag{9.3.6}$$

Then, using Lemma 9.3.3, we obtain that, for $0 < \alpha \leq \gamma(s)$ and $k \leq m$, $i < k < j$,

$$\begin{aligned} x_i x_k &\leq \tfrac{1}{s} x_k y_k + x_k y_i, \ \ x_j x_k \leq \tfrac{1}{s} x_k y_k + x_j y_k, \\ y_i y_k &\leq \tfrac{1}{s} x_k y_k + x_i y_k, \ \ y_j y_k \leq \tfrac{1}{s} x_k y_k + x_k y_j. \end{aligned} \tag{9.3.7}$$

Therefore, for any $i, k \leq m$,

$$x_i x_k + y_i y_k \leq \frac{2}{s} x_k y_k + x_k y_i + x_i y_k. \tag{9.3.8}$$

Lemma 9.3.9. *Let $0 < \alpha \leq \gamma(m)$ and let (X, d) be a semimetric space. Then, (X, d^α) is $(2m+1)$-gonal. Moreover, for $\alpha > \gamma(m)$, there exists a semimetric space (X, d) with $|X| = 2m+2$ (resp. $|X| = 2m+1$) for which (X, d^α) is not $(2m+2)$-gonal (resp. not $(2m+1)$-gonal.*

Proof. Consider a sequence of points as (9.3.5) and suppose that it satisfies (9.3.6). Given $k \leq m$, summing (9.3.8) (applied with $s := m$) over $i = 1, \ldots, m, i \neq k$, yields:

$$\sum_{i=1}^{m} x_i x_k + y_i y_k \leq -\frac{2}{m} x_k y_k + \sum_{i=1}^{m} x_k y_i + x_i y_k.$$

Summing now over $k = 1, \ldots, m$, we obtain:

$$\sum_{i,k=1}^{m} x_i x_k + y_i y_k \leq -\frac{2}{m} \sum_{k=1}^{m} x_k y_k + \sum_{i,k=1}^{m} x_k y_i + x_i y_k.$$

Dividing by 2 and adding the inequality:

$$\sum_{k=1}^{m} y_k y_{m+1} \leq \frac{1}{m} \sum_{k=1}^{m} x_k y_k + \sum_{k=1}^{m} x_k y_{m+1}$$

(which follows using Lemma 9.3.3), we obtain the desired inequality (9.3.4). Suppose now that $\alpha > \gamma(m)$. Consider as distance space (X, d) the graphic metric space of the complete bipartite graphs $K_{m+1,m+1}$ and $K_{m+1,m}$. Then, d^α is not $(2m+2)$-gonal in both cases. (This is immediate, chosing for the x_i's the points in one colour class and for the y_i's the points in the other colour class.) ∎

Lemma 9.3.10. *Let $0 < \alpha \leq s$ and let (X, d) be a semimetric space such that (X, d^α) is $(2m-1)$-gonal. Consider a sequence as in (9.3.5) with maximum repetition number greater than or equal to $\frac{2m}{s+1}$ (that is, at least one member of the sequence is repeated at least $\frac{2m}{s+1}$ times). Then, (X, d^α) satisfies the corresponding $(2m+1)$-gonal inequality (9.3.4).*

Proof. Let r denote the maximum repetition number in the sequence (9.3.5). By assumption, $r \geq \frac{2m}{s+1}$. We can assume that (9.3.6) holds. If some x_i coincides with some y_j, say, $x_m = y_{m+1}$, then the $(2m+1)$-gonal inequality follows from the fact (X, d^α) is $(2m-1)$-gonal (and, thus, satisfies the inequality associated with the sequence $x_1, \ldots, x_{m-1}, y_1, \ldots, y_m$). Hence, we can suppose that the two sets $\{x_1, \ldots, x_m\}$ and $\{y_1, \ldots, y_{m+1}\}$ are disjoint. Let k denote the smallest index i such that one of x_i or y_i is repeated r times in the sequence (9.3.5). Let us suppose that this is the case for x_k (the case with y_k is similar and omitted). By assumption, (X, d^α) satisfies the $(2m-1)$-gonal inequality associated to the sequence (9.3.5) with x_k and y_k omitted; that is,

$$\sum_{\substack{1 \leq i < j \leq m \\ i,j \neq k}} x_i x_j + \sum_{\substack{1 \leq i < j \leq m+1 \\ i,j \neq k}} y_i y_k \leq \sum_{\substack{1 \leq i \leq m,\ 1 \leq j \leq m+1 \\ i,j \neq k}} x_i y_j.$$

Therefore, we are done if we can show:

(a) $$\sum_{i \in I \cup J} x_i x_k + \sum_{1 \leq i \leq m+1,\ i \neq k} y_i y_k \leq \sum_{1 \leq i \leq m+1,\ i \neq k} x_k y_i + \sum_{1 \leq i \leq m} x_i y_k,$$

where $I := \{i \mid 1 \leq i < k,\ x_i \neq x_k\}$ and $J := \{i \mid k < i \leq m,\ x_i \neq x_k\}$. Note that $|I| + |J| + r = m$ and $I = \{1, \ldots, k-1\}$ by the choice of k. Let us evaluate the left-hand side of (a). Using (9.3.7), we have:

$$\begin{aligned}
\sum_{i \in I \cup J} x_i x_k + \sum_{i \neq k} y_i y_k &\leq \sum_{i \in I} (\frac{1}{s} x_k y_k + x_k y_i) + \sum_{i \in J} (\frac{1}{s} x_k y_k + x_i y_k) \\
&+ \sum_{i=1}^{k-1} (\frac{1}{s} x_k y_k + x_i y_k) + \sum_{i=k+1}^{m} (\frac{1}{s} x_k y_k + x_k y_i) \\
&= \frac{1}{s} x_k y_k (|I| + |J| + m) + \sum_{1 \leq i \leq m+1,\ i \neq k} x_k y_i + \sum_{i=1}^{m} x_i y_k - r x_k y_k.
\end{aligned}$$

But, $\frac{1}{s}(|I|+|J|+m)-r = \frac{2m-r}{s} - r \leq 0$, because $r \geq \frac{2m}{s+1}$. Therefore, (a) holds, which concludes the proof. ∎

Proof of Theorem 9.3.1. Let $0 < \alpha \leq \gamma(n-1)$. We show that d^α is $(2m+1)$-gonal for all $m \geq 1$, by induction on m. If $m = 1$, this is clear as d^α is a semimetric. Suppose that $m \geq 2$ and that d^α is $(2m-1)$-gonal. Clearly, a sequence of $2m+1$ points in X must contain an element which is repeated at least $\frac{2m+1}{n}$ times (as $|X| = n$). Applying Lemma 9.3.10 (with $s := n-1$), we deduce that d^α is $(2m+1)$-gonal. ∎

Corollary 9.3.11. *Let (X, d) be a semimetric space with $|X| = 5$, or 6. Then, (X, d^α) is ℓ_1-embeddable for all $0 < \alpha \leq \gamma(2)$. Moreover, $\gamma(2)$ is the largest such value of α; that is, for $\alpha > \gamma(2)$, there are distance spaces on 5 and 6 points for which d^α is not ℓ_1-embeddable.*

Proof. In the case $|X| = 5$, the result follows from the fact that a distance on 5 points is ℓ_1-embeddable if and only if it is 5-gonal (recall Remark 6.3.5 (i)). Consider now the case $|X| = 6$. If $\alpha \leq \gamma(2)$ then d^α is 5-gonal by Lemma 9.3.9 and, thus, 7-gonal by Lemma 9.3.10. (We use the fact that a distance on 6 points is ℓ_1-embeddable if and only if it is 7-gonal.) ∎

For $n \geq 3$, let us consider the parameters $g(n)$, $h(n)$, $c_1(n)$, and $c_2(n)$ which are defined as follows: $g(n)$ denotes the maximum $\alpha \in [0,1]$ for which (X, d^α) is n-gonal for every semimetric space (X, d) with $|X| = n$. Similarly, $h(n)$ (resp. $c_1(n)$, $c_2(n)$) denotes the maximum α for which (X, d^α) is hypermetric (resp. ℓ_1-embeddable, ℓ_2-embeddable) for every semimetric space (X, d) with $|X| = n$. Clearly,

$$c_2(n) \leq c_1(n) \leq h(n) \leq g(n) \text{ and } c_2(n) \geq \frac{1}{2}h(n).$$

We have shown in Theorem 9.3.1 that

$$h(n) \geq \gamma(n-1); \text{ implying } c_2(n) \geq \frac{1}{2}\gamma(n-1).$$

Using Lemma 9.3.9 (together with Lemma 6.1.10 in the case when n is even), one can give the exact value of the parameter $g(n)$.

Lemma 9.3.12. *$g(2m) = \gamma(m-1)$ for $m \geq 2$, and $g(2m+1) = \gamma(m)$ for $m \geq 1$.* ∎

This implies that $h(5) = h(6) = \gamma(2)$. The exact value of the parameter $c_1(n)$ is known in the cases $n \leq 6$, by Corollary 9.3.11. On the other hand, we have: $c_2(5) \geq \frac{1}{2}h(5) = \frac{1}{2}\gamma(2)$, but the exact value of $c_2(5)$ is not known. The value of $c_2(n)$ is known, however, in the case $n = 6$.

Lemma 9.3.13. $c_2(6) = \frac{1}{2}\gamma(2)$.

Proof. First, $c_2(6) \geq \frac{1}{2}h(6) = \frac{1}{2}\gamma(2)$. Equality is proved, by considering the path metric d of $K_{3,3}$. Indeed, if $\alpha > \frac{1}{2}\gamma(2)$, then $d^{2\alpha}$ is not 6-gonal and, thus, not of negative type. That is, d^α is not ℓ_2-embeddable. ∎

We summarize in Figure 9.3.14 the information known about the parameters $g(n)$, $h(n)$, $c_1(n)$, $c_2(n)$ for small values of n, $n \leq 6$. Note that $\gamma(2) = 0.5849...$, $\gamma(2)/2 = 0.2924....$

n	3	4	5	6
$c_2(n)$	1	1/2	?	$\gamma(2)/2$
$c_1(n)$	1	1	$\gamma(2)$	$\gamma(2)$
$h(n)$	1	1	$\gamma(2)$	$\gamma(2)$
$g(n)$	1	1	$\gamma(2)$	$\gamma(2)$

Figure 9.3.14

To conclude, let us mention the following upper bounds from Deza and Maehara [1990] for $c_2(n)$, obtained by considering the path metric of the complete bipartite graph $K_{m,n}$ for suitable m, n.

$$c_2(2m) \leq \frac{1}{2}\gamma(m-1), \; c_2(2m+1) \leq \frac{1}{2}\gamma\left(\frac{2m(m+1)}{2m+1} - 1\right).$$

Deza and Maehara [1990] conjecture that equality holds in the above inequalities. This conjecture is confirmed for $c_2(n)$ with $n = 4, 6$. (More information about metric transforms of graphs can be found in Maehara [1986].)

Chapter 10. Lipschitz Embeddings

Let (X, d) be a finite semimetric space on $n := |X|$ points. In general, (X, d) cannot be *isometrically* embedded into some ℓ_1-space. However, (X, d) admits always an embedding into some ℓ_1-space, where the distances are preserved up to a multiplicative factor whose size is of the order $\log n$. This result is due to Bourgain [1985]. We present this result, together with an interesting application due to Linial, London and Rabinovich [1994] for approximating multicommodity flows. We also present a generalization of the negative type condition for Lipschitz ℓ_2-embeddings.

10.1 Embeddings with Distortion

Definition 10.1.1. *Let (X, d) and (X', d') be two distance spaces and let $C \geq 1$. A mapping $\varphi : X \longrightarrow X'$ is called a* Lipschitz embedding *of (X, d) into (X', d') with* distortion C *if*

$$\frac{1}{C} d(x, y) \leq d'(\varphi(x), \varphi(y)) \leq d(x, y)$$

holds for all $x, y \in X$.

Hence, a Lipschitz embedding with distortion $C = 1$ is just an isometric embedding. Bourgain [1985] shows that every semimetric space on n points admits a Lipschitz embedding into ℓ_1 with distortion[1] $O(\log_2(n))$.

Theorem 10.1.2. *Let (X, d) be a finite semimetric space with $|X| = n$. There exist vectors $u_x \in \mathbb{R}^N$ $(x \in X)$ satisfying:*

$$\| u_x - u_y \|_1 \leq d(x, y) \leq c_0 \log_2(n) \| u_x - u_y \|_1$$

for all $x, y \in X$. Here, $N = \sum_{p=1}^{\lfloor \log_2 n \rfloor} \binom{n}{2^p}$ $(< 2^n)$, and $c_0 > 0$ is a constant.

[1] The Lipschitz embedding described in Theorem 10.1.2 has distortion $c_0 \log_2 n$, where $c_0 < 92$. Garg [1995b] presents a slight variation of it, for which he can show a better constant for the distortion. Namely, set $H(n) := 1 + \frac{1}{2} + \ldots + \frac{1}{n}$ and, for $x \in X$, let u'_x be the vector indexed by all proper subsets A of X with components $u'_x(A) := \frac{d(x,A)}{H(n)|A|\binom{n}{|A|}}$. Then, $\| u'_x - u'_y \|_1 \leq d(x, y) \leq 2H(n) \| u'_x - u'_y \|_1$ for all $x, y \in X$.

Proof. For each integer s $(1 \leq s \leq n)$ let $\mathcal{P}_s$ denote the family of subsets $A \subseteq X$ with $|A| = s$. Set $\mathcal{P} := \bigcup_{p=1}^{\lfloor \log_2 n \rfloor} \mathcal{P}_{2^p}$; so $|\mathcal{P}| = N$. For each $x \in X$, we define the vector $u_x \in \mathbb{R}^N$ indexed by the sets $A \in \mathcal{P}$ and defined by

$$u_x(A) := \frac{1}{\lfloor \log_2 n \rfloor \binom{n}{|A|}} d(x, A), \quad \text{for all } A \in \mathcal{P}.$$

We remind that $d(x, A) := \min_{y \in A} d(x, y)$. We show that the vectors u_x $(x \in X)$ satisfy the conditions of Theorem 10.1.2. We first check that

$$d(x, y) \geq \| u_x - u_y \|_1 \quad \text{for all } x, y \in X.$$

Indeed, using the fact that $|d(x, A) - d(y, A)| \leq d(x, y)$, we obtain

$$\begin{aligned} \| u_x - u_y \|_1 &= \sum_{A \in \mathcal{P}} |u_x(A) - u_y(A)| = \frac{1}{\lfloor \log_2 n \rfloor} \sum_{A \in \mathcal{P}} \frac{|d(x, A) - d(y, A)|}{\binom{n}{|A|}} \\ &\leq \frac{1}{\lfloor \log_2 n \rfloor} \sum_{A \in \mathcal{P}} \frac{d(x, y)}{\binom{n}{|A|}} = d(x, y). \end{aligned}$$

Let $x, y \in X$ be fixed. We now show that

$$d(x, y) \leq c_0 \lfloor \log_2 n \rfloor \| u_x - u_y \|_1 \tag{10.1.3}$$

for some constant c_0. For $z_0 \in X$ and $\rho > 0$, let $B(z_0, \rho) := \{z \in X \mid d(z_0, z) < \rho\}$ denote the open ball with center z_0 and radius ρ. We define a sequence of radii ρ_t in the following way: $\rho_0 := 0$ and, for $t \geq 1$, ρ_t denotes the smallest scalar ρ for which $|B(x, \rho)| \geq 2^t$ and $|B(y, \rho)| \geq 2^t$. We define ρ_t as long as $\rho_t < \frac{1}{2} d(x, y)$. Let t^* denote the largest such index. We let $\rho_{t^*+1} := \frac{1}{2} d(x, y)$. Observe that, for $t = 1, 2, \ldots, t^*, t^* + 1$, the balls $B(x, \rho_t)$ and $B(y, \rho_t)$ are disjoint.

Let $t \in \{1, \ldots, t^* + 1\}$ and let ρ be such that $\rho_{t-1} < \rho < \rho_t$. Note that

$$|d(x, A) - d(y, A)| \geq \rho - \rho_{t-1}$$

holds for any subset $A \subseteq X$ such that $A \cap B(x, \rho) = \emptyset$ and $A \cap B(y, \rho_{t-1}) \neq \emptyset$. Hence, for any integer s $(1 \leq s \leq n)$,

$$\begin{aligned} &\tfrac{1}{|\mathcal{P}_s|} \sum_{A \in \mathcal{P}_s} |d(x, A) - d(y, A)| \\ &\geq (\rho - \rho_{t-1}) \tfrac{|\{A \in \mathcal{P}_s \mid A \cap B(x, \rho) = \emptyset, A \cap B(y, \rho_{t-1}) \neq \emptyset\}|}{|\mathcal{P}_s|} = (\rho - \rho_{t-1}) \mu_t, \end{aligned} \tag{10.1.4}$$

after setting

$$\mu_t := \frac{|\{A \in \mathcal{P}_s \mid A \cap B(x, \rho) = \emptyset, A \cap B(y, \rho_{t-1}) \neq \emptyset\}|}{|\mathcal{P}_s|}.$$

The key argument is now to observe that, if we choose s in such a way that $s \approx \frac{n}{10 \cdot 2^t}$, then the quantity μ_t is bounded below by an absolute constant μ_0 (not depending on t, x or y). More precisely, let us choose $s = s_t = 2^p$, where p is the smallest integer such that $\frac{n}{10 \cdot 2^t} \leq 2^p$, i.e., $p = \lceil \log_2(\frac{n}{10 \cdot 2^t}) \rceil$. Let us assume for a moment that the relation $\mu_t \geq \mu_0$ holds for all t, for some constant μ_0. We show how to conclude the proof of (10.1.3).

As the values of s_t corresponding to $t = 1, \ldots, t^* + 1$ are all distinct, by applying (10.1.4) with $s = s_t$, we deduce that

$$\frac{1}{|\mathcal{P}_{s_t}|} \sum_{A \in \mathcal{P}_{s_t}} |d(x, A) - d(y, A)| \geq (\rho - \rho_{t-1})\mu_0.$$

Letting $\rho \longrightarrow \rho_t$, the same relation holds with $\rho = \rho_t$, i.e.,

$$\frac{1}{|\mathcal{P}_{s_t}|} \sum_{A \in \mathcal{P}_{s_t}} |d(x, A) - d(y, A)| \geq (\rho_t - \rho_{t-1})\mu_0.$$

Summing the above relation over $t = 1, \ldots, t^* + 1$, we obtain:

$$\lfloor \log_2 n \rfloor \parallel u_x - u_y \parallel_1 \geq \sum_{1 \leq t \leq t^*+1} (\rho_t - \rho_{t-1})\mu_0 = \rho_{t^*+1}\mu_0 = \frac{\mu_0}{2} d(x, y).$$

This shows that (10.1.3) holds with $c_0 := \frac{2}{\mu_0}$. (Note that $c_0 < 91.26$.)

Let us return to the evaluation of the quantity μ_t. Recall that $s = s_t = 2^p$, where $p = \lceil \log_2(\frac{n}{10 \cdot 2^t}) \rceil$; hence, $\frac{n}{10 \cdot 2^t} \leq s < \frac{2n}{10 \cdot 2^t}$. As $\rho_{t-1} < \rho < \rho_t$, we have $|B(x, \rho)| < 2^t$ or $|B(y, \rho)| < 2^t$. We can suppose without loss of generality that $|B(x, \rho)| < 2^t$. Set $\beta_x := |B(x, \rho)|$ and $\beta_y := |B(y, \rho_{t-1})|$; then, $\beta_x < 2^t$ and $\beta_y \geq 2^{t-1}$. We have:

$$\begin{aligned}
\mu_t &= \frac{\binom{n-\beta_x}{s} - \binom{n-\beta_x-\beta_y}{s}}{\binom{n}{s}} \\
&= \prod_{i=0}^{\beta_x-1} (1 - \frac{s}{n-i}) - \prod_{i=0}^{\beta_x+\beta_y-1} (1 - \frac{s}{n-i}) \\
&= \prod_{i=0}^{\beta_x-1} (1 - \frac{s}{n-i}) \left(1 - \prod_{i=\beta_x}^{\beta_x+\beta_y-1} (1 - \frac{s}{n-i}) \right).
\end{aligned}$$

We show that

$$\prod_{i=0}^{\beta_x-1} (1 - \frac{s}{n-i}) \geq e^{-\frac{8}{10}} \text{ and } \prod_{i=\beta_x}^{\beta_x+\beta_y-1} (1 - \frac{s}{n-i}) \leq e^{-\frac{1}{20}},$$

which implies that $\mu_t \geq e^{-\frac{8}{10}}(1 - e^{-\frac{1}{20}}) =: \mu_0$. Indeed,

$$\begin{aligned}
\prod_{i=0}^{\beta_x-1} (1 - \frac{s}{n-i}) \geq (1 - \frac{s}{n-2^t+1})^{\beta_x} &= \exp(\beta_x \log(1 - \tfrac{s}{n-2^t+1})) \\
&\geq \exp(-\beta_x \tfrac{2s}{n-2^t+1})
\end{aligned}$$

(we have used the fact that $\log(1-u) \geq -2u$ for $0 \leq u \leq \frac{1}{2}$). Now, the latter quantity is greater than or equal to $\exp(-\frac{8}{10})$, as $s < \frac{2n}{10 \cdot 2^t}$, $\beta_x < 2^t$, and

$\frac{n}{n-2^t+1} \leq 2$. On the other hand,

$$\prod_{i=\beta_x}^{\beta_x+\beta_y-1} (1-\frac{s}{n-i}) \leq (1-\frac{s}{n-\beta_x})^{\beta_y} = \exp(\beta_y \log(1-\frac{s}{n-\beta_x})) \leq \exp(-\beta_y \frac{s}{n-\beta_x})$$

(we have used the fact that $\log(1-u) \leq -u$ for $0 \leq u \leq 1$). The latter quantity is less than or equal to $\exp(-\frac{1}{20})$, as $\beta_y \geq 2^{t-1}$, $s \geq \frac{n}{10 \cdot 2^t}$, and $\frac{n}{n-\beta_x} \geq 1$. ∎

Theorem 10.1.2 extends, in fact, easily to the case of ℓ_p-spaces for $p \geq 1$, as observed in Linial, London and Rabinovich [1994].

Theorem 10.1.5. *Let $1 \leq p < \infty$. Let (X, d) be a semimetric space with $|X| = n$. There exist vectors $v_x \in \mathbb{R}^N$ ($x \in X$) such that*

$$\| v_x - v_y \|_p \leq d(x, y) \leq c_0 \log_2 n \| v_x - v_y \|_p$$

for all $x, y \in X$. Here, $N = \sum_{p=1}^{\lfloor \log_2 n \rfloor} \binom{n}{2^p}$ ($< 2^n$), and $c_0 > 0$ is a constant.

Proof. The vectors can be constructed by slightly modifying the vectors u_x from the proof of Theorem 10.1.2. We remind that

$$u_x(A) = \lambda_A d(x, A)$$

where $\lambda_A := \frac{1}{\lfloor \log_2 n \rfloor \binom{n}{|A|}}$ for $A \in \mathcal{P}$. Define $v_x \in \mathbb{R}^N$ by setting

$$v_x(A) := (\lambda_A)^{\frac{1}{p}} d(x, A) \quad \text{for } A \in \mathcal{P}.$$

Observe that $\sum_{A \in \mathcal{P}} \lambda_A = 1$. From this follows that $\| v_x - v_y \|_p \leq d(x, y)$. By convexity of the function $x \longmapsto x^p$ ($x \geq 0$) we obtain that

$$\sum_{A \in \mathcal{P}} \lambda_A |d(x, A) - d(y, A)| \leq \left(\sum_{A \in \mathcal{P}} \lambda_A |d(x, A) - d(y, A)|^p \right)^{\frac{1}{p}}.$$

This implies that $\| u_x - u_y \|_1 \leq \| v_x - v_y \|_p$. Thus, $d(x, y) \leq c_0 \log_2 n \| v_x - v_y \|_p$. ∎

We have just seen that, for each $p \geq 1$, every semimetric on n points can be embedded with distortion $O(\log n)$ into ℓ_p^N, where $N < 2^n$. In fact, Linial, London and Rabinovich [1994] show that the above proofs can be modified so as to yield an embedding into $\ell_p^{O((\log n)^2)}$ with distortion $O(\log n)$. Hence, the embedding takes place in a space of dimension $O((\log n)^2)$ instead of the dimension $N = \sum_p \binom{n}{2^p}$ (exponential in n) demonstrated in Theorems 10.1.2 and 10.1.5. For this, instead of the vectors u_x, v_x ($x \in X$) (constructed in the above two theorems) which are indexed by *all* subsets $A \subseteq X$ whose cardinality is a power of 2, one considers their projections in a $O((\log n)^2)$-dimensional subspace.

Namely, for each cardinality $s \leq |X| = n$ which is a power of 2 one randomly chooses $O(\log n)$ subsets $A \subseteq X$ of cardinality s; then, one restricts the coordinates of u_x, v_x to be indexed by these $O((\log n)^2)$ subsets of X. Moreover, if $1 \leq p \leq 2$, every semimetric on n points can be embedded into $\ell_p^{O(\log n)}$ with distortion $O(\log n)$; this is shown in Linial, London and Rabinovich [1994], using earlier results by Johnson and Lindenstrauss [1984] and Bourgain [1985]. Such embeddings can be constructed in random polynomial time. For a semimetric d on n points, let $C(d)$ denote the smallest distortion with which d can be embedded into some ℓ_2-space. We will see below (see Proposition 10.2.1 and the remarks thereafter) that, for every $\epsilon > 0$, one can find in polynomial time an embedding of d into ℓ_2^n with distortion $C(d) + \epsilon$.

The following result is a variant of Theorem 10.1.2 due to Linial, London and Rabinovich [1994], that we will use later in this section for approximating multicommodity flows.

Theorem 10.1.6. *Let (X,d) be a finite semimetric space and let $Y \subseteq X$ with $k := |Y|$. There exist vectors $w_x \in \mathbb{R}^K$ $(x \in X)$ satisfying*

$$\begin{array}{ll} \| w_x - w_y \|_1 \leq d(x,y) & \text{for all } x,y \in X, \\ d(x,y) \leq c_0 \log_2(k) \| w_x - w_y \|_1 & \text{for all } x,y \in Y. \end{array}$$

Here, $K = \sum_{p=1}^{\lfloor \log_2 k \rfloor} \binom{k}{2^p}$ $(< 2^k)$ and $c_0 > 0$ is a constant.

Proof. The proof is analogous to that of Theorem 10.1.2 but, now, we construct a Lipschitz ℓ_1-embedding using only subsets of Y. Namely, we define $w_x \in \mathbb{R}^K$ by setting

$$w_x(A) := \frac{1}{\lfloor \log_2 k \rfloor \binom{k}{|A|}} d(x,A) \quad \text{for } A \subseteq Y, |A| = 2^p, \ p \leq \log_2 k.$$

Then, $\| w_x - w_y \|_1 \leq d(x,y)$ holds for all $x,y \in X$. Moreover, by the proof of Theorem 10.1.2, the relation: $d(x,y) \leq c_0 \log_2 k \| w_x - w_y \|_1$ holds for all $x,y \in Y$. ∎

In fact, for the purpose of finding good approximations in polynomial time for multicommodity flows, we will need the following strengthening of Theorem 10.1.6, whose proof can be found in Linial, London and Rabinovich [1994].

Theorem 10.1.7. *Let (X,d) be a finite semimetric space and let $Y \subseteq X$ with $k := |Y|$. There exist vectors $w_x \in \mathbb{R}^{O(n^2)}$ $(x \in X)$ satisfying*

$$\begin{array}{ll} \| w_x - w_y \|_1 \leq d(x,y) & \text{for all } x,y \in X, \\ d(x,y) \leq c_0 \log_2(k) \| w_x - w_y \|_1 & \text{for all } x,y \in Y \end{array}$$

(where $c_0 > 0$ is a constant). Moreover, the vectors w_x $(x \in X)$ can be found in polynomial time. ∎

10.2 The Negative Type Condition for Lipschitz Embeddings

Recall from Theorem 6.2.2 that isometric ℓ_2-embeddability can be characterized in terms of the negative type inequalities. In fact, this characterization extends to Lipschitz ℓ_2-embeddings, as observed in Linial, London and Rabinovich [1994].

Proposition 10.2.1. *Let $C \geq 1$ and let (X,d) be a finite semimetric space with $X = \{0,1,\ldots,n\}$. The following assertions are equivalent.*

(i) *$(X,\sqrt{d})$ has an ℓ_2-embedding with distortion C, i.e., there exist $u_i \in \mathbb{R}^N$ $(i \in X)$ such that*

$$\frac{1}{C^2} d_{ij} \leq (\| u_i - u_j \|_2)^2 \leq d_{ij} \quad \text{for } i,j \in X. \tag{10.2.2}$$

(ii) *There exists a positive semidefinite $n \times n$ matrix $A = (a_{ij})$ satisfying:*

$$\begin{cases} \frac{1}{C^2} d_{ij} \leq a_{ii} + a_{jj} - 2a_{ij} \leq d_{ij} & \text{for } 1 \leq i \neq j \leq n, \\ \frac{1}{C^2} d_{0i} \leq a_{ii} \leq d_{0i} & \text{for } 1 \leq i \leq n. \end{cases} \tag{10.2.3}$$

(iii) *For every $b \in \mathbb{Z}^X$ with $\sum_{i \in X} b_i = 0$, d satisfies:*

$$\frac{1}{C^2} \sum_{i,j \in X \mid b_i b_j > 0} b_i b_j d_{ij} + \sum_{i,j \in X \mid b_i b_j < 0} b_i b_j d_{ij} \leq 0. \tag{10.2.4}$$

Before giving the proof, let us make some observations. First, note that (10.2.4) in the case $C = 1$ is nothing but the usual negative type condition. Hence, (10.2.4) is a generalization of the negative type condition for Lipschitz embeddings. Proposition 10.2.1 has some algorithmic consequences. In particular, one can evaluate in polynomial time (with an arbitrary precision) the smallest distortion C^* with which $(X,\sqrt{d})$ can be embedded into an ℓ_2-space. Indeed, $C^* = \frac{1}{\sqrt{\lambda^*}}$, where

$$\begin{array}{rll} \lambda^* = \max & \lambda & \\ & \lambda d_{ij} \leq a_{ii} + a_{jj} - 2a_{ij} \leq d_{ij} & (1 \leq i \neq j \leq n) \\ & \lambda d_{0i} \leq a_{ii} \leq d_{0i} & (i = 1,\ldots,n) \\ & A \succeq 0 & \\ & \lambda \geq 0. & \end{array}$$

This optimization program can be solved using, for instance, the ellipsoid algorithm (cf. Grötschel, Lovász and Schrijver [1988]). By the results mentioned earlier, $C^* = O(\log n)$.

Proof of Proposition 10.2.1. We set $V_n = X \setminus \{0\} = \{1, \ldots, n\}$.
(i) $\Longrightarrow$ (ii) We can suppose without loss of generality that $u_0 = 0$. Set $a_{ij} := u_i^T u_j$ for all $i, j \in V_n$. Then, the matrix (a_{ij}) satisfies (ii).
(ii) $\Longrightarrow$ (i) As A is positive semidefinite, it is the Gram matrix of some vectors $u_1, \ldots, u_n$. Then, the vectors $u_0 := 0, u_1, \ldots, u_n$ satisfy (10.2.2).
(ii) $\Longrightarrow$ (iii) Let A be a matrix satisfying (ii). We construct a distance D on X by setting $D_{0i} := a_{ii}$ for $i \in V_n$ and $D_{ij} := a_{ii} + a_{jj} - 2a_{ij}$ for $i \neq j \in V_n$. Then, D is of negative type as A is positive semidefinite. Hence, given $b \in \mathbb{Z}^X$ with $\sum_{i \in X} b_i = 0$, we have: $\sum_{i,j \in X} b_i b_j D_{ij} \leq 0$. By assumption (ii), $b_i b_j D_{ij} \geq \frac{1}{C^2} b_i b_j d_{ij}$ if $b_i b_j > 0$, and $b_i b_j D_{ij} \geq b_i b_j d_{ij}$ if $b_i b_j < 0$. This shows that (10.2.4) holds.
(iii) $\Longrightarrow$ (ii) We suppose that (iii) holds. This implies that, for every positive semidefinite matrix $B = (b_{ij})_{i,j \in X}$ such that $Be = 0$ (e denoting the all-ones vector),

$$\frac{1}{C^2} \sum_{i,j \in X | b_{ij} > 0} b_{ij} d_{ij} + \sum_{i,j \in X | b_{ij} < 0} b_{ij} d_{ij} \leq 0. \tag{10.2.5}$$

Let us suppose that (ii) does not hold. Then, the two cones PSD_n (the cone of positive semidefinite $n \times n$ matrices) and $K := \{A = (a_{ij}) \mid A \text{ satisfies } (10.2.3)\}$ are disjoint. Therefore, there exists a hyperplane separating PSD_n and K. In other words, there exists a symmetric matrix Z and a scalar α satisfying: $\langle Z, A \rangle > \alpha$ for all $A \in \mathrm{PSD}_n$ and $\langle A, Z \rangle \leq \alpha$ for all $A \in K$, where

$$\langle Z, A \rangle := \sum_{1 \leq i,j \leq n} z_{ij} a_{ij}.$$

Hence, $\alpha < 0$, which implies that Z is positive semidefinite. Moreover, as the inequality $\langle Z, A \rangle \leq \alpha$ is valid for the cone K, it can be expressed as a nonnegative linear combination of the inequalities defining K. Therefore, there exist two symmetric matrices (λ_{ij}) and (μ_{ij}) with nonnegative entries such that

$$\langle Z, A \rangle = \sum_{i \neq j \in V_n} (\lambda_{ij} - \mu_{ij})(a_{ii} + a_{jj} - 2a_{ij}) + \sum_{i \in V_n} (\lambda_{ii} - \mu_{ii}) a_{ii}$$

for all A, and

$$\sum_{i \neq j \in V_n} \lambda_{ij} d_{ij} - \mu_{ij} \frac{d_{ij}}{C^2} + \sum_{i \in V_n} \lambda_{ii} d_{0i} - \mu_{ii} \frac{d_{0i}}{C^2} \leq \alpha. \tag{10.2.6}$$

We define the symmetric matrix B indexed by X by

$$\begin{aligned}
&b_{00} := \sum_{j \in V_n} \lambda_{jj} - \mu_{jj}, \; b_{ii} := \sum_{j \in V_n} \lambda_{ij} - \mu_{ij} \quad \text{for } i \in V_n, \\
&b_{0i} := -\lambda_{ii} + \mu_{ii} \quad \text{for } i \in V_n, \\
&b_{ij} := -\lambda_{ij} + \mu_{ij} \quad \text{for } i \neq j \in V_n.
\end{aligned}$$

By construction, $Be = 0$ and B is positive semidefinite (as $\sum_{i,j \in V_n} b_{ij} a_{ij} = \langle Z, A \rangle \geq 0$ for all $A \in \text{PSD}_n$). Observe now that, as $C \geq 1$,

$$\min(b_{ij}, \frac{b_{ij}}{C^2}) \geq \frac{\mu_{ij}}{C^2} - \lambda_{ij} \ (i \neq j \in V_n), \ \min(b_{0i}, \frac{b_{0i}}{C^2}) \geq \frac{\mu_{ii}}{C^2} - \lambda_{ii} \ (i \in V_n).$$

These relations together with (10.2.6) yield:

$$\sum_{i,j \in X | b_{ij} > 0} b_{ij} \frac{d_{ij}}{C^2} + \sum_{i,j \in X | b_{ij} < 0} b_{ij} d_{ij} \geq -\alpha > 0,$$

which contradicts the assumption (10.2.5). ∎

10.3 An Application for Approximating Multicommodity Flows

We present here an application of the above results on Lipschitz ℓ_1-embeddings with distortion to the approximation of multicommodity flows.

We start with recalling several definitions. Let $G = (V, E)$ be a connected graph and let $(s_1, t_1), \ldots, (s_k, t_k)$ be k distinct pairs of nodes in V, called *terminal pairs* (or *commodity pairs*). We are given for each edge $e \in E$ a number $C_e \in \mathbb{R}_+$, called the *capacity* of the edge e and, for each terminal pair (s_h, t_h), a number $D_h \geq 0$ called its *demand.* For $h = 1, \ldots, k$, let $\mathcal{P}_h$ denote the set of paths joining the terminals s_h and t_h in G; set

$$\mathcal{P} := \bigcup_{1 \leq h \leq k} \mathcal{P}_h.$$

A *multicommodity flow* is a function $f : \mathcal{P} \longrightarrow \mathbb{R}_+$. The multicommodity flow f is said to be *feasible* for the instance (G, C, D) if it satisfies the following capacity and demand constraints:

$$\sum_{P \in \mathcal{P} | e \in P} f_P \leq C_e \quad \text{for all } e \in E,$$
$$\sum_{P \in \mathcal{P}_h} f_P \geq D_h \quad \text{for } h = 1, \ldots, k.$$

A classical problem in the theory of networks is to find a feasible multicommodity flow (possibly satisfying some further cost constraints). We are interested here in the following variant of the problem, known as the *concurrent flow problem*:

> *Determine the largest scalar λ^* for which there exists a feasible multicommodity flow for the instance $(G, C, \lambda^* D)$.*

So, in this problem, one tries to satisfy the largest possible fraction λD_h of demand for each terminal pair (s_h, t_h), while respecting the capacity constraints.

Let us call the maximum such λ^* the *max-flow*. The max-flow can be computed by solving the following linear programming problem:

$$(10.3.1) \qquad \begin{array}{lll} \lambda^* = \max & \lambda & \\ & \sum_{P\in\mathcal{P}|e\in P} f_P \leq C_e & (e \in E) \\ & \sum_{P\in\mathcal{P}_h} f_P \geq \lambda D_h & (h = 1, \ldots, k) \\ & f_P \geq 0 & (P \in \mathcal{P}). \end{array}$$

An upper bound for the max-flow λ^* can be easily formulated in terms of cuts. Given a subset $S \subseteq V$, recall that $\delta_G(S)$ denotes the cut in G determined by S, which consists of the edges $e \in E$ having one endnode in S and the other endnode in $V \setminus S$. Then, the quantity

$$\text{cap}(S) := \sum_{e\in\delta_G(S)} C_e$$

denotes the capacity of the cut $\delta_G(S)$. Furthermore, let

$$\text{dem}(S) := \sum_{h\in H_S} D_h$$

denote its demand, where H_S consists of the indices h for which the terminal pair (s_h, t_h) is separated by the partition $(S, V \setminus S)$ (i.e., such that $s_h \in S$ and $t_h \in V \setminus S$, or vice versa).

Lemma 10.3.2. *We have:* $\lambda^* \leq \min_{S\subseteq V} \frac{\text{cap}(S)}{\text{dem}(S)}$.

Proof. We use the formulation of λ^* from (10.3.1). Let S be a subset of V. Then,

$$\begin{aligned} \text{cap}(S) &= \sum_{e\in\delta_G(S)} C_e \geq \sum_{e\in\delta_G(S)} \sum_{P\in\mathcal{P}|e\in P} f_P = \sum_{P\in\mathcal{P}} \sum_{e\in\delta_G(S)\cap P} f_P \\ &= \sum_{P\in\mathcal{P}} |\delta_G(S) \cap P| f_P = \sum_{1\leq h\leq k} \sum_{P\in\mathcal{P}_h} |\delta_G(S) \cap P| f_P \\ &\geq \sum_{h\in H_S} \sum_{P\in\mathcal{P}_h} f_P \geq \sum_{h\in H_S} \lambda^* D_h = \lambda^* \text{dem}(S). \end{aligned}$$

∎

The quantity: $\min_{S\subseteq V} \frac{\text{cap}(S)}{\text{dem}(S)}$ is called the *min-cut*. The inequality from Lemma 10.3.2 says that

$$\text{max-flow} \quad \leq \quad \text{min-cut}.$$

This inequality is, in general, strict. However, it is an equality in some special cases. In particular, it is an equality for the cases $k = 1, 2$ of one or two commodity pairs. In the case $k = 1$, this follows from the well-known max-flow min-cut theorem by Ford and Fulkerson [1962]. In the case $k = 2$ this is the max-flow min-cut theorem for two commodities, proved by Hu [1963]. These two results can also be derived from some properties of ℓ_1-embeddings; see Corollary 10.3.7 below.

For an arbitrary number k of commodity pairs, the max-flow is in general less than the min-cut. However, the gap between the max-flow and the min-cut cannot be too big. The first result in this direction is due to Leighton and Rao [1988], who proved that the ratio

$$\rho := \frac{\min_{S \subseteq V} \frac{\mathrm{cap}(S)}{\mathrm{dem}(S)}}{\lambda^*}$$

is $O(\log n)$ in the case when all demands are one and there is a commodity between any pair of vertices, i.e., $k = \binom{n}{2}$. Several improvements of this result have been proposed thereafter, in particular, by Klein, Agrawal, Ravi and Rao [1990], Plotkin and Tardos [1993], Garg, Vazirani and Yannakakis [1993]. The best result is due to Linial, London and Rabinovich [1994], who proved that $\rho = O(\log k)$ for arbitrary demands and capacities and an arbitrary number k of commodity pairs; see Theorem 10.3.3[2] below. This result is based on the results from Theorems 10.1.2, 10.1.6 and 10.1.7.

Theorem 10.3.3. *We have:* $\min_{S \subseteq V} \frac{\mathrm{cap}(S)}{\mathrm{dem}(S)} \leq (c_0 \log k)\lambda^*$, *where* $c_0 > 0$ *is a constant. Moreover, a subset* S *for which* $\frac{\mathrm{cap}(S)}{\mathrm{dem}(S)} \leq (c_0 \log k)\lambda^*$ *can be found in polynomial time.*

Proof. We remind that λ^* is the optimum value of the linear program (10.3.1). Applying the linear programming duality theorem, we obtain:

$$\text{(10.3.4)} \qquad \begin{array}{lll} \lambda^* = \min & \sum_{e \in E} C_e z_e & \\ & \sum_{e \in P} z_e \geq y_h & (P \in \mathcal{P}_h, h = 1, \ldots, k) \\ & \sum_{1 \leq h \leq k} D_h y_h \geq 1 & \\ & y_h \geq 0 & (h = 1, \ldots, k) \\ & z_e \geq 0 & (e \in E). \end{array}$$

Let (y, z) be an optimum solution to the program (10.3.4). We can suppose that $z_e \leq \sum_{f \in P} z_f$ for every edge $e = ij$ of G and every path P from i to j in G. (For, if not, replace z by z' where, for an edge $e = ij$ of G, $z'_e := \min(z_e, \sum_{f \in P} z_f$ for P path from i to j in G).) Let d_z denote the shortest path

[2] A randomized version of Theorem 10.3.3 appeared in the earliest version of Linial, London and Rabinovich [1994]. This randomized version was independently obtained by Aumann and Rabani [1997]. Garg [1995a] has also derandomized the randomized version from Linial, London and Rabinovich [1994] and, thus, obtained Theorem 10.3.3.

metric of the weighted graph (G, z) (where, for $i, j \in V$, $d_z(i,j)$ is defined as the smallest value of $\sum_{e \in P} z_e$ taken over all paths P from i to j in G). Then,

$$\lambda^* = \sum_{e \in E} C_e z_e \geq \frac{\sum_{ij \in E} C_{ij} d_z(i,j)}{\sum_{h=1}^k D_h d_z(s_h, t_h)}$$

because $\sum_{h=1}^k D_h d_z(s_h, t_h) \geq \sum_{h=1}^k D_h y_h \geq 1$. On the other hand, given a semimetric d on V, set $z_e := \frac{d(i,j)}{\sum_{h=1}^k D_h d(s_h,t_h)}$ for an edge $e = ij$ in G and $y_h := \frac{d(s_h,t_h)}{\sum_{h=1}^k D_h d(s_h,t_h)}$ for $h = 1, \ldots, k$. Then, (y, z) is feasible for the program (10.3.4), which implies that

$$\frac{\sum_{ij \in E} C_{ij} d(i,j)}{\sum_{h=1}^k D_h d(s_h, t_h)} = \sum_{e \in E} C_e z_e \geq \lambda^*.$$

This shows, therefore, that λ^* can be reformulated as

$$\text{(10.3.5)} \qquad \lambda^* = \min_{d \text{ semimetric on } V} \frac{\sum_{ij \in E} C_{ij} d(i,j)}{\sum_{h=1}^k D_h d(s_h, t_h)}.$$

Let d be a semimetric on V. If d happens to be a cut semimetric $\delta(S)$, then

$$\frac{\sum_{ij \in E} C_{ij} d(i,j)}{\sum_{h=1}^k D_h d(s_h, t_h)} = \frac{\text{cap}(S)}{\text{dem}(S)}.$$

More generally, if d is isometrically ℓ_1-embeddable, then $d = \sum_S \lambda_S \delta(S)$ for some $\lambda_S \geq 0$, which implies that

$$\frac{\sum_{ij \in E} C_{ij} d(i,j)}{\sum_{h=1}^k D_h d(s_h, t_h)} = \frac{\sum_S \lambda_S \text{cap}(S)}{\sum_S \lambda_S \text{dem}(S)} \geq \min_S \frac{\text{cap}(S)}{dem(S)}.$$

In general, d is not isometrically ℓ_1-embeddable. However, by Theorem 10.1.6, there exist a constant c_0 and vectors $w_i \in \mathbb{R}^K$ $(i \in V)$ such that

$$\text{(10.3.6)} \qquad \begin{cases} \| w_i - w_j \|_1 \leq d(i,j) & \text{for all } i, j \in V, \\ d(s_h, t_h) \leq (c_0 \log k) \| w_{s_h} - w_{t_h} \|_1 & \text{for } h = 1, \ldots, k. \end{cases}$$

(We apply Theorem 10.1.6 with $X = V$ and $Y = \{s_h, t_h \mid h = 1, \ldots, k\}$.) Therefore,

$$\frac{\sum_{ij \in E} C_{ij} d(i,j)}{\sum_{h=1}^k D_h d(s_h, t_h)} \geq \frac{\sum_{ij \in E} C_{ij} \| w_i - w_j \|_1}{(c_0 \log k) \sum_{h=1}^k D_h \| w_{s_h} - w_{t_h} \|_1} \geq \frac{1}{(c_0 \log k)} \min_S \frac{\text{cap}(S)}{\text{dem}(S)}$$

(using the argument above). This shows that $\lambda^* \geq \frac{1}{c_0 \log k} \min_S \frac{\text{cap}(S)}{\text{dem}(S)}$.

A subset S realizing $\frac{\text{cap}(S)}{\text{dem}(S)} \leq (c_0 \log k)\lambda^*$ can be constructed in the following way. In view of (10.3.5), the max-flow λ^* can be found by solving the linear program:

$$\begin{aligned}\lambda^* = \min \quad & \sum_{ij\in E} C_{ij} d(i,j) \\ & \sum_{h=1}^{k} D_h d(s_h,t_h) = 1 \\ & d \text{ semimetric on } V,\end{aligned}$$

which can be clearly done in polynomial time. Let d be an optimum solution. By Theorem 10.1.7, we can find vectors $w_i = (w_i(r))_{r=1}^K \in \mathbb{R}^K$ ($i \in V$) satisfying (10.3.6) and with $K = O(n^2)$. Determine the coordinate index $r \in K$ for which the quantity

$$\frac{\sum_{ij\in E} C_{ij} |w_i(r) - w_j(r)|}{\sum_{h=1}^k D_h |w_{s_h}(r) - w_{t_h}(r)|}$$

is minimum. The semimetric $d^{(r)}$ on V defined by

$$d^{(r)}(i,j) := |w_i(r) - w_j(r)| \text{ for } i,j \in V,$$

is ℓ_1-embeddable by construction. Hence, $d^{(r)} = \sum_{A\in\mathcal{A}} \lambda_A \delta(A)$ where $\lambda_A > 0$ for $A \in \mathcal{A}$. Such a decomposition can be easily found; recall, e.g., Proposition 4.2.2 and its proof. Then, a set $A_0 \in \mathcal{A}$ realizing $\min_{A\in\mathcal{A}} \frac{\text{cap}(A)}{\text{dem}(A)}$ satisfies the inequality: $\frac{\text{cap}(A_0)}{\text{dem}(A_0)} \le (c_0 \log k)\lambda^*$. Indeed, we have:

$$\begin{aligned}\lambda^* &= \frac{\sum_{ij\in E} C_{ij} d(i,j)}{\sum_{h=1}^k D_h d(s_h,t_h)} \ge \frac{\sum_{ij\in E} C_{ij} \|w_i - w_j\|_1}{(c_0 \log k)\sum_{h=1}^k \|w_{s_h} - w_{t_h}\|_1} \\ &\ge \frac{\sum_{ij\in E} C_{ij} |w_i(r) - w_j(r)|}{(c_0 \log k)\sum_{h=1}^k D_h |w_{s_h}(r) - w_{t_h}(r)|} = \frac{\sum_{A\in\mathcal{A}} \lambda_A \text{cap}(A)}{(c_0 \log k)\sum_{A\in\mathcal{A}} \lambda_A \text{dem}(A)} \\ &\ge \frac{1}{c_0 \log k} \frac{\text{cap}(A_0)}{\text{dem}(A_0)}.\end{aligned}$$

∎

Corollary 10.3.7. *For the case of $k \le 2$ commodity pairs, max-flow = min-cut, i.e.,*

$$\lambda^* = \min_{S\subseteq V} \frac{\text{cap}(S)}{\text{dem}(S)}.$$

Proof. We use the proof of Theorem 10.3.3. Let d be a semimetric on V which is an optimum solution for the program (10.3.5). Suppose first that $k = 1$. Define the semimetric d' on V by setting

$$d'(i,j) := |d(i,s_1) - d(j,s_1)| \text{ for } i,j \in V.$$

Then, $d(i,j) \ge d'(i,j)$ for all $i,j \in V$ and $d'(s_1,t_1) = d(s_1,t_1)$. Hence,

$$\lambda^* = \frac{\sum_{ij\in E} C_{ij} d(i,j)}{\sum_{h=1}^k D_h d(s_h,t_h)} \ge \frac{\sum_{ij\in E} C_{ij} d'(i,j)}{\sum_{h=1}^k D_h d'(s_h,t_h)}.$$

As d' is isometrically ℓ_1-embeddable (by construction), the latter quantity is greater than or equal to $\min_S \frac{\text{cap}(S)}{\text{dem}(S)}$. This shows that $\lambda^* = \min_S \frac{\text{cap}(S)}{\text{dem}(S)}$. Suppose now that $k = 2$. We define a semimetric d' on V by setting

$$d'(i,j) := \max(|d(i,s_1) - d(j,s_1)|, |d(i,s_2) - d(j,s_2)|) \text{ for } i,j \in V.$$

Then, d' is isometrically ℓ_∞^2-embeddable (by construction), $d'(i,j) \leq d(i,j)$ for all $i,j \in V$, and $d'(s_h,t_h) = d(s_h,t_h)$ for $h = 1,2$. We can now conclude the proof as in the previous case, because the semimetric d' is again ℓ_1-embeddable (since ℓ_∞^2 is an isometric subspace of ℓ_1^2 by relation (3.1.6)). ∎

Chapter 11. Dimensionality Questions for ℓ_1-Embeddings

Given a distance space (X,d) which is ℓ_1-embeddable, a natural question is to determine the smallest dimension m of an ℓ_1-space $\ell_1^m = (\mathbb{R}^m, d_{\ell_1})$ in which (X,d) can be embedded. A next question is whether there exists a finite point criterion for ℓ_1^m-embeddability, analogue to Menger's result for the Euclidean space; this is the question of finding the order of congruence of ℓ_1^m. We present in this chapter several results related to these questions. Unfortunately fairly little is known. For instance, the order of congruence of ℓ_1^m is known only for $m \leq 2$ and it is not even known whether ℓ_1^m has a finite order of congruence for $m \geq 3$. More precisely, a distance space (X,d) is ℓ_1^1- (resp. ℓ_1^2-embeddable) if and only if the same holds for every subspace of (X,d) on 4 (resp. on 6) points. As a consequence, one can recognize in polynomial time whether a distance is ℓ_1^m-embeddable for $m \leq 2$. On the other hand, the complexity of checking ℓ_1^m-embeddability is not known for $m \geq 3$. These results are presented in Section 11.1. A crucial tool for the proofs is the notion of 'totally decomposable distance' studied by Bandelt and Dress [1992]; Section 11.1.2 contains the facts about total decomposability that are needed for our treatment. Then we consider in Section 11.2 some bounds for the minimum ℓ_p-dimension of an arbitrary ℓ_p-embeddable distance space on n points.

11.1 ℓ_1-Embeddings in Fixed Dimension

Let us begin with several observations concerning the structure of ℓ_1^m-embeddable distances, where $m \geq 1$ is an integer. Let (V_n, d) be a distance space which is ℓ_1-embeddable. Then, d admits a decomposition as

$$(11.1.1) \qquad d = \sum_{\delta(S) \in \mathcal{C}} \lambda_S \delta(S),$$

where $\lambda_S > 0$ for all $\delta(S) \in \mathcal{C}$ and $\mathcal{C}$ is a family of cut semimetrics on V_n. When d is ℓ_1^m-embeddable then d has such a decomposition which has a special structure. We need some definitions in order to formulate it.

Given a family $\mathcal{C}$ of cut semimetrics on V_n, set

$$(11.1.2) \qquad \widetilde{\mathcal{C}} := \{S, \overline{S} := V_n \setminus S \mid \delta(S) \in \mathcal{C}\};$$

the members of $\widetilde{\mathcal{C}}$ are called the *shores* of the cut semimetrics in $\mathcal{C}$. The family $\mathcal{C}$

is said to be *nested* if the elements of $\widetilde{\mathcal{C}}$ can be ordered as $S_1, \ldots, S_m, \overline{S}_1, \ldots, \overline{S}_m$ in such a way that $S_1 \subset S_2 \subset \ldots \subset S_m$ (and, thus, $\overline{S}_m \subset \ldots \subset \overline{S}_2 \subset \overline{S}_1$). Given an integer $m \geq 1$, $\mathcal{C}$ is said to be *m-nested* if $\mathcal{C}$ can be partitioned into m nested subfamilies. (Hence, 1-nested means nested.)

We have the following (easy) characterization for ℓ_1^m-embeddability; it was in fact already contained in Proposition 4.2.2 but we recall it here for clarity.

Lemma 11.1.3. *Let (V_n, d) be a distance space. The following assertions are equivalent.*

(i) *(V_n, d) can be embedded into ℓ_1^m.*

(ii) *d admits a decomposition* (11.1.1) *where $\mathcal{C}$ is m-nested.*

(iii) *(V_n, d) can be isometrically embedded into the Cartesian product of m weighted paths.*

Proof. We first consider the case when $m = 1$.
(i) $\Longrightarrow$ (ii), (iii). Suppose that d is ℓ_1^1-embeddable. Then, there exist scalars $u_1, \ldots, u_n$ such that $d_{ij} = |u_i - u_j|$ for all $i, j \in V_n$. Up to a reordering of the elements of V_n, we can suppose that $u_1 \leq u_2 \leq \ldots \leq u_n$. Then,

$$d = \sum_{i=1}^{n-1} (u_{i+1} - u_i)\delta(\{1, \ldots, i\}).$$

Then, (ii) holds as the family $\{\delta(\{1, \ldots, i\}) \mid i = 1, \ldots, n-1\}$ is obviously nested. Consider the path $P = (1, \ldots, n)$ with weight $w_{i,i+1} := u_{i+1} - u_i$ on edge $(i, i+1)$ $(i = 1, \ldots, n-1)$. Then, the shortest path metric of the weighted path P coincides with d, which shows (iii).
(iii) $\Longrightarrow$ (ii). Let $P = (1, \ldots, n)$ be a path with nonnegative weight w_i on edge $(i, i+1)$ for $i = 1, \ldots, n-1$. Then, its path metric $d_{P,w}$ can be decomposed as $\sum_{i=1}^{n-1} w_i \delta(\{1, \ldots, i\})$.
(ii) $\Longrightarrow$ (i). Suppose now that $d = \sum_{i=1}^{p} \alpha_i \delta(S_i)$, where $\alpha_i > 0$ for all i and $S_1 \subset S_2 \ldots \subset S_p$. Define $u_1, \ldots, u_n \in \mathbb{R}$ by setting

$$\begin{array}{ll} u_i := 0 & \text{for } i \in S_1, \\ u_i := \sum_{h=1}^{k} \alpha_h & \text{for } i \in S_{k+1} \setminus S_k,\ k = 1, \ldots, m-1, \\ u_i := \sum_{h=1}^{p} \alpha_h & \text{for } i \in V_n \setminus S_p. \end{array}$$

Then, one can verify that $d_{ij} = u_j - u_i$ for all $i, j \in V_n$, which shows that d is ℓ_1^1-embeddable.
The result for $m \geq 2$ follows now easily using additivity, since d is ℓ_1^m-embeddable if and only if d can be decomposed as a sum $d = d_1 + \ldots + d_m$, where all d_i's are ℓ_1^1-embeddable. ∎

It is not clear how to use this result for devising a polynomial algorithm permitting to recognize ℓ_1^m-embeddable distances. The difficulty is that a distance in the cut cone CUT_n has many decompositions as (11.1.1) (in fact, an infinity, if it is not ℓ_1-rigid). However, as we will see in Section 11.1.3 below, in the cases $m = 1, 2$, it is enough to check one special decomposition (namely, the one corresponding to a 'total decomposition' of d). In particular, one can check embeddability in the space ℓ_1^m for $m \in \{1, 2\}$ in polynomial time. Moreover, one can determine the order of congruence of ℓ_1^2.

These results need the notion of total decomposability for their proofs. So we organize the rest of the section in the following manner. We summarize in Section 11.1.1 what is known about the order of congruence of the space ℓ_1^m as well as some results on the ℓ_1-dimension of some specific distance spaces. Then, after introducing in Section 11.1.2 the definitions and facts about totally decomposable distances that we need for our treatment, we present in Section 11.1.3 the proofs for the results concerning embeddability in the ℓ_1-space of dimension $m \leq 2$.

11.1.1 The Order of Congruence of the ℓ_1-Space

Let $f_p(m)$ denote the order of congruence of ℓ_p^m; that is, $f_p(m)$ is the smallest integer such that an arbitrary distance space (X, d) is ℓ_p^m-embeddable if and only if every subspace of (X, d) on $f_p(m)$ points is ℓ_p^m-embeddable. By convention, we set $f_p(m) = \infty$ if $f_p(m)$ does not exist. We remind from Theorem 3.2.2 that we may restrict our attention to finite distance spaces (X, d).

The study of the parameter $f_p(m)$ is motivated by the result of Menger in the case $p = 2$, quoted in Theorem 6.2.13. Namely, Menger [1928] showed that a distance space (X, d) embeds isometrically in the Euclidean space $(\mathbb{R}^m, d_{\ell_2})$ if and only if each subspace of (X, d) on $m + 3$ points embeds isometrically in $(\mathbb{R}^m, d_{\ell_2})$. In other words,

$$f_2(m) = m + 3 \quad \text{for each } m \geq 1.$$

Hence, we have the natural question of looking for analogues of Menger's theorem for the case of arbitrary ℓ_p-metrics and, in particular, in the case $p = 1$. This turns out to be a difficult question. Only the following partial results are known.

Since the spaces $(\mathbb{R}, d_{\ell_p})$ and $(\mathbb{R}, d_{\ell_2})$ are identical, we deduce from Menger's theorem that $f_p(1) = 4$ for all p. (See also Theorem 11.1.21 for a direct proof of the fact that $f_1(1) = 4$.) Malitz and Malitz [1992] show that

$$6 \leq f_1(2) \leq 11 \quad \text{and} \quad f_1(m) \geq 2m + 1 \text{ for all } m \geq 1.$$

These results are improved by Bandelt and Chepoi [1996a] who show that

$$f_1(2) = 6$$

and by Bandelt, Chepoi and Laurent [1997] who show that

$$f_1(m) \geq m^2 \text{ for } m \geq 3 \text{ odd}, \quad f_1(m) \geq m^2 - 1 \text{ for } m \geq 4 \text{ even}.$$

The equality: $f_1(2) = 6$ will be proved in Theorem 11.1.24 and the inequalities: $f_1(m) \geq m^2$ for m odd, $f_1(m) \geq m^2 - 1$ for m even, are given in Corollary 11.1.6 below. Malitz and Malitz [1992] conjecture that $f_1(m)$ is finite for all m.

In view of the above mentioned results, one can test in polynomial time whether a given finite distance space (X, d) embeds in $(\mathbb{R}^m, d_{\ell_1})$, when $m = 1$ or 2. (The case $m = 1$ is anyway easy, as d_{ℓ_1} and d_{ℓ_2} coincide when restricted to $\mathbb{R}$.) It would be very interesting if one could show that $f_1(m)$ is finite for all m. Indeed, this would imply the existence of a polynomial time algorithm for checking embeddability of a finite distance space in the space $(\mathbb{R}^m, d_{\ell_1})$, for any given m. The complexity of the embedding problem into $(\mathbb{R}^m, d_{\ell_1})$ for some fixed $m \geq 3$ is not known. We recall that, on the other hand, checking ℓ_1-embeddability of a finite distance space (i.e., embeddability into some $(\mathbb{R}^m, d_{\ell_1})$ for unrestricted m) is NP-complete (cf. Section 4.4).

We conclude this section with some remarks on the ℓ_1-dimension of some concrete examples of distance spaces. We remind that $m_{\ell_1}(X, d)$ denotes the smallest m such that (X, d) can be embedded into ℓ_1^m. This parameter can be easily formulated in the case of ℓ_1-rigid metrics. Indeed, if (X, d) is an ℓ_1-rigid distance space, then it has a unique decomposition as (11.1.1) and, therefore, $m_{\ell_1}(X, d)$ is equal to the smallest number of nested families needed to cover the set $\mathcal{C}$ of cut semimetrics entering in this decomposition. For instance, hypercube embeddable graphs have their path metric which is ℓ_1-rigid. In particular, trees are hypercube embeddable and, thus, ℓ_1-rigid. (See Proposition 19.1.2.) We now give some more details about trees as they will play some role here; justification for the facts quoted below can be found in Section 19.1.

Let $T = (V, E)$ be a tree. Every edge $e \in E$ determines a partition of the node set V into two sets S_e and $V \setminus S_e$. Then, the path metric of T can be decomposed as

$$d_T = \sum_{e \in E} \delta(S_e)$$

and this is its only $\mathbb{R}_+$-realization. More generally, if $w \in \mathbb{R}_+^E$ are weights assigned to the edges of T, then the path metric $d_{T,w}$ of the weighted tree can be decomposed in a unique way as

$$d_{T,w} = \sum_{e \in E} w_e \delta(S_e).$$

One can easily determine the ℓ_1-dimension of a tree in terms of its number of leaves[1]. The next result was already formulated by Hadlock and Hoffman [1978].

Proposition 11.1.4. *The following assertions are equivalent for a tree T.*

(i) *The path metric of T can be embedded into ℓ_1^m.*

(ii) *T can be covered by m paths.*

(iii) *T has at most $2m$ leaves.*

Therefore, if T has p leaves, then the ℓ_1-dimension of its path metric is equal to $\lceil \frac{p}{2} \rceil$.

---

[1] A node v of T is a *leaf* if its degree is equal to 1.

Proof. Let $T = (V, E)$ be a tree and, with the above notation, let $\mathcal{C}_T := \{\delta(S_e) \mid e \in E\}$ denote the family of cuts entering the decomposition of d_T. Then, (i) can be equivalently formulated as: (iv) The family $\mathcal{C}_T$ is m-nested.
We show the equivalence of (ii)-(iv). The implications (ii) $\Longrightarrow$ (iii), (iv) are clear. For (iv) $\Longrightarrow$ (iii), suppose that T has at least $2m+1$ leaves. Then, there are at least $2m+1$ minimal sets in $\widetilde{\mathcal{C}}$ (as each leaf v yields the minimal set $\{v\}$). Hence, $\mathcal{C}$ cannot be covered with m nested families.
The implication (iii) $\Longrightarrow$ (ii) can be shown by induction on $m \geq 1$. We leave the details to the reader. ∎

In order to give lower bounds for the order of congruence $f_1(m)$, we need as a tool the following result about the ℓ_1-dimension of a direct product of distance spaces.

Lemma 11.1.5. *Let (X_1, d_1), (X_2, d_2) be two distance spaces and let $(X, d) := (X_1 \times X_2, d_1 \oplus d_2)$ be their direct product. Then the following relation holds: $m_{\ell_1}(X, d) = m_{\ell_1}(X_1, d_1) + m_{\ell_1}(X_2, d_2)$.*

Proof. The inequality: $m_{\ell_1}(X, d) \leq m_{\ell_1}(X_1, d_1) + m_{\ell_1}(X_2, d_2)$ follows by additivity of the ℓ_1-distance. Conversely, suppose that (X, d) embeds in ℓ_1^m. Then, d can be decomposed as $d = \sum_{\delta(S) \in \mathcal{C}} \lambda_S \delta(S)$, where $\mathcal{C}$ is a family of cut semimetrics on X that is m-nested. By the proof of Proposition 7.5.2, we know that every cut semimetric in $\mathcal{C}$ is of the form $\delta(A \times X_2)$ or $\delta(X_1 \times B)$ for some $A \subset X_1$, $B \subset X_2$. Say, $d = \sum_{A \in \mathcal{A}} \lambda_A \delta(A \times X_2) + \sum_{B \in \mathcal{B}} \lambda_B \delta(X_1 \times B)$. Then, $d_1 = \sum_{A \in \mathcal{A}} \lambda_A \delta(A)$ and $d_2 = \sum_{B \in \mathcal{B}} \lambda_B \delta(B)$. Note that no two cut semimetrics $\delta(A \times X_2)$ and $\delta(X_1 \times B)$ can be put together in a nested family. Hence, the m nested families partitioning $\mathcal{C}$ yield a partition of the cut semimetrics $\delta(A)$ ($A \in \mathcal{A}$) in p nested families and a partition of the cut semimetrics $\delta(B)$ ($B \in \mathcal{B}$) in q nested families with $p + q = m$. Therefore, $p \geq m_{\ell_1}(X_1, d_1)$ and $q \geq m_{\ell_1}(X_2, d_2)$, which implies the converse inequality: $m_{\ell_1}(X, d) \geq m_{\ell_1}(X_1, d_1) + m_{\ell_1}(X_2, d_2)$. ∎

Corollary 11.1.6. *$f_1(m) \geq m^2$ for m odd; $f_1(m) \geq m^2 - 1$ for m even.*

Proof. We start with an observation. Let (X, d) denote the graphic metric space of the graph $G := K_{1,s} \times K_{1,t}$ (the Cartesian product of the two complete bipartite graphs $K_{1,s}$ and $K_{1,t}$) where $s, t \geq 3$. Its ℓ_1-dimension is $m_{\ell_1}(X, d) = \lceil \frac{s}{2} \rceil + \lceil \frac{t}{2} \rceil$ (applying Proposition 11.1.4 and Lemma 11.1.5, as $K_{1,s}$, $K_{1,t}$ are trees with s and t leaves, respectively). Say, G has node set $X := V_1 \times V_2$, where $V_1 := \{u_0, u_1, \ldots, u_s\}$ and $V_2 := \{v_0, v_1, \ldots, v_t\}$ (u_0, v_0 denoting the respective 'centers' of the stars $K_{1,s}$ and $K_{1,t}$). Then, d can be decomposed as

$$d = \sum_{i=1}^{s} \delta(S_i) + \sum_{j=1}^{t} \delta(T_j),$$

setting $S_i := \{u_i\} \times V_2$ ($i = 1, \ldots, s$) and $T_j := V_1 \times \{v_j\}$ ($j = 1, \ldots, t$). Clearly, no two cut semimetrics $\delta(S_i)$, $\delta(T_j)$ form a nested pair, while any two $\delta(S_i)$, $\delta(S_j)$ (or $\delta(T_i)$, $\delta(T_j)$) do form a nested pair. This shows again that $m_{\ell_1}(X,d) = \lceil \frac{s}{2} \rceil + \lceil \frac{t}{2} \rceil$. On the other hand, if we delete the element $x := (u_{i_0}, v_{j_0})$ from X (where $i_0 \in \{1,\ldots,s\}$ and $j_0 \in \{1,\ldots,t\}$ correspond to leaves in the two trees), then the two cut semimetrics $\delta(S_{i_0} \setminus \{x\})$ and $\delta(T_{j_0} \setminus \{x\})$ become a nested pair. Hence, one can now partition the cut semimetrics $\delta(S_i \setminus \{x\})$, $\delta(T_j \setminus \{x\})$ into $\lceil \frac{s}{2} \rceil + \lceil \frac{t}{2} \rceil - 1$ nested subfamilies, which shows that

$$\text{(a)} \qquad m_{\ell_1}(X \setminus \{x\}, d) \leq \left\lceil \frac{s}{2} \right\rceil + \left\lceil \frac{t}{2} \right\rceil - 1$$

for such x.

We can now proceed with the proof. We first show that $f_1(m) \geq m^2$ for m odd. For this, it suffices to construct a distance space (Y,d) such that $|Y| \geq m^2$, $m_{\ell_1}(Y,d) \geq m+1$ and $m_{\ell_1}(Y \setminus \{y\}, d) \leq m$ for all $y \in Y$. Namely, let (X,d) be the graphic metric space of the graph $K_{1,m} \times K_{1,m}$, with $m_{\ell_1}(X,d) = m+1$. Let $Y \subset X$ be a minimal subset of X such that $m_{\ell_1}(Y,d) = m+1$; then, $m_{\ell_1}(Y \setminus \{y\}, d) \leq m$ for all $y \in Y$. This distance space (Y,d) does the job as $|Y| \geq m^2$ in view of relation (a) above. One proceeds in the same way for showing the inequality: $f_1(m) \geq m^2 - 1$ for m even. Namely, one considers now for (X,d) the graphic metric space of $K_{1,m-1} \times K_{1,m+1}$ with $m_{\ell_1}(X,d) = m+1$ and (Y,d) is a subspace such that $m_{\ell_1}(Y,d) = m+1$, $m_{\ell_1}(Y \setminus \{y\}, d) \leq m$ for all $y \in Y$. ∎

To conclude let us mention that the exact value of the ℓ_1-dimension of the equidistant metric is not known. This amounts to determining the maximum number of equidistant points (with respect to the ℓ_1-distance) that can be placed in the m-dimensional space for given m. This maximum is at least $2m$, as the coordinate vectors $\pm e_i$ ($i = 1,\ldots,m$) form obviously an equidistant set of $2m$ points. Is it the best one can do ? So we have the following problem:

Problem 11.1.7. *Show that there are at most $2m$ equidistant points in the m-dimensional ℓ_1-space $(\mathbb{R}^m, d_{\ell_1})$.*

Bandelt, Chepoi and Laurent [1997] have settled Problem 11.1.7 in dimension $m \leq 3$, but the problem remains open in dimension $m \geq 4$. In other words, $m_{\ell_1}(\mathbb{1}_m) \leq \lceil \frac{m}{2} \rceil$ with equality if $m \leq 8$, but it is not known whether equality holds for $m \geq 9$.

11.1.2 A Canonical Decomposition for Distances

An ℓ_1-embeddable distance d on the set V_n can be decomposed as a nonnegative sum of cut semimetrics. In general, there is *not* unicity of such a decomposition. Bandelt and Dress [1992] develop a theory that permits to decompose d in a unique way. They achieve this by requiring some very specific properties for the cuts entering the decomposition. Moreover, the theory applies to an arbitrary distance d; for this, one has to allow a residual 'split-prime' term in the decomposition. We present here the main ideas and results from Bandelt and Dress [1992] that we need for our treatment. We do not give proofs as this would take too much space.

Let d be a distance on V_n. Given two subsets $A, B \subseteq V_n$, set

$$\alpha_d(A,B) := \frac{1}{2} \min_{a,a' \in A,\ b,b' \in B} \max(0, \quad d(a,b) + d(a',b') - d(a,a') - d(b,b'),$$
$$d(a,b') + d(a',b) - d(a,a') - d(b,b')).$$

(Here, the elements a, a' (or b, b') may coincide.) When $B = V_n \setminus A$, we also set

$$\alpha_d(A) := \alpha_d(A, V_n \setminus A).$$

The quantity $\alpha_d(A)$ is called the *isolation index* of the cut semimetric $\delta(A)$ (with respect to the distance d). Then, $\delta(A)$ is said to be a *d-split* if $\alpha_d(A) > 0$. Clearly, if $\delta(A)$ is a d-split then both sets A and $V_n \setminus A$ are d-convex. (Recall Definition 4.2.7.) A distance d is said to be *split-prime* if d has no d-split. In general, we let Σ_d denote the set of d-splits of d. We illustrate the definitions on some examples.

Example 11.1.8.

(i) Let $d = \delta(A)$ be a cut semimetric. Then, d has only one d-split, namely, $\delta(A)$ itself with isolation index 1.

(ii) Let d be a semimetric on 4 points. Then, d is ℓ_1-embeddable. Moreover, one can verify that $d = \sum_{\delta(S) \in \Sigma_d} \alpha_d(S)\delta(S)$.

(iii) Let $d := d_{T,w}$ be the path metric of a weighted tree $T = (V, E)$ with nonnegative edge weights w. Recall that d can be decomposed as $d = \sum_{e \in E} w_e \delta(S_e)$ where $S_e, V \setminus S_e$ denote the two components of the graph $T \backslash e$. Clearly, every cut semimetric $\delta(S_e)$ is a d-split with isolation index w_e and the cut semimetrics $\delta(S_e)$ $(e \in E)$ constitute all the d-splits.

(iv) Let d be the path metric of the complete bipartite graph $K_{2,3}$. Then, d is split-prime (as, for every set A, either A or its complement is not d-convex). In fact, $d(K_{2,3})$ is the only (up to multiple) split-prime semimetric on 5 points (Lemma 1 in Bandelt and Dress [1992]). There are examples of split-prime distances that are ℓ_1-embeddable. This is the case, for instance, for the path metric of the 3-dimensional hypercube $H(3,2)$, or for the metric on 5 points $d := \frac{1}{2}(\delta(\{1,4\}) + \delta(\{1,5\}) + \delta(\{2,4\}) + \delta(\{2,5\}))$ (taking value 2 on the pairs $(1,2)$ and $(4,5)$ and value 1 elsewhere). ∎

Bandelt and Dress [1992] (Theorem 2) show that every distance d on V_n can be decomposed as

(11.1.9) $$d = d_0 + \sum_{\delta(S) \in \Sigma_d} \alpha_d(S)\delta(S),$$

where d_0 is a split-prime distance on V_n (that is, there is no d_0-split) and the sum is taken over the set Σ_d of d-splits; d_0 is called the *split-prime residue* of

d. Moreover, d_0 is a semimetric whenever d is a semimetric. The decomposition (11.1.9) is clearly unique. It can be found in time $O(n^6)$.

The collection of d-splits has some specific property. We need one more definition in order to formulate it. Call a collection $\mathcal{C}$ of cut semimetrics on V_n *weakly compatible* if there does not exist four points $x_1, x_2, x_3, x_4 \in V_n$ and three cut semimetrics $\delta(A_1)$, $\delta(A_2)$, $\delta(A_3) \in \mathcal{C}$ whose restrictions on the set $X := \{x_1, x_2, x_3, x_4\}$ would induce the three distinct cut semimetrics where X is partitioned into two pairs. In other words, the family $\mathcal{C}$ is weakly compatible if and only if, for all $A, B, C \in \widetilde{\mathcal{C}}$, $A \cap B \cap C \neq \emptyset \Longrightarrow A \subseteq B \cup C$, or $B \subseteq A \cup C$, or $C \subseteq A \cup B$ (recall the definition of $\widetilde{\mathcal{C}}$ from (11.1.2)). One can verify that the set of d-splits of a distance d is weakly compatible. Conversely, we have (from Theorem 3 in Bandelt and Dress [1992]):

Fact 11.1.10. *Let $\mathcal{C}$ be a weakly compatible family of cut semimetrics on V_n and let $\lambda_S > 0$ be given scalars for $\delta(S) \in \mathcal{C}$. Then, $\mathcal{C}$ is the set of d-splits of the distance $d := \sum_{\delta(S)\in\mathcal{C}} \lambda_S \delta(S)$ and $\alpha_d(S) = \lambda_S$ for all $\delta(S) \in \mathcal{C}$.* ∎

As a consequence, every weakly compatible set of cut semimetrics on V_n is linearly independent and, thus, has cardinality $\leq \binom{n}{2}$. In fact, for a distance d having a nonzero split-prime residue d_0 in (11.1.9), the set $\Sigma_d \cup \{d_0\}$ is linearly independent.

A distance d on V_n is said to be *totally decomposable* if $d = \sum_{\delta(S)\in\Sigma_d} \alpha_d(S)\delta(S)$ holds. That is, if in the decomposition (11.1.9) there is no split-prime residue, i.e., $d_0 = 0$. Then, d is ℓ_1-embeddable. As mentioned in Example 11.1.8 (ii), every semimetric on 4 points is totally decomposable. In general, totally decomposable distances are characterized by the following 5-point criterion (Theorem 6 in Bandelt and Dress [1992]):

Fact 11.1.11. *A distance d on V_n is totally decomposable if and only if, for all $a, b, c, d, x \in V_n$, $\alpha_d(\{a,b\},\{c,d\}) = \alpha_d(\{a,b,x\},\{c,d\}) + \alpha_d(\{a,b\},\{c,d,x\})$.* ∎

As an application, one can check total decomposability in time $O(n^5)$.

Finally we introduce a notion of minor for distances. Let d be a distance on V_n, let d_0 be its split-prime residue, and let Σ_d be its set of d-splits. Given a subset $X \subseteq V_n$, a distance d' on X is said to be a *minor* of d if d' is of the form:

$$d' = \lambda_0 d'_0 + \sum_{\delta(S)\in\mathcal{C}} \lambda_S \delta(S \cap X), \tag{11.1.12}$$

where $\mathcal{C} \subseteq \Sigma_d$, $\lambda_S > 0$ for all $\delta(S) \in \mathcal{C}$, $\lambda_0 \geq 0$, and d'_0 denotes the restriction of d_0 to X. Then, the d'-splits are the nonzero cut semimetrics $\delta(S \cap X)$ for $\delta(S) \in \mathcal{C}$. (Here, $\delta(S \cap X)$ denotes the cut semimetric on X determined by the partition of X into $S \cap X$ and $\overline{S} \cap X$.) In other words, a minor d' of d is obtained by applying the following two operations: take a nonnegative combination of the d-splits and of the residue of d, and/or take the restriction to a subset X of the groundset of d. Total decomposability is obviously preserved by taking minors; the following is shown in Bandelt and Dress [1992]:

Fact 11.1.13. *A distance d is totally decomposable if and only if it does not have the path metric of the complete bipartite graph $K_{2,3}$ as a minor.* ∎

Examples of totally decomposable distances include path metrics of weighted trees and their isometric subspaces, known as tree metrics. In other words, a distance space (X, d) is called a *tree metric* if there exists a tree $T = (V, E)$ with edge weights $w \in \mathbb{R}_+^E$ and a mapping $f : X \longrightarrow V$ such that $d(x, y) = d_{T,w}(f(x), f(y))$ for $x, y \in X$.

Call two cut semimetrics $\delta(A)$ and $\delta(B)$ *crossing* if the four sets $A \cap B$, $A \cap \overline{B}$, $\overline{A} \cap B$, $\overline{A} \cap \overline{B}$ are nonempty and *cross-free* otherwise. That is, $\delta(A)$ and $\delta(B)$ are cross-free if two of the sets A, $\bar{A}$, B, $\bar{B}$ are comparable (for inclusion). Then, tree metrics admit the following characterization[2]:

Fact 11.1.14. *A distance d is a tree metric if and only if d is totally decomposable and any two d-splits are cross-free.* ∎

From this follows:

Fact 11.1.15. *Let d_1 and d_2 be two tree metrics. Then, their sum $d := d_1 + d_2$ is totally decomposable with set of d-splits $\Sigma_d = \Sigma_{d_1} \cup \Sigma_{d_2}$.* ∎

An important class of totally decomposable semimetrics is provided by the semimetrics that can be embedded into the space ℓ_1^m of dimension $m \leq 2$. Indeed, a semimetric that can be embedded into ℓ_1^1 is a tree metric (by Lemma 11.1.3). Hence, a distance that can be embedded into ℓ_1^2 is the sum of two tree metrics and, thus, is totally decomposable. This fact will play a central role for the recognition of ℓ_1^2-embeddable metrics, as we see in the next subsection.

11.1.3 Embedding Distances in the ℓ_1-Plane

We return here to the question of determining the order of congruence $f_1(m)$ of ℓ_1^m, the m-dimensional ℓ_1-space. It is known that $f_1(1) = 4$ (by Theorem 6.2.13; see also Theorem 11.1.21). It is not known whether $f_1(m) < \infty$ when $m \geq 3$. Bandelt and Chepoi [1996a] have computed the exact value of $f_1(2)$; namely, they show that $f_1(2) = 6$ (see Theorem 11.1.24 below).

The main results presented here are Theorems 11.1.21 and 11.1.24 which give several equivalent characterizations for ℓ_1^1- and ℓ_1^2-embeddability of a finite distance space; in particular, in terms of a list of forbidden minimal configurations. We follow essentially Bandelt and Chepoi [1996a] for the proofs.

[2] This is essentially a result of Buneman [1971] or, independently, Edmonds and Giles [1977], which shows how to represent cross-free families of cut semimetrics by trees.

An essential tool for these results is the theory of totally decomposable distance spaces, exposed in the previous subsection. We will use in particular the properties of totally decomposable distance spaces, mentioned above in Facts 11.1.10-11.1.15.

We start with several easy but crucial observations. As was observed in Lemma 11.1.3, a distance d can be embedded in the m-dimensional ℓ_1-space $(\mathbb{R}^m, d_{\ell_1})$ if and only if d has a decomposition $d = \sum_{\delta(S)\in\mathcal{C}} \lambda_S \delta(S)$ (with $\lambda_S > 0$ for all S), where $\mathcal{C}$ can be partitioned into m nested subfamilies. In the case $m = 1, 2$, it suffices, in fact, to check this property for the collection of d-splits.

Lemma 11.1.16. *A distance d on V_n is ℓ_1^1-embeddable if and only if d is totally decomposable and its set Σ_d of d-splits is nested. Then, d is the shortest path metric of a weighted path.*

Proof. This follows from Lemma 11.1.3 and the fact that a nested family of cut semimetrics is weakly compatible. ∎

Lemma 11.1.17. *A distance d on V_n is ℓ_1^2-embeddable if and only if d is totally decomposable and its set Σ_d of d-splits can be partitioned into two nested families. Then, d can be isometrically embedded into the Cartesian product of two weighted paths.*

Proof. The result follows using Fact 11.1.15 and Lemma 11.1.16. ∎

Lemma 11.1.18. *For $m \in \{1, 2\}$, if d is ℓ_1^m-embeddable, then so is every minor of d.*

Proof. Suppose that d is ℓ_1^1- or ℓ_1^2-embeddable and let d' be a minor of d, say, of the form (11.1.12). Then, d is totally decomposable, that is, its split-prime residue is equal to zero. Hence, d' too is totally decomposable. The result now follows using Lemmas 11.1.16 and 11.1.17. ∎

We group in Theorem 11.1.21 several equivalent characterizations for ℓ_1^1-embeddability. One of them is in terms of some distances that are forbidden as minors; they are the path metrics of the graphs $K_{2,3}$, K_3, and C_4. See Figure 11.1.19 where are displayed the embeddings in the ℓ_1-plane for the latter two distances.

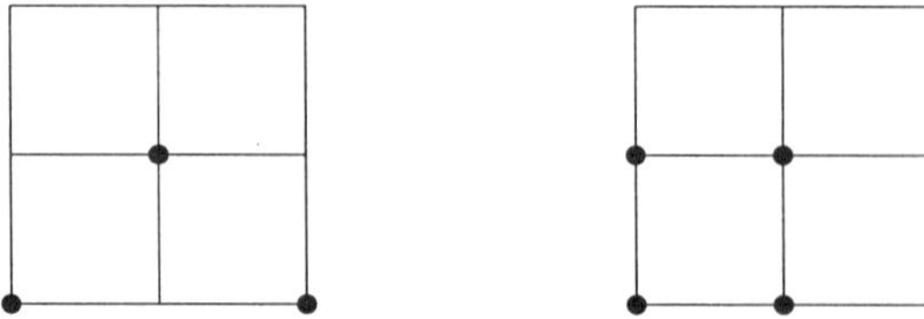

Figure 11.1.19: Embedding K_3 and C_4 in the ℓ_1-plane (up to scale)

We start with a characterization of nested families.

Lemma 11.1.20. *A family $\mathcal{C}$ of cut semimetrics is nested if and only if every subfamily $\mathcal{C}_0 \subseteq \mathcal{C}$ with cardinality $|\mathcal{C}_0| \leq 3$ is nested.*

Proof. We show the 'if' part by induction on the cardinality of $\mathcal{C}$. We suppose that $|\mathcal{C}| \geq 4$ and that every proper subset of $\mathcal{C}$ is nested; we show that $\mathcal{C}$ is nested. Let $\delta(S) \in \mathcal{C}$; then, the elements of $\mathcal{C} \setminus \{\delta(S)\}$ can be arranged as $\delta(A_1)$, $\ldots, \delta(A_m)$ in such a way that $A_1 \subset A_2 \subset \ldots \subset A_m$. If $S \subset A_1$ or $\overline{S} \subset A_1$ then $\mathcal{C}$ is nested. If $S, \overline{S} \not\subset A_i$ for every $i = 1, \ldots, m$, then $A_m \subset S$ (or $A_m \subset \overline{S}$) (because the three cut semimetrics $\delta(A_{m-1})$, $\delta(A_m)$, $\delta(S)$ form a nested family) and $\mathcal{C}$ is again nested. Else, let i be an index such that $S, \overline{S} \not\subset A_i$ and $S \subset A_{i+1}$. As $\delta(A_i)$, $\delta(A_{i+1})$ and $\delta(S)$ form a nested family, we obtain that $A_i \subset S \subset A_{i+1}$. This shows that $\mathcal{C}$ is nested. ∎

Theorem 11.1.21. *Let d be a distance on V_n. The following assertions are equivalent.*

(i) *(V_n, d) is ℓ_1^1-embeddable.*

(ii) *d is totally decomposable and its set Σ_d of d-splits is nested.*

(iii) *(X, d) is ℓ_1^1-embeddable for every subset $X \subseteq V_n$ such that $|X| \leq 4$.*

(iv) *d does not have as a minor the path metrics of the graphs $K_{2,3}$, K_3 and C_4.*

Proof. The equivalence (i) $\Longleftrightarrow$ (ii) holds by Lemma 11.1.16. The implication (i) $\Longrightarrow$ (iii) is obvious and (i) $\Longrightarrow$ (iv) follows from Lemma 11.1.18 and the fact that the path metrics of $K_{2,3}$, K_3 and $H(2,2)$ are not ℓ_1^1-embeddable.
We now show the implications (iv) $\Longrightarrow$ (ii) and (iii) $\Longrightarrow$ (ii). For this, we suppose that d does not satisfy (ii); we show that neither (iii) nor (iv) holds. If d is not totally decomposable, then d has $d(K_{2,3})$ as a minor (by Fact 11.1.15) and Σ_d is not weakly compatible. Hence, there exist three d-splits $\delta(A)$, $\delta(B)$, $\delta(C)$ such that $A \cap B \cap C \neq \emptyset$, $A \not\subset B \cup C$, $B \not\subset A \cup C$ and $C \not\subset A \cup B$. Let $x_1 \in A \cap B \cap C$, $x_2 \in A \setminus B \cup C$, $x_3 \in B \setminus A \cup C$, and set $X := \{x_1, x_2, x_3\}$. Then, (X, d) is not ℓ_1^1-embeddable because it has the distance $\delta(A \cap X) + \delta(B \cap X) + \delta(C \cap X) = 2d(K_3)$ as a minor.

We can now suppose that d is totally decomposable and that Σ_d is not nested. If there are two crossing d-splits $\delta(A)$ and $\delta(B)$, then we can choose four elements $x_1 \in A \cap B$, $x_2 \in A \cap \overline{B}$, $x_3 \in \overline{A} \cap B$ and $x_4 \in \overline{A} \cap \overline{B}$. Setting $X := \{x_1, \ldots, x_4\}$, the distance $\delta(A \cap X) + \delta(B \cap X)$ is a minor of d that coincides with the path metric of $H(2,2)$; hence, (X, d) is not ℓ_1^1-embeddable. Suppose now that any two d-splits are cross-free. By Lemma 11.1.20, we can find three d-splits $\delta(A)$, $\delta(B)$ and $\delta(C)$ which do not form a nested family. We can suppose without loss of generality that $A \subset B$ (as $\delta(A)$ and $\delta(B)$ are cross-free). As $\delta(C)$ is cross-free with $\delta(A)$, we have $A \subset C$ (or $A \subset \overline{C}$) (as $C, \overline{C} \not\subset A$, else $\delta(A)$, $\delta(B)$, $\delta(C)$ would form a nested family). Let $x_1 \in A$, $x_2 \in B \setminus C$ and $x_3 \in C \setminus B$, and

$X := \{x_1, x_2, x_3\}$. Then, the metric $\frac{1}{2}(\delta(A \cap X) + \delta(B \cap X) + \delta(C \cap X))$ on X coincides with the path metric of K_3 and is a minor of d, and (X, d) is not ℓ_1^1-embeddable. ∎

We now turn to the characterization of ℓ_1^2-embeddability. We start with establishing an analogue of Lemma 11.1.20 for 2-nested families of cut semimetrics. The result[3] from Proposition 11.1.22 will play a central role in the proof of Theorem 11.1.24 below, which contains several equivalent characterizations for ℓ_1^2-embeddable distances.

Proposition 11.1.22. *Let $\mathcal{C}$ be a family of cut semimetrics. Then, $\mathcal{C}$ can be partitioned into two nested families if only if the same holds for every subset $\mathcal{C}_0$ of $\mathcal{C}$ with cardinality $|\mathcal{C}_0| \leq 5$.*

Proof. Let $\mathcal{C}$ be a family of cut semimetrics. An element $\delta(S) \in \mathcal{C}$ is said to be *extremal* in $\mathcal{C}$ if one of S or $\overline{S}$ is minimal in $\widetilde{\mathcal{C}}$ (i.e., if S is minimal or maximal in $\widetilde{\mathcal{C}}$). Hence, $\delta(A)$ is not extremal if $B \subset A \subset C$ for some $B, C \in \widetilde{\mathcal{C}}$. Then, we say that $\delta(A)$ *separates* $\delta(B)$ from $\delta(C)$.
We show the 'if' part in Proposition 11.1.22 by induction on the cardinality of $\mathcal{C}$. So we can suppose that $|\mathcal{C}| \geq 6$ and that every proper subset of $\mathcal{C}$ is 2-nested. Suppose, for a contradiction, that $\mathcal{C}$ is not 2-nested. We first show:

(a) There are at most four extremal elements in $\mathcal{C}$.

For, suppose that there are five extremal elements in $\mathcal{C}$. By the assumption, they can be partitioned into two nested families. Hence, at least three of them form a nested family, which contradicts the extremality assumption. Next, we show:

(b) For every extremal element $\delta(S) \in \mathcal{C}$, the family $\mathcal{C} \setminus \{\delta(S)\}$ has no new extremal element.

Indeed, suppose that $\delta(S)$ is an extremal element in $\mathcal{C}$ and that $\delta(T)$ is an extremal element in $\mathcal{C} \setminus \{\delta(S)\}$ but not in $\mathcal{C}$. Consider a partition of $\mathcal{C} \setminus \{\delta(S)\}$ into two nested families. Then, $\delta(S)$ can be added to the nested family containing $\delta(T)$, so that the new family remains nested. Hence, $\mathcal{C}$ is 2-nested, in contradiction with our assumption. This shows (b). From this we derive:

(c) For every nonextremal element $\delta(A) \in \mathcal{C}$, there exist four extremal elements such that $\delta(A)$ separates two of them from the other two.

This follows from the fact that A contains at least two minimal sets and is contained in at least two maximal sets from $\widetilde{\mathcal{C}}$. Indeed, if S is a minimal set from

[3] Proposition 11.1.22 is an analogue of Lemma 11.1.20 for 2-nested families of cut semimetrics; it was proved by Schrijver [1995]. We prefer to use this combinatorial result rather than the corresponding result given in Theorem B from Bandelt and Chepoi [1996a], in particular, because it is self-contained while Bandelt and Chepoi need the notion of median graphs. We thank Lex Schrijver for his proof of Proposition 11.1.22.

$\widetilde{\mathcal{C}}$ contained in A, then there exists a minimal set T in $\widetilde{\mathcal{C}} \setminus \{S, \overline{S}\}$ which is also contained in A. Now, T is also minimal in $\widetilde{\mathcal{C}}$ because $\delta(T)$ is extremal in $\mathcal{C}$ by (b). This shows (c).

Therefore, there are exactly four extremal elements in $\mathcal{C}$ (by (a) and (c)). Say, they are $\delta(S_i)$ for $i = 1,2,3,4$, where S_1, S_2, S_3, S_4 are minimal in $\widetilde{\mathcal{C}}$. By (c), every nonextremal cut semimetric $\delta(A)$ separates two of them from the other two. This makes three possibilities for such a separation. We first observe that not all three possibilities can occur simultaneously. For this, note that if $\delta(A)$ and $\delta(B)$ separate the extremal cut semimetrics in two distinct ways, then they are crossing. (Indeed, say $\delta(A)$ separates $\delta(S_1)$, $\delta(S_2)$ from $\delta(S_3)$, $\delta(S_4)$ and $\delta(B)$ separates $\delta(S_1)$, $\delta(S_3)$ from $\delta(S_2)$, $\delta(S_4)$. We can suppose that $S_1, S_2 \subset A \subset \overline{S}_3, \overline{S}_4$, $S_1, S_3 \subset B \subset \overline{S}_2, \overline{S}_4$. From this follows that $A \not\subset B, \overline{B}$ and $B, \overline{B} \not\subset A$, i.e., $\delta(A)$ and $\delta(B)$ are crossing.) Now, there cannot be three pairwise crossing elements in $\mathcal{C}$ as they would form a family that is not 2-nested. Hence, at most two possibilities can occur for the separation of the extremal cut semimetrics. We distinguish two cases.

Case 1: Every nonextremal cut semimetric $\delta(A)$ separates the extremal ones in the same way; say, it separates $\delta(S_1)$, $\delta(S_2)$ from $\delta(S_3)$, $\delta(S_4)$. Consider a partition of the nonextremal cut semimetrics into two nested families: $\{\delta(A_1), \ldots, \delta(A_m)\}$ and $\{\delta(B_1), \ldots, \delta(B_p)\}$, where $A_1 \subset \ldots \subset A_m$ and $B_1 \subset \ldots \subset B_p$. We can always add the $\delta(S_i)$'s to either of these two nested families so as to retain the property of being nested. Indeed, say $A_m \subset \overline{S}_3, \overline{S}_4$; then, $S_1, S_2 \subset A_1$. Either, $B_p \subset \overline{S}_3, \overline{S}_4$ and $S_1, S_2 \subset B_1$; then, $S_2 \subset B_1 \subset \ldots \subset B_p \subset \overline{S}_4$ and $S_1 \subset A_1 \subset \ldots \subset A_m \subset \overline{S}_3$. Or, $B_p \subset \overline{S}_1, \overline{S}_2$ and $S_3, S_4 \subset B_1$; then, we can add $\overline{S}_2, S_4$ to the chain $B_1 \subset \ldots \subset B_p$ and $S_1, \overline{S}_3$ to the chain $A_1 \subset \ldots \subset A_m$.

Case 2: Every nonextremal cut semimetric separates, either $\delta(S_1)$, $\delta(S_2)$ from $\delta(S_3)$, $\delta(S_4)$, or $\delta(S_1)$, $\delta(S_3)$ from $\delta(S_2)$, $\delta(S_4)$. Let $\delta(A)$ satisfy the first possibility and $\delta(B)$ the second one. Then, $\delta(A)$ and $\delta(B)$ are crossing. Consider again a partition of the nonextremal cut semimetrics into two nested families $\mathcal{C}_1$ and $\mathcal{C}_2$. Say, $\delta(A) \in \mathcal{C}_1$ and $\delta(B) \in \mathcal{C}_2$. Then, all elements of $\mathcal{C}_1$ (resp. $\mathcal{C}_2$) separate the same two pairs of extremal cut semimetrics as $\delta(A)$ (resp. $\delta(B)$). From this follows that both $\mathcal{C}_1 \cup \{\delta(S_2), \delta(S_3)\}$ and $\mathcal{C}_2 \cup \{\delta(S_1), \delta(S_4)\}$ are nested. Hence, $\mathcal{C}$ is 2-nested. This concludes the proof. ∎

The following distances are not ℓ_1^2-embeddable: the path metrics of the graphs $K_{2,3}$, K_5, C_5, C_6 (the circuits on 5 and 6 nodes), $K_2 \times K_3$ (the Cartesian product of K_2 and K_3) (see Figure 11.1.23), as well as the four distances d_1, d_2, d_3, and d_4 displayed in Figures 11.1.25-11.1.28. (We display there an embedding in the 3-dimensional space for each of the distances.) (One can verify that, for each of these distances, their set of d-splits cannot be partitioned into two nested families.) It turns out that these nine distances are the only minimal obstructions to ℓ_1^2-embeddability; this result is the contents of Theorem 11.1.24 below which was proved by Bandelt and Chepoi [1996a].

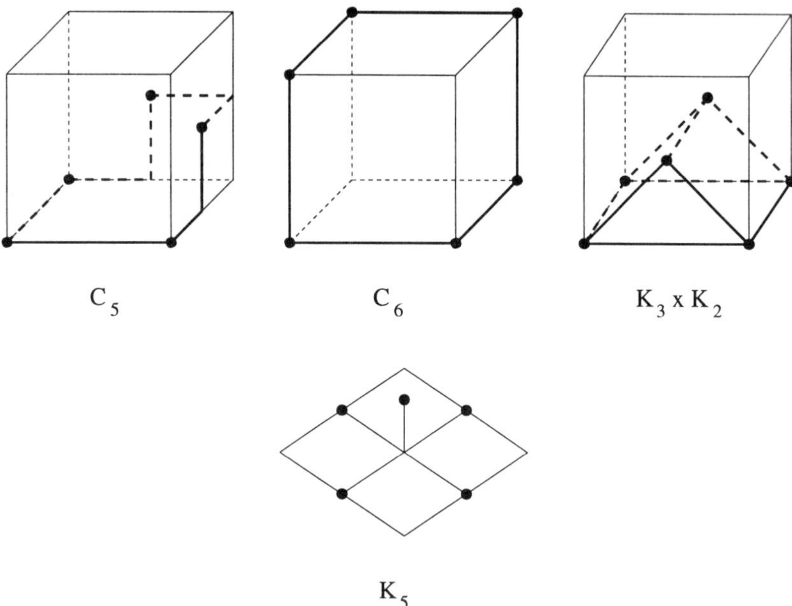

Figure 11.1.23: Embedding C_5, C_6, $K_2 \times K_3$, K_5 in the 3-dimensional ℓ_1-space

Theorem 11.1.24. *Let d be a distance on V_n. The following assertions are equivalent.*

(i) *(V_n, d) is ℓ_1^2-embeddable.*

(ii) *d is totally decomposable and its set Σ_d of d-splits can be partitioned into two nested subfamilies.*

(iii) *(X, d) is ℓ_1^2-embeddable for every subset $X \subseteq V_n$ with $|X| \leq 6$.*

(iv) *d does not have as a minor the path metrics of the graphs $K_{2,3}$, C_5, C_6, K_5, $K_2 \times K_3$, nor any of the metrics d_1, d_2, d_3, d_4 shown in Figures* 11.1.25-11.1.28.

Proof. Clearly, (i) $\Longleftrightarrow$ (ii) (by Lemma 11.1.17) and (i) $\Longrightarrow$ (iii), (iv). We show below the implications: (iii) $\Longrightarrow$ (ii) and (iv) $\Longrightarrow$ (ii). We start with two preliminary observations concerning an arbitrary distance d. We first show:

(a) If there exist three pairwise crossing d-splits, then the path metric of C_6 is a minor of d.

Indeed, suppose that $\delta(A), \delta(B), \delta(C) \in \Sigma_d$ are pairwise crossing. We can suppose without loss of generality that $A \cap B \cap C \neq \emptyset$. As Σ_d is weakly compatible, we deduce that, either $A \cap \overline{B} \cap \overline{C} = \emptyset$, or $\overline{A} \cap B \cap \overline{C} = \emptyset$, or $\overline{A} \cap \overline{B} \cap C = \emptyset$. We can suppose, for instance, that $A \cap \overline{B} \cap \overline{C} = \emptyset$. Then, $A \cap \overline{B} \cap C \neq \emptyset$ (as $A \cap \overline{B} \neq \emptyset$), $A \cap B \cap \overline{C} \neq \emptyset$, and $\overline{A} \cap \overline{B} \cap \overline{C} \neq \emptyset$. Using again the weak compatibility of Σ_d, we obtain that $\overline{A}\cap B\cap C=\emptyset$ and, then, that $\overline{A}\cap\overline{B}\cap C\neq\emptyset$, $\overline{A}\cap B\cap\overline{C}\neq\emptyset$. Pick an element in each of these six sets: $x_1\in A\cap B\cap C$, $x_2\in A\cap\overline{B}\cap C$, $x_3\in A\cap B\cap\overline{C}$, $x_4\in\overline{A}\cap\overline{B}\cap\overline{C}$, $x_5\in\overline{A}\cap\overline{B}\cap C$, and $x_6\in\overline{A}\cap B\cap\overline{C}$ and set $X:=\{x_1,\ldots,x_6\}$. Then, the distance $\delta(A\cap X)+\delta(B\cap X)+\delta(C\cap X)$ is a minor of d which coincides with the path metric of the 6-circuit $C_6=(x_1,x_3,x_6,x_4,x_5,x_2)$. Hence, (a) holds. Next, we show:

(b) If there exist four d-splits $\delta(A_i)(i=0,1,2,3)$ such that $\delta(A_0)$ and $\delta(A_i)$ are crossing for $i=1,2,3$ and A_1,A_2,A_3 are all minimal in $\{A_i,\overline{A}_i\mid i=1,2,3\}$, then the path metric of C_6 or of $K_2\times K_3$ is a minor of d.

Indeed, suppose that such d-splits exist. Then, $\delta(A_1)$, $\delta(A_2)$ and $\delta(A_3)$ are pairwise cross-free (else, we are done in view of (a)). Hence, $A_1\cap A_2=A_1\cap A_3=A_2\cap A_3=\emptyset$ (by minimality of A_1,A_2,A_3). Let $x_i\in A_0\cap A_i$ and $y_i\in\overline{A}_0\cap A_i$, for $i=1,2,3$ (such points exist by assumption) and set $X:=\{x_i,y_i\mid i=1,2,3\}$. Then, the distance $\delta(A_0\cap X)+\frac{1}{2}(\sum_{i=1}^{3}\delta(A_i\cap X))$ is a minor of d which coincides with the path metric of $K_2\times K_3$. This shows (b).

We can now proceed with the proof. We suppose that d does not satisfy (ii) and we show that both (iii) and (iv) are violated. If d is not totally decomposable, then we are done. Indeed, d has $d(K_{2,3})$ as a minor, which violates both (iii) and (iv). Suppose now that d is totally decomposable and that Σ_d is not 2-nested. By Proposition 11.1.22, there exists a subset $\mathcal{C}\subseteq\Sigma_d$ such that $|\mathcal{C}|\leq 5$ and $\mathcal{C}$ is not 2-nested. Choose such $\mathcal{C}$ with minimum cardinality. We distinguish three cases.

Case 1: $|\mathcal{C}|=3$. Then, any two members of $\mathcal{C}$ are crossing. By (a), we obtain that $d(C_6)$ is a minor of d; hence, (iii) and (iv) are violated.

Case 2: $|\mathcal{C}|=4$. Suppose first that every member of $\mathcal{C}$ is cross-free with at least another member of $\mathcal{C}$. Let G denote the graph on $\mathcal{C}$, where two elements of $\mathcal{C}$ are joined by an edge if they are cross-free. Then, G contains no matching of size 2 (else, $\mathcal{C}$ would be 2-nested). Moreover, the complement of G contains no triangle (by the minimality of $\mathcal{C}$). From this follows that G consists of a triangle. Hence, $\mathcal{C}=\{\delta(A_i)\mid i=0,1,2,3\}$, where $\delta(A_0)$ is crossing with $\delta(A_i)$ $(i=1,2,3)$ and the $\delta(A_i)$'s $(i=1,2,3)$ are pairwise cross-free. We claim that the set $\mathcal{A}:=\{A_i,\overline{A}_i\mid i=1,2,3\}$ has three minimal elements at least. (For, suppose that A_1 and A_2 are the only minimal elements of $\mathcal{A}$. Then, $A_1\subset\overline{A}_2$, $A_2\subset\overline{A}_1$ and, for instance, $A_1\subset A_3$, $A_2\subset\overline{A}_3$. This implies that $A_1\subset A_3\subset\overline{A}_2$. Hence, $\mathcal{C}$ could be covered by two nested families, a contradiction.) Hence, we can suppose that A_1, A_2 and A_3 are minimal elements of $\mathcal{A}$. Applying (b), we obtain that $d(K_2\times K_3)$ is a minor of d and, thus, (iii) and (iv) are violated.

Case 3: $|\mathcal{C}|=5$. Let H denote now the graph on $\mathcal{C}$, where two elements are joined by an edge if thay are crossing. We claim that the maximum degree of a node in H is ≤ 2. Suppose first that there is a node of degree 4 in H; say, $\delta(A_5)\in\mathcal{C}$ is crossing with the four other elements $\delta(A_i)$ $(i=1,2,3,4)$ of $\mathcal{C}$. The $\delta(A_i)$'s $(i=1,2,3,4)$ are pairwise noncrosssing and, for every $i=1,2,3,4$, the set $\mathcal{C}\setminus\{\delta(A_i)\}$ is 2-nested (by the minimality of $\mathcal{C}$). From this follows that the family

$\mathcal{C} \setminus \{\delta(A_5), \delta(A_i)\}$ is nested for every $i = 1, 2, 3, 4$. By Lemma 11.1.20, this implies that $\mathcal{C} \setminus \{\delta(A_5)\}$ is nested. Therefore, $\mathcal{C}$ can be covered by two nested families, a contradiction. Suppose now that H has a node of degree 3. Say, $\delta(A_5)$ is crossing with $\delta(A_i)$ $(i = 1, 2, 3)$ and cross-free with $\delta(A_4)$. The family $\mathcal{C} \setminus \{\delta(A_4)\}$ can be covered by two nested families. Therefore, $\{\delta(A_1), \delta(A_2), \delta(A_3)\}$ is nested and, thus, $\mathcal{C}$ can be covered by two nested families, a contradiction. So, we have shown that the maximum degree in H is ≤ 2. Therefore, H is either a circuit on 4 or 5 nodes, or a disjoint union of paths. For every $\delta(A_i)$ of degree at most 1 in H, we can select a point from A_i that does not belong to the other A_j's. Moreover, we can select a point in every nonempty intersection $A_j \cap A_k$, where at least one of $\delta(A_j)$ and $\delta(A_k)$ has degree 2 in H. Altogether we have selected a set X of five or six points such that the family $\{A_i \cap X, \overline{A}_i \cap X \mid i = 1, 2, 3, 4, 5\}$ has at least five minimal members. Hence, the family $\{\delta(A_i \cap X) \mid i = 1, 2, 3, 4, 5\}$ cannot be covered with two nested families and, thus, (X, d) is not ℓ_1^2-embeddable. We list below the possible configurations for the graph H together with the corresponding distance $d_H := \sum_{i=1}^{5} \delta(B_i)$ on X, setting $B_i := A_i \cap X$. In each case, we find one of the forbidden distances as a minor.

(i) When $H = C_5$, then $B_1 = \{1, 2\}$, $B_2 = \{2, 3\}$, $B_3 = \{3, 4\}$, $B_4 = \{4, 5\}$, and $B_5 = \{5, 1\}$. Hence, d_H is (up to a factor 2) the shortest path metric of C_5.

(ii) When H is the disjoint union of P_1 and two paths P_2, then $B_i = \{i\}$ for $i = 1, \ldots, 5$. Hence, d_H is (up to a factor 2) the path metric of the complete graph K_5.

The remaining cases are displayed in Figures 11.1.25-11.1.28. (We show an embedding of the distance in the 3-dimensional grid, as well as the sets B_i.) ∎

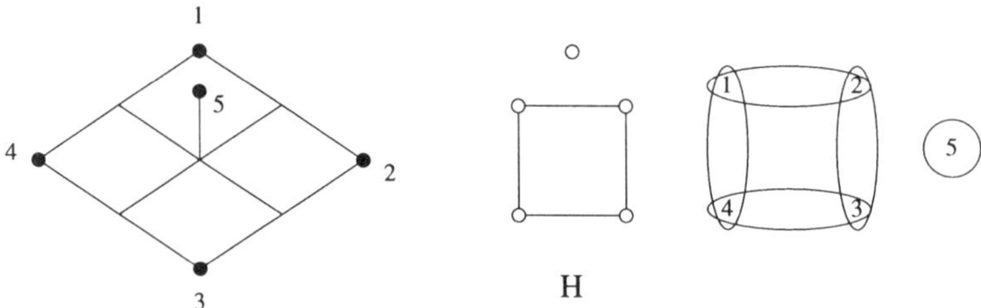

Figure 11.1.25: $H = C_4$, distance d_1

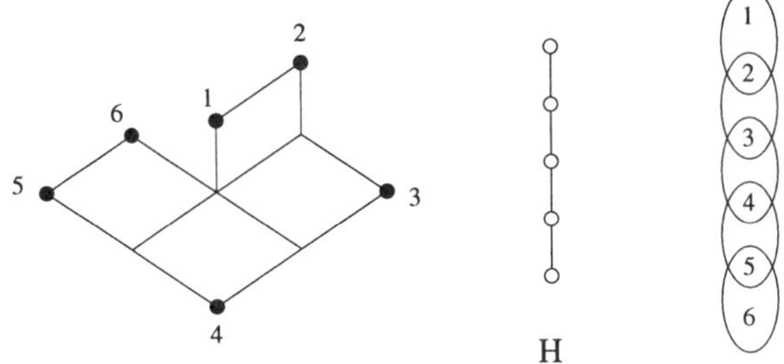

Figure 11.1.26: $H = P_5$, distance d_2

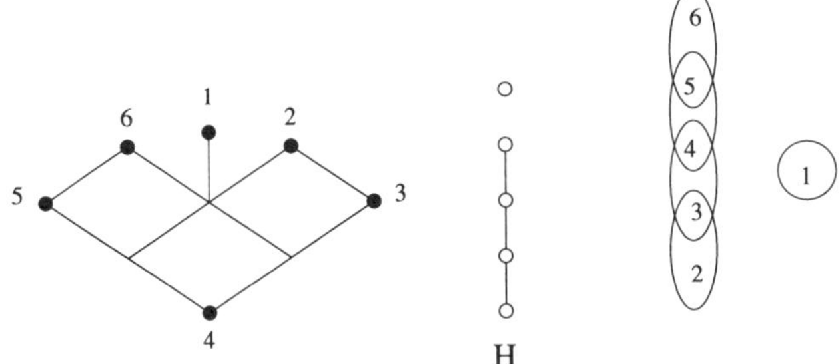

Figure 11.1.27: H is the disjoint union of P_1 and P_4, distance d_3

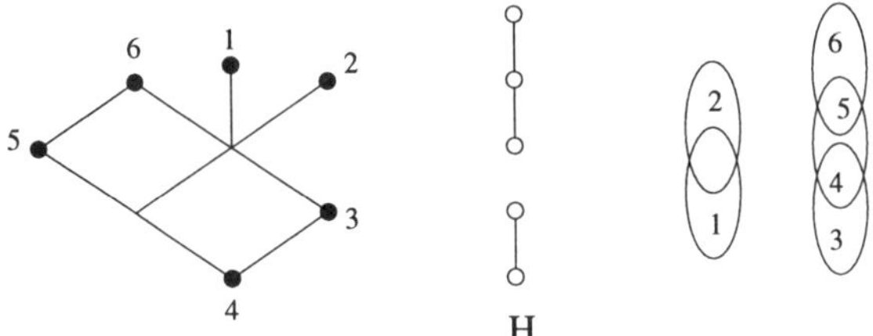

Figure 11.1.28: H is the disjoint union of P_2 and P_3, distance d_4

Finally, we mention without proof a result of Bandelt and Chepoi [1996b] concerning the characterization of the distance spaces that can be embedded into $(\mathbb{Z}^2, d_{\ell_1})$, the rectilinear 2-dimensional grid (or digital plane). Interestingly, this result is very similar in its formulation to the result for the ℓ_1-plane from Theorem 11.1.24. Namely,

Theorem 11.1.29. *Let (X, d) be a distance space where d is integer valued (and X arbitrary). The following assertions are equivalent.*

(i) *(X, d) is an isometric subspace of $(\mathbb{Z}^2, d_{\ell_1})$.*

(ii) *Every subspace (Y, d) of (X, d) with $|Y| \leq 6$ can be embedded into $(\mathbb{Z}^2, d_{\ell_1})$.*

(iii) (X, d) *satisfies the following parity condition*[4]*:* $d(x,y)+d(x,z)+d(y,z) \in 2\mathbb{Z}$ *for all* $x, y, z \in X$*, and every subspace* (Y, d) *of* (X, d) *with* $|Y| \leq 6$ *embeds in the rectilinear plane* $(\mathbb{R}^2, d_{\ell_1})$.

(iv) *Every finite subspace of* (X, d) *satisfies the parity condition, is totally decomposable and its collection of d-splits is 2-nested.* ∎

In particular, a distance space (X, d) embeds in the grid $(\mathbb{Z}^2, d_{\ell_1})$ if and only if it embeds in the plane $(\mathbb{R}^2, d_{\ell_1})$ and it satisfies the parity condition.

According to Malitz and Malitz [1992], one can test in $O(n^3)$ time whether a distance space on n points embeds in the ℓ_1-plane and construct such an embedding if one exists. Therefore, one can test embeddability in the 2-dimensional grid with the same time complexity. Moreover, Bandelt and Chepoi [1996b] show how to construct in time $O(n \log n)$ an embedding in the grid $\mathbb{Z}^2$ from an embedding in the plane $\mathbb{R}^2$ (if one exists).

11.2 On the Minimum ℓ_p-Dimension

We consider here the problem of evaluating the minimum ℓ_p-dimension $m_{\ell_p}(n)$ of an arbitrary ℓ_p-embeddable space on n points. We recall the definition of $m_{\ell_p}(n)$ from relation (3.1.2). That is, $m_{\ell_p}(n)$ is the smallest integer m such that any ℓ_p-embeddable space on n points can embedded in ℓ_p^m. The main results can be stated as follows.

As was already observed in relations (3.1.3) and (6.2.6),

$$m_{\ell_\infty}(n) \leq n-1 \quad \text{and} \quad m_{\ell_2}(n) = n-1$$

but, for general p, it is not immediate that $m_{\ell_p}(n)$ is finite. Wolfe [1967] showed that

$$m_{\ell_\infty}(n) \leq n-2$$

and Holsztysnki [1978] that

$$m_{\ell_\infty}(n) \geq \left\lfloor \frac{2}{3}n \right\rfloor \quad \text{for } n \geq 4.$$

Ball [1990] showed the existence of a constant c such that

$$m_{\ell_\infty}(n) \geq n - cn^{3/4}.$$

Witsenhausen proved that

$$m_{\ell_1}(n) \leq \binom{n}{2}.$$

Fichet [1988] and Ball [1990] extended (independently) the result for any $p \geq 1$ (see Proposition 11.2.3 below). In other words, every ℓ_p-embeddable distance

[4]The parity condition is an obvious necessary condition for embeddability in the hypercube or the ℓ_1-grid $\mathbb{Z}^m$ (for some $m \geq 1$); see relation (24.1.1).

on n points can embedded in ℓ_p^m, where $m = \binom{n}{2}$. The bound can be slightly improved to

$$m_{\ell_1}(n) \leq \binom{n}{2} - 1$$

as observed by Fichet [1994]. Ball [1990] proposes the following lower bounds for the minimum ℓ_p-dimension:

$$m_{\ell_p}(n) \geq \binom{n-1}{2} \text{ for } 1 < p < 2, n \geq 3, \text{ and}$$

$$m_{\ell_1}(n) \geq \binom{n-2}{2} \text{ for } n \geq 4.$$

(See Proposition 11.2.4 below.) In fact,

$$m_{\ell_1}(4) = m_{\ell_\infty}(4) = 2, \ m_{\ell_1}(5) = 3, \text{ and } m_{\ell_1}(6) = 6$$

(see Ball [1990] and Fichet [1994]). Ball [1990] made the following conjecture concerning the minimum ℓ_1-dimension:

Conjecture 11.2.1. $m_{\ell_1}(n) = \binom{n-2}{2}$ *for all* $n \geq 5$.

The upper bound: $m_{\ell_p}(n) \leq \binom{n}{2}$ is based on Carathéodory's theorem applied to the cut cone (if $p = 1$) or to the cone $\text{NOR}_n(p)$ (for $p \geq 1$). Let us recall the definition of $\text{NOR}_n(p)$. Given an integer $p \geq 1$, $\text{NOR}_n(p)$ consists of the distances d on V_n for which $d^{\frac{1}{p}}$ is ℓ_p-embeddable, i.e., for which there exist n vectors $v_1, \ldots, v_n \in \mathbb{R}^m$ ($m \geq 1$) such that

$$d_{ij} = (\| v_i - v_j \|_p)^p$$

for all $1 \leq i < j \leq n$. In the case $p = 1$, $\text{NOR}_n(1)$ coincides with the cut cone CUT_n (by Proposition 4.2.2). An element $d \in \text{NOR}_n(p)$ is said to be *linear* if $d^{\frac{1}{p}}$ is ℓ_p^1-embeddable, i.e., if there exist $x_1, \ldots, x_n \in \mathbb{R}$ such that $d_{ij} = |x_i - x_j|^p$ for all $1 \leq i < j \leq n$. For example, each cut semimetric belongs to $\text{NOR}_n(p)$ and is linear, i.e.,

$$\text{CUT}_n \subseteq \text{NOR}_n(p).$$

We collect in the next result a few (easy to verify) properties of the set $\text{NOR}_n(p)$.

Lemma 11.2.2.

(i) $\text{NOR}_n(p)$ *is a cone.*

(ii) *Let* $d \in \text{NOR}_n(p)$. *Then,* $d^{\frac{1}{p}}$ *is* ℓ_p^m*-embeddable if and only if* d *is the sum of* m *linear members of* $\text{NOR}_n(p)$. *In particular, if* d *lies on an extreme ray of* $\text{NOR}_n(p)$, *then* d *is linear.* ∎

Proposition 11.2.3. *$m_{\ell_1}(n) \leq \binom{n}{2} - 1$ and $m_{\ell_p}(n) \leq \binom{n}{2}$ for all $p \geq 1$.*

Proof. Consider first the case $p = 1$. We show that every semimetric $d \in \mathrm{CUT}_n$ can be written as a nonnegative combination of $\binom{n}{2} - 1$ linear members of CUT_n. This follows from Carathéodory's theorem if d lies on the boundary of CUT_n. Else, suppose that d lies in the interior of CUT_n. Let $\alpha > 0$ such that $d - \alpha\delta(1)$ lies on the boundary of CUT_n. Then, $d - \alpha\delta(1)$ can be written as a nonnegative combination of $\binom{n}{2} - 1$ cut semimetrics. This implies that d can be written as a nonnegative combination of $\binom{n}{2} - 1$ linear semimetrics (as $\delta(1)$ together with any other cut semimetric $\delta(S)$ form a nested family). We consider now the case $p \geq 1$. Let H denote the hyperplane in $\mathbb{R}^{E_n}$, which is defined by the equation $\sum_{1 \leq i < j \leq n} x_{ij} = 1$. Set

$$L := \{d \in \mathrm{NOR}_n(p) \mid d \in H \text{ and } d \text{ is linear}\}.$$

One can show that L is a compact set and that $\mathrm{NOR}_n(p) \cap H$ is a $(\binom{n}{2} - 1)$-dimensional convex set which coincides with the convex hull of L. Hence, Carathéodory's theorem implies that every member of $\mathrm{NOR}_n(p)$ can be written as the sum of $\binom{n}{2}$ linear members of $\mathrm{NOR}_n(p)$. This yields the result. ∎

Proposition 11.2.4.

(i) *$m_{\ell_1}(n) \geq \binom{n-2}{2}$ for $n \geq 4$.*

(ii) *$m_{\ell_p}(n) \geq \binom{n-1}{2}$ for $1 < p < 2$ and $n \geq 3$.*

Proof. (i) Set $m := \binom{n-2}{2}$. We exhibit a semimetric d on V_n which embeds in ℓ_1^m but not in ℓ_1^k if $k < m$. Set $d := \sum_{2 \leq r < s \leq n-1} \delta(\{1, r, s\})$; hence,

$$\begin{cases} d_{1n} & = \binom{n-2}{2}, \\ d_{1i} & = \binom{n-3}{2} \quad \text{for } 2 \leq i \leq n-1, \\ d_{ij} & = 2(n-4) \quad \text{for } 2 \leq i < j \leq n-1, \\ d_{in} & = n-3 \quad \text{for } 2 \leq i \leq n-1. \end{cases}$$

By construction, d embeds isometrically in ℓ_1^m. We show that d cannot be embedded in ℓ_1^k if $k < m$. For this, we consider the inequality of negative type (6.1.1) with $b := (-2, 1, \ldots, 1, -(n-4))$, i.e., the inequality

$$2(n-4)x_{1n} - 2\sum_{2 \leq i \leq n-1} x_{1i} - (n-4)\sum_{2 \leq i \leq n-1} x_{in} + \sum_{2 \leq i < j \leq n-1} x_{ij} \leq 0. \tag{11.2.5}$$

Let F denote the face of the cone $\mathrm{NOR}_n(1)$ ($=\mathrm{CUT}_n$) which is defined by the inequality (11.2.5). Clearly, the cut semimetrics $\delta(\{1, r, s\})$ $(2 \leq r < s \leq n-1)$ are the only cut semimetrics that lie on F. Moreover, they are linearly independent. Hence, F is a simplex face of $\mathrm{NOR}_n(1)$. Therefore, d is ℓ_1-rigid; that is, $d := \sum_{2 \leq r < s \leq n-1} \delta(\{1, r, s\})$ is its only $\mathbb{R}_+$-realization. No two cut semimetrics $\delta(\{1, r, s\})$ and $\delta(\{1, r', s'\})$ form a nested pair. Hence, the family $\{\delta(\{1, r, s\}) \mid$

$2 \le r < s \le n-1\}$ cannot be covered with less than m nested subfamilies. This shows that d is not ℓ_1^k-embeddable if $k < m$.
(ii) We only sketch the proof, which is along the same lines as for (i). Set $m := \binom{n-1}{2}$. Consider the vectors $v_1, \ldots, v_n \in \mathbb{R}^m$ defined by

$$(v_i)_{rs} = \begin{cases} 1 & \text{if } r = i \\ -1 & \text{if } s = i \\ 0 & \text{otherwise} \end{cases}$$

for $1 \le r < s \le n$. Define a distance d on V_n by setting $d_{ij} := \| v_i - v_j \|_p$ for $1 \le i < j \le n$. So d embeds in ℓ_p^m by construction. One can show that d does not embed in ℓ_p^k if $k < m$ by using, as in case (i), a special inequality which is valid for the cone $\mathrm{NOR}_n(p)$ and is satisfied at equality by d^p. Namely, one uses the inequality:

$$\sum_{1 \le i < j \le n} (\| u_i - u_j \|_p)^p - (n + 2^{p-1} - 2)(\| u_i \|_p)^p \le 0,$$

which holds for any set of n vectors $u_1, \ldots, u_n \in \mathbb{R}^h$ $(h \ge 1)$ if $1 \le p \le 2$ (Ball [1987]). ∎

Remark 11.2.6. Linial, London and Rabinovich [1994] define the *metric dimension* $\dim(G)$ of a connected graph G as the smallest integer m for which there exists a norm $\| \,.\, \|$ on $\mathbb{R}^m$ such that the graphic space (V, d_G) of the graph G can be isometrically embedded into the space $(\mathbb{R}^m, d_{\|.\|})$. The definition extends clearly to an arbitrary semimetric space. Hence, rather than looking only at embeddings in a fixed Banach ℓ_p-space, Linial, London and Rabinovich [1994] consider embeddings in an arbitrary normed space.

Actually, this notion of metric dimension is linked with ℓ_∞-embeddings in the following way. Let (V_n, d) be a semimetric space. Then, its metric dimension is equal to the minimum rank of a system of vectors $v_1, \ldots, v_n \in \mathbb{R}^k$ $(k \ge 1)$ providing an ℓ_∞-embedding of (V_n, d), i.e., such that $d_{ij} = \| v_i - v_j \|_\infty$ for all $1 \le i < j \le n$.

The metric dimension of several graphs is computed in Linial, London and Rabinovich [1994]. In particular, $\dim(K_n) = \lceil \log_2(n) \rceil$, $\dim(T) = O(\log_2(n))$ for a tree on n nodes (both being realized by an ℓ_∞-embedding), $\dim(C_{2n}) = n$ for a circuit on $2n$ nodes (realized by an ℓ_1-embedding), $\dim(K_{n\times 2}) \ge n-1$ for the cocktail party graph. It is also shown there that, if G is a graph on n nodes with metric dimension d, then each vertex has degree $\le 3^d - 1$, G has diameter $\ge \frac{1}{2}(n^{\frac{1}{d}} - 1)$, and there exists a subset S of $O(dn^{1-\frac{1}{d}})$ nodes whose deletion disconnects G and so that each connected component of $G \backslash S$ has no more than $(1 - \frac{1}{d} + o(1))n$ nodes.

Dewdney [1980] considers the question of embedding graphs isometrically into the ℓ_p-space (F^m, d_{ℓ_p}), where F is a field. He shows, in particular, that every connected graph G on n nodes can be isometrically embedded into the space $(\{0,1,2\}^{n-2}, d_{\ell_\infty})$. Moreover, computing the smallest m such that G embeds isometrically into $(\{0,1,2\}^m, d_{\ell_\infty})$ is an NP-hard problem. ∎

Chapter 12. Examples of the Use of the L_1-Metric

The L_1-metric is widely used in many areas, for instance, for the analysis of data structures, for the recognition of computer pictures, or for comparing random variables in probabilities. We provide here some (superficial) information on some areas of application of the L_1-metric. The importance of the L_1-metric is illustrated, in particular, by the great variety of names under which it is known. For example, the Manhattan metric, the taxi-cab metric, or the 4-metric are different names for the same notion, namely, the ℓ_1 distance in the plane; more terminology is given in Section 12.3.

12.1 The L_1-Metric in Probability Theory

Let $(\Omega, \mathcal{A}, \mu)$ be a probability space and let $X : \Omega \longrightarrow \mathbb{R}$ be a random variable belonging to $L_1(\Omega, \mathcal{A}, \mu)$; that is, such that $\int_\Omega |X(\omega)|\mu(d\omega) < \infty$. Let F_X denote the distribution function of X, i.e., $F_X(x) = \mu(\{\omega \in \Omega \mid X(\omega) \leq x\})$ for $x \in \mathbb{R}$; when it exists, its derivative F'_X is called the density of X. A great variety of metrics on random variables are studied in the monograph by Rachev [1991]; among them, the following are based on the L_1-metric:

- The usual L_1-metric between the random variables:

$$L_1(X,Y) = E(|X - Y|) = \int_\Omega |X(\omega) - Y(\omega)|\mu(d\omega).$$

- The Monge-Kantorovich-Wasserstein metric (i.e., the L_1-metric between the distribution functions):

$$k(X,Y) = \int_\mathbb{R} |F_X(x) - F_Y(x)|dx$$

- The total valuation metric (i.e., the L_1-metric between the densities when they exist):

$$\sigma(X,Y) = \frac{1}{2}\int_\mathbb{R} |F'_X(x) - F'_Y(x)|dx.$$

- The engineer metric (i.e., the L_1-metric between the expected values):

$$\mathrm{EN}(X,Y) = |E(X) - E(Y)|.$$

- The indicator metric:

$$i(X,Y) = E(1_{X \neq Y}) = \mu(\{\omega \in \Omega \mid X(\omega) \neq Y(\omega)\}).$$

In fact, the L_p-analogues ($1 \leq p \leq \infty$) of the above metrics, especially of the first two, are also used in probability theory.

Several results are known, establishing links among the above metrics. One of the main such results is the Monge-Kantorovich mass-transportation theorem which shows that the second metric $k(X,Y)$ can be viewed as a minimum of the first metric $L_1(X,Y)$ over all joint distributions of X and Y with fixed marginal. A relationship between the $L_1(X,Y)$ and the engineer metric $EN(X,Y)$ is given in Rachev [1991] as a solution of a moment problem. Similarly, a connection between the total valuation metric $\sigma(X,Y)$ and the indicator metric $i(X,Y)$ is given in Dobrushin's theorem on the existence and uniqueness of Gibbs fields in statistical physics. See Rachev [1991] for a detailed account on the above topics.

We mention another example of the use of the L_1-metric in probability theory, namely for Gaussian random fields. We refer to Noda [1987, 1989] for a detailed account. Let $B = (B(x) \mid x \in M)$ be a centered Gaussian system with parameter space M, $0 \in M$. The variance of the increment is denoted by

$$d(x,y) := E((B(x) - B(y))^2) \text{ for } x,y \in M.$$

When (M,d) is a metric space which is L_1-embeddable, the Gaussian system is called a Lévy's Brownian motion with parameter space (M,d). The case $M = \mathbb{R}^n$ and $d(x,y) =$
$\| x - y \|_2$ gives the usual Brownian motion with n-dimensional parameter. By Lemma 4.2.5, (M,d) is L_1-embeddable if and only if there exist a non negative measure space (H,ν) and a mapping $x \mapsto A_x \subseteq H$ with $\nu(A_x) < \infty$ for $x \in M$, such that $d(x,y) = \nu(A_x \triangle A_y)$ for $x,y \in M$. Hence, a Gaussian system admits a representation called of Chentsov type

$$B(x) = \int_{A_x} W(dh) \text{ for } x \in M$$

in terms of a Gaussian random measure based on the measure space (H,ν) with $d(x,y) = \nu(A_x \triangle A_y)$ if and only if d is L_1-embeddable.

This Chentsov type representation can be compared with the Crofton formula for projective metrics from Theorem 8.3.3. Actually both come naturally together in Ambartzumian [1982] (see parts A.8-A.9 of Appendix A there).

12.2 The ℓ_1-Metric in Statistical Data Analysis

A *data structure* is a pair (I,d), where I is a finite set, called *population*, and $d : I \times I \longrightarrow \mathbb{R}_+$ is a symmetric mapping with $d_{ii} = 0$ for $i \in I$, called *dissimilarity index*. A typical problem in statistical data analysis is to choose a "good representation" of a data structure; usually, "good" means a representation allowing to represent the data structure visually by a graphic display. Each sort of visual display corresponds, in fact, to a special choice of the dissimilarity index as a distance and the problem turns out to be the classical isometric embedding problem in special classes of metrics.

For instance, in hierarchical classification, the case when d is ultrametric corresponds to the possibility of having a representation of the data structure by a so-called indexed hierarchy (see Johnson [1967]). A natural extension is the case when d is the path metric of a weighted tree, i.e., when d satisfies the four point condition (cf. Section 20.4); then the data structure is called an *additive tree*. Data structures (I, d) for which d is ℓ_2-embeddable are considered in factor analysis and multidimensional scaling. These two cases together with cluster analysis are the main three techniques for studying data structures. The case when d is ℓ_1-embeddable is a natural extension of the ultrametric and ℓ_2 cases which has received considerable attention in the recent years.

An ℓ_p-approximation consists of minimizing the estimator $\| e \|_p$, where e is a vector or a random variable (representing an error, deviation, etc). The following criteria are used in statistical data analysis:

- the ℓ_2-norm, in the least square method; or its square,
- the ℓ_∞-norm, in the minimax method,
- the ℓ_1-norm, in the least absolute values (LAV) method.

In fact, the ℓ_1 criterion has also been increasingly used in the recent years.

The importance of the role played by the ℓ_1-metric in statistical data analysis can be seen, for instance, from the volumes by Dodge [1987b, 1992] and by van Cutsem [1994] of proceedings of conferences on the topic of statistical data analysis. We refer, in particular, to the papers by Crichtley and Fichet [1994], Dodge [1987a], Fichet [1987a, 1987b, 1992, 1994], Le Calve [1987], Vajda [1987] in those volumes.

12.3 The ℓ_1-Metric in Computer Vision and Pattern Recognition

The ℓ_p-metrics are also used in the new area called pattern recognition, or robot vision, or digital topology; see, e.g., Rosenfeld and Kak [1976], Horn [1986].

A computer picture is a subset of $\mathbb{Z}^n$ (or of a scaling $\frac{1}{m}\mathbb{Z}^n$ of $\mathbb{Z}^n$) which is called a *digital n-D-space* (or an *n-D m-quantized space*). Usually, pictures are represented in the digital plane $\mathbb{Z}^2$ or in the digital 3-D-space $\mathbb{Z}^3$. The points of $\mathbb{Z}^n$ are called the pixels.

Given a picture in $\mathbb{Z}^n$, i.e., a subset A of $\mathbb{Z}^n$, one way to define its volume vol(A) is by vol$(A) := |A|$, i.e., as the number of pixels contained in A. Then, the distance

$$d(A, B) := \text{vol}(A \triangle B)$$

is used in digital topology for evaluating the distance between pictures. It is a digital analogue of the symmetric difference metric used in convex geometry, where the distance between two convex bodies A and B in $\mathbb{R}^n$ is defined as the n-dimensional volume of their symmetric difference.

The above metric and other metrics on $\mathbb{Z}^n$ are used for studying analogues of classical geometric notions as volume, perimeter, shape complexity, etc., for computer pictures. The metrics on $\mathbb{Z}^n$ that are mainly used are the ℓ_1-, ℓ_∞-metrics, as well as the ℓ_2-metric after rounding to the nearest upper (or lower) integer.

When considered on $\mathbb{Z}^n$, the ℓ_1-metric is also called the *grid metric* and the ℓ_∞-metric is called the *lattice metric* (or *Chebyshev metric*, or *uniform metric*). More specific names are used in the case $n = 2$. Then, the ℓ_1-metric is also known as the *city-block metric* (or *Manhattan metric*, or *taxi-cab metric*, or *rectilinear metric*), or as the *4-metric* since each point of $\mathbb{Z}^2$ has exactly 4 closest neighbors in $\mathbb{Z}^2$ for the ℓ_1-metric. The reader may consult Krause [1986] for a leisurely account on the taxi-cab metric. Similarly, the ℓ_∞-metric on $\mathbb{Z}^2$ is called the *chessboard metric*, or the *8-metric* since each pixel has exactly 8 closest neighbors in $\mathbb{Z}^2$. Note indeed that the unit sphere $S^1_{\ell_1}$ (centered at the origin) for the ℓ_1-norm in $\mathbb{R}^2$ contains exactly 4 integral points while the unit sphere $S^1_{\ell_\infty}$ for the ℓ_∞-norm contains 8 integral points.

Observe also that the ℓ_1-metric, when considered on $\mathbb{Z}^n$, can be seen as the path metric of an (infinite) graph on $\mathbb{Z}^n$. Namely, consider the graph on $\mathbb{Z}^n$ where two lattice points are adjacent if their ℓ_1-distance is equal to 1; this graph is nothing but the usual grid. Then, the shortest path distance of two lattice points in the grid is equal to their ℓ_1-distance. Similarly, the ℓ_∞-metric on $\mathbb{Z}^n$ is the path metric of the graph on $\mathbb{Z}^n$ where adjacency is defined by the pairs at ℓ_∞-distance one. For $n = 2$, adjacency corresponds to the king move in chessboard terms; moreover, $(\mathbb{Z}^2, d_{\ell_\infty})$ is an isometric subspace of $(\frac{1}{2}\mathbb{Z}^2, d_{\ell_1})$ via the embedding given in relation (3.1.6).

There are some other useful metrics on $\mathbb{Z}^2$ which are obtained by combining the ℓ_1- and ℓ_∞-metrics. The following two examples, the octogonal and the hexagonal distances, are path metrics; hence, in order to define them, it suffices to describe the pairs of lattice points at distance 1, i.e., to describe their unit balls.

The Octogonal Distance d_{oct}. For each $(x, y) \in \mathbb{Z}^2$, its unit sphere $S^1_{oct}(x, y)$, centered at (x, y), is defined by

$$S^1_{oct}(x,y) = S^3_{\ell_1}(x,y) \cap S^2_{\ell_\infty}(x,y),$$

where $S^3_{\ell_1}(x, y)$ denotes the ℓ_1-sphere of radius 3 and $S^2_{\ell_\infty}(x, y)$ the ℓ_∞-sphere of radius 2, centered at (x, y). Hence, $S^1_{oct}(x, y)$ contains exactly 8 integral points; note that moving from (x, y) to its eight neighbors at distance 1 corresponds to the knight move in chessboard terms. Figure 12.3.1 shows the spheres $S^3_{\ell_1}$, $S^2_{\ell_\infty}$, and S^1_{oct}.

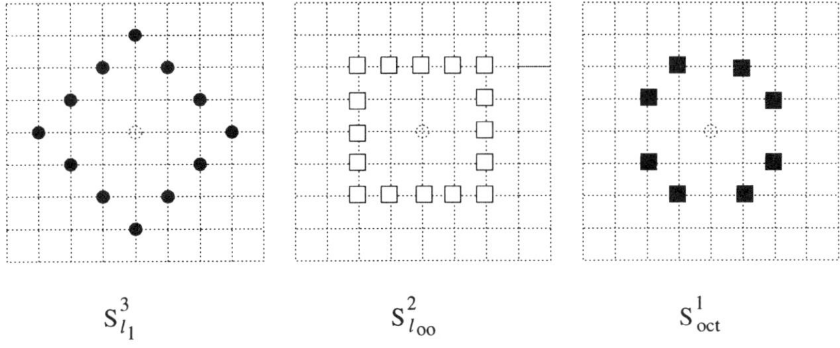

Figure 12.3.1

The Hexagonal Distance or 6-Metric d_{hex}. Its unit sphere $S^1_{hex}(x,y)$, centered at $(x,y) \in \mathbb{Z}^2$, is defined by

$$S^1_{hex}(x,y) = S^1_{\ell_1}(x,y) \cup \{(x-1,y-1),(x-1,y+1)\} \text{ for } x \text{ even},$$

$$S^1_{hex}(x,y) = S^1_{\ell_1}(x,y) \cup \{(x+1,y-1),(x+1,y+1)\} \text{ for } x \text{ odd}.$$

The unit sphere $S^1_{hex}(x,y)$ contains exactly 6 integral points. Figure 12.3.2 shows the unit spheres $S^1_{hex}(0,0)$ and $S^1_{hex}(1,-3)$. (In fact, the distance space $(\mathbb{Z}^2, d_{hex})$ embeds with scale 2 in the hexagonal grid A_2 (consisting of the vectors in $\mathbb{Z}^3$ with sum 0); see Luczak and Rosenfeld [1976] for details.)

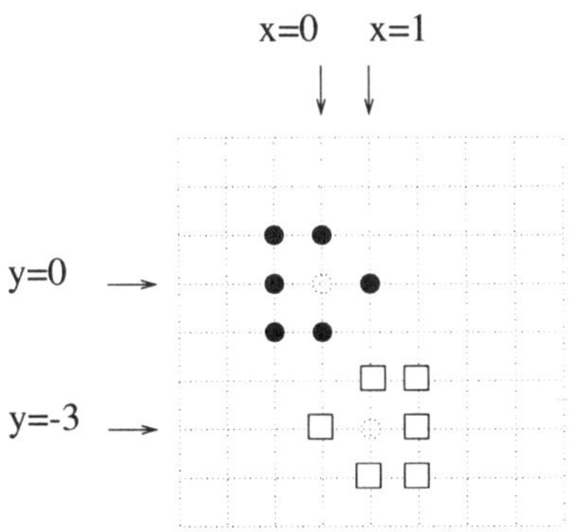

Figure 12.3.2

Several other modifications of the ℓ_1-metric on the plane have been considered; see, e.g., De Berg [1991] and references therein.

In practice, the subset $(\mathbb{Z}_k)^n := \{0,1,\dots,k-1\}^n$ is considered instead of the full space $\mathbb{Z}^n$. Note that $(\mathbb{Z}_2)^n$ is nothing but the vertex set of the n-dimensional hypercube and $((\mathbb{Z}_2)^n, d_{\ell_1})$ is the n-dimensional hypercube metric space. Note also that $(\mathbb{Z}_3)^2$ is the unit ball (centered at $(1,1)$) of the space $(\mathbb{Z}^n, d_{\ell_\infty})$. $(\mathbb{Z}_4)^n$

is known as the *tic-tac-toe board* (or *Rubik's n-cube*) and $(\mathbb{Z}_k)^2$, $(\mathbb{Z}_k)^3$ are called, respectively, the *k-grill* and the *k-framework*.

Other distances are used on $(\mathbb{Z}_k)^n$, in particular in coding theory, namely, the Hamming distance d_H and the *Lee distance* d_{Lee} defined by

$$d_{Lee}(x,y) = \sum_{1 \leq i \leq n} \min(|x_i - y_i|, k - |x_i - y_i|) \text{ for all } x, y \in (\mathbb{Z}_k)^n.$$

The metric space $(\mathbb{Z}_k, d_{Lee})$ can be seen as a discrete analogue of the elliptic metric space (which consists of the set of all the lines in $\mathbb{R}^2$ going through the origin and where the distance between two such lines is their angle).

The ℓ_1-distance and the Hamming distance coincide when restricted to $(\mathbb{Z}_2)^n$, i.e., the spaces $((\mathbb{Z}_2)^n, d_{\ell_1})$ and $((\mathbb{Z}_2)^n, d_H)$ are identical. Also, $(\mathbb{Z}_k, d_{\ell_1})$ coincides with the graphic metric space of the path P_k on k nodes, $(\mathbb{Z}_k, d_H)$ coincides with the graphic space of the complete graph K_k on k nodes, and $(\mathbb{Z}_k, d_{Lee})$ coincides with the graphic space of the circuit C_k on k nodes. Therefore, the spaces $((\mathbb{Z}_k)^n, d_{\ell_1})$, $((\mathbb{Z}_k)^n, d_H)$ and $((\mathbb{Z}_k)^n, d_{Lee})$ coincide with the graphic space of the Cartesian product G^n, where G is P_k, K_k and C_k, respectively. The following can be easily checked.

(i) P_k embeds isometrically in the $(k-1)$-dimensional hypercube (see Example 7.2.9), i.e., $(\mathbb{Z}_k, d_{\ell_1})$ is an isometric subspace of $((\mathbb{Z}_2)^{k-1}, d_{\ell_1})$ (simply, label each $x \in \mathbb{Z}_k$ by the binary string $1 \ldots 10 \ldots 0$ of length $k-1$ whose first x letters are equal to 1). Hence, $((\mathbb{Z}_k)^n, d_{\ell_1})$ is an isometric subspace of $((\mathbb{Z}_2)^{n(k-1)}, d_{\ell_1})$.

(ii) $((\mathbb{Z}_k)^n, d_H)$ is an isometric subspace of $((\mathbb{Z}_2)^{kn}, \frac{1}{2}d_{\ell_1})$ (label each $x \in \mathbb{Z}_k$ by the binary string of length k whose letters are all equal to 0 except the $(x+1)$th one equal to 1).

(iii) The even circuit C_{2k} embeds isometrically into the k-dimensional hypercube (see Example 7.2.9). Therefore, $((\mathbb{Z}_{2k})^n, d_{Lee})$ is an isometric subspace of $((\mathbb{Z}_2)^{nk}, d_{\ell_1})$. Also, $((\mathbb{Z}_{2k+1})^n, d_{Lee})$ is an isometric subspace of $((\mathbb{Z}_2)^{(2k+1)n}, \frac{1}{2}d_{\ell_1})$ (since the odd circuit C_{2k+1} embeds isometrically into the $(2k+1)$-dimensional halfcube).

More details about the ℓ_1-embeddings of the graphs P_k, C_k and K_k can be found in Parts III and IV.

Part II

Hypermetric Spaces: an Approach via Geometry of Numbers

Introduction

The central concept studied in Part II is hypermetricity. This is a natural strengthening of the notion of semimetric, which has many applications and connections. The main topics to which hypermetrics relate include ℓ_1- and ℓ_2-metrics in analysis, the cut cone and the cut polytope in combinatorial optimization, graphs with high regularity and, what will be our focus of interest here, quadratic forms, Delaunay polytopes and holes in lattices.

The notion of hypermetrics sheds a new light and gives a more ordered view on some well studied questions; for example, on equiangular sets of lines, on the graphs whose adjacency matrix has minimum eigenvalue -2, on the metric properties of regular graphs. For instance, the parameter characterizing the three layers composing the famous list from Bussemaker, Cvetković and Seidel [1976] of the 187 graphs with minimum eigenvalue -2 has now a more clear meaning: it comes from the radius of the Delaunay polytope associated with the graph metrics in each layer (see Section 17.2).

The links between hypermetrics and ℓ_1-,ℓ_2-metrics have been discussed in Section 6.3. Hypermetric inequalities, as valid inequalities for the cut cone and polytope, will be studied in Part V. In this second part, we focus on the connections existing between hypermetrics and geometry of numbers and, more precisely, with Delaunay polytopes and holes in lattices.

Our central objects here are hypermetric inequalities and hypermetric spaces; they have already been introduced in Section 6.1, but we recall the main definitions here. Given an integer vector $b \in \mathbb{Z}^n$ with $\sum_{i=1}^n b_i = 1$, the inequality

$$\text{(a)} \qquad \sum_{1 \leq i < j \leq n} b_i b_j x_{ij} \leq 0$$

is called a *hypermetric inequality*. When $b_i = b_j = 1 = -b_k$ and $b_h = 0$ for $h \neq i, j, k$ (for some distinct i, j, k), the inequality (a) is simply the triangle inequality:

$$x_{ij} - x_{ik} - x_{jk} \leq 0.$$

A distance space (X, d) is said to be *hypermetric* if d satisfies all hypermetric inequalities. As the hypermetric inequalities include the triangle inequalities, every hypermetric distance space is a semimetric space. The *hypermetric cone* HYP_n is the cone in $\mathbb{R}^{E_n}$ defined by the inequalities (a) for all $b \in \mathbb{Z}^n$ with $\sum_{i=1}^n b_i = 1$.

When $b \in \mathbb{Z}^n$ with $\sum_{i=1}^n b_i = 0$, the inequality (a) is called an *inequality of negative type*. The *negative type cone* NEG_n is the cone in $\mathbb{R}^{E_n}$ defined by the inequalities (a) for all $b \in \mathbb{Z}^n$ with $\sum_{i=1}^n b_i = 0$, and a distance space (X, d) is said to be of *negative type* if d satisfies all the negative type inequalities.

Many important semimetrics are hypermetric. In particular, all ℓ_1-semimetrics are hypermetric. More precisely, given a distance d, we have the following chain of implications (recall Theorem 6.3.1):

d is isometrically ℓ_2-embeddable $\Longrightarrow$ d is isometrically ℓ_1-embeddable $\Longrightarrow$ d is hypermetric $\Longrightarrow$ $\sqrt{d}$ is isometrically ℓ_2-embeddable

Moreover, if d is hypermetric then $\sqrt{d}$ has an ℓ_2-embedding on a sphere and, as we see below, this sphere corresponds to a hole in some lattice. The last property in the above chain of implications is well characterized. Namely, $\sqrt{d}$ is isometrically ℓ_2-embeddable if and only if d is of negative type or, equivalently, if and only if the image $\xi(d)$ of d under the covariance mapping ξ is a positive semidefinite matrix. Therefore, our central object: the hypermetric cone, is closely related to the positive semidefinite cone. (See Section 13.1 for details.)

A distance that we will use constantly here is the square of the Euclidean distance, namely the distance $d^{(2)}$ defined by

$$d^{(2)}(x,y) := (\| x - y \|_2)^2 = (x-y)^T(x-y)$$

for $x, y \in \mathbb{R}^n$. For convenience, we also denote $x^T x = (\| x \|_2)^2 = \sum_{i=1}^n (x_i)^2$ as x^2, for $x \in \mathbb{R}^n$. In this part we will use exclusively the ℓ_2-norm $\| \,.\, \|_2$. So, for simplicity, we sometimes omit the subscript and write $\| x \|$ instead of $\| x \|_2$.

In fact, the study of hypermetric distance spaces amouts to the study of holes in lattices, as we now briefly explain. Let L be a lattice. Blow up a sphere S in one of the interstices of L until it is held rigidly by lattices points. Then, there are no lattice points in the interior of the ball delimited by the sphere S and sufficiently many lattice points lie on S so that their convex hull is a full-dimensional polytope P. The sphere S is then called an empty sphere in L, its center is called a hole of L and the polytope P is called a Delaunay polytope. So the vertices of P are the lattice points lying on the boundary of the empty sphere S. Let $V(P)$ denote the set of vertices of P. Then, the distance space $(V(P), d^{(2)})$ (endowed with the square of the Euclidean distance) is called a Delaunay polytope space; such spaces are fundamental in our treatment.

Empty spheres in lattices have been intensively studied in the literature from the point of view of their centers (i.e., the holes of L). Hypermetricity provides a new way of studying empty spheres, namely from the point of view of the lattice points lying on their boundary, i.e., from the point of view of Delaunay polytope spaces. Indeed, Delaunay polytopes have the remarkable property (discovered by Assouad [1984]) that their Delaunay polytope spaces are hypermetric and, conversely, every hypermetric space can be realized as a subspace of a Delaunay polytope space (see Theorem 14.1.3). To each hypermetric space (X, d) corresponds an (essentially unique) Delaunay polytope P_d whose dimension is less than or equal to $|X| - 1$.

Hence, there is a connection between the members of the hypermetric cone HYP_n and the Delaunay polytopes of dimension $k \leq n - 1$.

An interesting application of this connection is for proving that the hypermetric cone is a polyhedral cone (see Theorem 14.2.1).

These two objects: hypermetric cone and Delaunay polytopes, have been studied for their own sake. For instance, the hypermetric cone HYP_n arises in connection with ℓ_1-metrics (recall Lemma 6.1.7); it forms a linear relaxation for the cut cone and, as such, its facial structure has been intensively investigated; results in this direction will be given in Chapter 28. On the other hand, Delaunay polytopes have been mostly studied in the literature from the classical point of view of geometry of numbers: holes, L-decomposition of the space, dual tiling by Voronoi polytopes, etc. The approach taken here is to study the metric structure of their sets of vertices. Moreover, taking advantage of the interplay with hypermetrics, we can transport and exploit some of the notions defined for the hypermetric cone to Delaunay polytopes and vice versa.

For instance, there is a natural notion of rank for hypermetrics (namely, the dimension of the smallest face of the hypermetric cone that contains a given hypermetric distance). We introduce the corresponding notion of rank for Delaunay polytopes. This notion of rank permits, for instance, to shed a new light on a classical notion studied by Voronoi; namely, the repartitioning polytopes which correspond to the facets of the hypermetric cone. The other extreme case for the rank, namely the case of rank 1 for the extreme rays of the hypermetric cone, corresponds to the class of extreme Delaunay polytopes. A Delaunay polytope P is extreme if and only if the only affine transformations T for which $T(P)$ is still a Delaunay polytope are the homotheties (see Corollary 15.2.4). Several examples of extreme Delaunay polytopes are presented in Chapter 16 arising, in particular, in root lattices or in sections of the Leech lattice Λ_{24} and of the Barnes-Wall lattice Λ_{16}.

Historically, Delaunay polytopes and the corresponding L-partitions of the space were introduced by G.F. Voronoi at the beginning of this century. The so-called empty sphere method was developed later by B.N. Delaunay[1], who showed that it yields the same partition of the space as Voronoi's L-partition. The topic has been studied extensively mainly by the Russian school, especially by B.N. Delaunay, E.P. Baranovskii, S.S. Ryshkov, and also by R.M. Erdahl from Canada. In dimensions 2 and 3, L-decompositions are used in computational geometry[2] under the name of Delaunay triangulations; actually, nonlattice

[1]We refer to the preface of the volume edited by Novikov et al. [1992] for a detailed historical account on the work of B.N. Delaunay.

[2]For information see, for instance, Chapter 13 in Edelsbrunner [1987] or the survey by

triangulations are also studied there. Delaunay polytopes are also used for the study of coverings in lattices (see Conway and Sloane [1988], Rogers [1964]); for instance, the *covering radius* of a lattice L is the maximum radius of an empty sphere in L, i.e., the radius of a deep hole in L. There is the following connection between Voronoi polytopes and Delaunay polytopes: The vertices of the Voronoi polytope at a lattice point u are the centers of the Delaunay polytopes that contain u as a vertex. Moreover, the two partitions of the space by Delaunay polytopes and by Voronoi polytopes are in combinatorial duality.

Within the list of references on this topic, the more relevant and fundamental ones include Voronoi's Deuxième mémoire [1908, 1909], the survey by Ryshkov and Baranovskii [1979], the papers by Erdahl and Ryshkov [1987, 1988] and the collection by Conway and Sloane [1988] of surveys on lattices and applications. The present treatment on hypermetric spaces is based, essentially, on the papers by Assouad [1984], Deza, Grishukhin and Laurent [1992, 1993], Deza and Grishukhin [1993]. Further relevant references will be given throughout the text. An earlier version of the material presented in Part II appeared in the survey by Deza, Grishukhin and Laurent [1995].

We now briefly describe the main results presented in Part II. Chapter 13 contains preliminaries on distance spaces, lattices and Delaunay polytopes. In Section 13.3, we give a short proof of Voronoi's result, which states that the number of distinct (up to affine equivalence) Delaunay polytopes in fixed dimension is finite.

We consider in Chapter 14 the connection existing between hypermetric spaces and Delaunay polytopes. In Section 14.1, this connection is described together with some first results showing how the polytope P_d inherits some of the properties of the hypermetric space (X, d), in particular, about subspaces (see Corollary 14.1.9) and ℓ_1-embeddability (see Proposition 14.1.10). In Section 14.2, the hypermetric cone is shown to be polyhedral. Several proofs are given; one of them is based on the above connection and Voronoi's finiteness result for the number of Delaunay polytopes in fixed dimension.

Section 14.3 describes all the Delaunay polytopes that can arise in root lattices; see, in particular, Figure 14.3.1 which lists the Delaunay polytopes in the irreducible root lattices together with their 1-skeleton graphs and radii. If P is a Delaunay polytope in a root lattice, then its edges are the pairs of vertices at squared distance 2, i.e., its 1-skeleton graph is determined by the metric structure of its Delaunay polytope space (see Proposition 14.3.3). As an application, we give a characterization of the connected strongly even distance spaces that are hypermetric or ℓ_1-embeddable (see Theorems 14.3.6 and 14.3.7).

In Section 14.4, we group several results dealing with the radius of the sphere circumscribing Delaunay polytopes. We consider, in particular, the spherical t-extension operation which consists of adding a new point to a distance space at distance t from all the other points.

Fortune [1995].

The notion of rank for Delaunay polytopes is considered in detail in Chapter 15. If (X, d) is a hypermetric space with $|X| = n$, then $d \in \text{HYP}_n$ and the rank of (X, d) is defined as the dimension of the smallest (by inclusion) face of HYP_n that contains d. If P is a Delaunay polytope, then the Delaunay polytope space $(V(P), d^{(2)})$ is hypermetric and the rank of P is defined as the rank of the space $(V(P), d^{(2)})$. Then, P is said to be extreme if its rank is equal to 1. In Section 15.1, we consider several properties for this notion of rank, in particular, its invariance (see Theorem 15.1.8) and its additivity (see Proposition 15.1.10). We describe in Section 15.2 how the faces of the hypermetric cone relate to Delaunay polytopes; see, in particular, Figure 15.2.9. In particular, hypermetrics lying on the interior of the same face of the hypermetric cone correspond to affinely equivalent Delaunay polytopes (see Corollary 15.2.2), a geometric interpretation for the rank of a Delaunay polytope is given in Theorem 15.2.5, and Delaunay polytopes associated with facets of the hypermetric cone are described in Proposition 15.2.7.

We present in Section 15.3 some bounds on the number of vertices of a basic Delaunay polytope, i.e., whose set of vertices contains a base of the lattice it spans (see Proposition 15.3.1).

Chapter 16 is devoted to the study of the extreme Delaunay polytopes, which correspond to the extreme rays of the hypermetric cone. The extreme Delaunay polytopes in root lattices are characterized in Theorem 16.0.1; they are the segment α_1, the Schläfli polytope 2_{21} and the Gosset polytope 3_{21}. In Section 16.1, we derive bounds on the number of vertices of an extreme basic Delaunay polytope, which turn out to be closely related with known bounds on the cardinality of equiangular sets of lines. We also present a general construction for equiangular sets of lines from integral lattices (see Proposition 16.1.9). In Sections 16.2, 16.3 and 16.4, we describe examples of extreme Delaunay polytopes arising in sections of the root lattice E_8, of the Leech lattice Λ_{24} and of the Barnes-Wall lattice Λ_{16}. Section 16.5 contains results on the construction of perfect lattices from extreme Delaunay polytopes.

Chapter 17 applies the notion of hypermetricity to graphs. Given a graph G, two distances can be defined: its shortest path metric d_G or its truncated distance d^*_G (with distance 1 on an edge and distance 2 on a non-edge). The graph G is said to be hypermetric if its path metric is hypermetric. A characterization of the hypermetric graphs and of the ℓ_1-graphs is given in Theorem 17.1.1; see also Theorems 17.1.8 and 17.1.9 for a refined result for the class of suspension graphs.

The connected regular graphs whose truncated distance is hypermetric are considered in Section 17.2; see Proposition 17.2.1 for several equivalent characterizations, one of them is that the minimum eigenvalue of their adjacency matrix is greater than or equal to -2. The graphs with minimum eigenvalue -2 are well studied. Those that are not line graphs nor cocktail-party graphs belong to the well-known list of 187 graphs from Bussemaker, Cvetković and Seidel [1976]. This list is partitioned into three layers, each of them being characterized by a parameter which is directly related to the radius of the Delaunay polytopes associated with the graphs in the layer.

We consider in Section 17.3 extreme hypermetric graphs, i.e., the graphs whose path metric lies on an extreme ray of the hypermetric cone. In fact, all of them are isometric subgraphs of the Gosset graph or of the Schläfli graph. See Proposition 17.3.4 for their characterization.

Chapter 13. Preliminaries on Lattices

We group in this chapter definitions and preliminary results about lattices and Delaunay polytopes. One of the results of Voronoi that will play a central role in our study concerns the finiteness of the number of types of Delaunay polytopes in given dimension. We give a proof of this result in Section 13.3.

13.1 Distance Spaces

Distance spaces have been introduced in Chapter 3. We give here some additional definitions that are needed in this chapter. Let (X, d) be a distance space. Then, $d_{\min}$ denotes the minimum nonzero value taken by d. The distance space (X, d) is said to be *connected* if the graph with vertex set X and whose edges are the pairs (i, j) with $d(i, j) = d_{\min}$, is connected. The distance space (X, d) is said to be *strongly even* if $d(i, j) \in 2\mathbb{Z}$ for all $i, j \in X$ and $d_{\min} = 2$.

A *representation* of the distance space (X, d) is a mapping

$$i \in X \mapsto v_i \in \mathbb{R}^n$$

(where $n \geq 1$) such that

$$d(i,j) = (v_i - v_j)^2 \quad \text{for } i, j \in X. \tag{13.1.1}$$

In other words, it is an isometric embedding of (X, d) into the space $(\mathbb{R}^n, d^{(2)})$, endowed with $d^{(2)}$, the square of the Euclidean distance. Hence, (X, d) has a representation if and only if $(X, \sqrt{d})$ is isometrically ℓ_2-embeddable. Clearly, every translation of a representation of (X, d) is again a representation of (X, d). Hence, we can always assume that a given element $i_0 \in X$ is represented by the zero vector. The representation $(v_i \mid i \in X)$ is said to be *spherical* if all v_i's lie on a sphere. The next result summarizes how the property of having a spherical representation relates to the hypermetric and negative type conditions. Two partial converses to the implications (ii) $\Longrightarrow$ (iv) and (i) $\Longrightarrow$ (ii) in Proposition 13.1.2 will be given in Propositions 14.4.1 and 14.4.4, respectively.

Proposition 13.1.2. *Let (X, d) be a distance space. Consider the assertions:*

(i) *(X, d) is hypermetric.*

(ii) *(X, d) has a spherical representation.*

(iii) (X,d) *has a representation.*

(iv) (X,d) *is of negative type.*

Then, (i) $\Longrightarrow$ (ii) $\Longrightarrow$ (iii) $\Longleftrightarrow$ (iv) *holds.*

Proof. (i) $\Longrightarrow$ (ii) will be shown in Proposition 14.1.2, (iii) $\Longleftrightarrow$ (iv) follows from Theorem 6.2.2, and (ii) $\Longrightarrow$ (iii) is trivial. ∎

We remind from Section 2.4 that PSD_n denotes the positive semidefinite cone, which consists of the vectors $p = (p_{ij})_{1 \le i \le j \le n} \in \mathbb{R}^{\binom{n+1}{2}}$ for which the symmetric matrix $(p_{ij})_{i,j=1}^n$ (setting $p_{ji} = p_{ij}$) is positive semidefinite. As mentioned in Lemma 2.4.2, positive semidefinite matrices can be characterized in terms of Gram matrices.

The covariance mapping, which has been introduced in Section 5.2, will play a crucial role here; so, we now recall its definition[1]. The covariance mapping is the mapping $\xi : \mathbb{R}^{\binom{n+1}{2}} \longrightarrow \mathbb{R}^{\binom{n+1}{2}}$ defined by $p = \xi(d)$, for $d = (d_{ij})_{0 \le i < j \le n}$, $p = (p_{ij})_{1 \le i \le j \le n}$, with

$$\left\{ \begin{array}{ll} p_{ii} = d_{0i} & \text{for } 1 \le i \le n, \\ p_{ij} = \frac{1}{2}(d_{0i} + d_{0j} - d_{ij}) & \text{for } 1 \le i < j \le n. \end{array} \right. \tag{13.1.3}$$

It is easy to verify that

$$d \in \mathrm{HYP}_{n+1} \Longleftrightarrow \sum_{1 \le i,j \le n} b_i b_j p_{ij} - \sum_{1 \le i \le n} b_i p_{ii} \ge 0 \quad \text{for all } b \in \mathbb{Z}^n, \tag{13.1.4}$$

$$d \in \mathrm{NEG}_{n+1} \Longleftrightarrow \sum_{1 \le i,j \le n} b_i b_j p_{ij} \ge 0 \quad \text{for all } b \in \mathbb{Z}^n. \tag{13.1.5}$$

where $p = \xi(d)$ is considered as a symmetric $n \times n$ matrix by setting $p_{ji} = p_{ij}$ for all i,j. Therefore,

$$\xi(\mathrm{NEG}_{n+1}) = \mathrm{PSD}_n, \tag{13.1.6}$$

$$\mathrm{HYP}_{n+1} \subseteq \mathrm{NEG}_{n+1}, \quad \text{i.e., } \xi(\mathrm{HYP}_{n+1}) \subseteq \mathrm{PSD}_n. \tag{13.1.7}$$

(These facts were already mentioned in Corollary 6.1.4 and in (6.1.15).) We also remind from (5.2.3) and (6.1.6) that

$$\xi(\mathrm{CUT}_{n+1}) = \mathrm{COR}_n \text{ and } \mathrm{CUT}_{n+1} \subseteq \mathrm{HYP}_{n+1}.$$

As the correlation cone COR_n is generated by the vectors $(x_i x_j)_{1 \le i \le j \le n}$ (for $x \in \{0,1\}^n$), its polar $(\mathrm{COR}_n)^\circ$ consists of the quadratic forms that are nonpositive

[1] We find it more convenient to now denote the distinguished point by 0 rather than by $n+1$ as was done earlier.

on binary variables. A well-known (easy) fact is that the polar $(\mathrm{PSD}_n)^\circ$ of PSD_n consists of the negative semidefinite quadratic forms, i.e.,

$$(\mathrm{PSD}_n)^\circ = -\mathrm{PSD}_n.$$

Hence, we have the following chain of inclusions:

$$\text{(13.1.8)} \qquad \mathrm{COR}_n \subseteq \xi(\mathrm{HYP}_{n+1}) \subseteq \mathrm{PSD}_n, \ (\mathrm{COR}_n)^\circ \supseteq (\xi(\mathrm{HYP}_{n+1}))^\circ \supseteq -\mathrm{PSD}_n.$$

This shows that our central object, namely the hypermetric cone (or, to be more precise, the polar of its image under the covariance mapping) is a subcone of the cone of quadratic forms that are nonpositive on binary variables and contains the cone $-\mathrm{PSD}_n$ of the quadratic forms that are nonpostive on integer (or real) variables.

We will frequently use in this chapter the graphic metric spaces attached to the following graphs:

- the complete graph K_n, the circuit C_n, the path P_n (on n nodes),
- the cocktail-party graph $K_{n\times 2}$ (i.e., K_{2n} with a perfect matching deleted),
- the hypercube graph $H(n,2)$ (i.e., the graph whose nodes are the vectors $x \in \{0,1\}^n$ with two nodes x, y adjacent if $d_{\ell_1}(x,y) = 1$),
- the half-cube graph $\frac{1}{2}H(n,2)$ (i.e., the graph whose nodes are the vectors $x \in \{0,1\}^n$ with $\sum_{1\le i\le n} x_i$ even and two nodes x, y are adjacent if $d_{\ell_1}(x,y) = 2$).

13.2 Lattices and Delaunay Polytopes

We give here several definitions related to lattices and Delaunay polytopes. More information can be found, e.g., in Cassels [1959], Conway and Sloane [1988], Lagarias [1995].

13.2.1 Lattices

A subset L of $\mathbb{R}^k$ is called a *lattice* (or *point lattice*) if L is a discrete subgroup of $\mathbb{R}^k$, i.e., if there exists a ball of radius $\beta > 0$ centered at each lattice point which contains no other lattice point. A subset $V := \{v_1, \ldots, v_m\}$ of L is said to be *generating* (resp. a *basis*) for L if, for every $v \in L$, there exist some integers (resp. a unique system of integers) $b_1, \ldots, b_m$ such that

$$v = \sum_{1\le i\le m} b_i v_i.$$

Every lattice has a basis; all bases have the same cardinality, called the *dimension* of L. Let $L \subseteq \mathbb{R}^k$ be a lattice of dimension k. Given a basis B of L, let M_B

denote the $k \times k$ matrix whose rows are the members of B. If B_1 and B_2 are two bases of L, then

$$M_{B_1} = AM_{B_2}$$

where A is an integer matrix with determinant $\det(A) = \pm 1$ (such a matrix is called a *unimodular matrix*). Therefore, the quantity $|\det(B)|$ does not depend on the choice of the basis in L; it is called the *determinant* of L and is denoted by $\det(L)$.

Given a finite set $V \subseteq \mathbb{R}^k$, its integer hull $\mathbb{Z}(V)$ is clearly a lattice whenever all vectors in V are rational valued.

Given a vector $a \in \mathbb{R}^k$, the translate

$$L' := L + a = \{v + a \mid v \in L\}$$

of a lattice L is called an *affine lattice.* A subset $V' := \{v_0, v_1, \dots, v_m\}$ of L' is called an *affine generating set* for L' (resp. an *affine basis* of L') if, for every $v \in L'$, there exist some integers (resp. a unique system of integers) $b_0, b_1, \dots, b_m$ such that

$$\sum_{0 \leq i \leq m} b_i = 1 \text{ and } v = \sum_{0 \leq i \leq m} b_i v_i.$$

Clearly, V' is an affine generating set (resp. an affine basis) of L' if and only if the set $V := \{v_1 - v_0, \dots, v_m - v_0\}$ is a (linear) generating set (resp. basis) of the lattice L.

For simplicity, we will use the same word "lattice" for denoting both a usual lattice (i.e., containing the zero vector) and an affine lattice (i.e., the translate of a lattice). We also often omit to precise whether we consider linear or affine bases (or generating sets).

Let L be a lattice. The quantity:

$$t := \min((u - v)^2 \mid u, v \in L, u \neq v)$$

is called the *minimal norm* of L. This terminology of minimal "norm" is classical in the theory of lattices, although it actually denotes the square of the Euclidean norm. In particular, if $0 \in L$, then

$$t = \min(u^2 \mid u \in L, u \neq 0).$$

The *minimal vectors* of L are then the vectors $v \in L$ with $v^2 = t$. Their set is denoted as $L_{\min}$ and the polytope $\mathrm{Conv}(L_{\min})$ is known as the *contact polytope* of L. Note that $\frac{\sqrt{t}}{2}$ coincides with the packing radius of L.

Let L be a lattice. Then, L is said to be *integral* if $u^T v \in \mathbb{Z}$ for all $u, v \in L$. L is said to be an *even lattice* if L is integral and $u^2 \in 2\mathbb{Z}$ for each $u \in L$. L is called a *root lattice* if L is integral and L is generated by a set of vectors v with $v^2 = 2$; then, each $v \in L$ with $v^2 = 2$ is called a *root* of L. Observe that, in a root lattice L,

(13.2.1) $u^T v \in \{0, -1, 1\}$ for all roots u, v of L such that $u \neq \pm v$.

(This follows from the fact that $(u-v)^2 = 4-2u^T v > 0$ and $(u+v)^2 = 4+2u^T v > 0$.) The *dual* L^* of a lattice $L \subseteq \mathbb{R}^k$ is defined as

$$L^* := \{x \in \mathbb{R}^k \mid x^T u \in \mathbb{Z} \text{ for all } u \in L\}.$$

If L is an integral lattice, then $L \subseteq L^*$ holds. L is said to be *self-dual* if $L = L^*$ holds. L is said to be *unimodular* if $\det(L) = \pm 1$. Hence, an integral unimodular lattice is self-dual. For example, the root lattice E_8 and the Leech lattice Λ_{24} (introduced later in the text) are even and unimodular and, therefore, self-dual. For every k-dimensional lattice $L \subseteq \mathbb{R}^k$,

$$(L^*)^* = L.$$

Let L_1 and L_2 be two orthogonal lattices, i.e., such that $u_1^T u_2 = 0$ for all $u_1 \in L_1$, $u_2 \in L_2$. Their *direct sum* $L_1 \oplus L_2$ is defined by

$$L_1 \oplus L_2 := \{u_1 + u_2 \mid u_1 \in L_1, u_2 \in L_2\}.$$

L is called *irreducible* if $L = L_1 \oplus L_2$ implies $L_1 = \{0\}$ or $L_2 = \{0\}$, and *reducible* otherwise. A well-known result by Witt gives the classification of the irreducible root lattices; cf. Section 14.3.

13.2.2 Delaunay Polytopes

Let $L \subseteq \mathbb{R}^k$ be a k-dimensional lattice and let $S = S(c, r)$ be a sphere with center c and radius r in $\mathbb{R}^k$. Then, S is said to be an *empty sphere* in L if the following two conditions hold:

(i) $(v - c)^2 \geq r^2$ for all $v \in L$, and

(ii) the set $S \cap L$ has affine rank $k + 1$.

Then, the center of S is called a *hole*[2]. The polytope P, which is defined as the convex hull of the set $S \cap L$, is called a *Delaunay polytope*, or an *L-polytope*. See Figure 13.2.2 for an illustration.

Equivalently, a k-dimensional polytope P in $\mathbb{R}^k$ with set of vertices $V(P)$ is a Delaunay polytope if the following conditions hold:

(i) The set $L(P) := \mathbb{Z}_{af}(V(P)) = \{\sum_{v \in V(P)} b_v v \mid b \in \mathbb{Z}^{V(P)}, \sum_{v \in V(P)} b_v = 1\}$ is a lattice,

(ii) P is inscribed on a sphere $S(c, r)$ (i.e., $(v - c)^2 = r^2$ for all $v \in V(P)$), and

(iii) $(v - c)^2 \geq r^2$ for all $v \in L(P)$, with equality if and only if $v \in V(P)$.

[2]The terminology of 'empty sphere' is used mainly in the Russian literature and that of 'hole' in the English literature.

Another equivalent definition will be given in Proposition 14.1.4. Given a Delaunay polytope P, the distance space $(V(P), d^{(2)})$ is called a *Delaunay polytope space.*

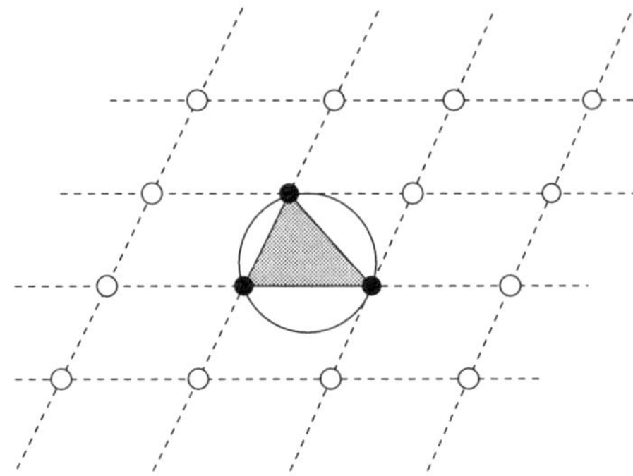

Figure 13.2.2: An empty sphere in a lattice and its Delaunay polytope

Let P be a Delaunay polytope and let L be a lattice such that $V(P) \subseteq L$. Then, P is said to be *generating in* L if $V(P)$ generates L, i.e., if $L = L(P)$. There are examples of lattices for which none of their Delaunay polytopes is generating; this is the case for the root lattice E_8, the Leech lattice Λ_{24} and, more generally, for all even unimodular lattices (see Lemma 13.2.6). However, when we say that P is an *Delaunay polytope in* L, we will always mean that P is generating in L, i.e., we suppose that $L = L(P)$.

A subset $B \subseteq V(P)$ is said to be *basic* if it is an affine basis of the lattice $L(P)$. Then, P is said to be *basic* if $V(P)$ contains a basic set, i.e., if $V(P)$ contains an affine basis of $L(P)$. Actually, we do not know an example of a nonbasic Delaunay polytope. We formulate this as an open problem for further reference.

Problem 13.2.3. *Is every Delaunay polytope basic ?*

The answer is positive for Delaunay polytopes having a small corank (cf. Proposition 15.2.12) and for concrete examples mentioned later in Part II. Some further information about this problem will be given in Section 27.4.3 for Delaunay polytopes arising in the context of binary matroids. The property of being basic will be useful on several occasions; for instance, for formulating upper bounds on the number of vertices of extreme Delaunay polytopes (cf. Section 15.3) or for the study of perfect lattices (cf. Section 16.5).

For instance, the n-dimensional cube $\gamma_n = [0,1]^n$ is a Delaunay polytope in the integer lattice $\mathbb{Z}^n$. As other example, we have the central object of the book, namely, the cut polytope $\mathrm{CUT}_n^\square$, which is a Delaunay polytope in the cut lattice $\mathcal{L}_n$ (cf. Example 13.2.5 below). Note that both γ_n and $\mathrm{CUT}_n^\square$ are basic, γ_n is centrally symmetric, while $\mathrm{CUT}_n^\square$ is asymmetric (see the definition in Lemma 13.2.7).

Two Delaunay polytopes have the same *type* if they are affinely equivalent, i.e., if $P' = T(P)$ for some affine bijection T.

Given a lattice point $v \in L$, the set of all the Delaunay polytopes in L that admit v as a vertex is called the *star* of L at v. Clearly, the stars at distinct lattice points are all identical (up to translation). The lattice L is called *general* if all the Delaunay polytopes of its star are simplices (which, in general, cannot be obtained from one another by translation or orthogonal transformation), and L is called *special* otherwise.

Two k-dimensional lattices L, L' are said to be *z-equivalent* if there exists an affine bijection T such that $L' = T(L)$ and such that T brings the star of L on the star of L'; one also says that L and L' have the same *type*. For example, in dimension 2, there are two distinct types of lattices: the triangular lattice which is general, and the square lattice which is special. (See Figure 13.2.4 for an illustration.)

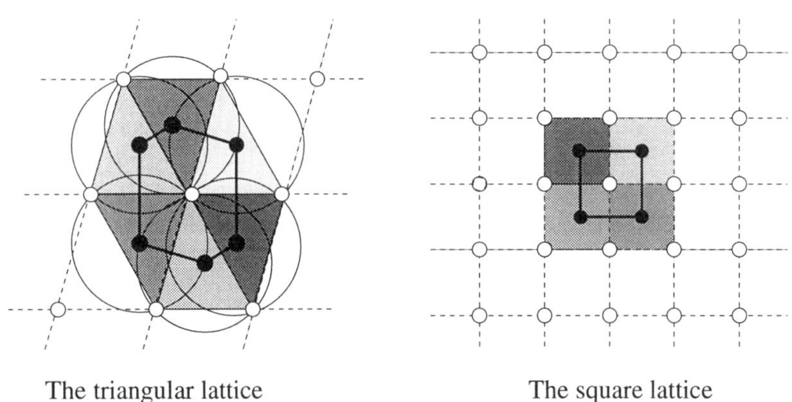

Figure 13.2.4: The star of Delaunay polytopes and the Voronoi polytope (The Delaunay polytopes are shaded and the Voronoi polytope is drawn with thick lines)

Example 13.2.5. Delaunay polytopes in the cut lattice. Let $\mathcal{L}_n$ denote the cut lattice, which is the sublattice of $\mathbb{Z}^{\binom{n}{2}}$ generated by all cut semimetrics on n points. One can easily verify that $\text{CUT}_n^{\square}$ is a (asymmetric) Delaunay polytope in $\mathcal{L}_n$. Other examples of Delaunay polytopes in the cut lattice $\mathcal{L}_n$ are described by Deza and Grishukhin [1995b]. In particular, using a result of Baranovskii [1992], they describe all symmetric Delaunay polytopes in $\mathcal{L}_n$. Moreover, they analyze in detail the Delaunay polytopes in $\mathcal{L}_n$ for small n.

If $n = 2$, then $\mathcal{L}_2 = \mathbb{Z}$ and $\text{CUT}_2^{\square} = \gamma_1$. In the case $n = 3$, then $\mathcal{L}_3 = D_3$ ($\simeq A_3$) is the unique 3-dimensional root lattice (the face-centered cubic lattice) (see Section 14.3) and $\text{CUT}_3^{\square} = \alpha_3$ is a regular 3-dimensional simplex. In the case $n = 4$, $\mathcal{L}_4 = \sqrt{2}D_6^+ = \Lambda_6\{3\}$, where D_6^+ is a union of the root lattice D_6 with a translated copy of it and $\Lambda_6\{3\}$ is an integral laminated lattice of minimal norm 3 (cf. Chap. 4, p.119 and Chap. 6, p.179 in Brouwer, Cohen and Neumaier [1989]). Moreover, Deza and Grishukhin [1995b] give a detailed description of the star of Delaunay polytopes in $\mathcal{L}_4$; it contains 588 distinct Delaunay polytopes which are grouped into four types: the cut polytope $\mathrm{CUT}_4^{\square}$, the simplex α_6, the cross-polytope β_6, and a 'twisted' cross-polytope. ∎

We conclude with recalling the connection existing between Delaunay polytopes and Voronoi polytopes. If L is a lattice in $\mathbb{R}^k$ and $u_0 \in L$, the *Voronoi polytope* at u_0 is the set $P_v(u_0)$ consisting of all the points $x \in \mathbb{R}^k$ that are at least as close to u_0 than to any other lattice point, i.e.,

$$P_v(u_0) := \{x \in \mathbb{R}^k : \| x - u_0 \| \leq \| x - u \| \text{ for all } u \in L\}.$$

The vertices of the Voronoi polytope $P_v(u_0)$ are precisely the centers of the Delaunay polytopes in L that contain u_0 as a vertex, i.e. of the Delaunay polytopes of the star of L at u_0. (Cf. Figure 13.2.4.)

The Voronoi polytopes $P_v(u)$ $(u \in L)$ form a normal (i.e., face-to-face) tiling of the space $\mathbb{R}^k$; this tiling is sometimes called the *Voronoi-Dirichlet tiling*. Another normal tiling is provided by the elementary cells $\{u + \sum_{i=1}^k b_i v_i \mid 0 \leq b_i \leq 1 \text{ for } 1 \leq i \leq k\}$ for $u \in L$, where $(v_1, \ldots, v_k)$ is a basis of L. Hence, the Voronoi polytopes and the elementary cells have the same volume, equal to $\det(L)$. Another normal partition of the space, called *L-decomposition*, is provided by the Delaunay polytopes in L. However, different types of Delaunay polytopes may occur in this partition; in particular, if L is special, then some of them are not simplices. For instance, if L is a general lattice of dimension 2, then the normal partition of $\mathbb{R}^2$ by the Delaunay polytopes in L is a Delaunay triangulation of the plane.

Given a k-dimensional lattice L, the two normal partitions of the space by the Voronoi polytopes and by the Delaunay polytopes in L are in combinatorial duality. Namely, there is a one-to-one correspondence $F \mapsto F^*$ between the faces F of one partition and the faces F^* of the other partition in such a way that:

(i) F and F^* are orthogonal,

(ii) if F has dimension h, then F^* has dimension $k - h$, and

(iii) if $F_1 \subseteq F_2$, then $F_2^* \subseteq F_1^*$.

13.2.3 Basic Facts on Delaunay Polytopes

We group here several basic properties on the symmetry, the number of vertices, and the volume of Delaunay polytopes. We start with an observation from Erdahl [1992] on generating Delaunay polytopes in even lattices.

Lemma 13.2.6. *Let P be a generating Delaunay polytope in an even lattice L. Then, the center of the sphere circumscribing P belongs to the dual lattice L^*. Therefore, an even unimodular lattice contains no generating Delaunay polytope.*

Proof. We can suppose that the origin is a vertex of P. Let c denote the center of the sphere S circumscribing P. Since L is generated by $V(P)$, it suffices to check that $c^T v \in \mathbb{Z}$ for each $v \in V(P)$, for showing that $c \in L^*$. For $v \in V(P)$, $(c-v)^2 = c^2$, i.e., $2c^T v = v^2$, implying that $c^T v \in \mathbb{Z}$ since v^2 is even. If L is even unimodular, then $c \in L^* = L$, contradicting the fact that S is an empty sphere in L. ∎

Let S be a sphere with center c. For $x \in S$, its *antipode* on S is the point $x^* := 2c - x$. It is immediate to see that:

Lemma 13.2.7. *For a Delaunay polytope P, one of the following assertions* (i) *or* (ii) *holds.*

(i) $v^* \in V(P)$ *for all* $v \in V(P)$.

(ii) $v^* \notin V(P)$ *for all* $v \in V(P)$. ∎

In case (i), we say that P is *centrally symmetric*[3] and, in case (ii), that P is *asymmetric.*

Proposition 13.2.8. *Every Delaunay polytope P in $\mathbb{R}^k$ has at most 2^k vertices.*

Proof. Without loss of generality, we can suppose that the origin is a vertex of P. Let $\{v_1, \ldots, v_k\}$ be a basis of the lattice $L = L(P)$. We consider the following equivalence relation on L: For $u, v \in L$, set $u \sim v$ if $u + v \in 2L$. Clearly, every vertex of P is in relation by $\sim$ with one of the elements $\sum_{i \in I} v_i$ for $I \subseteq \{1, \ldots, k\}$. On the other hand, no two vertices of P are in relation by $\sim$. Indeed, if $u \sim v$ for $u, v \in V(P)$, then $\frac{u+v}{2} \in L$, contradicting the fact that the sphere circumscribing P is empty in L. This shows that P has at most 2^k vertices. ∎

Let u, v, w be vertices of a Delaunay polytope P. One can check that

$$(u-w)^2 \leq (u-v)^2 + (v-w)^2. \tag{13.2.9}$$

This is the triangle inequality, expressing the fact that the Delaunay polytope space $(V(P), d^{(2)})$ is a semimetric space. Actually, we will see in Proposition 14.1.2 that every Delaunay polytope space is hypermetric, which is a much stronger property. The inequality (13.2.9) means that the points u, v, w form a triangle with no obtuse angles. The problem of determining the maximum cardinality of a set points in $\mathbb{R}^k$, any three of which form a triangle with no obtuse angle, was first posed by Erdös [1948, 1957], who conjectured that this maximum cardinality is 2^k. This conjecture was proved by Danzer and Grünbaum [1962]. Therefore, the inequality (13.2.9) is already sufficient for proving the upper bound 2^k on the number of vertices of a Delaunay polytope in $\mathbb{R}^k$.

[3] This coincides with the definition given earlier for centrally symmetric sets, up to a translation of the center of the sphere circumscribing P to the origin.

The following upper bound on the volume of a Delaunay polytope was observed by Lovász [1994].

Proposition 13.2.10. *Let P be a Delaunay polytope in a lattice L with volume* $\text{vol}(P)$. *Then,* $\text{vol}(P) \leq \det(L)$.

Proof. The bound $\text{vol}(P) \leq \det(L)$ follows from the fact that the polytopes $P+u$ ($u \in L$) form a packing, i.e., that their interiors are pairwise disjoint. ∎

13.2.4 Construction of Delaunay Polytopes

Clearly, every face of a Delaunay polytope is again a Delaunay polytope (in the space affinely spanned by that face). We present here some further methods for constructing Delaunay polytopes; namely, by taking suitable sections of the sphere of minimal vectors in a lattice, by direct product and by pyramid or bipyramid extension. We then give the complete classification of Delaunay polytopes in dimension $k \leq 4$.

Construction by Sectioning the Sphere of Minimal Vectors in a Lattice. Let L be a lattice in $\mathbb{R}^k$ with $0 \in L$ and let L_{min} be the set of minimal vectors of L. Given noncollinear vectors $a, b \in \mathbb{R}^k$ and some nonzero scalars α, β, set

$$V_a := \{v \in L_{\text{min}} \mid v^T a = \alpha\} \text{ and } V_b := \{v \in L_{\text{min}} \mid v^T b = \beta\}.$$

The following construction, taken from Deza, Grishukhin and Laurent [1992], can be easily checked; it will be applied on several occasions in Sections 16.2, 16.3 and 16.4.

Lemma 13.2.11. *If the sets V_a and $V_a \cap V_b$ are not empty, then the polytopes* $\text{Conv}(V_a)$ *and* $\text{Conv}(V_a \cap V_b)$ *are Delaunay polytopes.* ∎

Direct Product. Let L_i be a lattice in $\mathbb{R}^{k_i}$ and let P_i be a Delaunay polytope in L_i centered at the origin whose circumscribed sphere has radius r_i, for $i = 1, 2$. Then,

$$L := L_1 \times L_2 = \{(v_1, v_2) \mid v_1 \in L_1, v_2 \in L_2\}$$

is a lattice in $\mathbb{R}^k$ ($k = k_1 + k_2$) and

$$P := P_1 \times P_2 = \{(v_1, v_2) \mid v_1 \in P_1, v_2 \in P_2\}$$

is a Delaunay polytope in L whose circumscribed sphere is centered in the origin and has radius $r = \sqrt{r_1^2 + r_2^2}$. Therefore, the direct product of two Delaunay polytopes is again a Delaunay polytope. The direct product of P and a segment α_1 is called the *prism with base P*.

Call a Delaunay polytope *reducible* if it is the direct product of two other nontrivial (i.e., not reduced to a point) Delaunay polytopes and *irreducible* otherwise. Note that irreducible Delaunay polytopes arise in irreducible lattices.

Pyramid and Bipyramid. Let P be a polytope and let v be a point that does not lie in the affine space spanned by P, then

$$\mathrm{Pyr}_v(P) := \mathrm{Conv}(P \cup \{v\})$$

is called the *pyramid with base* P and *apex* v. Under some conditions, the pyramid of a Delaunay polytope is still a Delaunay polytope.

Namely, let P be a Delaunay polytope with radius r, suppose that v is at squared distance t from all the vertices of P and that $t > 2r^2$. Then, the pyramid $\mathrm{Pyr}_v(P)$ is a Delaunay polytope with radius $R = \frac{t}{2\sqrt{t-r^2}}$ (see Proposition 14.4.6).

Moreover, if P is centrally symmetric and if $t = 2r^2$, then the *bipyramid*

$$\mathrm{Bipyr}_v(P) := \mathrm{Conv}(P \cup \{v, v^*\})$$

is a Delaunay polytope with radius r, where v^* is the antipode of v on the sphere circumscribing $\mathrm{Pyr}_v(P)$ (see Proposition 14.4.6).

The Layerwise Construction. The following *layerwise construction* for Delaunay polytopes is described in Ryshkov and Erdahl [1989]. In fact, rather than a construction, it is a way of visualizing a given k-dimensional Delaunay polytope in a lattice L as the convex hull of its sections by the $(k-1)$-dimensional layers composing L.

Let L be a k-dimensional lattice and let $(v_1, \ldots, v_k)$ be a basis of L. Then, $L_0 := \mathbb{Z}(v_1, \ldots, v_{k-1})$ is a $(k-1)$-dimensional sublattice of L and $L = \bigcup_{a \in \mathbb{Z}}(L_0 + av_k)$. The layers $L_0 + av_k$ ($a \in \mathbb{Z}$) are affine translates of L_0 lying in parallel hyperplanes.

Let P be a k-dimensional Delaunay polytope, let L denote the lattice generated by $V(P)$, and let S be the sphere circumscribing P. Let F be a facet of P and let H denote the hyperplane spanned by F. Then, $L_0 := L \cap H$ is a $(k-1)$-dimensional sublattice of L and L is composed by the layers $L_0 + av$ ($a \in \mathbb{Z}$) for some $v \in L - L_0$. Therefore, $P = \mathrm{Conv}(\bigcup_{a \in \mathbb{Z}}(S \cap (L_0 + av)))$, where $S \cap L_0$ is the set of vertices of F and, for $a \in \mathbb{Z}$, $S \cap (L_0 + av)$ is empty or is the set of vertices of a face of a Delaunay polytope in L_0. So, we have the following result:

Proposition 13.2.12. *For each k-dimensional Delaunay polytope P, there exists a $(k-1)$-dimensional lattice L_0, an integer $p \geq 1$, and a sequence $F_0, F_1, \ldots, F_p$ of polytopes that are faces of Delaunay polytopes in L_0 (where $\dim(F_0) = k-1$, but $F_1, \ldots, F_p$ may be empty) such that $P = \mathrm{Conv}(\bigcup_{0 \leq a \leq p}(F_a + av))$, where v is a vector not lying in the space spanned by L_0.* ∎

For instance, the pyramid construction can be viewed as the above layerwise construction with $p = 1$, with a facet on the layer L_0 and a single point on the layer $L_0 + v$.

Let $p(k)$ denote the smallest number p of polytopes $F_1, \ldots, F_p$ in Proposition 13.2.12 needed for constructing any k-dimensional Delaunay polytope.

Given a lattice L, if P is a Delaunay polytope in L which is a simplex, then its volume is an integer multiple of $\frac{\det(L)}{k!}$ (this can be checked by induction on the dimension). This integer is called the *relative volume* of the simplex P. The maximum relative volume of all simplices that are Delaunay polytopes in any k-dimensional lattice is denoted by $p_0(k)$.

It is shown in Ryshkov and Erdahl [1989] that $p(k) = p_0(k)$ holds. In particular, $p(2) = p(3) = p(4) = 1$, $p(5) = 2$ and $\lfloor \frac{k-1}{2} \rfloor \leq p(k) \leq k!$.

There is a Delaunay polytope of dimension 6, namely the Schläfli polytope 2_{21}, for which the integer p (from Proposition 13.2.12) satisfies $p > 1$. In fact, for 2_{21}, $p = 2$, i.e., three layers are needed to obtain 2_{21} from its 5-dimensional sections. We mention two ways of visualizing 2_{21} via the layerwise construction. In the first construction, L_0 is the root lattice D_5 and the layers L_0, $L_0 + v$, $L_0 + 2v$ carry, respectively, $F_0 = \beta_5$, $F_1 = h\gamma_5$ (the 5-dimensional half-cube) and F_2 which is a single point. In the second construction, L_0 is the root lattice A_5 and the layers carry, respectively, $F_0 = \alpha_5$, $F_1 = J(6,2)$ and $F_2 = \alpha_5$. We refer to Coxeter [1973] for a description of all faces of 2_{21}.

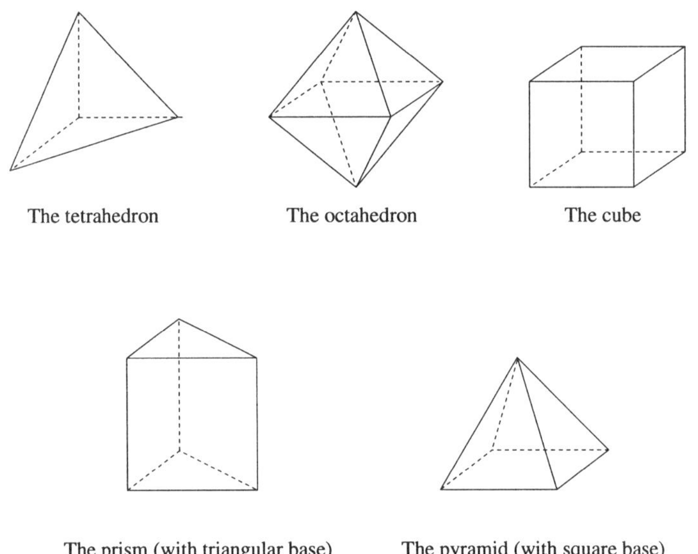

Figure 13.2.13: The five types of Delaunay polytopes in dimension 3

Delaunay Polytopes in Dimension $k \leq 4$. Examples of Delaunay polytopes include the simplex α_k, the cross-polytope β_k, and the hypercube γ_k in every dimension $k \geq 1$. Indeed, $\alpha_k = \text{Pyr}(\alpha_{k-1})$, $\beta_k = \text{Bipyr}(\beta_{k-1})$ and $\gamma_k = \gamma_{k-1} \times \gamma_1$ for $k \geq 2$ and $\alpha_1 = \beta_1 = \gamma_1$ is trivially a Delaunay polytope. We remind that every k-dimensional simplex with no obtuse angles is a Delaunay polytope which is affinely equivalent to α_k; similarly, every k-dimensional parallepiped (with square angles) is a Delaunay polytope which is affinely equivalent to γ_k.

In fact, all the types of Delaunay polytopes of dimension $k \leq 4$ are known. They have been classified by Erdahl and Ryshkov [1987]; we summarize this classification below.

(i) There is only one type of Delaunay polytope of dimension $k = 1$, namely, the segment $\alpha_1 = \beta_1 = \gamma_1$.

(ii) There are two types of Delaunay polytopes of dimension $k = 2$, namely, the triangle (with no obtuse angles) α_2 and the rectangle $\beta_2 = \gamma_2$. (Recall Figure 13.2.4.)

(iii) There are five types of Delaunay polytopes of dimension $k = 3$. They are the tetrahedron α_3, the octahedron β_3, the cube γ_3, the prism with triangular base (i.e., $\alpha_2 \times \alpha_1$) and the pyramid with square base (i.e., $\text{Pyr}(\gamma_2)$). (See Figure 13.2.13.)

(iv) There are 19 types of Delaunay polytopes of dimension $k = 4$. They are described in Tables V and VII from Erdahl and Ryshkov [1987]. Among them, 13 can be obtained from the Delaunay polytopes of dimension 1, 2 or 3 by applying the direct product, pyramid and bipyramid constructions, as indicated below.
- Using the pyramid construction, we obtain the pyramids with base α_3 (this gives α_4), with base β_3, with base γ_3, with base the triangular prism, and with base the squared base pyramid.
- Using the bipyramid construction, we obtain the bipyramids with base β_3 (this gives β_4) and with base γ_3.
- By taking the direct product of the 3-dimensional Delaunay polytopes with α_1, we obtain the prisms with base α_3, with base β_3, with base γ_3 (this gives γ_4), with base the triangular prism, and with base the squared base pyramid.
- By taking the direct product of two 2-dimensional Delaunay polytopes, we obtain $\alpha_2 \times \alpha_2$. (Indeed, $\alpha_2 \times \gamma_2$ and $\gamma_2 \times \gamma_2$ have already been mentioned.)
In addition, we have the repartitioning polytope $P^0_{2,2}$ (associated with the pentagonal facet; see Section 15.2) which is one more Delaunay polytope of dimension 4; it is the polytope A in the Table VI from Erdahl and Ryshkov [1987]. The remaining five Delaunay polytopes are those numbered 4, 5, 6, 9 and 13 in Table V from Erdahl and Ryshkov [1987].

The number of combinatorial types of Voronoi polytopes is also known in small dimension $k \leq 4$; for $k = 2$, this number is 2 (rectangles and hexagons being the two possibilities; recall Figure 13.2.4), it is 5 for $k = 3$ and 52 for $k = 4$. It is conjectured to be about 75000 for $k = 5$. (See Waldschmidt, Moussa, Luck and Itzykson [1992], p.479.)

13.2.5 Additional Notes

Lattices and Positive Semidefinite Matrices. Let $p \in \text{PSD}_n$ and $(p_{ij})_{i,j=1}^n$ be the corresponding positive semidefinite matrix (setting $p_{ji} = p_{ij}$). By Lemma 2.4.2, there exist n vectors $v_1, \ldots, v_n \in \mathbb{R}^k$ ($1 \leq k \leq n$) such that $p_{ij} = v_i^T v_j$ for all $i, j = 1, \ldots, n$, where k is the rank of the system $(v_1, \ldots, v_n)$ and of the matrix $(p_{ij})_{i,j=1}^n$. So, $k = n$ if p is positive definite (i.e., if p lies in the interior of PSD_n) and $k < n$ otherwise. Set $L := \mathbb{Z}(v_1, \ldots, v_n)$. Sometimes, L is a lattice. This is the case, in particular, if p is positive definite.

There is a many-to-one correspondence between the positive definite matrices $p \in \text{PSD}_n$ and the lattices in $\mathbb{R}^n$. Indeed, the action on p of the group $GL(n, \mathbb{Z})$ of integral unimodular transformations produces distinct bases of the same lattice L. The following was proved by Voronoi [1908, 1909] (we follow Erdahl and Ryshkov [1988] for the exposition): The action of $GL(n, \mathbb{Z})$ induces a partition of the cone PSD_n into disjoint relatively open convex subcones, called the *L-type domains*, of dimension $1, 2, \ldots, \binom{n+1}{2}$, and having the following properties:

(i) On each of these subcones the affine structure of the L-decompositions of corresponding lattices is constant, i.e., the lattices corresponding to the points of a given subcone are all z-equivalent.

(ii) Subcones of dimension $\binom{n+1}{2}$ correspond to general lattices, i.e. having simplicial L-decompositions. These L-type domains are polyhedral.

(iii) A subcone of dimension less than $\binom{n+1}{2}$ is a relatively open face of two or more L-type domains. If such a cone makes contact with the boundary of an L-type domain, then it is necessarily a face of that domain. The lattice corresponding to a quadratic form on such a face is special, i.e., it has among its Delaunay polytopes some that are not simplices.

Voronoi [1908, 1909] showed that, in any given dimension k, the number of distinct (up to z-equivalence) k-dimensional lattices is finite. Therefore, many of the L-type domains correspond to z-equivalent lattices.

Delaunay Polytopes and Empty Ellipsoids. As we will see in Section 14.1, the study of the hypermetric spaces on n points amounts to the study of the Delaunay polytopes of dimension $k \leq n-1$. It is also closely related to that of empty ellipsoids. Indeed, empty ellipsoids, which arise as the solution sets of the quadratic functions that are nonnegative on integer variables, are nothing but affine images of empty spheres in lattices.

There is a sequence of papers by Erdahl [1974, 1987, 1992] and by Erdahl and Ryshkov [1987, 1988] studying the set of integer solutions of equations of the form:

(13.2.14) $$f(x) := a_0 + \sum_{1 \leq i \leq n} a_i x_i + \sum_{1 \leq i,j \leq n} a_{ij} x_i x_j = 0,$$

where $a_0, a_1, \ldots, a_n \in \mathbb{R}$, $a_{ij} = a_{ji} \in \mathbb{R}$, and f satisfies the condition:

(13.2.15) $$f(x) \geq 0 \text{ for all } x \in \mathbb{Z}^n.$$

The set of integer solutions of $f(x) = 0$ is called the *root figure* of f and is denoted by R_f. From relation (13.2.15), the matrix $A_f := (a_{ij})_{1 \leq i,j \leq n}$ is positive semidefinite and the region $\{x \in \mathbb{R}^n \mid f(x) < 0\}$ is free of integral points. Set

$$E_f := \{x \in \mathbb{R}^n \mid f(x) = 0\}.$$

Suppose first that A_f is positive definite. Then, the set E_f is an ellipsoid whose interior is free of integral points; the ellipsoid E_f is said to be *empty* in $\mathbb{Z}^n$. Hence, the root figure R_f consists of the integral points lying on E_f and, thus, is finite. In fact, the root figure R_f is affinely equivalent to the set of vertices V of a Delaunay polytope, with $\dim(V) = \dim(R_f) \leq n$. Moreover, every finite root figure arises in this way. (See Erdahl [1992].)

Suppose now that A_f has a nonzero kernel[4] V. Erdahl [1992] shows that V contains a basis composed only of integral vectors. Then, the set E_f contains infinite directions, as $f(x+v)=f(x)$ holds for all $x \in \mathbb{R}^n$, $v \in V$. If W is a subspace complement of V in $\mathbb{R}^n$, then $E_f = V + E'$ where $E' := E_f \cap W$ is an ellipsoid in W. Hence, E_f can be seen as a "cylinder" with axis V and ellipsoidal section; we say that E_f is a *degenerate ellipsoid with axis* V. In this case, the root figure R_f is also infinite. In fact, every infinite root figure arises from the finite ones by a simple construction; essentially, every infinite root figure is of the form $R+\Gamma$ where R is a finite root figure and Γ is a sublattice of $\mathbb{Z}^n$ (see Theorem 2.1 in Erdahl [1992]).

Therefore, the study of the root figures amounts to the classification of the Delaunay polytopes of dimension $k \leq n$.

Finally, consider the cone:

$$Q^+(\mathbb{Z}^n) := \quad \{a=(a_0,a_1,\ldots,a_n,a_{ij}, 1\leq i\leq j\leq n) \mid a_0 + \sum_{1\leq i\leq n} a_i x_i + \sum_{1\leq i,j\leq n} a_{ij}x_ix_j \geq 0 \text{ for all } x\in\mathbb{Z}^n\},$$

i.e., each member $a \in Q^+(\mathbb{Z}^n)$ corresponds to a function f_a satisfying (13.2.14) and (13.2.15). Erdahl [1992] shows that every $a \in Q^+(\mathbb{Z}^n)$ lying on an extreme ray of $Q^+(\mathbb{Z}^n)$ satisfies one of the following:

(i) f_a is constant (i.e., $a_1 = \ldots = a_n = a_{ij} = 0$),

(ii) $f_a(x) = (\sum_{1\leq i\leq n} \alpha_i x_i + \eta)^2$ where $(\alpha_1,\ldots,\alpha_n)$ is not proportional to an integer vector,

(iii) f_a is perfect which, in the terminology of Erdahl [1992], means that the dimension of the set $\{b \in Q^+(\mathbb{Z}^n) \mid R_{f_a} \subseteq R_{f_b}\}$ is equal to 1.

Clearly, the hypermetric cone HYP_{n+1} is (via the covariance mapping) a section of the cone $Q^+(\mathbb{Z}^n)$, as

$$\xi(\mathrm{HYP}_{n+1}) = \{a \in Q^+(\mathbb{Z}^n) \mid a_0 = 0 \text{ and } a_i = -a_{ii} \text{ for } i=1,\ldots,n\}.$$

Note that the notion of root figure corresponds to that of annullator, used in Sections 14.2 and 15.1. Moreover, there is the following link between the perfect elements of $Q^+(\mathbb{Z}^n)$ and the extreme rays of HYP_{n+1}: For $d \in \mathrm{HYP}_{n+1}$, d lies on an extreme ray of HYP_{n+1} if and only if $\xi(d)$ is a perfect element of $Q^+(\mathbb{Z}^n)$.

13.3 Finiteness of the Number of Types of Delaunay Polytopes

Recall that two lattices L, L' are z-equivalent if there exists an affine bijection T such that $L' = T(L)$ and T brings the star of L on the star of L'. (Note that any k-dimensional lattice is *affinely equivalent* to $\mathbb{Z}^k$!) Voronoi [1908, 1909] proved that the number of distinct, up to z-equivalence, k-dimensional lattices is finite. This implies obviously that the number of distinct, up to affine equivalence, k-dimensional Delaunay polytopes is finite. In other words, the number of types of k-dimensional Delaunay polytopes is finite. (Recall that two Delaunay polytopes have the same type if they are affinely equivalent.)

[4]The *kernel* of a matrix A is the set Ker A consisting of all vectors x such that $Ax = 0$.

We give here a direct proof of the finiteness of the number of types of Delaunay polytopes in $\mathbb{R}^k$ since Voronoi's original proof is very involved; it is taken from Deza, Grishukhin and Laurent [1993].

Let γ be a type of Delaunay polytopes of dimension k. A subset $B \subseteq \mathbb{R}^k$ is called a *representative basis* of γ if there exist a Delaunay polytope P of type γ and a lattice $L \subseteq \mathbb{R}^k$ which contains the set $V(P)$ of vertices of P and admits B as a basis. (Note that L may be larger than the lattice $L(P)$ generated by the set of vertices of P.)

Suppose that P has N vertices and let Q_P denote the $N \times k$ matrix whose rows are the vectors $v \in V(P)$. Let M_B denote the $k \times k$ matrix whose rows are the members of B. Then, there exists an integer $N \times k$ matrix Y_γ such that

$$Q_P = Y_\gamma M_B. \tag{13.3.1}$$

If P' is another Delaunay polytope of type γ, i.e., if $P' = T(P)$ for some affine bijection T, then the relation

$$Q_{T(P)} = Y_\gamma M_{T(B)}$$

holds. Hence, the matrix Y_γ characterizes the type γ (once a representative basis B has been chosen). The next result shows that, for each type γ, one can choose a representative basis B in such a way that the matrix Y_γ has a very special form, which will imply that there are only finitely many possibilities for Y_γ.

Proposition 13.3.2. *Let γ be a type of Delaunay polytopes of dimension k. One can choose a representative basis B of γ in such a way that the matrix Y_γ satisfies the following relations:*

(i) *There exists a $k \times k$ submatrix $D = (\alpha_{ij})_{1 \leq i,j \leq k}$ of Y_γ which is lower triangular and satisfies: $0 \leq \alpha_{ij} < \alpha_{ii}$ for all $1 \leq j < i \leq k$.*

(ii) *$p = |\det(D)|$ is the maximum possible value of the absolute value of the determinant of any $k \times k$ submatrix of Y_γ.*

(iii) *$p \leq k!$.*

For the proof, we need the following classical result about lattices from Cassels [1959].

Proposition 13.3.3. *Let L, L' be two k-dimensional lattices in $\mathbb{R}^k$ such that $L' \subseteq L$. For every basis $\{a_1, \ldots, a_k\}$ of L', there exists a basis $\{b_1, \ldots, b_k\}$ of L such that*

$$a_i = \alpha_{i1} b_1 + \ldots + \alpha_{ii} b_i,$$

for $i = 1, \ldots, k$, where $(\alpha_{ij})_{1 \leq i,j \leq k}$ are integers satisfying

$$0 \leq \alpha_{ij} < \alpha_{ii}$$

for all $1 \leq j < i \leq k$. ∎

Proof of Proposition 13.3.2. Let P be a Delaunay polytope of type γ with N vertices and let L be a lattice in $\mathbb{R}^k$ containing the set of vertices $V(P)$ of P. Let V_0 be a subset of $V(P)$ of size k and let Q_0 denote the $k \times k$ submatrix of Q_P whose rows are the members of V_0. We choose V_0 in such a way that $|\det(Q_0)|$ is largest possible. We can suppose that Q_0 is the submatrix of Q_P formed by its first k rows. The lattice $L' := \mathbb{Z}(V_0)$ is a sublattice of L and admits V_0 as a basis. Applying Proposition 13.3.3, we deduce the existence of a basis B of L such that

$$Q_0 = DM_B,$$

where D is a lower triangular integer matrix satisfying Proposition 13.3.3 (ii). Let us choose B as representative basis for the type γ. Then, as $Q_P = (Q_P M_B^{-1})M_B$, by comparing with relation (13.3.1), we obtain that $Q_P(M_B)^{-1}$ coincides with the integer matrix Y_γ. Note that the matrix $Q_P(M_B)^{-1}$ is an $N \times k$ matrix whose first k rows form the matrix D with

$$p = |\det(D)| = \frac{|\det(Q_0)|}{|\det(M_B)|} = \frac{|\det(Q_0)|}{\det(L)}.$$

Hence, by the choice of Q_0, the absolute value of the determinant of any $k \times k$ submatrix of $Y_\gamma = Q_P(M_B)^{-1}$ is less than or equal to p. Therefore, Y_γ satisfies the conditions (i),(ii) of Proposition 13.3.2.

Finally, we check (iii). Let Δ denote the simplex whose vertices are the members of V_0, i.e., the rows of Q_0. Then, Δ is contained in P and, thus, $\mathrm{vol}(\Delta) \leq \mathrm{vol}(P)$. But,

$$\mathrm{vol}(\Delta) = \frac{|\det(Q_0)|}{k!} = \frac{p\ \det(L)}{k!} \quad \text{and} \quad \mathrm{vol}(P) \leq \det(L)$$

from Proposition 13.2.10. This implies that $p \leq k!$. ∎

We can now show the finiteness of the number of types of Delaunay polytopes in $\mathbb{R}^k$.

Theorem 13.3.4. *The number of types of Delaunay polytopes in* $\mathbb{R}^k$ *is finite.*

Proof. Every type γ of Delaunay polytopes in $\mathbb{R}^k$ with N vertices is characterized by an $N \times k$ integer matrix Y_γ satisfying Proposition 13.3.2 (i)-(iii). It suffices to show that there is only a finite number of such matrices. For this, we show that, for fixed p, there is only a finite number of matrices satisfying Proposition 13.3.2 (i)-(ii).

Let Y be an $N \times k$ integer matrix satisfying Proposition 13.3.2 (i),(ii). Suppose that D is the upper $k \times k$ submatrix of Y. Then, the upper $k \times k$ submatrix of YD^{-1} is the identity matrix. Let r_{ih} be a nonzero entry of YD^{-1}, where $k+1 \leq i \leq N$ and $1 \leq h \leq k$. Let C denote the matrix obtained from

D by replacing its h-th row by the i-th row of Y. By Proposition 13.3.2 (ii), $|\det(C)| \leq p$. On the other hand, $|\det(CD^{-1})| = |r_{ih}|$, implying that

$$|r_{ih}| = \frac{|\det(C)|}{p} \in \left\{0, \frac{1}{p}, \ldots, \frac{p-1}{p}, 1\right\}.$$

Since YD^{-1} is an $N \times k$ matrix with $N \leq 2^k$ (from Proposition 13.2.8), we deduce that, for fixed p and k, there is only a finite number of such matrices YD^{-1}. Now, D is a $k \times k$ integer matrix with $p = \alpha_{11} \ldots \alpha_{kk}$ and satisfying Proposition 13.3.2 (i); therefore, there is only a finite number of such matrices D. Consequently, there is a finite number of possibilities for Y. ∎

Chapter 14. Hypermetrics and Delaunay Polytopes

In this chapter we establish the fundamental connection existing between hypermetric spaces and Delaunay polytopes. Hence, for a hypermetric distance space (X, d), one may speak of its associated Delaunay polytope P_d; the case when (X, d) is ℓ_1-embeddable corresponding to the case when P_d can be embedded in a parallepiped. As an application of this connection, one can show polyhedrality of the hypermetric cone; several proofs for this fact are given in Section 14.2. As another application (and using the classification of the irreducible root lattices), one can characterize the graphs whose shortest path metric is hypermetric or ℓ_1-embeddable. Such graphs arise essentially from cocktail-party graphs, half-cube graphs, and a single graph on 56 nodes (the Gosset graph) by taking Cartesian products and isometric subgraphs (see Section 14.3). We group in Section 14.4 several results concerning spherical representations of distance spaces and the radius of Delaunay polytopes.

14.1 Connection between Hypermetrics and Delaunay Polytopes

In this section, we establish the fundamental connection existing between hypermetric spaces and Delaunay polytopes. This connection was discovered by Assouad [1982, 1984]; it is stated in Theorem 14.1.3, whose proof is based on the next two propositions.

Proposition 14.1.1. *Let $c, v_0 := 0, v_1, \ldots, v_n \in \mathbb{R}^k$ be vectors satisfying*

(i) $\| v_i - c \| = \| c \|$ *for* $1 \leq i \leq n$,

(ii) $\| \sum_{1 \leq i \leq n} b_i v_i - c \| \geq \| c \|$ *for all* $b \in \mathbb{Z}^n$.

Then, the set $L := \mathbb{Z}(v_1, \ldots, v_n)$ is a lattice.

Proof. For $b \in \mathbb{Z}^n$, set $v(b) := \sum_{1 \leq i \leq n} b_i v_i$; then, $v(b) \pm v_i \in L$. Hence, (ii) yields

$$(v_i \pm v(b) - c)^2 \geq c^2, \quad \text{i.e.,} \ (v_i - c)^2 + (v(b))^2 \pm 2(v_i - c)^T v(b) \geq c^2$$

which, together with (i), implies

(a) $$(v(b))^2 \geq 2|(v_i - c)^T v(b)| \text{ for } 1 \leq i \leq n.$$

Consider the units vectors

$$e_i := \frac{v_i - c}{\| c \|} \text{ for } i = 0, 1, \ldots, n, \quad \text{and } e(b) := \frac{v(b)}{\| v(b) \|}.$$

Set

$$\beta := \min\{\max(e_i^T u \mid 1 \leq i \leq n) \mid u \in \mathbb{R}^k, \| u \| = 1\}.$$

We show that $\beta > 0$. Then, (a) will imply that $\| v(b) \| \geq 2\beta \| c \|$ for all $b \in \mathbb{Z}^n$ such that $v(b) \neq 0$; in other words, the open ball centered at the origin with radius $2\beta \| c \|$ contains no other lattice point besides the origin, which shows that L is a lattice. Suppose that $\beta = 0$. Then, we can find a sequence $(u_p)_{p\geq 1}$ of unit vectors of $\mathbb{R}^k$ such that $|e_i^T u_p| \leq \frac{1}{p}$ for any $1 \leq i \leq n$, $p \geq 1$. By the compactness of the unit sphere, we can suppose that the sequence $(u_p)_{p\geq 1}$ admits a limit u when p goes to infinity (replacing, if necessary, $(u_p)_{p\geq 1}$ by a subsequence). Therefore, $\| u \| = 1$, while $e_i^T u = 0$ for $i = 1, \ldots, n$, implying that $u = 0$ since the vectors $v_1, \ldots, v_n$ span $\mathbb{R}^k$. We have a contradiction. This shows that $\beta > 0$. ∎

Proposition 14.1.2. *Let (X, d) be a distance space, $X = \{0, 1, \ldots, n\}$. The following assertions are equivalent.*

(i) *(X, d) is hypermetric.*

(ii) *(X, d) has a representation $i \in X \mapsto v_i \in \mathbb{R}^k$ (where $k \leq n$) on a sphere S which is empty in the set*

$$L_{af}(X, d) := \{\sum_{i \in X} b_i v_i \mid b \in \mathbb{Z}^X \text{ and } \sum_{i \in X} b_i = 1\}.$$

Proof. (i) $\Longrightarrow$ (ii) Since (X, d) is of negative type, we know from Proposition 13.1.2 that (X, d) has a representation $v_0, v_1, \ldots, v_n \in \mathbb{R}^k$ for some $1 \leq k \leq n$. Moreover, the system $(v_0, \ldots, v_n)$ has rank k and we can suppose without loss of generality that $v_0 = 0$. We first show that the vectors $v_0, v_1, \ldots, v_n$ lie on a sphere, i.e., that there exists $c \in \mathbb{R}^k$ such that

$$\text{(a)} \qquad 2c^T v_i = v_i^2 \text{ for } 1 \leq i \leq n.$$

If $k = n$, then the vectors $v_1, \ldots, v_n$ are linearly independent and, therefore, the system (a) admits a unique solution c. Suppose that $k \leq n - 1$. Let M denote the $n \times k$ matrix whose rows are the vectors $v_1, \ldots, v_n$, let U denote the subspace of $\mathbb{R}^n$ spanned by the columns of M and set $f := (v_1^2, \ldots, v_n^2)^T$. The system (a) has a solution if and only if $f \in U$ or, equivalently, if $f^T g = 0$ for each $g \in U^\perp$ ($U^\perp$ is the orthogonal complement of U in $\mathbb{R}^k$). Take $g \in U^\perp$, let $b \in \mathbb{Z}^n$ such that $|g_i - b_i| < 1$ for $i = 1, \ldots, n$ and set $\delta := g - b$; so δ belongs to the unit cube. Set $p := \xi(d)$. Then, $p_{ij} = v_i^T v_j$ for $1 \leq i < j \leq n$. Using relation (13.1.4), we deduce that

$$\sum_{1 \leq i,j \leq n} b_i b_j p_{ij} - \sum_{1 \leq i \leq n} b_i p_{ii} \geq 0, \quad \text{i.e.,}$$

(b) $$(\sum_{1\le i\le n} b_i v_i)^2 - \sum_{1\le i\le n} b_i v_i^2 \ge 0.$$

Therefore,

$$f^T b = \sum_{1\le i\le n} b_i v_i^2 \le (\sum_{1\le i\le n} b_i v_i)^2 = (b^T M)^2 = (g^T M - \delta^T M)^2 = (\delta^T M)^2,$$

since $g \in U^{\perp}$. Hence, $f^T b \le (\delta^T M)^2$, implying that $f^T g \le f^T \delta + (\delta^T M)^2$. This implies that $f^T g = 0$; otherwise, the left hand side of the latter inequality could be made arbitrarily large while its right hand side is bounded. Note that the solution c to the system (a) is unique since $(v_1, \dots, v_n)$ has full rank k.

From (a) and (b), we deduce that

$$(\sum_{i\in X} b_i v_i - c)^2 \ge c^2$$

for all $b \in \mathbb{Z}^X$ with $\sum_{i\in X} b_i = 1$. This shows that the sphere S with center c and radius $\| c \|$ is empty in $L_{af}(X,d)$.
(ii) $\Longrightarrow$ (i) Let $b \in \mathbb{Z}^X$ with $\sum_{i\in X} b_i = 1$. Then,

$$\begin{aligned}
&\sum_{i,j\in X} b_i b_j d(i,j) = \sum_{i,j\in X} b_i b_j (v_i - v_j)^2 \\
&= \sum_{i,j\in X} b_i b_j (v_i - c + c - v_j)^2 = \sum_{i,j\in X} b_i b_j (2r^2 - 2(v_i - c)^T (v_j - c)) \\
&= 2r^2 - 2(\sum_{i\in X} b_i (v_i - c))^2 = 2(r^2 - (\sum_{i\in X} b_i v_i - c)^2) \le 0
\end{aligned}$$

since the sphere S is empty in $L_{af}(X,d)$. This shows that (X,d) is hypermetric. ∎

As a consequence of Propositions 14.1.1 and 14.1.2, we have the next theorem which summarizes the connection existing between hypermetrics and Delaunay polytopes.

Theorem 14.1.3. *Let (X,d) be a hypermetric space, $|X| = n+1$. There exist a k-dimensional Delaunay polytope P_d in $\mathbb{R}^k$, for some $1 \le k \le n$, and a mapping*

$$f_d : i \in X \mapsto v_i \in V(P_d)$$

which is generating, which means that the set $\{v_i \mid i \in X\}$ generates the set of vertices $V(P_d)$ of P_d, and such that

$$d(i,j) = (v_i - v_j)^2 \text{ for } i,j \in X.$$

Moreover, the pair (P_d, f_d) is unique, up to translation and orthogonal transformation. ∎

We refer to P_d as the *Delaunay polytope associated with the hypermetric space* (X,d), the lattice

$$\mathbb{Z}_{af}(V(P_d)) = \{ \sum_{v\in V(P_d)} b_v v \mid b \in \mathbb{Z}^{V(P_d)} \text{ and } \sum_{v\in V(P_d)} b_v = 1\}$$

is denoted as L_d and the sphere circumscribing P_d as S_d.

As another application of Propositions 14.1.1 and 14.1.2, we obtain the following characterization for Delaunay polytopes.

Proposition 14.1.4. *Let P be a polytope of dimension k in $\mathbb{R}^k$. Then, P is a Delaunay polytope if and only if the following assertions hold.*

(i) *The distance space $(V(P), d^{(2)})$ is hypermetric.*

(ii) *If Q is a polytope of dimension k in $\mathbb{R}^k$ such that*

 (iia) $V(P) \subseteq V(Q)$,
 (iib) $\mathbb{Z}_{af}(V(P)) = \mathbb{Z}_{af}(V(Q))$, *and*
 (iic) *the distance space $(V(Q), d^{(2)})$ is hypermetric,*

 then P and Q coincide. ∎

For instance, take for Q a square (2-dimensional hypercube) and for P the triangle having as vertices three of the vertices of Q. Then, P satisfies (i), but not (ii). Indeed, the triangle P is not a Delaunay polytope as it has a right angle. Recall that a triangle is a Delaunay polytope if and only if it has no obtuse angle. On the other hand, there exist pairs of Delaunay polytopes (P,Q) of the same dimension in $\mathbb{R}^k$ and satisfying (iia), (iic), but not (iib). Such an example is given in Section 16.4 for the Barnes-Wall lattice (see the pair (Q,P) there).

Note that we may assume to be dealing with hypermetric distances taking nonzero distances between distinct points (i.e., with metrics). Indeed, let (X,d) be a hypermetric space with $d(i_0,j_0) = 0$ for two distinct points $i_0, j_0 \in X$, and let $(X' := X \setminus \{j_0\}, d)$ denote its subspace on $X \setminus \{j_0\}$. Then, both (X,d) and (X',d) have the same associated Delaunay polytope P (simply representing j_0 by the same vertex of P as i_0).

We would like to emphasize the following fact, since it will be often used in the sequel.

Proposition 14.1.5. *Let (X,d) be a hypermetric space with representation $(v_i \mid i \in X)$ in the set of vertices of its associated Delaunay polytope P_d. Given $b \in \mathbb{Z}^X$ with $\sum_{i\in X} b_i = 1$,*

$$\sum_{i,j\in X} b_i b_j d(i,j) = 0 \iff \sum_{i\in X} b_i v_i \text{ is a vertex of } P_d.$$

Proof. This follows from the equality

$$\sum_{i,j \in X} b_i b_j d(i,j) = 2(r^2 - (\sum_{i \in X} b_i v_i - c)^2),$$

stated in the proof of Proposition 14.1.2 (ii) $\Longrightarrow$ (i). ∎

Example 14.1.6. Consider the cut semimetric $\delta(S)$ for some subset $S \subseteq X$. It is obviously hypermetric. Its associated Delaunay polytope is the segment $\alpha_1 = [0,1]$ and a representation of the hypermetric space $(X, \delta(S))$ is

$$i \in S \mapsto v_i := 1, \ i \in X \setminus S \mapsto v_i := 0.$$

∎

Example 14.1.7. Let (X,d) be a semimetric space. Then, d lies in the interior of the hypermetric cone HYP(X) if and only if its associated Delaunay polytope is the simplex $\alpha_{|X|-1}$ of dimension $|X|-1$ (since, by Proposition 14.1.5, d satisfies no nontrivial hypermetric equality if and only if $|V(P_d)| = \dim(P_d)+1$). ∎

We conclude this section with two additional properties concerning the connection between hypermetrics and Delaunay polytopes. A first observation is that, if (Y,d) is a subspace of the hypermetric space (X,d), then its associated Delaunay polytope is embedded in the Delaunay polytope associated to (X,d).

Lemma 14.1.8. *Let P be a Delaunay polytope with set of vertices $V(P)$ and let X be a subset of $V(P)$. Let P_X denote the Delaunay polytope associated with the hypermetric space $(X, d^{(2)})$. Then, $V(P_X) \subseteq V(P)$ with equality if and only if X is a generating subset of $V(P)$.*

Proof. Let L_X denote the sublattice of L generated by X and let A_X denote the affine space generated by X. Let S be the circumscribed sphere to P, so S is an empty sphere in L. The sphere $S_X = S \cap A_X$ is empty in L_X. Hence, $P_X = \text{Conv}(S_X \cap L_X)$ is a Delaunay polytope and it is the Delaunay polytope associated with the hypermetric space $(X, d^{(2)})$. Therefore, $V(P_X) = S_X \cap L_X$ is indeed contained in $V(P) = S \cap L$. It is easy to see that $V(P_X) = V(P)$ if and only if X generates the lattice L. ∎

In particular, every face of a Delaunay polytope is a Delaunay polytope. For instance, every 2-dimensional face of a Delaunay polytope is a rectangle or a triangle with no obtuse angles.

Corollary 14.1.9. *Let (X,d) be a hypermetric space and (Y,d) be a subspace of (X,d), i.e., $Y \subseteq X$. Let P_X and P_Y denote the Delaunay polytopes associated, respectively, with (X,d) and (Y,d). Then, $V(P_Y) \subseteq V(P_X)$ holds.* ∎

There are some properties of the hypermetric space (X,d) which are inherited by its associated Delaunay polytope. This is the case for hypercube or

ℓ_1-embeddability as shown in Proposition 14.1.10 below; (i) is proved in Assouad [1982] and (ii) in Deza and Grishukhin [1993]. Another such property is the notion of rank and extremality as will be seen in Section 15.1.

Proposition 14.1.10. *Let (X, d) be a hypermetric space and let P_d be its associated Delaunay polytope with set of vertices $V(P_d)$.*

(i) *Then, (X, d) is isometrically ℓ_1-embeddable if and only if P_d can be embedded in a parallepiped, i.e., if $(V(P_d), d^{(2)})$ is isometrically ℓ_1-embeddable.*

(ii) *Moreover, if d is rational valued, then (X, d) is ℓ_1-embeddable if and only if P_d can be embedded in a hypercube with side length λ for some $\lambda > 0$; the smallest such λ is the minimum scale of both spaces (X, d) and $(V(P_d), d^{(2)})$.*

Proof. (i) The set of vertices of a parallepiped endowed with the distance $d^{(2)}$ is clearly ℓ_1-embeddable. Conversely, suppose that (X, d) is ℓ_1-embeddable. Then,

$$d = \sum_{1 \leq h \leq m} \lambda_h \delta(S_h)$$

where $\lambda_1, \ldots, \lambda_m > 0$ and $S_1, \ldots, S_m \subset X$. Set $e'_h := \sqrt{\lambda_h} e_h$, where e_h is the h-th unit vector in $\mathbb{R}^m$, for $1 \leq h \leq m$. Let L denote the lattice in $\mathbb{R}^m$ generated by $(e'_1, \ldots, e'_m)$. It is easy to check that the sphere S with center $c := \frac{1}{2} \sum_{1 \leq h \leq m} e'_h$ and radius $\| c \|$ is empty in L. For $i = 1, \ldots, n$, set

$$I_i := \{h \in \{1, \ldots, m\} \mid i \in S_h\}.$$

So, $h \in I_i$ if and only if $i \in S_h$. Therefore,

$$d(i,j) = \sum_{1 \leq h \leq m,\ |S_h \cap \{i,j\}| = 1} \lambda_h = \sum_{1 \leq h \leq m,\ h \in I_i \Delta I_j} \lambda_h.$$

From this, we deduce that the mapping

$$i \in X \mapsto v_i = \sum_{h \in I_i} e'_h \in \mathbb{R}^m$$

is a representation of (X, d); indeed,

$$d^{(2)}(v_i, v_j) = (\sum_{h \in I_i \Delta I_j} e'_h)^2 = \sum_{h \in I_i \Delta I_j} \lambda_h = d(i,j).$$

This shows that the Delaunay polytope P_d associated with (X, d) is embedded in the parallepiped spanned by e'_h, $1 \leq h \leq m$. The assertion (ii) can be derived in the same way. ∎

14.2 Polyhedrality of the Hypermetric Cone

The hypermetric cone HYP_{n+1} is defined by infinitely many inequalities. Hence, a natural question is whether a finite subset of them suffices for describing HYP_{n+1} or, in other words, whether the cone HYP_{n+1} is polyhedral. The answer is yes, as stated in the next theorem proved by Deza, Grishukhin and Laurent [1993].

Theorem 14.2.1. *For any $n \geq 2$, the hypermetric cone HYP_{n+1} is polyhedral.*

We present three different proofs for this result. In each of them, the image $\xi(\mathrm{HYP}_n)$ of the hypermetric cone HYP_{n+1} under the covariance mapping ξ is considered instead of the cone HYP_{n+1} itself. The cone $\xi(\mathrm{HYP}_{n+1})$ is shown to be polyhedral in the following three ways: either by showing that it has a finite number of faces, or by showing that is can be decomposed as a finite union of polyhedral cones, or by showing that it coincides with a larger cone defined by a finite subset of its inequalities. Recall from (13.1.4) that $\xi(\mathrm{HYP}_{n+1})$ is the cone defined by the inequalities

$$(14.2.2) \qquad \sum_{1\leq i,j\leq n} b_i b_j p_{ij} - \sum_{1\leq i\leq n} b_i p_{ii} \geq 0$$

for all $b \in \mathbb{Z}^n$.

The first proof was given by Deza, Grishukhin and Laurent [1993]. It is based on the connection existing between faces of the hypermetric cone and types of Delaunay polytopes and it uses as essential tool the fact that the number of types of Delaunay polytopes in any given dimension is finite.

The second proof relies on the fact that the cone $\xi(\mathrm{HYP}_{n+1})$ can be decomposed as a finite union of L-type domains. It uses two results of Voronoi; the first one concerns the finiteness of the number of types of lattices in any given dimension and the second one concerns properties of the partition of the cone PSD_n into L-type domains (in particular, the fact that each L-type domain is a polyhedral cone).

The third proof is due to Lovász [1994]. It consists of proving directly that, among all the inequalities (14.2.2) defining $\xi(\mathrm{HYP}_{n+1})$, only a finite subset of them is necessary; namely, one shows that the inequalities (14.2.2) with b bounded in terms of n are sufficient for the description of $\xi(\mathrm{HYP}_{n+1})$. For each $p \in \xi(\mathrm{HYP}_{n+1})$, the set

$$E_p := \{x \in \mathbb{R}^n \mid \sum_{1\leq i,j\leq n} x_i x_j p_{ij} - \sum_{1\leq i\leq n} x_i p_{ii} = 0\}$$

is an ellipsoid (possibly degenerate with infinite directions) whose interior is free of integral points. The key argument consists of showing that if p lies on the boundary of the cone $\xi(\mathrm{HYP}_{n+1})$, then E_p contains an integral point b which is distinct from 0 and the unit vectors and is "short", which means that all its components are bounded by a constant depending only on n, e.g., $\max_{1\le i\le n} |b_i| \le n!$. A better upper bound than $n!$ is given in Proposition 14.2.4.

FIRST PROOF. Let $d \in \mathrm{HYP}_{n+1}$ and let $p := \xi(d)$. The *annullator* $\mathrm{Ann}(p)$ of p is defined as

$$\mathrm{Ann}(p) = \{b \in \mathbb{Z}^n \mid b \neq 0, e_1, \ldots, e_n \text{ and } \sum_{1\le i,j\le n} b_i b_j p_{ij} - \sum_{1\le i\le n} b_i p_{ii} = 0\},$$

where $e_1, \ldots, e_n$ denote the unit vectors in $\mathbb{R}^n$. Let $F(p)$ denote the smallest face of $\xi(\mathrm{HYP}_{n+1})$ containing p, i.e.,

$$F(p) = \xi(\mathrm{HYP}_{n+1}) \cap \bigcap_{b\in\mathrm{Ann}(p)} H_b,$$

where H_b denotes the hyperplane in $\mathbb{R}^{\binom{n+1}{2}}$ defined by the equation

$$\sum_{1\le i,j\le n} b_i b_j p_{ij} - \sum_{1\le i\le n} b_i p_{ii} = 0.$$

Clearly, showing that $\xi(\mathrm{HYP}_{n+1})$ is polyhedral amounts to showing that the number of its distinct faces is finite or, equivalently, that the number of distinct annullators $\mathrm{Ann}(p)$ (for $p \in \xi(\mathrm{HYP}_{n+1})$) is finite.

Let P_d denote the Delaunay polytope associated with d, let L_d be the associated lattice and let $i \in X \mapsto v_i \in V(P_d)$ be the representation of (X, d) on the sphere S_d with center c circumscribing P_d. We can assume that $v_0 = 0$; then, $L_d = \mathbb{Z}(v_1, \ldots, v_n)$. For $v \in L_d$, set

$$Z(v) := \{b \in \mathbb{Z}^n \mid v = \sum_{1\le i\le n} b_i v_i\}.$$

Then, from Proposition 14.1.5,

$$\text{(a)} \qquad \mathrm{Ann}(p) \cup \{0, e_1, \ldots, e_n\} = \bigcup_{v\in V(P_d)} Z(v).$$

Suppose that the polytope P_d has type γ. Let B be a representative basis of the type γ and let Y_γ be the integer matrix characterizing the type γ, as defined in Proposition 13.3.2. Then,

$$Q_{P_d} = Y_\gamma M_B,$$

where Q_{P_d} denotes the matrix whose rows are the vectors $v \in V(P_d)$ and M_B denotes the matrix whose rows are the vectors of B. Let Q denote the $n \times k$ matrix whose rows are the vectors v_i for $1 \le i \le n$. So, Q may have repeated rows and every row of Q is a row of Q_{P_d}. Then,

$$Q = Y M_B,$$

for some integer matrix Y. Let y_v ($v \in V(P_d)$) denote the rows of Y_γ. Then the rows of Y are the vectors y_{v_i} for $1 \le i \le n$. Note that

$$v = \sum_{1\le i\le n} b_i v_i \iff y_v = \sum_{1\le i\le n} b_i y_{v_i}.$$

Therefore, for each $v \in V(P_d)$,

$$Z(v) = \{b \in \mathbb{Z}^n \mid y_v = \sum_{1 \leq i \leq n} b_i y_{v_i}\}.$$

Hence, for $v \in V(P_d)$, $Z(v)$ depends only on $(y_v, y_{v_1}, \ldots, y_{v_n})$. Using (a), we deduce that $\text{Ann}(p)$ is entirely determined by the matrix Y_γ and the subsystem $(y_{v_1}, \ldots, y_{v_n})$ of its rows. In other words, for each $d \in \text{HYP}_{n+1}$, the annullator $\text{Ann}(\xi(d))$ is completely determined by a pair (γ, θ), where γ is a type of Delaunay polytopes in $\mathbb{R}^k$ with $k \leq n$, and θ is a mapping from $\{1, \ldots, n\}$ to the set of rows of Y_γ. As Y_γ has $|V(P_d)| \leq 2^k$ rows (by Proposition 13.2.8), the number of such mappings θ is finite. Moreover, the number of types of Delaunay polytopes in given dimension is finite (from Theorem 13.3.4). Therefore, we deduce that the number of distinct annullators $\text{Ann}(\xi(d))$ (for $d \in \text{HYP}_{n+1}$) is finite. This shows that $\xi(\text{HYP}_{n+1})$ is a polyhedral cone. ∎

Second proof. We use the results from Section 13.2.5 concerning the partition of the cone PSD_n into L-type domains. Recall from relation (13.1.7) that $\xi(\text{HYP}_{n+1})$ is a subcone of the cone PSD_n. More precisely, the following holds. Let $v_1, \ldots, v_n \in \mathbb{R}^k$ with rank k (where $1 \leq k \leq n$), set $L := \mathbb{Z}(v_1, \ldots, v_n)$, and define $p \in \text{PSD}_n$ by setting $p_{ij} = v_i^T v_j$ for $1 \leq i, j \leq n$. Then, by the results of Section 14.1,

(b) $p \in \xi(\text{HYP}_{n+1})$ if and only if L is a lattice and $v_1, \ldots, v_n$ are all vertices of the same Delaunay polytope in the star of L at the origin.

Moreover, if $p \in \xi(\text{HYP}_{n+1})$ then the whole L-type domain containing p is entirely contained in $\xi(\text{HYP}_{n+1})$. Hence, $\xi(\text{HYP}_{n+1})$ is a union of L-type domains. In fact, this union is *finite*. Indeed, $\xi(\text{HYP}_{n+1})$ contains only finitely many L-type domains whose associated lattices are all z-equivalent (by (b) and since, in a given lattice L, there are only finitely many ways of choosing a set of n vectors $v_1, \ldots, v_n$ that are all vertices of the same Delaunay polytope in the star of L). Finally, the number of distinct (up to z-equivalence) lattices of given dimension is finite. Therefore, $\xi(\text{HYP}_{n+1})$ is a finite union of L-type domains. But, a fundamental property is that each L-type domain is a polyhedral cone. Hence, $\xi(\text{HYP}_{n+1})$ is a polyhedral cone. ∎

Third proof. For $p \in \xi(\text{HYP}_{n+1})$, set

$$E_p := \{x \in \mathbb{R}^n \mid \sum_{1 \leq i,j \leq n} x_i x_j p_{ij} - \sum_{1 \leq i \leq n} x_i p_{ii} = 0\}$$

and let V_p denote the kernel of the matrix $(p_{ij})_{i,j=1}^n$. If p is rational valued then V_p admits a basis composed of integral vectors and, therefore, $x + v \in E_p$ for all $x \in E_p$ and $v \in V_p$. Hence, if the matrix (p_{ij}) has a nonzero kernel, then E_p is a degenerate ellipsoid with axis V_p. As in the first proof, set

$$\text{Ann}(p) := E_p \cap (\mathbb{Z}^n \setminus \{0, e_1, \ldots, e_n\}),$$

where $e_1, \ldots, e_n$ are the unit vectors. By definition of $\xi(\mathrm{HYP}_{n+1})$, there are no integral points lying in the interior of the region delimited by the ellipsoid E_p. Note moreover that the vectors $0, e_1, \ldots, e_n$ lie on E_p. The key idea of the proof consists of showing that, for each rational valued p lying on the boundary of $\xi(\mathrm{HYP}_{n+1})$, there exists $b \in \mathrm{Ann}(p)$ which is "short", which means here that $\max_{1 \leq i \leq n} |b_i| \leq n!$. If we can show this fact, then we obtain that $\xi(\mathrm{HYP}_{n+1})$ coincides with the cone defined by the inequalities (14.2.2) for all b such that $\max_{1 \leq i \leq n} |b_i| \leq n!$ (indeed, the latter is a cone containing $\xi(\mathrm{HYP}_{n+1})$ and whose boundary fully contains the boundary of $\xi(\mathrm{HYP}_{n+1})$, which means that the two cones coincide). Assume that p is rational valued and lies on the boundary of $\xi(\mathrm{HYP}_{n+1})$ and let $k' := \dim(V_p)$. We show that

(c) there exists $b \in \mathrm{Ann}(p)$ such that $\max_{1 \leq i \leq n} |b_i| \leq (n-k')!$.

Suppose first that $k' = 0$, i.e., that the matrix (p_{ij}) is positive definite. Then, E_p is a (nondegenerate) ellipsoid in $\mathbb{R}^n$ and $\mathrm{Ann}(p) \neq \emptyset$ as p lies on the boundary of $\xi(\mathrm{HYP}_{n+1})$. Let $b \in \mathrm{Ann}(p)$ and suppose, e.g., that $|b_1|, \ldots, |b_{n-1}| \leq |b_n|$. Let Δ denote the n-dimensional simplex spanned by the $n+1$ vectors $0, e_1, \ldots, e_{n-1}, b$ (which all lie on E_p) and let $\mathrm{vol}(\Delta)$ denote its volume. Then,

$$\mathrm{vol}(\Delta) = \frac{|b_n|}{n!} \quad \text{and} \quad \mathrm{vol}(\Delta) \leq 1;$$

the latter inequality holds since E_p is an ellipsoid whose interior is free of integral points, which implies that

$$\mathrm{vol}(\Delta) \leq \mathrm{vol}(\mathrm{Conv}(\mathbb{Z}^n \cap E_p)) \leq 1$$

(the last inequality is the analogue of the inequality $\mathrm{vol}(P) \leq \det(L)$ from Proposition 13.2.10, if P is a Delaunay polytope in a lattice L). Therefore, $|b_n| \leq n!$, i.e., (c) holds.

Suppose now that $\dim(V_p) = k' \geq 1$. Then, E_p is a degenerate ellipsoid with axis V_p. We show that V_p contains an integral vector whose components are all less than or equal to $(n-k')!$. For this, let $a^1, \ldots, a^{k'}$ be k' linearly independent vectors of $V_p \cap \mathbb{Z}^n$. For $i = 1, \ldots, n$, let M_i denote the $k' \times k'$ matrix whose rows are the vectors $(a_1^h, \ldots, a_{k'-1}^h, a_i^h)$ for $h = 1, \ldots, k'$, and let D_i denote the determinant of M_i. Then, $D_1 = \ldots = D_{k'-1} = 0$ and we can suppose without loss of generality that $D_{k'} \neq 0$. We can also suppose that $|D_{k'+1}|, \ldots, |D_n| \leq |D_{k'}|$. Set

$$b := (D_1, \ldots, D_n)^T.$$

For $h \in \{1, \ldots, k'\}$, let M'_h denote the $(k'-1) \times (k'-1)$ matrix whose rows are the vectors $(a_1^r, \ldots, a_{k'-1}^r)$ for $r \in \{1, \ldots, k'\} \setminus \{h\}$ and set $\gamma_h := (-1)^{h+k'} \det(M'_h)$. By developing the determinant D_i with respect to its last column, we obtain that $D_i = \sum_{1 \leq h \leq k'} \gamma_h a_i^h$, i.e.,

$$b = \sum_{1 \leq h \leq k'} \gamma_h a^h.$$

This shows that $b \in V_p$. Let g denote the g.c.d. of $D_1, \ldots, D_n$. Then, $\frac{1}{g}b$ is an integral vector belonging to V_p. We show that $\frac{1}{g}b$ satisfies (c), i.e., that $\frac{|D_{k'}|}{g} \leq (n-k')!$.

Set $W := \{x \in \mathbb{R}^n \mid x_1 = \ldots = x_{k'} = 0\}$. Then, $\mathbb{R}^n = V_p + W$. Let π denote the projection from $\mathbb{R}^n$ on W along V_p. So, if $x = y + z$ is the unique decomposition of $x \in \mathbb{R}^n$ with $y \in V_p$, $z \in W$, then $\pi(x) = z$. Then, $L := \pi(\mathbb{Z}^n)$ is a lattice in $\mathbb{R}^{n-k'}$ which contains $\mathbb{Z}^{n-k'}$ as a sublattice. Moreover, $\pi(E_p)$ is a (nondegenerate) ellipsoid in $\mathbb{R}^{n-k'}$ which is empty in L, i.e., no lattice point of L is lying in the interior of $\pi(E_p)$. Let Δ denote the $(n-k')$-dimensional simplex spanned by the $n-k'+1$ vectors $0, e_{k'+1}, \ldots, e_n$, which all belong to $\pi(E_p) \cap L$. Then, $\text{vol}(\Delta) = \frac{1}{(n-k')!}$ and $\text{vol}(\Delta) \leq \det(L)$ since the ellipsoid $\pi(E_p)$ is empty in L. Hence,

$$\frac{1}{\det(L)} \leq (n-k')!.$$

Consider the quotient set $L/\mathbb{Z}^{n-k'}$. Its cardinality is equal to the index of $\mathbb{Z}^{n-k'}$ in L, i.e., to $\frac{1}{\det(L)}$. Therefore,

$$|L/\mathbb{Z}^{n-k'}| = \frac{1}{\det(L)} \leq (n-k')!.$$

On the other hand, as shown below, $|L/\mathbb{Z}^{n-k'}| \geq \frac{|D_n|}{g}$, which implies that the vector $\frac{1}{g}b$ is "short". We show

$$|L/\mathbb{Z}^{n-k'}| \geq \frac{|D_n|}{g}.$$

For this, consider the vectors $(0_{k'-1}, \alpha, 0_{n-k'})$, whose components are all equal to 0 except the k'-th one equal to α, where α is an integer such that $0 \leq \alpha < \frac{|D_{k'}|}{g}$. We show that the projections of these $\frac{|D_{k'}|}{g}$ vectors belong to distinct residue classes in the quotient set $L/\mathbb{Z}^{n-k'}$. For suppose not. Let α, α' be integers with $0 \leq \alpha < \alpha' < \frac{|D_{k'}|}{g}$ and such that the two vectors $(0_{k'-1}, \alpha, 0_{n-k'})$, $(0_{k'-1}, \alpha', 0_{n-k'})$ belong to the same class. Then, there exist $z \in \mathbb{Z}^{n-k'}$ and scalars $\beta_1, \ldots, \beta_{k'} \in \mathbb{R}$ such that

$$(0_{k'-1}, \alpha' - \alpha, 0_{n-k'}) = \sum_{1 \leq h \leq k'} \beta_h a^h + (0_{k'}, z), \text{ i.e.,}$$

$$\begin{cases} \sum_{1 \leq h \leq k'} \beta_h a_i^h = 0 & \text{for } i = 1, \ldots, k'-1, \\ \sum_{1 \leq h \leq k'} \beta_h a_{k'}^h = \alpha_0 := \alpha' - \alpha, & \\ \sum_{1 \leq h \leq k'} \beta_h a_i^h = -z_i & \text{for } i = k'+1, \ldots, n. \end{cases}$$

The first k' equations of the above system imply that $\beta_h = \frac{\gamma_h \alpha_0}{D_{k'}}$ for $h = 1, \ldots, k'$. Using the last $n-k'$ equations and the fact that $\sum_{1 \leq h \leq k'} \gamma_h a_i^h = b_i = D_i$, we obtain that $\alpha_0 D_i = -z_i D_{k'}$, i.e.,

$$\alpha_0 \frac{D_i}{g} = -z_i \frac{D_{k'}}{g}$$

for each $i = k'+1, \ldots, n$. As the integers $\frac{D_{k'+1}}{g}, \ldots, \frac{D_n}{g}$ are relatively prime, we deduce that $\frac{D_{k'}}{g}$ divides α_0. This yields a contradiction as $0 < \alpha_0 < \frac{|D_{k'}|}{g}$. ∎

Remark 14.2.3. Let us make more precise the connections existing between the notions used in the first proof (empty sphere S_d, lattice L_d, etc) and in the third proof (ellipsoid E_p, etc). We use the notation from both proofs. Let $d \in \mathrm{HYP}_{n+1}$ and $p := \xi(d)$ with rank k and such that $p_{ij} = v_i^T v_j$, where $v_1, \ldots, v_n \in \mathbb{R}^k$. The vectors $v_1, \ldots, v_n$ lie on the sphere S_d which is empty in the lattice $L_d = \mathbb{Z}(v_1, \ldots, v_n)$, and $P_d = \mathrm{Conv}(S_d \cap L_d)$ is the Delaunay polytope associated to d. Let V_p denote the kernel of (p_{ij}) with dimension $k' = n-k$ and let W denote its complement in $\mathbb{R}^n$ considered in the third proof. Clearly,

$$b \in \mathrm{Ann}(p) \Longleftrightarrow \sum_{1 \le i \le n} b_i v_i \in V(P_d),$$

$$b \in V_p \Longleftrightarrow \sum_{1 \le i \le n} b_i v_i = 0.$$

Let $\varphi : \mathbb{R}^n \mapsto \mathbb{R}^k$ denote the linear mapping defined by

$$\varphi(b) = \sum_{1 \le i \le n} b_i v_i, \quad \text{for } b \in \mathbb{R}^n.$$

Hence, the kernel of φ is V_p and, thus, the spaces W and $\mathbb{R}^k$ are in bijection via φ. In particular, the ellipsoid $\pi(E_p)$ is in one-to-one correspondence with the sphere S_d via φ. Also, φ maps the integral lattice $\mathbb{Z}^n$ onto the lattice L_d. ∎

An immediate consequence of Lovász's proof (presented above as the third proof) is that the inequalities (14.2.2) for $b \in \mathbb{Z}^n$ with

$$\max_{1 \le i \le n} |b_i| \le n!$$

are sufficient for describing the cone $\xi(\mathrm{HYP}_{n+1})$.

Avis and Grishukhin [1993] show that the inequalities (14.2.2) for $b \in \mathbb{Z}^n$ with

$$\max_{1 \le i \le n} |b_i| \le (n-1)!$$

are sufficient for describing $\xi(\mathrm{HYP}_{n+1})$. (They give, in fact, the bound $\frac{2^{n-2}(n-1)!}{n+1}$, but their bound can be improved to $(n-1)!$ using the upper bound on the volume of a Delaunay polytope from Proposition 13.2.10.) More precisely, they show that, if the inequality (14.2.2) defines a facet of the cone $\xi(\mathrm{HYP}_{n+1})$, then $\max_{1 \le i \le n} |b_i| \le (n-1)!$. Their proof uses the fact that the facets of the hypermetric cone correspond to a very special class of Delaunay polytopes, namely, the repartitioning polytopes (described in Section 15.2).

The following tighter bound is due to Lovász [1994]; its proof is based on Lemma 14.2.5.

Proposition 14.2.4. *The cone* $\xi(\mathrm{HYP}_{n+1})$ *is defined by the inequalities* (14.2.2) *for* $b \in \mathbb{Z}^n$ *with*

$$\max_{1 \le i \le n} |b_i| \le \frac{2^n}{\binom{2n}{n}} n!.$$

Lemma 14.2.5. *Let* E *be an ellipsoid in* $\mathbb{R}^n$ *and let* L *be an* n*-dimensional lattice in* $\mathbb{R}^n$*. Suppose that* E *is empty in* L*, i.e., that there are no lattice points lying in the interior of the region delimited by* E*. Let* $x_0 = 0, x_1, \ldots, x_n$ *be affinely independent points of* $E \cap L$ *and let* Δ *denote the simplex they span. Then,*

$$\mathrm{vol}(\Delta) \le \frac{2^n}{\binom{2n}{n}} \det(L).$$

Proof. Consider the difference body $\Delta - \Delta := \{x - y \mid x, y \in \Delta\}$. It is a centrally symmetric convex body around the origin and one can check that the only lattice point lying in the interior of $\Delta - \Delta$ is the origin. Hence, by Minkowski's theorem (see, e.g., Siegel [1989]),

$$\mathrm{vol}(\Delta - \Delta) \le 2^n \det(L).$$

On the other hand, it is shown in Rogers and Shephard [1957] that

$$\mathrm{vol}(\Delta - \Delta) = \binom{2n}{n} \mathrm{vol}(\Delta).$$

This yields the result. ∎

Proof of Proposition 14.2.4. Let us now call a vector b "short" if

$$|b_i| \le \frac{2^n}{\binom{2n}{n}} n!$$

holds for all $1 \le i \le n$. Using the bound on the volume of a simplex from Lemma 14.2.5, the third proof of Theorem 14.2.1 can be adapted so as to show the existence of a short vector in $\mathrm{Ann}(p)$ for each p lying on the boundary of $\xi(\mathrm{HYP}_{n+1})$. (One also uses the fact that

$$\frac{2^k}{\binom{2k}{k}} k! \le \frac{2^n}{\binom{2n}{n}} n!$$

for all $1 \le k \le n$.) ∎

Remark 14.2.6. From Proposition 14.2.4, a (rough) upper bound on the number of facets of the cone $\xi(\mathrm{HYP}_{n+1})$ (or of the hypermetric cone HYP_{n+1}) is given by $(2B_n + 1)^n$, where

$$B_n := \frac{2^n n!}{\binom{2n}{n}}.$$

Therefore, the problem of testing whether a given distance d is hypermetric is in co-NP. It is not known whether testing hypermetricity is NP-hard. But the following complexity results are proved by Avis and Grishukhin [1993]; they will be treated in detail in Section 28.3.

(i) Given an integral distance d and an integer m, does d satisfy all $(2m+1)$-gonal hypermetric inequalities ? This problem is co-NP-complete.

(ii) Given an integral distance d. Is d hypermetric ? If not, give the smallest k such that d violates a $(2k+1)$-gonal inequality. This problem is NP-hard. ∎

14.3 Delaunay Polytopes in Root Lattices

In this section, we group several results on Delaunay polytopes in root lattices. First, we recall the description of the irreducible root lattices and of their Delaunay polytopes. We show that, if P is a Delaunay polytope in a root lattice, then its 1-skeleton graph is completely determined by the metric structure of P (see Proposition 14.3.3). Then, we see that Delaunay polytopes in root lattices arise in a natural way from hypermetric spaces that are connected and strongly even (see Proposition 14.3.5). As a consequence, we obtain a characterization of the connected strongly even distance spaces that are hypermetric, or isometrically ℓ_1-embeddable (see Theorems 14.3.6 and 14.3.7).

Let P be a Delaunay polytope which is generating in a root lattice L. If L is reducible, then $L = L_1 \oplus L_2$ where L_1 and L_2 are root lattices. Hence, $P = P_1 \times P_2$ where P_i is a Delaunay polytope in L_i, for $i = 1, 2$. Therefore, it suffices to describe the Delaunay polytopes that are generating in some irreducible root lattice.

The irreducible root lattices have been classified by Witt (see, for instance, Brouwer, Cohen and Neumaier [1989]). They are A_n $(n \geq 0)$, D_n $(n \geq 4)$, and E_n $(n = 6, 7, 8)$. We recall their description below; we will consider in more detail the lattices E_6, E_7 and E_8 in Section 16.2.

For each of the lattices A_n, D_n and E_n, we recall some information about its roots (i.e., its minimal vectors) and about its empty spheres (i.e., its holes). For more details we refer, for instance, to Brouwer, Cohen and Neumaier [1989], Conway and Sloane [1988, 1991].

Case of A_n, $n \geq 0$.

• $A_n = \{x \in \mathbb{Z}^{n+1} \mid \sum_{0 \leq i \leq n} x_i = 0\}$.

• The roots of A_n are the $n(n+1)$ vectors $e_i - e_j$, $0 \leq i \neq j \leq n$, where e_i denote the i-th unit vector in $\mathbb{R}^{n+1}$.

• There are $\lfloor \frac{n+1}{2} \rfloor$ types of empty spheres in A_n. Their centers are

$$c_a = (\frac{a}{n+1}, \ldots, \frac{a}{n+1}; -\frac{n+1-a}{n+1}, \ldots, -\frac{n+1-a}{n+1}),$$

where $\frac{a}{n+1}$ is repeated $n+1-a$ times and $-\frac{n+1-a}{n+1}$ is repeated a times, with corresponding radius $r_a = \sqrt{\frac{a(n+1-a)}{n+1}}$, for $1 \leq a \leq \lfloor \frac{n+1}{2} \rfloor$. The case $a = \lfloor \frac{n+1}{2} \rfloor$ corresponds to a deep hole, i.e., a hole with maximum radius.

• The Delaunay polytope circumscribed by the empty sphere with center c_a and radius r_a has for vertices the following $\sum_b \binom{n+1-a}{b}\cdot\binom{a}{b}$ vectors $(1^b, 0^{n+1-a-b}; (-1)^b, 0^{a-b})$ for $0 \leq b \leq a, n+1-a$, where the first b ones are chosen among the $n+1-a$ positions of the entries $\frac{a}{n+1}$ of c_a and the last b minus ones are chosen among the a positions of the entries $-\frac{n+1-a}{n+1}$ of c_a. Its 1-skeleton graph is the Johnson graph[1] $J(n+1, a)$.

Case of D_n, $n \geq 4$.

• $D_n = \{x \in \mathbb{Z}^n \mid \sum_{1\leq i\leq n} x_i \in 2\mathbb{Z}\}$.

• The roots of D_n are the $2n(n-1)$ vectors $\pm e_i \pm e_j$ for $1 \leq i \neq j \leq n$.

• There are two types of empty spheres in D_n, namely, an empty sphere S_1 with center $c_1 = (0, \ldots, 0, 1)$ and radius $r_1 = 1$, and an empty sphere S_2 with center $c_2 = (\frac{1}{2}, \ldots, \frac{1}{2})$ and radius $r_2 = \frac{\sqrt{n}}{2}$.

• The Delaunay polytope circumscribed by the sphere S_1 has for vertices the $2n$ vectors $(0, \ldots, 0)$, $(0, \ldots, 0, 2)$ and $(0, \ldots, 0, \pm 1, 0, \ldots, 0, 1)$ where the component ± 1 is in one of the first $n-1$ positions. This is the cross-polytope β_n whose 1-skeleton graph is the cocktail-party graph $K_{n\times 2}$.

• The Delaunay polytope circumscribed by the second sphere S_2 has for vertices the 2^{n-1} vectors $x \in \{0,1\}^n$ with $\sum_{1\leq i\leq n} x_i \in 2\mathbb{Z}$. This is the half-cube $h\gamma_n$ whose 1-skeleton graph is the half-cube graph $\frac{1}{2}H(n,2)$. It corresponds to a deep hole in D_n.
Note that, for $n = 4$, β_4 and $h\gamma_4$ are congruent (i.e., coincide up to orthogonal transformation and translation).

Case of E_8.

• $E_8 = \{x \in \mathbb{R}^8 \mid x \in \mathbb{Z}^8 \cup (\frac{1}{2} + \mathbb{Z})^8$ and $\sum_{1\leq i\leq 8} x_i \in 2\mathbb{Z}\}$, i.e., E_8 is the lattice generated by D_8 and $\frac{1}{2}\sum_{1\leq i\leq 8} e_i$. The lattice E_8 is even and unimodular; hence, $E_8^* = E_8$.

• The roots of E_8 are the 240 vectors $\pm e_i \pm e_j$ and $\frac{1}{2}(\pm e_1 \pm \ldots \pm e_n)$, where there is an even number of minus signs in a root of the second kind.

• There are two types of empty spheres in E_8, namely, the sphere S_1 with center $c_1 = (1, 0^7)$ and radius $r_1 = 1$, and the sphere S_2 with center $c_2 = (\frac{5}{6}, \frac{1}{6}^7)$ and radius $r_2 = \sqrt{\frac{8}{9}}$.

• The Delaunay polytope circumscribed by the sphere S_1 has for vertices the following 16 vectors (0^8), $(2, 0^7)$, $(1, 0, \ldots, 0, \pm 1, 0, \ldots, 0)$, where ± 1 is in one of the last seven positions. This is the cross-polytope β_8 whose 1-skeleton graph is $K_{8\times 2}$. It corresponds to a deep hole in E_8.

• The Delaunay polytope circumscribed by the sphere S_2 has for vertices the following 9 vectors (0^8), $(\frac{1}{2}, \ldots, \frac{1}{2})$ and $(1, 0, \ldots, 0, 1, 0, \ldots, 0)$, where the second

[1] For $1 \leq t \leq n$, the *Johnson graph* $J(n,t)$ is defined as the graph with node set $\{A \subseteq \{1, \ldots, n\} : |A| = t\}$ and with edges the pairs (A, B) with $|A \cap B| = t - 1$.

1 is in one of the last seven positions. This is the simplex α_8 with 1-skeleton graph K_9.
We point out that, as E_8 is an even unimodular lattice, none of its Delaunay polytopes is generating (by Lemma 13.2.6).

Case of E_7.

• The root lattice E_7 consists of the vectors of E_8 that are orthogonal to a given minimal vector v_0 of E_8. If we choose $v_0 = (\frac{1}{2}, \ldots, \frac{1}{2})$, then $E_7 = \{x \in E_8 \mid \sum_{1 \leq i \leq 8} x_i = 0\}$. Another choice for v_0 could be $v_0 = (1, 1, 0^6)$; we will work with this second definition of E_7 in Section 16.2 (in fact, we shall use there for E_7 the following affine translate $\{x \in E_8 \mid x^T v_0 = x_1 + x_2 = 1\}$).

• There are two types of empty spheres in E_7, namely, the sphere S_1 with center $c_1 = (\frac{3}{4}^2, -\frac{1}{4}^6)$ and radius $r_1 = \sqrt{\frac{3}{2}}$, and the sphere S_2 with center $c_2 = (\frac{7}{8}, -\frac{1}{8}^7)$ and radius $r_2 = \sqrt{\frac{7}{8}}$.

• The Delaunay polytope circumscribed by the sphere S_1 has for vertices the 56 vectors $c_1 \pm (\frac{3}{4}^2, -\frac{1}{4}^6)$. This is the Gosset polytope 3_{21} whose 1-skeleton graph is the Gosset graph G_{56}. It corresponds to a deep hole in E_7.

• The Delaunay polytope circumscribed by the sphere S_2 has for vertices the 8 following vectors (0^8) and $(1, 0, \ldots, 0, -1, 0, \ldots, 0)$, where -1 is in one of the last seven positions. This is the 7-dimensional simplex α_7 with 1-skeleton graph K_8.

Case of E_6.

• The root lattice E_6 consists of the vectors of E_8 that are orthogonal to two nonorthogonal given minimal vectors v_0 and w_0 of E_8. If we choose $v_0 = (1, 1, 0^6)$ and $w_0 = (-\frac{1}{2}^8)$, then $E_6 = \{x \in E_8 \mid x_1 + x_2 = x_3 + \ldots + x_8 = 0\}$. (In Section 16.2, we select differently v_0 and w_0 and we consider an affine translate as E_6.)

• There is only one type of empty sphere in E_6. Its radius is $\sqrt{\frac{4}{3}}$ and it circumscribes the Delaunay polytope whose vertices are the following 27 vectors $(\frac{1}{2}, -\frac{1}{2}, \frac{5}{6}, -\frac{1}{6}^5)$, $(-\frac{1}{2}, \frac{1}{2}, \frac{5}{6}, -\frac{1}{6}^5)$ where $\frac{5}{6}$ is in one of the last six positions, and $(0, 0, -\frac{2}{3}^2, \frac{1}{3}^4)$ where the two $-\frac{2}{3}$'s are in the last six positions. This is the Schläfli polytope 2_{21} whose 1-skeleton graph is the Schläfli graph G_{27}. So the star of E_6 contains only copies of 2_{21} and of its image under central symmetry.

We summarize in Figure 14.3.1 some information about the Delaunay polytopes P arising in the irreducible root lattices. For each irreducible root lattice L, we list all the Delaunay polytopes P arising in the star of L. Note that P is always generating in L (i.e., $V(P)$ generates L) with the exception of the polytopes α_8 and β_8 that are *not* generating in E_8. For each Delaunay polytope P, we present its 1-skeleton graph, denoted by $H(P)$ and called a *Delaunay polytope graph*. The last column gives the square r^2 of the radius r of the sphere circumscribing P. As $d_{H(P)}(u, v) = \frac{1}{2}(u - v)^2$ for all $u, v \in V(P)$ (by the next Proposition 14.3.3), the graphic metric space $(V(P), d_{H(P)})$ of the graph $H(P)$ is hypermetric and its associated L-polytope has radius $\frac{r}{\sqrt{2}}$.

lattice L	Delaunay polytope P	Delaunay polytope graph $H(P)$	squared radius r^2
A_n $(n \geq 0)$	see description above	$J(n+1,t)$ for $1 \leq t \leq \lfloor \frac{n+1}{2} \rfloor$	$\frac{t(n+1-t)}{n+1}$
D_n $(n \geq 4)$	β_n $h\gamma_n$	$K_{n\times 2}$ $\frac{1}{2}H(n,2)$	1 $n/4$
E_8	α_8 β_8	K_9 $K_{8\times 2}$	8/9 1
E_7	α_7 3_{21}	K_8 G_{56}	7/8 3/2
E_6	2_{21}	G_{27}	4/3

Figure 14.3.1: Delaunay polytopes in the irreducible root lattices

Remark 14.3.2. We group here several observations about the graphs $J(n,t)$, $\frac{1}{2}H(n,2)$, $K_{n\times 2}$, K_n, the Schläfli graph G_{27}, and the Gosset graph G_{56} occurring in Figure 14.3.1.

(i) There are some isomorphisms among them, namely, $J(n,1) = K_n$, $\frac{1}{2}H(2,2) = K_2$, $\frac{1}{2}H(3,2) = K_4$, $K_{3\times 2} = J(4,2)$, $K_{4\times 2} = \frac{1}{2}H(4,2)$. Note that $J(n,2)$ coincides with the line graph $L(K_n)$ of K_n, which is also called the *triangular graph* and denoted by $T(n)$. The half-cube graph $\frac{1}{2}H(5,2)$ is also known as the *Clebsch graph.*

(ii) $J(n,t)$ is an isometric subgraph of $\frac{1}{2}H(n,2)$ and of $J(n+1,t)$; $\frac{1}{2}H(n,2)$ is an isometric subgraph of $\frac{1}{2}H(n+1,2)$; $K_{n\times 2}$ is an isometric subgraph of $K_{(n+1)\times 2}$. Also, $\nabla\frac{1}{2}H(5,2)$ is an isometric subgraph of G_{27}; $\frac{1}{2}H(6,2)$, $K_{6\times 2}$, $J(8,2)$ and ∇G_{27} are isometric subgraphs of G_{56}. Hence, G_{27} is an isometric subgraph of G_{56} (as G_{27} has diameter 2). In fact, $J(5,2)$ (resp. $\frac{1}{2}H(5,2)$, G_{27}) is the subgraph of $\frac{1}{2}H(5,2)$ (resp. of G_{27}, G_{56}) induced by the neighborhood of one of its nodes.

(iii) $J(n,t)$, $K_{n\times 2}$ $(n \geq 2)$, $\frac{1}{2}H(n,2)$ are ℓ_1-graphs, but G_{27}, G_{56} are not ℓ_1-graphs. ∎

We consider in the next result an interesting property for a Delaunay polytope P in a root lattice. Namely, a geometric feature of P is entirely determined by the metric structure of P: its 1-skeleton graph consists of the pairs of vertices at squared distance 2. This was proved in Deza and Grishukhin [1993].

Proposition 14.3.3. *Let P be a generating Delaunay polytope in a root lattice. Let $G(P)$ denote the graph with set of vertices $V(P)$ and with edges the pairs (u,v) for which $d^{(2)}(u,v) = 2$, for $u,v \in V(P)$, and let $d_{G(P)}$ denote its path metric. Then,*

$$d^{(2)}(u,v) = 2d_{G(P)}(u,v)$$

holds for all $u,v \in V(P)$, i.e., the Delaunay polytope space $(V(P), d^{(2)})$ coincides with the space $(V(P), 2d_{G(P)})$. Moreover, $G(P)$ coincides with the 1-skeleton graph $H(P)$ of P, i.e., two vertices u,v form an edge of P if and only if $d^{(2)}(u,v) = 2$.

Proof. Let $u,v \in V(P)$ such that $d_{G(P)}(u,v) = 2$. Let (u,u_1,v) be a path in $G(P)$ from u to v, i.e., $(u-u_1)^2 = (u_1-v)^2 = 2$ and $(u-v)^2 > 2$. Observe that

$$(u_1-u)^T(u_1-v) = 0. \tag{14.3.4}$$

Indeed, $(u_1-u)^T(u_1-v) \geq 0$ since any three vertices of P form a triangle with no obtuse angles. Using relation (13.2.1), we obtain that $(u_1-u)^T(u_1-v) = 0, 1$. Moreover, $(u-v)^2 = 4-2(u_1-u)^T(u_1-v) > 2$, implying that $(u_1-u)^T(u_1-v) = 0$ and, thus, $(u-v)^2 = 4 = 2d_{G(P)}(u,v)$.

Consider now $u,v \in V(P)$ such that $d_{G(P)}(u,v) = k \geq 2$. Let $(u_0 = u, u_1, \ldots, u_k = v)$ be a shortest path from u to v in $G(P)$. Then, $u - v = \sum_{1\leq i\leq k} r_i$, where $r_i := u_i - u_{i-1}$ is a root (i.e., $r_i^2 = 2$) for $1 \leq i \leq k$. So this path corresponds to the sequence of roots $(r_1, \ldots, r_k)$. Consider the subpath (u_{i-1}, u_i, u_{i+1}). Applying (14.3.4), we deduce that $r_i^T r_{i+1} = 0$ holds. So, any two consecutive roots in the sequence $(r_1, \ldots, r_k)$ are orthogonal.
Note that $w := u_{i-1} + u_{i+1} - u_i$ is also a vertex of P since $w \in L$ and w also lies on the sphere circumscribing P. Hence, $(u_0, u_1, \ldots, u_{i-1}, w, u_{i+1}, \ldots, u_k)$ is another shortest path from u to v; it corresponds to the sequence of roots $(r_1, \ldots, r_{i-1}, r_{i+1}, r_i, r_{i+2}, \ldots, r_k)$. By the above argument, $r_i^T r_{i+2} = (r_{i-1})^T r_{i+1} = 0$. After iteration, we obtain that any two roots r_i, r_j, $i \neq j$, are orthogonal. Therefore,

$$(u-v)^2 = \sum_{1\leq i\leq k} r_i^2 = 2k = 2d_{G(P)}(u,v)$$

holds. Moreover, $u-v$ is the diagonal of the k-cube spanned by $r_1, \ldots, r_k$, whose vertices all are vertices of P. Therefore, u,v do not form an edge of P.

It is easy to see that, conversely, any two vertices u,v of P with $(u-v)^2 = 2$ form an edge of P. ∎

We now see that Delaunay polytopes in root lattices arise in a natural way from connected strongly even hypermetric spaces. Recall that a distance space (X,d) is strongly even if $d(i,j)$ is an even integer for all $i,j \in X$ and $d(i,j) = 2$ for some $i,j \in X$.

Proposition 14.3.5. *Let (X, d) be a connected strongly even distance space. If (X, d) is hypermetric with associated Delaunay polytope P_d generating the lattice L_d, then L_d is a root lattice.*

Proof. Let $i \in X \mapsto v_i \in V(P_d)$ denote a representation of (X, d) in L_d. As the distance space (X, d) is connected and strongly even, the lattice L_d is generated by the set $\{v_i - v_j \mid i, j \in X \text{ and } d(i,j) = 2\}$. Hence, L_d is a root lattice. ∎

As an application, we can characterize the connected strongly even distance spaces which are hypermetric, or ℓ_1-embeddable. The following Theorems 14.3.6 and 14.3.7 are due, respectively, to Terwiliger and Deza [1987] and Deza and Grishukhin [1993]. An application to graphs will be formulated in Section 17.1.

Theorem 14.3.6. *Let (X, d) be a connected strongly even distance space. The following assertions are equivalent.*

(i) *(X, d) is hypermetric.*

(ii) *$(X, \frac{1}{2}d)$ is an isometric subspace of a direct product of half-cube graphs $\frac{1}{2}H(n, 2)$ ($n \geq 7$), cocktail-party graphs $K_{n\times 2}$ ($n \geq 7$), and copies of the Gosset graph G_{56}.*

Proof. (i) $\Longrightarrow$ (ii) From Proposition 14.3.5, the Delaunay polytope P_d associated with (X, d) is generating in a root lattice. Therefore, from Proposition 14.3.3, The Delaunay polytope space $(V(P_d), \frac{d^{(2)}}{2})$ coincides with the graphic space $(V(P_d), d_{H(P_d)})$ which, using Figure 14.3.1, is a direct product of Johnson graphs, cocktail-party graphs, half-cube graphs, copies of G_{27} and G_{56}. The result now follows using Remark 14.3.2. The implication (ii) $\Longrightarrow$ (i) is obvious. ∎

Theorem 14.3.7. *Let (X, d) be a connected strongly even distance space. The following assertions are equivalent.*

(i) *(X, d) is isometrically ℓ_1-embeddable.*

(ii) *$(X, \frac{1}{2}d)$ is an isometric subspace of a product of half-cube graphs and cocktail-party graphs.* ∎

Theorem 14.3.7 can be proved in the same way as Theorem 14.3.6, using Proposition 14.1.10 and the fact that the graphs G_{27} and G_{56} are not ℓ_1-graphs. In the (main) subcase of graphic metric spaces, another proof was given by Shpectorov [1993]; it is elementary (it does not use Delaunay polytopes) but longer. We will present the latter proof in Chapter 21.

14.4 On the Radius of Delaunay Polytopes

We present here several results which give, in some cases, a more precise information on the radius of Delaunay polytopes. The first result is a partial converse to the implication (ii) $\Longrightarrow$ (iv) from Proposition 13.1.2; it gives explicitly the value of the radius of the spherical representation of a distance space (X,d) of negative type when $\sum_{i\in X} d(i,j)$ is a constant. This result was already formulated in Theorem 6.2.18; we repeat it here for convenience.

Proposition 14.4.1. *Suppose that (X,d) is of negative type and that the sum $\sum_{i\in X} d(i,j)$ does not depend on $j \in X$. Then, (X,d) has a spherical representation, on a sphere whose center is the center of mass of the representation and whose radius r is given by*

$$r^2 = \frac{1}{2|X|} \sum_{j\in X} d(i,j). \tag{14.4.2}$$

∎

An example of distance space with constant sum $\sum_{i\in X} d(i,j)$ is the graphic metric space $(V(G), d_G)$, where G is a distance regular graph or a regular graph of diameter 2; see Section 17.2. Proposition 14.4.3 below is a specification of Proposition 14.4.1 to hypermetric spaces, and Proposition 14.4.4 is a partial converse to the implication (i) $\Longrightarrow$ (ii) from Proposition 13.1.2. Both results are given in Deza and Grishukhin [1993].

Proposition 14.4.3. *Let (X,d) be a hypermetric space, let P_d be its associated Delaunay polytope and let r denote the radius of its circumscribed sphere S_d. If $\sum_{i\in X} d(i,j)$ does not depend on $j \in X$, then the radius r is given by relation* (14.4.2).

Proof. From Proposition 14.4.1, we can suppose that X lies on a sphere S with center the center of mass of X and with radius r given by (14.4.2). On the other hand, S_d is a minimal dimension sphere containing X. Hence, $S_d \subseteq S$ holds. The affine space spanned by S_d contains X and thus its center of mass, i.e., the center of S. Therefore, S and S_d have the same radius. ∎

Proposition 14.4.4. *Let (X,d) be a connected strongly even distance space. Suppose that (X,d) has a representation on a sphere with radius r such that $r^2 < 2$. Then, (X,d) is hypermetric.*

Proof. Let $(v_i \mid i \in X)$ be a representation of (X,d) on a sphere S. Up to translation, we can suppose that $v_i = 0$ for some index $i \in X$. From Proposition 14.3.5, $L(X,d) := \mathbb{Z}(v_i \mid i \in X)$ is a root lattice. We show that the sphere S is empty in $L(X,d)$ which, by Proposition 14.1.2, implies that (X,d) is hypermetric. Let H be the affine space spanned by $\{v_i \mid i \in X\}$. We can suppose that S lies in H (else replace S by $S \cap H$). Let ℓ be the line in $\mathbb{R}^{n+1}$ orthogonal to H going through the center of S, and let q be a point on ℓ such that $(q - v_i)^2 = 2$ for all $i \in X$. Note that $(q-v)^2 < 2$ for each point v lying in the interior of the ball delimited by S. Note also that $(q-v_i)^T(q-v_j) \in \{0,-1,1\}$ for all $i \neq j \in X$.

Indeed, $(v_i - v_j)^2$ is even since (X, d) is strongly even and

$$(v_i - v_j)^2 = 4 - 2(q - v_i)^T(q - v_j) \leq 4r^2 < 8,$$

implying that $(v_i - v_j)^2 \in \{2, 4, 6\}$. Therefore, $L' := \mathbb{Z}(q - v_i \mid i \in X)$ is a root lattice and, in particular, $a^2 \geq 2$ for each $a \in L'$, $a \neq 0$. Suppose now that some point $v \in L(X, d)$ lies in the interior of the ball delimited by S. Then, $(v - q)^2 < 2$, yielding a contradiction with the fact that $v - q \in L'$. ∎

We now present some results on spherical t-extensions of hypermetric spaces. Given a distance space (X, d), we remind that its spherical t-extension $(X' := X \cup \{i_0\}, \mathrm{sph}_t(d))$ is defined by

$$\begin{array}{ll} \mathrm{sph}_t(d)(i,j) = d(i,j) & \text{for } i,j \in X, \\ \mathrm{sph}_t(d)(i,i_0) = t & \text{for } i \in X. \end{array}$$

The next lemma characterizes the parameters[2] t for which $\mathrm{sph}_t(d)$ belongs to the negative type cone and Proposition 14.4.6 below specifies values of t for which the spherical t-extension operation preserves hypermetricity. Both results are taken from Grishukhin [1992b].

Lemma 14.4.5. *Let (X, d) be a distance space. Then, $\mathrm{sph}_t(d) \in \mathrm{NEG}_{n+1}$ if and only if (X, d) has a spherical representation with radius r and $r^2 \leq t$. Moreover,*

(i) *If $r^2 < t$, then $\mathrm{sph}_t(d)$ has a spherical representation with radius $R := \frac{t}{2\sqrt{t-r^2}}$.*

(ii) *If $t = r^2$, then $\mathrm{sph}_t(d)$ has no spherical representation.*

Proof. If $\mathrm{sph}_t(d) \in \mathrm{NEG}_{n+1}$, then $\mathrm{sph}_t(d)$ has a representation $i \in X' \mapsto v_i$ with $(v_i - v_{i_0})^2 = t$. Hence, the v_i's ($i \in X$) lie on the sphere S_0 with center v_{i_0} and radius $\sqrt{t}$. Let H denote the affine space spanned by $(v_i \mid i \in X)$. Therefore, the v_i's ($i \in X$) lie on the sphere $S_d = S_0 \cap H$ whose radius r is less than or equal to $\sqrt{t}$.
Conversely, consider a representation $(v_i \mid i \in X)$ of (X, d) on a sphere S_d with radius r such that $r^2 \leq t$. Consider the line orthogonal to the affine space H spanned by $(v_i \mid i \in X)$ and going through the center of S_d. Choose a point v_{i_0} on this line which is at squared distance $t - r^2$ from H. Then, $(v_i \mid i \in X')$ is a representation of $\mathrm{sph}_t(d)$, which shows that $\mathrm{sph}_t(d) \in \mathrm{NEG}_{n+1}$.
Suppose $r^2 < t$. Let S denote the sphere of dimension one higher than that of S_d, which contains S_d and v_{i_0}. So, $S_d = S \cap H$ and the radius of S is $R := \frac{t}{2\sqrt{t-r^2}}$.
On the other hand, if $r^2 = t$, then v_{i_0} is the center of $S_0 = S_d$ and the representation of $\mathrm{sph}_t(d)$ is not spherical. ∎

Proposition 14.4.6. *Let (X, d) be a hypermetric space and let r denote the radius of the sphere S_d circumscribing P_d.*

[2]Note that the quantity $\min(\sqrt{t} \mid \mathrm{sph}_t(d) \in \mathrm{NEG}_{n+1})$ is considered in approximation theory where it is called the Chebyshev radius of (X, d) (Chebyshev [1947]).

(i) *Suppose that* $t \geq 2r^2$. *Then* $\mathrm{sph}_t(d)$ *is hypermetric, its radius is* $R := \frac{t}{2\sqrt{t-r^2}}$, *with* $t \geq 2R^2$ *(and* $R \geq r$, *with equality if and only if* $t = 2r^2$*). Therefore,* $\mathrm{sph}_t^m(d)$ *is hypermetric for any integer* $m \geq 1$. *Let* P *be the Delaunay polytope associated with* $\mathrm{sph}_t(d)$. *If* $t > 2r^2$, *or if* $t = 2r^2$ *and* P_d *is asymmetric, then* P *is a pyramid with base* P_d. *If* P_d *is centrally symmetric, then* P *is a bipyramid with base* P_d.

(ii) *If* $r^2 < t < 2r^2$ *and if* P_d *is centrally symmetric, then* $\mathrm{sph}_t(d)$ *is not hypermetric.*

Proof. We use the same notation as in the proof of Lemma 14.4.5.
(i) Let L_d be the lattice spanned by $V(P_d)$ and let L denote the lattice generated by L_d and v_{i_0}. So, L consists of layers which are translates of L_d, the distance between consecutive layers being $h = \sqrt{t-r^2}$. By assumption, $t \geq 2r^2$, implying that $h \geq R = \frac{t}{2\sqrt{t-r^2}}$. This shows that the sphere S is empty in L. Therefore, $\mathrm{sph}_t(d)$ is hypermetric and its associated Delaunay polytope P has radius R. If $t > 2r^2$, then P is the pyramid with base P_d and apex v_{i_0}. If $t = 2r^2$, then one checks easily that the antipode $v_{i_0}^*$ of v_{i_0} on the sphere S belongs to L if and only if P_d is centrally symmetric. Therefore, if P_d is centrally symmetric, then P is the bipyramid with base P_d and apex v_{i_0} and, if P_d is asymmetric, then P is the pyramid with base P_d and apex v_{i_0}. Note that $t > 2R^2$ follows from $R = \frac{t}{2h}$ and $h > R$.
(ii) Let $v \in V(P_d)$ and let v^* be its antipode on the sphere S_d. Then, $w = v + v^* - v_{i_0}$ belongs to L. We show that w lies inside S, which implies that $\mathrm{sph}_t(d)$ is not hypermetric. Indeed,

$$(v-v^*)^2 = 4r^2,\ (v-v_{i_0})^2 = (v^*-v_{i_0})^2 = t,\ (v-c)^2 = (v^*-c)^2 = (v_{i_0}-c)^2 = R^2,$$

from which we deduce that $(w-c)^2 = R^2 + 2t - 4r^2 < R^2$. ∎

We finally mention (without proof) a result from Deza and Grishukhin [1996b] relating the covering radius of a lattice to the radius of its symmetric Delaunay polytopes.

Proposition 14.4.7. *Let* L *be a* k*-dimensional lattice in* $\mathbb{R}^k$ *with covering radius* $\rho(L)$*; that is,* $\rho(L)$ *is the maximum radius of (the sphere circumscribing) a Delaunay polytope in* L. *Let* R *denote the maximum radius of a symmetric Delaunay polytope in* L *(setting* $R := 0$ *if none exists) and let* r *denote the maximum radius of a proper symmetric face of a Delaunay polytope in* L. *If* $\frac{4}{3}r^2 \leq R^2$, *then* $\rho(L) = R$. *Otherwise,* $R^2 \leq (\rho(L))^2 \leq \frac{4}{3}r^2$. ∎

We now present some examples of applications (taken, in particular, from Avis and Maehara [1994] and Grishukhin [1992b]).

Example 14.4.8. Consider the complete bipartite graph $K_{1,n}$ and its suspension graph $\nabla K_{1,n}$. Then, their path metrics can be expressed as $d(K_{1,n}) =$

$\mathrm{sph}_1(d)$, $d(\nabla K_{1,n}) = \mathrm{sph}_1(\mathrm{sph}_1(d))$, where $d := 2d(K_n)$ takes value 2 on all pairs of $\{1, \ldots, n\}$. The distance d is hypermetric with radius $r = \sqrt{\frac{n-1}{n}}$ (by Proposition 14.4.3). Hence, using Lemma 14.4.5, we obtain that $d(K_{n,1})$ has a spherical representation with radius $R = \frac{1}{2\sqrt{1-r^2}} = \sqrt{\frac{n}{4}}$. Therefore, $d(\nabla K_{1,n})$ is of negative type if and only if $n \leq 4$. Moreover, $d(\nabla K_{n,1})$ has no spherical representation if $n = 4$ and, for $n = 2, 3$, $d(\nabla K_{1,n})$ has a spherical representation with radius $\frac{1}{2\sqrt{1-R^2}} = \frac{1}{\sqrt{4-n}}$. ∎

Example 14.4.9. Consider the graph $K_n \setminus P_3$ for $n \geq 4$ (where P_3 denotes the path on 3 nodes). Let d denote the distance on 3 points with two values equal to 2 and one equal to 1. Clearly, $d(K_4 \setminus P_3) = \mathrm{sph}_1(d)$ and $d(K_n \setminus P_3) = \mathrm{sph}_1^{n-3}(d)$ for $n \geq 4$. One can easily verify that d is hypermetric with radius $r_3^2 = \frac{4}{7}$ and with associated Delaunay polytope α_2. Set

$$(r_{n+1})^2 = \frac{1}{4(1 - r_n^2)}$$

for $n \geq 3$. Then,

$$r_4^2 = \frac{7}{12},\ r_5^2 = \frac{3}{5},\ r_6^2 = \frac{5}{8},\ r_7^2 = \frac{2}{3},\ r_8^2 = \frac{3}{4},\ r_9^2 = 1.$$

From Lemma 14.4.5, we obtain that $d(K_n \setminus P_3)$ has a spherical representation with radius r_n if $n \leq 9$ and that $d(K_{10} \setminus P_3)$ is of negative type but has no spherical representation. Fom Proposition 14.4.4 applied to $2d(K_n \setminus P_3)$, we obtain that $d(K_n \setminus P_3)$ is hypermetric if $n \leq 8$. It is known that the Delaunay polytope associated with $2d(K_7 \setminus P_3)$ is the Schläfli polytope 2_{21} and that the Delaunay polytope associated with $2d(K_8 \setminus P_3)$ is the Gosset polytope 3_{21} (see Section 16.2). Hence, Proposition 14.4.6 yields that $d(K_9 \setminus P_3)$ is not hypermetric (since 3_{21} is centrally symmetric). Finally, note that $d(K_n \setminus P_3)$ is ℓ_1-embeddable if $n \leq 6$ but not for $n = 7$. This latter statement will be justified later in Remark 19.2.9. ∎

At this point let us observe as a curiosity an analogy between the minimum ℓ_1-size and the radius of the Delaunay polytope associated to an ℓ_1-embeddable distance. Let (X, d) be a distance space with $|X| = n$ and consider the iterated spherical t-extension $\mathrm{sph}_t^m(d)$ of d. Part of Proposition 14.4.6 can be rephrased as follows:

(a) Suppose $d \in \mathrm{HYP}_n$. Then, $\mathrm{sph}_t^m(d) \in \mathrm{HYP}_{n+m}$ for all $m \geq 1$ if $t \geq \frac{1}{2}(\mathrm{diam}(P_d))^2$, where $\mathrm{diam}(P_d)$ denotes the diameter of the sphere circumscribing the Delaunay polytope P_d associated with d.

Compare (a) with Lemma 7.3.3, which states:

(b) Suppose $d \in \mathrm{CUT}_n$. Then, $\mathrm{sph}_t^m(d) \in \mathrm{CUT}_{n+m}$ for all $m \geq 1$ if $t \geq \frac{1}{2}s_{\ell_1}(d)$, where $s_{\ell_1}(d)$ denotes the minimum ℓ_1-size of d.

Observe, moreover, that the limit value of $(\mathrm{diam}(P_{\mathrm{sph}_t^m(d)}))^2$ in (a) and of $s_{\ell_1}(\mathrm{sph}_t^m(d))$ in (b) is equal to $2t$ when m goes to infinity.

Example 14.4.10. Set $d := d(K_n)$. Then, d is hypermetric with radius $\sqrt{\frac{n-1}{2n}}$ (by Proposition 14.4.3). Then,

$$\mathrm{sph}_t(d) \in \mathrm{NEG}_{n+1} \Longleftrightarrow t \geq \frac{n-1}{2n}$$

(by Lemma 14.4.5 (i)) and it is easy to check that

$$\mathrm{sph}_t(d) \in \mathrm{MET}_{n+1} \text{ (or } \mathrm{CUT}_{n+1}, \text{ or } \mathrm{HYP}_{n+1}) \Longleftrightarrow t \geq \frac{1}{2}.$$

■

Example 14.4.11. Set $d := d(K_{n\times 2})$. Then, d is hypermetric with radius $\frac{1}{\sqrt{2}}$ (by Proposition 14.4.3). One can check that, for each $m \geq 1$,

$$\mathrm{sph}_t^m(d) \in \mathrm{MET}_{n+m} \text{ (or } \mathrm{CUT}_{n+m}, \text{ or } \mathrm{HYP}_{n+m}) \Longleftrightarrow t \geq 1.$$

(Apply Proposition 14.4.6 (i) and relations (a),(b) above, after noting that $s_{\ell_1}(d) = 2$.) ■

Chapter 15. Delaunay Polytopes: Rank and Hypermetric Faces

There is a natural notion of rank for hypermetric spaces. Namely, if (X, d) is a hypermetric space, then its rank $\text{rk}(X, d)$ is defined as the dimension of the smallest face of the cone $\text{HYP}(X)$ that contains d. The extremal cases when the rank of (X, d) or its corank is equal to 1 correspond, respectively, to the cases when d lies on an extreme ray or on a facet of $\text{HYP}(X)$. Correspondingly, the rank $\text{rk}(P)$ of a Delaunay polytope P is defined as the rank of its Delaunay polytope space $(V(P), d^{(2)})$. Delaunay polytopes of rank 1 are called extreme; they are associated to hypermetrics lying on an extreme ray of the hypermetric cone. This notion of rank for a Delaunay polytope P has the following geometric interpretation: It coincides with the number of degrees of freedom one has when deforming P in such a way that the deformed polytope remains a Delaunay polytope; a precise formulation can be found in Theorem 15.2.5.

Extreme Delaunay polytopes have a highly rigid geometric structure; indeed, their only affine transforms which are still Delaunay polytopes are their homothetic transforms (see Corollary 15.2.4). The first example of an extreme Delaunay polytope is the segment α_1, associated with the cut semimetrics. Other examples are known, such as the Schläfli polytope 2_{21}, the Gosset polytope 3_{21} constructed from the root lattice E_8 and some others constructed from the Leech lattice Λ_{24} and the Barnes-Wall lattice Λ_{16}; they will be described in the next Chapter 16.

Delaunay polytopes of corank 1, which correspond to facets of the hypermetric cone, are well understood; they are the repartitioning polytopes considered by Voronoi (see Section 15.2).

In Section 15.1, we study several properties for the notion of rank of a Delaunay polytope; in particular, its invariance under taking generating subspaces (see Theorem 15.1.8) and its additivity with respect to the direct product of polytopes (see Proposition 15.1.10). We present in Section 15.3 some bounds for the rank of a Delaunay polytope in terms of its number of vertices (see Proposition 15.3.1). We also investigate in detail in Section 15.2 the links between faces of the hypermetric cone and their associated Delaunay polytopes.

15.1 Rank of a Delaunay Polytope

We consider here the notion of rank for a Delaunay polytope; we follow essentially Deza, Grishukhin and Laurent [1992], where this notion was introduced and studied. Let (X,d) be a hypermetric space. We define its *annullator*[1] $\mathrm{Ann}(X,d)$ by

$$\mathrm{Ann}(X,d) = \{b \in \mathbb{Z}^X \mid b \neq e_i \text{ for } i \in X, \sum_{i\in X} b_i = 1, \sum_{i,j\in X} b_i b_j d(i,j) = 0\}.$$

Let $\mathcal{S}(X,d)$ denote the system which consists of the equations:

$$\sum_{i,j\in X} b_i b_j x(i,j) = 0 \quad \text{for } b \in \mathrm{Ann}(X,d);$$

that is, $\mathcal{S}(X,d)$ consists of the hypermetric inequalities that are satisfied at equality by d. Let $F(X,d)$ (or $F(d)$) denote the smallest (by inclusion) face of the hypermetric cone $\mathrm{HYP}(X)$ that contains d. Hence,

$$F(X,d) = \mathrm{HYP}(X) \cap \bigcap_{b\in \mathrm{Ann}(X,d)} H_b,$$

where H_b denotes the hyperplane in $\mathbb{R}^{\binom{|X|}{2}}$ defined by the equation:

$$\sum_{i,j\in X} b_i b_j x(i,j) = 0.$$

The dimension of $F(X,d)$ is equal to the rank of the solution set to the system $\mathcal{S}(X,d)$.

Definition 15.1.1.

(i) *The* rank $\mathrm{rk}(X,d)$ *of a hypermetric space* (X,d) *is defined as the dimension of the smallest face* $F(X,d)$ *of* $\mathrm{HYP}(X)$ *that contains* d. *Its* corank *is defined as* $\binom{|X|}{2} - \mathrm{rk}(X,d)$.

(ii) *The* rank $\mathrm{rk}(P)$ *of a Delaunay polytope* P *is defined as the rank of the Delaunay polytope space* $(V(P), d^{(2)})$, *i.e.*, $\mathrm{rk}(P) := \mathrm{rk}(V(P), d^{(2)})$. *A Delaunay polytope of rank* 1 *is called* extreme.

Hence, $\mathrm{rk}(X,d) = 1$ if d lies on an extreme ray of the hypermetric cone; $\mathrm{rk}(X,d) = \binom{|X|}{2}$ if d lies in the interior of $\mathrm{HYP}(X)$, i.e., $F(X,d) = \mathrm{HYP}(X)$; and $\mathrm{rk}(X,d) = \binom{|X|}{2} - 1$ if $F(X,d)$ is a facet of $\mathrm{HYP}(X)$.

In fact, the rank of a hypermetric space is an invariant of its associated Delaunay polytope; namely, $\mathrm{rk}(X,d) = \mathrm{rk}(P_d)$ holds (see Corollary 15.1.9). In order to prove this invariance result, we need to investigate some properties of the system $\mathcal{S}(X,d)$ of hypermetric equalities satisfied by d.

We first observe that a hypermetric space and any gate 0-extension of it have the same rank. This means that we may consider only metrics rather than semimetrics. Let (X,d) be a distance space and $i_0 \in X$, $j_0 \notin X$. Let $(X' :=$

[1]This notion was already used in the proof of Theorem 14.2.1.

$X \cup \{j_0\}, d')$ be its gate 0-extension, defined by $d'(i_0, j_0) = 0$, $d'(i, j_0) = d(i_0, i)$ for $i \in X$, and $d'(i,j) = d(i,j)$ for $i, j \in X$.

Lemma 15.1.2. *Let (X', d') be a gate 0-extension of the hypermetric space (X, d). Then,*

$$\mathrm{rk}(X', d') = \mathrm{rk}(X, d).$$

Proof. Suppose that (X', d') is defined as above. We show that the solution sets of the systems $\mathcal{S}(X, d)$ and $\mathcal{S}(X', d')$ have the same rank. Since $\mathcal{S}(X, d)$ is a subsystem of the system $\mathcal{S}(X', d')$, it suffices to check that each additional variable $x(i, j_0)$ ($i \in X$) in the system $\mathcal{S}(X', d')$ can be expressed in terms of the variables $x(i,j)$ ($i, j \in X$). As $d(i_0, j_0) = 0$, the triangle equalities $x(i_0, i) - x(j_0, i) - x(i_0, j_0) = 0$ and $x(j_0, i) - x(i_0, i) - x(i_0, j_0) = 0$ belong to the system $\mathcal{S}(X', d')$. This implies that the equality $x(j_0, i) = x(i_0, i)$ ($i \in X$) follows from $\mathcal{S}(X', d')$. ∎

Let $P \subseteq \mathbb{R}^k$ be a k-dimensional Delaunay polytope with set of vertices $V(P)$ and let $V \subseteq V(P)$ be a generating subset of $V(P)$. For $w \in V(P)$, every $a \in \mathbb{Z}^V$ such that $w = \sum_{v \in V} a_v v$ and $\sum_{v \in V} a_v = 1$ is called an *affine realization* of w *in the set* V. Proposition 14.1.5 implies that, for $b \in \mathbb{Z}^V$ with $\sum_{v \in V} b_v = 1$,

$$b \in \mathrm{Ann}(V, d^{(2)}) \iff \sum_{v \in V} b_v v \in V(P).$$

In other words, there is a one-to-one correspondence between

(i) the equations of $\mathcal{S}(V, d^{(2)})$, and

(ii) the affine realizations of the vertices of P in the set V.

In particular, if B is a basic set in $V(P)$, then each vertex has a unique affine realization in the set B and, therefore, $\mathcal{S}(B, d^{(2)})$ is a system of $|V(P) \setminus B| = |V(P)| - k - 1$ equations in $\binom{k+1}{2}$ variables. Therefore,

$$\binom{k+2}{2} - |V(P)| \le \mathrm{rk}(B, d^{(2)}) \le \binom{k+1}{2}. \tag{15.1.3}$$

Lemma 15.1.4. *Let V be a generating subset of $V(P)$ and let $c \in \mathbb{Z}^V$ such that $\sum_{v \in V} c_v = 0$ and $\sum_{v \in V} c_v v = 0$. The following equations:*

$$\sum_{v \in V} c_v x(u, v) = 0 \quad \text{for } u \in V, \tag{15.1.5}$$

$$\sum_{u,v \in V} c_u c_v x(u, v) = 0 \tag{15.1.6}$$

are implied by the system $\mathcal{S}(V, d^{(2)})$.

Proof. Let $u \in V$ and $c' \in \mathbb{Z}^V$ be defined by $c'_u := c_u + 1$, $c'_v := c_v$ for $v \in V \setminus \{u\}$. Hence, c' is an affine realization of u in V. Therefore, the equation:

$$\sum_{v,w \in V} c'_v c'_w x(v,w) = 0$$

belongs to the system $\mathcal{S}(V, d^{(2)})$. It can be rewritten as

$$\sum_{v,w \in V} c_v c_w x(v,w) + 2 \sum_{v \in V} c_v x(u,v) = 0.$$

By multiplying the above equality by c_u and summing over $u \in V$, we obtain (15.1.6). Hence, the equation (15.1.6) follows from $\mathcal{S}(V, d^{(2)})$ and, thus, the equations (15.1.5) as well. ∎

For each $w \in V(P)$, let $a^w \in \mathbb{Z}^V$ be a given affine realization of w in the set V. Let $\mathcal{S}'(V, d^{(2)})$ denote the system consisting of the equations (15.1.5) and (15.1.6) together with the hypermetric equations:

$$\sum_{u,v \in V} a^w_u a^w_v x(u,v) = 0 \quad \text{for} \quad w \in V(P).$$

Lemma 15.1.7. *The systems $\mathcal{S}(V, d^{(2)})$ and $\mathcal{S}'(V, d^{(2)})$ have the same solutions.*

Proof. It remains only to show that each equation of $\mathcal{S}(V, d^{(2)})$ follows from the system $\mathcal{S}'(V, d^{(2)})$. Let $w \in V(P)$ and let b be another affine realization of w in V. Then, we can apply (15.1.6) with $c := a^w - b$, which yields

$$\sum_{u,v \in V} (a^w_u - b_u)(a^w_v - b_v) x(u,v) = 0, \quad \text{i.e.,}$$

$$\sum_{u,v \in V} a^w_u a^w_v x(u,v) - 2 \sum_{u,v \in V} a^w_u b_v x(u,v) + \sum_{u,v \in V} b_u b_v x(u,v) = 0.$$

We check that the first two terms in the above equality are equal to zero, which will imply that the equation:

$$\sum_{u,v \in V} b_u b_v x(u,v) = 0$$

follows from the system $\mathcal{S}'(V, d^{(2)})$. Indeed, the first term is equal to 0, since it corresponds to an equation of $\mathcal{S}'(V, d^{(2)})$. On the other hand, for $u \in V$, the equation:

$$\sum_{v \in V} a^w_v x(u,v) = \sum_{v \in V} b_v x(u,v)$$

follows from (15.1.5). Hence, the second term is equal to

$$-2 \sum_{u,v \in V} a^w_u a^w_v x(u,v) = 0. \; ∎$$

Theorem 15.1.8. *Let P be a Delaunay polytope and let V be a generating subset of $V(P)$. Then,*

$$\text{rk}(V, d^{(2)}) = \text{rk}(V(P), d^{(2)}).$$

Proof. We show that the solution sets to the systems $\mathcal{S}(V, d^{(2)})$ and $\mathcal{S}(V(P), d^{(2)})$ have the same rank. Since $\mathcal{S}(V, d^{(2)})$ is a subsystem of the system $\mathcal{S}(V(P), d^{(2)})$, it suffices to check that each variable $x(w, w')$ ($(w, w') \in (V \times (V(P) \setminus V)) \cup (V(P) \setminus V)^2$) can be expressed in terms of the variables $x(u, v)$ ($u, v \in V$).

Let $w, w' \in V(P) \setminus V$ and let a, a' denote affine realizations of w, w' in V, respectively. We show that the following equations (a) and (b) are implied by $\mathcal{S}(V(P), d^{(2)})$:

(a) $$x(w, u) = \sum_{v \in V} a_v x(u, v) \quad \text{for} \quad u \in V,$$

(b) $$x(w, w') = \sum_{u,v \in V} a_u a'_v x(u, v).$$

For this, let $b, b' \in \mathbb{Z}^{V(P)}$ be defined by $b_w := -1$, $b_v := a_v$ for $v \in V$ and $b_v := 0$ for $v \in V(P) \setminus (V \cup \{w\})$, $b'_{w'} := -1$, $b'_v := a'_v$ for $v \in V$ and $b'_v := 0$ for $v \in V(P) \setminus (V \cup \{w'\})$. We can apply Lemma 15.1.4. From (15.1.5), we obtain:

$$\sum_{v \in V(P)} b_v x(u, v) = 0 \quad \text{for all } u \in V,$$

which implies (a). Applying (15.1.6) for b, b' and $b + b'$, we obtain:

$$\sum_{u,v \in V(P)} b_u b_v x(u, v) = 0, \quad \sum_{u,v \in V(P)} b'_u b'_v x(u, v) = 0,$$

$$\sum_{u,v \in V(P)} (b_u + b'_u)(b_v + b'_v) x(u, v) = 0.$$

This implies:

$$\sum_{u,v \in V(P)} b_u b'_v x(u, v) = 0.$$

Expressing b' in terms of a' and b in terms of a in the latter equality, we obtain:

$$x(w, w') - \sum_{u \in V} a_u x(u, w') - \sum_{v \in V} a'_v x(w, v) + \sum_{u,v \in V} a_u a'_v x(u, v) = 0.$$

The second and third terms in the above equality are equal to $x(w, w')$ by (a). This shows that (b) holds. ∎

Corollary 15.1.9. *Let (X, d) be a hypermetric space with associated Delaunay polytope P_d. Then,*

$$\text{rk}(X, d) = \text{rk}(P_d).$$

Proof. Let $V \subseteq V(P_d)$ representing (X,d). Then, $\text{rk}(X,d) = \text{rk}(V,d^{(2)})$ by Lemma 15.1.2 and $\text{rk}(V,d^{(2)}) = \text{rk}(P_d)$ by Theorem 15.1.8 as V generates $V(P_d)$. ∎

We conclude this section with an additivity property of the rank of a Delaunay polytope.

Proposition 15.1.10. *Let P_1 and P_2 be Delaunay polytopes. Their direct product $P_1 \times P_2$ is a Delaunay polytope with rank*

$$\text{rk}(P_1 \times P_2) = \text{rk}(P_1) + \text{rk}(P_2).$$

∎

For instance, $\text{rk}(\gamma_k) = k$ for the hypercube γ_k, since γ_k is the direct product $(\gamma_1)^k$ and $\text{rk}(\gamma_1) = 1$.

15.2 Delaunay Polytopes Related to Faces

15.2.1 Hypermetric Faces

We show that hypermetrics that lie in the interior of the same face of the hypermetric cone are associated with affinely equivalent Delaunay polytopes; therefore, one can speak of *the* Delaunay polytope associated with a face of the hypermetric cone. This result is presented in Theorem 15.2.1 below; it was proved by Deza, Grishukhin and Laurent [1992].

Let T be an affine bijection of $\mathbb{R}^k$. For $u, v \in \mathbb{R}^k$, set

$$d_T(u,v) := (T(u) - T(v))^2.$$

Theorem 15.2.1. *Let $P \subseteq \mathbb{R}^k$ be a Delaunay polytope and let V be a generating subset of $V(P)$. Let T be an affine bijection of $\mathbb{R}^k$. Let F denote the smallest face of the hypermetric cone* HYP(V) *that contains $(V, d^{(2)})$.*

(i) *If $T(P)$ is a Delaunay polytope, then d_T lies in the interior of F, i.e., $F(d_T) = F$.*

(ii) *Let $d \in$ HYP(V). If d lies in the interior of F, then the Delaunay polytope P_d associated with d is affinely equivalent to P.*

Proof. (i) Let $b \in \mathbb{Z}^V$ with $\sum_{v \in V} b_v = 1$. We show that

$$\sum_{u,v \in V} b_u b_v (T(u) - T(v))^2 = 0 \Longleftrightarrow \sum_{u,v \in V} b_u b_v (u - v)^2 = 0.$$

For this, let c, r (resp. c_T, r_T) denote the center and radius of the sphere circumscribing P (resp. $T(P)$). Then,

$$\sum_{u,v \in V} b_u b_v (T(u) - T(v))^2 = 2(r_T)^2 - 2(\sum_{u \in V} b_u T(u) - c_T)^2,$$

which is equal to 0 if and only if $\sum_{u\in V} b_u T(u)$ is a vertex of $T(P)$. On the other hand,

$$\sum_{u,v\in V} b_u b_v (u-v)^2 = 2r^2 - 2(\sum_{u\in V} b_u u - c)^2$$

is equal to 0 if and only if $\sum_{u\in V} b_u u$ is a vertex of P. The result now follows as

$$\sum_{u\in V} b_u T(u) = T(\sum_{u\in V} b_u u)$$

is a vertex of $T(P)$ if and only if $\sum_{u\in V} b_u u$ is a vertex of P, since T is an affine bijection. This shows that d_T lies in the interior of the face F.
(ii) Let P_d be the Delaunay polytope associated with d and let $T : V \longrightarrow V(P_d)$ be a generating mapping such that $d(u,v) = (T(u) - T(v))^2$ for $u, v \in V$. The mapping T is one-to-one since $d(u,v) \neq 0$ for $u \neq v \in V$. We show that T can be extended to an affine bijective mapping of the space spanned by V, mapping $V(P)$ to $V(P_d)$.

First, we verify that T preserves the affine dependencies on V, i.e., that if $c \in \mathbb{Z}^V$ with $\sum_{v\in V} c_v = 0$, then

$$\sum_{v\in V} c_v v = 0 \Longleftrightarrow \sum_{v\in V} c_v T(v) = 0.$$

Since all the vectors $v \in V$ lie on a sphere and the same holds for their images $T(v)$, we have:

(a) $$\sum_{u,v\in V} c_u c_v d^{(2)}(u,v) = \sum_{u,v\in V} c_u c_v (u-v)^2 = -2\left(\sum_{v\in V} c_v v\right)^2,$$

(b) $$\sum_{u,v\in V} c_u c_v d(u,v) = \sum_{u,v\in V} c_u c_v (T(u) - T(v))^2 = -2\left(\sum_{v\in V} c_v T(v)\right)^2.$$

By assumption, $F(d) = F$, i.e., the systems $\mathcal{S}'(V,d)$ and $\mathcal{S}'(V,d^{(2)})$ have the same sets of solutions (using Lemma 15.1.7). This implies that the quantities in (a) and (b) are simultaneously equal to zero.

We now check that, for $b \in \mathbb{Z}^V$ with $\sum_{v\in V} b_v = 1$,

$$\sum_{v\in V} b_v v \text{ is a vertex of } P \Longleftrightarrow \sum_{v\in V} b_v T(v) \text{ is a vertex of } P_d.$$

But, by Proposition 14.1.5, $\sum_{v\in V} b_v v \in V(P)$ if and only if $d^{(2)}$ satisfies the equation $\sum_{u,v\in V} b_u b_v x(u,v) = 0$ and $\sum_{v\in V} b_v T(v) \in V(P_d)$ if and only if d satisfies the same equation. Therefore, we can extend T to the space spanned by V by setting

$$T(\sum_{v\in V} b_v v) := \sum_{v\in V} b_v T(v);$$

T is affine bijective and maps P on P_d. ∎

Corollary 15.2.2. *Let (X, d) and (X, d') be two hypermetric spaces with associated Delaunay polytopes P_d and $P_{d'}$. Let $F(d)$ and $F(d')$ denote the smallest faces of* HYP(X) *that contain d and d', respectively. If $F(d) = F(d')$, then P_d and $P_{d'}$ are affinely equivalent.*

Proof. If $d(i,j) \neq 0$ for all $i \neq j \in X$, then (X, d) is isomorphic to a subspace $(V, d^{(2)})$ of $(V(P_d), d^{(2)})$, where V is a generating subset of $V(P_d)$. As d' lies in the interior of $F(d)$, Theorem 15.2.1 (ii) implies that $P_{d'}$ is affinely equivalent to P_d. Otherwise, note that $d(i,j) = 0$ if and only if $d'(i,j) = 0$ as d, d' satisfy the same triangle equalities since $F(d) = F(d')$. Hence, X can be partitioned into $X_1 \cup \ldots \cup X_m$ for $m < |X|$ in such a way that, for $i \neq j \in X$, $d(i,j) = d'(i,j) = 0$ if and only if $i, j \in X_h$ for some $h \in \{1, \ldots, m\}$. Let $x_1 \in X_1, \ldots, x_m \in X_m$ and denote by d_0 (resp. d'_0) the projection of d (resp. d') on $X_0 := \{x_1, \ldots, x_m\}$. One can easily see that d_0,d'_0 lie in the interior of the same face of HYP(X_0) which, by the above argument, implies that P_d and $P_{d'}$ are affinely equivalent (as d_0 is associated to the same Delaunay polytope P_d as d and d'_0 is associated to $P_{d'}$). ∎

Remark 15.2.3. The implication from Corollary 15.2.2 is strict, in general. Indeed, it is easy to construct examples of hypermetrics $d, d' \in$ HYP(X) whose associated Delaunay polytopes are affinely equivalent, but lying on distinct faces of HYP(X), i.e., such that $F(d) \neq F(d')$. For instance, any two distinct cut semimetrics $\delta(S)$, $\delta(T)$ are associated to the same Delaunay polytope, namely α_1, but they lie on distinct faces as each cut semimetric lies on an extreme ray of HYP(X). Another example can be easily obtained by taking a generating subset $V := \{v_1, \ldots, v_n\}$ of a Delaunay polytope P and considering the hypermetrics $d, d' \in$ HYP$_n$ defined by $d(i,j) = (v_i - v_j)^2$, $d'(i,j) = (v_{\sigma(i)} - v_{\sigma(j)})^2$ for $i, j \in \{1, \ldots, n\}$, where σ is a permutation of $\{1, \ldots, n\}$. ∎

Corollary 15.2.4. *Let P be a Delaunay polytope in $\mathbb{R}^k$. Then, P is extreme if and only if the only (up to orthogonal transformation and translation) affine bijective transformations T of $\mathbb{R}^k$ for which $T(P)$ is a Delaunay polytope are the homotheties*[2].

Proof. Suppose first that rk$(P) = 1$, i.e., that $(V(P), d^{(2)})$ lies on an extreme ray of HYP$(V(P))$. Assume that $T(P)$ is a Delaunay polytope. By Theorem 15.2.1 (i), $d_T = \lambda^2 d^{(2)}$ for some scalar λ. Hence, $(T(u) - T(v))^2 = \lambda^2 (u - v)^2$ for all $u, v \in V(P)$. It is not difficult to see that, up to translation, $\lambda^{-1}T$ is an orthogonal transformation.
Suppose now that the only affine bijective transformations T for which $T(P)$ is a Delaunay polytope are the homotheties. Let $d \in$ HYP$(V(P))$ with $F(d) = F(d^{(2)})$. By Theorem 15.2.1 (ii), the Delaunay polytope P_d associated to d is of the form λP, where $\lambda > 0$, implying that $d = \lambda^2 d^{(2)}$. This shows that

[2] A *homothety* is a mapping $T : \mathbb{R}^k \longrightarrow \mathbb{R}^k$ such that $T(x) = \lambda x$ for all $x \in \mathbb{R}^k$, for some $\lambda \neq 0$.

$(V(P), d^{(2)})$ lies on an extreme ray of HYP$(V(P))$, i.e., rk$(P) = 1$. ∎

Let P be a k-dimensional Delaunay polytope in $\mathbb{R}^k$. Let $\mathcal{T}(P)$ denote the set of affine bijections T of $\mathbb{R}^k$ (taken up to translations and orthogonal transformations) for which $T(P)$ is again a Delaunay polytope. If P has rank 1 then, by Corollary 15.2.4, $\mathcal{T}(P)$ consists only of homotheties and, hence, its dimension is also equal to 1. This fact can be generalized for arbitrary ranks, as the next result from Laurent [1996a] shows.

Theorem 15.2.5. *For a Delaunay polytope P, the dimension of the set $\mathcal{T}(P)$ (in the topological sense) is equal to the rank of P.*

Proof. We can suppose without loss of generality that P is a k-dimensional Delaunay polytope in $\mathbb{R}^k$ which contains the origin as a vertex. Let V denote the set of vertices of P, let d_P denote the associated distance on V defined by

$$d_P(u, v) := (u - v)^2 \quad \text{for all } u, v \in V,$$

and let F_P denote the smallest face of the hypermetric cone HYP(V) that contains d_P. Denote by $\mathcal{T}_0(P)$ the set of nonsingular $k \times k$ matrices A for which the polytope $A(P) := \{Ax \mid x \in P\}$ is again a Delaunay polytope. Clearly, the sets $\mathcal{T}(P)$ and $\mathcal{T}_0(P)/\text{OA}(k)$ coincide. (Here, $\mathcal{T}_0(P)/\text{OA}(k)$ denotes the quotient set of $\mathcal{T}_0(P)$ by the set OA(k) of orthogonal matrices of order k.) We show that the dimension of the set $\mathcal{T}_0(P)/\text{OA}(k)$ is equal to the rank of P. For this we need to introduce an intermediary cone $\mathcal{C}_P$.

As the hypermetric cone HYP(V) is polyhedral, we can suppose that it is defined by the hypermetric inequalities $\sum_{u,v \in V, u<v} b_u b_v x_{uv} \leq 0$ for $b \in \mathcal{B}$, where $\mathcal{B}$ is a finite subset of $\{b \in \mathbb{Z}^V \mid \sum_{u \in V} b_u = 1\}$. Let $\mathcal{A}$ denote the subset of $\mathcal{B}$ corresponding to the hypermetric equalities defining the face F_P; that is, $b \in \mathcal{A}$ if $\sum_{u,v \in V} b_u b_v (u - v)^2 = 0$. Let us now introduce the cone $\mathcal{C}_P$ which consists of the symmetric $k \times k$ positive semidefinite matrices M satisfying:

(a) $$\sum_{u,v \in V} b_u b_v (u - v)^T M (u - v) \leq 0 \quad \text{for } b \in \mathcal{B} \setminus \mathcal{A},$$

(b) $$\sum_{u,v \in V} b_u b_v (u - v)^T M (u - v) = 0 \quad \text{for } b \in \mathcal{A}.$$

Then the relative interior of the cone $\mathcal{C}_P$ consists of the positive definite matrices M satisfying (b), and (a) with strict inequality. We claim that

(c) The dimension of the cone $\mathcal{C}_P$ is equal to the rank of P.

Indeed, set $s := \text{rk}(P)$ and $t := \dim(\mathcal{C}_P)$. Let $d_1, \ldots, d_s$ be linearly independent points lying in the relative interior of the face F_P. Applying Theorem 15.2.1 (ii) we obtain $A_1, \ldots, A_s \in \mathcal{T}_0(P)$ such that $d_i(u, v) = (u - v)^T A_i^T A_i (u - v)$ for all $u, v \in V$ and $i = 1, \ldots, s$. Then, $A_1^T A_1, \ldots, A_s^T A_s$ are linearly independent members of the cone $\mathcal{C}_P$, which shows $s \leq t$. Conversely, let $M_1, \ldots, M_t$ be linearly independent and lying in the relative interior of $\mathcal{C}_P$. As each M_h is positive definite, it is of the form $A_h^T A_h$ for some $k \times k$ nonsingular matrix A_h. Set $d_h(u,v) := (u-v)^T M_h (u-v) = (A_h u - A_h v)^2$ for $u, v \in V$ and $h = 1, \ldots, t$. Then, the points $d_1, \ldots, d_t$ lie in the relative interior of the face F_P. Moreover, they are linearly independent, which shows that $t \leq s$. Hence, $s = t$ and (c) holds. (Indeed, suppose $\sum_{h=1}^t \lambda_h d_h = 0$ for some scalars λ_h. Then, $(u-v)^T (\sum_{h=1}^t \lambda_h M_h)(u-v) = 0$ for all $u, v \in V$. As V is full-dimensional and contains the origin, this implies easily that $\sum_{h=1}^t \lambda_h M_h = 0$ and, thus, $\lambda_h = 0$ for all h.)

We can now formulate a homeomorphism (=bicontinuous bijection) between the set $\mathcal{T}_0(P)/\mathrm{OA}(k)$ and the relative interior of the cone $\mathcal{C}_P$. Namely, consider the mapping

$$\theta : A \mapsto A^T A.$$

Then, θ establishes clearly a one-to-one correspondence between $\mathcal{T}_0(P)/\mathrm{OA}(k)$ and the relative interior of $\mathcal{C}_P$. The result now follows as θ is a homeomorphism between, on the one hand, the quotient set of the set of $k \times k$ matrices by the equivalence relation: $A \sim B$ if $A^T A = B^T B$ and, on the other hand, the set of $k \times k$ positive semidefinite matrices. ∎

15.2.2 Hypermetric Facets

Following Avis and Grishukhin [1993], we now describe the Delaunay polytopes which are associated with the facets of the hypermetric cone.

Let Δ_1, Δ_2 be two simplices lying in affine spaces that intersect in one point which belongs to $\Delta_1 \cap \Delta_2$. Then, their convex hull $P := \mathrm{Conv}(\Delta_1 \cup \Delta_2)$ is called a *repartitioning polytope*. This polytope was studied by Voronoi [1908, 1909]. There is only one affine dependency among the vertices of Δ_1 and Δ_2; namely,

$$\sum_{v \in V_1} b_v v = \sum_{v \in V_2} b_v v,$$

where $\sum_{v \in V_1} b_v = \sum_{v \in V_2} b_v = 1$, $b_v \geq 0$ for $v \in V_1 \cup V_2$, and V_i denotes the set of vertices of Δ_i, $i = 1, 2$. Set

$$V_0 := \{v \in V_1 \cup V_2 \mid b_v = 0\}.$$

Then,

$$P_1 := \mathrm{Conv}(V_1 \cup V_2 \setminus V_0)$$

is also a repartitioning polytope, with the same affine dependency between its vertices as P and

$$P = \prod_{v \in V_0} \mathrm{Pyr}_v(P_1).$$

We denote the repartitioning polytope P by $P^m_{p,q}$, where $m = |V_0|$, $p+1 = |V_1 \setminus V_0|$ and $q+1 = |V_2 \setminus V_0|$. Hence, $P^m_{p,q}$ has $m+p+q+2$ vertices (if $p, q \geq 1$) and its dimension is $m+p+q$. Note that $P^m_{p,q}$ does not denote a concrete polytope, but a class of affinely equivalent repartitioning polytopes.

We now show that the Delaunay polytope associated with a facet of the hypermetric cone is a repartitioning polytope. Let $b \in \mathbb{Z}^X$ with $\sum_{i\in X} b_i = 1$ and suppose that the hypermetric equation:

$$\sum_{i,j\in X} b_i b_j x(i,j) = 0 \tag{15.2.6}$$

defines a facet of HYP(X). Let $d \in$ HYP(X) and suppose that d lies in the interior of this facet, i.e., that (15.2.6) is the only hypermetric equality satisfied by d. In particular, $d(i,j) > 0$ for distinct i,j (else, d would satisfy $2(|X|-2)$ triangle equalities).

Proposition 15.2.7. *Let P_d be the Delaunay polytope associated with d lying in the interior of the facet defined by* (15.2.6). *Then, P_d is a repartitioning polytope $P^m_{p,q}$ where $m = |\{i \mid b_i = 0\}|$, $p+1 = |\{i \mid b_i > 0\}|$ and $q = |\{i \mid b_i < 0\}|$. Moreover, P_d is basic.*

Proof. Let $(v_i \mid i \in X)$ denote the representation of d on $V(P_d)$. From Proposition 14.1.5, the equality (15.2.6) is equivalent to the point

$$v_0 := \sum_{i\in X} b_i v_i \tag{15.2.8}$$

being a vertex of P_d. From Proposition 14.1.5 again and the fact that (15.2.6) is the only hypermetric equality satisfied by d, we deduce that $v_0 \notin \{v_i \mid i \in X\}$, $V(P_d) = \{v_i \mid i \in X\} \cup \{v_0\}$ and the set $\{v_i \mid i \in X\}$ is affinely independent. Hence, P_d has $|X|+1$ vertices and $\sum_{v\in V(P_d)} b_v v = 0$ is the only affine dependency between the vertices of P_d, after setting $b_{v_i} := b_i$ for $i \in X$ and $b_{v_0} := -1$. Set

$$V_0 := \{v \in V(P_d) \mid b_v = 0\},\ V_+ := \{v \in V(P_d) \mid b_v > 0\},$$

$$V_- := \{v \in V(P_d) \mid b_v < 0\},\ m := |V_0|, p+1 := |V_+| \text{ and } q+1 := |V_-|.$$

Then, $P_1 := \mathrm{Conv}(V_+ \cup V_-)$ is a repartitioning polytope $P^0_{p,q}$ and the polytope $P_d := \prod_{v\in V_0} \mathrm{Pyr}_v(P_1)$ is a repartitioning polytope $P^m_{p,q}$. ∎

As we see in Example 15.2.10 (iii) below, there exist distinct hypermetric facets for which the b_i's have the same numbers of positive and negative components; hence, they correspond to repartitioning polytopes with the same parameters p and q. For this reason, we also denote the repartitioning polytope associated with the hypermetric facet (15.2.6) by $P^m_{p,q}(b)$. Note that the matrix Y_γ characterizing the type of the repartitioning polytope $P^m_{p,q}(b)$ is of the form $\left[\frac{I_n}{b_1\ldots b_n}\right]$ (recall Section 13.3).

We summarize in Figure 15.2.9 the main facts we know about the connections between faces of the hypermetric cone and their associated Delaunay polytopes.

For the first two equivalences, see Examples 14.1.6 and 14.1.7 and, for the last four equivalences, see, respectively, Propositions 14.1.10, 15.2.7, Theorem 15.1.8 and Corollary 15.2.2.

hypermetric d		Delaunay polytope P_d
d is a cut semimetric	$\Longleftrightarrow$	$P_d = \alpha_1$
$F(d) = \text{HYP}_{n+1}$	$\Longleftrightarrow$	$P_d = \alpha_n$
$d \in \text{CUT}_{n+1}$	$\Longleftrightarrow$	$V(P)$ is contained in the set of vertices of a parallepiped
$F(d)$ is a facet	$\Longleftrightarrow$	P_d is a repartitioning polytope
$F(d)$ is an extreme ray	$\Longleftrightarrow$	P_d is extreme
$F(d) = F(d')$	$\Longrightarrow$	$P_d, P_{d'}$ are affinely equivalent

Figure 15.2.9: Hypermetric faces and Delaunay polytopes

Example 15.2.10.

(i) Let (15.2.6) be the triangle equality $x(1,2) - x(1,3) - x(2,3) \leq 0$, i.e., $b_1 = b_2 = 1, b_3 = -1$ and $b_i = 0$ otherwise. Then, (15.2.8) reads $v_0 = v_1 + v_2 - v_3$ and $V_+ = \{v_1, v_2\}$, $V_- = \{v_3, v_0\}$. Therefore, the Delaunay polytope associated with a triangle facet is $P^0_{1,1}$ or, more precisely, $P^0_{1,1}(1,1,-1)$, a rectangle whose diagonals are the segments $[v_1, v_2]$ and $[v_0, v_3]$.

(ii) Let (15.2.6) be a pentagonal facet, i.e., $b_1 = b_2 = b_3 = 1$, $b_4 = b_5 = -1$. Then, (15.2.8) reads $v_0 = v_1 + v_2 + v_3 - v_4 - v_5$. Therefore, the Delaunay polytope associated with the pentagonal facet is $P^0_{2,2}$ or, more precisely, $P^0_{2,2}(1,1,1,-1,-1)$, the convex hull of two intersecting triangles.

(iii) Set $b_1 := (2,2,2,1,1,1,-2,-2,-2,-1,-1)$, $b_2 := (1,1,1,1,1,1,-1,-1,-1,-1,-1)$. Then, (15.2.6) defines a facet for both b_1 and b_2. Hence, these two facets are associated with repartitioning polytopes with the same parameters $p = q = 5$ (with, of course, distinct affine dependencies (15.2.8) between their vertices). ∎

Remark 15.2.11. We group here several observations on hypermetric facets.

(i) For $n \leq 6$, $\text{HYP}_n = \text{CUT}_n$ and all the facets of HYP_n are known (see Section 30.6). Namely, for $n = 3,4$, the facets of HYP_n are defined by the triangle inequalities. For $n = 5$, they are defined by the triangle inequalities and the pentagonal inequalities (6.1.9). For $n = 6$, they are defined by the triangle inequalities, the pentagonal inequalities, and the inequalities $\sum_{1\le i<j\le 6} b_ib_jx_{ij} \le 0$ for $b = (2,1,1,-1,-1,-1)$ and $(-2,-1,1,1,1,1)$ (up to permutation of the components). (The description in the case $n = 6$ was obtained independently by Baranovskii [1971] and Avis [1989].)

(ii) When $n = 7$, the hypermetric inequality $\sum_{1\le i<j\le 7} b_ib_jx_{ij} \le 0$ for $b = (3,1,1,1,-1,-2,-2)$ defines a facet of HYP_7, but not of CUT_7. Indeed, there are precisely 19 linearly independent cut semimetrics satisfying this hypermetric inequality at equality. An additional hypermetric distance d satisfying equality can be obtained in the following manner: Consider the graph $\overline{G_9}$ shown in Figure 16.2.4 (labeling its nodes as $1,2,3,4,5,6,7$ corresponding to degrees $3,2,2,2,5,1,1$) and set $d_{ij} := 2$ if ij is an edge of $\overline{G_9}$, $d_{ij} := 1$ if ij is not an edge in $\overline{G_9}$. This distance d together with 19 cut semimetrics form a set of 20 linearly independent vectors satisfying the hypermetric equality; this shows that it defines a facet of HYP_7.

In fact, Baranovskii [1995] describes all the facets of the cone HYP_7. They are the hypermetric facets for CUT_7 (see their list in Section 30.6) together with the facets defined by the inequalities $\sum_{1\le i<j\le 7} b_ib_jx_{ij} \le 0$, for $b = (3,1,1,1,-1,-2,-2)$, $(-3,1,1,1,1,-2,2)$, $(3,-1,-1,-1,1,-2,2)$, and $(-3,1,1,-1,-1,2,2)$ (up to permutation).

(iii) There is an easy way of constructing new hypermetric facets from given ones, namely, using the so-called 'switching' operation. This operation will be described in detail in Section 26.3; we indicate here how it acts on hypermetric inequalities. Given $b \in \mathbb{Z}^n$ and a subset $A \subset V_n := \{1,\ldots,n\}$, define the vector $b^A \in \mathbb{Z}^n$ by $b_i^A := -b_i$ if $i \in A$ and $b_i^A := b_i$ if $i \in V_n \setminus A$. If $\sum_{i=1}^n b_i = 1$ and $b(A) = 0$, then the inequality $\sum_{1\le i<j\le n} b_i^Ab_j^Ax_{ij} \le 0$ is again a hypermetric inequality. In fact,

(a) The hypermetric inequality $\sum_{1\le i<j\le n} b_ib_jx_{ij} \le 0$ defines a facet of $\text{HYP}_n \Longleftrightarrow$ its switching $\sum_{1\le i<j\le n} b_i^Ab_j^Ax_{ij} \le 0$ defines a facet of HYP_n.

See for an example the facets cited in (ii) above. We briefly sketch the proof for assertion (a).

Consider the mapping $r_{\delta(A)} : \mathbb{R}^{\binom{n}{2}} \longrightarrow \mathbb{R}^{\binom{n}{2}}$ where $y = r_{\delta(A)}(x)$ is defined by $y_{ij} := 1 - x_{ij}$ if $\delta(A)_{ij} = 1$ and $y_{ij} := x_{ij}$ if $\delta(A)_{ij} = 0$. As HYP_n is a polyhedral cone, we can suppose that HYP_n is defined by the hypermetric inequalities $\sum_{1\le i<j\le n} b_ib_jx_{ij} \le 0$ for $b \in \mathcal{B}$, where $\mathcal{B}$ is a finite subset of $\{b \in \mathbb{Z}^n \mid \sum_{i=1}^n b_i = 1\}$. Set

$$\mathcal{B}' := \{b^A \mid b \in \mathcal{B}, A \subseteq V_n\}.$$

For $d \in \text{HYP}_n$, set

$$\alpha_d := \min \frac{(\sum_{i=1}^n b_i)^2 - 1}{4\sum_{1\le i<j\le n} b_ib_jd_{ij}}$$

where the minimum is taken over all $b \in \mathcal{B}'$ for which $\sum_{1\le i<j\le n} b_ib_jd_{ij} > 0$, setting $\alpha_d := 1$ if there is no such b. One can easily verify that $r_{\delta(A)}(\alpha_d d) \in$

HYP_n for all $d \in \mathrm{HYP}_n$. We can now show (a). If $\sum_{1\leq i<j\leq n} b_i b_j x_{ij} \leq 0$ defines a facet of HYP_n, then we can find $\binom{n}{2}$ affinely independent vectors $d_1 := 0, d_2, \ldots, d_{\binom{n}{2}} \in \mathrm{HYP}_n$ satisfying the equality $\sum_{1\leq i<j\leq n} b_i b_j x_{ij} = 0$. Then, the vectors $r_{\delta(A)}(\alpha_{d_i} d_i)$ $(i = 1, \ldots, \binom{n}{2})$ are affinely independent and satisfy the equality $\sum_{1\leq i<j\leq n} b_i^A b_j^A x_{ij} = 0$. This shows that (a) holds. ∎

We conclude this section with an observation on Delaunay polytopes with small corank. We recall Problem 13.2.3, which asks whether every Delaunay polytope is basic. This is indeed the case for simplices and repartitioning polytopes, i.e., for Delaunay polytopes associated with hypermetrics with corank 0 and 1. We extend this fact to the case of hypermetrics with corank 2 and 3.

Proposition 15.2.12. *Let P be a k-dimensional Delaunay polytope and let V be a generating subset of $V(P)$. If*

$$\binom{|V|}{2} - \mathrm{rk}(V, d^{(2)}) \leq 3,$$

then P is basic.

Proof. We show that V is affinely independent, which implies that P is basic. Suppose, for contradiction, that $\sum_{v\in C} b_v v = 0$ is an affine dependency with $C \subseteq V$ and $b_v \neq 0$ for $v \in C$. By Lemma 15.1.4, the equations $\sum_{v\in C} b_v x(u,v) = 0$ (for $u \in V$) follow from the system $\mathcal{S}(V, d^{(2)})$. One can check that the matrix of the subsystem $\sum_{v\in C} b_v x(u,v) = 0$ (for $u \in C$) has full rank $|C|$. Since the corank of $(V, d^{(2)})$ is equal to the rank of the matrix of the system $\mathcal{S}(V, d^{(2)})$, we deduce that $\mathrm{corank}(V, d^{(2)}) \geq |C|$, implying that $|C| \leq 3$. Hence, $C = \{v_1, v_2, v_3\}$ and, for instance, v_3 belongs to the segment $[v_1, v_2]$. So we have a triangle with an obtuse angle, yielding a contradiction. ∎

15.3 Bounds on the Rank of Basic Delaunay Polytopes

In this section, we present some bounds for the rank of a basic Delaunay polytope; they are taken from Deza, Grishukhin and Laurent [1992]. Recall that a Delaunay polytope P is basic if its set of vertices $V(P)$ contains a basis of the lattice generated by $V(P)$.

Proposition 15.3.1. *Let P be a basic k-dimensional Delaunay polytope. Then,*

$$\binom{k+2}{2} - |V(P)| \leq \mathrm{rk}(P) \leq \binom{k+1}{2}, \tag{15.3.2}$$

(15.3.3) $$\mathrm{rk}(P) \geq \binom{k+1}{2} - \frac{|V(P)|}{2} + 1 \text{ if } P \text{ is centrally symmetric.}$$

Proof. (15.3.2) follows immediately from relation (15.1.3) and Theorem 15.1.8. We show (15.3.3). Let B be a basic set in $V(P)$. For each $w \in V(P)$, let a^w denote the affine realization of w in B and let $h(w)$ denote the corresponding hypermetric equality of the system $\mathcal{S}(B, d^{(2)})$,

$$h(w) := \sum_{u,v \in B} a_u^w a_v^w x(u, v).$$

Let $v \in B$. Since $w^* = v + v^* - w$, the affine realization a^{w^*} of w^* in B is given by $a^{w^*} = e_v + a^{v^*} - a^w$, where e_v is the v-th unit vector in $\mathbb{R}^B$. Hence,

$$\begin{aligned} h(w^*) &= h(v^*) + h(w) + 2\sum_{u' \in B} a_{u'}^{v^*} x(v, u') - 2\sum_{u' \in B} a_{u'}^{w} x(v, u') \\ &\quad -2 \sum_{u,u' \in B} a_{u'}^{v^*} a_u^w x(u, u') \\ &= h(w) + \sum_{u \in B} a_u^w \left(h(v^*) - 2x(v,u) + 2\sum_{u' \in B} a_{u'}^{v^*} (x(v,u') - x(u,u')) \right). \end{aligned}$$

If $w \in B$, then $h(w)$ is zero and, thus, the above relation implies:

(15.3.4) $$h(w^*) = h(v^*) - 2x(v, w) + 2\sum_{u' \in B} a_{u'}^{v^*} (x(v, u') - x(w, u')).$$

We deduce from (15.3.4) that, for each $w \in V(P)$,

(15.3.5) $$h(w^*) = h(w) + \sum_{u \in B} a_u^w h(u^*).$$

Relation (15.3.5) applied to $w = v^*$ yields

(15.3.6) $$0 = h(v^*) + \sum_{u \in B} a_u^{v^*} h(u^*).$$

We show that the system $\mathcal{S}(B, d^{(2)})$ can be reduced to a system of $\frac{|V(P)|}{2} - 1$ equations, which implies that the rank of its solution set is greater than or equal to $\binom{k+1}{2} - \frac{|V(P)|}{2} + 1$. Clearly, the basis B contains at most one pair of antipodal points. For a set A, we set $A^* := \{a^* \mid a \in A\}$. Suppose first that B contains no pair of antipodal points. Then, $V(P) = B \cup B^* \cup A \cup A^*$, for some $A \subseteq V(P) \setminus B$. By (15.3.5), each equation $h(a^*) = 0$ $(a \in A)$ follows from the equations $h(u) = 0$ for $u \in A \cup B^*$. In view of (15.3.6), one of the equations $h(b^*) = 0$ $(b \in B)$ follows from the others. Therefore, the system $\mathcal{S}(B, d^{(2)})$ reduces to $|A| + |B^*| - 1 = \frac{|V(P)|}{2} - 1$ equations. Suppose now that B contains one antipodal pair, i.e., $B = B' \cup \{v, v^*\}$ with $|B'| = k - 1$. Then, $V(P) = B \cup (B')^* \cup A \cup A^*$ for some $A \subseteq V(P) \setminus B$. Hence, $\mathcal{S}(B)$ reduces again to $|A| + |(B')^*| = \frac{|V(P)}{2} - 1$ equations. ∎

For example, the k-dimensional simplex α_k has $k+1$ vertices; hence both inequalities in (15.3.2) hold at equality for α_k. It is easy to check that the rank of the k-dimensional cross-polytope β_k is

$$\operatorname{rk}(\beta_k) = \binom{k+1}{2} - k + 1.$$

Hence, β_k realizes equality in the bound (15.3.3).

The next lemma will be useful for computing the rank of Delaunay polytopes.

Lemma 15.3.7. *Let P be a basic k-dimensional centrally symmetric Delaunay polytope and let $B := \{v_0, v_1, \ldots, v_k\}$ be a basic set in $V(P)$. Let H denote the affine space spanned by $B_1 := \{v_1, \ldots, v_k\}$ and set $P_1 := P \cap H$. If P_1 is an asymmetric Delaunay polytope and if there exists $w \in V(P) \setminus H$ such that $w \notin \{v_1^*, \ldots, v_k^*\}$ and $w - v_0 \notin H$, then $\operatorname{rk}(P_1) = \operatorname{rk}(P)$ holds.*

Proof. The set B_1 is basic in $V(P_1) = V(P) \cap H$. Hence, $\operatorname{rk}(P_1)$ is equal to the rank of the solution set to the system $\mathcal{S}(B_1, d^{(2)})$. In order to show that $\operatorname{rk}(P) = \operatorname{rk}(P_1)$, it suffices to check that each variable $x(v_0, v_i)$ $(1 \leq i \leq k)$ can be expressed in terms of the variables $x(v_i, v_j)$ $(1 \leq i, j \leq k)$ in the system $\mathcal{S}(B, d^{(2)})$. Let $a, b \in \mathbb{Z}^{k+1}$ denote the affine realizations of w, v_0^* in B. We have $a_0 \neq 0, 1$ since $w \notin H$ and $w - v_0 \notin H$. Moreover, $b_0 \neq -1$, else the center $\frac{v_0+v_0^*}{2}$ of P would lie in H contradicting the fact that P_1 is asymmetric. Using relation (15.3.4) (applied to $v = v_0$ and $w = v_i$), we deduce that

$$h(v_i^*) = h(v_0^*) - 2x(v_0, v_i) + 2 \sum_{0 \leq j \leq k} b_j (x(v_0, v_j) - x(v_i, v_j)).$$

Set

$$h_i := -2 \sum_{1 \leq j \leq k} b_j x(v_i, v_j)$$

for $1 \leq i \leq k$. Then,

$$h(v_i^*) = h(v_0^*) - 2x(v_0, v_i)(b_0 + 1) + h_i + 2 \sum_{0 \leq j \leq k} b_j x(v_0, v_j).$$

By subtracting the above relations with indices i and 1, we obtain that the equation:

$$\text{(a)} \qquad x(v_0, v_i) = x(v_0, v_1) + \frac{h_i - h_1}{2(b_0 + 1)}$$

follows from $\mathcal{S}(B, d^{(2)})$. Consider now the equation $h(w) = 0$, i.e.,

$$0 = \sum_{1 \leq i < j \leq k} a_i a_j x(v_i, v_j) + \sum_{1 \leq i \leq k} a_i a_0 x(v_0, v_i).$$

Using (a), it can be rewritten as

$$0 = \sum_{1 \le i < j \le k} a_i a_j x(v_i, v_j) + a_0(1 - a_0)x(v_0, v_1) + \frac{a_0}{2(b_0 + 1)} \sum_{1 \le i \le k} a_i(h_i - h_1).$$

Therefore, $x(v_0, v_1)$ and, thus, each $x(v_i, v_0)$ can be expressed in terms of $x(v_i, v_j)$ (for $1 \le i < j \le k$). ∎

Chapter 16. Extreme Delaunay Polytopes

In this chapter, we consider extreme Delaunay polytopes, i.e., Delaunay polytopes with rank 1. A geometric characterization of extreme Delaunay polytopes has been given in Corollary 15.2.4. Extreme Delaunay polytopes are of particular interest since they correspond to the extreme rays of the hypermetric cone. More precisely, if $d \in \mathrm{HYP}_n$ lies on an extreme ray of HYP_n, then its associated Delaunay polytope P_d is an extreme Delaunay polytope of dimension $k \leq n-1$. Conversely, if P is a k-dimensional extreme Delaunay polytope then, for each generating subset V of its set of vertices, the hypermetric space $(V, d^{(2)})$ lies on an extreme ray of the hypermetric cone $\mathrm{HYP}(V)$. Moreover, by taking gate 0-extensions of $(V, d^{(2)})$, we obtain extreme rays of the cone HYP_n for any $n \geq |V|$. In particular, if P is basic, then each basic subset of $V(P)$ yields an extreme ray of the hypermetric cone HYP_{k+1} and, thus, of HYP_n for $n \geq k+1$. Therefore, finding all extreme rays of the hypermetric cone HYP_n yields the question of finding all extreme Delaunay polytopes of dimension $k \leq n-1$.

The only basic extreme Delaunay polytope of dimension $k \leq 5$ is the segment α_1, of dimension 1. Indeed, it is known that the only extreme rays of the hypermetric cone HYP_n (for $n \leq 6$) are the cut semimetrics with associated Delaunay polytope α_1 (see Deza [1960] for $n \leq 5$ and Avis [1989] for $n = 6$; this result is also implicit in Baranovskii [1971, 1973]). Actually, it is announced in Erdahl [1992] that α_1 is the only extreme Delaunay polytope of dimension $k \leq 5$, i.e., the assumption about "basic" can be dropped.

For $n \geq 7$, the hypermetric cone has extreme rays which are not generated by cut semimetrics. Indeed, there exists a basic extreme Delaunay polytope of dimension 6, namely, the Schläfli polytope 2_{21}. Further examples of extreme Delaunay polytopes are presented in Sections 16.2, 16.3 and 16.4. The extreme Delaunay polytopes occurring in root lattices can be easily characterized. Namely,

Theorem 16.0.1. *Let P be a generating Delaunay polytope in a root lattice. Then, P is extreme if and only if P is the segment α_1, the Schläfli polytope 2_{21} or the Gosset polytope 3_{21}.*

Proof. Let L denote the lattice generated by $V(P)$. By assumption, L is a root lattice and L is irreducible by Proposition 15.1.10. Hence, P is one of the Delaunay polytopes from Figure 14.3.1. Therefore, P is equal to α_1, 2_{21} or 3_{21} since the other polytopes are not extreme as their Delaunay polytopes spaces are ℓ_1-spaces. ∎

We start the chapter with formulating in Section 16.1 lower bounds for the number of vertices of an extreme basic Delaunay polytope. It turns out that they relate with some known upper bounds for sets of equiangular lines. We refer to Deza, Grishukhin and Laurent [1992] for details on the topics treated in this chapter.

16.1 Extreme Delaunay Polytopes and Equiangular Sets of Lines

In this section, we present bounds on the number of vertices of a basic extreme Delaunay polytope and we compare them with some known bounds for the cardinality of equiangular sets of lines. We also present some constructions of equiangular sets of lines by taking sections of the sphere of minimal vectors in a lattice. The next result follows immediately from Proposition 15.3.1.

Theorem 16.1.1. *Let P be a k-dimensional basic Delaunay polytope. If P is extreme, then*

(16.1.2) $$|V(P)| \geq \frac{k(k+3)}{2} \quad \text{if } P \text{ is asymmetric},$$

(16.1.3) $$|V(P)| \geq k(k+1) \quad \text{if } P \text{ is centrally symmetric.}$$

∎

Let $N_p(k)$ denote the maximum number of points in a spherical two-distance set of dimension k and let $N_\ell(k)$ denote the maximum number of lines in an equiangular set of lines of dimension k. There is a striking analogy between the lower bounds (16.1.2), (16.1.3) and the following known upper bounds (16.1.4), (16.1.5) for $N_p(k)$ and $N_\ell(k)$ (see Lemmens and Seidel [1973]):

(16.1.4) $$N_p(k) \leq \frac{k(k+3)}{2},$$

(16.1.5) $$N_\ell(k) \leq \frac{k(k+1)}{2}.$$

Recall that equiangular sets of lines and spherical two-distance sets are in one-to-one correspondence. In particular, each the two bounds (16.1.4), (16.1.5) can be deduced from the other.

The bound (16.1.5) was given by Gerzon who proved, furthermore, that if equality holds in (16.1.5), then $k+2 = 4, 5$ or $k+2 = q^2$ for some odd integer $q \geq 3$ (see Lemmens and Seidel [1973]). The first case of equality in (16.1.5) is $N_\ell = 28$ for $q = 3, k = 7$; it is well-known that an equiangular set of 28 lines can be constructed from the Gosset polytope 3_{21} (see Section 16.2). Also, the set of vertices of the Schläfli polytope 2_{21} is a spherical two-distance set in $\mathbb{R}^6$, realizing equality in (16.1.4). The next case of equality is $N_\ell = 276$ for $q = 5, k = 23$. Neumaier [1987] has shown how to construct a set of 276 equiangular lines using the Leech lattice Λ_{24}. In Section 16.3, we shall see that an extreme centrally symmetric Delaunay polytope of dimension 23 and with 552 vertices can be constructed from this set of lines, also that a suitable section of it is an extreme asymmetric Delaunay polytope of dimension 22 and with 275 vertices. The next cases of equality in (16.1.5) are $N_\ell = 1128$ for $q = 7, k = 47$, and $N_\ell = 3160$ for $q = 9, k = 79$; but it is not known whether such sets of equiangular lines exist in these two cases.

On the other hand, we shall see in Section 16.4 some examples of extreme Delaunay polytopes realizing equality in the bound (16.1.2) or (16.1.3), but not arising from some spherical two-distance set or from some equiangular set of lines. Also, we shall have examples of extreme Delaunay polytopes that do not realize equality in the bound (16.1.2) or (16.1.3).

Let $N_\ell(k, \alpha)$ denote the maximum number of lines in an equiangular set of lines of dimension k and with common angle $\arccos \alpha$; so, $N_\ell(k) = \max_\alpha N_\ell(k, \alpha)$. The following results can be found in Lemmens and Seidel [1973]: If $N_\ell(k, \alpha) > 2k$ then $\frac{1}{\alpha}$ is an odd integer. Moreover, if $k < \frac{1}{\alpha^2}$, then

$$(16.1.6) \qquad N_\ell(k, \alpha) \leq \frac{k(1-\alpha^2)}{1-k\alpha^2}.$$

When $k = \binom{n}{2}$ and $\alpha = \frac{1}{n-1}$, the upper bound in (16.1.6) is equal to n^2. For n even, Deza and Grishukhin [1996a] propose a method for constructing a set of equiangular lines meeting almost the bound. More precisely,

Proposition 16.1.7. *If $n \equiv 0$ (resp. $n \equiv 2$) (modulo 4) and if there exists a Hadamard matrix*[1] *of order n (resp. order $n-2$), then one can construct n^2 lines (resp. $n(n-2)$ lines) in dimension $\binom{n}{2}$ with common angle* $\arccos(\frac{1}{n-1})$.

Proof. In order to construct N lines with common angle $\arccos(\frac{1}{n-1})$, it suffices to find N vectors with norms $n-1$ and mutual inner products ± 1. The construction goes as follows. Consider first the case when $n \equiv 0 \pmod 4$. Let A be a Hadamard matrix of order n. Rescale A so that its first column is the all-ones vector. Deleting the first column, we obtain a matrix whose rows are n vectors $u_1, \ldots, u_n$ in $\{\pm 1\}^{n-1}$ with norms $n-1$ and pairwise inner products -1. For each $i = 1, \ldots, n$, consider the cut semimetric $\delta(i)$, which is a 0,1-vector of length $\binom{n}{2}$ with $n-1$ units. What we now do is to place copies of the vectors $u_1, \ldots, u_n$ on every $\delta(i)$ (for $i = 1, \ldots, n$). In this way, we obtain $N = n^2$ vectors in $\mathbb{R}^{\binom{n}{2}}$ with norms $n-1$ and pairwise inner products ± 1. The reasoning is similar in the case $n \equiv 2 \pmod 4$. Namely, we add the all-ones vector as a new column to

[1] A *Hadamard matrix* of order m is an $m \times m$ ± 1 matrix A such that $A^T A = mI$. Then, $m \equiv 0 \pmod 4$ if $m > 2$.

a Hadamard matrix A of order $n-2$; the rows of this extended matrix provide $n-2$ vectors of norm $n-1$ and pairwise inner products 1. Placing copies of them on every $\delta(i)$ yields $n(n-2)$ vectors in $\mathbb{R}^{\binom{n}{2}}$ with norms $n-1$ and inner products ± 1. ∎

We now present a general construction for equiangular sets of lines by taking a suitable section of the sphere of minimal vectors in an integral lattice.

Let L be a lattice with minimal norm t and let $L_{\min}$ be its set of minimal vectors. Given $a \in L$, $a \neq 0$, set $V := \{u \in L_{\min} \mid 2u^T a = a^2\}$. Hence, all $u \in V$ lie on a sphere with center $\frac{a}{2}$. By Lemma 13.2.11, if $V \neq \emptyset$, then the polytope $P := \text{Conv}(V)$ is a Delaunay polytope. Moreover, P is centrally symmetric.

The following properties can be easily checked: $V \neq \emptyset$ if and only if $a = a_1 + a_2$ for some $a_1, a_2 \in L_{\min}$ and, then, $a_1, a_2 \in V$. If $V \neq \emptyset$, then $|V| = 1$ if and only if $a^2 = 4t$. If $|V| \geq 2$ then, for all $u, v \in V$ such that $v \neq u, a-u$, we have

$$\text{(16.1.8)} \qquad \frac{a^2 - t}{2} \leq u^T v \leq \frac{t}{2}.$$

(This follows from the fact that $(u-v)^2 \geq t$ and $(u+v-a)^2 \geq t$.) This implies that $t \leq a^2 \leq 2t$ if $|V| \geq 3$.

Since P is centrally symmetric, we can arrange its vertices into pairs of antipodal vertices. Each such pair determines a line going through $\frac{a}{2}$ and with direction $2u-a$, for $u \in V$. Let $\mathcal{L}$ denote this set of lines and let $V' := \{\sqrt{2}(u - \frac{a}{2}) \mid u \in V\}$ denote the set of their directions. Note that $u'^2 = 2t - \frac{a^2}{2}$ for $u' \in V'$, and $u'^T v' = 2u^T v - \frac{a^2}{2}$ for $u', v' \in V'$. Therefore, if L is an integral lattice, then u'^2, $u'^T v'$ are integers with the same parity as $\frac{a^2}{2}$. Note also that, from relation (16.1.8), we have that $-(t - \frac{a^2}{2}) \leq u'^T v' \leq (t - \frac{a^2}{2})$ for $u', v' \in V'$, $v' \neq u', -u'$. Using the above observations, we obtain the following result from Deza and Grishukhin [1995a].

Proposition 16.1.9. *Let $\mathcal{L}$ denote the set lines determined by the diagonals of the polytope $P = \text{Conv}(V)$ (defined as above). The following assertions hold.*

(i) *If $a^2 = 2t$, then the lines in $\mathcal{L}$ are pairwise orthogonal.*

(ii) *Suppose $a^2 = 2t-2$, $t \geq 2$ and L is an integral lattice. Then, $\mathcal{L}$ is an equiangular set of lines with common angle $\arccos(\frac{1}{t+1})$ (resp. $\arccos(0) = \frac{\pi}{2}$) if t is even (resp. odd).*

(iii) *Suppose $a^2 = 2t-4$, $t \geq 4$ and L is an integral lattice. If t is odd, then $\mathcal{L}$ is equiangular with common angle $\arccos(\frac{1}{t+2})$ and, if t is even, then there are two possible angles between the lines of $\mathcal{L}$, namely $\arccos(\frac{2}{t+2})$ and $\arccos(0) = \frac{\pi}{2}$.* ∎

We give an illustration of the above construction in the case (ii) when $a^2 = 2t-2$, $t = 2$ and L is a root lattice (see Deza and Grishukhin [1995a] for details). If L is an irreducible root lattice, we indicate what is the Delaunay polytope P produced by the construction, the number of lines in the equiangular set $\mathcal{L}$ of its diagonals and the dimension in which $\mathcal{L}$ occurs.

- for $L = A_{n-1}$, $P = \beta_{n-1}$, $|\mathcal{L}| = n-1$, in dimension $n-1$,
- for $L = D_n$, $P = \alpha_1 \times \beta_{n-2}$, $|\mathcal{L}| = 2(n-2)$, in dimension $n-1$,
- for $L = E_6$, the 1-skeleton graph of P is $J(6,3)$, $|\mathcal{L}| = 10$, in dimension 5,
- for $L = E_7$, $P = \frac{1}{2}H(6,2)$, $|\mathcal{L}| = 16$, in dimension 6,

- for $L = E_8$, $P = 3_{21}$, $|\mathcal{L}| = 28$, in dimension 7.

Note that, in dimensions 5 and 6, the maximum cardinality of an equiangular set of lines is equal to 10 and 16, respectively; so the two examples above from E_6 and E_7 are optimum.

16.2 The Schläfli and Gosset Polytopes are Extreme

In this section, we show that the Schläfli polytope 2_{21} and the Gosset polytope 3_{21} are extreme. The proof uses the treatment for the notion of rank developed in Section 15.1. The main steps of the proof are:

(i) Find an affine basis B; so $|B| = 7$ for 2_{21} and $|B| = 8$ for 3_{21} (thus, showing that both $2_{21}, 3_{21}$ are basic Delaunay polytopes).

(ii) Using the affine decomposition of each nonbasic vertex in B, find the explicit description of the system $\mathcal{S}(B, d^{(2)})$ (it consists of $27 - 7 = 20$ equations for 2_{21} and of $\frac{56}{2} - 1 = 27$ equations for 3_{21}).

(iii) Show that the solution set to the system $\mathcal{S}(B, d^{(2)})$ has rank 1.

For this, we need an explicit description of the polytopes $2_{21}, 3_{21}$. We refer, for instance, to Brouwer, Cohen and Neumaier [1989], Conway and Sloane [1988, 1991] for a detailed account of the facts about E_6, E_7, E_8 mentioned below. The lattice E_8 is defined by

$$E_8 = \{x \in \mathbb{R}^8 \mid x \in \mathbb{Z}^8 \cup (\frac{1}{2} + \mathbb{Z})^8 \text{ and } \sum_{1 \le i \le 8} x_i \in 2\mathbb{Z}\}.$$

Let V_8 denote the set of minimal vectors of E_8. Then V_8 consists of
- the 112 vectors $(\pm 1^2, 0^6)$ and
- the 128 vectors $\left(\pm \frac{1}{2}^8\right)$ that have an even number of minus signs.

So, $|V_8| = 240$ and $v^T v = 2$ for $v \in V_8$. The set V_8 lies on the sphere S_8 with center 0 and radius $\sqrt{2}$.

Let $v_0 = (1, 1, 0^6)$ be a given minimal vector. One can check that $v^T v_0 = 0, \pm 1$ for all $v \in V_8, v \neq \pm v_0$. The lattice E_7 is defined by

$$E_7 = \{x \in E_8 \mid x^T v_0 = 1\}.$$

Let H_7 denote the hyperplane defined by the equation $x^T v_0 = 1$; then, $S_7 = S_8 \cap H_7$ is the 7-dimensional sphere with center $\frac{v_0}{2}$ and radius $\sqrt{\frac{3}{2}}$. Set

$$V_7 := \{x \in V_8 \mid x^T v_0 = 1\}.$$

Then, V_7 consists of
- the 12 vectors $(1, 0, \pm 1, 0^5)$,
- the 12 vectors $(0, 1, \pm 1, 0^5)$ and
- the 32 vectors $\left(\frac{1}{2}, \frac{1}{2}, \pm \frac{1}{2}^6\right)$ with an even number of minus signs.

So, $|V_7| = 56$ and V_7 lies on the sphere S_7. By Lemma 13.2.11, the polytope $\mathrm{Conv}(V_7)$ is a Delaunay polytope; it is known as the *Gosset polytope* and is denoted by 3_{21}. Observe that the 56 points of V_7 are partitioned into 28 pairs of antipodal points (with respect to the sphere S_7, i.e., the antipode of v is $v^* = v_0 - v$). So, the polytope 3_{21} is centrally symmetric.

Let $w_0 = \left(\frac{1}{2}\right)^8 \in V_7$, so $w_0^* = \left(\frac{1}{2}, \frac{1}{2}, -\frac{1}{2}^6\right)$. One can check that $v^T w_0 = 0, 1$ for all $v \in V_7, v \neq w_0$ and $v \neq w_0^*$. Then, the lattice E_6 is defined by

$$E_6 = \{x \in E_7 \mid x^T w_0 = 1\}.$$

Note that, if v^* is the antipode of $v \in V_7$, then $v^T w_0 + (v^*)^T w_0 = v_0^T w_0 = 1$ and, thus, $v^T w_0 = 1$ if and only if $(v^*)^T w_0 = 0$. Let H_6 denote the hyperplane defined by the equation $x^T w_0 = 1$; then, $S_6 = S_7 \cap H_6 = S_8 \cap H_7 \cap H_6$ is the 6-dimensional sphere with center $\frac{v_0 + w_0}{3}$ and radius $\sqrt{\frac{4}{3}}$. Set

$$V_6 := \{x \in V_7 \mid x^T w_0 = 1\}$$

and $V_6^* = \{v^* \mid v \in V_6\}$. Hence, $V_7 = V_6 \cup V_6^* \cup \{w_0, w_0^*\}$. The set V_6 consists of
- the 6 vectors $(1, 0, 1, 0^5)$,
- the 6 vectors $(0, 1, 1, 0^5)$ and
- the 15 vectors $\left(\frac{1}{2}, \frac{1}{2}, -\frac{1}{2}^2, \frac{1}{2}^4\right)$.

Hence, $|V_6| = 27$ and V_6 lies on the sphere S_6. The polytope $\mathrm{Conv}(V_6)$ is a Delaunay polytope (by Lemma 13.2.11). It is known as the *Schläfli polytope* and is denoted by 2_{21}; it is asymmetric.

Remark 16.2.1.

(i) The 28 lines determined by the diagonals of 3_{21} form a 7-dimensional set of equiangular lines with common angle $\arccos(\frac{1}{3})$; this can be seen directly or as an application of Proposition 16.1.9 (ii).

(ii) For $u, v \in V_6, v \neq u, u^T v \in \{0, 1\}$ and thus $d^{(2)}(u, v) = (u - v)^2 = 4$ (if $u^T v = 0$) or 2 (if $u^T v = 1$). Therefore, the 27 vertices of 2_{21} form a 6-dimensional spherical two-distance set of points.

(iii) The graph whose nodes are the vertices of 2_{21} and with edges the pairs (u, v) of vertices at the smallest distance $d^{(2)}(u, v) = 2$, is called the *Schläfli graph* and is denoted by G_{27}. The graph whose nodes are the vertices of 3_{21} and with edges the pairs (u, v) of vertices with $d^{(2)}(u, v) = 2$ is called the *Gosset graph* and is denoted by G_{56}. From Proposition 14.3.3, G_{27} (resp. G_{56}) is the 1-skeleton graph of 2_{21} (resp. of 3_{21}). ∎

We now show that the polytopes 2_{21} and 3_{21} are extreme. This result was proved in Deza, Grishukhin and Laurent [1992] and, independently, in Erdahl [1992].

Theorem 16.2.2. *The Schläfli polytope 2_{21} and the Gosset polytope 3_{21} are basic extreme Delaunay polytopes.*

Proof. We denote the vectors of V_6 by $u_i := (1,0,1_i,0^5)$, $v_i := (0,1,1_i,0^5)$ (where the first two coordinates are fixed and the second 1 stays in the $(2+i)$–th position) for $1 \leq i \leq 6$, and $u_{ij} := \left(\frac{1}{2},\frac{1}{2},\left(-\frac{1}{2}\right)_i,\left(-\frac{1}{2}\right)_j,\frac{1}{2}^4\right)$ (where the two $-\frac{1}{2}$'s stay in the $(2+i)$–th and $(2+j)$–th positions) for $1 \leq i < j \leq 6$. Setting $t := 2$, we have

$$(16.2.3)\qquad \begin{cases} d(u_i,u_j) & = d(v_i,v_j) = t \text{ for } i \neq j,\\ d(u_i,v_j) & = \begin{cases} t & \text{if } i = j,\\ 2t & \text{if } i \neq j,\end{cases}\\ d(u_i,u_{kl}) & = d(v_i,u_{kl}) = \begin{cases} t & \text{if } i \notin \{k,l\},\\ 2t & \text{if } i \in \{k,l\},\end{cases}\\ d(u_{ij},u_{kl}) & = \begin{cases} t & \text{if } |\{i,j\}\cap\{k,l\}| = 1,\\ 2t & \text{if } |\{i,j\}\cap\{k,l\}| = 0.\end{cases}\end{cases}$$

One can check that the set

$$B_6 := \{u_{12},u_{24},u_{34},u_{35},u_{15},u_6,v_6\} =: \{1,2,3,4,5,6,7\}$$

is an affine basis of E_6. The affine decompositions of the points of $V_6 \setminus B_6$ in B_6 give the following system of 20 equations in the 21 variables $d(i,j)$ $(1 \leq i < j \leq 7)$ (the indices are taken modulo 5):

$$\begin{cases} d(i,6)+d(i+1,6)-d(i,i+1) & = 0 \quad \text{for } 1 \leq i \leq 5,\\ d(i,7)+d(i+1,7)-d(i,i+1) & = 0 \quad \text{for } 1 \leq i \leq 5,\\ d(i,i+2)+d(i,i+3)-d(i+2,i+3) & = 0 \quad \text{for } 1 \leq i \leq 5,\\ d(6,7)+\sum\limits_{\substack{i<j\\ i,j\in\{k,k+1,k+2\}}} d(i,j) - \sum\limits_{i\in\{k,k+1,k+2\}} (d(i,6)+d(i,7)) & = 0 \quad \text{for } 1 \leq k \leq 5.\end{cases}$$

The equalities of the first, second and fourth lines correspond to the representations of v_i, u_i and u_{k6} in B_6, respectively. The equalities of the third line correspond to the representations of $u_{45}, u_{25}, u_{23}, u_{13}$ and u_{14} in B_6. (For example, the equality $d(1,6)+d(2,6)-d(1,2) = 0$ comes from the affine decomposition $v_5 = u_{12}+u_{34}-u_6$ of v_5 in B_6.) One can verify that the solution set to the system $\mathcal{S}(B_6,d^{(2)})$ described above is precisely given by (16.2.3) and thus has rank 1. Therefore, $\mathrm{rk}(2_{21}) = \mathrm{rk}(B_6,d^{(2)}) = 1$, showing that 2_{21} is extreme.

We now turn to the case of 3_{21}. The set $B_7 := B_6 \cup \{w_0\}$ is clearly an affine basis of E_7. Indeed, $V_7 = V_6 \cup V_6^* \cup \{w_0,w_0^*\}$, $v_0 = u_{12}+u_{34}+u_{56}-w_0$ and, for $v \in V_6$, $v^* = v_0 - v = u_{12}+u_{34}+u_{56}-w_0-v$ is thus affinely decomposable in B_7. Since $w_0^T v = 1$ for all $v \in B_6$, we have that $d^{(2)}(w_0,v) = 2$ for $v \in B_6$. From Lemma 15.3.7 (applied to $P = 3_{21}$, $P_1 = 2_{21}$, $H = H_6$ and $w = u_{13}^*$), we deduce that $\mathrm{rk}(2_{21}) = \mathrm{rk}(3_{21})$, implying that 3_{21} is extreme.

Note that the system $\mathcal{S}(B_7,d^{(2)})$ consists of the system $\mathcal{S}(B_6,d^{(2)})$ together with the following seven equations:

$$\begin{cases} d(i,8)+d(i+1,8)-d(i,i+1) & = 0 \quad \text{for } 1 \leq i \leq 5,\\ d(1,2)+d(1,3)+d(2,3)+d(k,8) - \sum\limits_{i=1,2,3} (d(i,k)+d(i,8)) & = 0 \quad \text{for } k = 6,7.\end{cases}$$

These equations correspond to the decomposition of v^* in B_7 (for $v \in B_6$). ∎

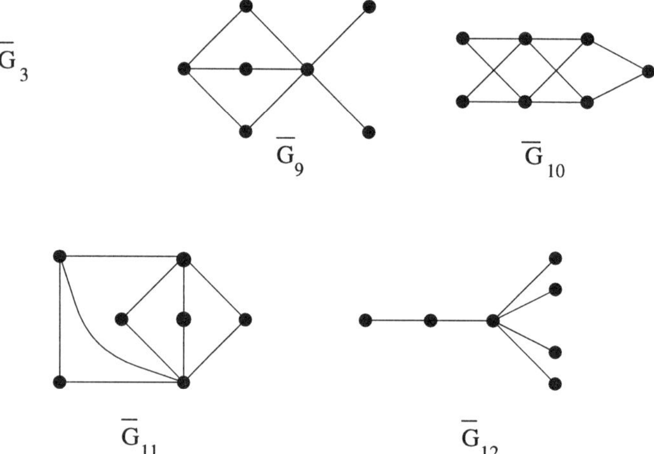

Figure 16.2.4: Class $q = 8$

Since 2_{21} is extreme and basic, each basic set $B \subseteq V(2_{21})$ yields an extreme ray of the hypermetric cone HYP_7. We have constructed in the proof of Theorem 16.2.2 the basic set B_6. It is interesting to know how many distinct (up to permutation) extreme rays of HYP_7 arise in this way from 2_{21}. Actually, we believe that all the extreme rays of HYP_7, other than those generated by the cut semimetrics, arise from 2_{21}.

For each basic subset $B \subseteq V(2_{21}) = V_6$, let $G_{27}[B]$ denote the subgraph of the Schläfli graph G_{27} induced by B; its set of nodes is B and its edges are the pairs of points at the smallest distance 2. $G_{27}[B]$ is called a *basic subgraph* of G_{27}. For instance, for the basic set B_6 defined above, $G_{27}[B_6]$ is $K_7 \setminus C_5$ (where C_5 is the circuit on the nodes $(u_{12}, u_{34}, u_{15}, u_{24}, u_{35})$).

By a direct inspection of the 7-vertices subgraphs of the Schläfli graph, we found that there are in total 26 distinct basic subsets in 2_{21}. Eight of them are connected with Theorem 17.1.9; namely, they are the graphs G_i $(1 \leq i \leq 8)$ where $G_1 = \nabla B_9$ (so, $G_1 = G_{27}[B_6]$), $G_2 = \nabla H_2$, $G_3 = \nabla H_1$, $G_4 = \nabla B_8$, $G_5 = \nabla B_7$, $G_6 = \nabla H_4$, $G_7 = \nabla H_3$ and $G_8 = \nabla B_5$. The graphs B_i $(1 \leq i \leq 8)$ and H_i $(1 \leq i \leq 4)$ will be shown in Figures 17.1.3 and 17.1.6, respectively.

We show in Figures 16.2.4, 16.2.5, 16.2.6 and 16.2.7 the 26 basic subgraphs of G_{27}. Actually, we depict there the complements $\overline{G_i}$ of the graphs G_i since they appear to be simpler to draw. Hence, in Figures 16.2.4-16.2.7, an edge means a pair of points at the largest distance 4. The 26 basic graphs G_i $(1 \leq i \leq 26)$ are partitioned into five classes indexed by some integer $q \in \{8, 11, 12, 14, 15\}$. In fact, all basic graphs of the same class are switching equivalent and the invariant[2]

[2]The graph $\overline{G}_{18}$ was incorrectly assigned to the class $q = 12$ in Deza, Grishukhin and Laurent [1992]. It belongs, in fact, to the class $q = 11$ as indicated here.

q of each switching class is the number of odd triples, i.e., the triples of nodes carrying an odd number of edges. (See Deza, Grishukhin and Laurent [1992] for more details about the occurrence of switching here.) Finally, note that one obtains at least 26 distinct extreme rays for HYP_8 from the Gosset polytope 3_{21}. Indeed, each basic set of 2_{21} can be augmented to a basic set of 3_{21}. We do not know about the classification of all other basic sets of 3_{21}.

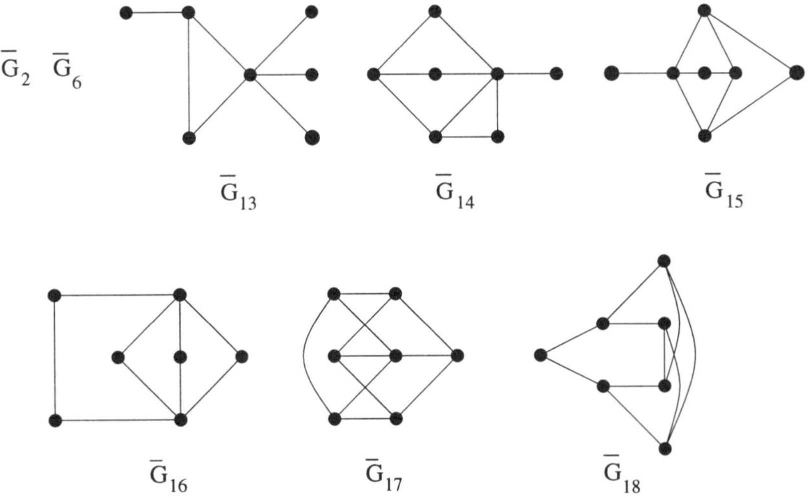

Figure 16.2.5: Class $q = 11$

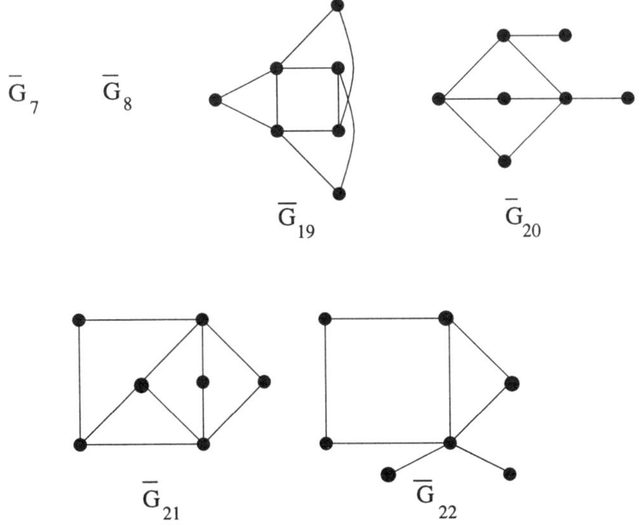

Figure 16.2.6: Class $q = 12$

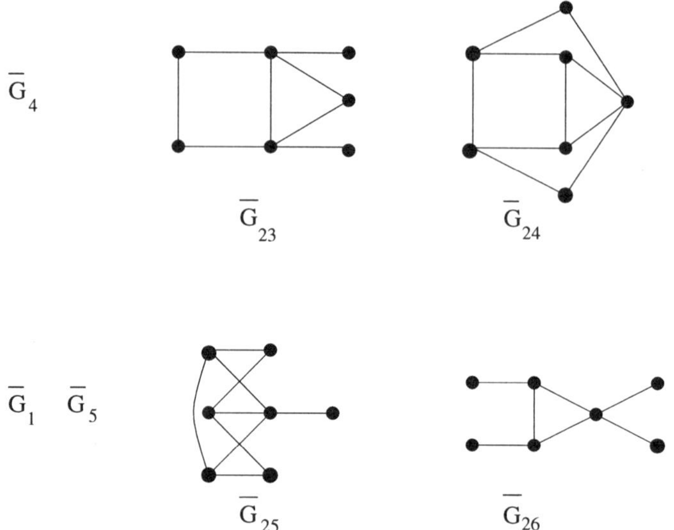

Figure 16.2.7: Class $q = 14$ (for the first three graphs $\overline{G}_4, \overline{G}_{23}, \overline{G}_{24}$) and $q = 15$ (for the remaining four graphs)

16.3 Extreme Delaunay Polytopes in the Leech Lattice Λ_{24}

In this section, we describe two extreme Delaunay polytopes coming from the Leech lattice Λ_{24}; they are taken from Deza, Grishukhin and Laurent [1992]. These polytopes have dimension 22, 23 and they are constructed by taking two consecutive suitable sections of the sphere of minimal vectors of Λ_{24}, precisely in the same way as the Gosset polytopes $3_{21}, 2_{21}$ were constructed from the lattice E_8 in Section 16.2. We refer to Conway and Sloane [1988] for a precise description of the Leech lattice Λ_{24}; we only recall some facts that we need for our treatment.

The *Leech lattice* Λ_{24} is a 24-dimensional lattice in $\mathbb{R}^{24}$. For convenience, the coordinates of the vectors $x \in \mathbb{R}^{24}$ are indexed by the elements of the set $I := \{\infty, 0, 1, \ldots, 22\}$. For $i \in I$, let e_i denote the i-th unit vector in $\mathbb{R}^{24}$. For a subset S of I, set $e_S := \sum_{i \in S} e_i$.

Let $\mathcal{B}_{24}$ denote the family of blocks of the Steiner system $S(5, 8, 24)$ defined on the set I; hence, $|\mathcal{B}_{24}| = 759$. Set

$$\mathcal{B}_{23} := \{B \setminus \{\infty\} \mid B \in \mathcal{B}_{24} \text{ with } \infty \in B\};$$

so $\mathcal{B}_{23}$ is the family of blocks of the Steiner system $S(4, 7, 23)$ defined on the set $\{0, 1, \ldots, 22\}$ and $|\mathcal{B}_{23}| = 253$. In $\mathcal{B}_{23}$, there are exactly 176 blocks that do not contain a given point and there are exactly 77 blocks that do contain a given point.

The Leech lattice Λ_{24} is generated by the vectors $e_I - 4e_\infty$ and $2e_B$ for all blocks $B \in \mathcal{B}_{24}$. Let V denote the set of minimal vectors of Λ_{24}; so, $x^Tx = 32$ for $x \in V$. (Note that, in the usual definition, all vectors are scaled by a factor of $\frac{1}{\sqrt{8}}$ and the minimal norm is 4; we choose to omit this factor in order to make the notation easier.) The set V consists of the following vectors:

(I) $(\pm 4^2, 0^{22})$ ($1104 = 2 \times 24 \times 23$ such vectors),

(II) $(\pm 2^8, 0^{16})$, where the positions of the nonzero components form a block of $\mathcal{B}_{24}$ and there is an even number of minus signs ($2^7 \times 759$ such vectors),

(III) $(\mp 3, \pm 1^{23})$, where the ∓ 3 may be in any position, but the upper signs are taken on the coordinates of a codeword of the Golay code $\mathcal{C}_{24}$. (Recall that the codewords of $\mathcal{C}_{24}$ which have exactly 8 nonzero coordinates are precisely the blocks of $\mathcal{B}_{24}$.)

Set $c := (5, 1^{23})$ and $a_0 := (4, 4, 0^{22})$; so $c, a_0 \in \Lambda_{24}$, $c^Tc = 48$ and $a_0 \in V$. Set

$$V_{23} := \{v \in V \mid v^Tc = 24\} \text{ and } V_{22} := \{v \in V \mid v^Tc = 24 \text{ and } v^Ta_0 = 16\}.$$

By Lemma 13.2.11, the polytopes

$$P_{23} := \mathrm{Conv}(V_{23}),\ P_{22} := \mathrm{Conv}(V_{22})$$

are Delaunay polytopes. The polytope P_{23} is centrally symmetric and P_{22} is asymmetric. In fact, the set V_{22} is a spherical two-distance set (indeed, the distances between the points of V_{22} take the two values 32 or 48). Moreover, the 276 lines defined by the 276 pairs of antipodal vertices of the polytope P_{23} are equiangular (with common angle $\arccos(\frac{1}{5})$).

Theorem 16.3.1.

(i) *The polytope P_{23} is a basic centrally symmetric extreme Delaunay polytope of dimension* 23 *with* 552 *vertices, hence realizing equality in the bound* (16.1.3).

(ii) *The polytope P_{22} is a basic asymmetric extreme Delaunay polytope of dimension* 22 *with* 275 *vertices, hence realizing equality in the bound* (16.1.2).

∎

16.4 Extreme Delaunay Polytopes in the Barnes-Wall Lattice Λ_{16}

In this section, we describe some examples of extreme Delaunay polytopes constructed from the Barnes-Wall lattice Λ_{16}; they are taken from Deza, Grishukhin and Laurent [1992]. (See Conway and Sloane [1988] for a precise description of Λ_{16}.) The *Barnes-Wall lattice* Λ_{16} is a 16-dimensional lattice in $\mathbb{R}^{16}$. Its set V of minimal vectors consists of:

(I) 480 vectors of the form $(\pm 2^2, 0^{14})$, where there are two nonzero components equal to 2 or -2,

(II) 3840 vectors of the form $(\pm 1^8, 0^8)$, where the positions of the ± 1's form one of the 30 codewords of weight 8 of the first order Reed-Muller code and there are an even number of minus signs.

Hence, $|V| = 4320$ and $v^T v = 8$ for $v \in V$. (Note that in the usual definition, the minimal norm is 4 and all vectors should be scaled by a factor $\frac{1}{\sqrt{2}}$; we omit this factor in order to make the notation easier.)

c_{12}	001111	1111	0000	00
c_{13}	010111	0010	1010	01
c_{14}	011011	0100	0011	10
c_{15}	011101	0001	0100	11
c_{16}	011110	1000	1101	00
c_{23}	100111	0010	0101	10
c_{24}	101011	0100	1100	01
c_{25}	101101	0001	1011	00
c_{26}	101110	1000	0010	11
c_{34}	110011	1001	0110	00
c_{35}	110101	1100	0001	01
c_{36}	110110	0101	1000	10
c_{45}	111001	1010	1000	10
c_{46}	111010	0011	0001	01
c_{56}	111100	0110	0110	00

Figure 16.4.1: Codewords of weight 8 in the Reed-Muller code

We show in Figure 16.4.1 a list of 15 codewords of weight 8 of the first order Reed-Muller code; the other 15 codewords of weight 8 are obtained by taking their complements.

Set $a := (2^6, 0^{10}) \in \Lambda_{16}$ (the six 2's are in the first six positions which are precisely the first six positions distinguished in Figure 16.4.1). Let S denote the sphere with center $\frac{a}{2}$ and radius $\sqrt{6}$. Then, S is an empty sphere in Λ_{16} corresponding to a deep hole (i.e., with maximum radius). Hence, the polytope

$$P := \{v \in \Lambda_{16} \mid (v - \frac{a}{2})^2 = 6\}$$

is a Delaunay polytope. P is centrally symmetric, has dimension 16 and has 512 vertices. One can check that the vertices of P lie in the parallel layers: $x^T a \in \{0, 8, 12, 16, 24\}$. For $\alpha \in \mathbb{R}$, let H^α denote the hyperplane with equation $x^T a = \alpha$ and set $V^\alpha := V(P) \cap H^\alpha$. Then,

$$V(P) = V^0 \cup V^8 \cup V^{12} \cup V^{16} \cup V^{24},$$

with $V^0 = \{0\}$, $V^{24} = (V^0)^* = \{a\}$, $V^{16} = (V^8)^*$, $|V^8| = |V^{16}| = 135$, and $|V^{12}| = 240$.

Moreover, for $\alpha = 8, 12, 16$, the section

$$P^\alpha := P \cap H^\alpha = \text{Conv}(V^\alpha)$$

by the hyperplane H^α is a Delaunay polytope of dimension 15 in the lattice $\Lambda_{16} \cap H^\alpha$. The polytopes P^8 and P^{16} are asymmetric with 135 vertices, but their sets of vertices are not spherical two-distance sets (indeed, there are three possible distances: 8,12,16 between the vertices). The polytope P^{12} is centrally symmetric with 240 vertices, but the 120 lines defined by its 120 pairs of antipodal vertices are not equiangular (there are two possible angles: arccos(0),arccos($\frac{1}{3}$)). Note that P^{16} has radius $\frac{4}{\sqrt{3}}$ and that P^{12} has radius $\sqrt{6}$ ($> \frac{4}{\sqrt{3}}$). This shows that P^{16} does not correspond to a deep hole of the lattice $\Lambda_{16} \cap H^{16}$. Finally, set

$$Q := \text{Conv}(V^0 \cup V^8 \cup V^{16} \cup V^{24}).$$

So, Q has $2 \times 135 + 2 = 272$ vertices and its dimension is 16. Q is a Delaunay polytope in the lattice $\Lambda_{16} \cap \{x \mid x^T a = 0 \pmod 8\}$. The polytopes P, P^8, P^{16} and Q can be verified to be basic.

Theorem 16.4.2.

(i) *P is a centrally symmetric extreme Delaunay polytope of dimension* 16 *with* 512 *vertices.*

(ii) *P^8 and P^{16} are asymmetric extreme Delaunay polytopes of dimension* 15*, each having* 135 *vertices.*

(iii) *P^{12} is not extreme.*

(iv) *Q is a centrally symmetric extreme Delaunay polytope of dimension* 16 *with* 272 *vertices, hence realizing equality in the bound* (16.1.3). ∎

Finally, let us consider the section of the sphere of minimal vectors by the hyperplane H^4. In this way, one obtains the Delaunay polytope

$$Q' := \text{Conv}(x \in \Lambda_{16} \mid x^T x = 8 \text{ and } x^T a = 4).$$

Q' is a 15-dimensional polytope with 1080 vertices. Consider the vertex $c := (2, 0, .., 0, 2)$ of Q'. Then, the distances $d^{(2)}(c, v)$ from the other vertices v to c take the values $8, 12, 16, 20, 24$; in fact, value 8 (respectively, 12,16,20,24) is taken for 119 (respectively, 336, 427,176,21) vertices of Q'. Therefore, the set of the 119 vertices that are at distance 8 from c forms a 14-dimensional asymmetric Delaunay polytope which realizes equality in the bound (16.1.2). However, this polytope is not extreme. On the other hand, the polytope Q' is extreme.

We summarize in Figure 16.4.3 the results from this section about the Delaunay polytopes constructed from the Barnes-Wall lattice Λ_{16}. (The second column indicates the dimension, the fourth column indicates whether the polytope is asymmetric (A) or centrally symmetric (CS), and the fifth column indicates whether equality holds in the bounds (16.1.2) or (16.1.3).) Recall that $a = (2^6, 0^{10})$, $c = (2, 0^{14}, 2)$, S denotes the sphere with center $\frac{a}{2}$ and radius $\sqrt{6}$, and H^α denotes the hyperplane $x^T a = \alpha$.

Delaunay polytope	dim.	number of vertices	sym. ?	equality in bound ?	extreme ?
$P = \mathrm{Conv}(S \cap \Lambda_{16})$	16	512	CS	No	Yes
$P^8 = \mathrm{Conv}(S \cap \Lambda_{16} \cap H^8)$	15	135	A	Yes	Yes
$P^{16} = \mathrm{Conv}(S \cap \Lambda_{16} \cap H^{16})$	15	135	A	Yes	Yes
$P^{12} = \mathrm{Conv}(S \cap \Lambda_{16} \cap H^{12})$	15	240	CS	Yes	No
$Q = \mathrm{Conv}(S \cap \Lambda_{16}$ $\cap \{x \mid x^T a = 0, 8, 16, 24\})$	16	272	CS	Yes	Yes
$\mathrm{Conv}(x \in \Lambda_{16} \mid x^T x = 8,$ $a^T x = 4, c^T x = 8)$	14	119	A	Yes	No
$Q' = \mathrm{Conv}(x \in \Lambda_{16} \mid$ $x^T x = 8, a^T x = 4)$	15	1080	A	No	Yes

Figure 16.4.3: Delaunay polytopes in the Barnes-Wall lattice

16.5 Extreme Delaunay Polytopes and Perfect Lattices

Let L be a k-dimensional lattice (containing the origin) with minimal norm t and set $L_{\min} := \{v \in L \mid v^2 = t\}$. Let $(v_1, \ldots, v_k)$ be a basis of L and, for each $v \in L_{\min}$, let $v = \sum_{i=1}^k b_i^v v_i$ denote its decomposition in the basis, with $b^v \in \mathbb{Z}^k$. Let $\mathcal{S}_L$ denote the system composed by the equations

$$\sum_{1 \le i \le j \le k} b_i^v b_j^v x_{ij} = t \text{ for } v \in L_{\min}$$

in $\binom{k+1}{2}$ variables. The lattice L is said to be *perfect* if the system $\mathcal{S}_L$ has full rank $\binom{k+1}{2}$; that is, if it has a unique solution which is then given by

$$x_{ij} = 2v_i^T v_j(\text{ for } 1 \le i < j \le k),\ x_{ii} = v_i^2(\text{ for } 1 \le i \le k).$$

Perfect lattices are important since they include the lattices with the locally most dense packings (see, for instance, Ryshkov and Baranovskii [1979]). If L is an affine lattice, i.e., if L is the translate of a lattice L_0, we say that L is perfect if L_0 is perfect.

The notion of perfect lattice is closely related to the notion of extreme Delaunay polytope as the following Propositions 16.5.1, 16.5.2 and 16.5.3 show; all three results are taken from Grishukhin [1993].

Proposition 16.5.1. *Let P be a Delaunay polytope with radius r, let L_0 denote the lattice generated by the set of vertices $V(P)$ of P and let t denote its minimal norm. Suppose that P is a basic extreme Delaunay polytope, that there exist $u, v \in V(P)$ with $(u-v)^2 = t$ and that $t \geq \frac{4}{3}r^2$. Then, there exists w not lying on the affine space spanned by P such that $(w-v)^2 = t$ for all $v \in V(P)$ and such that the lattice L generated by $L_0 \cup \{w\}$ is perfect.*

Proof. We can suppose without loss of generality that the origin is a vertex of P. By Lemma 14.4.5, the spherical t-extension of the space $(V(P), d^{(2)})$ has a spherical representation. Let w denote the vector representating the extension point. So, $(w-v)^2 \geq t$ for all $v \in L_0$ with equality if $v \in V(P)$. Let L denote the lattice generated by $L_0 \cup \{w\}$. Then, $L = \bigcup_{a \in \mathbb{Z}} L_a$, where $L_a := (L_0 + aw)$ are the layers composing L. The distance between two consecutive layers is $h = \sqrt{t - r^2}$.

We check that the minimal norm of L is equal to t, i.e., that $v^2 \geq t$ for all $v \in L$, $v \neq 0$. This is obvious if v lies in L_0. If v lies in a layer L_a which is not consecutive to the layer L_0, then $\| v \| \geq 2h$, i.e., $v^2 \geq 4h^2 = 4(t - r^2) \geq t$ since $t \geq \frac{4}{3}r^2$. If v lies in a layer consecutive to L_0, say $v = u - w$ where $u \in L_0$, then $v^2 \geq t$.

Since P is basic, we can find a basis $(v_1, \ldots, v_k)$ of L_0 composed of vertices of P. Then, $(w, v_1, \ldots, v_k)$ is a basis of L. So, the system $\mathcal{S}_L$ is composed by the equations:

$$\sum_{0 \leq i \leq j \leq k} b_i b_j x_{ij} = t, \quad \text{where } (b_0 w + \sum_{1 \leq i \leq k} b_i v_i)^2 = t \quad \text{with } b \in \mathbb{Z}^{k+1}.$$

We show that $\mathcal{S}_L$ has full rank. Let x denote a solution of $\mathcal{S}_L$.

Since $w, w - v_1, \ldots, w - v_k \in L_{\min}$, we deduce that the equations:

$$x_{00} = t, \; x_{00} + x_{ii} - x_{0i} = t \; (1 \leq i \leq k)$$

belong to $\mathcal{S}_L$. Therefore, $x_{00} = t$ and $x_{ii} = x_{0i}$ for $i = 1, \ldots, k$. Let $v \in V(P)$, $v = \sum_{1 \leq i \leq k} b_i^v v_i$ with $b^v \in \mathbb{Z}^k$. Then, $v - w \in L_{\min}$, implying the equation:

$$x_{00} - \sum_{1 \leq i \leq k} b_i^v x_{0i} + \sum_{1 \leq i \leq j \leq k} b_i^v b_j^v x_{ij} = t$$

of $\mathcal{S}_L$. Hence, x satisfies

$$\text{(a)} \qquad \sum_{1 \leq i \leq k} ((b_i^v)^2 - b_i^v) x_{ii} + \sum_{1 \leq i < j \leq k} b_i^v b_j^v x_{ij} = 0$$

for each $v \in V(P)$.

By assumption, P is an extreme Delaunay polytope; that is, the system $\mathcal{S}(V(P), d^{(2)})$, composed by the equations:

$$\text{(b)} \qquad \sum_{1 \leq i \leq k} (1 - \sum_{1 \leq j \leq k} b_j^v) b_i^v d_{0i} + \sum_{1 \leq i < j \leq k} b_i^v b_j^v d_{ij} = 0$$

for all $v \in V(P)$, has rank $\binom{k+1}{2} - 1$.

Set $d_{0i} := x_{ii}$ for $1 \leq i \leq k$ and $d_{ij} := x_{ii} + x_{jj} - 2x_{ij}$ for $1 \leq i < j \leq k$. Then, since x satisfies (a), we deduce that d satisfies (b). Therefore, d is uniquely determined up to multiple. This implies that x too is uniquely determined up to multiple. The fact that there exist $u, v \in V(P)$ with $u - v \in L_{\min}$ permits to fix the multiple. Hence, $\mathcal{S}_L$ has a unique solution x. This shows that L is perfect. ∎

Note that Proposition 16.5.1 still holds if we replace the assumption: $t \geq \frac{4}{3}r^2$ by the assumption: $t \geq r^2$ and t is the minimal norm of L.

As we saw in Lemma 13.2.11, every section of the contact polytope by a hyperplane not containing the origin is a Delaunay polytope. Hence, Proposition 16.5.1 can be reformulated as follows.

Proposition 16.5.2. *Let L be a k-dimensional lattice with minimal norm t and let P be a Delaunay polytope obtained by taking a section of the contact polytope of L by a hyperplane not containing the origin. If P is basic and extreme and if P contains two vertices u, v with $(u - v)^2 = t$, then L is perfect.* ∎

For example, the root lattice E_8 and the Leech lattice Λ_{24} are perfect. This can be seen by applying Proposition 16.5.2; for E_8, take $t = 2$ and $P = 3_{21}$ whose squared radius is $\frac{3}{2}$, and for Λ_{24}, take $t = 32$ and $P = P_{23}$ whose squared radius is 24 (see Sections 16.2 and 16.3). Another example of perfect lattice is the lattice $\Lambda_{16} \cap \{x \mid x^T a = 0 \mod (8)\}$, where Λ_{16} is the Barnes-Wall lattice and a is a minimal vector of it; apply Proposition 16.5.2 with the polytope P^{16} (see Section 16.4).

The following result can also be checked.

Proposition 16.5.3. *Let P be an extreme basic Delaunay polytope with radius r and let L' denote the lattice generated by the set of vertices of P and the center of P (L' is known as the* centered lattice*). If L' has minimal norm r^2, then L' is perfect.* ∎

For instance, the Schläfli polytope 2_{21} is an extreme basic Delaunay polytope in E_6. The lattice generated by $V(2_{21})$ and its center is the dual lattice E_6^* which is indeed perfect.

Chapter 17. Hypermetric Graphs

We group in this chapter several results concerning hypermetricity of distance spaces arising from graphs.

There are essentially two ways of constructing a distance space from a graph. The most classical construction of a distance space from a connected graph G is by considering the graphic metric space $(V(G), d_G)$ where d_G is the path metric of G, with $d_G(u,v)$ denoting the smallest length of a path connecting the nodes $u, v \in V(G)$. If $(V(G), d_G)$ is hypermetric (resp. isometrically ℓ_1-embeddable, of negative type), we say that G is a *hypermetric graph* (resp. an *ℓ_1-graph*, a *graph of negative type*).

Another distance space which can be constructed from a graph G is the space $(V(G), d_G^*)$, where d_G^* is the *truncated distance* of G defined by

$$\begin{cases} d_G^*(i,j) = 1 & \text{if } ij \in E(G), i \neq j, \\ d_G^*(i,j) = 2 & \text{if } ij \notin E(G), \ i \neq j, \\ d_G^*(i,i) = 0 & \text{for all } i \in V(G). \end{cases}$$

If G has diameter[1] ≤ 2, then these two notions of path metric and truncated distance coincide. This is the case, for instance, for suspension graphs. In fact, the graphs whose suspension is of negative type form a class of graphs which has received a lot of attention in the literature; indeed, they are the graphs whose adjacency matrix has minimum eigenvalue greater than or equal to -2.

For a graph $G = (V, E)$ on n nodes we remind that its *adjacency matrix* A_G is the $n \times n$ symmetric matrix with zero diagonal entries and whose (i,j)-th entry is equal to 1 if i, j are adjacent in G and to 0 otherwise, for distinct $i, j \in V$. We let $\lambda_{\min}(A_G)$ denote the smallest eigenvalue of matrix A_G.

We consider here questions related to hypermetricity and ℓ_1-embeddability for either of the two distances d_G and d_G^* attached to a graph G. In Sections 17.1 and 17.2 we present a number of results dealing with the problem of characterizing the graphs whose path metric or truncated distance is hypermetric or ℓ_1-embeddable. We are interested, in particular, in finding 'good' characterizations; that is, characterizations leading to polynomial time recognition algorithms. Such results are known for several classes of graphs. We focus in Section 17.1 on suspension graphs and on bipartite graphs equipped with the truncated

[1]The *diameter* of a graph G is defined as the largest distance (with respect to the shortest path metric) between two nodes of G.

distance. Section 17.2 deals with graphs having some regularity properties. Finally, Section 17.3 studies the graphs G for which either of the two distances d_G and d_G^* lies on an extreme ray of the hypermetric cone.

17.1 Characterizing Hypermetric and ℓ_1-Graphs

Characterizing Hypermetricity and ℓ_1-Embeddability for Path Metrics. We start with a characterization of the graphs whose path metric is hypermetric or isometrically ℓ_1-embeddable, which follows directly from Theorems 14.3.6 and 14.3.7. (In fact, the characterization of ℓ_1-graphs will also follow from the results of Shpectorov [1993] exposed in Chapter 21.)

Theorem 17.1.1. *Let G be a connected graph. Then,*

(i) *G is hypermetric if and only if G is an isometric subgraph of a Cartesian product of half-cube graphs, cocktail-party graphs and copies of the Gosset graph G_{56}.*

(ii) *G is an ℓ_1-graph if and only if G is an isometric subgraph of a Cartesian product of half-cube graphs and cocktail-party graphs.* ∎

Several characterizations of the hypercube embeddable graphs will be given in Chapter 19; they are good characterizations, in the sense that they permit to recognize whether a graph is an isometric subgraph of a hypercube in polynomial time. The result from Theorem 17.1.1 (ii) does not yield, a priori, a good characterization for ℓ_1-graphs. However, the proof method developed by Shpectorov [1993] permits to recognize ℓ_1-graphs in polynomial time (see Corollary 21.1.9). No good characterization is known yet for hypermetric graphs (recall Remark 14.2.6). We state this as an open problem.

Problem 17.1.2. *What is the complexity of the problem of testing whether (the path metric of) a graph is hypermetric ?*

Problem 17.1.2 is solved for the class of suspension graphs. Indeed, for these graphs, some refined characterizations for hypermetricity and ℓ_1-embeddability are known that lead, in particular, to polynomial-time recognition algorithms (cf. Theorems 17.1.8 and 17.1.9). A suspension graph has diameter ≤ 2 and thus its path metric coincides with its truncated distance. The graphs whose truncated distance is hypermetric or ℓ_1-embeddable are not well understood in general. Good characterizations are, however, available for some subclasses; for instance, for bipartite graphs (cf. Theorem 17.1.13) and for graphs with regularity properties (cf. Section 17.2).

A Good Characterization of Hypermetricity and ℓ_1-Embeddability for Suspension Graphs. We consider here in detail suspension graphs. A first observation is that, for a graph G, its suspension ∇G is hypermetric (resp. an ℓ_1-graph) if and only if ∇H is hypermetric (resp. an ℓ_1-graph) for each connected component H of G. Indeed, the path metric of ∇G arises as the 1-sum of the path metrics of $\nabla H_1, \ldots, \nabla H_m$, if $H_1, \ldots, H_m$ are the connected components of G. (Recall Section 7.6.) Hence, we can restrict our attention to the case when G is a connected graph.

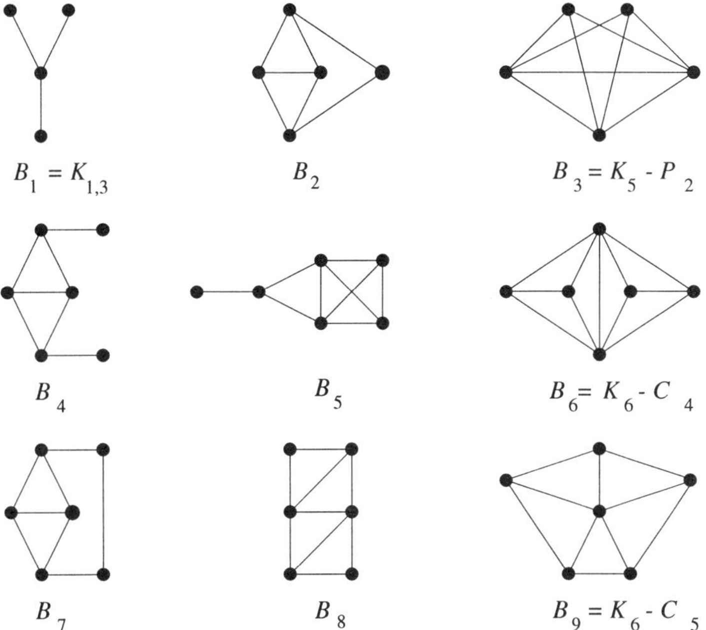

Figure 17.1.3: The excluded configurations for line graphs

We start with a characterization of the suspension graphs that are of negative type, which was obtained by Assouad and Delorme [1980]. (Compare the results from Propositions 17.1.4 and 17.2.1.)

Proposition 17.1.4. *Let G be a graph. Then, its suspension ∇G is of negative type if and only if $\lambda_{\min}(A_G) \geq -2$ holds.*

Proof. We use Proposition 13.1.2, so we show that $\lambda_{\min}(A_G) \geq -2$ if and only if the space $(V(\nabla G), d_{\nabla G})$ has a representation. Let i_0 denote the apex node of ∇G and suppose that G has n nodes. If $\lambda_{\min}(A_G) \geq -2$, then the matrix $A_G + 2I$ is positive semidefinite. Hence, there exist n vectors $u_1, \ldots, u_n \in \mathbb{R}^m$ (for some m) such that

$$\begin{cases} (u_i)^2 = 2 & \text{for } i = 1, \ldots, n, \\ u_i^T u_j = 1 & \text{if } ij \in E(G), \\ u_i^T u_j = 0 & \text{otherwise.} \end{cases}$$

Then, the mapping

$$i \in V(G) \mapsto u_i, \ i_0 \mapsto u_0 := 0,$$

provides a representation of $(V(\nabla G), 2d_{\nabla G})$. Indeed, $(u_i - u_j)^2 = 2$ if $ij \in E(\nabla G)$ and $(u_i - u_j)^2 = 4$ otherwise. All the above arguments can be reversed, stating the converse implication: If ∇G is of negative type, then $\lambda_{\min}(A_G) \geq -2$. ∎

We recall that $L(H)$ denotes the line graph of a graph H. It is easy to see that the suspension $\nabla L(H)$ of any line graph is an ℓ_1-graph. Indeed, if we label the apex node by the zero vector and each edge $e := ij \in E(H)$ by the vector $\frac{e_i+e_j}{2}$ (e_i denoting the i-th unit vector in the space $\mathbb{R}^{V(H)}$), then we obtain an ℓ_1-embedding of $\nabla L(H)$. This shows, moreover, that $2d(\nabla L(H))$ is hypercube embeddable. Line graphs have been characterized by Beineke [1970] by means of excluded subgraphs. Namely,

Theorem 17.1.5. *A graph G is a line graph if and only if G does not contain as an induced subgraph any of the nine graphs B_i $(1 \leq i \leq 9)$ shown in Figure 17.1.3.* ∎

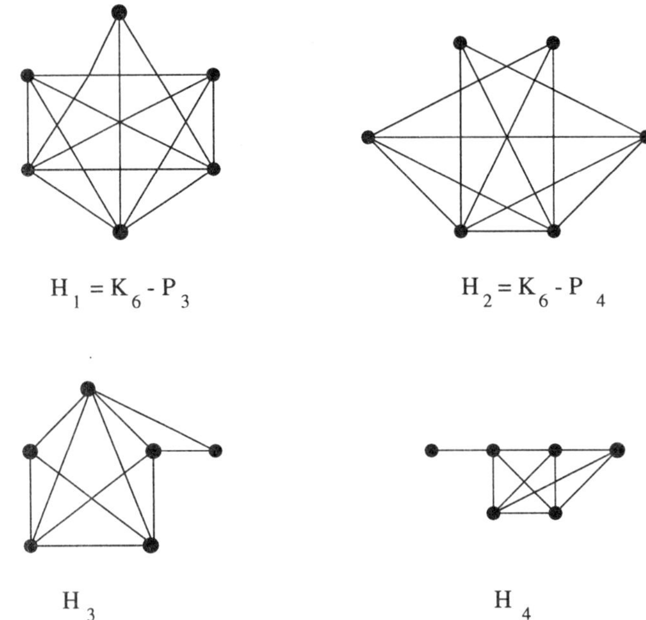

Figure 17.1.6: Four graphs whose suspensions are not ℓ_1-graphs

Remark 17.1.7. The following can be checked.

(i) ∇B_i is not an ℓ_1-graph for all $1 \leq i \leq 9$ except $i = 3$; in fact, ∇B_1, B_2 are not 5-gonal and ∇B_4, ∇B_6 are not 7-gonal.

(ii) For each of the graphs H_i $(1 \leq i \leq 4)$ shown in Figure 17.1.6, ∇H_i is not an ℓ_1-graph.

In other words, if d denotes the path metric of any of the graphs ∇B_i ($1 \leq i \leq 9, i \neq 3$) or of ∇H_i ($1 \leq i \leq 4$), there exists an inequality $v^T x \leq 0$ defining a facet of CUT_7 which is violated by d, i.e., such that $v^T d > 0$. See Deza and Laurent [1992a] for an explicit description of such inequalities. ∎

Let G be a connected graph and suppose that its suspension ∇G is hypermetric. Let H denote the 1-skeleton graph of the Delaunay polytope associated with the space $(V(\nabla G), 2d_{\nabla G})$. Then, ∇G is an induced subgraph of H and H is one of the Delaunay polytope graphs shown in Figure 14.3.1. Therefore, if ∇G is an ℓ_1-graph then, by Proposition 14.1.10, $H \neq G_{27}$, G_{56} and, thus, H is one of $J(m,t)$, $\frac{1}{2}H(m,2)$ and $K_{m\times 2}$. More precisely, we have the following results, due to Assouad and Delorme [1980, 1982].

Theorem 17.1.8. *Let G be a connected graph. Then, the following assertions are equivalent.*

(i) *∇G is an ℓ_1-graph.*

(ii) *G does not contain as an induced subgraph any of the graphs from the family*

$$\mathcal{F} := \{B_1, B_2, B_4, B_5, B_6, B_7, B_8, B_9, H_1, H_2, H_3, H_4\}.$$

(iii) *G is a line graph or G is an induced subgraph of a cocktail-party graph.*

Proof. The implication (i) $\Longrightarrow$ (ii) follows from the fact that the suspensions of the graphs from $\mathcal{F}$ are not ℓ_1-graphs. The implication (iii) $\Longrightarrow$ (i) is clear. We now show that (ii) $\Longrightarrow$ (iii) holds. Let G be a connected graph that does not contain any member of $\mathcal{F}$ as an induced subgraph. If G does not contain B_3 as an induced subgraph, then G is a line graph by Theorem 17.1.5. Hence, we can suppose that B_3 is an induced subgraph of G; say, $B_3 = G[Y]$ is the subgraph of G induced by the subset of nodes Y, $|Y| = 5$. We show that G is an induced subgraph of a cocktail-party graph. For this consider the following property (P):

(P) For each subset $Z \subseteq V(G)$ such that $Y \subseteq Z$ and for each $i \in V(G) \setminus Z$, if $G[Z]$ is an induced subgraph of a cocktail-party graph and if $G[Z \cup \{i\}]$ is connected, then $G[Z \cup \{i\}]$ is also an induced subgraph of a cocktail-party graph.

We show that (P) holds by induction on $|Z|$.

Case 1. We show that (P) holds for $Z = Y$. Let $i \in V(G) \setminus Y$ such that the graph $G[Y \cup \{i\}]$ is connected. So, $G[Y \cup \{i\}]$ is a connected graph on six nodes containing $B_3 = K_5 \setminus P_2$ as an induced subgraph. By direct inspection, one can check that there are eleven connected graphs on six nodes containing B_3 as an induced subgraph. Among them, we find H_1, H_2, H_3, H_4; we also find two graphs containing B_2 and three graphs containing B_1; these cases are excluded since G

does not contain any member of $\mathcal{F}$. The remaining two graphs are $K_6 \setminus P_2$ and $\nabla\nabla K_{2\times 2}$ which are, respectively, induced subgraphs of $K_{5\times 2}$ and $K_{4\times 2}$. Hence, the property (P) holds for $Z = Y$.

Consider now Z such that $Y \subseteq Z \subseteq V(G)$, $|Z| \geq 6$ and $G[Z]$ is an induced subgraph of a cocktail-party graph, and let $i \in V(G) \setminus Z$ such that $G[Z \cup \{i\}]$ is connected. Set $Y := \{y_1, y_2, y_3, y_4, y_5\}$ where, for instance, y_1 and y_2 are not adjacent in G and, thus, every other pair of nodes of Y is adjacent in G.

Case 2. Let $s, t \in Z$ such that s and t are not adjacent in G. We show that i is adjacent to both s and t. Since $G[Z]$ is contained in a cocktail-party graph, every other node of Z is adjacent to both s and t. Let $u \in Z$ be a node which is adjacent to i. Then, i is adjacent to at least one of s or t (else, $G[\{u, s, t, i\}]$ would be a induced subgraph B_1 of G). Hence, for $U := \{s, t, y_3, y_4, y_5\}$, $G[U]$ is B_3 and $G[U \cup \{i\}]$ is connected. By Case 1, we deduce that $G[U \cup \{i\}]$ is an induced subgraph of a cocktail-party graph, which implies that i is adjacent to both s and t.

Case 3. Let $s, t \in Z$ such that s and t are adjacent in G. We show that i is adjacent to at least one of s or t. If there exists $r \in Z$ which is not adjacent to s then, by Case 2, i is adjacent to both r and s. Similarly, if there exists $r \in Z$ which is not adjacent to t, then i is adjacent to t. Else, each $r \in Z$ is adjacent to both s and t. Let $r \in Z$ which is adjacent to i. We can find a set U such that $|U| = 5$, $r, s, t \in U$ and $G[U] = B_3$. Therefore, $G[U \cup \{i\}]$ is an induced subgraph of a cocktail-party graph, which implies that i is adjacent to at least one of s or t.

We deduce from Cases 2 and 3 that $G[Z \cup \{i\}]$ is an induced subgraph of a cocktail-party graph. So, we have shown that (P) holds. ∎

Theorem 17.1.9. *Let G be a connected graph. The following assertions are equivalent.*

(i) *∇G is a hypermetric graph, but not an ℓ_1-graph.*

(ii) *G is an induced subgraph of the Schläfli graph G_{27} and G contains as an induced subgraph one of the graphs of the family*

$$\mathcal{F}_0 := \mathcal{F} \setminus \{B_1, B_2, B_4, B_6\} = \{B_5, B_7, B_8, B_9, H_1, H_2, H_3, H_4\}.$$

Proof. (i) $\Longrightarrow$ (ii) By Theorem 17.1.8, if ∇G is not an ℓ_1-graph, then G contains as an induced subgraph one of the members of $\mathcal{F}$ and, in fact, of $\mathcal{F}_0$ since ∇B_1, ∇B_2, ∇B_4, ∇B_6 are not hypermetric (recall Remark 17.1.7). Let P denote the Delaunay polytope associated with the hypermetric space $(V(\nabla G), 2d_{\nabla G})$ and let H denote its 1-skeleton graph. By Proposition 14.3.5, P is a generating Delaunay polytope in a root lattice. Thus, P is a direct product of Delaunay polytopes from Figure 14.3.1 and H is a direct product of Delaunay polytopes graphs from Figure 14.3.1. In fact, since the graph G is connected, H is not a direct product, i.e., H is one of the Delaunay polytope graphs from Figure

14.3.1. Now, H is G_{27} or G_{56} since all the other Delaunay polytope graphs are ℓ_1-graphs. Therefore, ∇G is an isometric subgraph of G_{56} and, thus, G is an isometric subgraph of G_{27}. The implication (ii) $\Longrightarrow$ (i) is clear. ∎

Corollary 17.1.10. *Let G be a connected graph on n nodes.*

(i) *If $n \geq 37$, then ∇G is an ℓ_1-graph if and only if ∇G is 5-gonal and of negative type.*

(ii) *If $n \geq 28$, then ∇G is an ℓ_1-graph if and only if ∇G is hypermetric.* ∎

A Good Characterization of Hypermetricity and ℓ_1-Embeddability for Truncated Distances of Bipartite Graphs. We consider now, more generally, ℓ_1-embeddability and hypermetricity for truncated distances of graphs. For bipartite graphs, Assouad and Delorme [1982] have obtained several equivalent characterizations, leading to a polynomial-time recognition algorithm; they are formulated in Theorem 17.1.13 below.

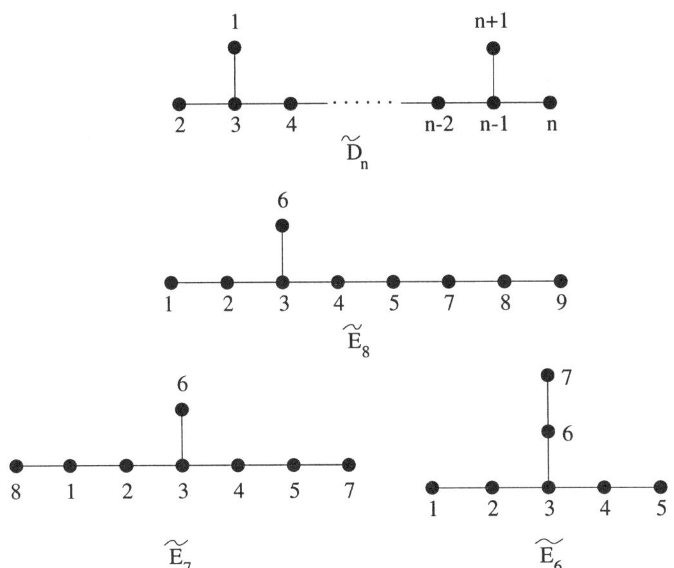

Figure 17.1.11: The graphs[2] $\tilde{D}_n$, $\tilde{E}_8$, $\tilde{E}_7$, $\tilde{E}_6$

[2]The graphs $\tilde{D}_n$, $\tilde{E}_8$, $\tilde{E}_7$, $\tilde{E}_6$ and $\tilde{A}_n := C_{n+1}$ arise, in fact, as the Dynkin diagrams of the root lattices (cf. Brouwer, Cohen and Neumaier [1989]).

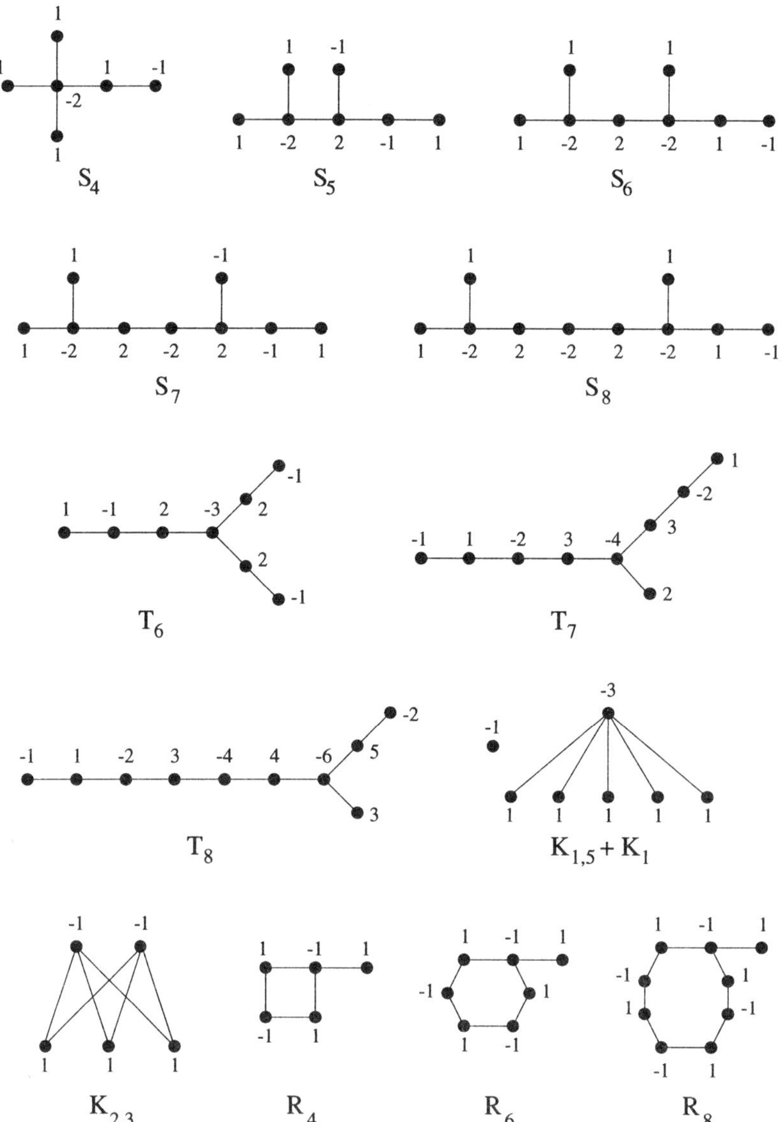

Figure 17.1.12: The family $\mathcal{F}_1$

Theorem 17.1.13. *Let G be a bipartite graph and let d_G^* denote its truncated distance. The following assertions are equivalent.*

(i) *d_G^* is ℓ_1-embeddable.*

(ii) *d_G^* is hypermetric.*

(iii) *d_G^* is 31-gonal.*

(iv) *G does not contain as an induced subgraph any of the thirteen graphs from the family $\mathcal{F}_1$ shown in Figure* 17.1.12.

(v) *G is a star (i.e., $G = K_{1,n}$ for some $n \geq 1$), or every connected component of G is an induced subgraph of an even circuit C_{2n} ($n \geq 2$) or of one of the graphs $\tilde{D}_n$ ($n \geq 4$), $\tilde{E}_8$, $\tilde{E}_7$, $\tilde{E}_6$ shown in Figure* 17.1.11. ∎

The proof of Theorem 17.1.13 relies on several lemmas.

Lemma 17.1.14. *Let $G_1(V_1, E_1)$ and $G_2(V_2, E_2)$ be two graphs with disjoint node sets and let G denote their clique* 0*-sum, with node set $V_1 \cup V_2$ and edge set $E_1 \cup E_2$. If the truncated distance $d^*_{G_i}$ of G_i admits an ℓ_1-embedding of size ≤ 4 for $i = 1, 2$, then the truncated distance d^*_G of G also admits an ℓ_1-embedding of size ≤ 4.*

Proof. Say, $d^*_{G_1} = \sum_{S \subseteq V_1} \alpha_S \delta(S)$ where $\alpha_S \geq 0$ for all S and $\sum_S \alpha_S = 4$ (if $\sum_S \alpha_S \leq 4$ one can make this sum equal to 4 by introducing the empty cut with coefficient $4 - \sum_S \alpha_S$ in the decomposition). Similarly, $d^*_{G_2} = \sum_{T \subseteq V_2} \beta_T \delta(T)$ where $\beta_T \geq 0$ for all T and $\sum_T \beta_T = 4$. Then, one can verify that d admits the following decomposition:

$$d^*_G = \frac{1}{8} \sum_{S \subseteq V_1} \sum_{T \subseteq V_2} \alpha_S \beta_T [\delta(S \cup T) + \delta(S \cup (V_2 \setminus T))]$$

with size $\frac{1}{4} \sum_{S,T} \alpha_S \beta_T = 4$. ∎

We need a result concerning generalized line graphs. These graphs are defined in the following manner. Let H be a graph with, say, node set $\{v_1, \ldots, v_n\}$ and let $a_1, \ldots, a_n$ be nonnegative integers. For every node v_i of H consider a cocktail-party graph $K_{a_i \times 2}$ (all being defined on disjoint node sets). Then, the *generalized line graph* $L(H; a_1, \ldots, a_n)$ is obtained by juxtaposing the line graph $L(H)$ of H, the n cocktail-party graphs $K_{a_i \times 2}$, and adding for every node $e := v_i v_j$ of $L(H)$ (corresponding to an edge e in H) edges between e and all nodes of $K_{a_i \times 2}$ and $K_{a_j \times 2}$. Hence we obtain usual line graphs when $a_1 = \ldots = a_n = 0$. Line graphs and cocktail-party graphs have an ℓ_1-embeddable truncated distance; Assouad and Delorme observed that this extends to generalized line graphs of bipartite graphs.

Lemma 17.1.15. *If H is a bipartite graph, then the truncated distance of any generalized line graph of H admits an ℓ_1-embedding of size ≤ 4.*

Proof. Let $V = \{v_1, \ldots, v_n\} = S \cup T$ denote the bipartition of H. Given integers $a_1, \ldots, a_n \geq 0$, we consider the generalized line graph $L(H; a_1, \ldots, a_n)$ and we let $Y_{v_i} \cup Y'_{v_i}$ denote the node set on which the cocktail-party graph $K_{a_i \times 2}$ is defined. Let $V^{(\alpha)}$, $Y^{(\alpha)}_{v_i}$ ($\alpha = 1, 2, 3, 4$) be four disjoint copies of the sets V and Y_{v_i} ($i = 1, \ldots, n$). We now consider two cocktail-party graphs G_1 and G_2,

G_1 being defined on the node set $\bigcup_{\alpha=1,2}(V^{(\alpha)} \cup \bigcup_{i=1}^{n} Y_{v_i}^{(\alpha)})$ and G_2 being defined on $\bigcup_{\alpha=3,4}(V^{(\alpha)} \cup \bigcup_{i=1}^{n} Y_{v_i}^{(\alpha)})$ (with the obvious pairing for opposite nodes). We now construct an isometric embedding of $L(H;a)$ equipped with the truncated distance into the Cartesian product of G_1 and G_2. Namely, to a node v_iv_j of $L(H)$ (where $v_i \in S$ and $v_j \in T$), assign the pair $(v_i^{(1)}, v_j^{(3)})$; to a node $y \in Y_{v_i}$ (resp. $y \in Y'_{v_i}$) with $v_i \in S$, associate the pair $(v_i^{(1)}, y^{(3)})$ (resp. $(v_i^{(1)}, y^{(4)})$); to a node $y \in Y_{v_i}$ (resp. $y \in Y'_{v_i}$) with $v_i \in T$, associate the pair $(y^{(1)}, v_i^{(3)})$ (resp. $(y^{(2)}, v_i^{(3)})$). The result now follows since a cocktail-party graph is ℓ_1-embeddable with size 2, which implies that the Cartesian product of two cocktail-party graphs has an ℓ_1-embedding of size 4. ∎

Lemma 17.1.16. *For each of the graphs C_{2n} $(n \geq 2)$, $\tilde{D}_n$ $(n \geq 4)$, $\tilde{E}_6$, $\tilde{E}_7$, $\tilde{E}_8$, the truncated distance admits an ℓ_1-embedding of size ≤ 4.*

Proof. For C_{2n} and $\tilde{D}_n$, the result follows by applying Lemma 17.1.15. This is obvious for C_{2n} as $C_{2n} = L(C_{2n})$. On the other hand, $\tilde{D}_n$ can be obtained as an induced subgraph of the line graph $L(C_{2m};a)$ of some even circuit choosing a with all zero components except two equal to 1. We now indicate explicit ℓ_1-embeddings for the truncated distances of $\tilde{E}_6$, $\tilde{E}_7$, $\tilde{E}_8$ (using the node labelings shown in Figure 17.1.11), of respective sizes 3,4,4:

$$\begin{aligned} 2d^*_{\tilde{E}_6} &= \delta(\{1,2,5\}) + \delta(\{1,4,5\}) + \delta(\{1,2,7\}) + \delta(\{4,5,7\}) + \delta(\{1,6,7\}) \\ &\quad + \delta(\{5,6,7\}); \end{aligned}$$

$$\begin{aligned} 2d^*_{\tilde{E}_7} &= \delta(\{1,2,5,7,8\}) + \delta(\{1,2,3,4,8\}) + \delta(\{1,6,8\}) + \delta(\{2,3,6\}) \\ &\quad + \delta(\{1,2\}) + \delta(\{4,5\}) + \delta(\{7\}) + \delta(\{8\}); \end{aligned}$$

$$\begin{aligned} 2d^*_{\tilde{E}_8} &= \delta(\{1,2,5,7\}) + \delta(\{1,2,3,4\}) + \delta(\{1,6\}) + \delta(\{2,3,6\}) \\ &\quad + \delta(\{1,2,8,9\}) + \delta(\{4,5\}) + \delta(\{9\}) + \delta(\{7,8\}). \end{aligned}$$

∎

Proof of Theorem 17.1.13. The implications (i) $\Longrightarrow$ (ii) $\Longrightarrow$ (iii) are obvious. The implication (iii) $\Longrightarrow$ (iv) relies on the fact that each of the graphs $G \in \mathcal{F}_1$ has a nonhypermetric truncated distance. That is, there exists an integer vector b with $\sum_i b_i = 1$ such that $Q(b)^T d^*_G > 0$. Explicit values for b are indicated for each graph of $\mathcal{F}_1$ in Figure 17.1.12. Note that $\sum_i |b_i| \leq 31$ with equality for graph T_8.
The implication (v) $\Longrightarrow$ (i) is immediate. Indeed, d^*_G is obviously ℓ_1-embeddable if G is a star and otherwise the assertion follows from Lemmas 17.1.14 and 17.1.16.
We now show the last implication (iv) $\Longrightarrow$ (v). Let G be a bipartite graph

not containing any of the thirteen graphs from Figure 17.1.12 as an induced subgraph. We first claim that

(a) every connected component of G is a tree or an even circuit.

Indeed suppose that H is a connected component of G which is not a tree nor a circuit. Let C be an induced circuit in H of minimum length; thus, $|C| \geq 4$ and is even. There exists a node x outside C adjacent to some node of C. In fact, if $|C| \geq 6$, then x is adjacent to exactly one node of C (by minimality of $|C|$); then we find R_6, R_8 or T_7 as an induced subgraph of H if $|C| = 6, 8$ or ≥ 10, respectively. If $|C| = 4$ and x is adjacent to only one node of C, then we find R_4 and, if $|C| = 4$ and x is adjacent to 2 nodes of C, then we find $K_{2,3}$. Thus, (a) holds.

If G has a node of degree ≥ 5, then G is connected (else, we find $K_{1,5} + K_1$ as an induced subgraph of G) and, moreover, G is a star (else, we find S_4).

We can now suppose that G has maximum degree ≤ 4. Let H be a connected component of G with maximum degree Δ, which is a tree; we show that H is an induced subgraph of $\tilde{D}_n$, $\tilde{E}_6$, $\tilde{E}_7$, or $\tilde{E}_8$. If $\Delta \leq 2$ then H is a path and, thus, is contained in $\tilde{D}_n$. If $\Delta = 4$ then $H = K_{1,4} = \tilde{D}_4$ (else we find S_4 as induced subgraph). We now suppose that $\Delta = 3$. Let x be a node of degree 3 in H, let L_1, L_2, L_3 denote the connected components of $H\backslash x$ and, for $i = 1, 2, 3$, let m_i denote the longest geodesic distance from x to a node in L_i. Say, $1 \leq m_1 \leq m_2 \leq m_3$. Observe that $m_3 \geq 3 \Longrightarrow m_1 = 1$ (because of T_6), $m_2 \geq 3 \Longrightarrow m_3 = 3$ (because of T_7), $m_1 = 1$ and $m_3 \geq 6 \Longrightarrow m_2 = 1$ (because of T_8). We distinguish two cases.

Case 1: There exists a node x of degree 3 in H for which $m_2 \geq 2$. Then, in view of the above observations, the only possibilities for (m_1, m_2, m_3) are the following sequences: (1,2,2), (2,2,2), (1,3,3), (1,2,3), (1,2,4) and (1,2,5). Now H has no node outside $L_1 \cup L_2 \cup L_3$ (else one would find one of the forbidden induced subgraphs) and, thus, H is an induced subgraph of $\tilde{E}_6$, $\tilde{E}_7$ or $\tilde{E}_8$.

Case 2: Every node of degree 3 in H has $m_2 = 1$. Then, we obtain that H is an induced subgraph of some $\tilde{D}_n$. ∎

On the other hand, there is no finite point criterion, analogue to Theorem 17.1.13, for truncated distances of general graphs. Indeed, for every $n \geq 2$, there exists a graph G on $2n + 1$ nodes whose truncated distance is not hypermetric while the truncated distance of any proper induced subgraph of G is ℓ_1-embeddable. We present in Example 17.1.17 below an example of such graph, taken from Assouad and Delorme [1982]. We state a preliminary result for generalized line graphs.

Example 17.1.17. Let G_n denote the generalized line graph $L(C_{2n-1}, a)$, where C_{2n-1} is the circuit on $2n - 1$ nodes and $a := (1, 0, \ldots, 0)$. Figure 17.1.18 shows the graph G_4 on 9 nodes. The truncated distance of G_n is not hypermetric (as it violates the pure hypermetric inequality where the ± 1 coefficients are assigned as indicated in Figure 17.1.18). On the other hand, the truncated distance of any proper induced subgraph of G_n is ℓ_1-embeddable. Indeed, if we delete a node of degree 2 on the circuit, we find the generalized line graph of a path and, if we delete a node of degree 4, we find a tree which is a subtree of some D_n. In both cases, we find a graph whose truncated distance is ℓ_1-embeddable (recall Lemmas 17.1.15 and 17.1.16). Finally if we delete one of the two remaining nodes of degree 2, an ℓ_1-embedding can be very easily constructed. ∎

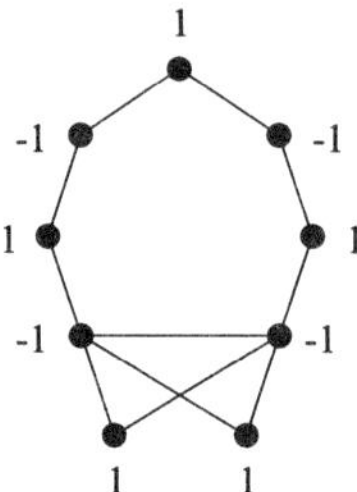

Figure 17.1.18: A minimal graph whose truncated distance is not hypermetric

Finally, let us mention that Assouad and Delorme [1980, 1982] have characterized the graphs whose truncated distance is hypercube embeddable at any given scale λ; they show the existence of an integer $n(\lambda)$ such that λd^*_G is hypercube embeddable whenever this holds for any induced subgraph of G on at most $n(\lambda)$ nodes. This result will be discussed in detail in Section 24.4.

17.2 Hypermetric Regular Graphs

In this section we consider regular graphs and some subclasses such as distance-regular graphs and strongly regular graphs (defined later). Recall that a graph G is said to be *regular* if all its nodes have the same degree, called the *valency* of G. We aim again at understanding which such graphs enjoy metric properties such as hypermetricity, ℓ_1-embeddability, being of negative type, etc. We address this question for both the path metric and the truncated distance.

Properties of the Truncated Distance of a Regular Graph. We group here several results on the hypermetricity of the truncated distance space of a regular graph. They will apply, in particular, to the usual path metric of strongly regular graphs, i.e., distance-regular graphs of diameter 2.

Given a graph G on n nodes, we denote by D^*_G the symmetric $n \times n$ matrix whose (i,j)-th entry is equal to $d^*_G(i,j)$, for all $i,j \in V(G)$. We recall that A_G denotes the adjacency matrix of G.

The first result is from Deza and Grishukhin [1993] and gives several equivalent characterizations for the regular graphs whose truncated distance is hypermetric.

Proposition 17.2.1. *Let G be a connected regular graph on n nodes with valency k. The following assertions are equivalent.*

(i) *d^*_G is of negative type.*

(ii) *the distance space $(V(G), 2d^*_G)$ has a spherical representation with radius r satisfying $r^2 < 2$.*

(iii) *d^*_G is hypermetric.*

(iv) *∇G is of negative type.*

(v) *$\lambda_{\min}(A_G) \geq -2$.*

(vi) *D^*_G has exactly one positive eigenvalue.*

Moreover, if d^*_G is hypermetric, then the radius r of the Delaunay polytope associated with the space $(V(G), 2d^*_G)$ is given by

$$r^2 = 2 - \frac{k+2}{n}. \tag{17.2.2}$$

Proof. (i) $\Longrightarrow$ (ii) Note that $\sum_{i \in V(G)} 2d^*_G(i,j) = 2(2n-2-k)$ is a constant. Hence, by Proposition 14.4.1, $(V(G), 2d^*_G)$ has a spherical representation whose radius r is given by relation (14.4.2). Therefore, $r^2 = 2 - \frac{k+2}{n}$ and, thus, $r^2 < 2$.
The implication (ii) $\Longrightarrow$ (iii) follows from Proposition 14.4.4.
(iii) $\Longrightarrow$ (iv) By Proposition 14.4.3, the radius of the Delaunay polytope associated with $(V(G), 2d^*_G)$ is given by (17.2.2). Since $(V(\nabla G), 2d_{\nabla G})$ is the spherical 2-extension of the space $(V(G), 2d^*_G)$, we deduce from Lemma 14.4.5 that $(V(\nabla G), d_{\nabla G})$ is of negative type.
The equivalence (iv) $\Longleftrightarrow$ (v) follows from Proposition 17.1.4.
(v) $\Longrightarrow$ (vi) Let $\lambda_1 = k, \lambda_2, \ldots, \lambda_n \geq -2$ denote the eigenvalues of the adjacency matrix A_G of G. Note that

$$D^*_G = 2J - (A_G + 2I),$$

where J is the $n \times n$ matrix of all ones. The vector of all ones is a common eigenvector of A_G and D^*_G for the eigenvalues k and $2n-2-k$, respectively. One checks easily that the other eigenvalues of D^*_G are $-\lambda_2 - 2, \ldots, -\lambda_n - 2$ which are all nonpositive. Hence, $2n - 2 - k$ is the only positive eigenvalue of D^*_G.
The implication (vi) $\Longrightarrow$ (v) follows by reversing the arguments used above for the implication (v) $\Longrightarrow$ (vi). Using the obvious implication (iv) $\Longrightarrow$ (i), we obtain the equivalence of (i)-(vi). ∎

Proposition 17.2.1 applies, in particular, to the regular graphs of diameter 2; then, the two distances d_G and d^*_G coincide. However, without the regularity assumption, the equivalence of (i)-(vi) does not hold. For instance, $K_9 \setminus P_3$ has diameter 2, is not regular, satisfies (v) but not (iii) (recall Example 14.4.9).

Let G be a connected regular graph with $\lambda_{\min}(A_G) \geq -2$. Hence, its truncated distance d^*_G is hypermetric. Let P^*_G denote the Delaunay polytope associated with the space $(V(G), 2d^*_G)$ and let H^*_G denote its 1-skeleton graph. By Proposition 14.3.3, $(V(G), d^*_G)$ is an isometric subspace of the graphic metric space $(V(H^*_G), d_{H^*_G})$. By Proposition 14.3.5, P^*_G is a Delaunay polytope in a root lattice and, thus, H^*_G is a direct product of some of the Delaunay polytope graphs shown in Figure 14.3.1. The next result from Deza and Grishukhin [1993] shows that, if H^*_G is a nontrivial direct product, then it can only be the direct product of two complete graphs.

A bipartite graph B with bipartition $V_1 \cup V_2$ of its set of nodes is said to be *semiregular* if all nodes in V_1 (resp. all nodes in V_2) have the same degree.

Lemma 17.2.3. *Let G be a connected regular graph on n nodes with valency k. Suppose that $\lambda_{\min}(A_G) \geq -2$ and let H^*_G denote the 1-skeleton graph of the Delaunay polytope P^*_G associated with $(V(G), 2d^*_G)$. If H^*_G is a nontrivial direct product, then $H^*_G = K_{n_1} \times K_{n_2}$ for some $n_1, n_2 \geq 1$, G is the line graph of a bipartite semiregular graph and*

$$n = \frac{n_1 + n_2}{n_1 n_2}(k+2).$$

Proof. Suppose that H is the nontrivial direct product $H_1 \times H_2$. By the assumption, $(V(G), d^*_G)$ is an isometric subspace of the graphic metric space $(V(H), d_H)$. Let

$$f : i \in V(G) \mapsto f(i) = (f_1(i), f_2(i)) = (i_1, i_2) \in V(H_1) \times V(H_2)$$

denote this isometric embedding. For $i \in V(G)$, set

$$V_1(i) := \{j \in V(G) \mid f_1(i) = f_1(j)\}, \; V_2(i) := \{j \in V(G) \mid f_2(i) = f_2(j)\}.$$

If i, j are adjacent in G, then $j \in V_1(i) \cup V_2(i)$. Conversely, we check that, if $|V_1(i)|, |V_2(i)| > 1$, then both $V_1(i)$ and $V_2(i)$ induce a complete graph in G.

For this, let $j \in V_1(i)$ and $h \in V_2(i)$ with $j \neq i$, $h \neq i$. Then,

$$2 \geq d^*_G(j,h) = d_{H_1}(j_1, h_1) + d_{H_2}(j_2, h_2) = d_{H_1}(i_1, h_1) + d_{H_2}(j_2, i_2)$$

(since $i_1 = j_1$ and $i_2 = h_2$) which is equal to $d^*_G(i,h) + d^*_G(i,j) \geq 2$. This implies that $d^*_G(i,h) = d^*_G(i,j) = 1$, i.e., both h and j are adjacent to i. One deduces easily that any two nodes in V_1, or in V_2, are adjacent.

Therefore, if $|V_1(i)|, |V_2(i)| > 1$, then $|V_1(i)| + |V_2(i)| = k+2$. For $j \in V_1(i)$,

$$V_1(i) = V_1(j), \; k+2 \leq |V_1(j)| + |V_2(j)|,$$

implying that $|V_2(i)| \leq |V_2(j)|$ and, thus, $|V_1(j)|, |V_2(j)| > 1$, yielding

$$k+2 = |V_1(j)| + |V_2(j)| \;\text{ and, thus, } |V_2(j)| = |V_2(i)|.$$

Therefore, since G is connected, there exist integers $p, q \geq 1$ such that $|V_1(i)| = p$, $|V_2(i)| = q$ for all $i \in V(G)$.

Let B denote the bipartite graph with node bipartition $V_1 \cup V_2$, where

$$V_1 := f_1(V(G)) \subseteq V(H_1) \text{ and } V_2 := f_2(V(G)) \subseteq V(H_2),$$

and two nodes $i_1 \in V_1$, $i_2 \in V_2$ are adjacent in B if $(i_1, i_2) = f(i)$ for some node $i \in V(G)$. So each node of V_1 (resp. of V_2) has valency p (resp. q), i.e., B is semiregular. It is immediate to see that G is the line graph of B.

We now show that H_1 and H_2 are complete graphs. Set $n_1 := |V_1|$, $n_2 := |V_2|$ and $n := |V(G)|$. Let r denote the radius of the Delaunay polytope P_G^*; r is given by relation (17.2.2). So,

$$r^2 = 2 - \frac{k+2}{n} = \frac{n_1 - 1}{n_1} + \frac{n_2 - 1}{n_2}.$$

Let r_m denote the radius of the Delaunay polytope whose 1-skeleton graph is the graph H_m, for $m = 1, 2$. Then, $r^2 = r_1^2 + r_2^2$ holds. We use the following observation: For each Delaunay polytope P in a root lattice, its radius r satisfies

$$r^2 \geq \frac{|V(P)| - 1}{|V(P)|}$$

with equality if and only if P is a simplex. Therefore,

$$r_m^2 \geq \frac{|V(H_m)| - 1}{|V(H_m)|} \geq \frac{n_m - 1}{n_m},$$

since $|V(H_m)| \geq n_m$, for $m = 1, 2$. But,

$$r^2 = r_1^2 + r_2^2 = \frac{n_1 - 1}{n_1} + \frac{n_2 - 1}{n_2},$$

from which we deduce that

$$r_m^2 = \frac{n_m - 1}{n_m}, \ |V(H_m)| = n_m$$

and, thus, H_m is the complete graph K_{n_m} for $m = 1, 2$. ∎

Corollary 17.2.4. *Let G be a connected regular graph on n nodes with valency k and such that $\lambda_{\min}(A_G) \geq -2$. Then, one of the following assertions holds.*

(i) *G is the line graph of a bipartite semiregular graph and $n = \frac{n_1 n_2}{n_1 + n_2}(k+2)$, for some $n_1, n_2 \geq 1$.*

(ii) *G is the line graph of a regular graph and $n = \frac{m}{4}(k+2)$ for some $m \geq 3$.*

(iii) *$G = K_{m \times 2}$ and $n = k + 2$.*

(iv) *G is an induced subgraph of the Gosset graph G_{56} and $n = 2(k+2)$.*

(v) *G is an induced subgraph of the Schläfli graph G_{27} and $n = \frac{3}{2}(k+2)$.*

(vi) *G is an induced subgraph of the Clebsch graph $\frac{1}{2}H(5,2)$ and $n = \frac{3}{2}(k+2)$.*

Proof. Let H_G^* denote the 1-skeleton graph of the Delaunay polytope P_G^* associated with the hypermetric space $(V(G), 2d_G^*)$. If H_G^* is a direct product, then we have (i) by Lemma 17.2.3. So we now suppose that H_G^* is one of the Delaunay polytope graphs from Figure 14.3.1. We know that the radius r of P_G^* satisfies $r^2 = 2 - \frac{k+2}{n} < 2$.

• If $H_G^* = J(m,t)$ for some $t \geq 1$, $n \geq 2t$, then $r^2 = \frac{t(m-t)}{m} < 2$ implying that $t = 1,2,3$. If $H_G^* = J(m,1) = K_m$, then $G = H_G^* = K_m$ is the line graph of the bipartite semiregular graph $K_{1,m}$; hence, $m = n$ and we have (i). If $H_G^* = J(m,2) = L(K_m)$, then G is a line graph. Since G is regular, one can check that G is the line graph of a regular graph or a bipartite semiregular graph. Since $r^2 = \frac{2(m-2)}{m}$, we deduce that $n = \frac{m}{4}(k+2)$. So, we have (i) or (ii). If $H_G^* = J(m,3)$, then $m = 6,7,8$. If $H_G^* = J(6,3)$, then G is an induced subgraph of G_{56} and $r^2 = \frac{3}{2} = 2 - \frac{k+2}{n}$, yielding $n = 2(k+2)$, i.e., we have (iv). The cases $m = 7,8$ are excluded. Indeed, one can check that every subgraph K of $J(m,3)$ $(m = 7,8)$ such that K is not contained in $J(6,3)$ nor in $J(n,2)$ and such that no pair of nodes of K is at distance 3 in $J(m,3)$ has strictly less than $\frac{m(k+2)}{9-m}$ nodes.

• If $H_G^* = K_{m\times 2}$, then we have (iii).

• If $H_G^* = \frac{1}{2}H(m,2)$ for some $m \geq 4$, then $r^2 = \frac{m}{4} < 2$, implying that $m = 4,5,6,7$. If $m = 4$, then $H_G^* = K_{4\times 2}$ and, thus, we have (iii). If $m = 5$, then $r^2 = \frac{5}{4}$ yielding $n = \frac{4}{3}(k+2)$ and, thus, we have (vi). If $m = 6$, then $r^2 = \frac{3}{2}$ yielding $n = 2(k+2)$ and, thus, we have (iv) since $\frac{1}{2}H(6,2)$ is an isometric subgraph of G_{56}. The case $m = 7$ is excluded (similarly to the exclusion above of the cases $J(7,3)$ and $J(8,3)$; indeed, there is no k-regular subgraph of $\frac{1}{2}H(7,2)$ on $n = 4(k+2)$ nodes which is not contained in $\frac{1}{2}H(6,2)$ or $J(7,2)$ and does not contain a pair of vertices at distance 3).

• If $H_G^* = G_{56}$, then we have (iv) and, if $H_G^* = G_{27}$, then we have (v). ∎

Remark 17.2.5. Under the assumptions of Corollary 17.2.4, the only possibilities for the 1-skeleton graph H_G^* of the Delaunay polytope P_G^* associated with the hypermetric space $(V(G), 2d_G^*)$ are $H_G^* = K_{n_1} \times K_{n_2}$, $J(m,1)$, $J(m,2)$, $J(6,3)$, $K_{m\times 2}$, $\frac{1}{2}H(5,2)$, $\frac{1}{2}H(6,2)$, G_{27} and G_{56}. In particular, if G is not a line graph nor a cocktail-party graph, then H_G^* is one of $J(6,3)$, $\frac{1}{2}H(5,2)$, $\frac{1}{2}H(6,2)$, G_{27} or G_{56}. Note that the radius r of the Delaunay polytope P_G^* satisfies $r^2 = \frac{5}{4}$ for $\frac{1}{2}H(5,2)$, $r^2 = \frac{4}{3}$ for G_{27} and $r^2 = \frac{3}{2}$ for $\frac{1}{2}H(6,2)$, $J(6,3)$ and G_{56}. ∎

The graphs for which $\lambda_{\min}(A_G) \geq -2$ have been intensively studied in the literature.

Clearly, $\lambda_{\min}(A_G) \geq -2$ for every line graph G (indeed, if $G = L(H)$, then $2I + A_G = N^T N$ is positive semidefinite, where N is the node-edge incidence matrix of H). Moreover, $\lambda_{\min}(A_G) = -2$ if G is a cocktail-party graph. More generally, $\lambda_{\min}(A_G) \geq -2$ for all generalized line graphs.

In fact, generalized line graphs constitute (up to a finite number of exceptions) the only connected graphs with $\lambda_{\min}(A_G) \geq -2$. More precisely, Cameron,

Goethals, Seidel and Shult [1976] show the following results: If G is a connected graph on $n > 36$ nodes satisfying $\lambda_{\min}(A_G) \geq -2$, then G is a generalized line graph. Moreover, if G is connected regular with $n > 28$ nodes and $\lambda_{\min}(A_G) \geq -2$, then G is a line graph or a cocktail-party graph.

This leads us to consider the class $\mathcal{L}_{BCS}$, consisting of the connected regular graphs with $\lambda_{\min}(A_G) \geq -2$ and which are not line graphs nor cocktail-party graphs. Bussemaker, Cvetković and Seidel [1976] have completely described the graphs in $\mathcal{L}_{BCS}$. The class $\mathcal{L}_{BCS}$ consists of 187 graphs, each of them has $n \leq 28$ nodes and valency $k \leq 16$. The graphs in $\mathcal{L}_{BCS}$ are partitioned into three layers $\mathcal{L}_1$, $\mathcal{L}_2$, $\mathcal{L}_3$ depending on the value of the quantity $\frac{n}{k+2}$, where n is the number of nodes and k the valency of a graph in $\mathcal{L}_{BCS}$. The layer $\mathcal{L}_1$ (resp. $\mathcal{L}_2$, $\mathcal{L}_3$) consists of the graphs $G \in \mathcal{L}_{BCS}$ for which $\frac{n}{k+2} = 2$ (resp. $\frac{n}{k+2} = \frac{3}{2}$, $\frac{n}{k+2} = \frac{4}{3}$).

Our approach permits to shed a new light on the parameter $\frac{k+2}{n}$ characterizing each layer of $\mathcal{L}_{BCS}$. Namely, the parameter $\frac{k+2}{n}$ is nothing but the quantity $2 - r^2$, where r is the radius of the Delaunay polytope associated with the hypermetric space $(V(G), 2d^*_G)$ for any graph $G \in \mathcal{L}_{BCS}$. Therefore, each layer in $\mathcal{L}_{BCS}$ is characterized by a quantity directly derived from the hypermetricity of its graphs.

We summarize below several facts about the class $\mathcal{L}_{BCS}$ and its three layers.

- The first layer $\mathcal{L}_1$ consists of 163 graphs (the graphs NN1-163 in Bussemaker, Cvetković and Seidel [1976]); it is characterized by $\frac{n}{k+2} = 2$. For each graph $G \in \mathcal{L}_{BCS}$, the Delaunay polytope P^*_G associated with the hypermetric space $(V(G), 2d^*_G)$ has radius $\frac{3}{2}$ and its 1-skeleton graph is $\frac{1}{2}H(6,2)$, $J(6,3)$, or G_{56}. Hence, each graph $G \in \mathcal{L}_1$ is an induced subgraph of G_{56} and thus has diameter 2 or 3. Therefore, the graphs of $\mathcal{L}_1$ with diameter 2 are hypermetric with Delaunay polytope graph $\frac{1}{2}H(6,2)$, $J(6,3)$, or G_{56}.

- The second layer $\mathcal{L}_2$ consists of 21 graphs (the graphs NN164-184 in Bussemaker, Cvetković and Seidel [1976]) including the Schläfli graph G_{27} (which is N184). It is characterized by the value $\frac{n}{k+2} = \frac{3}{2}$. For each $G \in \mathcal{L}_2$, the Delaunay polytope P^*_G is 2_{21} with radius r, $r^2 = \frac{4}{3}$. Hence, each $G \in \mathcal{L}_2$ is an isometric subgraph of G_{27} and thus has diameter 2 and is hypermetric.

- The third layer $\mathcal{L}_3$ consists of 3 graphs; they are the Clebsch graph $\frac{1}{2}H(5,2)$ (N187 in Bussemaker, Cvetković and Seidel [1976]) and two of its regular subgraphs (the graphs NN185,186). $\mathcal{L}_2$ is characterized by the value $\frac{n}{k+2} = \frac{4}{3}$. For each graph $G \in \mathcal{L}_3$, $P^*_G = h\gamma_5$ with radius r, $r^2 = \frac{5}{4}$, with 1-skeleton graph $\frac{1}{2}H(5,2)$. Therefore, each graph of $\mathcal{L}_3$ is an isometric subgraph of $\frac{1}{2}H(5,2)$ and thus has diameter 2 and is an ℓ_1-graph with Delaunay polytope graph $\frac{1}{2}H(5,2)$.

Properties of the Path Metric of a Distance-Regular Graph. We conclude this section with some results on hypermetric distance-regular graphs. A graph G is said to be *distance-regular* if there exist integers b_m, c_m $(m > 0)$ such that for any two nodes $i, j \in V(G)$ at distance $d_G(i,j) = m$ there are exactly c_m nodes at distance 1 from i and distance $m - 1$ from j, and there are b_m

nodes at distance 1 from i and distance $m+1$ from j. Hence, G is regular with valency b_0 and there are k_m nodes at distance m from any node $i \in V(G)$, where $k_0 = 1$, $k_1 = 1$, $k_{m+1} = \frac{k_m b_m}{c_{m+1}}$, $m \geq 0$. Let μ denote the number of common neighbors of two nodes at distance 2, i.e., $\mu = c_2$. A *strongly regular* graph is a distance-regular graph of diameter 2.

Note that the quantity $\sum_{i \in V(G)} d_G(i,j) = \sum_{m \geq 0} m k_m$ does not depend on $j \in X$ for a distance-regular graph. Hence, we can derive the following result (e.g., from Theorem 6.2.18).

Proposition 17.2.6. *For a distance-regular graph G, the following are equivalent.*

(i) *G is of negative type.*

(ii) *The graphic metric space $(V(G), d_G)$ has a spherical representation.*

(iii) *The distance matrix D_G has exactly one positive eigenvalue.* ∎

Koolen and Shpectorov [1994] have completely classified the distance-regular graphs of negative type. We present below their classification; we refer, e.g., to Brouwer, Cohen and Neumaier [1989] for the description of the graphs not defined here.

Theorem 17.2.7. *Let G be a distance-regular graph. Then, G is of negative type if and only if one of the following holds.*

(i) *$\mu = 2n-2$ and G is a cocktail-party graph $K_{n\times 2}$.*

(ii) *$\mu = 10$ and G is the Gosset graph G_{56}.*

(iii) *$\mu = 8$ and G is the Schläfli graph G_{27}.*

(iv) *$\mu = 6$ and G is a half-cube graph $\frac{1}{2}H(n,2)$ $(n \geq 4)$.*

(v) *$\mu = 4$ and G is one of the three Chang graphs.*

(vi) *$\mu = 4$ and G is a Johnson graph $J(n,d)$ $(d \geq 2)$.*

(vii) *$\mu = 2$ and G is a Hamming graph $H(n,d)$ $(= (K_d)^n)$ $(n,d \geq 2)$.*

(viii) *$\mu = 2$ and G is a Doob graph (including the Shrikhande graph).*

(ix) *$\mu = 2$ and G is the icosahedron graph.*

(x) *$\mu = 1$ and G is the dodecahedron graph.*

(xi) *$\mu = 1$ and G is the Petersen graph.*

(xii) *$\mu = 1$ and G is a circuit C_n.*

(xiii) *$\mu = 1$ and G is a double-odd graph DO_{2n+1}.*

(xiv) *$\mu = 0$ and G is a complete graph K_n.* ∎

In fact, all the graphs listed in Theorem 17.2.7 are hypermetric. Therefore,

Corollary 17.2.8. *A distance-regular graph is hypermetric if and only if it is of negative type, i.e., if it is one of the graphs* (i)-(xiv). ∎

Therefore, the metric hierarchy from Theorem 6.3.1 partially collapses for distance-regular graphs; we will see in Theorem 19.2.8 that it also does for connected bipartite graphs.

Note that all the graphs listed in Theorem 17.2.7, with the exception of G_{27}, G_{56} and the three Chang graphs, are ℓ_1-graphs and, if we exclude moreover $K_{n\times 2}$ for $n \geq 5$, all of them are isometric subgraphs of a half-cube graph. Hence,

Corollary 17.2.9. *Let G be a distance-regular graph.*

(i) *G is an ℓ_1-graph if and only if G is one of the graphs from* (i), (iv), *or* (vi)-(xiii).

(ii) *G is an isometric subgraph of a half-cube graph if and only if G is one of the graphs from* (iv), (vi)-(xiv).

(iii) *G is an isometric subgraph of a hypercube graph if and only if G is a double-odd graph DO_{2n+1}, a hypercube $H(n,2)$, or an even circuit C_{2n}.*

(iv) *Suppose that G be a strongly regular graph. Then, G is hypermetric if and only if G is $K_n \times K_n$, $J(n,2)$, $K_{n\times 2}$, $\frac{1}{2}H(5,2)$, G_{27}, the 5-circuit C_5, the Petersen graph, the Shrikhande graph, or one of the three Chang graphs.* ∎

The assertion (ii) was obtained by Shpectorov [1996]; (iii) can be found in Koolen [1990] and in Weichsel [1992]; and (iv) in Koolen [1990] and in Deza and Grishukhin [1993]. Further results concerning the distance-regular graphs satisfying some subclass of hypermetric inequalities (such as, for instance, the pentagonal inequalities, or the 6-gonal inequalities) can be found in Koolen [1990, 1994].

17.3 Extreme Hypermetric Graphs

In this section, we consider extreme hypermetrics arising graphs, i.e., the graphs G whose path metric d_G (or whose truncated distance d^*_G) lies on an extreme ray of the hypermetric cone. We remind that G is said to be hypermetric when its path metric is hypermetric. All the results presented here are taken from Deza and Grishukhin [1993].

Let G be a hypermetric graph. Let P_G denote the Delaunay polytope associated with the hypermetric space $(V(G), 2d_G)$ and let H_G denote its 1-skeleton graph. Hence, P_G is a Delaunay polytope in a root lattice and G is an isometric subgraph of H_G. Moreover, G is an extreme hypermetric if and only if P is an extreme Delaunay polytope (by Theorem 15.1.8). By Theorem 16.0.1, the only extreme Delaunay polytopes in a root lattice are the segment α_1, the

Schläfli polytope 2_{21} and the Gosset polytope 3_{21}. Therefore, if G is an extreme hypermetric graph distinct from K_2, then we are in one of the following two situations:

(i) Either $H_G = G_{56}$, i.e., G is an isometric subgraph of G_{56} which is generating (i.e., $V(G)$ viewed as subset of the set of vertices $V(3_{21})$ of 3_{21} generates $V(3_{21})$); we then say that G is an extreme hypermetric graph of *Type I.*

(ii) Or $H_G = G_{27}$, i.e., G is an isometric subgraph of G_{27} which is generating (i.e., $V(G)$ generates $V(2_{21})$); we then say that G is an extreme hypermetric graph of *Type II.*

A generating subset in G_{27} has at least 7 elements. We found that there are 26 distinct (up to permutation) generating subsets in G_{27} with 7 elements (i.e., basic subsets of 2_{21}; see Section 16.2). For $B \subseteq V(G_{27})$, recall that $G_{27}[B]$ denotes the subgraph of G_{27} induced by B. Note that $G_{27}[B]$ is an isometric subgraph of G_{27} if and only if $G_{27}[B]$ has diameter 2 and, in this case, $G_{27}[B]$ is a hypermetric graph. Among the 26 basic subsets B of G_{27} (whose graphs $G_{27}[B]$ are shown in Figures 16.2.4, 16.2.5, 16.2.6 and 16.2.7), the graph $G_{27}[B]$ has diameter 2 for twelve of them, namely for the graphs G_i for $1 \leq i \leq 8$, G_{16}, G_{18}, G_{24}, and G_{26}. Hence, these twelve graphs are extreme hypermetric graphs on 7 nodes with Delaunay polytope graph G_{27}. For each of these twelve graphs G_i, their suspension ∇G_i (for $1 \leq i \leq 8$, $i = 16, 18, 24, 26$) is an extreme hypermetric graph on 8 nodes with Delaunay polytope graph G_{56}.

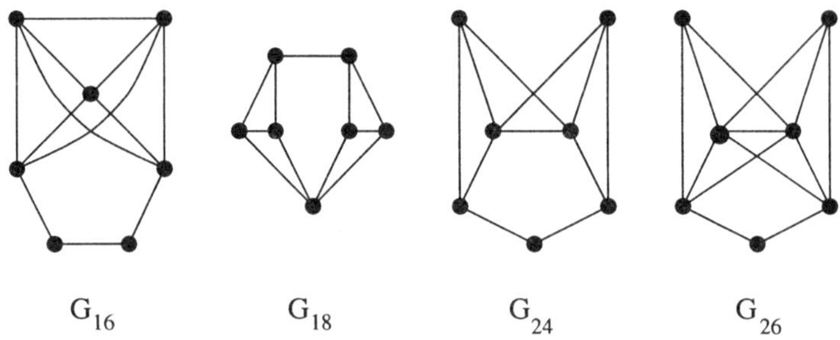

Figure 17.3.1

We recall that $G_1 = \nabla B_9$, $G_2 = \nabla H_2$, $G_3 = \nabla H_1$, $G_4 = \nabla B_8$, $G_5 = \nabla B_7$, $G_6 = \nabla H_4$, $G_7 = \nabla H_3$ and $G_8 = \nabla B_5$, where the graphs B_i ($1 \leq i \leq 8$) and H_i ($1 \leq i \leq 4$) are shown in Figures 17.1.3 and 17.1.6. We show in Figure 17.3.1 the graphs G_{16}, G_{18}, G_{24} and G_{26} (their complements are shown in Figures 16.2.5, 16.2.6 and 16.2.7).

Lemma 17.3.2. *Let H be a maximal (by inclusion) Delaunay polytope graph which is a proper isometric subgraph of G_{56}. Then, H is one of the following graphs:*

(i) $H = J(8,2)$.

(ii) $H = K_{6\times 2} \times K_2$.

(iii) $H = \frac{1}{2}H(6,2)$.

(iv) $H = G_{27}$.

Proof. We know that H is a direct product of the Delaunay polytope graphs from Figure 14.3.1. Let r denote the radius of the Delaunay polytope whose 1-skeleton graph is H. Then, $r^2 \leq \frac{3}{2}$, since H is contained in G_{56}. (We remind that, for two graphs H_1 and H_2, $H_1 \hookrightarrow H_2$ means that H_1 is an isometric subgraph of H_2.)

• If $H = J(n,t)$, then $r^2 = \frac{t(n-t)}{n} \leq \frac{3}{2}$, implying that $t = 1,2,3$. Then, H is not maximal except for $J(8,2)$. Indeed, if $t = 1$, then $n \leq 7$ and $K_n \hookrightarrow J(8,2)$; if $t = 2$, then $n \leq 8$ and $J(n,2) \hookrightarrow J(8,2)$; if $t = 3$, then $n = 6$ and $J(6,3) \hookrightarrow \frac{1}{2}H(6,2)$.

• If $H = K_{n\times 2}$, then $n \leq 6$ and $K_{n\times 2} \hookrightarrow K_{6\times 2} \times K_2$.

• If $H = \frac{1}{2}H(n,2)$, then $r^2 = \frac{n}{4} \leq \frac{3}{2}$, implying that $n \leq 6$ and thus $H \hookrightarrow \frac{1}{2}H(6,2)$.

Else $H = G_{27}$ or H is a direct product. Suppose that $H = H_1 \times H_2$. Denote by r_1, r_2 the radius of the Delaunay polytope whose 1-skeleton graph is H_1, H_2, respectively. Then, $r^2 = r_1^2 + r_2^2 \leq \frac{3}{2}$. Looking at the radii of the Delaunay polytopes from Figure 14.3.1, it is easy to see that the only possibility is $H_1 = K_{6\times 2}$, $H_2 = K_2$ ($r_1^2 = 1$, $r_2^2 = \frac{1}{2}$) (for instance, for $H_1 = H_2 = K_4$, $r_1^2 = r_2^2 = \frac{3}{4}$ but $K_4 \times K_4 \hookrightarrow J(8,2)$). ∎

Lemma 17.3.3. *Let H be a maximal (by inclusion) Delaunay polytope graph which is a proper isometric subgraph of G_{27}. Then, one of the following holds.*

(i) $H = J(6,2)$.

(ii) $H = K_{5\times 2}$.

(iii) $H = \frac{1}{2}H(5,2)$.

(iv) $H = K_6$.

Proof. The proof is similar to that of Lemma 17.3.2. We use the fact that the radius r of the Delaunay polytope whose 1-skeleton graph is H satisfies $r^2 \leq \frac{4}{3}$. It is easily seen that H cannot be a direct product.

• If $H = J(n,t)$, then $r^2 = \frac{t(n-t)}{n} \leq \frac{4}{3}$, implying that $t = 1,2$ and $n \leq 6$. Hence, we have (i) or (iv).

• If $H = K_{n\times 2}$, then $n \leq 5$ (because $K_{6\times 2}$ is not contained in G_{27}) and thus $H \hookrightarrow K_{5\times 2}$.

• If $H = \frac{1}{2}H(n,2)$, then $r^2 = \frac{n}{4} \leq \frac{4}{3}$, implying that $n \leq 5$ and thus $H \hookrightarrow \frac{1}{2}H(5,2)$. ∎

We deduce the following characterization for extreme hypermetric graphs.

Proposition 17.3.4. *Let G be a connected graph distinct from K_2. Then, G is an extreme hypermetric graph if and only if one of the following assertions hold.*

(I) *G is an isometric subgraph of G_{56} and G is not an isometric subgraph of $J(8,2)$, $K_{6\times 2} \times K_2$, $\frac{1}{2}H(6,2)$ or G_{27}.*

(II) *G is an isometric subgraph of G_{27} and G is not an isometric subgraph of $K_{5\times 2}$, $J(6,2)$, K_6 or $\frac{1}{2}H(5,2)$.* ∎

Observe that all the excluded graphs in Proposition 17.3.4 are ℓ_1-graphs. In other words, every isometric subgraph of G_{56} is either an extreme hypermetric graph or an ℓ_1-graph.

As an application of Proposition 17.3.4, we obtain that:

- Every isometric subgraph of G_{27} on $n \geq 17$ nodes is extreme.
- Every induced subgraph of G_{27} on $n \geq 20$ nodes is extreme (since deleting 7 nodes from G_{27} preserves the diameter 2 because $\mu(G_{27}) = 8$).
- Every isometric subgraph of G_{56} on $n \geq 33$ nodes is extreme.
- Every induced subgraph of G_{56} on $n \geq 47$ nodes is extreme (since $\mu(G_{56}) = 10$).
- If G is a connected graph of diameter 2, then its suspension ∇G is an extreme hypermetric graph of Type I if and only if G is an extreme hypermetric graph of Type II.

Let us finally collect some properties for the extreme hypermetric spaces arising from the graphs $G \in \mathcal{L}_{BCS}$.

As we saw in Section 17.2, if G is a connected regular graph with $\lambda_{\min}(A_G) \geq -2$, then its truncated distance d^*_G is hypermetric. Let P^*_G denote the Delaunay polytope associated with $(V(G), 2d^*_G)$ and let H^*_G denote its 1-skeleton graph.

Suppose that G belongs to the class $\mathcal{L}_{BCS}$; that is, G is connected regular with $\lambda_{\min}(A_G) \geq -2$ and G is not a line graph nor a cocktail-party graph. By Remark 17.2.5, H^*_G is one of $J(6,3)$, $\frac{1}{2}H(5,2)$, $\frac{1}{2}H(6,2)$, G_{27} or G_{56}. Since $(V(G), d^*_G)$ is an isometric subspace of $(V(H^*_G), d_{H^*_G})$ which, in turn, is an isometric subspace of $(V(G_{56}), d_{G_{56}})$, we deduce that G does not contain any pair of nodes at distance 3 in G_{56}; in particular, if G is an induced subgraph of $\frac{1}{2}H(6,2)$, then G has at most $n \leq 16$ nodes. Hence,

Proposition 17.3.5. *Let G be a graph of $\mathcal{L}_{BCS}$. If G is not an induced subgraph of $\frac{1}{2}H(6,2)$, then d^*_G is extreme hypermetric. In particular, if G has $n \geq 17$ nodes, then d^*_G is extreme hypermetric.* ∎

Proposition 17.3.6.

(i) *A graph $G \in \mathcal{L}_{BCS}$ is extreme hypermetric if and only if it has diameter* 2 *and it is not an induced subgraph of* $\frac{1}{2}H(6,2)$.

(ii) *Every extreme hypermetric graph which is regular and has diameter* 2 *belongs to* $\mathcal{L}_{BCS}$.

(iii) *Let G be an extreme hypermetric graph from $\mathcal{L}_{BCS}$; then, G is of Type I (resp. of Type II) if and only if G belongs to the layer $\mathcal{L}_1$ (resp. $\mathcal{L}_2$).*

(iv) *Every graph from $\mathcal{L}_{BCS}$ on $n \geq 17$ and with valency $k \geq 9$ is an extreme hypermetric graph. They are the* 29 *graphs in layer $\mathcal{L}_1$ numbered* NN135 – 163 *and the* 8 *graphs in layer $\mathcal{L}_2$ numbered* NN177 – 184.

(v) *All the nine maximal (by inclusion) graphs of $\mathcal{L}_{BCS}$ are extreme hypermetric graphs; they are the Schläfli graph G_{27} (numbered* N184*), the three Chang graphs* NN161 – 163, *and the five graphs* NN148 – 152 *on* 22 *nodes.*

∎

Part III

Isometric Embeddings of Graphs

Introduction

In Part III, we study various embeddability properties of graphs. A metric space can be attached to any connected graph in the following way. Let $G = (V, E)$ be a connected graph. Its *path metric* d_G is the metric defined on V by letting $d_G(a, b)$ denote the length of a shortest path joining a to b in G, for all nodes $a, b \in V$. Then, (V, d_G) is a metric space, called the *graphic metric space* associated with G. The *distance matrix* of G is the matrix $D_G := (d_G(a, b))_{a,b \in V}$.

We have seen in Part I a hierarchy of metric properties that a given distance space may enjoy; in particular, isometric embeddability into the hypercube and into the Banach ℓ_1-, ℓ_2-spaces, hypermetricity, and the negative type condition. We study here the graphs whose path metric enjoys some of these properties. Accordingly, a graph G is called an *ℓ_1-graph*, a *hypercube embeddable graph*, a *hypermetric graph*, a *graph of negative type*, if its path metric d_G is isometrically ℓ_1-embeddable, hypercube embeddable, hypermetric, of negative type, respectively.

Given two connected graphs G and H, we write

$$G \hookrightarrow H$$

and say that G is an *isometric subgraph* (or, *distance-preserving* subgraph) of H if there exists a mapping

$$\sigma : V(G) \longrightarrow V(H)$$

such that

$$d_H(\sigma(a), \sigma(b)) = d_G(a, b)$$

for all nodes $a, b \in V(G)$. We will consider in Part III in particular the cases when the host graph H is a hypercube (see Chapter 19), a Hamming graph or, more generally, a Cartesian product of irreducible graphs (see Chapter 20).

Several other weaker types of embeddings of graphs have been considered in the literature. For instance, one may consider the graphs G that can be embedded into H as an induced subgraph; such embeddings are called topological embeddings and will not be considered here. An even weaker notion of embedding consists of asking which graphs G can be embedded into H as a (partial) subgraph, i.e., requiring only that the edges be preserved; see Section 19.3 where the case of the hypercube as host graph H is briefly discussed.

The theory of isometric embeddings of graphs is a rich theory, with many applications. The main goal is to try to embed graphs isometrically into some other simpler graphs. Research in this area was probably motivated by a problem in communication theory posed by Pierce [1972]. In a telephone network one wishes to be able to establish a connection between two terminals A and B without B knowing that a message is on its way. The idea is to let the message be preceded by some "address" of B, permitting to decide at each node of the network in which direction the message should proceed. Namely, the message will proceed to the next node if its Hamming distance to the destination node B is shorter. The most natural way of devising such a scheme is by labeling the nodes by binary strings, which amounts to try to embed the graph in a hypercube. Unfortunately, not all graphs can be embedded into hypercubes. We study in detail in Chapter 19 the hypercube embeddable graphs. We present their basic structural characterization, due to Djokovic (Theorem 19.1.1), and some other equivalent characterizations (Theorems 19.2.1, 19.2.5 and 19.2.8).

The notion of isometric embedding into hypercubes can be relaxed in several ways.

First, one may consider isometric embeddings into squashed hypercubes as in Graham and Pollack [1971]. Namely, one tries to label the nodes by sequences using the symbols "$0, 1, *$", with the distance between $x, y \in \{0, 1, *\}$ being equal to 1 if $\{x, y\} = \{0, 1\}$ and to 0 otherwise. It turns out that every connected graph on n nodes can be isometrically embedded into the squashed hypercube of dimension $n-1$ (Winkler [1983]). (Note that the squashed hypercube is *not* a semimetric space.)

One may also consider isometric embeddings into arbitrary Cartesian products. In fact, every connected graph admits a unique canonical isometric embedding into a Cartesian product whose factors are irreducible (Graham and Winkler [1985]). This result together with some applications is presented in Chapter 20.

Another way of relaxing isometric embeddings into hypercubes is to look for isometric embeddings into hypercubes up to scale, i.e., to consider ℓ_1-graphs; such embeddings were first considered in Blake and Gilchrist [1973]. Chapter 21 contains results on ℓ_1-graphs; among them, a polynomial time algorithm for recognizing ℓ_1-graphs and a structural characterization for isometric subgraphs of half-cube graphs.

Chapter 18. Preliminaries on Graphs

We introduce here several notions about graphs and embeddings, that we will need in Part III. We start with defining a certain relation θ on the edge set of a graph, which leads to the notion of isometric dimension of the graph.

Let $G = (V, E)$ be a graph. Each edge (a, b) of G induces a partition of the node set V of G into

$$V = G(a, b) \cup G(b, a) \cup G_{=}(a, b),$$

where

$$\text{(18.0.1)} \qquad \begin{cases} G(a,b) & := \{x \in V \mid d_G(x,a) < d_G(x,b)\}, \\ G(b,a) & := \{x \in V \mid d_G(x,b) < d_G(x,a)\}, \\ G_{=}(a,b) & := \{x \in V \mid d_G(x,a) = d_G(x,b)\}. \end{cases}$$

Clearly, if G is a bipartite graph, then $G_{=}(a, b) = \emptyset$ for each edge (a, b) of G.

The following relation θ, defined on the edge set of a graph, was first introduced by Djokovic [1973]. It plays a crucial role in the theory of isometric embeddings of graphs. Given two edges $e = (a, b)$ and $e' = (a', b')$ of G, let

$$\text{(18.0.2)} \qquad e\ \theta\ e' \ \text{ if } \ d_G(a', a) - d_G(a', b) \neq d_G(b', a) - d_G(b', b).$$

In other words, e' is in relation by θ with e if the edge e' "cuts" the partition $V = G(a, b) \cup G(b, a) \cup G_{=}(a, b)$ induced by the edge e, i.e., if the endpoints of e' belong to distinct sets in this partition. The relation θ is clearly reflexive and symmetric, but not transitive in general. For instance, θ is not transitive if G is the complete bipartite graph $K_{2,3}$. Actually, the relation θ is transitive precisely when the graph G can be isometrically embedded into $(K_3)^m$ for some $m \geq 1$ (see Corollary 20.1.3). The transitive closure of θ is denoted by θ^*. The number of equivalence classes of θ^* is called the *isometric dimension* of G and is denoted by $\dim_I(G)$. As will be seen in Chapter 20, every connected graph G can be embedded in a canonical way in a Cartesian product of $\dim_I(G)$ irreducible graphs.

We now recall further definitions needed in Part III. Given two sequences $x = (x_1, \ldots, x_m)$, $y = (y_1, \ldots, y_m) \in \mathbb{R}^m$, their *Hamming distance* $d_H(x, y)$ is defined by

$$d_H(x, y) = |\{i \in \{1, \ldots, m\} \mid x_i \neq y_i\}|.$$

Given two graphs G and H, their *Cartesian product* is the graph $G \times H$ with node set $V(G) \times V(H)$ and whose edges are the pairs $((a,x),(b,y))$ with $a, b \in V(G)$, $x, y \in V(H)$ and, either $(a,b) \in E(G)$ and $x = y$, or $a = b$ and $(x,y) \in E(H)$. The Cartesian product $H_1 \times \ldots \times H_k$ of k graphs $H_1, \ldots, H_k$ is also denoted as $\prod_{h=1}^{k} H_h$. A *Hamming graph* is a Cartesian product of complete graphs, i.e., of the form $\prod_{j=1}^{m} K_{q_j}$ for some integers $q_1, \ldots, q_m, m \geq 1$. Note that the graphic metric space of the Hamming graph $\prod_{j=1}^{m} K_{q_j}$ coincides with the Hamming distance space $(\prod_{j=1}^{m} \{0, 1, \ldots, q_j - 1\}, d_H)$. The *m-hypercube graph* is the graph $H(m,2)$ with node set $\{0,1\}^m$ and whose edges are the pairs $(x,y) \in \{0,1\}^m \times \{0,1\}^m$ with $d_H(x,y) = 1$; $H(m,2)$ has 2^m nodes and $m2^{m-1}$ edges. Hence, $H(m,2)$ is isomorphic to the Hamming graph $(K_2)^m$ and its graphic metric space coincides with the space $(\{0,1\}^m, d_H)$. Equivalently, given a finite set Ω, the $|\Omega|$-hypercube graph, also denoted as $H(\Omega)$, can be defined as the graph whose node set is the set of all subsets of Ω and whose edges are the pairs (A,B) of subsets of Ω such that $|A \triangle B| = 1$. The *half-cube graph* $\frac{1}{2}H(m,2)$ is the graph whose node set is the set of all subsets of even cardinality of $\{1, \ldots, m\}$ and with edges the pairs (A,B) such that $|A \triangle B| = 2$. The *cocktail-party graph* $K_{m \times 2}$ is the complete multipartite graph with m parts, each of size 2. Hence, $K_{m \times 2}$ is the graph on $2m$ nodes $v_1, \ldots, v_{2m}$ whose edges are all pairs of nodes except the m pairs (v_i, v_{i+m}) for $i = 1, \ldots, m$.

A connected graph G is said to be *hypercube embeddable* if its nodes can be labeled by binary vectors in such a way that the distance between two nodes coincides with the Hamming distance between their labels (or, equivalently, if its path metric d_G can be decomposed as a nonnegative integer combination of cut semimetrics). In other words, G is hypercube embeddable if G is an isometric subgraph of $(K_2)^m$ for some $m \geq 1$. Then, the smallest integer m such that G can be isometrically embedded into $H(m,2)$ is denoted by $m_h(G)$. Note that $m_h(G)$ coincides with the notion of minimum h-size $s_h(d_G)$ for the path metric of G (introduced in Section 4.3). The graph G is said to be an *ℓ_1-graph* if its path metric d_G is ℓ_1-embeddable. Equivalently, by Proposition 4.3.8, G is an ℓ_1-graph if G is hypercube embeddable, up to scale; then the smallest integer η such that ηd_G is hypercube embeddable is called the *minimum scale* of G. G is an *ℓ_1-rigid graph* if its path metric d_G is ℓ_1-rigid (see Section 4.3 for the definition of ℓ_1-rigidity). Observe that the Cartesian product $G \times H$ is an ℓ_1-rigid graph if and only if both graphs G and H are ℓ_1-rigid (see Section 7.5).

More generally, a graph G is an isometric subgraph of a Hamming graph if and only if its path metric d_G can be decomposed as a nonnegative integer combination of multicut semimetrics (recall Proposition 4.2.9). If $(S_1, \ldots, S_t)$ is a partition of V_n, then the multicut semimetric $\delta(S_1, \ldots, S_t)$ can be decomposed in the following way:

$$\delta(S_1, \ldots, S_t) = \frac{1}{2} \sum_{1 \le i \le t} \delta(S_i).$$

This implies that, for every isometric subgraph G of a Hamming graph, $2d_G$ is a nonnegative integer combination of cut semimetrics, i.e., $2d_G$ is hypercube embeddable. In other words, every isometric subgraph of a Hamming graph is an ℓ_1-graph with scale ≤ 2 or, equivalently, is an isometric subgraph of a half-cube graph. We summarize in the figure below the links existing between the various embeddings just discussed. In fact, as we will see here, each of the following graph properties can be checked in polynomial time.

G is hypercube embeddable
$\Longrightarrow$ G is an isometric subgraph of a Hamming graph
$\Longrightarrow$ G is an isometric subgraph of a half-cube graph
$\Longrightarrow$ G is an ℓ_1-graph

Chapter 19. Isometric Embeddings of Graphs into Hypercubes

We study in this chapter the graphs that can be isometrically embedded into hypercubes. We give several equivalent characterizations for these graphs in Theorems 19.1.1, 19.2.1, 19.2.5 and 19.2.8. As an application, one can recognize in polynomial time whether a graph can be isometrically embedded in a hypercube. Hypercube embeddable graphs admit, in fact, an essentially unique embedding in a hypercube; two formulations for the dimension of this hypercube are given in Propositions 19.1.2 and 19.2.12.

19.1 Djokovic's Characterization

We recall that, given a graph $G = (V, E)$, a subset $U \subseteq V$ is said to be d_G-convex or, simply, *convex* if it is closed under taking shortest paths.

We now state the main result of this section, which is a structural characterization of the hypercube embeddable graphs, due to Djokovic [1973]. Recall the definition of the set $G(a, b)$ from relation (18.0.1).

Theorem 19.1.1. *Let G be a connected graph. The following assertions are equivalent.*

(i) *G can be isometrically embedded into a hypercube.*

(ii) *G is bipartite and $G(a, b)$ is convex for each edge (a, b) of G.*

Proof. (i) $\Longrightarrow$ (ii) If G is hypercube embeddable, then its path metric d_G satisfies

$$d_G(a, b) + d_G(a, c) + d_G(b, c) \equiv 0 \pmod 2$$

for all nodes a, b, c of G, which means that G is bipartite. Let us now check the convexity of $G(a, b)$ for all adjacent nodes a, b. Let (a, b) be an edge of G, let $x, y \in G(a, b)$ and $z \in V$ lying on a shortest path from x to y. Consider a hypercube embedding of G in which node a is labeled by $\emptyset$, node b is labeled by a singleton $\{1\}$, and nodes x, y, z are labeled by the sets X, Y, Z. Then, $1 \notin X, Y$ since $x, y \in G(a, b)$, and $|X \triangle Y| = |X \triangle Z| + |Y \triangle Z|$ since $d_G(x, y) = d_G(x, z) + d_G(z, y)$. This implies that $1 \notin Z$, i.e., that $z \in G(a, b)$. This shows that the set $G(a, b)$ is convex.
(ii) $\Longrightarrow$ (i) We first show that, given two edges $e = (a, b), e' = (a', b')$ of G, $e\theta e'$ if and only if the two bipartitions of V into $G(a, b) \cup G(b, a)$ and $G(a', b') \cup G(b', a')$

are identical. Suppose, for instance, that $a' \in G(a,b)$ and $b' \in G(b,a)$. We show that $G(a,b) = G(a',b')$. For this, it suffices to check that $G(a,b) \subseteq G(a',b')$. Let $x \in G(a,b)$. If $x \in G(b',a')$, then b' lies on a shortest path from x to a'. By the convexity of $G(a,b)$, this implies that $b' \in G(a,b)$, yielding a contradiction. Therefore, the relation θ is transitive. Let $\overline{E} := E/\theta$ denote the set of equivalence classes of the relation θ. For $e \in E$, let $\overline{e}$ denote the equivalence class of e in $\overline{E}$. So, all edges (a,b) of a common equivalence class correspond to the same bipartition $G(a,b) \cup G(b,a)$ of V. Fix a node x_0 of G. For each node $x \in V$, let $A(x)$ denote the set of all $\overline{e} \in \overline{E}$ for which x and x_0 belong to distinct sets of the bipartition $V = G(a,b) \cup G(b,a)$, if (a,b) is an edge of $\overline{e}$. In particular, $A(x_0) = \emptyset$. We show that this labeling provides a hypercube embedding of G, i.e., that

$$|A(x) \triangle A(y)| = d_G(x,y)$$

holds for all nodes $x, y \in V$. Let $x, y \in V$ and $m := d_G(x,y)$. Let $P := (x_0 = x, x_1, \ldots, x_m = y)$ be a shortest path in G from x to y, with edges $e_i = (x_{i-1}, x_i)$ for $i = 1, \ldots, m$. We claim that

$$A(x) \triangle A(y) = \{\overline{e}_1, \ldots, \overline{e}_m\}.$$

Clearly, each $\overline{e}_i$ belongs to $A(x) \triangle A(y)$. Indeed if, for instance, $x_0 \in G(x_{i-1}, x_i)$, then $\overline{e}_i \in A(y) \setminus A(x)$ since $x \in G(x_{i-1}, x_i)$ and $y \in G(x_i, x_{i-1})$. Conversely, let $e = (a,b) \in E$ such that $\overline{e} \in A(x) \triangle A(y)$. We can suppose, for instance, that $\overline{e} \in A(y) \setminus A(x)$ with $x_0, x \in G(a,b)$ and $y \in G(b,a)$. Let i be the largest index from $\{1, \ldots, p\}$ for which $x_{i-1} \in G(a,b)$. Then, $e_i \theta e$, which shows that $\overline{e} = \overline{e}_i$. Therefore, we have shown that $|A(x) \triangle A(y)| = d_G(x,y)$ holds for all nodes $x, y \in V$. This shows that G can be isometrically embedded into the hypercube of dimension $\dim_I(G) := |\overline{E}|$. ∎

The following result from Deza and Laurent [1994a] will also be a consequence of Theorem 20.3.1.

Proposition 19.1.2. *If G is hypercube embeddable, then G is ℓ_1-rigid; in particular, G has a unique (up to equivalence) isometric embedding into a hypercube whose dimension is $m_h(G) = \dim_I(G)$.*

Proof. Suppose that G is hypercube embeddable. We show that G is ℓ_1-rigid. Then, this will imply that G has a unique hypercube embedding and, therefore, that $m_h(G) = \dim_I(G)$. We keep the notation from the proof of Theorem 19.1.1. For each $\overline{e} \in \overline{E}$ with $e = (a,b)$, let $S_{\overline{e}}$ denote the one of the two sets $G(a,b)$ and $G(b,a)$ that does not contain the fixed node x_0. From the fact that $d_G(x,y) = |A(x) \triangle A(y)|$ for all nodes $x, y \in V$, we deduce that d_G can be decomposed as

$$d_G = \sum_{\overline{e} \in \overline{E}} \delta(S_{\overline{e}}).$$

Let F_G denote the smallest face of the cut cone CUT_n (n is the number of nodes of G) that contains d_G. We claim that F_G is a simplex face of CUT_n of dimension

$\dim_I(G)$. Clearly, all the cut semimetrics $\delta(S_{\overline{e}})$ ($\overline{e} \in \overline{E}$) belong to F_G and they are linearly independent. We show that every cut semimetric $\delta(S)$ lying on F_G is of the form $\delta(S_{\overline{e}})$ for some $\overline{e} \in \overline{E}$. If this is the case, then we have indeed shown that F_G is a simplex face of CUT_n of dimension $|\overline{E}| = \dim_I(G)$. Let S be a subset of V such that $\delta(S) \in F_G$. Then, $\delta(S)$ satisfies the same triangle equalities as d_G. As the graph G is connected, we can find an edge $e = (a, b)$ such that $a \in S$ and $b \in V \setminus S$. Suppose, for instance, that $x_0 \in G(b, a)$, i.e., $S_{\overline{e}} = G(a, b)$. As d_G satisfies the triangle equality $d_G(x_0, a) = d_G(x_0, b) + d_G(a, b)$, we deduce that $\delta(S)$ satisfies the equality $\delta(S)(x_0, a) = \delta(S)(x_0, b) + \delta(S)(a, b)$, which implies that $x_0 \in V \setminus S$. We claim that $S = G(a, b)$ holds. If $x \in G(a, b)$, then $d_G(x, b) = d_G(x, a) + d_G(a, b)$ from which we deduce that $\delta(S)(x, b) = \delta(S)(x, a) + \delta(S)(a, b)$, implying that $x \in S$. In the same way, $G(b, a)$ is contained in $V \setminus S$, which implies that $S = G(a, b)$. ∎

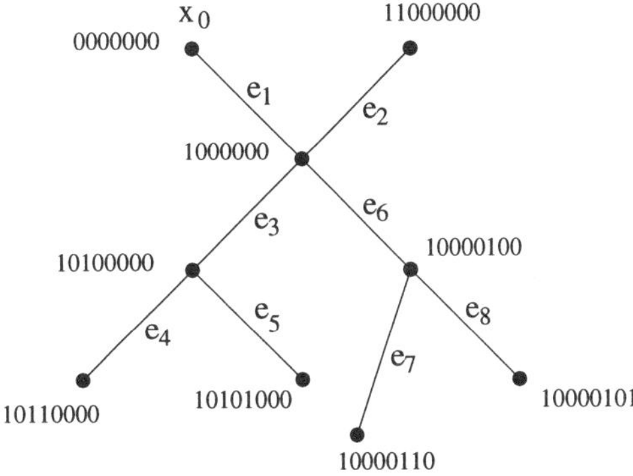

Figure 19.1.3: Embedding a tree in the hypercube

Example 19.1.4. Case of trees.
Let T be a tree on n nodes and with edge set E. Then, T embeds isometrically into the $(n-1)$-hypercube, i.e., $\dim_I(T) = n-1$. The hypercube embedding of T can be easily constructed, as follows from the proof of Theorem 19.1.1. Namely, choose a node x_0 in T and label each node x of T by the set $A(x)$ consisting of the edges of T lying on the path from x_0 to x. We give in Figure 19.1.3 an example of a tree together with its hypercube embedding. Alternatively, the path metric d_T of the tree T can be decomposed as $d_T = \sum_{e \in E} \delta(S_e)$ where, for an edge $e \in E$, S_e is the connected component in $T \backslash e$ that does not contain the specified node x_0.

The distance matrix of a tree has some remarkable properties. In particular, its determinant depends only on the number of nodes of the tree. Namely, let T be a tree on n nodes with distance matrix D_T. Then, $\det(D_T) = (-1)^{n-1}(n-1)2^{n-2}$ (Graham and Pollack [1971]). (To see it, label the nodes of T as $a_1, \ldots, a_n$ in such a way that a_n is adjacent only to a_{n-1}. In the matrix D_T, subtract the $(n-1)$-column to the n-th one and the $(n-1)$-row to the n-th one. Iterating this process brings D_T into the form of an $n \times n$ symmetric matrix having all entries equal to 0 except the $(1,i)$- and $(i,1)$ entries equal to 1 and the (i,i)-entries equal to -2, for $i = 2, \ldots, n$.) Graham and Lovász [1978] show, more generally, how the coefficients of the characteristic polynomial of D_T can be expressed in terms of the number of occurrences of certain forests in T. ∎

Remark 19.1.5. As an immediate consequence of Theorem 19.1.1, one can test in polynomial time whether a graph G is hypercube embeddable. Moreover, the minimum dimension $m_h(G)$ of a hypercube containing G as an isometric subgraph can also be computed in polynomial time, since it coincides with the isometric dimension $\dim_I(G)$ of G (by Proposition 19.1.2).

Suppose G is a graph on n nodes with m edges. Aurenhammer and Hagauer [1991] present an algorithm for testing whether G is hypercube embeddable that runs in time $O(n^2 \log n)$ using $O(n^2)$ space. Their algorithm is based on the characterization from Corollary 20.1.3 (iv); that is, it consists of checking whether G is bipartite and whether the relation θ is transitive. Feder [1992] proposes another algorithm running in time $O(mn)$ and using $O(m)$ space; its space complexity is better than in the previous algorithm as one can easily check that $m \leq \frac{1}{2} n \log_2 n$ if G is hypercube embeddable. Feder's algorithm is based on the results of Section 20.1; that is, it consists of checking whether all the factors in the canonical metric representation of G are isomorphic to K_2. ∎

19.2 Further Characterizations

We start with presenting two further characterizations for hypercube embeddable graphs, due respectively to Avis [1981] and to Roth and Winkler [1986].

Theorem 19.2.1. *Let G be a connected graph. Then, G is hypercube embeddable if and only if G is bipartite and d_G satisfies the following 5-gonal inequality:*

$$d(i_1,i_2) + d(i_1,i_3) + d(i_2,i_3) + d(i_4,i_5) - \sum_{\substack{h=1,2,3 \\ k=4,5}} d(i_h,i_k) \leq 0 \qquad (19.2.2)$$

for all nodes $i_1, \ldots, i_5 \in V$.

Proof. If G is hypercube embeddable, then its path metric d_G is ℓ_1-embeddable and, therefore, satisfies the 5-gonal inequality by Theorem 6.3.1. Suppose now that G is bipartite and not hypercube embeddable. Then, by Theorem 19.1.1, there exists an edge (a,b) of G for which the set $G(a,b)$ is not convex. Hence, there exist $x,y \in G(a,b)$ and $z \in G(b,a)$ such that $d_G(x,z)+d_G(z,y) = d_G(x,y)$. Consider the inequality (19.2.2) for the nodes $i_1 = x, i_2 = y, i_3 = b, i_4 = a$, and $i_5 = z$. One computes easily that the left hand side of (19.2.2) takes the value 2, which shows that d_G violates some 5-gonal inequality. ∎

	x	y	a	b	s
x	0	1	$n+1$	n	1
y	1	0	n	$n+1$	2
a	$n+1$	n	0	1	n
b	n	$n+1$	1	0	$n+1$
s	1	2	n	$n+1$	0

Figure 19.2.3: The distance space $A(n)$ on the 5 points of $\{x, y, a, b, s\}$

	x	y	a	b	r	s
x	0	1	$m+1$	m	p	$p+1$
y	1	0	m	$m+1$	$p+1$	$p+2$
a	$m+1$	m	0	1	$n+1$	n
b	m	$m+1$	1	0	n	$n+1$
r	p	$p+1$	$n+1$	n	0	1
s	$p+1$	$p+2$	n	$n+1$	1	0

Figure 19.2.4: The distance space $B(m, n, p)$ on the 6 points of $\{x, y, a, b, r, s\}$

Theorem 19.2.5. *Let G be a connected bipartite graph. Then, G is hypercube embeddable if and only if the space (V, d_G) does not contain as an isometric subspace any of the spaces $A(n)$ or $B(m, n, p)$, whose distance matrices are shown in Figures* 19.2.3 *and* 19.2.4, *respectively.*

Proof. Suppose that G is not hypercube embeddable. Then, by Theorem 19.1.1, there exists an edge (a, b) of G for which $G(a, b)$ is not closed. Let P be an isometric path in G connecting two nodes of $G(a, b)$ such that P meets $G(b, a)$ and P has minimal length with respect to these properties. Say, $P = (y, x, z_1, \ldots, z_k, r, s)$, where $y, s \in G(a, b)$ and $x, r \in G(b, a)$. Set $m = d_G(x, b)$, $n = d_G(r, b)$ and $p = d_G(x, r)$ (hence, $p = k + 1$). One can check that the distances between the points a, b, x, y, r, s are entirely determined by the parameters m, n, p. Namely, if both points x and r coincide, then $p = 0$, $m = n$, and the 5-point subspace $(\{x, y, a, b, s\}, d_G)$ of (V, d_G) coincides with the space $A(n)$, whose distance matrix is shown in Figure 19.2.3. If the points x and r are distinct, then the 6-point subspace $(\{x, y, a, b, r, s\}, d_G)$ coincides with the space $B(m, n, p)$, whose distance matrix is shown in Figure 19.2.4.
Conversely, if (V, d_G) contains $A(n)$ or $B(m, n, p)$ as an isometric subspace, then G is not hypercube embeddable, by Theorem 19.2.1. Indeed, both $A(n)$ and $B(m, n, p)$ violate the 5-gonal inequality; namely, they violate the inequality (19.2.2) for $\{i_1, i_2, i_3\} = \{b, y, s\}$ and $\{i_4, i_5\} = \{x, a\}$. ∎

Recall that, for a finite distance space (X, d), the following chain of implications holds (see Proposition 4.3.8, Theorems 6.2.16 and 6.3.1).

(X,d) is hypercube embeddable
$\Longrightarrow$ (X,d) is ℓ_1-embeddable
$\Longrightarrow$ (X,d) is hypermetric
$\Longrightarrow$ (X,d) is of negative type
$\Longrightarrow$ the distance matrix of (X,d) has exactly one positive eigenvalue.

Figure 19.2.6: The metric hierarchy

In addition, We recall (from Lemma 6.1.14 and Theorem 6.2.2) the following equivalences.

(X,d) is of negative type
$\Longleftrightarrow$ $(X,\sqrt{d})$ is ℓ_2-embeddable
$\Longleftrightarrow$ the matrix $(d(x,x_0)+d(y,x_0)-d(x,y))_{x,y\in X\setminus\{x_0\}}$ is positive semidefinite, where x_0 is a given element of X.

Figure 19.2.7: Characterizing ℓ_2-embeddability

Roth and Winkler [1986] show that, for the graphic metric spaces of bipartite graphs, the metric hierarchy from Figure 19.2.6 collapses. (Blake and Gilchrist [1973] had earlier observed that connected bipartite ℓ_1-graphs are hypercube embeddable.)

Theorem 19.2.8. *Let G be a connected bipartite graph. The following assertions are equivalent.*

(i) *G is hypercube embeddable.*

(ii) *G is an ℓ_1-graph.*

(iii) *G is hypermetric.*

(iv) *G is of negative type.*

(v) *The distance matrix of G has exactly one positive eigenvalue.*

Proof. It suffices to show that, if G is not hypercube embeddable, then its distance matrix D_G has at least two positive eigenvalues. Suppose that G is not hypercube embedable. By Theorem 19.2.5, (V,d_G) contains as an isometric subspace a space C which is one of the forbidden subspaces $A(n)$ or $B(m,n,p)$. In other words, the distance matrix D_C of C is a principal submatrix of D_G. Clearly, D_C has at least one positive eigenvalue since its trace is equal to 0. If we can show that D_C has at least two positive eigenvalues then, by applying Lemma 2.4.4, we deduce that the number of positive eigenvalues of D_G is greater than or equal to the number of positive eigenvalues of D_C and, therefore, that D_G has at least two positive eigenvalues.
Consider first the case when C is of the form $A(n)$. One can check that the determinant of D_C is equal to $-8n(n+1)$. Hence, D_C is nonsingular and has at least two positive eigenvalues (indeed, if D_C would have only one positive eigenvalue, then its determinant would be positive).

Suppose now that C is of the form $B(m,n,p)$. One can check that the determinant of D_C is equal to

$$4(4mnp + 2mp + 2np + 2mn - m^2 - n^2 - p^2),$$

which can be rewritten as

$$16mnp+4(n+p-m)(p+m-n)+4(p+m-n)(m+n-p)+4(m+n-p)(n+p-m).$$

As m,n,p are the distances between pairs of nodes of G, we deduce from the triangle inequality that each of the quantities into parentheses in the above expression is nonnegative. Hence, the determinant of D_C is positive. This implies that D_C is nonsingular and has at least two positive eigenvalues (else, its determinant would be negative). ∎

Remark 19.2.9. All the implications in the metric hierarchy from Figure 19.2.6 are strict for general (nonbipartite) graphs. We present here a unified set of counterexamples for the converse implications, proposed by Avis and Maehara [1994], which is based on the graph $K_n \backslash P_3$ (with P_3 denoting the path on 3 nodes). This remark is, therefore, a continuation of Example 14.4.9. The metric $d(K_n \backslash P_3)$ was considered in Example 14.4.9 for disproving some implications between the properties of being hypermetric or of negative type or of having a spherical representation. The treatment there was based on features of Delaunay polytopes, while we use here conditions involving linear inequalities.

- The path metrics of $K_4 \backslash P_3$, $K_5 \backslash P_3$, and $K_6 \backslash P_3$ are ℓ_1-embeddable (since $2d_G$ is hypercube embeddable), but not hypercube embeddable (since they contain three points at pairwise distances one).

- The path metrics of $K_7 \backslash P_3$, $K_8 \backslash P_3$ are hypermetric, but not ℓ_1-embeddable. (Hint: The inequality:

$$5x_{12} + 5x_{13} + 3x_{23} - 3 \sum_{j=4,5,6,7} x_{1j} - 2 \sum_{j=4,5,6,7} (x_{2j} + x_{3j}) + \sum_{4 \leq i < j \leq 7} x_{ij} \leq 0$$

is valid for the cut cone CUT_7 (this is the clique-web inequality:

$$\mathrm{CW}_7^1(3,2,2,-1,-1,-1,-1)^T x \leq 0$$

which defines a facet of CUT_7; see Chapter 29). But, the path metric of $K_7 \backslash P_3$ violates this inequality if P_3 is the path $(2,1,3)$ in the complete graph K_7 on the nodes 1,2,3,4,5,6,7. Hence, $K_7 \backslash P_3$ is not an ℓ_1-graph.)

- The path metrics of $K_9 \backslash P_3$, $K_{10} \backslash P_3$ are of negative type, but not hypermetric. (Hint: The path metric of $K_9 \backslash P_3$ violates the hypermetric inequality:

$$Q_9(3,2,2,-1,-1,-1,-1,-1,-1)^T x \leq 0$$

if P_3 is the path $(2,1,3)$. To see that the path metric of $K_{10} \backslash P_3$ is of negative type, one can use Theorem 6.2.16. Namely, it suffices to check that the bordered matrix $M = \begin{pmatrix} D & 1 \\ 1 & 0 \end{pmatrix}$ has exactly one positive eigenvalue, where D denotes the distance matrix of $K_{10}\backslash P_3$. One can indeed check that the eigenvalues of M are $4+\sqrt{41}$, 0, $4-\sqrt{41}$ and -1 (with multiplicity 8).)

• The distance matrix of $K_{11}\backslash P_3$ has exactly one positive eigenvalue, but $K_{11}\backslash P_3$ is not of negative type; the distance matrix of $K_n\backslash P_3$ has two positive eigenvalues for all $n \geq 12$. (Hint: $K_{11}\backslash P_3$ is not of negative type since it violates the negative type inequality:

$$Q_{11}\left(\frac{24}{7}, \frac{16}{7}, \frac{16}{7}, -1, -1, -1, -1, -1, -1, -1, -1\right)^T x \leq 0$$

if P_3 is the path $(2,1,3)$.) Another example of a graph which is not of negative type but whose distance matrix has one positive eigenvalue is given in Example 19.2.11 below. ∎

We saw in Figure 19.2.7 two characterizations for the distance spaces of negative type. Winkler [1985] proposes yet another characterization for the graphs of negative type, exposed in Theorem 19.2.10 below. Let G be a graph. Consider an orientation G' of G which has, for each edge (a,b) of G, exactly one of the arcs (a,b) or (b,a). Given two arcs $e=(a,b)$ and $e'=(a',b')$ of G', set

$$\langle e, e'\rangle := \frac{1}{2}(d_G(a,b') - d_G(a,a') - d_G(b,b') + d_G(b,a')).$$

(This is the same definition as the one considered later in (21.2.13) up to a factor 2.) Observe that, if d_G is of negative type and $u_a \in \mathbb{R}^m$ $(a \in V)$ are vectors satisfying $d_G(a,b) = (\| u_a - u_b \|_2)^2$ for all $a, b \in V$, then $\langle e, e'\rangle$ coincides with the scalar product $(u_b - u_a)^T(u_{b'} - u_{a'})$.

Theorem 19.2.10. *Let G be a connected graph on $n+1$ nodes and let G' be an arbitrary orientation of G. Let T be a spanning tree in G, with corresponding arcs $e_1, \ldots, e_n$ in G'. The following assertions are equivalent.*

(i) *d_G is of negative type.*

(ii) *The $n \times n$ matrix $(\langle e_i, e_j\rangle)_{i,j=1,\ldots,n}$ is positive semidefinite.*

Proof. Let $V = \{a_0, a_1, \ldots, a_n\}$ denote the set of nodes of G. By definition, d_G is of negative type if and only if

$$\sum_{0 \leq r < s \leq n} d_G(a_r, a_s) x_r x_s \leq 0 \text{ for all } x \in U := \{x \in \mathbb{R}^{n+1} \mid \sum_{0 \leq r \leq n} x_r = 0\}.$$

For each node $a_r \in V$, set

$$A(a_r) := \{i \in \{1, \ldots, n\} \mid \text{ the arc } e_i \text{ ends in } a_r\},$$

$$B(a_r) := \{i \in \{1, \ldots, n\} \mid \text{ the arc } e_i \text{ begins in } a_r\}.$$

For $y \in \mathbb{R}^n$, define $x \in \mathbb{R}^{n+1}$ by setting

$$x_r = \sum_{i \in A(v_r)} y_i - \sum_{i \in B(v_r)} y_i$$

for $r = 0, 1, \ldots, n$. One can check that $\sum_{0 \le r \le n} x_r = 0$, i.e., $x \in U$, and $x = (0, \ldots, 0)$ implies that $y = (0, \ldots, 0)$. Hence, we have found a 1-1 linear correspondence between the spaces $\mathbb{R}^n$ and U. We check that, under this correspondence,

$$\sum_{1 \le i,j \le n} \langle e_i, e_j \rangle y_i y_j = - \sum_{0 \le r < s \le n} d_G(a_r, a_s) x_r x_s.$$

Indeed, $d_G(a_r, a_s)$ appears in $\sum_{1 \le i,j \le n} \langle e_i, e_j \rangle y_i y_j$ with the coefficient

$$\sum_{(i,j) \in A(v_r) \times B(v_s)} y_i y_j + \sum_{(i,j) \in B(v_r) \times A(v_s)} y_i y_j$$
$$- \sum_{(i,j) \in A(v_r) \times A(v_s)} y_i y_j - \sum_{(i,j) \in B(v_r) \times B(v_s)} y_i y_j,$$

which is equal to $-x_r x_s$. This shows the equivalence of (i) and (ii). ∎

Example 19.2.11. Consider the graph $G_n := K_{n+1} \backslash K_{n-1}$ with node set $\{a_0, \ldots, a_n\}$ and with edges the pairs (a_0, a_1), (a_0, a_i) and (a_1, a_i) for $i = 2, 3, \ldots, n$. Then, the path metric of G_n is of negative type if and only if $n \le 5$. (To see it, consider the oriented spanning tree T with arcs $e_1 = (a_0, a_1), \ldots, e_n = (a_0, a_n)$. The matrix $(\langle e_i, e_j \rangle)_{i,j=1,\ldots,n}$ has all its entries equal to 0 except the diagonal entries equal to 1 and the $(1,i)$- and $(i,1)$-entries equal to $\frac{1}{2}$ for $i = 2, \ldots, n$. Its determinant is equal to $\frac{5-n}{4}$.) Note that, for $n \ge 6$, G_n provides a counterexample to the converse of the last implication from Figure 19.2.6, since the distance matrix of G_n has exactly one positive eigenvalue, but G_n is not of negative type. (Indeed, the eigenvalues of the distance matrix of G_n are $2n-1, -1, -2$ with respective multiplicities $1, 1, n-1$.) ∎

Finally, let us mention another formulation for the isometric dimension of a hypercube embeddable graph in terms of the number of negative eigenvalues of its distance matrix, due to Graham and Winkler [1985].

Proposition 19.2.12. *Let G be a graph with distance matrix D_G and let $n_+(D_G)$, $n_-(D_G)$ denote the number of positive and negative eigenvalues of D_G. If G is hypercube embeddable, then $\dim_I(G) = n_-(D_G)$ and $n_+(D_G) = 1$ hold.*

Proof. Suppose that $\dim_I(G) = k$. Let σ be an isometric embedding of G into the k-hypercube $H(k, 2)$; denote by $\sigma(a) = (a_1, \ldots, a_k) \in \{0,1\}^k$ the image of each node $a \in V$ under this embedding. For $h = 1, \ldots, k$, set

$$X_h := \{a \in V \mid a_h = 0\}, \; Y_h := \{a \in V \mid a_h = 1\} = V \setminus X_h.$$

Then,

$$\begin{aligned}\sum_{a,b\in V} d_G(a,b)x_a x_b &= \sum_{a,b\in V}(\sum_{1\le h\le k}|a_h-b_h|)x_a x_b = \sum_{1\le h\le k}(\sum_{a\in X_h} x_a)(\sum_{b\in Y_h} x_a)\\ &= \tfrac{k}{4}(\sum_{a\in V} x_a)^2 - \frac{1}{4}\sum_{1\le h\le k}(\sum_{a\in X_h} x_a - \sum_{b\in Y_h} x_b)^2\end{aligned}$$

(where the last equality is obtained using the identity $xy = \frac{1}{4}((x+y)^2-(x-y)^2))$). Hence, the quadratic form

$$\sum_{a,b\in V} d_G(a,b)x_a x_b$$

can be written as the sum of one "positive" square and k "negative" squares. By Sylvester's law of inertia, this implies that $n_+(D_G) \le 1$ and $n_-(D_G) \le k$. On the other hand, $n_+(D_G) \ge 1$ since D_G has trace zero. Hence, $n_+(D_G) = 1$ and the rank of D_G satisfies $\text{rank}(D_G) = n_+(D_G) + n_-(D_G) \le k+1$. We show that $\text{rank}(D_G) = k+1$. This will imply that $n_-(D_G) = k$, thus stating the result.
We can suppose without loss of generality that a given node $a^{(0)}$ of G receives the label $\sigma(a^{(0)}) := (0,\dots,0)$ in the hypercube embedding. We claim that there exist k nodes $a^{(1)},\dots,a^{(k)}$ of G whose labels $\sigma(a^{(1)}),\dots,\sigma(a^{(k)})$ are linearly independent. For this, it suffices to check that the system $\{\sigma(a) \mid a\in V\} \subseteq \{0,1\}^k$ has full dimension k. Suppose for contradiction that, say, the k-th coordinate can be expressed in terms of the others, i.e., there exist scalars $\lambda_1,\dots,\lambda_{k-1}$ such that $a_k = \sum_{1\le j\le k-1}\lambda_j a_j$ for all $a\in V$. Then, $a_k = b_k$ holds for any two adjacent nodes a,b in G. This implies that $a_k = 0$ holds for each node $a\in V$, by considering a shortest path from $a^{(0)}$ to a. So, one could have embedded G into the $(k-1)$-hypercube, contradicting the fact that $\dim_I(G) = k$. We now claim that the submatrix

$$M := (d_G(a^{(i)},a^{(j)}))_{i,j=0,\dots,k}$$

is nonsingular. This will imply that $\text{rank}(D_G) \ge k+1$ and, therefore, $\text{rank}(D_G) = k+1$. For $i = 0,1,\dots,k$, set

$$u^{(i)} := 2\sigma(a^{(i)}) - e,$$

where $e = (1,\dots,1)^T$. As the vectors $u^{(i)}$ are ± 1-valued, we have

$$\begin{aligned}d_G(a^{(i)},a^{(j)}) &= \sum_{1\le h\le k}|a_h^{(i)} - a_h^{(j)}| = \frac{1}{2}\sum_{1\le h\le k}|u_h^{(i)} - u_h^{(j)}|\\ &= \tfrac{1}{2}\sum_{1\le h\le k}(1 - u_h^{(i)}u_h^{(j)}) = \frac{k}{2} - \frac{1}{2}(u^{(i)})^T u^{(j)}.\end{aligned}$$

Therefore,

$$M = \frac{k}{2}J - \frac{1}{2}\text{Gram}(u^{(0)},u^{(1)},\dots,u^{(k)}),$$

where J denotes the all-ones matrix and $\text{Gram}(u^{(0)},u^{(1)},\dots,u^{(k)})$ denotes the Gram matrix of the vectors $u^{(0)},u^{(1)},\dots,u^{(k)}$. One can easily check that

$$\begin{aligned}\det(M) = \ & (-2)^{-(k+1)}\Big(\det(\text{Gram}(u^{(0)},u^{(1)},\dots,u^{(k)}))\\ & -k\ \det(\text{Gram}(u^{(1)}-u^{(0)},u^{(2)}-u^{(0)},\dots,u^{(k)}-u^{(0)}))\Big).\end{aligned}$$

But, $\det(\text{Gram}(u^{(0)}, u^{(1)}, \ldots, u^{(k)})) = 0$ since the vectors $u^{(0)}, u^{(1)}, \ldots, u^{(k)}$ are linearly dependent, and $\det(\text{Gram}(u^{(1)} - u^{(0)}, u^{(2)} - u^{(0)}, \ldots, u^{(k)} - u^{(0)})) \neq 0$ since the vectors $u^{(1)} - u^{(0)}, u^{(2)} - u^{(0)}, \ldots, u^{(k)} - u^{(0)}$ are linearly independent. Therefore, $\det(M) \neq 0$. ∎

19.3 Additional Notes

We mention here some remarks on possible relaxations of the notion of isometric embeddability into the hypercube. First, one may consider isometric embeddings into the squashed hypercube; second, one may consider embeddings as a subgraph (not necessarily isometric) into the hypercube; finally, the notion may be extended to hypergraphs.

Isometric Embedding into Squashed Hypercubes. As we just saw, not every graph can be isometrically embedded into a hypercube. For this reason, Graham and Pollak [1971] considered isometric embeddings into squashed hypercubes. Let d_* denote the distance defined on the set $B_* = \{0, 1, *\}$ by setting

$$d_*(x,y) = \begin{cases} 1 \text{ if } \{x,y\} = \{0,1\} \\ 0 \text{ otherwise} \end{cases}$$

for $x, y \in B_*$. Hence, the symbol $*$ is at distance 0 from the other symbols; it is also called the "don't care" symbol. The distance d_* can be extended to B_*^m by setting

$$d_*((x_1, \ldots, x_m), (y_1, \ldots, y_m)) = \sum_{1 \leq i \leq m} d_*(x_i, y_i).$$

The distance space (B_*^m, d_*) is called the *squashed m-hypercube.* It contains the usual m-hypercube as a subspace. Each element $(x_1, \ldots, x_m) \in B_*^m$ can be thought of as representing a face of the m-dimensional hypercube, namely, the face consisting of all $y \in \{0,1\}^m$ such that $y_i = x_i$ for all i such that $x_i \in \{0, 1\}$. A nice property of squashed hypercubes is that every connected graph can be isometrically embedded in some squashed hypercube. Indeed, let G be a connected graph with node set $\{1, \ldots, n\}$. Set

$$m := \sum_{1 \leq i < j \leq n} d_G(i,j).$$

For $1 \leq i < j \leq n$, let D_{ij} be pairwise disjoint subsets of $\{1, \ldots, m\}$ with $|D_{ij}| = d_G(i,j)$. Label each node i by the m-tuple $(i_1, \ldots, i_m) \in B_*^m$ by setting

$$i_k = \begin{cases} 0 \text{ if } k \in \bigcup_{h=i+1}^{n} D_{ih} \\ 1 \text{ if } k \in \bigcup_{h=1}^{i-1} D_{ih} \\ \ \text{otherwise.} \end{cases}$$

Then, $d_*((i_1, \ldots, i_m), (j_1, \ldots, j_m)) = |D_{ij}| = d_G(i,j)$. This shows that G can be isometrically embedded into the squashed m-hypercube. Let $r(G)$ denote the smallest dimension of a squashed hypercube in which G can be embedded. Winkler [1983] showed that $r(G) \leq n-1$ for each graph on n nodes. On the other hand, $r(G) \geq \max(n_+(D_G), n_-(D_G))$, where $n_+(D_G), n_-(D_G)$ denote the number of positive and negative eigenvalues of the distance matrix D_G of G (Graham and Pollack [1972]). For instance, $r(K_n) = n-1$ since $n_-(D_{K_n}) = n-1$. The following provides an isometric embedding of K_3 into the squashed 2-hypercube:

$$1 \mapsto (0,0), \; 2 \mapsto (0,1), \; 3 \mapsto (1,*).$$

Nonisometric Embedding of Graphs into Hypercubes. Another relaxation of the notion of hypercube embeddable graphs is that of cubical graphs. A graph G is said to be *cubical* if G is a subgraph of some hypercube $H(m,2)$, i.e., there exists an injective mapping from the node set of G to the node set of $H(m,2)$ which maps edges of G to edges of $H(m,2)$. Clearly, every cubical graph is bipartite and every hypercube embeddable graph is cubical. We show below an example of a graph which is cubical but not hypercube embeddable.

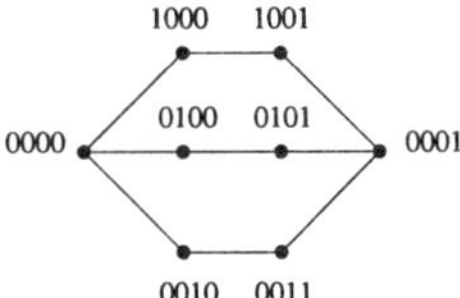

The structure of the minimal noncubical graphs has been studied in Garey and Graham [1975], where some constructions of such graphs are presented. For instance, $K_{2,3}$ and odd circuits are minimal noncubical graphs. Recall from Remark 19.1.5 that one can check in polynomial time whether a graph G is hypercube embeddable and, moreover, the minimum dimension $m_h(G)$ of a hypercube containing G as an isometric subgraph can be computed in polynomial time. On the other hand, it has been proved that deciding whether a graph G is cubical is an NP-complete problem (Afrati, Papadimitriou and Papageorgiou [1985, 1989]; Krumme, Venkataraman and Cybenko [1986]). Moreover, for G cubical, computing the minimum dimension of a hypercube containing G as a subgraph is also a difficult problem. For instance, each tree is cubical (in fact, a tree on n nodes can be isometrically embeded into an $(n-1)$-hypercube). But, given a tree T and an integer m, it is NP-complete to decide whether T is a subgraph of the m-hypercube (Wagner and Corneil [1990]). The problem of determining the minimum dimension of a hypercube containing a tree has been long studied (see, e.g., Havel and Liebl [1972, 1973]). Along the same lines, given a graph G and integers m, k, it is NP-complete to decide whether G is a subgraph of $(K_m)^k$ (Wagner and Corneil [1993]).

Isometric Embeddings into Cube-Hypergraphs. The characterization of the graphs that can be isometrically embedded into the hypercube has been extended in the context of uniform hypergraphs by Burosch and Ceccherini [1995].

A t-uniform hypergraph $H = (V, \mathcal{E})$ consists of a set V of vertices and a set $\mathcal{E}$ of (hyper)edges each having the same cardinality $t \geq 2$. A semimetric d_H can be defined on V in the following way: Construct the graph G_H with vertex set V and with two nodes $x, y \in V$ being adjacent if they are contained in a common edge of H. Then, we take for d_H the path metric of the graph G_H. A hypergraph $H = (V, \mathcal{E})$ is said to be isometrically embeddable into another hypergraph $H' = (V', \mathcal{E}')$ if the graph G_H can be isometrically embedded into $G_{H'}$. As hypergraph analogue of the hypercube, Burosch and Ceccherini [1995] considers the following t-uniform hypergraph $Q(n,t)$: its vertex set is $\{0, 1, \ldots, t-1\}^n$ and an edge consists of all the vectors $x \in \{0, 1, \ldots, t-1\}^n$ whose coordinates are fixed on $n-1$ positions with the last coordinate being free in $\{0, 1, \ldots, t-1\}$. Hence, two vectors x, y belong to a common edge of $Q(n,t)$ if and only if their Hamming distance is 1. In other words, the graph structure $G_{Q(n,t)}$ underlying the hypergraph $Q(n,t)$ is the Hamming graph $H(n,t)$. Burosch and Ceccherini characterize the t-uniform hypergraphs that can be embeded into the cube-hypergraph $Q(n,t)$; their characterization is a direct extension of the corresponding results in the graph case, namely, of the results by Djokovic (Theorem 19.1.1) and by Graham and Winkler (Corollary 20.1.3 (iv)).

Chapter 20. Isometric Embeddings of Graphs into Cartesian Products

We have characterized in the previous chapter the graphs that can be isometrically embedded into a hypercube. The hypercube is the simplest example of a Cartesian product of graphs; indeed, the m-hypercube is nothing but $(K_2)^m$. We consider here isometric embeddings of graphs into arbitrary Cartesian products. It turns out that every graph can be isometrically embedded in a canonical way into a Cartesian product whose factors are "irreducible", i.e., cannot be further embedded into Cartesian products. We present two applications of this result, for finding the prime factorization of a graph, and for showing that the path metric of every bipartite graph can decomposed in a unique way as a nonnegative combination of primitive semimetrics.

20.1 Canonical Metric Representation of a Graph

Let G, $H_1, \ldots, H_k$ be graphs. An isometric embedding of G into the Cartesian product $\prod_{1\leq i\leq k} H_i$ is said to be *irredundant* if each factor H_h is a connected graph on at least two nodes, and if each vertex of every factor H_h appears as a coordinate in the image of at least one node of G. Clearly, any isometric embedding into a Cartesian product can be made irredundant by discarding the factors consisting of an isolated node and the unused nodes in each factor. An irredundant isometric embedding of G into a Cartesian product is also called a *metric representation* of G. Two isometric embeddings of G into Cartesian products are said to be *equivalent* if there is a bijection between the factors of one and the factors of the other, together with isomorphisms between the corresponding factors for which the obvious diagram commutes. A graph G is said to be *irreducible* if all its metric representations are equivalent to the trivial embedding of G into itself.

Examples of irreducible graphs include: the complete graph K_n $(n \geq 2)$, odd circuits C_{2n+1} $(n \geq 1)$, the half-cube $\frac{1}{2}H(n,2)$ $(n \geq 2)$, the cocktail-party graph $K_{n\times 2}$ $(n \geq 3)$, the Petersen graph P_{10}, the Gosset graph G_{56}, the Schläfli graph G_{27}, etc. Actually, it is observed in Graham and Winkler [1985] that the probability that a random graph (with edge probability 1/2) on n nodes is irreducible goes to 1 as $n \longrightarrow \infty$.

The following theorem is the main result of this section; it is due to Graham and Winkler [1985]; see also Winkler [1987b] and Graham [1988].

Theorem 20.1.1. *Every connected graph G has a unique metric representation*

$$G \hookrightarrow \prod_{1\le h\le k} G_h$$

in which each factor G_h is irreducible; it is called the canonical metric representation *of G. Moreover, $k = \dim_I(G)$ and, if*

$$G \hookrightarrow \prod_{1\le i\le m} H_i$$

is another metric representation of G, then there exist a partition $(S_1, \ldots, S_m)$ of $\{1, \ldots, k\}$ and metric representations

$$H_i \hookrightarrow \prod_{h\in S_i} G_h,$$

for $i \in \{1, \ldots, m\}$, for which the obvious diagram commutes.

Theorem 20.1.1 is an essential result in the metric theory of graphs, which has many applications; we will present a number of them, in particular, in Sections 20.2, 20.3 and in Chapter 21. The crucial tool for constructing the canonical metric representation of a graph G is Djokovic's relation θ, introduced earlier in (18.0.2). The factors in the canonical metric representation correspond, in fact, to the equivalence classes of the transitive closure θ^* of θ. Let us mention some useful rules for computing them:
- Any two edges on an odd isometric circuit are in relation by θ.
- Let $C = (a_1, \ldots, a_{2m})$ be an isometric even circuit in G. Call the two edges $e_i := (a_i, a_{i+1})$ and $e_{m+i} := (a_{m+i}, a_{m+i+1})$ (where the indices are taken modulo m) *opposite* on C if $d_G(a_i, a_{m+i}) = d_G(a_{i+1}, a_{m+i+1}) = m$. Clearly, if e_i and e_{m+i} are opposite on C, then e_i and e_{m+i} are in relation by θ.
It is observed in Lomonosov and Sebö [1993] that, if G is a bipartite graph, then two edges are in relation by θ if and only if they are opposite on some even circuit of G.

The following lemma is crucial for the proof of Theorem 20.1.1.

Lemma 20.1.2. *Let $E_1, \ldots, E_k$ denote the equivalence classes of the transitive closure θ^* of the relation θ, defined in relation* (18.0.2). *Given two nodes a, b of G, let P be a shortest path from a to b, and let Q be another path joining a to b in G. Then, for all $h = 1, \ldots, k$,*

$$|E(P) \cap E_h| \le |E(Q) \cap E_h|.$$

Proof. Set $P = (x_0 = a, x_1, \ldots, x_p = b)$. For any index $h \in \{1, \ldots, k\}$ and any node x of G, set

$$f_h(x) := \sum_{i\in\{1,\ldots,p\} | (x_{i-1},x_i)\in E_h} (d_G(x, x_i) - d_G(x, x_{i-1})).$$

Hence, $f_h(a) = |E(P) \cap E_h|$ and $f_h(b) = -|E(P) \cap E_h|$. Let (x, y) be an edge of G. We claim

$$f_h(x) = f_h(y) \text{ if } (x, y) \notin E_h.$$

Indeed,

$$f_h(x) - f_h(y) = \sum_{i|(x_{i-1},x_i) \in E_h} (d_G(x, x_i) - d_G(y, x_i)) - (d_G(x, x_{i-1}) - d_G(y, x_{i-1}))$$

is equal to 0, since the edge (x, y) is not in relation by θ with any of the edges of E_h. On the other hand,

$$|f_h(x) - f_h(y)| \leq 2 \text{ if } (x, y) \in E_h.$$

Indeed, by the above argument,

$$\begin{aligned} |f_h(x) - f_h(y)| &= |\sum_{1 \leq j \leq k} (f_j(x) - f_j(y))| \\ &= |(d_G(x, b) - d_G(x, a)) - (d_G(y, b) - d_G(y, a))| \\ &\leq |d_G(x, b) - d_G(x, a)| + |d_G(y, b) - d_G(y, a)| \leq 2. \end{aligned}$$

As $f_h(a) = |E(P) \cap E_h|$ and $f_h(b) = -|E(P) \cap E_h|$, when moving along the nodes of the path Q, the function $f_h(.)$ changes in absolute value by $2|E(P) \cap E_h|$. But, on an edge of $E \setminus E_h$, the function $f_h(.)$ remains unchanged and, on an edge of E_h, $f_h(.)$ increases by at most 2. This implies that the path Q must contain at least $|E(P) \cap E_h|$ edges from E_h. ∎

Proof of Theorem 20.1.1. As in Lemma 20.1.2, let $E_1, \ldots, E_k$ denote the equivalence classes of the transitive closure θ^* of the relation θ. For each $h = 1, \ldots, k$, let G_h denote the graph obtained from G by contracting the edges of $E \setminus E_h$. In other words, for constructing G_h, one identifies any two nodes of G that are joined by a path containing no edge from E_h. This defines a surjective mapping σ_h from $V(G)$ to $V(G_h)$ and a mapping

$$\sigma : V(G) \longrightarrow \prod_{1 \leq h \leq k} V(G_h)$$

by setting $\sigma(v) := (\sigma_1(v), \ldots, \sigma_k(v))$ for each node v of G. We show that the mapping σ provides the required metric representation of G. For this, we have to check that σ is an irredundant isometric embedding and that each factor G_h is irreducible. Take two nodes a, b of G and a shortest path P from a to b in G. We show

$$d_G(a, b) = \sum_{1 \leq h \leq k} d_{G_h}(\sigma_h(a), \sigma_h(b)).$$

Indeed, for each h, $d_{G_h}(\sigma_h(a), \sigma_h(b))$ is the minimum value of $|E(Q) \cap E_h|$ taken over all paths Q joining a and b; hence, by Lemma 20.1.2, $d_{G_h}(\sigma_h(a), \sigma_h(b)) = |E(P) \cap E_h|$. Therefore,

$$\sum_{1 \leq h \leq k} d_{G_h}(\sigma_h(a), \sigma_h(b)) = \sum_{1 \leq h \leq k} |E(P) \cap E_h| = |E(P)| = d_G(a, b).$$

This shows that σ is an isometric embedding of G into $\prod_{h=1}^k G_h$. Moreover, using again Lemma 20.1.2, the endpoints of an edge of E_h are not identified when constructing G_h. Hence, each factor G_h has at least two nodes. Therefore, the embedding σ is irredundant since the mappings σ_h are surjective. Consider now another metric representation

$$G \hookrightarrow \prod_{1 \leq j \leq m} H_j$$

of G and denote by $(x_1, \ldots, x_m)$ the image of a node x of G. If $e = (x, y)$ is an edge of G corresponding to an edge in the j-th factor H_j, i.e., $(x_j, y_j) \in E(H_j)$ and $x_i = y_i$ for all $i \in \{1, \ldots, m\} \setminus \{j\}$, then each edge f in relation by θ with e is also an edge in H_j. Therefore, each factor H_j "contains" exactly the edges of $\bigcup_{i \in J} E_i$ for some nonempty set J of indices. In particular, $m \leq k$ holds. This implies that each factor G_h is irreducible (else, one would have a metric representation of G with more than k factors). Therefore, $G \hookrightarrow G_1 \times \ldots \times G_k$ is the canonical metric representation of G. This concludes the proof. ∎

Corollary 20.1.3. *Let G be a connected graph.*

(i) *G is irreducible if and only if* $\dim_I(G) = 1$.

(ii) *If G has n nodes, then* $\dim_I(G) \leq n - 1$, *with equality if and only if G is a tree.*

(iii) *G embeds isometrically into $(K_3)^m$ for some $m \geq 1$ if and only if the relation θ is transitive.*

(iv) *G embeds isometrically into $(K_2)^m$ for some $m \geq 1$ if and only if G is bipartite and θ is transitive.*

Proof. (i) follows immediately from Theorem 20.1.1.
(ii) Set $k := \dim_I(G)$ and let T be a spanning tree in G. We claim that T contains at least one edge from each equivalence class E_h. Indeed, if e is an edge from $E \setminus E(T)$ belonging to the class E_h then, by Lemma 20.1.2, T must contain at least one edge from E_h. Therefore, $n - 1 = |E(T)| \geq k$ holds. If there are two edges $e, f \in E$ in relation by θ, let T be a spanning tree containing both e and f; then, $k \leq n - 2$ holds. This shows that equality $k = n - 1$ holds only if G is a tree.
(iii) Note that G embeds isometrically into $(K_3)^m$ if and only if each factor G_h in the canonical representation of G is K_2 or K_3 (see Remark 20.1.10). On the other hand, G_h is K_2 or K_3 if and only if E_h consists of all the edges that are cut by the partition of V into $G(a,b) \cup G(b,a) \cup G_=(a,b)$, where $(a,b) \in E_h$, in which case θ is transitive.
The assertion (iv) follows from (iii) since $G_=(a,b) = \emptyset$ for each edge (a,b) when G is bipartite. ∎

One can easily check that, for G bipartite, the relation θ is transitive if and only if $G(a,b)$ is convex for all adjacent nodes a, b of G. Hence, Corollary 20.1.3 (iv) implies the characterization of hypercube embeddable graphs stated in Theorem 19.1.1. In particular, if G is hypercube embeddable with isometric dimension $\dim_I(G) = k$, then $G \hookrightarrow (K_2)^k$ is the canonical metric representation of G. Lomonosov and Sebö [1993] give the following additional information.

Proposition 20.1.4. *If G is a bipartite graph, then all the factors $G_1, \ldots, G_k$ of its canonical metric representation are bipartite graphs.*

Proof. Suppose, for contradiction, that a factor G_h of the canonical metric representation of G is not bipartite. Then, there exists a circuit C of G such that $|E(C) \cap E_h|$ is odd. Choose such a circuit C of minimal length. As G is bipartite, C has even length, say $C = (a_1, a_1, \ldots, a_{2m})$. Consider the pairs (a_i, a_{m+i}) (where the indices are taken modulo m) of diametrally opposed nodes of C. If $d_G(a_i, a_{m+i}) = d_G(a_{i+1}, a_{m+i+1}) = m$, then $d_G(a_{m+i}, a_{i+1}) - d_G(a_{m+i}, a_i) = -1$ and $d_G(a_{m+i+1}, a_{i+1}) - d_G(a_{m+i+1}, a_i) = 1$, which implies that the edges (a_i, a_{i+1}) and (a_{m+i}, a_{m+i+1}) are in relation by θ. Hence, there exists a pair (a_i, a_{m+i}) for which $d_G(a_i, a_{m+i}) < m$ (otherwise, any two oposite edges of C are in relation by θ, implying that $|E(C) \cap E_h|$ is even). Let P be a shortest path from a_i to a_{m+i} in G. Suppose that only the endnodes of P are on C. The endnodes of P partition C into two paths which, together with P, form two circuits C_1 and C_2. As C_1 and C_2 have smaller length than C, we deduce that both $|E(C_1) \cap E_h|$ and $|E(C_2) \cap E_h|$ are even. This implies that $|E(C) \cap E_h|$ is even, yielding a contradiction. The reasoning is the same if P meets C in other nodes than its endnodes. ∎

Remark 20.1.5. Let G be a graph on n nodes with m edges. Its canonical metric representation $G \hookrightarrow G_1 \times \ldots \times G_k$ can be found in polynomial time; more precisely, in time $O(mn)$ using $O(m)$ space. Indeed, it can be obtained in the following way:

(i) Compute the relation θ and determine the equivalence classes $E_1, \ldots, E_k$ of its transitive closure θ^*.

(ii) For each $h = 1, \ldots, k$, construct the graph G_h from G by contracting the edges of $E \setminus E_h$.

Step (ii) can be easily executed in time $O(nm)$ using $O(m)$ space. (Indeed, given an equivalence class E_h, the graph G_h can be constructed as follows: Delete the edges from E_h in G and compute the connected components in the resulting graph; they are precisely the vertices of G_h. There is an edge between two vertices of G_h if there is an edge between the two corresponding components. Finally, we know from Corollary 20.1.3 (ii) that there are at most $n-1$ equivalence classes.) Step (i) can obviously be executed in $O(m^2)$ time. Feder [1992] shows how to execute Step (i) in $O(mn)$ time. For this, he considers another relation θ_1 contained in θ and such that θ and θ_1 have the same transitive closure, i.e., $\theta^* = \theta_1^*$. Namely, given a spanning tree T in G, let

$$e \;\theta_1\; e' \Longleftrightarrow e \;\theta\; e' \text{ and } T \cap \{e, e'\} \neq \emptyset.$$

The relation θ_1 can be computed in time $O(mn)$ as it suffices to compare every edge of T with the edges of G.

That the relations θ and θ_1 have the same transitive closure follows from Lemma 20.1.6 below, which shows that the statement from Lemma 20.1.2 remains valid for the relation θ_1. Indeed, suppose that θ_1^* has k_1 equivalence classes: $F_1, \ldots, F_{k_1}$. Then, $k \leq k_1$ as each class of θ^* is a union of classes of θ_1^*. On the other hand, the arguments used in the proof of Theorem 20.1.1 show that $G \hookrightarrow \prod_{h=1}^{k_1} H_h$ is a metric representation of G, where H_h is obtained from G by contracting the edges of $E \setminus F_h$. This shows that $k_1 \leq k$. Therefore, $k = k_1$; that is, θ^* and θ_1^* coincide. ∎

The following lemma was proved by Feder [1992].

Lemma 20.1.6. Let $F_1, \ldots, F_{k_1}$ denote the equivalence classes of the transitive closure θ_1^* of the relation θ_1. Given two nodes a, b of G, let P be a shortest path from a to b, and let Q be another path joining a to b in G. Then, for all $h = 1, \ldots, k_1$,

$$|E(P) \cap F_h| \leq |E(Q) \cap F_h|.$$

Proof. Let $P_T := (a := z_0, z_1, \ldots, z_t := b)$ denote the path joining a and b in the tree T and set $Q := (y_0 := a, y_1, \ldots, y_q := b)$. For $i \in \{1, \ldots, q\}$ and $j \in \{1, \ldots, t\}$, set

$$\mu_{ij} := d_G(z_{j-1}, y_i) - d_G(z_{j-1}, y_{i-1}) - d_G(z_j, y_i) + d_G(z_j, y_{i-1}).$$

Observe that $\mu_{ij} = 0$ if the edges (z_{j-1}, z_j) and (y_{i-1}, y_i) are not in relation by θ. Then,

$$\begin{aligned}
&\sum_{i|(y_{i-1},y_i)\in E(Q)\cap F_h} d_G(a, y_i) - d_G(a, y_{i-1}) - d_G(b, y_i) + d_G(b, y_{i-1}) \\
&= \sum_{j|(z_{j-1},z_j)\in E(P_T)} \sum_{i|(y_{i-1},y_i)\in E(Q)\cap F_h} \mu_{ij} \\
&= \sum_{j|(z_{j-1},z_j)\in E(P_T)\cap F_h} \sum_{i|(y_{i-1},y_i)\in E(Q)\cap F_h} \mu_{ij} \\
&= \sum_{j|(z_{j-1},z_j)\in E(P_T)\cap F_h} \sum_{i|(y_{i-1},y_i)\in E(Q)} \mu_{ij} \\
&= \sum_{j|(z_{j-1},z_j)\in E(P_T)\cap F_h} d_G(z_{j-1}, b) - d_G(z_{j-1}, a) - d_G(z_j, b) + d_G(z_j, a).
\end{aligned}$$

Setting $P := (x_0 := a, x_1, \ldots, x_p := b)$, we deduce that

$$\begin{aligned}
&\sum_{i|(x_{i-1},x_i)\in E(P)\cap F_h} d_G(a, x_i) - d_G(a, x_{i-1}) - d_G(b, x_i) + d_G(b, x_{i-1}) \\
&= \sum_{i|(y_{i-1},y_i)\in E(Q)\cap F_h} d_G(a, y_i) - d_G(a, y_{i-1}) - d_G(b, y_i) + d_G(b, y_{i-1}).
\end{aligned}$$

The result now follows as the first term is equal to $2|E(P) \cap F_h|$ while the second term is less than or equal to $2|E(Q) \cap F_h|$. ∎

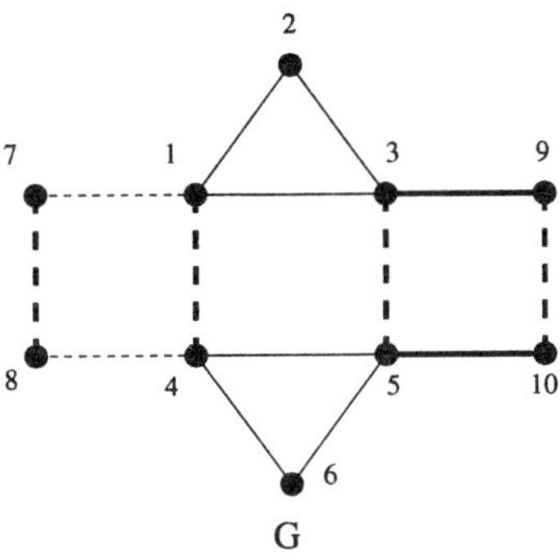

Figure 20.1.7: G

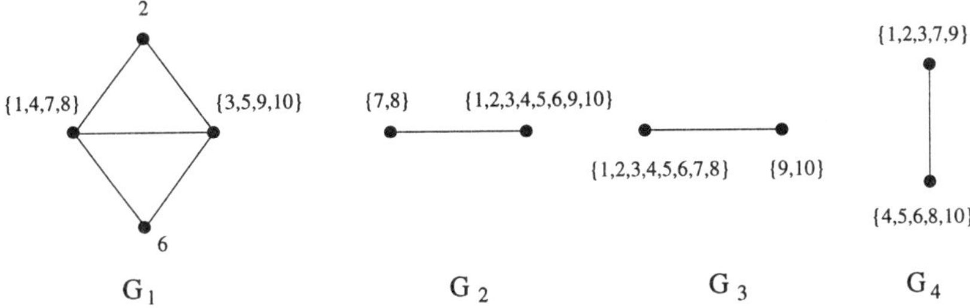

Figure 20.1.8: Factors in the canonical metric representation of G

Example 20.1.9. Let G be the graph from Figure 20.1.7. The relation θ^* has four equivalence classes:

$$E_1 = \{12, 13, 23, 45, 46, 56\}, \; E_2 = \{17, 48\},$$

$$E_3 = \{39, 5\ 10\}, \; E_4 = \{14, 35, 78, 9\ 10\}$$

(where we denote an edge (i,j) by the string ij). The edges of E_1,E_2,E_3, E_4 are represented by plain, dotted, dark, and dark dotted edges, respectively. Hence, $G \hookrightarrow G_1 \times G_2 \times G_3 \times G_4$ is the canonical metric representation of G, where G_1, G_2, G_3, G_4 are the graphs indicated in Figure 20.1.8. (The set associated to each node in the factor G_h is the set of nodes of G that have been identified during the construction of G_h.) ∎

Remark 20.1.10. Isometric embedding into Hamming graphs.
We recall that a graph can be isometrically embedded into a Hamming graph (i.e., a Cartesian product of complete graphs) if its nodes can be labeled by sequences of nonnegative integers in such a way that the shortest path distance between two nodes coincides with the Hamming distance between the corresponding sequences. It follows from Theorem 20.1.1 that a graph G embeds isometrically

into a Hamming graph if and only if each factor G_h in the canonical metric representation of G is a complete graph. (Indeed, let

$$\alpha : G \hookrightarrow \prod_{1 \leq i \leq m} K_{q_i}$$

be an isometric embedding of G into a Hamming graph. We may assume that α is irredundant since deleting a node from a complete graph yields another complete graph. Therefore, as complete graphs are irreducible, α is the canonical metric representation of G.) Therefore, the embedding into a Hamming graph is unique (Winkler [1984]). Moreover, one can recognize whether a graph G is an isometric subgraph of a Hamming graph in polynomial time. For this, it suffices to determine the canonical metric representation of G and to check whether all its factors are complete graphs. This can be done in time $O(mn)$ using $O(m)$ space (using Feder's algorithm mentioned in Remark 20.1.5). Wilkeit [1990] has proposed earlier an algorithm with running time $O(n^3)$, which yields moreover a structural characterization for isometric subgraphs of Hamming graphs.

As an example, consider the graph H from Figure 20.1.11 (taken from Wilkeit [1990]). The relation θ^* has three equivalences classes:

$$E_1 = \{12, 34, 35, 45\},$$

$$E_2 = \{28, 37, 56\}, \text{ and } E_3 = \{14, 23, 78\}.$$

Hence, $H \hookrightarrow K_3 \times K_2 \times K_2$ is the canonical metric representation of H. We also indicate for the graph H in Figure 20.1.11 the sequences from $\{0, 1, 2\} \times \{0, 1\}^2$ providing the correct labeling of the nodes of H. Equivalently, the path metric of H can be decomposed as an integer sum of multicut semimetrics, namely,

$$\begin{aligned} d_H = \ & \delta(\{1,4\},\{2,3,7,8\},\{5,6\}) + \delta(\{1,2,3,4,5\},\{6,7,8\}) \\ & + \delta(\{1,2,8\},\{3,4,5,6,7\}). \end{aligned}$$

(Recall Proposition 4.2.9.) ∎

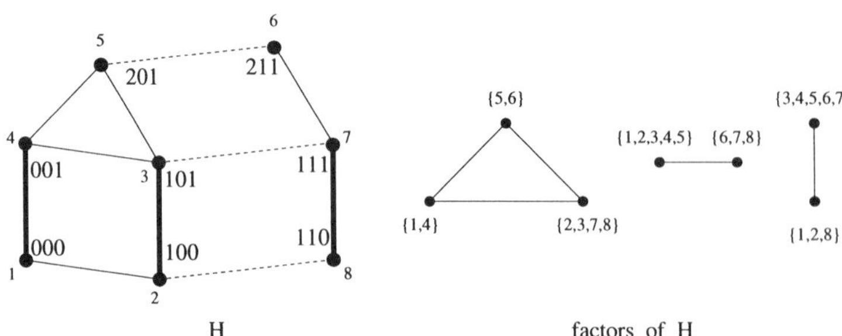

Figure 20.1.11: Embedding graph H in a Hamming graph

20.2 The Prime Factorization of a Graph

Let G be a connected graph. A *factorization* of G is a metric representation which is an isomorphism. G is said to be *prime* if G cannot be decomposed as the Cartesian product of two other graphs (each having at least two nodes). Sabidussi [1960] proved that every connected graph admits a unique prime factorization. Unicity is lost for disconnected graphs (see Zaretskii [1965]). The *graph factoring problem* can be stated as follows:

> *Given a connected graph G, decide whether G is prime. If not, find the prime factorization of G.*

This problem can be solved in time polynomial in the number of nodes (Feigenbaum, Hershberger and Schäffer [1985], Winkler [1987a]). We restrict ourselves to connected graphs since the graph factoring problem for disconnected graphs is at least as hard as the graph isomorphism problem. (Indeed, one can determine whether two graphs G and H are isomorphic by checking whether the graph consisting of two isolated nodes is a factor of the disjoint union of G and H.) The algorithm proposed by Feigenbaum, Hershberger and Schäffer [1985] is based on Sabidussi's original proof and is rather difficult; it runs in time $O(n^{4.5})$ where n is the number of nodes of the graph. Winkler [1987a] proposes an algorithm which is based on the canonical metric representation of graphs presented in Section 20.1, with running time $O(n^4)$. We describe briefly the main ideas of his algorithm.

Let G be the connected graph whose prime factorization is to be found. Let

$$\sigma : V(G) \longrightarrow \prod_{1 \le h \le k} V(G_h)$$

denote the canonical metric representation of G and set

$$\sigma(a) = (\sigma_1(a), \dots, \sigma_k(a))$$

for each node $a \in V$. Set $S := \{1, \dots, k\}$. For a subset T of S, let σ_T denote the mapping from V to $\prod_{h \in T} V(G_h)$ defined by

$$\sigma_T(a) = (\sigma_h(a) \mid h \in T)$$

for $a \in V$. A partition $(S_1, \dots, S_m)$ of S is said to be "*good*" if

$$\sigma(V) = \prod_{1 \le i \le m} \sigma_{S_i}(V).$$

If this is the case, then

$$G = \prod_{1 \le i \le m} \sigma_{S_i}(G)$$

gives a factoring of G. In particular, the prime factorization of G corresponds to a good partition of S. A subset T of S is said to be *complete* if

$$\sigma_T(V) = \prod_{h \in T} V(G_h).$$

A subset $T \subseteq S$ can be checked for completeness in polynomial time. If S itself is complete, then $G = G_1 \times \ldots \times G_k$ is the prime factorization of G. Otherwise, S is not complete. One can find a minimum incomplete subset T of S in polynomial time. (Indeed, check whether all $(k-1)$-subsets of S are complete. If yes, then S is minimal incomplete. Else, let T be an incomplete $(k-1)$-subset of S. Check all $(k-2)$-subsets of T, and so on.) The crucial fact is that, if T is minimal incomplete and if $(S_1, \ldots, S_m)$ is a good partition of S, then $T \subseteq S_i$ for some $i \in \{1, \ldots, m\}$. (If not, then $T \cap S_1, \ldots, T \cap S_m$ are complete, from which one deduces that T itself is complete.) Now, $\sigma_T(G)$ cannot be split in a factorization of G. Hence, we may consider the metric representation

$$G \hookrightarrow \sigma_T(G) \times \prod_{h \in S \setminus T} G_h$$

instead of the initial representation $G \hookrightarrow \prod_{1 \leq h \leq k} G_h$. The new representation has at most $k-1$ factors (since $|T| \geq 2$, as singletons are complete). We repeat the process with this new representation until we find a representation whose index set is complete. This final representation is the prime factorization of G.

Feder [1992] shows how this algorithm can be performed in $O(mn)$ steps using $O(m)$ space; a faster algorithm running in $O(m \log n)$ time and using $O(m)$ space is proposed by Aurenhammer, Hagauer and Imrich [1990].

20.3 Metric Decomposition of Bipartite Graphs

We recall that the semimetric cone MET_n is defined by

$$\mathrm{MET}_n = \{x \in \mathbb{R}^{\binom{n}{2}} \mid x_{ij} - x_{ik} - x_{jk} \leq 0 \text{ for all } i, j, k \in \{1, \ldots, n\}\}.$$

In other words, MET_n consists of all semimetrics on n points. A semimetric $d \in \mathrm{MET}_n$ is said to be *primitive* if d lies on an extreme ray of MET_n, i.e., if $d = d_1 + d_2$ with $d_1, d_2 \in \mathrm{MET}_n$ implies that $d_1 = \alpha_1 d$, $d_2 = \alpha_2 d$ for some $\alpha_1, \alpha_2 \geq 0$. Given $d \in \mathrm{MET}_n$, let $F(d)$ denote the smallest face of MET_n that contains d. Hence, $F(d)$ consists of all the vectors $y \in \mathrm{MET}_n$ that satisfy the same triangle equalities as d, i.e., such that $y_{ij} - y_{ik} - y_{jk} = 0$ whenever $d_{ij} - d_{ik} - d_{jk} = 0$. Then, $F(d)$ is a simplex face (i.e., the primitive semimetrics lying on $F(d)$ are linearly independent) if and only if d admits a unique decomposition as a sum of primitive semimetrics (recall Lemma 4.3.2).

Let G be a connected graph on n nodes. Its path metric d_G belongs to the semimetric cone MET_n. Hence, a natural question to ask is what are the possible decompositions of d_G as a sum of primitive semimetrics. Lomonosov and Sebö [1993] show that, if G is a bipartite graph, then d_G admits a unique such decomposition, i.e., d_G lies on a simplex face of MET_n. In fact, the primitive semimetrics entering the decomposition of d_G are gate 0-extensions of the path metrics of the factors in the canonical metric representation of G.

Theorem 20.3.1. *Let G be a connected bipartite graph on n nodes with isometric dimension $\dim_I(G) = k$. Let $F(d_G)$ denote the smallest face of the semimetric cone MET_n that contains d_G. Then, $F(d_G)$ is a simplex face of MET_n of dimension k.*

Proof. Let $E_1, \ldots, E_k$ denote the equivalence classes of the relation θ^* and let

$$G \hookrightarrow \prod_{1 \le h \le k} G_h$$

denote the associated canonical metric representation of G. For a node $a \in V$, denote by $(a_1, \ldots, a_k)$ its image under the canonical embedding. For $h = 1, \ldots, k$, let d_h denote the semimetric on V defined by

$$d_h(a, b) := d_{G_h}(a_h, b_h)$$

for $a, b \in V$. Then, d can be decomposed as

$$d = \sum_{1 \le h \le k} d_h.$$

The semimetrics $d_1, \ldots, d_k$ are clearly linearly independent and they belong to the face F_G. We show that $F(d_G)$ is generated by $\{d_1, \ldots, d_k\}$. For this, we show that each $x \in F_G$ is of the form $x = \sum_{1 \le h \le k} \alpha_h d_h$ for some scalars $\alpha_h \ge 0$. Let $x \in F_G$. By definition, this means that every triangle inequality which is satisfied at equality by d_G is also satisfied at equality by x. We claim that, if $e = (a, b)$ and $e' = (a', b')$ are edges of G, then

$$e \theta e' \implies x(a, b) = x(a', b').$$

Indeed, as G is bipartite, we can suppose that $a' \in G(a, b)$ and $b' \in G(b, a)$. One can easily check that d_G satisfies the following four triangle equalities:

$$d_G(a', b) = d_G(a', a) + d_G(a, b), \; d_G(a, b') = d_G(a, b) + d_G(b, b'),$$

$$d_G(a', b) = d_G(a', b') + d_G(b, b'), \text{ and } d_G(a, b') = d_G(a, a') + d_G(a', b').$$

Hence, x satisfies these four triangle equalities too. From the first two equalities, we obtain

$$x(a', b) - x(a, b') = x(a', a) - x(b, b'),$$

and the last two imply

$$x(a', b) - x(a, b') = x(b, b') - x(a, a').$$

Therefore, $x(a, a') = x(b, b')$ and $x(a, b') = x(a', b)$ which imply that $x(a, b) = x(a', b')$. Hence, there exist scalars $\alpha_1, \ldots, \alpha_k \ge 0$ such that $x(a, b) = \alpha_h$ for each edge (a, b) in the class E_h. We show

$$x = \sum_{1 \le h \le k} \alpha_h d_h.$$

Let $a, b \in V$ and let $P := (a_0 = a, a_1, \ldots, a_p = b)$ be a shortest path from a to b in G. Set $N_h := |E(P) \cap E_h|$ for $h = 1, \ldots, k$. Using the triangle equalities along P, one obtains

$$x(a,b) = \sum_{1 \leq i \leq p} x(a_{i-1}, a_i) = \sum_{1 \leq h \leq k} \alpha_h N_h.$$

As P contains N_h edges from E_h, by contracting the other edges of P, we obtain in the graph G_h a path from a_h to b_h of length N_h. This shows that $d_{G_h}(a_h, b_h) \leq N_h$. Let Q' be a shortest path from a_h to b_h in G_h. So, Q' arises from a path Q joining a to b in G. By Lemma 20.1.2, Q contains at least N_h edges from E_h. Therefore, $|Q'| \geq N_h$, implying that $d_{G_h}(a_h, b_h) = N_h$. Hence,

$$\sum_{1 \leq h \leq k} \alpha_h d_h(a,b) = \sum_{1 \leq h \leq k} \alpha_h N_h = x(a,b).$$

So, we have shown that $F(d_G)$ is generated by $\{d_1, \ldots, d_k\}$. Therefore, $F(d_G)$ is a simplex face of dimension k of MET_n. ∎

Corollary 20.3.2. *Let G be a connected bipartite graph. Then, its path metric d_G lies on an extreme ray of the semimetric cone MET_n if and only if $\dim_I(G) = 1$, i.e., if G is irreducible.* ∎

Corollary 20.3.2 is not valid for nonbipartite graphs. For instance, K_3 is irreducible, but its path metric lies in the interior of the semimetric cone MET_3.

20.4 Additional Notes

Several further aspects of the metric structure of graphs have received a considerable attention in the literature, leading to rich theories. For instance, distance-regular graphs, or strongly regular graphs, are defined by some invariance property of their path metric. The study of these graphs leads to a large and rich area of research, connected to algebraic graph theory. The monograph by Brouwer, Cohen and Neumaier [1989] is an excellent source of information on this topic. Let us only remind that some results along this line have been presented in Part II, especially in Chapter 17, where hypermetric graphs are considered. The papers by Koolen [1990, 1993, 1994], Koolen and Shpectorov [1994], Weichsel [1992] deal with the study of graphs with high regularity that have some specified metric properties such as hypermetricity, or some special cases of it (e.g., satisfying the pentagonal inequality, or the hexagonal inequality), etc. For instance, the distance-regular graphs that are hypercube embeddable are completely classified: they are the hypercubes, the even circuits, and the double-odd graphs (Koolen [1990], Weichsel [1992]). The distance-regular graphs that are of negative type (or, equivalently, hypermetric) are classified in Koolen and Shpectorov [1994]; see Theorem 17.2.7.

In this section we mention some further topics related to the metric structure of graphs. Among them, the study of graphs having specified metric properties (e.g., interval-regular graphs, geodetic graphs) and the question of embedding an arbitrary distance space into a (weighted) graph. There is a vast literature on these topics. So, we shall not attempt to give a detailed treatment; we only mention without proof some facts, results and references.

Interval-Regular Graphs and Geodetic Graphs. Let $G = (V, E)$ be a connected graph. For two nodes $x, y \in V$, let $\gamma(x, y)$ denote the number of shortest paths joining x to y in G. Set

$$I(x,y) := \{z \in V \mid d_G(x,y) = d_G(x,z) + d_G(z,y)\}.$$

Moreover, for $i = 0, 1, \ldots, d_G(x, y)$, set

$$N_i(x,y) := \{z \in I(x,y) \mid d_G(x,z) = i\}$$

and

$$N_{-1}(x,y) := \{z \in V \mid d_G(x,z) = 1 \text{ and } d_G(z,y) = d_G(x,z) + 1\}.$$

Then, G is distance-regular if the numbers $|N_1(x, y)|$ and $|N_{-1}(x, y)|$ depend only on $d_G(x, y)$. The graph G is said to be *interval-regular* if $|N_1(x, y)| = d_G(x, y)$ for all nodes $x, y \in V$ (see Mulder [1980, 1982]); G is said to be *uniformely geodetic* (in Cook and Pryce [1983]) (or *F-geodetic* in Ceccherini and Sappa [1986]) if $\gamma(x, y)$ depends only on $d_G(x, y)$. Every distance-regular graph is uniformely geodetic (Cook and Pryce [1983]), and every Hamming graph is interval-regular (since the subgraph induced by the interval $I(x, y)$ is isomorphic to the $d_G(x, y)$-hypercube). See, e.g., Scapellato [1990] and Koolen [1993] for more information on uniformely geodetic graphs; Koolen [1993] characterizes the uniformely geodetic bipartite graphs.

Several characterizations of the hypercube are known. Foldes [1977] shows that the hypercube is the only connected bipartite graph for which $\gamma(x, y) = d_G(x, y)!$ holds for any pair of nodes. Ceccherini and Sappa [1986] show that a connected bipartite graph G is isomorphic to a hypercube if and only if the Cartesian product $G \times K_2$ is uniformely geodetic.

Interval-regular graphs are linked to hypercubes in the following way: Mulder [1982] shows that a connected graph G is interval-regular if and only if, for any two nodes x, y, the subgraph of G consisting of the edges connecting two consecutive levels $N_i(x, y)$ and $N_{i+1}(x, y)$ ($i = 0, 1, \ldots, d_G(x, y) - 1$) is isomorphic to the $d_G(x, y)$-hypercube. Equivalently, G is interval-regular if and only if $\gamma(x, y) = d_G(x, y)!$ for all nodes $x, y \in V$.

Hamming graphs can be characterized in terms of interval-regular graphs in the following way (Bandelt and Mulder [1991]): A connected graph G is a Hamming graph if and only if G is an interval-graph, G does not contain $K_{1,1,2}$ as an induced subgraph, and the only isometric odd circuits in G are triangles. More generally, Bandelt and Mulder [1991] characterize the connected graphs that can be decomposed as a Cartesian product where each factor is the suspension of a geodetic graph of diameter at most 2. (A geodetic graph is a graph in which there is exactly one shortest path joining any pair of nodes.)

Embedding Metrics into Graphs. We now consider the question of embedding metrics into graphs or, more generally, into weighted graphs. This topic has many applications in various areas, such as psychology (Cunningham [1978]) and biology (Penny, Foulds and Hendy [1982]).

Let $G = (V, E)$ be a graph and let $w_e \in \mathbb{R}_+$ $(e \in E)$ be nonnegative weights assigned to its edges. The path metric $d_{G,w}$ of the weighted graph (G, w) is defined by letting $d_{G,w}(x, y)$ denote the smallest value of $\sum_{e \in E(P)} w_e$, taken over all paths P joining x and y in G.

Given a finite metric space (X, d), one says that the weighted graph (G, w) *realizes* (X, d) if there exists a mapping $i \in X \mapsto x_i \in V$ such that

$$d(i, j) = d_{G,w}(x_i, x_j)$$

for all $i, j \in X$. The graph G may have more nodes than those corresponding to points of X. Every metric space can clearly be realized by some graph, namely, by the complete graph on $|X|$ nodes with weights $d(i, j)$ on its edges. Consider, for instance, the metric d on $X = \{1, 2, 3\}$ defined by $d(1, 2) = 4$, $d(1, 3) = 8$, $d(2, 3) = 6$. Then, d can be realized by the following two weighted graphs: K_3 and a tree with one auxiliary node.

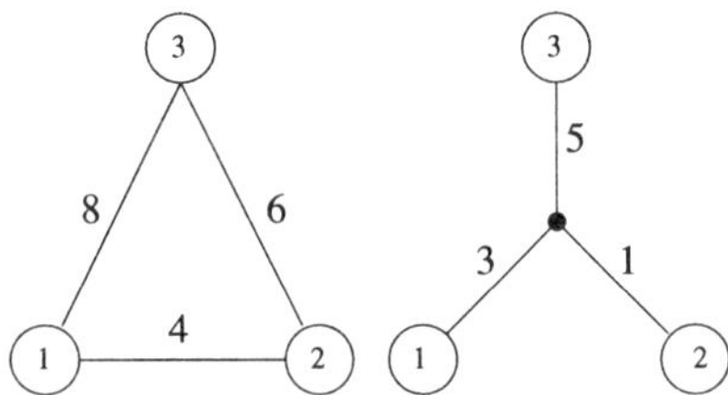

The objective is, therefore, to find a graph (G, w) realizing (X, d) whose total weight $\sum_{e \in E} w_e$ is as small as possible. The existence of an optimal realization, i.e., with minimum total weight among all possible realizations, was shown in Imrich, Simões-Pereira and Zamfirescu [1984]. But finding an optimal realization is an NP-hard problem even if the metric is assumed to be integer valued (Althöfer [1988], Winkler [1988]).

On the other hand, the metric spaces that can be realized by weighted trees are well characterized; such graphs have been introduced earlier under the name of tree metrics. Namely, (X, d) is realizable by a weighted tree, if and only if d satisfies the following condition, known as the *four-point condition*:

$$d(i, j) + d(r, s) \leq \max(d(i, r) + d(j, s), d(i, s) + d(j, r)),$$

i.e., the two largest of the three sums $d(i, j)+d(r, s)$, $d(i, r)+d(j, s)$, $d(i, s)+d(j, r)$ are equal, for all $i, j, r, s \in X$ (Buneman [1974]). Note that the four point condition implies the metric condition (by taking $r = s$). Moreover, if (X, d) is realizable by a tree, then there is only one such realization; it is optimal among all graph realizations, and it can be found in polynomial time (Hakimi and Yau [1964]).

The four point condition is closely related to another metric condition, namely, ultrametricity. Recall that a distance space (X, d) is said to be *ultrametric* if it satisfies

$$d(i,j) \leq \max(d(i,k), d(j,k))$$

for all $i, j, k \in X$. In other words, any three points form an isosceles triangle with the third side shorter than or equal to the other two. See Aschbacher, Baldi, Baum and Wilson [1987] for applications and references on ultrametrics. Clearly, every ultrametric space satisfies the four point condition. Actually, each tree metric can be characterized in terms of an associated ultrametric in the following way (Bandelt [1990]). Let (X, d) be a distance space, let $r \in X$, and let c be a constant such that $c \geq \max(d(i,j) \mid i, j \in X)$. Define the distance $d^{(c)}$ on $X \setminus \{r\}$ by setting

$$d^{(c)}(i,j) := c + \frac{1}{2}(d(i,r) + d(j,r) - d(i,j))$$

for $i \neq j \in X$. Then, d is a tree metric if and only if $d^{(c)}$ is ultrametric. Ultrametrics have also a tree-like representation, which is used in classification theory, in particular, in taxonomy (see Gordon [1987] and references therein for details). Let $T = (V, E)$ be a tree and $w_e \in \mathbb{R}_+$ $(e \in E)$ be nonnegative weights on its edges. Let $r \in V$ be a specified node (a root) of T and let $X = \{x_1, \ldots, x_k\}$ denote the set of leaves (nodes of degree 1) of T other than r. We assume that $d_{T,w}(r, x) = h$ for all $x \in X$, for some constant h, called the height of T. Then, T is also called a *dendrogram*, or *indexed hierarchy*. The height $h(v)$ of a node v of T is defined as the length of a shortest path joining v to some leaf of X. Then, one can define a metric space (X, d_X) on X by letting $d_X(x, y)$ denote the height of the first predecessor of x and y. The metric space (X, d_X) is ultrametric and every ultrametric arises in this way. Moreover, the distance spaces (X, d_X) and $(X, \frac{1}{2}d_{T,w})$ coincide.

Chapter 21. ℓ_1-Graphs

We study in this chapter ℓ_1-graphs, i.e., the graphs whose path metric can be isometrically embedded into an ℓ_1-space. Such graphs can be characterized as the isometric subgraphs of Cartesian products of two types of elementary graphs, namely, half-cube graphs and cocktail-party graphs. This result has already been established in Part II, using the theory of Delaunay polytopes. We present here another proof due to Shpectorov [1993] which is elementary and, moreover, yields a polynomial time recognition algorithm. Section 21.4 contains additional results on ℓ_1-graphs; in particular, a characterization in terms of forbidden isometric subspaces for isometric subgraphs of half-cube graphs.

21.1 Results on ℓ_1-Graphs

As was recalled in Chapter 18, a graph G is an ℓ_1-graph if and only if it is hypercube embeddable, up to scale. A λ-*embedding* of G into the hypercube $H(\Omega)$ is any mapping

$$x \in V \mapsto X \subseteq \Omega$$

such that

$$\lambda d_G(x,y) = |X \triangle Y|$$

for all nodes x, y of G. If G has a λ-embedding into a hypercube, we also say that G is hypercube embeddable with scale λ. A 1-embedding in a hypercube is nothing but an isometric embedding in a hypercube.

Three classes of graphs play a crucial role in the theory of ℓ_1-graphs: complete graphs, cocktail-party graphs, and half-cube graphs. All of them are ℓ_1-graphs, so is any Cartesian product of them. Actually, we show below that any ℓ_1-graph arises as an isometric subgraph of such a Cartesian product. Complete graphs K_m $(m \geq 3)$ and half-cube graphs $\frac{1}{2}H(m,2)$ $(m \geq 3)$ have minimum scale 2. Note that

$$\frac{1}{2}H(2,2) = K_2,\ \frac{1}{2}H(3,2) = K_{4\times 2},$$

and K_m is an isometric subgraph of both $K_{m\times 2}$ and $\frac{1}{2}H(m,2)$. The following holds clearly.

Lemma 21.1.1. *A graph G is hypercube embeddable with scale 2 (that is, $2d_G$ is hypercube embeddable) if and only if G is an isometric subgraph of some half-*

cube graph. ∎

On the other hand, determining the minimum scale of cocktail-party graphs is a hard problem. This problem is considered in detail in Chapter 23; see also Section 7.4. We already know from Theorem 19.2.8 that every connected bipartite ℓ_1-graph is hypercube embeddable. For nonbipartite graphs we have the following observation (Blake and Gilchrist [1973]).

Lemma 21.1.2. *Let G be an ℓ_1-graph and suppose that G has a λ-embedding in a hypercube. If G is not a bipartite graph, then λ is an even integer. Therefore, the minimum scale of an ℓ_1-graph is equal to 1 or is even.*

Proof. Suppose that G is not bipartite. Let C be an odd circuit in G of minimal length. Then, C is an isometric subgraph of G. Say, $C = (a_1, \ldots, a_{2k+1})$. We can suppose that, in the λ-embedding of G in a hypercube, the nodes a_1, a_{k+1}, a_{k+2} are labeled by the sets $\emptyset$,A,B, respectively. Then, as $d_G(a_1, a_{k+1}) = d_G(a_1, a_{k+2}) = k$ and $d_G(a_{k+1}, a_{k+2}) = 1$, we have $\lambda = |A \triangle B|$ and $|A| = |B| = \lambda k$. Hence, $\lambda = 2\lambda k - 2|A \cap B|$. Therefore, λ is an even integer. ∎

We now present the main results of this chapter; they are due to Shpectorov [1993].

Theorem 21.1.3. *Let G be an ℓ_1-graph. Then, there exist a graph $\widehat{G}$ and an isometric embedding $\widehat{\sigma}$ from G into $\widehat{G}$ such that*

(i) *$\widehat{G} = \widehat{G}_1 \times \ldots \times \widehat{G}_k$, where each $\widehat{G}_h$ is isomorphic to a complete graph, a cocktail-party graph $K_{m \times 2}$ ($m \geq 3$), or a half-cube graph, and*

(ii) *if ψ is a λ-embedding of G into the hypercube, then there is a λ-embedding $\widehat{\psi}$ of $\widehat{G}$ into the same hypercube such that $\psi = \widehat{\psi}\widehat{\sigma}$.*

Corollary 21.1.4. *A connected graph G is an ℓ_1-graph if and only if all the factors in its canonical metric representation are ℓ_1-graphs.*

Corollary 21.1.5. *A connected graph G is an ℓ_1-graph if and only if G is an isometric subgraph of a Cartesian product of cocktail-party graphs and half-cube graphs.*

Corollary 21.1.6. *Let G be an ℓ_1-graph. Then, G is ℓ_1-rigid if and only if $\widehat{G}$ is ℓ_1-rigid.*

Corollary 21.1.7. *Every ℓ_1-rigid graph is an isometric subgraph of a half-cube graph and, therefore, its minimum scale η is equal to 1 or 2.*

Corollary 21.1.8. *Let G be an ℓ_1-graph on $n \geq 4$ nodes. Then its minimum scale η satisfies $\eta \leq n - 2$.*

Corollary 21.1.9. *Let G be a graph with n nodes and m edges. There exists an algorithm permitting to decide whether G is an ℓ_1-graph which runs in $O(mn)$ time using $O(n^2)$ space.*

We present in Section 21.2 a concrete construction of the graph $\widehat{G}$ from Theorem 21.1.3, using a specific λ-embedding of G. We group in Section 21.3 the proofs for Theorem 21.1.3 and Corollaries 21.1.4-21.1.9.

The result from Corollary 21.1.5 was already established in Part II (in Theorem 17.1.1 (ii)), as an application of the correspondence existing between Delaunay polytopes and hypermetrics. However, the proof method developed there did not permit to obtain further results such as the characterization of ℓ_1-rigidity and the fact that ℓ_1-graphs can be recognized in polynomial time. In contrast, the proof method presented here uses only elementary notions. It is, in a way, a continuation of the theory of canonical metric representations of graphs. Indeed, the essential step of the proof will be to show that each factor in the canonical metric representation of an ℓ_1-graph can be further embedded into a complete graph, a cocktail-party graph, or a half-cube graph.

In view of Corollaries 21.1.4 and 21.1.5, one can check whether a graph G is an ℓ_1-graph in the following way:

(i) Construct the canonical metric representation of G.

(ii) For each factor G_h in the canonical metric representation, check whether G_h is an isometric subgraph of a cocktail-party graph or of a half-cube graph.

Then, G is an ℓ_1-graph if and only if the answer is always positive in Step (ii). As was seen in Remark 20.1.5, Step (i) can be performed in $O(mn)$ time using $O(m)$ space. The main difficulty in Step (ii) consists of recognizing the isometric subgraphs of the half-cube graphs. The following result was already implicit in Shpectorov [1993]; full details about the algorithm are provided in Deza and Shpectorov [1996]. We will give the proof in Section 21.3.

Proposition 21.1.10. *Let G be a graph on n nodes with m edges. There exists an algorithm permitting to decide whether G is an isometric subgraph of some half-cube graph which runs in $O(mn)$ time using $O(n^2)$ space. The algorithm constructs an embedding if one exists.*

Suppose that each factor G_h in the canonical metric representation of G has n_h nodes and m_h edges, where $m_h \geq n_h - 1$ as G_h is connected. Then, $m \geq m_1 + \ldots + m_k$ and $n_h \leq n$ for all h. Therefore, $n_1m_1 + \ldots + n_km_k \leq nm$. Checking whether G_h is a subgraph of a cocktail-party graph can be done in $O(n_h^2)$ time and space (simply check that every node is adjacent to all other nodes except at most one). Therefore, the overall time complexity for checking whether G is an ℓ_1-graph is in $O(mn)$ and the space complexity in $O(n^2)$, as claimed in Corollary 21.1.9.

21.2 Construction of $\widehat{G}$ via the Atom Graph

In this section, we show how to construct the graph $\widehat{G}$ from Theorem 21.1.3, using a specific scale embedding of G. It will turn out that, in fact, $\widehat{G}$ does not depend on the choice of the scale embedding and that $\widehat{G}$ is an isometric extension of the canonical metric representation of G. The main tool for the construction of $\widehat{G}$ is the atom graph of G, as we explain below.

Let $G = (V, E)$ be an ℓ_1-graph. Let

$$\psi : x \in V \mapsto X \subseteq \Omega$$

be a λ-embedding of G into the hypercube $H(\Omega)$. We can suppose without loss of generality that $\Omega = \bigcup_{x \in V} X$ and that a given node $x_0 \in V$ is assigned to $\emptyset$. Set

$$E_0 := \{e = (x, y) \in E \mid d_G(x_0, x) \neq d_G(x_0, y)\}. \tag{21.2.1}$$

For an edge $e = (x, y) \in E_0$, we can suppose, e.g., that $x_0 \in G(x, y)$. One can easily check the following statements:

$$|X| = \lambda d_G(x_0, x) \quad \text{for all } x \in V, \tag{21.2.2}$$

$$|X \cap Y| = \frac{\lambda}{2}(d_G(x_0, x) + d_G(x_0, y) - d_G(x, y)) \quad \text{for all } x, y \in V. \tag{21.2.3}$$

For an edge $e = (x, y)$,

$$\begin{cases} |X \setminus Y| = |Y \setminus X| = \frac{\lambda}{2} & \text{if } e \notin E_0, \\ X \subseteq Y & \text{if } e \in E_0. \end{cases} \tag{21.2.4}$$

Call *atom* every set of the form $X \triangle Y$ corresponding to an edge $e = (x, y)$ of G, and *proper atom* every set of the form $Y \setminus X$ corresponding to an edge $e = (x, y) \in E_0$ (with $x_0 \in G(x, y)$). Atoms have cardinality λ and

$$\text{if } A, B \text{ are distinct proper atoms, then } |A \cap B| = 0, \frac{\lambda}{2}. \tag{21.2.5}$$

We define the *atom graph* $\Lambda(G)$ as the graph with node set the set of proper atoms of G and with two proper atoms A, B being adjacent if $|A \cap B| = \frac{\lambda}{2}$. Let $\Lambda_1, \ldots, \Lambda_k$ denote the connected components of $\Lambda(G)$. For $h = 1, \ldots, k$, let Ω_h denote the union of the proper atoms that are nodes of Λ_h. Hence, each proper atom is either contained in Ω_h, or is disjoint from Ω_h. Actually, the same property holds for all atoms, as we show in the next result.

Claim 21.2.6. *Let A be an atom of G. Then, for each $h = 1, \ldots, k$, either*

$A \subseteq \Omega_h$, or $A \cap \Omega_h = \emptyset$.

Proof. Let (x, y) be an edge of G corresponding to the atom A, i.e., $A = X \triangle Y$. We can suppose that the edge (x, y) does not belong to E_0. Hence, the node x_0 is at the same distance s from x and y. Let $(x_0, x_1, \ldots, x_s = x)$ and $(y_0 = x_0, y_1, \ldots, y_s = y)$ be shortest paths from x_0 to x and y in G. Hence,

$$X = \bigcup_{1 \leq i \leq s} B_i, \; Y = \bigcup_{1 \leq i \leq s} C_i,$$

where B_i is the proper atom $X_i \setminus X_{i-1}$, C_i is the proper atom $Y_i \setminus Y_{i-1}$, for $i = 1, \ldots, s$. We claim that

$$X \setminus Y \subseteq B_{i_0}$$

for some $i_0 \in \{1, \ldots, s\}$. Indeed, take $\alpha \in X \setminus Y$ and suppose, for instance, that $\alpha \in B_1$. Then, $B_1 \cap Y \neq B_1$. On the other hand, $B_1 \cap Y \neq \emptyset$, else $B_1 \subseteq X \setminus Y$ implying that $|X \setminus Y| \geq \lambda$, contradicting (21.2.4). As the cardinality of $B_1 \cap Y$ is a multiple of $\frac{\lambda}{2}$ by (21.2.5), we obtain that $|B_1 \cap Y| = \frac{\lambda}{2}$, $|B_1 \setminus Y| = \frac{\lambda}{2}$. Therefore, $X \setminus Y = B_1 \setminus Y \subseteq B_1$. Similarly,

$$Y \setminus X \subseteq C_{j_0}$$

for some $j_0 \in \{1, \ldots, s\}$. Furthermore, as the C_j's are pairwise disjoint, each B_i either coincides with some C_j, or meets exactly two of them, unless $B_i = B_{i_0}$ in which case B_i meets exactly one C_j. The symmetric statement holds for each C_i. This means that the subgraph of the atom graph $\Lambda(G)$ induced by the set $\{B_1, \ldots, B_s, C_1, \ldots, C_s\}$ consists of isolated nodes, cycles, and exactly one path whose endpoints are B_{i_0} and C_{j_0}. Let Λ_{h_0} be the connected component of $\Lambda(G)$ that contains this path. Then, $B_{i_0}, C_{j_0} \subseteq \Omega_{h_0}$, which implies that $A = X \triangle Y \subseteq \Omega_{h_0}$. Moreover, for $h \neq h_0$, B_{i_0}, C_{j_0} are disjoint from Ω_h, implying that A is disjoint from Ω_h. ∎

Let $\overline{G}_h$ denote the graph with node set $\{\overline{X} := X \cap \Omega_h \mid x \in V\}$ and with $(\overline{X}, \overline{Y})$ being an edge if $|\overline{X} \triangle \overline{Y}| = \lambda$. Set $\overline{G} := \prod_{1 \leq h \leq k} \overline{G}_h$.

Claim 21.2.7.

(i) *Each $\overline{G}_h$ is λ-embedded into the hypercube $H(\Omega_h)$ and its atom graph $\Lambda(\overline{G}_h)$ coincides with Λ_h.*

(ii) *$\overline{G}$ is λ-embedded into the hypercube $H(\Omega)$.*

(iii) *The mapping $x \in V \mapsto (X \cap \Omega_1, \ldots, X \cap \Omega_k)$ is an isometric embedding of G into $\overline{G}$.*

Proof. Let x, y be two nodes of G, giving the two nodes $\overline{X} = X \cap \Omega_h$, $\overline{Y} = Y \cap \Omega_h$ of $\overline{G}_h$. We show

$$|\overline{X} \triangle \overline{Y}| = \lambda d_{\overline{G}_h}(\overline{X}, \overline{Y}).$$

Set $s := d_G(x,y)$ and $t := d_{\overline{G}_h}(\overline{X},\overline{Y})$. Let $(y_0 = x, y_1, \ldots, y_s = y)$ be a shortest path from x to y in G. Then,

$$X \triangle Y = \sum_{1 \leq i \leq s} Y_i \setminus Y_{i-1}$$

is a disjoint union of atoms. Let t_h denote the number of atoms $Y_i \setminus Y_{i-1}$ that are contained in Ω_h. By Claim 21.2.6, we obtain

$$|\overline{X} \triangle \overline{Y}| = |(X \triangle Y) \cap \Omega_h| = t_h \lambda.$$

Moreover, we have found a path of length t_h joining $\overline{X}$ to $\overline{Y}$ in $\overline{G}_h$, which implies that $t_h \geq t$. Let $(\overline{Z}_0 = \overline{X}, \overline{Z}_1, \ldots, \overline{Z}_t = \overline{Y})$ be a shortest path joining $\overline{X}$ to $\overline{Y}$ in $\overline{G}_h$. So,

$$|\overline{X} \triangle \overline{Y}| = |(\overline{X} \triangle \overline{Z}_1) \triangle (\overline{Z}_1 \triangle \overline{Z}_2) \triangle \ldots \triangle (\overline{Z}_{t-1} \triangle \overline{Y})| \leq \sum_{1 \leq i \leq t} |\overline{Z}_{i-1} \triangle \overline{Z}_i| = t\lambda.$$

This implies that $t_h \leq t$ and, therefore, $t_h = t$. Hence, the graph $\overline{G}_h$ is λ-embedded into the hypercube $H(\Omega_h)$. One checks easily that its atom graph is Λ_h. Hence, (i) holds. Moreover,

$$(X \cap \Omega_1, \ldots, X \cap \Omega_k) \mapsto \bigcup_h (X \cap \Omega_h) = X$$

provides a λ-embedding of $\overline{G}_1 \times \ldots \times \overline{G}_k$ into the hypercube $H(\Omega)$, showing (ii). It also follows that

$$d_G(x,y) = \sum_{1 \leq h \leq k} d_{\overline{G}_h}(X \cap \Omega_h, Y \cap \Omega_h)$$

for all nodes $x, y \in V$. This shows (iii). ∎

We now show that each factor $\overline{G}_h$ can be further embedded into some graph $\widehat{G}_h$ which is isomorphic to a complete graph, a cocktail-party graph, or a half-cube graph. We first deal with the case when the atom graph $\Lambda(G)$ is connected, i.e., when $k = 1$. Then, the graph $\overline{G}$ is nothing but the graph G embedded into the hypercube $H(\Omega)$.

Claim 21.2.8. *If $\Lambda(G)$ is connected, then there exists a unique minimal graph $\widehat{G}$ containing G as an isometric subgraph and such that $\widehat{G}$ is isomorphic to a complete graph, a cocktail-party graph $K_{m \times 2}$ $(m \geq 3)$, or a half-cube graph. Moreover, $\widehat{G}$ is λ-embedded into the hypercube $H(\Omega)$.*

Proof. We distinguish three cases.
Case 1: *$\Lambda(G)$ is a complete graph.* Then, G itself is a complete graph and $\widehat{G} = G$. Indeed, each node x is adjacent to x_0 (else, X would be a disjoint union of the proper atoms corresponding to the edges of a shortest path from x_0 to

x). For two nodes $x, y \in V$, X and Y are adjacent proper atoms, implying that $|X \triangle Y| = \lambda$ and, therefore, x and y are adjacent in G. (In fact, $\Lambda(K_n) = K_{n-1}$.)

Case 2: *$\Lambda(G)$ is not a complete graph, but is an induced subgraph of a cocktail-party graph.* Let A, B be two proper atoms at distance 2 in $\Lambda(G)$. Each other proper atom C is adjacent to both A and B, which implies that $C \subseteq A \cup B$. Hence, for each node $x \in V$, X is contained in the 2λ-element set $A \cup B$. We claim that G is an induced subgraph of a cocktail-party graph. Indeed, any two nonadjacent nodes in G are necessarily at distance 2 since $|X \triangle Y| \leq 2\lambda$ for all $x, y \in V$. Moreover, each node x is adjacent to all other nodes except maybe one, which is then labeled by the complement of X. Then, we take for $\widehat{G}$ the cocktail-party graph $K_{m\times 2}$, obtained by adding an "opposite" node labeled by the complement of X for each node x which is adjacent to all other nodes in G. Hence, $\widehat{G}$ is λ-embedded into the same hypercube $H(\Omega)$. Moreover, $m \geq 3$. Otherwise, G would be a subgraph of $K_{2\times 2}$, which implies that G is P_3 or C_4 in which cases $\Lambda(G)$ consists of two isolated nodes.

Case 3: *$\Lambda(G)$ is not an induced subgraph of a cocktail-party graph.* We show that G can isometrically embedded into a half-cube graph. First, we claim the existence of distinct proper atoms A, B, C, D satisfying

$$\begin{cases} A \cap C = \emptyset, \\ A \cap D = \emptyset, \\ C \text{ is adjacent to } D \text{ in } \Lambda(G), \\ B \text{ is adjacent to } A \text{ and } C \text{ in } \Lambda(G). \end{cases}$$

Indeed, let A, C be two proper atoms at distance 2 in $\Lambda(G)$ and let B be a proper atom adjacent to A and C. Suppose for contradiction that, for each proper atom D, D is adjacent to A if and only if D is adjacent to C. If D is adjacent to A and C, then $D \subseteq A \cup C$. If D' is adjacent to A and C and D is adjacent to D', then D meets A or C and, thus, D is adjacent to both A and C, implying $D \subseteq A \cup C$. By connectivity of the atom graph $\Lambda(G)$, we deduce that each proper atom D is contained in $A \cup C$. Therefore, if D, D' are disjoint proper atoms, then D' is the complement of D. This shows that each proper atom is adjacent to all other proper atoms except at most one, contradicting the assumption that $\Lambda(G)$ is not a subgraph of a cocktail-party graph.
Let us call a *half* each set of the form $A \cap B$ or $A \setminus B$, where A, B are adjacent proper atoms. Each half has cardinality $\frac{\lambda}{2}$ and each proper atom is the disjoint union of two halves. We claim that

(21.2.9) distinct halves are disjoint.

If (21.2.9) holds then, for each node $x \in V$, X can be uniquely expressed as a disjoint union of halves. Indeed, if $(x_0, x_1, \dots, x_s = x)$ is a shortest path from x_0 to x in G, then $X = \cup_{1 \leq i \leq s} X_i \setminus X_{i-1}$ where each proper atom $X_i \setminus X_{i-1}$ is the union of two halves; this set of halves does not depend on the choice of the shortest path. This gives an isometric embedding of G into the half-cube graph $\widehat{G}$ defined on the set of halves. By construction, $\widehat{G}$ is λ-embedded into the hypercube $H(\Omega)$.

We now show that (21.2.9) holds. As $\Lambda(G)$ is connected, we can order the proper atoms $A_1, A_2, \ldots, A_p$ in such a way that each A_j ($j \geq 2$) is adjacent to at least one A_s, $s < j$. We suppose that $A_1 = A, A_2 = B, A_3 = C, A_4 = D$. We show by induction on $j \geq 4$ that the distinct halves that are created by the first j proper atoms $A_1, \ldots, A_j$ are pairwise disjoint.

Consider first the case $j = 4$. By construction, the halves $H_1 = A \setminus B$, $H_2 = A \cap B$, $H_3 = B \cap C$, $H_4 = C \setminus B$ are disjoint. Consider the half $C \cap D$. Since $B \cap D = B \cap C \cap D = (C \cap D) \cap H_3$ has cardinality 0 or $\frac{\lambda}{2}$, we obtain that $C \cap D$ is equal to H_3 or H_4. The half $H_5 = D \setminus C$ is disjoint from H_1, H_2, H_3, H_4.

We suppose now that all halves in the set $\mathcal{H}$ of the halves created by the first $j-1$ ($j \geq 5$) proper atoms are pairwise disjoint. Call two halves $H, H' \in \mathcal{H}$ neighboring if $H \cup H'$ is a proper atom A_s for some $s < j$. This defines a graph structure on $\mathcal{H}$, for which $\mathcal{H}$ is connected. Suppose that A_j is adjacent to A_s, for $s < j$, and let $A_s = X_1 \cup X_2$ with $X_1, X_2 \in \mathcal{H}$. Suppose that $A_j \cap A_s$ is not equal to X_1, nor to X_2. Set $\alpha = |A_j \cap X_1|$ and $\beta = |A_j \cap X_2$, where $\alpha, \beta > 0$ and $\alpha + \beta = \frac{\lambda}{2}$. If Y_1, Y_2 are two neighboring halves and $|A_j \cap Y_1| = \alpha$ or β, then $|A_j \cap Y_2| = \beta$ or α, respectively. By connectivity of $\mathcal{H}$, we deduce that $A_j \cap Y_1| = \alpha$, $|A_j \cap Y_2| = \beta$, or vice versa, for every pair (Y_1, Y_2) of neighboring halves. Now, $|A_j \cap (H_1 \cup H_2 \cup H_3 \cup H_4 \cup H_5)| = 2(\alpha + \beta) + \alpha$ or $2(\alpha + \beta) + \beta$, which is greater than λ, yielding a contradiction. Therefore, $A_j \cap A_s$ is equal to X_1 or X_2. Hence, $A_j \setminus A_s$ is either a half from $\mathcal{H}$ or a new half disjoint from all halves in $\mathcal{H}$. This concludes the induction, and the proof of Claim 21.2.8. ∎

Claim 21.2.10. *If $\Lambda(G)$ has k connected components $\Lambda_1, \ldots, \Lambda_k$, then there exists a unique minimal graph $\widehat{G}$ containing G as an isometric subgraph and such that*

$$\widehat{G} = \prod_{1 \leq h \leq k} \widehat{G}_h,$$

where each factor $\widehat{G}_h$ is isomorphic to a complete graph, a cocktail-party graph $K_{m \times 2}$ ($m \geq 3$), or a half-cube graph. Moreover, $\widehat{G}$ is λ-embedded into the hypercube $H(\Omega)$.

Proof. As the atom graph $\Lambda(\overline{G}_h) = \Lambda_h$ is connected, we can apply Claim 21.2.8. Hence, for each $h = 1, \ldots, k$, there exists a unique minimal graph $\widehat{G}_h$ which contains $\overline{G}_h$ as an isometric subgraph and is isometric to a complete graph, a cocktail-party graph $K_{m \times 2}$ ($m \geq 3$), or a half-cube graph. Therefore,

$$G \hookrightarrow \widehat{G} = \prod_{1 \leq h \leq k} \widehat{G}_h,$$

providing a minimal graph $\widehat{G}$ satisfying Claim 21.2.10. Moreover, $\widehat{G}$ is λ-embedded into $H(\Omega)$ as each factor $\widehat{G}_h$ is λ-embedded into $H(\Omega_h)$ and the sets Ω_h are disjoint subsets of Ω. ∎

Remark 21.2.11. Each of the graphs $\overline{G}_h$ is irreducible since it is an isometric subgraph of a complete graph, a cocktail-party graph on at least 6 nodes, or a half-cube graph, which are all irreducible graphs. As the embedding $G \hookrightarrow \overline{G}$ is clearly irredundant, we deduce from Theorem 20.1.1 that the metric representation

$$G \hookrightarrow \overline{G} = \prod_{1 \leq h \leq k} \overline{G}_h$$

is, in fact, the canonical metric representation of G (which explains why we denoted the number of connected components of the atom graph by k, the letter used in Chapter 20 for denoting the isometric dimension of G). In particular, the graph $\overline{G}$ whose construction depends, a priori, on the choice of the scale embedding of G into a hypercube does not, in fact, depend on the specific embedding. Hence, the graph $\widehat{G}$ too does not depend on the specific embedding. ∎

Remark 21.2.12. One can also verify directly that the graph $\overline{G}$ does not depend on the specific scale embedding of G. Indeed, the atom graph can be defined in an abstract way, not using the specific embedding. Given two edges $e = (x, y)$, $e' = (x', y')$ of G, set

$$\langle e, e' \rangle := d_G(y', x) - d_G(y', y) - d_G(x', x) + d_G(x', y). \tag{21.2.13}$$

The quantity $\langle e, e' \rangle$ takes the values $0, \pm 1, \pm 2$, depending to which sets of the partition $V = G(x, y) \cup G(y, x) \cup G_{=}(x, y)$ (defined in (18.0.1)) the nodes x', y' belong. Observe that $\langle e, e' \rangle \neq 0$ if and only if e, e' are in relation by θ (defined in (18.0.2)). In the case when both e, e' belong to the set E_0 (recall (21.2.1)) with, say, $x_0 \in G(x, y) \cap G(x', y')$, then

$$\langle e, e' \rangle = \frac{1}{\lambda} |(Y \setminus X) \cap (Y' \setminus X')| \in \{0, 1, 2\},$$

if $x \mapsto X$ is a λ-embedding of G into a hypercube. In particular, $\langle e, e' \rangle = 2$ if and only if the edges e, e' correspond to the same proper atom $Y \setminus X = Y' \setminus X'$. For $e, e' \in E_0$, set

$$e \sim e' \text{ if } \langle e, e' \rangle = 2.$$

The relation $\sim$ is an equivalence relation on E_0. Clearly, the set of equivalence classes of E_0 under $\sim$ is in bijection with the set of proper atoms. One can define a graph $\mathcal{E}$ on the set of equivalence classes by letting two classes $\overline{e}$, $\overline{e}'$ be adjacent if $\langle e, e' \rangle = 1$ (the value of $\langle e, e' \rangle$ does not depend on the choice of e in the class $\overline{e}$ and of e' in the class $\overline{e}'$). The graph $\mathcal{E}$ clearly coincides with the atom graph $\Lambda(G)$. Let $\mathcal{E}_1, \ldots, \mathcal{E}_k$ denote the connected components of $\mathcal{E}$. Hence, each edge $e \in E_0$ is assigned to a node in one of the $\mathcal{E}_h$'s. We now see how to assign the other edges of G to some component $\mathcal{E}_h$. Let $e = (x, y)$ be an edge that does not belong to E_0, i.e., such that $d_G(x_0, x) = d_G(x_0, y) =: s$. Let $(x_0, x_1, \ldots, x_s = x)$ and $(y_0 = x_0, y_1, \ldots, y_s = y)$ be shortest paths joining x_0 to x and y in G and set $e_i := (x_{i-1}, x_i), f_i := (y_{i-1}, y_i)$ for $i = 1, \ldots, s$. Consider the subgraph of $\mathcal{E}$

induced by the set $\{\overline{e}_i, \overline{f}_i : i = 1, \ldots, s\}$. An analogue of Claim 21.2.6 shows that this graph consists of isolated nodes, cycles, and exactly one path. Moreover, the component $\mathcal{E}_h$ containing this path depends only on the edge (x, y) (not on the choice of the shortest paths from x_0 to x and y). This permits us to partition the edge set E of G into $E_1 \cup \ldots \cup E_k$, where E_h consists of the edges that are assigned to $\mathcal{E}_h$ by the above procedure. Then, let $\mathcal{G}_h$ denote the graph obtained by contracting the edges from $E \setminus E_h$. The graph $\mathcal{G}_h$ coincides with the graph $\overline{G}_h$ (up to renumbering of the factors).
So we have shown how to construct the graph $\overline{G}$ in an abstract way, not depending on the specific scale embedding of G. ∎

21.3 Proofs

Proof of Theorem 21.1.3. The existence of a graph $\widehat{G}$ satisfying Theorem 21.1.3 (i) follows from Claim 21.2.10. We prove the second part of Theorem 21.1.3. Let $\psi : x \mapsto X$ be a λ-embedding of G into a hypercube $H(\Omega)$. Suppose first that ψ assigns the given node x_0 to $\emptyset$. Using ψ, by the construction of Section 21.2, we obtain a graph $\widehat{G_\psi}$ which is λ-embedded into $H(\Omega)$ and is isomorphic to $\widehat{G}$ (by Remark 21.2.11). This gives the λ-embedding $\widehat{\psi}$ such that $\psi = \widehat{\psi}\widehat{\sigma}$. Suppose now that ψ assigns the set X_0 to the node x_0. Consider the λ-embedding $x \mapsto X \triangle X_0$, denoted by $\psi \triangle X_0$, of G into $H(\Omega)$. As $\psi \triangle X_0$ maps x_0 to $\emptyset$, we obtain $(\widehat{\psi \triangle X_0}) \triangle X_0$ for the embedding $\widehat{\psi}$. ∎

Proof of Corollaries 21.1.4 and 21.1.5. If G is an ℓ_1-graph, then the graph $\overline{G}$ is an ℓ_1-graph (by Claim 21.2.7) and $\overline{G}$ coincides with the canonical metric representation of G (by Remark 21.2.11). This shows Corollary 21.1.4. Corollary 21.1.5 is an immediate consequence of Corollary 21.1.4. ∎

Proof of Corollaries 21.1.6 and 21.1.7. The implication: $\widehat{G}$ is ℓ_1-rigid $\Longrightarrow$ G is ℓ_1-rigid follows from Theorem 21.1.3 (ii). Conversely, suppose that G is ℓ_1-rigid. We show that $\widehat{G}$ is ℓ_1-rigid. Consider a scale embedding $\widehat{\psi}_i$ of $\widehat{G}$ in the hypercube Ω_i, for $i = 1, 2$. We can suppose that Ω_1 and Ω_2 have the same cardinality (if not, add some redundant elements). We can also suppose that $\widehat{\psi}_1$ and $\widehat{\psi}_2$ have the same scale λ. (If, for $i = 1, 2$, $\widehat{\psi}_i$ has scale λ_i, then replace $\widehat{\psi}_i$ by $\widehat{\psi}'_i$, where $\widehat{\psi}'_1$ is the $\lambda_1\lambda_2$-embedding constructed from $\widehat{\psi}_1$ by replacing the elements of Ω_1 by disjoint sets each of cardinality λ_2 and $\widehat{\psi}'_2$ is the $\lambda_1\lambda_2$-embedding constructed from $\widehat{\psi}_2$ in the same way.) Then, $\psi_i := \widehat{\psi}_i\widehat{\sigma} : x \mapsto X_i$ is a λ-embedding of G into the hypercube $H(\Omega_i)$, for $i = 1, 2$. As G is ℓ_1-rigid, any two isometric ℓ_1-embeddings of G are equivalent (recall the definition from Chapter 18). It is not difficult to see that this implies the existence of a bijection $\alpha : \Omega_1 \longrightarrow \Omega_2$ and of a set $A \subseteq \Omega_2$ such that $X_2 = \varphi(X_1)$ for each node x of G where, for a subset $Z \subseteq \Omega_1$, we set

$$\varphi(Z) = \sigma(Z) \triangle A.$$

Using ψ_i, by the construction of Section 21.2, we obtain the graph $\widehat{G}_{\psi_i}$, which is λ-embedded into the hypercube $H(\Omega_i)$ via $\widehat{\psi}_i$. By the minimality of the graphs $\widehat{G}_{\psi_1}$,$\widehat{G}_{\psi_2}$ (see Claim 21.2.10), we deduce that φ establishes the equivalence of the embeddings $\widehat{\psi}_1$ and $\widehat{\psi}_2$. Hence, $\widehat{G}$ is ℓ_1-rigid. This shows Corollary 21.1.6. Corollary 21.1.7 now follows easily. Indeed, if G is ℓ_1-rigid, then $\widehat{G}$ is ℓ_1-rigid, which implies that each factor $\widehat{G}_h$ is ℓ_1-rigid (as a product of graphs is ℓ_1-rigid if and only if each factor is ℓ_1-rigid; see Proposition 7.5.2). Therefore, each $\widehat{G}_h$ is one of the following graphs: K_2, K_3, $K_{3\times 2}$, or $\frac{1}{2}H(m,2)$ for $m \geq 5$, which are all hypercube embeddable with scale 2. Therefore, G is hypercube embeddable with scale 2, i.e., G is an isometric subgraph of a half-cube graph. ∎

Proof of Corollary 21.1.8. Suppose G has n nodes. If some factor $\widehat{G}_h$ is a cocktail-party graph $K_{m\times 2}$, then $m < n$. Hence, by Lemma 7.4.4, $\widehat{G}_h$ is hypercube embeddable with scale 2^{k-1}, if $2^{k-1} < n-1 \leq 2^k$. All other factors are also hypercube embeddable with scale 2^{k-1} since $k \geq 2$ as $n \geq 4$. Hence, G is hypercube embeddable with scale 2^{k-1}, which implies that its minimum scale η satisfies: $\eta \leq 2^{k-1} < n-1$. ∎

Proof of Proposition 21.1.10. Let $G=(V,E)$ be a connected graph on n nodes and m edges. All the ingredients for constructing an algorithm permitting to recognize whether G is an isometric subgraph of a half-cube have been essentially given earlier, especially in Remark 21.2.12 (taking $\lambda = 2$). We describe the main steps of the proof.

Let x_0 be a given node of G. Given two edges e, e' in G, we remind the definition of the quantity $\langle e, e'\rangle$ from relation (21.2.13); recall that $\langle e, e'\rangle \in \{0, \pm 1, \pm 2\}$. Moreover, if $e := (x,y)$ and $e' := (x', y')$ are such that

$$d_G(x_0,y) = d_G(x_0,x)+1, \; d_G(x_0,y') = d_G(x_0,x')+1$$

(i.e., if e, e' belong to the set E_0 defined in (21.2.1)), then $\langle e, e'\rangle \geq 0$ in the case when G is an isometric subgraph of a half-cube graph. A "trick" used in Deza and Shpectorov [1996] in order to reduce the complexity is to consider a spanning tree rather than the entire set E_0. The algorithm consists of the following steps (i)-(v).

(i) Let $T=(V,E_T)$ be a spanning tree in G such that

$$d_T(x_0,x) = d_G(x_0,x) \text{ for all nodes } x \text{ in } G.$$

(Such a tree can be constructed in $O(m)$ time using a breadth first search algorithm.)

(ii) For any two edges $e=(x,y)$, $e'=(x',y') \in E_T$ (such that $x_0 \in G(x,y) \cap G(x',y')$) check whether $\langle e, e'\rangle \in \{0,1,2\}$. If not, then G cannot be isometrically embedded in a half-cube graph.

(iii) Define a relation $\sim$ on E_T by letting

$$e \sim e' \text{ if } \langle e, e'\rangle = 2.$$

Verify that this relation is an equivalence relation on E_T. If not, then G cannot be isometrically embedded in a half-cube graph.

(iv) Define a graph Σ on the set of equivalence classes of $(E_T, \sim)$, where there is an edge between two classes $\overline{e}$ and $\overline{e}'$ if $\langle e, e'\rangle = 1$. Check that this graph is well defined, i.e., that the value of $\langle e, e'\rangle$ does not depend on the choice of the elements e, e' in the classes. If not, then G cannot be isometrically embedded in a half-cube graph.

(v) Check whether Σ is a line graph. (This can be done in $O(m')$ time if Σ has m' edges (Lehot [1975]); note that $m' < n^2$.) If not, then G cannot be isometrically embedded in a half-cube graph.

Let H_Σ be a graph whose line graph is Σ. One can then construct an isometric embedding of G into a half-cube graph in the following way.

We label each node x of G by a set X in a recursive manner. First, label x_0 by $X_0 := \emptyset$. Let $e := (x, y)$ be an edge of G such that $x_0 \in G(x, y)$ and suppose that x has been already labeled by X. The equivalence class $\overline{e}$ (in $(E_T, \sim)$) is a node of Σ and, thus, corresponds to an edge $p(\overline{e})$ in H_Σ, $p(\overline{e})$ being a two-element set. We label the node y by the set $Y := X \cup p(\overline{e})$. Let us check that this labeling $x \mapsto X$ provides an isometric embedding of G into a half-cube graph.

Observe that

$$|p(\overline{e}) \cap p(\overline{e}')| = \langle e, e'\rangle$$

for all edges $e, e' \in E_T$. In particular, $p(\overline{e}) \cap p(\overline{e}') = \emptyset$ if and only if the edges e and e' are not in relation by θ. We first show that

$$|X| = 2d_G(x_0, x) \quad \text{for all } x \in V.$$

For this, let $(x_0, x_1, \ldots, x_p := x)$ be the path from x_0 to x in T, where $p := d_G(x_0, x)$. Let e_i denote the edge (x_{i-1}, x_i) for $i = 1, \ldots, p$. By the construction, x is labeled by the set

$$X := p(\overline{e}_1) \cup \ldots \cup p(\overline{e}_p).$$

Hence, $|X| = 2p = 2d_G(x_0, x)$, as the $p(\overline{e}_i)$'s are pairwise disjoint since the edges $e_1, \ldots, e_p$ are not in relation by θ. We now show that

$$|Y \triangle Y'| = 2d_G(y, y') \quad \text{for all } y, y' \in V.$$

We prove it by induction on the quantity $s := d_G(x_0, y) + d_G(x_0, y')$. We can suppose that y and y' are distinct from x_0. Let x be the predecessor of y on the path from x_0 to y in T, and let x' be the predecessor of y' on the path from x_0 to y' in T. Set $e := (x, y)$ and $e' := (x', y')$. By the induction assumption, we have

$$|X \triangle X'| = 2d_G(x, x'),\ |X \triangle Y'| = 2d_G(x, y'),\ |X' \triangle Y| = 2d_G(y, x')$$

or, equivalently,

$$\text{(a)} \qquad \begin{cases} |X \cap X'| = d_G(x_0, x) + d_G(x_0, x') - d_G(x, x'), \\ |X \cap Y'| = d_G(x_0, x) + d_G(x_0, y') - d_G(x, y'), \\ |X' \cap Y| = d_G(x_0, y) + d_G(x_0, x') - d_G(y, x'). \end{cases}$$

By the construction, $Y = X \cup p(\overline{e})$, $Y' = X' \cup p(\overline{e}')$ and, thus,

(b) $$|Y \cap Y'| = |X \cap Y'| + |X' \cap Y| - |X \cap X'| + |p(\overline{e}) \cap p(\overline{e}')|.$$

On the other hand,

$$|Y \triangle Y'| = |Y| + |Y'| - 2|Y \cap Y'| = 2d_G(x_0, y) + 2d_G(x_0, y') - 2|Y \cap Y'|.$$

Therefore, using (a), (b) and (21.2.13), we obtain

$$|Y \triangle Y'| = 2(d_G(x, y') + d_G(y, x') - d_G(x, x') - |p(\overline{e}) \cap p(\overline{e}')|) = 2d_G(y, y'),$$

which concludes the proof. ∎

21.4 More about ℓ_1-Graphs

We group here several additional facts and results on structural properties of ℓ_1-graphs.

We saw in Proposition 19.1.2 that every hypercube embeddable graph is ℓ_1-rigid. Hence, we have the following chain of implications:

G is an isometric subgraph of a hypercube $\Longrightarrow$ G is ℓ_1-rigid $\Longrightarrow$ G is an isometric subgraph of a half-cube graph

Several classes of graphs were shown to be ℓ_1-rigid in Deza and Laurent [1994a]; among them, the half-cube graph $\frac{1}{2}H(n,2)$ for $n \neq 3, 4$, the Johnson graph $J(n,d)$ for $d \neq 1$, the Petersen graph, the Shrikhande graph, the dodecahedron, the icosahedron, any weighted circuit. The method of proof is analogue to that of Proposition 19.1.2; namely, one shows that the path metric of the graph in question lies on a simplex face of the corresponding cut cone. This question of ℓ_1-rigidity is further investigated for other classes of graphs in Chepoi, Deza and Grishukhin [1996], Deza and Grishukhin [1996c], Deza, Deza and Grishukhin [1996].

An interesting fact is that, if an ℓ_1-graph G is not ℓ_1-rigid, then this is essentially due to the fact that complete graphs on at least four nodes are not ℓ_1-rigid. Indeed, it follows from Theorem 21.1.3 that any ℓ_1-embedding of G arises from an ℓ_1-embedding of its extension $\widehat{G}$. As $\widehat{G}$ is a Cartesian product of complete graphs, cocktail-party graphs and half-cube graphs, the variety of ℓ_1-embeddings of $\widehat{G}$ follows from the variety of ℓ_1-embeddings of its factors. But the half-cube graph is ℓ_1-rigid unless it coincides with K_4 or $K_{4\times 2}$. Moreover, any ℓ_1-embedding of $K_{n\times 2}$ arises from some ℓ_1-embedding of K_n, since the path metric of $K_{n\times 2}$ can be constructed from the path metric of K_n via the antipodal operation (recall Section 7.4). Therefore, the variety of ℓ_1-embeddings of $\widehat{G}$ and, hence, that of G, arises from the variety of ℓ_1-embeddings of the complete graph;

we will study in detail in Chapter 23 the variety of embeddings of the complete graph. In fact, we will see in Proposition 21.4.4 below that, if an ℓ_1-graph is not ℓ_1-rigid, then it must contain a clique on at least 4 nodes.

We have seen in Chapter 19 several structural characterizations for isometric subgraphs of hypercubes. We present below in Theorem 21.4.2 a structural characterization for isometric subgraphs of half-cube graphs. This result is due to Chepoi, Deza and Grishukhin [1996] and it can be seen as an analogue of Theorem 19.2.5 for isometric subgraphs of hypercubes. On the other hand, no result of this type is known for ℓ_1-graphs in general. Thus, the following problem is open.

Problem 21.4.1. *Find a structural characterization for ℓ_1-graphs (e.g., in terms of forbidden isometric subspaces).*

Such a characterization exists for some classes of graphs. For instance, we gave in Theorem 17.1.8 a structural characterization for the graphs with a universal node that are ℓ_1-graphs. We presented this result in Part II since its proof relies on the techniques of hypermetrics and Delaunay polytopes. In particular, we saw in Corollary 17.1.10 that, if G is a graph on $n \geq 28$ (resp. $n \geq 37$) nodes, then its suspension ∇G is an ℓ_1-graph if and only if ∇G is hypermetric (resp. ∇G satisfies the 5-gonal inequalities and is of negative type).

Let us now turn to the study of isometric subgraphs of half-cube graphs (that is, ℓ_1-graphs with scale 2). We need a definition. Given an integer $k \geq 0$, let $T_k = (X, d_k)$ denote the distance space on $X := \{a_0, a_1, a_2, a_3, a_4, b_0, b_1, b_2, b_3, b_4\}$ defined by

$$\begin{aligned}
&d_k(a_i, a_j) = d_k(b_i, b_j) := 1 \quad \text{for all } 0 \leq i < j \leq 4,\\
&d_k(a_0, b_0) := k + 2,\\
&d_k(a_i, b_i) := k \quad \text{for } i = 1, 2, 3, 4,\\
&d_k(x, y) := k + 1 \quad \text{elsewhere, i.e., on the pairs } (a_0, b_i), (b_0, a_i) \ (i = 1, 2, 3, 4)\\
&\qquad\qquad\qquad\qquad \text{and } (a_i, b_j) \ (i \neq j \in \{1, 2, 3, 4\}).
\end{aligned}$$

In the case $k = 0$, the distance space T_0 coincides (up to gate 0-extension) with the graphic metric space of the graph $K_6 \setminus e$, which has minimum scale 4 (recall Section 7.4). As $d_k = d_0 + k\delta(\{a_0, a_1, a_2, a_3, a_4\})$, every T_k is hypercube embeddable with scale 4. In fact, no T_k is hypercube embeddable with scale 2 and the distance spaces T_k turn out to be the only obstructions for isometric embeddability in half-cube graphs.

Theorem 21.4.2. *Let $G = (V, E)$ be an ℓ_1-graph. Then, G is an isometric subgraph of some half-cube graph if and only if its graphic metric space (V, d_G) does not contain T_k $(k \geq 0)$ as an isometric subspace.*

Proof. We start with verifying that the distance space $T_k = (X, d_k)$ $(k \geq 0)$ is not hypercube embeddable with scale 2. This is known for T_0. Suppose that T_k

admits a hypercube embedding with scale 2 and let $k \geq 1$ be the smallest index for which this is true. Consider a decomposition

$$2d_k = \sum_{\delta(S)\in\mathcal{C}_k} \delta(S)$$

where $\mathcal{C}_k$ is a collection of (not necessarily distinct) cut semimetrics on X. All cut semimetrics in $\mathcal{C}_k$ are d_k-convex by Lemma 4.2.8. Note that the only d_k-convex cut semimetric $\delta(S)$ with the property that $a_1 \in S$ and $b_1 \notin S$ is $\delta(S_0) := \delta(\{a_0, a_1, a_2, a_3, a_4\})$. Hence this cut semimetric occurs $2k$ times in $\mathcal{C}_k$. Now, $2d_k - 2\delta(S_0)$ coincides with $2d_{k-1}$ and admits a decomposition as a sum of cut semimetrics (as $k \geq 1$). This shows that T_{k-1} too admits a scale 2 hypercube embedding, yielding a contradiction.

Conversely, let $G = (V, E)$ be an ℓ_1-graph that is not an isometric subgraph of a half-cube graph. We show that (V, d_G) contains some T_k as an isometric subspace. By Corollary 21.1.5, G is an isometric subgraph of a Cartesian product of half-cube graphs and cocktail-party graphs. This product contains some $K_{m\times 2}$ with $m \geq 5$ (else, G would have scale 2). Say, G is an isometric subgraph of the graph $\Gamma := K_{m\times 2} \times H$, where $m \geq 5$ and H is a product of half-cube and cocktail-party graphs. Moreover, we can suppose that $K_{m\times 2}$ contains a subgraph $K_{m+1} \setminus e$ such that, for every node v of $K_{m+1} \setminus e$, the set $\{v\} \times V(H)$ contains at least one node of G (for, if not, one could have replaced $K_{m\times 2}$ by a smaller cocktail-party graph or by a complete graph). Denote by K the set of nodes of $K_{m+1} \setminus e$ forming a clique of size m. For every node v of $K_{m\times 2}$ we call the set $\{v\} \times V(H)$ a *fiber* in the product Γ. The following can be easily observed:

(a) Every union of fibers of the form: $\bigcup_{v\in K_0}\{v\} \times V(H)$ where $K_0 \subseteq K$, is convex (with respect to the path metric of Γ).

We claim:

(b) The set $V(G) \cap (\bigcup_{v\in K}\{v\} \times V(H))$ contains a clique of size m meeting each fiber $\{v\} \times V(H) (v \in K)$ in exactly one node.

For this, let $C \subseteq V(G) \cap (\bigcup_{v\in K}\{v\} \times V(H))$ be a clique of maximum size (whose elements all have the same H-coordinate). Suppose that $C \cap (\{w\} \times V(H)) = \emptyset$ for some $w \in K$. Note that every node $x \in V(G) \cap (\{w\} \times V(H))$ is at the same distance from all nodes in C; choose such x for which this distance is minimum. For every node $y \in C$ consider a shortest path in G from x to y; this path is entirely contained in the union of the two fibers containing x and y (by (a)). Say, this path is of the form $(x, \tilde{y}, \ldots, y)$. The node $\tilde{y}$ does not belong to the same fiber as x (by the minimality assumption on x). Thus, $\tilde{y}$ belongs to the fiber of y. Now, the nodes x and $\tilde{y}$ (for $y \in C$) form a clique of larger size than C. This shows that (b) holds.

Let $C \subseteq V(G)$ be a clique of size m meeting each fiber $\{v\} \times V(H)$ for $v \in K$. Denote by w, w' the two nonadjacent nodes in $K_{m+1} \setminus e$ with $w \in K$, $w' \notin K$. Let s denote the node of C lying in the fiber $\{w\} \times V(H)$. By assumption, the fiber $\{w'\} \times V(H)$ also contains some node of G. Every such node is at the same distance $k \geq 1$ from all nodes in $C \setminus \{s\}$; choose such a node t for which the distance k is minimum. Thus, $d_G(s,t) = k+1$ and $d_G(t,y) = k$ for all $y \in C\setminus\{s\}$. Consider a shortest path in G from t to $y \in C \setminus \{s\}$; it is of the form $(t, y', \ldots, y)$ where y' belongs to the same fiber as y by the minimality assumption on k. Therefore, the nodes t, y' (for $y \in C\setminus\{s\}$) form a clique and $d_G(y,y') = k-1$ for $y \in C \setminus \{s\}$. As $|C \setminus \{s\}| \geq 4$ we have found within $\{s,t\} \cup \{y,y' \mid y \in C \setminus \{s\}\}$ an isometric subspace of (V, d_G) of the form T_{k-1}. This concludes the proof. ∎

Corollary 21.4.3. *Let G be an ℓ_1-graph and suppose that G does not have a clique of cardinality 5. Then, G is an isometric subgraph of a half-cube graph.* ∎

Combining facts about ℓ_1-graphs established earlier in this chapter and in the proof of Theorem 21.4.2, one can show the following refinement of the result from Corollary 21.4.3.

Proposition 21.4.4. *Let G be an ℓ_1-graph that does not contain a clique of size 4, then G is ℓ_1-rigid (and, thus, is an isometric subgraph of a half-cube graph).*

Proof. Suppose that G is an ℓ_1-graph which is not ℓ_1-rigid. Following the notation from Section 21.2, let

$$G \hookrightarrow \overline{G} = \prod_{h=1}^{k} \overline{G}_h \hookrightarrow \widehat{G} = \prod_{h=1}^{k} \widehat{G}_h,$$

where $\prod_{h=1}^{k} \overline{G}_h$ is the canonical metric representation of G and each factor $\widehat{G}_h$ is a complete graph K_m ($m \geq 2$), a cocktail-party graph $K_{m\times 2}$ ($m \geq 2$), or a half-cube graph $\frac{1}{2}H(m,2)$ ($m \geq 5$). By Corollary 21.1.6, $\widehat{G}$ is not ℓ_1-rigid. Therefore, some factor $\widehat{G}_h$ is not ℓ_1-rigid and, thus, is K_m or $K_{m\times 2}$ with $m \geq 4$. This implies that $\overline{G}_h$ contains a clique of size $m \geq 4$ (see the proof of Claim 21.2.8). Now, we have that G is an isometric subgraph of $\overline{G}_h \times H$, where $H := \prod_{h' \neq h} \overline{G}_{h'}$. One can easily verify that the statement from relation (b) in the proof of Theorem 21.4.2 remains valid in the present situation. Thus, we have found a clique of size $m \geq 4$ in G. ∎

In particular, we find again that bipartite ℓ_1-graphs are ℓ_1-rigid but we also find, for instance, that tripartite ℓ_1-graphs are ℓ_1-rigid.

Let us now return to general ℓ_1-graphs. A question of interest is to classify ℓ_1-graphs within some restricted classes of graphs. Such a classification is known, for instance, for distance-regular graphs (recall Corollary 17.2.9). This question is studied in several papers for other classes of graphs, that we now mention briefly.

The graphs G for which both G and its complement $\overline{G}$ are ℓ_1-graphs are studied in Deza and Huang [1996a] and ℓ_1-embeddability of graphs related with some designs is considered in Deza and Huang [1996b].

Recently, Deza and Grishukhin [1996c] classify ℓ_1-graphs within polytopal graphs (that is, 1-skeleton graphs of polytopes) of a variety of well-known polytopes, such as semiregular, regular-faced polytopes, zonotopes, Delaunay polytopes of dimension ≤ 4, and several generalizations of prisms and antiprisms.

Deza, Deza and Grishukhin [1996] study the class of polytopal graphs arising from fullerenes. Fullerenes and their duals have many applications in chemistry, computer graphics, microbiology, architecture, etc. For instance, they occur as carbon molecules, spherical wavelets, virus capsids and geodesic domes. They can be defined in the following way. A fullerene F_n is a simple 3-dimensional polytope with n vertices that are arranged in 12 pentagons and $\frac{n}{2} - 10$ hexagons. Such polytopes can be constructed for every even $n \geq 20$, except $n = 22$. Among other results, Deza, Deza and Grishukhin [1996] show that any hypermetric fullerene graph is ℓ_1-embeddable (and thus has scale 2, by Proposition 21.4.5 below) and give an infinite family of fullerenes F_{20a^2} (a integer) which are ℓ_1-embeddable with an icosahedral group of symmetries.

Related work is made by Deza and Stogrin [1996] who study ℓ_1-embeddability of the (infinite) graphs arising as skeletons of plane tilings. In particular, they classify such ℓ_1-graphs for all semiregular and 2-uniform partitions of the plane, their duals and regular partitions of the hyperbolic plane.

Several authors study ℓ_1-graphs within the class of planar graphs. First, as an application of Corollary 21.4.3, we obtain:

Proposition 21.4.5. *Every planar ℓ_1-graph has scale* 2 *(that is, is an isometric subgraph of a half-cube graph).* ∎

Prisăkar, Soltan and Chepoi [1990] show that every planar graph satisfying the following conditions (i)-(iv) is an ℓ_1-graph. Suppose a plane drawing of G is given; then the conditions on G are: (i) every edge belongs to at most two faces; (ii) each interior face of G has at least five nodes; (iii) any two interior faces intersect in at most one edge; (iv) each interior node has degree ≥ 4.

Deza and Tuma [1996] characterize ℓ_1-graphs within the class of subdivided wheels. A *wheel* is the graph obtained from a circuit C by adding a new node adjacent to all nodes on C. For instance, K_4 is a wheel; Figure 31.3.12 (a) shows a wheel where C has length 6. Let us call *subdivided wheel* the graph which is obtained from a wheel by replacing some edges of the circuit C by paths. See Figure 21.4.7; there, the dotted lines indicate paths and the indication 'even' or 'odd' on a face means that the circuit bounding it has an even or odd length. The following result is proved in Deza and Tuma [1996].

Theorem 21.4.6. *A subdivided wheel is an ℓ_1-graph if and only if it is not one of the graphs shown in Figure* 21.4.7 (a),(b),(c). *Moreover, it is ℓ_1-rigid if and only if it is distinct from K_4.* ∎

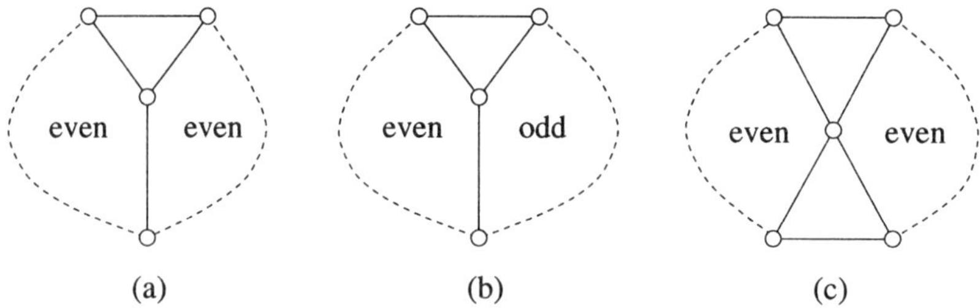

Figure 21.4.7: Subdivided wheels that are not ℓ_1-graphs

Chepoi, Deza and Grishukhin [1996] further study ℓ_1-graphs within the class of planar graphs satisfying the following two conditions: (i) each face of G is bounded by an isometric cycle; (ii) two interior faces meet in at most one edge (assuming a plane embedding is given). In particular, they show that every outerplanar[1] graph is an ℓ_1-graph. Another example of graphs satisfying (i), (ii) consists of *triangulations*; that is, planar graphs admitting a plane embedding in which every interior face is bounded by a cycle of length 3. Bandelt and Chepoi [1996c] show that every triangulation with the property that all interior vertices have degree larger than 5 is an ℓ_1-graph and, moreover, is ℓ_1-rigid. Moreover, by Proposition 21.4.4, every triangulation which is an ℓ_1-graph and does not have K_4 as an induced subgraph is ℓ_1-rigid.

[1]An *outerplanar graph* is a planar graph admitting a plane embedding in which all nodes lie on one face.

Part IV

Hypercube Embeddings and Designs

Introduction

In Part IV, we study the question of embedding semimetrics isometrically in the hypercube and, in particular, its link with the theory of designs.

Let $t \geq 1$ be an integer. A very simple metric is the *equidistant metric* on n points, denoted by $2t\mathbb{1}_n$, which takes the same value $2t$ on each pair of points. The metric $2t\mathbb{1}_n$ is obviously hypercube embeddable. Indeed, a hypercube embedding of $2t\mathbb{1}_n$ can be obtained by labeling the points by pairwise disjoint sets, each of cardinality t. A basic result established in Chapter 22 is that, for n large enough (e.g., for $n \geq t^2+t+3$), this embedding is essentially the unique hypercube embedding of $2t\mathbb{1}_n$. Moreover, for $n \geq t^2$, the existence of another hypercube embedding of $2t\mathbb{1}_n$ depends solely on the existence of a projective plane of order t. In Chapter 23, we further investigate how various hypercube embeddings of $2t\mathbb{1}_n$ arise from designs. We then consider in Chapter 24 some other classes of metrics for which we are able to characterize hypercube embeddability. Typically, these metrics have a small range of values so that one can still take advantage of the knowledge available for their equidistant submetrics. For instance, one can characterize the hypercube embeddable metrics with values in the set $\{a, 2a\}$, or $\{a, b, a+b\}$ (if two of $a, b, a+b$ are odd), where a, b are given integers. Moreover, this characterization yields a polynomial time algorithm for checking hypercube embeddability of such metrics. We recall that, for general semimetrics, it is an NP-hard problem to check whether a given semimetric is hypercube embeddable. Several additional results related to the notion of hypercube embeddability are grouped in Chapter 25, namely, on cut lattices, quasi h-points and Hilbert bases of cuts.

We now recall some definitions and terminology that we use in this part. Let d be a distance on the set $V_n := \{1, \ldots, n\}$. Then, d is said to be *hypercube embeddable* if there exist vectors $u_i \in \{0,1\}^m$ (for some $m \geq 1$) ($i \in V_n$) such that

(a) $$d(i,j) = \| u_i - u_j \|_1$$

for all $i, j \in V_n$. Let M denote the $n \times m$ matrix whose rows are the vectors $u_1, \ldots, u_n$; M is the *realization matrix* of the embedding $u_1, \ldots, u_n$ of d. Any matrix arising as the realization matrix of some hypercube embedding of d is called an *h-realization matrix* of d. Each vector u_i can be seen as the incidence vector of a subset A_i of $\{1, \ldots, m\}$. Hence,

(b) $$d(i,j) = |A_i \triangle A_j|$$

for all $i, j \in V_n$. We also say that the sets $A_1, \ldots, A_n$ form an *h-labeling* of d.

Clearly, if M is an h-realization matrix of d, we can assume without loss of generality that a row of M is the zero vector. This amounts to assuming that one of the points is labeled by the empty set in the corresponding h-labeling of d.

Let $\mathcal{B}$ denote the collection of subsets of V_n whose incidence vectors are the columns of M; $\mathcal{B}$ is a multiset, i.e., it may contain several times the same member. Then, (a) is equivalent to

$$\text{(c)} \qquad d = \sum_{B \in \mathcal{B}} \delta(B).$$

This shows again (recall Proposition 4.2.4) that hypercube embeddable semimetrics are exactly the semimetrics that can be decomposed as a nonnegative integer combination of cut semimetrics. If (c) holds, we also say that $\sum_{B \in \mathcal{B}} \delta(B)$ is a $\mathbb{Z}_+$-*realization* of d. It will be convenient to use both representations (a) (or (b)) and (c) for a hypercube embeddable semimetric d. So, we shall speak of a hypercube embedding (or of an h-labeling of d), and of a $\mathbb{Z}_+$-realization of d. This amounts basically to looking either to the rows, or to the columns of the matrix M.

Recall that d is said to be *h-rigid* if d has a unique $\mathbb{Z}_+$-realization or, equivalently, if d has a unique (up to a certain equivalence) hypercube embedding. Equivalent embeddings were defined in Section 4.3; we remind the definition below. Let $\mathcal{B}$ be a collection of subsets of V_n satisfying (c). If we apply the following operations to $\mathcal{B}$:

(i) delete or add to $\mathcal{B}$ the empty set or the full set V_n,

(ii) replace some $B \in \mathcal{B}$ by its complement $V_n \setminus B$,

then we obtain a new set family $\mathcal{B}'$ satisfying again (c). We then say that $\mathcal{B}$ and $\mathcal{B}'$ are *equivalent* (or define *equivalent* hypercube embeddings), as they yield the same $\mathbb{Z}_+$-realization of d.

Chapter 22. Rigidity of the Equidistant Metric

In this chapter we study h-rigidity of the equidistant metric $2t\mathbb{1}_n$ for n, t integers, $n \geq 3$, $t \geq 1$. As was already mentioned, $2t\mathbb{1}_n$ is hypercube embeddable. Indeed, a hypercube embedding of $2t\mathbb{1}_n$ is obtained by labeling the n points by pairwise disjoint sets, each of cardinality t. This embedding is called the *star embedding* of $2t\mathbb{1}_n$; it corresponds to the following $\mathbb{Z}_+$-realization:

$$2t\mathbb{1}_n = \sum_{1 \leq i \leq n} t\delta(\{i\}), \tag{22.0.1}$$

called the *star realization* of $2t\mathbb{1}_n$. The word "star" is used since each cut semimetric $\delta(\{i\})$ takes nonzero values on the pairs (i, j) for $j \in \{1, \ldots, n\} \setminus \{i\}$, i.e., on the edges of a star. Let us call a hypercube embedding of $2t\mathbb{1}_n$ *nontrivial* if it is not equivalent to the star embedding, i.e., if it provides a $\mathbb{Z}_+$-realization of $2t\mathbb{1}_n$ distinct from the star realization (22.0.1). Hence, the metric $2t\mathbb{1}_n$ is h-rigid if it has no nontrivial hypercube embedding.

For $n = 3$, the equidistant metric $2t\mathbb{1}_3$ is h-rigid. It is, in fact, ℓ_1-rigid since the cut cone CUT_3 is a simplex cone (indeed, CUT_3 is generated by the three linearly independent vectors $\delta(\{i\})$ for $i = 1, 2, 3$). For $n = 4$, the metric $2t\mathbb{1}_4$ is not h-rigid. Indeed, besides the star realization from (22.0.1), $2t\mathbb{1}_4$ admits the following $\mathbb{Z}_+$-realization:

$$2t\mathbb{1}_4 = t(\delta(\{1,2\}) + \delta(\{1,3\}) + \delta(\{1,4\})); \tag{22.0.2}$$

$2\mathbb{1}_4$ has no other $\mathbb{Z}_+$-realization. In fact, $2t\mathbb{1}_n$ is not ℓ_1-rigid for any $n \geq 4$ as, for instance,

$$2t\mathbb{1}_n = \frac{t}{n-2} \sum_{1 \leq i < j \leq n} \delta(\{i, j\})$$

is a decomposition of $2t\mathbb{1}_n$ as a nonnegative sum of cut semimetrics which is distinct from the decomposition in relation (22.0.1). On the other hand, as we see below, if n is large with respect to t, then $2t\mathbb{1}_n$ is h-rigid.

There is an easy way of constructing nontrivial hypercube embeddings for the metric $2t\mathbb{1}_n$, using (partial) projective planes. Let us first recall their definition.

A (finite) *projective plane of order* t, commonly denoted by $PG(2, t)$, consists of a pair $(X, \mathcal{L})$, where X is the set of *points* and $\mathcal{L}$ is a collection of subsets of X, called *lines*, satisfying:

(i) each line $L \in \mathcal{L}$ has cardinality $t+1$,

(ii) each point of X belongs to $t+1$ lines, and

(iii) any two distinct points of X belong to exactly one common line.

Hence, there is the same number t^2+t+1 of lines and points in $PG(2,t)$. Moreover,

(iv) any two distinct lines in $\mathcal{L}$ intersect in exactly one point.

In the case $t=1$ there are three points and three lines which are the possible pairs of points. We shall consider projective planes of order $t \geq 2$. It is well-known that projective planes exist of any order t that is a prime power. It is one of the major open problems in combinatorics to determine what are the possible values for the order of a projective plane. It is known that no projective plane exists with order 6 or 10; the case $t=10$ was settled using computer by Lam, Thiel and Swierzc [1989]. (See, e.g., Hall [1967] for more information on projective planes.)

A *partial projective plane of order* t is a pair $P=(X,\mathcal{L})$ satisfying (i) and (iv) above. We assume that every point lies on at least one line and P is said to be *trivial* if there exists a point lying on every line. It is easy to see that, if P is nontrivial, then $|\mathcal{L}| \leq t^2+t+1$ with equality if and only if P is a projective plane (of order t); moreover, every point is on at most $t+1$ lines. Clearly, a nontrivial partial projective plane of order $t=1$ has $|\mathcal{L}| \leq 3$ lines.

Hall [1977] shows that every nontrivial partial projective plane with sufficiently many lines can be extended to a projective plane. Namely,

Theorem 22.0.3. *Let $P=(X,\mathcal{L})$ be a nontrivial partial projective plane of order $t \geq 2$ with $|\mathcal{L}| \geq t^2-1$ lines. Then, P can be extended to a projective plane $P_1=(X_1,\mathcal{L}_1)$ of order t; that is, $X \subseteq X_1$ and $\mathcal{L} \subseteq \mathcal{L}_1$.* ∎

Let $P=(X,\mathcal{L})$ be a partial projective plane of order t with $n:=|\mathcal{L}|$ lines. Then, P yields a hypercube embedding of $2t\mathbb{1}_n$; namely, by labeling the elements of V_n by the lines in $\mathcal{L}$. Note that this embedding is nontrivial precisely if P is nontrivial. Therefore, the existence of a nontrivial partial projective plane of order t ensures that the metric $2t\mathbb{1}_n$ is not h-rigid. (Recall that $n \leq t^2+t+1$ in this case.) It ensures, moreover, that the metric $2t\mathbb{1}_{n+1}$ is not h-rigid. Indeed, let Z be a set of cardinality $t-1$ disjoint from X. If we label the elements of V_n by the sets $L \cup Z$ for $L \in \mathcal{L}$ and the remaining element of $V_{n+1} \setminus V_n$ by the empty set, then we obtain a nontrivial hypercube embedding of $2t\mathbb{1}_{n+1}$.

In fact, for n large enough, the existence of a nontrivial hypercube embedding of $2t\mathbb{1}_n$ depends only on the existence of a (partial) projective plane of order t with a suitable number of lines. The following two theorems were proved by Hall [1977].

Theorem 22.0.4. *Let $n \geq t^2 \geq 4$. The metric $2t\mathbb{1}_n$ is not h-rigid if and only if $n \leq t^2+t+2$ and there exists a projective plane of order t.*

Theorem 22.0.5. *Let $n \geq \frac{1}{2}(t+2)^2$ with $t \geq 2$. If $2t\mathbb{1}_n$ is not h-rigid then there exists a nontrivial partial projective plane of order t with $n-1$ lines.*

In fact, Theorem 22.0.4 can be derived easily from Theorems 22.0.3 and 22.0.5, as we indicate below. As an application of Theorem 22.0.5 we have the following Theorems 22.0.6 and 22.0.7 which were obtained earlier, respectively, by Deza [1973b] and van Lint [1973]. (The case $t = 1$ in Theorem 22.0.6 is not covered by Theorem 22.0.5 but can be very easily checked directly.) Another case of h-rigidity for the metric $2t\mathbb{1}_n$ will be given later in Corollary 23.2.5.

Theorem 22.0.6. *If $n \geq t^2+t+3$ and $t \geq 1$, then $2t\mathbb{1}_n$ is h-rigid, i.e., the only $\mathbb{Z}_+$-realization of $2t\mathbb{1}_n$ is the star realization from* (22.0.1). ∎

Theorem 22.0.7. *Let $n = t^2+t+2$ with $t \geq 3$. If the metric $2t\mathbb{1}_n$ is not h-rigid, then there exists a projective plane of order t.* ∎

Proof of Theorem 22.0.4. Let $n \geq t^2 \geq 4$ and suppose that $2t\mathbb{1}_n$ is not h-rigid. We show that there exists a projective plane of order t. If $t \geq 5$, this follows immediately from Theorems 22.0.3 and 22.0.5, as $t^2 \geq \frac{1}{2}(t+2)^2$; on the other hand, projective planes are known to exist for any order $t \in \{2,3,4\}$. We now verify that $n \leq t^2+t+2$. For this, note that $n \geq t^2+t+3$ implies $n \geq \frac{1}{2}(t+2)^2$; then a nontrivial partial projective plane of order t with $n-1$ lines exists (by Theorem 22.0.5), which implies that $n-1 \leq t^2+t+1$, yielding a contradiction. Conversely, it is immediate to verify that a projective plane of order t yields a nontrivial hypercube embedding of $2t\mathbb{1}_n$ when $n \geq t^2$. ∎

We now turn to the proof of Theorem 22.0.5. We will use the following notation. Let M be a binary $n \times m$ matrix which is an h-realization matrix of $2t\mathbb{1}_n$. Without loss of generality, we can suppose that the first row of M is the zero vector. Then, every other row of M has $2t$ units and any two rows (other than the first one) have t units in common. Given a nonzero row u_0 of M, $M \oplus u_0$ denotes the matrix obtained from M by replacing each row by its sum modulo 2 with u_0. Clearly, $M \oplus u_0$ is again an h-realization matrix of $2t\mathbb{1}_n$ with one zero row. Moreover, $M \oplus u_0$ provides a nontrivial hypercube embedding of $2t\mathbb{1}_n$ if and only if the same holds for M. We start with a preliminary result on the number r of units in the columns of M; the inequality $r(n-r) \leq nt$ in Lemma 22.0.8 was proved by Deza [1973b] and the equality case was analyzed by Hall [1977].

Lemma 22.0.8. *Let r denote the number of units in a column of M. Then, $r(n-r) \leq nt$, implying that*

$$\min(r, n-r) \leq \frac{1}{2}(n - \sqrt{n^2-4nt}).$$

Moreover, $r = \frac{n}{2} = 2t$ if $r(n-r) = nt$.

Proof. Let w be a column of M, let r denote the number of 1's in w, and let ρ denote the number of columns of M identical to w. Let M' denote the $n \times (m-\rho)$ denote the submatrix obtained from M by deleting these ρ columns, and let d' denote the distance on n points defined by letting d'_{ij} denote the Hamming distance between the i-th and j-th rows of M'. We can suppose that the first $n-r$ entries of w are equal to 0 and its last r entries are equal to 1. Then,

$$\begin{cases} d'_{ij} = 2t & \text{if } 1 \leq i < j \leq n-r, \text{ or } n-r+1 \leq i < j \leq n, \\ d'_{ij} = 2t - \rho & \text{if } 1 \leq i \leq n-r < j \leq n. \end{cases}$$

Consider the inequality of negative type:

$$\text{(a)} \quad \begin{array}{l} Q_n(-r,\ldots,-r,n-r,\ldots,n-r)^T x := \\ \sum\limits_{1 \leq i<j \leq n-r} r^2 x_{ij} + \sum\limits_{n-r+1 \leq i<j \leq n} (n-r)^2 x_{ij} - \sum\limits_{1 \leq i \leq n-r<j \leq n} r(n-r) x_{ij} \leq 0. \end{array}$$

As d' is hypercube embeddable by construction, d' satisfies the above inequality (recall Section 6.1). We deduce from it that $\rho r(n-r) \leq nt$, which implies

$$r(n-r) \leq nt.$$

From the latter relation follows immediately that

$$\min(r, n-r) \leq \frac{1}{2}(n - \sqrt{n^2 - 4nt}).$$

Suppose now that $r(n-r) = nt$. Then, $\rho = 1$. For each column v of M distinct from w, let a_v (resp. b_v) denote its number of units in the first $n-r$ rows (resp. in the last r rows). From the fact that d' satisfies the inequality (a) at equality, we deduce that

$$r a_v = (n-r) b_v.$$

Note that

$$\sum_v a_v(n-r-a_v) = \sum_{1 \leq i<j \leq n-r} d'_{ij} = 2t\binom{n-r}{2},$$

$$\sum_v b_v(r-b_v) = \sum_{n-r+1 \leq i<j \leq n} d'_{ij} = 2t\binom{r}{2},$$

where the sums are taken over all columns v of M distinct from w. As

$$\sum_v \frac{a_v}{n-r}(1 - \frac{a_v}{n-r}) = \sum_v \frac{b_v}{r}(1 - \frac{b_v}{r}),$$

we obtain that $\frac{2t}{(n-r)^2}\binom{n-r}{2} = \frac{2t}{r^2}\binom{r}{2}$, which yields $r = \frac{n}{2}$. Finally, $r(n-r) = \frac{n^2}{4} = nt$, implying $n = 4t$. ∎

Proof of Theorem 22.0.5. Note first that we can clearly assume that $n > \frac{1}{2}(t+2)^2$ when $t = 2$ (as a projective plane of order 2 does exist). Let M be an h-realization matrix of $2t\mathbb{1}_n$ providing a nontrivial hypercube embedding of $2t\mathbb{1}_n$. We assume that the first row of M is the zero vector. Let τ denote the maximum value taken by $\min(r, n-r)$, where r is the number of units in a column of M. Hence, each column of M has either $\leq \tau$ units, or $\geq n-\tau$ units. Call a column *heavy* if it has $\geq n-\tau$ units and *light* otherwise. We know from Lemma 22.0.8 that $\tau(n-\tau) \leq nt$, i.e., $\tau \leq \frac{1}{2}(n-\sqrt{n^2-4nt})$. Note that

$$\text{(a)} \qquad \frac{1}{2}(n-\sqrt{n^2-4nt}) < t+2 \Longleftrightarrow n > \frac{1}{2}(t+2)^2$$

and equality holds simultaneously in both inequalities. We will use this fact later in the proof.

Let us suppose for a contradiction that there does not exist a nontrivial partial projective plane of order t with $n-1$ lines. We claim:

$$\text{(22.0.9)} \qquad \text{If } \tau \leq t+1, \text{ then } n \leq \frac{1}{2}(t^2+3t+4),$$

$$\text{(22.0.10)} \qquad n \leq \frac{1}{2}(t+2)^2.$$

We first show that (22.0.9) holds. For this, suppose that $\tau \leq t+1$ and $n > \frac{1}{2}(t^2+3t+4)$. Then,

(c) there are at most $t+1$ heavy columns.

For, suppose that there are at least $t+2$ heavy columns in M. Consider the submatrix Y of M consisting of $t+2$ heavy columns. Then, the total number f of units in Y satisfies: $f \geq (t+2)(n-\tau) \geq (t+2)(n-t-1)$ (by counting units per columns). We now count the units in Y per rows. If Y has an all-ones row then every other nonzero row has at most t units and, thus, $f \leq t+2+t(n-2) = (t+2)(t+1)+t(n-t-4)$. Else, at most $t+2$ rows in Y have $t+1$ units and, thus, $f \leq (t+2)(t+1)+t(n-t-3)$. In both cases, we obtain: $(t+2)(n-t-1) \leq (t+2)(t+1)+t(n-t-3)$, which contradicts the fact that $n > \frac{1}{2}(t^2+3t+4)$. This shows (c). Next, we claim:

(d) Every nonzero row in M has at least $t-1$ units in the heavy columns.

Indeed, suppose that a nonzero row u_0 of M has at most $t-2$ units in the heavy columns. Then, the matrix $M \oplus u_0$ (defined as mentioned above by replacing each row in M by its sum modulo 2 with u_0) is an h-realization matrix of $2t\mathbb{1}_n$ having $t+2$ heavy columns, thus contradicting (c).

Hence, M has $t-1$, t, or $t+1$ heavy columns. Suppose first that M has $t-1$ heavy columns. Then, by (d), each nonzero row of M has units in all heavy columns. Therefore, the submatrix of M restricted to the light columns (and deleting the first zero row) is the incidence matrix of a nontrivial partial projective plane of order t with $n-1$ lines, in contradiction with our assumption. Suppose now that M has t heavy columns. Then, there is a row u_0 of M having exactly $t-1$ units in the heavy columns (else, M would correspond to the star embedding of $2t\mathbb{1}_n$). But, in this case, the matrix $M \oplus u_0$ has $t+2$ heavy columns, contradicting (c). Finally, suppose that there are $t+1$ heavy columns. If some row u_0 of M has $t+1$ units (resp. $t-1$ units) in the heavy columns, then $M \oplus u_0$ has $t-1$ (resp. $t+3$) heavy columns; the first case has just been excluded above and the second one is forbidden by (c). Hence, every nonzero row of M has t units in the heavy columns. Let x denote the binary vector indexed by the columns of M and having ones precisely in the positions of the heavy columns. Then, $M \oplus x$ is the incidence matrix of a nontrivial partial projective plane of order t with n lines. We have again a contradiction with our assumption. Therefore, relation (22.0.9) holds.

We now verify (22.0.10). Indeed, if $n > \frac{1}{2}(t+2)^2$, then $\tau \leq t+1$ by (a) and thus, using (22.0.9), $n \leq \frac{1}{2}(t^2+3t+4) < \frac{1}{2}(t+2)^2$, a contradiction.

Therefore, we have obtained that $n = \frac{1}{2}(t+2)^2$ and $t \geq 3$. By (22.0.9), this implies that $\tau \geq t+2$. On the other hand, $\tau \leq t+2$ by (a). Hence, $\tau = t+2$. Thus, $\tau(n-\tau) = nt$ which, by Lemma 22.0.8, implies that $\tau = \frac{n}{2} = 2t$. That is, $t = 2$, while we had assumed that $t \geq 3$. This concludes the proof. ∎

Consider, for instance, the case $t = 6$. By Theorem 22.0.4, the metric $12\mathbb{1}_n$ is h-rigid if $n \geq 36$ (as $PG(2,6)$ does not exist). It is, in fact, h-rigid for all $n \geq 33$ as stated in the next result, proved by Hall, Jansen, Kolen and van Lint [1977].

Proposition 22.0.11. *The equidistant metric $12\mathbb{1}_n$ is h-rigid for all $n \geq 33$.* ∎

The h-rigidity result from Theorem 22.0.6 was extended by Deza, Erdös and Frankl [1978] to the class of metrics of the form $\sum_{1\leq i\leq n} t_i\delta(\{i\})$ for $t_1,\ldots,t_n \in \mathbb{Z}_+$; the case $t_1 = \ldots = t_n = t$ corresponding to the case of the equidistant metric $2t\mathbb{1}_n$.

Theorem 22.0.12. *Let $t_1,\ldots,t_n$ be positive integers. If n is large with respect to $\max(t_1,\ldots,t_n)$, then the metric $\sum_{1\leq i\leq n} t_i\delta(\{i\})$ is h-rigid.* ∎

Chapter 23. Hypercube Embeddings of the Equidistant Metric

We study in this chapter how to construct various hypercube embeddings for the equidistant metric $2t\mathbb{1}_n$ from designs. A $\mathbb{Z}_+$-realization of $2t\mathbb{1}_n$ consists of a family $\mathcal{B}$ of (not necessarily distinct) subsets of V_n such that

$$\sum_{B\in\mathcal{B}} \delta(B) = 2t\mathbb{1}_n.$$

Given $i_0 \in V_n$, we can suppose without loss of generality that $i_0 \not\in B$ for all $B \in \mathcal{B}$ (replacing if necessary B by $V_n \setminus B$). Then, $\mathcal{B}$ is a collection of subsets of V_{n-1} satisfying:

(i) each point of V_{n-1} belongs to $2t$ members of $\mathcal{B}$, and

(ii) any two distinct points of V_{n-1} belong to t common members of $\mathcal{B}$.

Such a set family $\mathcal{B}$ is known as a $(2t, t, n-1)$-design. Therefore, the hypercube embeddings of $2t\mathbb{1}_n$ are nothing but special classes of designs. We review in Section 23.1 some known results on designs and we state precisely the link with hypercube embeddings of the equidistant metric in Section 23.2. Results on the minimum h-size of the equidistant metric are grouped in Section 23.3. We describe all the hypercube embeddings of $2t\mathbb{1}_n$ for small n or t in Section 23.4. Much of the exposition in this chapter follows Deza and Laurent [1993c].

23.1 Preliminaries on Designs

23.1.1 (r, λ, n)-Designs and BIBD's

Let $\mathcal{B}$ be a collection of (not necessarily distinct) subsets of V_n. The sets $B \in \mathcal{B}$ are called *blocks*. Let r, k, λ be positive integers. Consider the following properties:

(i) Each point of V_n belongs to r blocks.

(ii) Any two distinct points of V_n belong to λ common blocks.

(iii) Each block has cardinality k.

Clearly, if (ii),(iii) hold, then (i) holds with

(23.1.1) $$r = \lambda\frac{n-1}{k-1}$$

and the total number b of blocks in $\mathcal{B}$ (counting multiplicities) is given by

$$\text{(23.1.2)} \qquad b = \frac{rn}{k} = \lambda\frac{n(n-1)}{k(k-1)}.$$

The multiset $\mathcal{B}$ is called a (r, λ, n)-*design* if (i),(ii) hold with $0 < \lambda < r$. $\mathcal{B}$ is said to be *trivial* if $\mathcal{B}$ consists of the following blocks: V_n repeated λ times and, for each $i \in V_n$, the block $\{i\}$ repeated $r - \lambda$ times. In fact, if n is large with respect to r and λ, then every (r, λ, n)-design is trivial (this follows, e.g., from the rigidity results of Chapter 22).

The multiset $\mathcal{B}$ is called a (n, k, λ)-*BIBD* if (i),(ii),(iii) hold with $\lambda > 0$, $1 < k < n-1$. (BIBD stands for balanced incomplete block design.) A (n, k, λ)-BIBD is said to be *symmetric* if $r = k$ holds or, equivalently, the number of blocks b is equal to the number n of points.

Let $\mathcal{B}$ be a (n, k, λ)-BIBD. Then, the collection

$$\mathcal{B}^* := \{V_n \setminus B \mid B \in \mathcal{B}\}$$

is a $(n, k' := n-k, \lambda' := b - 2r + \lambda)$-BIBD, called the *dual* of $\mathcal{B}$. (Note that $1 < k' < n-1$ and $(n-1)(b-2r+\lambda) = (b-r)(n-k-1)$, which permits to check that $\lambda' > 0$.) If $\mathcal{B}$ is symmetric, then $\mathcal{B}^*$ too is symmetric. For instance, the dual of $PG(2,t)$ is a symmetric (t^2+t+1, t^2, t^2-t)-BIBD.

The following result is due to Ryser [1963].

Theorem 23.1.3. *Let $\mathcal{B}$ be a (r, λ, n)-design with b blocks. Then, $b \geq n$ holds, with equality if and only if $\mathcal{B}$ is a symmetric (n, r, λ)-BIBD.*

Proof. Let A denote the incidence matrix of $\mathcal{B}$, i.e., A is the $n \times b$ matrix with entries $a_{i,B} = 1$ if $i \in B$ and $a_{i,B} = 0$ if $i \notin B$, for $i \in V_n$, $B \in \mathcal{B}$. Suppose that $b < n$. Let M denote the $n \times n$ matrix obtained by adding $n - b$ zero columns to A. Then,

$$MM^T = \lambda J + (r - \lambda)I,$$

where J is the all-ones matrix and I the identity matrix. One can check that the eigenvalues of MM^T are $r + (n-1)\lambda$ and $r - \lambda$ (with multiplicity $n-1$), which shows that M is nonsingular. This contradicts the fact that M has a zero column. Hence, we have shown that $b \geq n$. Suppose now that $b = n$. We show that each block of $\mathcal{B}$ has cardinality r. From the above argument, the matrix A is an $n \times n$ matrix satisfying

$$\text{(a)} \qquad AA^T = \lambda J + (r-\lambda)I, \quad \text{and} \quad AJ = rJ.$$

Hence,

$$A^{-1}J = r^{-1}J \ \text{ and } AA^TJ = (\lambda n + r - \lambda)J,$$

implying

$$\text{(b)} \qquad A^TJ = (\lambda n + r - \lambda)r^{-1}J, \text{ i.e., } JA = (\lambda n + r - \lambda)r^{-1}J.$$

Therefore,

$$JAJ = (\lambda n + r - \lambda)r^{-1}nJ.$$

But, $JAJ = rnJ$ from (a), which implies

$$\text{(c)} \qquad r - \lambda = r^2 - \lambda n.$$

Substituting (c) in (b), we obtain $JA = rJ$. This shows that each block of $\mathcal{B}$ has size r. Hence, $\mathcal{B}$ is a symmetric (n, r, λ)-BIBD. ∎

Clearly, from (23.1.2), a necessary condition for the existence of a (n, k, λ)-BIBD is the following divisibility condition:

$$\text{(23.1.4)} \qquad k-1 \text{ is a divisor of } \lambda(n-1) \text{ and } k(k-1) \text{ is a divisor of } \lambda n(n-1).$$

This condition is, in some cases, already sufficient for the existence of a (n, k, λ)-BIBD. A few such cases are described in Theorem 23.1.5 below; assertion (i) is proved in Wilson [1975], (ii) in Hanani [1975], and (iii) in Mills [1990].

Theorem 23.1.5.

(i) *Suppose that* (23.1.4) *holds and that* n *is large with respect to* k *and* λ. *Then, there exists a* (n, k, λ)-BIBD.

(ii) *For* $k \leq 5$, *a* (n, k, λ)-BIBD *exists whenever* (23.1.4) *holds with the single exception:* $n = 15, k = 5, \lambda = 2$. *For* $k = 6$, $\lambda \geq 2$, *a* $(n, 6, \lambda)$-BIBD *exists whenever* (23.1.4) *holds with the single exception:* $n = 21, \lambda = 2$.

(iii) *For* $k = 6, \lambda = 1$, *a* $(n, 6, 1)$-BIBD *exists whenever* (23.1.4) *holds with the possible exception of* 95 *undecided cases (including* $n = 46, 51, 61, 81, 141, \ldots, 5391, 5901$*).* ∎

Two important cases of parameters for a symmetric BIBD are:

(i) The $(t^2 + t + 1, t + 1, 1)$-BIBD, which is nothing but the projective plane of order t, denoted by $PG(2, t)$.

(ii) The $(4t-1, 2t, t)$-BIBD, also known as the *Hadamard design* of order $4t-1$.

Recall that Hadamard designs are in one-to-one correspondence with Hadamard matrices. Namely, a *Hadamard matrix* is an $n \times n$ ± 1-matrix A such that $AA^T = nI$. Its order n is equal to 1, 2 or $4t$ for some $t \geq 1$. We can suppose without loss of generality that all entries in the first row and in the first column of A are equal to 1. Replace each -1 entry of A by 0 and delete its first row and column. We obtain a $(4t - 1) \times (4t - 1)$ binary matrix whose columns are the incidence vectors of the blocks of a Hadamard design of order $4t - 1$.

It is conjectured that Hadamard matrices of order $4t$ exist for all $t \geq 1$. This was proved for $t \leq 106$. (For more information on Hadamard matrices see, e.g., Geramita and Seberry [1979] and Wallis [1988].)

Remark 23.1.6. The parameters (k, λ) with $3 \leq k \leq 15$ for which there exists a symmetric (n, k, λ)-BIBD (then, $n = 1 + \frac{k(k-1)}{\lambda}$, by (23.1.1)) have been completely classified (with the exception of $k = 13, \lambda = 1$ corresponding to the question of existence of $PG(2, 12)$) (see Biggs and Ito [1981]). Besides the parameters corresponding to a projective plane, or to a Hadamard design, or to a dual of them, a symmetric (n, k, λ)-BIBD exists if and only if (n, k, λ) is one of the following list: $(16, 6, 2)$, $(37, 9, 2)$, $(25, 9, 3)$, $(16, 10, 6)$ (which is dual to the case $(16, 6, 2)$), $(56, 11, 2)$, $(31, 10, 3)$, $(45, 12, 3)$, $(79, 13, 2)$, $(40, 13, 4)$, $(71, 15, 3)$ and $(36, 15, 6)$. ∎

A useful notion is that of extension of a design. Let $\mathcal{B}$ be a collection of subsets of V_n and let $i_0 \notin V_n$. Given an integer s, the *s-extension* of $\mathcal{B}$ is the collection $\mathcal{B}'$ whose blocks are the blocks of $\mathcal{B}$ together with the block $\{i_0\}$ repeated s times.

23.1.2 Intersecting Systems

Let $\mathcal{A}$ be a collection of subsets of a finite set and let r, λ be positive integers. Then, $\mathcal{A}$ is called a *(r, λ)-intersecting system* if $|A| = r$ for all $A \in \mathcal{A}$ and $|A \cap B| = \lambda$ for all distinct $A, B \in \mathcal{A}$. The maximum cardinality of a (r, λ)-intersecting system consisting of subsets of V_b is denoted by $f(r, \lambda; b)$.

$\mathcal{A}$ is called a *Δ-system* with *center* K and *parameters* (r, λ) if $|K| = \lambda$, $|A| = r$ for all $A \in \mathcal{A}$, and $A \cap B = K$ for all distinct $A, B \in \mathcal{A}$. Clearly, if $\mathcal{A}$ consists of subsets of V_b, then $|\mathcal{A}| \leq \frac{b-\lambda}{r-\lambda}$.

Remark 23.1.7. (r, λ, n)-designs and (r, λ)-intersecting systems are basically the same objects. Namely, let M be a $n \times b$ binary matrix, let $\mathcal{B}$ denote the family of subsets of V_n whose incidence vectors are the columns of M, and let $\mathcal{A}$ denote the family of subsets of V_b whose incidence vectors are the rows of M. Then, $\mathcal{B}$ is a (r, λ, n)-design if and only if $\mathcal{A}$ is a (r, λ)-intersecting system of cardinality n. Moreover, $\mathcal{B}$ is trivial if and only if $\mathcal{A}$ is a Δ-system. As a reformulation of Theorem 23.1.3, $f(r, \lambda; b) \leq b$ holds. These two terminologies of (r, λ, n)-designs and intersecting systems are commonly used in the literature; this is why we present them both here. Moreover, we will need intersecting systems in Section 24.3.

Note also that partial projective planes of order t (with n lines) and $(t+1, 1)$-intersecting systems (with n members) are exactly the same notions and they correspond to $(t + 1, 1, n)$-designs (trivial partial projective planes or designs corresponding to Δ-systems). ∎

By the above remark, intersecting systems arise as the h-labelings of the equidistant metric. Namely,

Proposition 23.1.8. *There is a one-to-one correspondence between the h-labelings of the equidistant metric $2t\mathbb{1}_n$ and the $(2t, t)$-intersecting systems of*

cardinality $n-1$.

Proof. Indeed, in any h-labeling of $2t\mathbb{1}_n$, we may assume that one of the points is labeled by $\emptyset$ and then the sets labeling the remaining $n-1$ points are the members of a $(2t,t)$-intersecting system. ∎

Hence, Theorem 23.1.9 below from Deza [1973b] follows as a reformulation of Theorem 22.0.6.

Theorem 23.1.9. *Let* $t \geq 1$ *be an integer and let* $\mathcal{A}$ *be a* $(2t,t)$*-intersecting system. If* $|\mathcal{A}| \geq t^2+t+2$*, then* $\mathcal{A}$ *is a* Δ*-system.* ∎

As an application of Theorem 23.1.9, Deza [1974] proved the following result, solving a conjecture of Erdös and Lovász.

Theorem 23.1.10. *Let* $t \geq 1$ *be an integer and let* $\mathcal{A}$ *be a collection of subsets of a finite set such that* $|A \cap B| = t$ *for all* $A \neq B \in \mathcal{A}$*. Set* $k := \max(|A| : A \in \mathcal{A})$*. If* $|\mathcal{A}| \geq k^2-k+2$*, then* $\mathcal{A}$ *is a* Δ*-system.* ∎

We conclude with an easy application, that will be needed later.

Lemma 23.1.11. *Let* $k,t \geq 1$ *be integers such that* $t < k^2+k+1$ *and let* $\mathcal{A}$ *be a* $(k+t,t)$*-intersecting system. If* $|\mathcal{A}| \geq k^2+k+3$*, then* $\mathcal{A}$ *is a* Δ*-system.*

Proof. Let $A_1 \in \mathcal{A}$ and set $\mathcal{A}' := \{A\Delta A_1 \mid A \in \mathcal{A}\setminus\{A_1\}\}$. One checks easily that $\mathcal{A}'$ is a $(2k,k)$-intersecting system with $|\mathcal{A}'| \geq k^2+k+2$. By Theorem 23.1.9, $\mathcal{A}'$ is a Δ-system. Let K denote its center, $|K| = k$. Let $A \in \mathcal{A}, A \neq A_1$. Set $\alpha := |A_1 \cap K|$, then $|A_1 \cap ((A\Delta A_1) \setminus K)| = k - \alpha$ since $A_1 \cap (A\Delta A_1) = A_1 \setminus A$ has cardinality k. If $\alpha \leq k-1$, then

$$k+t = |A_1| \geq \alpha + |\mathcal{A}'|(k-\alpha) \geq \alpha + (k^2+k+2)(k-\alpha),$$

implying $t \geq (k-\alpha)(k^2+k+1)$, contradicting the assumption on t. Hence, $\alpha = k$, i.e., $A_1 \setminus A = K$ and, thus, $A_1 \cap A = A_1 \setminus K$. This shows that $\mathcal{A}$ is a Δ-system. ∎

23.2 Embeddings of $2t\mathbb{1}_n$ and Designs

Let $t,n \geq 1$ be integers. Every $\mathbb{Z}_+$-realization of $2t\mathbb{1}_n$ is of the form

$$2t\mathbb{1}_n = \sum_{B\in\mathcal{B}} \delta(B), \tag{23.2.1}$$

where $\mathcal{B}$ is a collection of (not necessarily distinct) subsets of V_n. Let $k \geq 1$ be an integer. The realization (23.2.1) is said to be *k-uniform* if $|B| = k, n-k$ for all $B \in \mathcal{B}$. It is very easy to construct $\mathbb{Z}_+$-realizations of the equidistant metric from designs.

For instance, let $\mathcal{B}$ be a (r, λ, n)-design. Then,

$$\sum_{B \in \mathcal{B}} \delta(B) = 2(r - \lambda)\mathbb{1}_n.$$

Moreover, if $r \geq 2\lambda$, then the $(r - 2\lambda)$-extension of $\mathcal{B}$ yields a $\mathbb{Z}_+$-realization of $2(r - \lambda)\mathbb{1}_n$, namely,

$$\sum_{B \in \mathcal{B}} \delta(B) + (r - 2\lambda)\delta(\{i_0\}) = 2(r - \lambda)\mathbb{1}_{n+1},$$

where $V_{n+1} \setminus V_n = \{i_0\}$. In particular, each $(t + 1, 1, n)$-design yields a $\mathbb{Z}_+$-realization of $2t\mathbb{1}_n$ and its $(t - 1)$-extension yields a realization of $2t\mathbb{1}_{n+1}$. Also, the 0-extension of a $(2t, t, n - 1)$-design gives a $\mathbb{Z}_+$-realization of $2t\mathbb{1}_n$.

If $\mathcal{B}$ is a (n, k, λ)-BIBD, then (23.2.1) is a $\mathbb{Z}_+$-realization of the equidistant metric $2\lambda\frac{n-k}{k-1}\mathbb{1}_n$. In particular, if $\mathcal{B}$ is a Hadamard design of order $4t - 1$, then (23.2.1) is a $\mathbb{Z}_+$-realization of $2t\mathbb{1}_{4t-1}$ and the 0-extension of $\mathcal{B}$ yields a $\mathbb{Z}_+$-realization of $2t\mathbb{1}_{4t}$. If $\mathcal{B}$ is $PG(2, t)$, then (23.2.1) is a $\mathbb{Z}_+$-realization of $2t\mathbb{1}_{t^2+t+1}$ and the $(t - 1)$-extension of $\mathcal{B}$ yields a $\mathbb{Z}_+$-realization of $2t\mathbb{1}_{t^2+t+2}$.

The next result makes precise the correspondence between $\mathbb{Z}_+$-realizations of the equidistant metric and designs. The first assertion (i) is nothing but a reformulation of Proposition 23.1.8 (using the link between intersecting systems and designs, explained in Remark 23.1.7).

Proposition 23.2.2.

(i) *There is a one-to-one correspondence between the $\mathbb{Z}_+$-realizations of $2t\mathbb{1}_n$ and the $(2t, t, n - 1)$-designs.*

(ii) *For $k \neq \frac{n}{2}$, there is a one-to-one correspondence between the k-uniform $\mathbb{Z}_+$-realizations of $2t\mathbb{1}_n$ and the $(n, k, \frac{t(k-1)}{n-k})$-BIBD's.*

Proof. (i) follows by assuming that all blocks $B \in \mathcal{B}$ do not contain a given point i_0 of V_n (replacing, if necessary, B by $V_n \setminus B$).
(ii) It is immediate to check that (23.2.1) holds if $\mathcal{B}$ is a $(n, k, \frac{t(k-1)}{n-k})$-BIBD. Suppose now that (23.2.1) holds, with $|B| = k$ for all $B \in \mathcal{B}$, and $k \neq \frac{n}{2}$. By taking the scalar product of both sides of (23.2.1) with the all-ones vector, we obtain that the number b of blocks satisfies

$$b = \frac{tn(n-1)}{k(n-k)}.$$

We show that each point belongs to the same number of blocks. For this, let r denote the number of blocks that contain the point 1 and denote by a_i the number of blocks containing both points 1 and i, for $i = 2, \ldots, n$. Then, $\sum_{2 \leq i \leq n} a_i = r(k-1)$. Counting in two ways the total number of units in the incidence matrix of $\mathcal{B}$ (summing over the columns or over the rows), we obtain

$$bk = r + \sum_{2 \leq i \leq n} (2t - r + 2a_i),$$

implying $r = t\frac{n-1}{n-k}$. Hence, any two points of V_n belong to $r - t = t\frac{k-1}{n-k}$ common blocks. Therefore, $\mathcal{B}$ is a $(n, k, \frac{t(k-1)}{n-k})$-BIBD. ∎

It is convenient for further reference to reformulate Theorem 22.0.5 in terms of designs. As can be easily verified, the above proof of Theorem 22.0.5 permits to formulate the following sharper statement.

Theorem 23.2.3. *Suppose that $n \geq \frac{1}{2}(t+2)^2$ with $t \geq 3$, or that $n > \frac{1}{2}(t+2)^2$ with $t = 2$. Let $\mathcal{B}$ be a family of subsets of V_n for which* (23.2.1) *holds and defines a nontrivial hypercube embedding of $2t\mathbb{1}_n$. Then $\mathcal{B}$ is equivalent, either to a nontrivial $(t+1, 1, n)$-design, or to the $(t-1)$-extension of a nontrivial $(t+1, 1, n-1)$-design.* ∎

Take, for instance, $t = 3$ and $n = 12$ $(< \frac{1}{2}(t+2)^2)$. Then, $6\mathbb{1}_{12}$ has a $\mathbb{Z}_+$-realization which is *not* of the form indicated in Theorem 23.2.3; such a realization can be obtained from the 1-extension of a (5,2,11)-design. The following result of McCarthy and Vanstone [1977] will yield a new case of h-rigidity for the equidistant metric.

Theorem 23.2.4. *Let α, t be positive integers such that $t > 2\alpha^2 + 3\alpha + 2$ (i.e., $\alpha < \frac{\sqrt{8t-7}-3}{4}$). Suppose that $PG(2,t)$ does not exist. Then, for $n \geq t^2 - \alpha$, each $(t+1, 1, n)$-design is trivial.* ∎

Corollary 23.2.5. *Suppose that $PG(2,t)$ does not exist. If $n > t^2 + 1 - \frac{\sqrt{8t-7}-3}{4}$, then the metric $2t\mathbb{1}_n$ is h-rigid.*

Proof. Let $\mathcal{B}$ be a family of subsets of V_n for which (23.2.1) holds and defines a nontrivial hypercube embedding of $2t\mathbb{1}_n$. By Theorem 23.2.3, $\mathcal{B}$ is equivalent to a $(t+1, 1, n)$-design or to the $(t-1)$-extension of a $(t+1, 1, n-1)$-design (as $n \geq \frac{1}{2}(t+2)^2$). By Theorem 23.2.4, such designs are trivial. Hence, $\mathcal{B}$ yields the star realization of $2t\mathbb{1}_n$, a contradiction. ∎

23.3 The Minimum h-Size of $2t\mathbb{1}_n$

Recall that the minimum h-size of $2t\mathbb{1}_n$ is defined as the smallest cardinality of a multiset $\mathcal{B} \subseteq 2^{V_n}$ satisfying (23.2.1); it is denoted by $s_h(2t\mathbb{1}_n)$. (The notion of minimum h-size was introduced in Section 4.3.) The following result is a reformulation of Ryser's result on the number of blocks of a $(2t, t, n-1)$-design.

Theorem 23.3.1.

(i) $s_h(2t\mathbb{1}_n) \geq n - 1$, *with equality if and only if $n = 4t$ and there exists a Hadamard matrix of order $4t$.*

(ii) *Suppose* $n \neq 4t$. *If* $n = 2t + \lambda + \frac{t(t-1)}{\lambda}$ *for some integer* $\lambda \geq 1$ *and if there exists a symmetric* $(n, \lambda + t, \lambda)$*-BIBD, then* $s_h(2t\mathbb{1}_n) = n$.

Proof. (i) By Proposition 23.2.2, the minimum h-size of $2t\mathbb{1}_n$ is equal to the minimum number of blocks in a $(2t, t, n-1)$-design, which is greater than or equal to $n-1$, by Theorem 23.1.3. If $s_h(2t\mathbb{1}_n) = n-1$, then there exists a $(2t, t, n-1)$-design $\mathcal{B}$ with $n-1$ blocks. Applying again Theorem 23.1.3, we deduce that $\mathcal{B}$ is a symmetric $(4t-1, 2t, t)$-design, i.e., a Hadamard design of order $4t-1$. The assertion (ii) can be easily checked. ∎

As an application of Theorem 23.3.1 and Remark 23.1.6, we deduce that $s_h(2t\mathbb{1}_n) = n$ for the following parameters (t, n): (7,37), (6,25), (9,56), (7,31), (9,45), (11,79), (9,40), (12,71). Note also that $s_h(2t\mathbb{1}_n) = n-1$ for $(t,n) = (9,36), (4,16)$.

The implication from Theorem 23.3.1 (ii) is, in fact, an equivalence in the cases $\lambda = 1$ (i.e., $n = t^2 + t + 1$) and $\lambda = t$ (i.e., $n = 4t - 1$).

Proposition 23.3.2.

(i) $s_h(2t\mathbb{1}_{t^2+t+1}) = t^2 + t + 1$ *if and only if there exists a projective plane of order* t.

(ii) $s_h(2t\mathbb{1}_{4t-1}) = 4t - 1$ *or, equivalently,* $s_h(2t\mathbb{1}_{4t}) = 4t - 1$ *if and only if there exists a Hadamard design of order* $4t - 1$.

(iii) *Suppose* $PG(2,t)$ *exists. Then,* $s_h(2t\mathbb{1}_{t^2+t+2}) = t^2 + 2t$ *if* $t \geq 3$ *and* $s_h(2t\mathbb{1}_{t^2+t+2}) = t^2 + t + 1$ *if* $t = 1, 2$.

(iv) *Suppose* $PG(2,t)$ *does not exist. If* $n > t^2 + 1 - \frac{\sqrt{8t-7}-3}{4}$, *then* $s_h(2t\mathbb{1}_n) = nt$.

Proof. (i) follows from Theorems 22.0.4 and 23.3.1.
(ii) Suppose $\mathcal{B}$ is a block family yielding a $\mathbb{Z}_+$-realization of $2t\mathbb{1}_{4t-1}$ with $|\mathcal{B}| = 4t-1$. Then, $|\mathcal{B}| = \frac{n(n-1)t}{\lceil\frac{n}{2}\rceil\lfloor\frac{n}{2}\rfloor}$ $(n = 4t-1)$, which implies that all blocks of $\mathcal{B}$ have size $2t$. Hence, $\mathcal{B}$ is a Hadamard design of order $4t-1$. The remaining of (ii) follows from Theorem 23.3.1.
(iii) For the case $t = 1, 2$, use Theorem 23.3.1. Suppose $t \geq 3$ and set $n := t^2 + t + 2$. The $(t-1)$-extension of $PG(2,t)$ yields a $\mathbb{Z}_+$-realization of $2t\mathbb{1}_n$ of size $t^2 + t$, implying $s_h(2t\mathbb{1}_n) \leq t^2 + 2t$. Let $\mathcal{B}$ be a block family yielding a $\mathbb{Z}_+$-realization of $2t\mathbb{1}_n$. We show that $|\mathcal{B}| \geq t^2 + 2t$. This is obvious if $\mathcal{B}$ corresponds to the star realization. Otherwise, we can use Theorem 23.2.3. Either, $\mathcal{B}$ is equivalent to a nontrivial $(t+1, 1, n)$-design; then, its $(t-1)$-extension yields a $\mathbb{Z}_+$-realization of $2t\mathbb{1}_{t^2+t+3}$ distinct from the star realization, in contradiction with Theorem 22.0.6. Or, $\mathcal{B}$ is equivalent to the $(t-1)$-extension of a $(t+1, 1, n-1)$-design and, then, $|\mathcal{B}| \geq n - 1 + t - 1 = t^2 + 2t$. This shows that $s_h(2t\mathbb{1}_n) = t^2 + 2t$. Finally, (iv) is a reformulation of Corollary 23.2.5. ∎

Set

$$a_n^t := \left\lceil \frac{n(n-1)t}{\lfloor \frac{n}{2} \rfloor \lceil \frac{n}{2} \rceil} \right\rceil = \left\lceil 4t - \frac{2t}{\lceil \frac{n}{2} \rceil} \right\rceil.$$

By taking the scalar product of both sides of (23.2.1) with the all-ones vector, we obtain the following bounds:

$$a_n^t \leq s_h(2t\mathbb{1}_n) \leq nt.$$

The equality

$$s_h(2t\mathbb{1}_n) = nt$$

holds if and only if the star realization (22.0.1) is the only $\mathbb{Z}_+$-realization of $2t\mathbb{1}_n$, i.e., if $2t\mathbb{1}_n$ is h-rigid. This is the case, for instance, if $n \geq t^2 + t + 3$ (by Theorem 22.0.6). Several other results about classes of parameters n, t for which $2t\mathbb{1}_n$ is h-rigid have been given in Chapter 22. A natural question is what are the parameters n, t for which the equality

$$s_h(2t\mathbb{1}_n) = a_n^t$$

holds. If $2t\mathbb{1}_n$ admits a $\mathbb{Z}_+$-realization $\sum_S \lambda_S \delta(S)$ where $\lambda_S > 0$ only if $\delta(S)$ is an equicut (i.e., satisfies $|S| = \lfloor \frac{n}{2} \rfloor, \lceil \frac{n}{2} \rceil$), then the equality $s_h(2t\mathbb{1}_n) = a_n^t$ holds. For instance, $s_h(4\mathbb{1}_7) = a_7^2 = 7$ and $s_h(4\mathbb{1}_8) = a_8^2 = 7$, as each of $4\mathbb{1}_7$ and $4\mathbb{1}_8$ has a $\mathbb{Z}_+$-realization using only equicuts (see Proposition 23.4.4).

Clearly (from Theorem 23.3.1), the equality $s_h(2t\mathbb{1}_n) = a_n^t$ can occur only if $n \leq 4t$. The case $n = 4t$ is well understood: equality holds if and only if there exists a Hadamard matrix of order $4t$. The following conjectures are posed by Deza and Laurent [1993c].

Conjecture 23.3.3. *Suppose that $n \leq 4t$ and that there exists a Hadamard matrix of order $4t$. Then, $s_h(2t\mathbb{1}_n) = a_n^t$.*

Conjecture 23.3.4. *Suppose that $n \leq 4t$ and that there exist Hadamard matrices of suitable orders. Then, $s_h(2t\mathbb{1}_n) = a_n^t$.*

Conjecture 23.3.4 is obviously weaker than Conjecture 23.3.3 (the word "suitable" remains to be defined in an appropriate way). We refer to Deza and Laurent [1993c] for partial results related to these conjectures. In particular, the following results are proved there.

Proposition 23.3.5.

(i) *Conjecture* 23.3.3 *holds for all n, t such that $n \leq 4t$, and $\frac{2t}{3} < \lceil \frac{n}{2} \rceil$ or* $\min(n, t) \leq 20$.

(ii) *Conjecture* 23.3.4 *holds for all n, t such that n is even and satisfies $2\sqrt{2t} \leq n \leq 4t$. (It suffices to assume the existence of Hadamard matrices of orders $2n$, $4n$, and n (if $\frac{n}{2}$ is even) and $n + 2$ (if $\frac{n}{2}$ is odd).)* ∎

Corollary 23.3.6. *If* $n \leq 4t \leq 80$ *then* $s_h(2t\mathbb{1}_n) = a_n^t$. ∎

Example 23.3.7. As an example, let us consider the minimum h-size of the metric $2t\mathbb{1}_n$ for $t = 6$ and $n \geq 31$. We have:

(i) $s_h(12\mathbb{1}_n) = 6n$ for all $n \geq 33$ (by Proposition 22.0.11),

(ii) $s_h(12\mathbb{1}_{32}) = 67$ and $s_h(12\mathbb{1}_{31}) \leq 62$.

Indeed, let $\mathcal{B}$ be a block design on V_{32} for which (23.2.1) holds and defines a nontrivial hypercube embedding of $12\mathbb{1}_{32}$. By Theorem 23.2.3, $\mathcal{B}$ is equivalent to a (7,1,32)-design, or to the 5-extension of a (7,1,31)-design. Each (7,1,32)-design is trivial (as its 5-extension yields a $\mathbb{Z}_+$-realization of the h-rigid metric $12\mathbb{1}_{33}$). It is shown in McCarthy, Mullin, Schellenberg, Stanton and Vanstone [1976, 1977] that the unique nontrivial (7,1,31)-design is the block family obtained by taking the blocks of $PG(2,5)$ together with the 31 singletons; it yields a $\mathbb{Z}_+$-realization of $12\mathbb{1}_{31}$ of size $31 + 31 = 62$. Its 5-extension yields a $\mathbb{Z}_+$-realization of $12\mathbb{1}_{32}$ of size $62 + 5 = 67$. This shows that $s_h(12\mathbb{1}_{32}) = 67$. ∎

23.4 All Hypercube Embeddings of $2t\mathbb{1}_n$ for Small n, t

We list all the $\mathbb{Z}_+$-realizations of the equidistant metric $2t\mathbb{1}_n$ in the following cases: $t = 1$, $t = 2$, $n = 4$, and we give partial information in the case $n = 5$. The results are taken from Deza and Laurent [1993c].

Let t, n be positive integers. For each integer s such that $t - \lfloor \frac{t}{n-3} \rfloor \leq s \leq t$, we have the following $\mathbb{Z}_+$-realization of $2t\mathbb{1}_n$:

$$2t\mathbb{1}_n = (t-(n-3)(t-s))\delta(\{n\}) + \sum_{1 \leq i \leq n-1} (t-s)\delta(\{i,n\}) + s\delta(\{i\}). \tag{23.4.1}$$

Its size is equal to $(n-3)s + 3t$ and (23.4.1) coincides with the star realization (22.0.1) for $s = t$.

Proposition 23.4.2. **(Case $n = 4$)** *The metric* $2t\mathbb{1}_4$ *has* $t+1$ $\mathbb{Z}_+$*-realizations, given by* (23.4.1) *for* $0 \leq s \leq t$.

Proof. This follows from the fact that the restriction to V_3 of any $\mathbb{Z}_+$-realization of $2t\mathbb{1}_4$ coincides with the star realization of $2t\mathbb{1}_3$. ∎

Proposition 23.4.3. **(Case $t = 1$)** *For* $n \neq 4$, (22.0.1) *is the only* $\mathbb{Z}_+$*-realization of the metric* $2\mathbb{1}_n$ *and, for* $n = 4$, $2\mathbb{1}_4$ *has two* $\mathbb{Z}_+$*-realizations: the star realization* (22.0.1) *and* (23.4.1) *for* $s = 0$*, namely,*

$$2\mathbb{1}_4 = \textstyle\sum_{1 \leq i \leq 4} \delta(\{i\}) = \delta(\{1,4\}) + \delta(\{2,4\}) + \delta(\{3,4\}). \quad ∎$$

Proposition 23.4.4. (Case $t = 2$)

(i) *For $n \geq 9$, (22.0.1) is the only $\mathbb{Z}_+$-realization of $4\mathbb{1}_n$.*

(ii) *For $n = 4$, $4\mathbb{1}_4$ has three $\mathbb{Z}_+$-realizations: (22.0.1) and (23.4.1) for $s = 0, 1$, namely,*

$$4\mathbb{1}_4 = 2(\sum_{1 \leq i \leq 4} \delta(\{i\})) = 2(\sum_{1 \leq i \leq 3} \delta(\{i, 4\})) = \sum_{1 \leq i \leq 4} \delta(\{i\}) + \sum_{1 \leq i \leq 3} \delta(\{i, 4\}).$$

(iii) *For $n = 5$, $4\mathbb{1}_5$ has (up to permutation) three $\mathbb{Z}_+$-realizations: the star realization (22.0.1), (23.4.1) for $s = 1$, i.e.,*

$$4\mathbb{1}_5 = \sum_{1 \leq i \leq 4} \delta(\{i, 5\}) + \delta(\{i\}), \text{ and}$$

$$4\mathbb{1}_5 = \delta(\{5\}) + \sum_{1 \leq i < j \leq 4} \delta(\{i, j\}).$$

(iv) *For $n = 6$, $4\mathbb{1}_6$ has (up to permutation) three $\mathbb{Z}_+$-realizations: the star realization (22.0.1),*

$$\begin{aligned} 4\mathbb{1}_6 = {} & \delta(\{2\}) + \delta(\{3\}) + \delta(\{4, 6\}) + \delta(\{5, 6\}) + \delta(\{1, 4\}) \\ & + \delta(\{1, 5\}) + \delta(\{1, 2, 6\}) + \delta(\{1, 3, 6\}), \text{ and} \end{aligned}$$

$$\begin{aligned} 4\mathbb{1}_6 = {} & \delta(\{1, 2\}) + \delta(\{3, 4\}) + \delta(\{5, 6\}) + \delta(\{1, 3, 6\}) \\ & + \delta(\{2, 4, 6\}) + \delta(\{1, 4, 5\}) + \delta(\{2, 3, 5\}) + \delta(\{1, 3, 6\}). \end{aligned}$$

(v) *For $n = 7$, $4\mathbb{1}_7$ has (up to permutation) three $\mathbb{Z}_+$-realizations: the star realization (22.0.1),*

$$\begin{aligned} 4\mathbb{1}_7 = {} & \delta(\{7\}) + \delta(\{1, 2\}) + \delta(\{3, 4\}) + \delta(\{5, 6\}) \\ & + \delta(\{1, 3, 6\}) + \delta(\{2, 4, 6\}) + \delta(\{1, 4, 5\}) + \delta(\{2, 3, 5\}), \text{ and} \end{aligned}$$

$$\begin{aligned} 4\mathbb{1}_7 = {} & \delta(\{1, 2, 7\}) + \delta(\{3, 4, 7\}) + \delta(\{5, 6, 7\}) + \delta(\{1, 3, 6\}) \\ & + \delta(\{2, 4, 6\}) + \delta(\{1, 4, 5\}) + \delta(\{2, 3, 5\}). \end{aligned}$$

(vi) *For $n = 8$, $4\mathbb{1}_8$ has (up to permutation) three $\mathbb{Z}_+$-realizations: the star realization (22.0.1),*

$$\begin{aligned} 4\mathbb{1}_8 = {} & \delta(\{8\}) + \delta(\{1, 2, 7\}) + \delta(\{3, 4, 7\}) + \delta(\{5, 6, 7\}) \\ & + \delta(\{1, 3, 6\}) + \delta(\{2, 4, 6\}) + \delta(\{1, 4, 5\}) + \delta(\{2, 3, 5\}), \text{ and} \end{aligned}$$

$$\begin{aligned} 4\mathbb{1}_8 = {} & \delta(\{1, 2, 7, 8\}) + \delta(\{3, 4, 7, 8\}) + \delta(\{5, 6, 7, 8\}) + \delta(\{1, 3, 6, 8\}) \\ & + \delta(\{2, 4, 6, 8\}) + \delta(\{1, 4, 5, 8\}) + \delta(\{2, 3, 5, 8\}) \end{aligned}$$

(corresponding to a Hadamard design). ∎

It seems a rather difficult task to list all the $\mathbb{Z}_+$-realizations of the metric $2t\mathbb{1}_n$ in the case $n = 5$. Note that we already have the realizations (23.4.1) for $t - \lfloor \frac{t}{2} \rfloor \leq s \leq t$. For t odd and $t \geq 3$, we also have

$$\begin{aligned} 2t\mathbb{1}_5 = \ & \tfrac{t-1}{2}(\delta(\{5\}) + \delta(\{1,2\}) + \delta(\{1,3\}) + \delta(\{2,4\}) + \delta(\{3,4\})) \\ & + \delta(\{1,5\}) + \delta(\{2\}) + \delta(\{3\}) + \delta(\{4,5\}) \\ & + \tfrac{t+1}{2}\delta(\{1,4\}) + \tfrac{t-3}{2}\delta(\{2,3\}). \end{aligned} \tag{23.4.5}$$

The following is also a $\mathbb{Z}_+$-realizations of $2t\mathbb{1}_5$:

$$\begin{aligned} 2t\mathbb{1}_5 = \ & p\delta(\{5\}) + q\delta(\{1\}) + (s-q)\delta(\{1,5\}) + \alpha \sum_{2\leq i\leq 4} \delta(\{i,5\}) \\ & + (s-\alpha) \sum_{2\leq i\leq 4} \delta(\{i\}) + \beta \sum_{2\leq i\leq 4} \delta(\{1,i\}) \\ & + (t-s-\beta) \textstyle\sum_{2\leq i\leq 4} \delta(\{1,i,5\}), \end{aligned} \tag{23.4.6}$$

where α, β, p, q, s are integers satisfying

$$\begin{cases} 0 \leq s \leq t, \\ 0 \leq \alpha \leq \min(s, \frac{t}{2}), \\ \max(0, s-2\alpha, \frac{t-3\alpha}{2}) \leq p \leq \min(t-2\alpha, \frac{t-3\alpha+s}{2}), \\ \beta = t - 2\alpha - p, \\ q = 3\alpha + 2p - t. \end{cases}$$

Let $\lambda(s,t,\alpha,p)$ denote the realization from (23.4.6).

For $t = 3$, the feasible parameters for (23.4.6) are (s, α, p) =(1,0,2), (1,1,0), (2,0,2), (2,1,0), (2,1,1), (3,0,3), (3,1,1), (3,0,3), and (3,1,1). Note, however, that $\lambda(3,3,0,3)$ coincides with the star realization (22.0.1); $\lambda(3,3,1,1)$ reads

$$6\mathbb{1}_5 = \delta(\{5\}) + \sum_{1\leq i\leq 4} \delta(\{i,5\}) + 2\delta(\{i\}) \tag{23.4.7}$$

(this is (23.4.1) in the case $t = 3, n = 5, s = 2$); $\lambda(2,3,0,2)$ is a permutation of (23.4.7); and $\lambda(2,3,1,1)$ coincides with $\lambda(1,3,0,2)$ (up to permutation).

Proposition 23.4.8. (Case $t = 3, n = 5$) *The metric $6\mathbb{1}_5$ has five distinct (up to permutation) $\mathbb{Z}_+$-realizations: the star realization (22.0.1), (23.4.7), (23.4.5) (with $t = 3$), and (23.4.6) for the parameters (s, α, p) =(2,1,1), (2,1,0), (1,1,0) which read, respectively,*

$$6\mathbb{1}_5 = \delta(\{5\}) + 2\delta(\{1\}) + \sum_{2\leq i\leq 4} \delta(\{i,5\}) + \delta(\{i\}) + \delta(\{1,i,5\}),$$

$$6\mathbb{1}_5 = 2\delta(\{1,5\}) + \sum_{2\leq i\leq 4} \delta(\{i,5\}) + \delta(\{i\}) + \delta(\{1,i\}),$$

$$6\mathbb{1}_5 = \delta(\{1,5\}) + \textstyle\sum_{2\leq i\leq 4} \delta(\{i,5\}) + \delta(\{1,i\}) + \delta(\{1,i,5\}). \qquad \blacksquare$$

Chapter 24. Recognition of Hypercube Embeddable Metrics

In this chapter we consider the following problem, called the *hypercube embeddability problem*:

Given a distance d on V_n, test whether d is hypercube embeddable.

When restricted to the class of path metrics of connected graphs, this is the problem of testing whether a graph can be isometrically embedded into a hypercube. Such graphs have a good characterization and can be recognized in polynomial time. This topic has been discussed in detail in Chapter 19.

The hypercube embeddability problem is NP-hard for general distances; it is, in fact, NP-complete for the class of distances with values in the set $\{2, 3, 4, 6\}$ (see Theorem 24.1.8).

However, the hypercube embeddability problem can be shown to be solvable in polynomial time for several classes of metrics. This is the case, typically, for metrics having a restricted range of values. For instance, the hypercube embeddability problem is polynomial-time solvable for the class of distances with range of values $\{1, 2, 3\}$, or $\{3, 5, 8\}$ or, more generally, $\{x, y, x+y\}$ where x, y are two positive integers not both even. This class is discussed in Section 24.3. We consider in Section 24.4 the distances taking two values of the form $a, 2a$ where $a \geq 1$ is an integer. Note that such distances can be seen as scale multiples of truncated distances of graphs. For this class of distances, hypercube embeddability can be characterized by a finite point criterion and, thus, recognized in polynomial time.

We also consider generalized bipartite metrics, which are the metrics d on V_n for which there exists a subset $S \subseteq V_n$ such that $d(i,j) = 2$ for all $i \neq j \in S$ and for all $i \neq j \in V_n \setminus S$. The hypercube embeddable generalized bipartite metrics can also be recognized in polynomial time; see Section 24.2.

The basic idea which is used for characterizing the hypercube embeddable metrics within the above classes is the existence of equidistant submetrics, which are h-rigid if they are defined on sufficiently many points (by the results of Chapter 22).

We present in Section 24.5 the following result of Karzanov [1985]: Let d be a metric whose extremal graph is K_4, C_5, or a union of two stars; then, d is hypercube embeddable if and only if d satisfies the parity condition (24.1.1).

Let us point out that, in contrast with the results from Section 24.3, no characterization is known for the hypercube embeddable metrics taking three values, all of them even. For instance, the complexity of the hypercube embeddability problem for the class of distances with range of values $\{2,4,6\}$ is not known. The case of distances taking two even values a, b is also unsettled; Section 24.4 solves only the case when $b = 2a$.

We start in Section 24.1 with some preliminary results.

24.1 Preliminary Results

Let d be a distance on the set V_n. A first easy observation is that we may assume that no pair of distinct points is at distance 0. Indeed, if $d(i,j) = 0$ for some distinct $i, j \in V_n$, then d is hypercube embeddable if and only if its restriction to the set $V_n \setminus \{j\}$ is hypercube embeddable (as the points i and j should be labeled by the same set in any hypercube embedding of d).

If d is hypercube embeddable, then

$$(24.1.1) \qquad d(i,j) + d(i,k) + d(j,k) \in 2\mathbb{Z} \text{ for all } i,j,k \in V_n.$$

(Indeed, if $A_1, \ldots, A_n$ are sets forming an h-labeling of d, then $d(i,j) + d(i,k) + d(j,k) = 2(|A_i| + |A_j| + |A_k| - |A_i \cap A_j| - |A_i \cap A_k| - |A_j \cap A_k|) \in 2\mathbb{Z}$.) The condition (24.1.1) is known as the *parity condition*; it was first introduced in Deza [1960]. This condition expresses the fact that each hypercube embeddable distance d on V_n can be decomposed as an *integer* combination of cut semimetrics, i.e., belongs to the cut lattice $\mathcal{L}_n$ (indeed, (24.1.1) characterizes membership in $\mathcal{L}_n$; see Proposition 25.1.1). As an application, we deduce that each hypercube embeddable distance has some bipartite structure, namely, the set of pairs at an odd distance forms a complete bipartite graph.

Lemma 24.1.2. *Let d be a distance on V_n. If d satisfies the parity condition* (24.1.1), *then V_n can be partitioned into $V_n = S \cup T$ in such a way that $d(i,j)$ is even if $i,j \in S$ or if $i,j \in T$, and $d(i,j)$ is odd if $i \in S, j \in T$.* ∎

This simple fact will be central in our treatment. For instance, the generalized bipartite metrics, considered in Section 24.2, have only one even distance equal to 2, i.e., they satisfy $d(i,j) = 2$ for $i \neq j \in S$, $i \neq j \in T$, for some bipartition (S,T) of V_n.

Obviously, every hypercube embeddable distance d on V_n is ℓ_1-embeddable, i.e., belongs to the cut cone CUT_n. In other words, d can be decomposed as a *nonnegative* combination of cut semimetrics. Hence, we have the implication:

$$d \text{ is hypercube embeddable } \Longrightarrow d \in \mathrm{CUT}_n \text{ and } d \text{ satisfies (24.1.1)}$$

In general, this implication is strict. But, for some classes of distances, this implication turns out to be an equivalence; this is the case, for instance, for the distances with range of values $\{1,2\}$, or $\{1, 2\alpha, 2\alpha+1\}$ ($\alpha \geq 2$) (see Propositions 24.1.9 and 24.1.10), or for the distances considered in Proposition 24.3.27

or in Theorem 24.5.1. This is also the case for the distances on $n \le 5$ points, as the next result by Deza [1960, 1982] shows.

Theorem 24.1.3. *Let d be a distance on $n \le 5$ points. Then, d is hypercube embeddable if and only if $d \in \mathrm{CUT}_n$ and d satisfies the parity condition* (24.1.1). ∎

The parity condition (24.1.1) also suffices to characterize hypercube embeddability for the distances that lie on specified facets of CUT_n, in particular, on several classes of simplex facets; see relation (31.8.3). We will consider in Chapter 25 the quasi h-points, which are the distances d that belong to CUT_n and satisfy (24.1.1) but are not hypercube embeddable.

Each valid inequality for the cut cone yields therefore a necessary condition for hypercube embeddability. It turns out that the hypermetric inequalities will play a crucial role for the characterization of certain classes of hypercube embeddable distances; see Propositions 24.1.9, 24.3.10, 24.3.11, 24.3.27. Hypermetric inequalities have been introduced in Section 6.1 and studied in detail in Part II; we recall here the main definitions in order to make the chapter self-contained. Let d be a distance on V_n and let $k \ge 1$ be an integer. Then, d is said to be $(2k+1)$-*gonal* if, for all (not necessarily distinct) points $i_1, \ldots, i_k, i_{k+1}, j_1, \ldots, j_k \in V_n$, the following inequality holds:

$$\text{(24.1.4)} \qquad \sum_{1 \le r < s \le k+1} d(i_r, i_s) + \sum_{1 \le r < s \le k} d(j_r, j_s) - \sum_{\substack{1 \le r \le k+1 \\ 1 \le s \le k}} d(i_r, j_s) \le 0.$$

Equivalently, d is $(2k+1)$-gonal if, for all $b \in \mathbb{Z}^n$ with $\sum_{1 \le i \le n} b_i = 1$ and $\sum_{1 \le i \le n} |b_i| = 2k+1$,

$$\text{(24.1.5)} \qquad \sum_{1 \le i < j \le n} b_i b_j d(i,j) \le 0.$$

Moreover, d is said to be *hypermetric* if d is $(2k+1)$-gonal for all $k \ge 1$. The inequality (24.1.4) is called the $(2k+1)$-gonal inequality.

We now recall the link existing between hypercube embeddable distances and intersection patterns. A vector $p \in \mathbb{R}^{V_n \cup E_n}$ is called an *intersection pattern* if there exist a set Ω and n subsets $A_1, \ldots, A_n$ of Ω such that

$$\text{(24.1.6)} \qquad p_{ij} = |A_i \cap A_j| \text{ for all } 1 \le i \le j \le n.$$

Hypercube embeddable distances are in one-to-one correspondence with intersection patterns, via the covariance mapping (see Sections 5.2 and 5.3 and, in particular, Proposition 5.3.5). Namely, for a distance d on V_{n+1}, d is hypercube embeddable if and only if its image $p = \xi(d)$ under the covariance mapping is an intersection pattern (indeed, the sets $A_1, \ldots, A_n, A_{n+1} = \emptyset$ form an h-labeling of

d if and only if $A_1, \ldots, A_n$ satisfy (24.1.6)). Recall that $p = \xi(d)$ is defined by

$$\begin{cases} p_{ii} = d(i, n+1) & \text{for } 1 \leq i \leq n, \\ p_{ij} = \frac{1}{2}(d(i, n+1) + d(j, n+1) - d(i,j)) & \text{for } 1 \leq i < j \leq n. \end{cases}$$

An early reference on intersection patterns is Kelly [1968]; in particular, Kelly characterizes there the intersection patterns of order $n \leq 4$, thus, providing another proof for Theorem 24.1.3. The above correspondence permits, for instance, to obtain the following result of Chvátal [1980] (which will also follow from the treatment in Section 24.4).

Proposition 24.1.7. *The hypercube embeddability problem is polynomial for the class of distances with range of values* $\{2, 4\}$ *and having a point at distance* 2 *from all other points.*

Proof. Let d be a distance on V_{n+1} such that $d(i, n+1) = 2$ for all $i \in V_n$ and $d(i,j) \in \{2, 4\}$ for all $i \neq j \in V_n$. Its image $p = \xi(d)$ satisfies $p_{ii} = 2$ for all $i \in V_n$ and $p_{ij} \in \{0, 1\}$ for all $i \neq j \in V_n$. Let H denote the graph on V_n with edges the pairs (i,j) such that $p_{ij} = 1$. Then, d is hypercube embeddable if and only if p is an intersection pattern which, in turn, is equivalent to H being a line graph. The result now follows from the fact that line graphs can be recognized in polynomial time (Beineke [1970]). ∎

The hypercube embeddability problem is hard for general metrics; this was shown by Chvátal [1980] in the context of intersection patterns.

Theorem 24.1.8. *The hypercube embeddability problem is NP-complete for the class of distances having a point at distance* 3 *from all other points and with distances between those points belonging to* $\{2, 4, 6\}$.

Proof. We sketch the proof. Let d be a distance as in the theorem. Hence, its image $p = \xi(d)$ satisfies $p_{ii} = 3$ for all $i \in V_n$ and $p_{ij} \in \{0, 1, 2\}$ for all $i \neq j \in V_n$. Let H denote the multigraph with node set V_n and having p_{ij} parallel edges between nodes i and j. It is easy to see that d is hypercube embeddable, i.e., p is an intersection pattern, if and only if the edge set of H can be partitioned into cliques in such a way that each node belongs to at most three of these cliques. Chvátal [1980] shows that the problem of testing whether a 4-regular graph is 3-colourable (which is NP-complete) can be polynomially reduced to the above edge partitioning problem for H. ∎

There are some classes of distances for which hypercube embeddability is very easy to characterize. Here are two examples taken, respectively, from Assouad and Deza [1980] and Deza and Laurent [1995a]. The first example will be contained in Proposition 24.3.10.

Proposition 24.1.9. *Let* d *be a distance on* V_n *with values in* $\{1, 2\}$. *The* *following assertions are equivalent.*

(i) *d is hypercube embeddable.*

(ii) *d is 5-gonal and satisfies the parity condition* (24.1.1).

(iii) *d is the path metric of the complete bipartite graphs $K_{1,n-1}$ or $K_{2,2}$, or $d = 2d(K_n)$.* ∎

Proposition 24.1.10. *Let d be a metric on V_n with range of values $\{1, 2\alpha, 2\alpha+1\}$, for some integer $\alpha \geq 2$. Then, d is hypercube embeddable if and only if d satisfies the parity condition* (24.1.1).

Proof. Suppose that d satisfies (24.1.1). Hence, the set of pairs at odd distance forms a complete bipartite graph $K_{S,T}$ for some bipartition (S,T) of V_n. As $\alpha \geq 2$, the pairs at distance 1 form a matching, say, $d(i_1, j_1) = \ldots = d(i_k, j_k) = 1$ for $i_1, \ldots, i_k \in S$ and $j_1, \ldots, j_k \in T$. Then,

$$d = \delta(S) + \sum_{1 \leq h \leq k} \alpha\delta(\{i_h, j_h\}) + \sum_{\substack{i \in S \setminus \{i_1, \ldots, i_k\} \\ j \in T \setminus \{j_1, \ldots, j_k\}}} \alpha\delta(\{i\}),$$

showing that d is hypercube embeddable. ∎

The case $\alpha = 1$, i.e., the case of the distances with range of values $\{1, 2, 3\}$, is significantly more complicated and will be treated in Section 24.3.

We close this section with a result on the number of distinct hypercube embeddings of a given distance. Given a hypercube embedable distance d on V_n and an integer $s \geq 0$, let $N_n(d, s)$ denote the number of distinct $\mathbb{Z}_+$-realizations $d = \sum_S \lambda_S \delta(S)$ (with $\lambda_S \in \mathbb{Z}_+$) of d with size $\sum_S \lambda_S = s$. Set

$$M_n(x) := \sum N(d, s)$$

where the sum is taken over all $s \in \mathbb{Z}_+$ and over all distances d on V_n satisfying $\sum_{1 \leq i < j \leq n} d(i,j) = x$. It is shown in Deza, Ray-Chaudhuri and Singhi [1990] that the function $x \in \mathbb{Z}_+ \mapsto M_n(x)$ is quasipolynomial. In other words, there exist an integer $t \geq 1$ and polynomials $f_0, f_1, \ldots, f_{t-1}$ such that

$$M_n(x) = f_i(x) \text{ if } x \equiv i \pmod{t}, \text{ for } 0 \leq i \leq t-1.$$

In particular, $M_n(x)$ is bounded by a polynomial in x. Therefore, the number of distinct $\mathbb{Z}_+$-realizations of d is bounded by a polynomial in $x = \sum_{1 \leq i < j \leq n} d(i,j)$.

24.2 Generalized Bipartite Metrics

Let d be a metric on V_n such that $d(i,j) = 2$ for all $i \neq j \in S$ and $i \neq j \in T$, for some bipartition (S,T) of V_n. Such a metric is called a *generalized bipartite metric*. The $|S| \times |T|$ matrix D with entries $d(i,j)$ for $i \in S, j \in T$ is called the *(S,T)-distance matrix* of d. For instance, the path metric of a complete bipartite graph is a generalized bipartite metric. In this section, we prove the following result, which is due to Deza and Laurent [1995a].

Theorem 24.2.1. *The hypercube embeddability problem is polynomial-time solvable for the class of generalized bipartite metrics.*

We start with an easy observation.

Lemma 24.2.2. *Let d be a generalized bipartite metric with bipartition (S,T). If d is hypercube embeddable, then there exists an integer α such that $d(i,j) \in \{\alpha, \alpha+2, \alpha+4\}$ for all $i \in S, j \in T$.*

Proof. Let α, β denote the smallest and largest value taken by $d(i,j)$ for $i \in S, j \in T$; say $\alpha = d(i,j)$, $\beta = d(i',j')$ for $i,i' \in S$, $j,j' \in T$. Using the triangle inequality, we obtain $\beta = d(i',j') \leq d(i',i) + d(i,j) + d(j,j') \leq 4 + \alpha$. Moreover, α, β have the same parity by (24.1.1). ∎

We will see below what are the possible configurations for the pairs at distance $\alpha, \alpha+2, \alpha+4$.

Set $s := |S|$ and $t := |T|$. Let d_S (resp. d_T) denote the restriction of d to the set S (resp. T). Then, $d_S = 2\mathbb{1}_s$ and $d_T = 2\mathbb{1}_t$ are equidistant metrics. Recall (from Proposition 23.4.3) that the equidistant metric $2\mathbb{1}_n$ is h-rigid if $n \neq 4$ and that $2\mathbb{1}_4$ has exactly two $\mathbb{Z}_+$-realizations, namely, its star realization: $2\mathbb{1}_4 = \sum_{1\leq i\leq 4} \delta(\{i\})$, and an additional realization:

$$2\mathbb{1}_4 = \delta(\{1,2\}) + \delta(\{1,3\}) + \delta(\{1,4\}),$$

called here the *special realization.*

The proof of Theorem 24.2.1 is based on the following simple observation. Let

$$d = \sum_{A\subseteq V_n} \lambda_A \delta(A)$$

be a $\mathbb{Z}_+$-realization of d. Then, its projection on S:

$$\sum_{A\subseteq V_n} \lambda_A \delta(A\cap S),$$

is a $\mathbb{Z}_+$-realization of d_S. Hence, if $s \neq 4$, then it must coincide with the star realization of $2\mathbb{1}_s$ and, if $s = 4$, it must coincide with the star realization or with the special realization of $2\mathbb{1}_4$. The same holds for d_T.

The following definitions will be useful in the sequel. A $\mathbb{Z}_+$-realization of d is called a *star-star* realization if both its projections on S and on T are the star realizations of $2\mathbb{1}_s$ and $2\mathbb{1}_t$, respectively. A realization of d is called a *star-special* realization if its projection on S is the star realization of $2\mathbb{1}_s$, but $t = 4$ and its projection on T is the special realization of $2\mathbb{1}_4$. Finally, a realization of d is called a *special-special* realization if $s = t = 4$ and both its projections on S and T are the special realization of $2\mathbb{1}_4$.

	A'	C'	B'	D'
A	$(f+2)J_a$ $-2I_a$	$f+2$	f	f
C	$f+2$	$f+2$	f	f
B	f	f	$(f-2)J_b$ $+2I_b$	$f-2$
D	f	f	$f-2$	$f-2$

Figure 24.2.3

We now analyze the structure of the hypercube embeddable generalized bipartite metrics admitting a star-star realization.

Proposition 24.2.4. *Let d be a generalized bipartite metric with bipartition (S,T). Then, d admits a star-star realization if and only if there exist a partition $\{A,B,C,D\}$ of S and a partition $\{A',B',C',D'\}$ of T (with possibly empty members) with $|A|=|A'|$ and $|B|=|B'|$ and there exist one-to-one mappings $\sigma : A \longrightarrow A'$ and $\tau : B \longrightarrow B'$ and an integer $f \geq |B|+|D|+|D'|$ such that the values $d(i,j)$ are given by*

$$(24.2.5) \quad \begin{cases} f & \text{if } (i,j) \in ((A\cup C)\times(B'\cup D'))\cup((B\cup D)\times(A'\cup C')) \\ & \quad \cup\{(k,\sigma(k)) \mid k\in A\}\cup\{(k,\tau(k)) \mid k \in B\}, \\ f+2 & \text{if } (i,j)\in((A\cup C)\times(A'\cup C'))\setminus\{(k,\sigma(k)) \mid k\in A\}, \\ f-2 & \text{if } (i,j)\in((B\cup D)\times(B'\times D'))\setminus\{(k,\tau(k)) \mid k\in B\}. \end{cases}$$

Proof. Let d be a generalized bipartite metric admitting a star-star realization: $d = \sum_{U\in\mathcal{U}} \delta(U)$, where $\mathcal{U}$ is a collection (allowing repetition) of nonempty subsets of V. Hence, $|U\cap S| \in \{0,s,1,s-1\}$ and $|U\cap T|\in\{0,t,1,t-1\}$ for all $U\in\mathcal{U}$. We can suppose without loss of generality that $|U\cap S|\in\{0,1\}$ for all $U\in\mathcal{U}$. Let M denote the matrix whose columns are the incidence vectors of the members of $\mathcal{U}$. Combining the above mentioned two possibilities for $U\cap S$ with the four possibilities for $U\cap T$, we obtain that M has the form shown in Figure 24.2.6. Hence the sets A,B,C,D and A',B',C',D' form the desired partitions of S and T. We can now compute $d(i,j)$ for $(i,j)\in S\times T$ and verify that they satisfy relation (24.2.5), after setting $f := |B|+|D|+|D'|+m$.
Conversely, suppose that d is defined by (24.2.5). Set $A=\{x_1,\dots,x_n\}$ and

$B = \{y_1, \ldots, y_n\}$. One can easily check that d satisfies:

$$d = \sum_{1\leq i\leq |A|} \delta(\{x_i, \sigma(x_i)\}) + \sum_{1\leq i\leq |B|} \delta(T\setminus\{\tau(y_i)\}\cup\{y_i\}) + \sum_{x\in C\cup C'} \delta(\{x\}) + \sum_{x\in D} \delta(T\cup\{x\}) + \sum_{x\in D'} \delta(T\setminus\{x\}) + (f-|B|-|D|-|D'|)\delta(T).$$

This realization is clearly a star-star realization. ∎

Figure 24.2.3 shows the (S,T)-distance matrix of the metric d defined by (24.2.5). We use the following notation in Figures 24.2.3 and 24.2.6: I_a denotes the $a \times a$ identity matrix, J_a the $a \times a$ all-ones matrix, and a block marked, say, with f, has all its entries equal to f. As a rule, we denote the cardinality of a set by the same lower case letter; e.g., $a = |A|, a' = |A'|$, etc.

	a	b	c	d	c'	d'	m
A	I_a	0	0	0	0	0	0
B	0	I_b	0	0	0	0	0
C	0	0	I_c	0	0	0	0
D	0	0	0	I_d	0	0	0
A'	I_a	1	0	1	0	1	1
B'	0	$J_b - I_b$	0	1	0	1	1
C'	0	1	0	1	$I_{c'}$	1	1
D'	0	1	0	1	0	$J_{d'} - I_{d'}$	1

Figure 24.2.6

It is fairly clear that the description from Proposition 24.2.4 permits to test in polynomial time whether a generalized bipartite metric has a star-star realization and to find it (if one exists) (see Deza and Laurent [1995a] for details). Actually, this can be done in $O(n^2)$ if the metric is on n points.

One can check whether a generalized bipartite metric has a star-special realization in the following way. Suppose $|T| = 4$. Let $z' \in T$ and let d' denote the restriction of d to the set $V \setminus \{z'\}$. If d has a star-special realization then d' has a star-star realization. We see easily that there are $O(1)$ possible star-star realizations for d' and all of them can be found in polynomial time. One then checks whether one of them can be extended to a star-special realization of d. (If a star-star realization of d' is as in Figure 24.2.6, there is a unique way to complete it to a star-special realization of d, namely, by adjoining the following row as a last row to Figure 24.2.6.)

	a	b	c	d	c'	d'	m
z'	1	0	0	1	1	0	1

Finally, a generalized bipartite metric d has a special-special realization if and only if, for some $m \in \mathbb{Z}_+$, the (S,T)-distance matrix of the semimetric $d - m\delta(T)$ is one of the nine matrices from Figure 24.2.7 (up to permutation on S and T). (This fact can be checked, using a characterization of the generalized bipartite metrics admitting a special-special realization analogous to that of Proposition 24.2.4, see Deza and Laurent [1995a].)

$$\begin{array}{ccc}
\begin{pmatrix}3&1&1&1\\1&3&1&1\\1&1&3&1\\1&1&1&3\end{pmatrix} & \begin{pmatrix}0&2&2&2\\2&0&2&2\\2&2&0&2\\2&2&2&0\end{pmatrix} & \begin{pmatrix}1&1&1&3\\1&3&3&3\\1&3&3&3\\3&3&3&5\end{pmatrix}\\
\begin{pmatrix}4&4&4&2\\4&2&2&2\\4&2&2&2\\2&2&2&0\end{pmatrix} & \begin{pmatrix}3&1&1&1\\1&3&1&1\\3&3&3&1\\3&3&1&3\end{pmatrix} & \begin{pmatrix}0&2&2&2\\2&0&2&2\\2&2&2&4\\2&2&4&2\end{pmatrix}\\
\begin{pmatrix}2&2&2&4\\2&2&2&4\\2&2&2&4\\4&4&4&6\end{pmatrix} & \begin{pmatrix}4&4&4&2\\4&4&4&2\\4&4&4&2\\2&2&2&0\end{pmatrix} & \begin{pmatrix}3&3&3&1\\3&3&3&1\\3&3&3&1\\5&5&5&3\end{pmatrix}
\end{array}$$

Figure 24.2.7

Example 24.2.8. Given an integer $k \geq 5$, let d_{2k} denote the metric defined on $2k$ points by: $d_{2k}(i, i+k) = 4$ for any $1 \leq i \leq k$ and $d_{2k}(i,j) = 2$ for all other pairs $(i,j), 1 \leq i \neq j \leq 2k$. Hence, d_{2k} is a generalized bipartite metric with bipartition $(\{1,2\ldots,k\},\{k+1,k+2,\ldots,2k\})$. Note that $\frac{1}{2}d_{2k}$ is the path metric of the cocktail-party graph $K_{k\times 2}$. The metric d_{2k} can be obtained from the equidistant metric using the full antipodal extension operation, as $d_{2k} = \text{Ant}_4(2\mathbb{1}_k)$. It is an easy exercise to verify, for instance using the above procedure, that d_{2k} is not hypercube embeddable (another proof of this fact has been given in Section 7.2, see Example 7.2.7). On the other hand, one verifies easily that d_{2k} belongs to the cut cone CUT_{2k} and to the cut lattice $\mathcal{L}_{2k}$ (see also Example 7.2.7). ∎

The same technique could be used for testing hypercube embeddability for other metrics than generalized bipartite metrics. Let d be a semimetric on V_n. Suppose that there exists a bipartition (S,T) of V such that the projections d_S and d_T of d on S and T are of the form:

$$(24.2.9) \qquad d_S = \sum_{x \in S} \alpha_x \delta(\{x\}), \quad d_T = \sum_{x \in T} \beta_x \delta(\{x\})$$

for some positive integers α_x, β_x. From Theorem 22.0.12, we know that d_S and d_T are h-rigid if $|S|$ is big enough with respect to $\max_{x \in S} \alpha_x$ and if $|T|$ is big enough with respect to $\max_{x \in T} \beta_x$. So, theoretically, one could use the same technique as the one used in Proposition 24.2.4 for studying hypercube embeddability of these metrics. However, a precise analysis of the structure of the distance matrix of such metrics seems to be technically much more involved than in the case considered above where all α_x, β_x are equal to 1.

The next simplest case to consider after the case of generalized bipartite metrics would be the class of metrics d for which $d(x,y) = 4$ for $x \neq y \in S$ and $d(x,y) = 2$ for $x \neq y \in T$ (i.e., all α_x's are equal to 2 and all β_x's to 1). One can characterize h-embeddability of these metrics by a similar reasoning as was applied to generalized bipartite metrics and, as a consequence, recognize them in polynomial time. Indeed, the metric $4\mathbb{1}_n$ is rigid for $n = 3$ and $n \geq 9$ and $4\mathbb{1}_n$ has exactly three $\mathbb{Z}_+$-realizations: its star realization and two special ones, for each $n \in \{4,5,6,7,8\}$ (cf. Proposition 23.4.4).

Another relatively simple case is when one of the sets S or T is small. Deza and Laurent [1995a] give a complete characterization of the hypercube embeddable metrics satisfying (24.2.9) in the case $|T| \leq 2$.

24.3 Metrics with Few Values

In this section, we consider the distances taking two values with distinct parities, and the distances taking three values, not all even and one of them being the sum of the other two. Namely, given $a, b \in \mathbb{Z}_+$, we consider the following classes of distances d:

(a) d takes the values $2a, b$, with b odd,

(b) d takes the values $a, b, a+b$, with a, b odd,

(c) d takes the values $2a, b, 2a+b$, with b odd and $b < 2a$, and

(d) d takes the values $2a, b, 2a+b$, with b odd and $2a < b$.

Laurent [1994] shows that the hypercube embeddability problem can be solved in polynomial time within each of these classes.

Theorem 24.3.1. *For fixed a, b, the hypercube embeddability problem within each of the classes* (a), (b), (c), (d) *can be solved in polynomial time.*

We sketch the proof of Theorem 24.3.1 in the rest of the section. It turns out that each of the classes (a), (b), (c), (d) has to be treated separately. The instance $a = b = 1$ of the class (c) was considered by Avis [1990], who showed that the hypercube embeddable distances with range of values $\{1,2,3\}$ can be recognized in polynomial time. The proof for the class (c) is essentially the same as in the subcase $a = b = 1$.

The basic steps of the proof are as follows. Let d be a distance on V_n from one of the classes (a), (b), (c), or (d). One first checks whether d satisfies the parity condition (24.1.1). If not, then d is not hypercube embeddable. Otherwise, let

(S,T) be the partition of V_n provided by Lemma 24.1.2 with, say, $|S| \geq |T|$. Set $n(a,b) := a^2 + a + 3$ if d belongs to the classes (a), (c), or (d), and $n(a,b) := (\frac{a+b}{2})^2 + \frac{a+b}{2} + 4$ if d belongs to the class (b).
If $n < 2n(a,b) - 1$, then one can test directly whether d is hypercube embeddable, for instance, by brute force enumeration (the number of operations in this step depends only on a, b, but may be exponential in a, b).
If $n \geq 2n(a,b) - 1$, then $|S| \geq n(a,b)$. Hence, the restriction of d to the set S is an h-rigid equidistant metric. Therefore, the points of S should be labeled by the star embedding (or an equivalent of it) in any h-labeling of d. For the classes (a), (b), (c), (d), this information enables us to completely characterize the hypercube embeddable distances on $n \geq 2n(a,b) - 1$ points by a set of conditions that can be checked in polynomial time.

We present below these characterizations for the classes (a), (b) and (c); see Propositions 24.3.8, 24.3.16, 24.3.18, and 24.3.27. We do not present here the results on the characterization of hypercube embeddability for the class (d), as they involve too many technical details.

We also have some partial results for the characterization of the hypercube embeddable distances on n points, for n arbitrary. See Propositions 24.3.9, 24.3.10, 24.3.11, and 24.3.17.

As mentioned above, we refer to Laurent [1994] for the study of the class (d). Characterization of hypercube embeddability within the class (d) needs many technical conditions. In some subcases, one needs conditions involving the existence of some designs, namely, of intersecting systems with prescribed parameters. Consider, for instance, the distance d from Figure 24.3.2. One can check that, if $|S| \geq a^2 + a + 3$, then d is hypercube embeddable if and only if $|U| \leq f(2a, a; a+b)$, i.e., if there exists a $(2a,a)$-intersecting system on V_{a+b} of cardinality $|U|$.

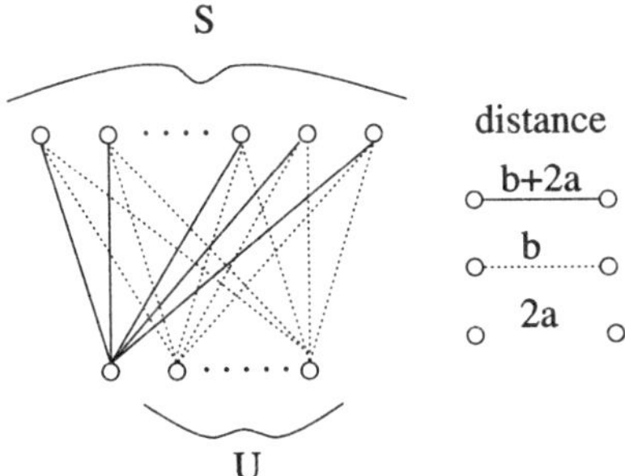

Figure 24.3.2

24.3.1 Distances with Values $2a, b$ (b odd)

Let a, b be positive integers with b odd. Let d be a distance on V_n with range of values $\{2a, b\}$. Suppose that d is a semimetric and satisfies the parity condition (24.1.1). Then, $b \geq a$ and let (S,T) be the partition of V_n provided by Lemma 24.1.2. Then an h-labeling of d consists of two set families $\mathcal{A}$ and $\mathcal{B}$ such that

$$(24.3.3)\qquad \begin{cases} \mathcal{A} \text{ is a } (b, b-a)\text{-intersecting system,} \\ \mathcal{B} \text{ is a } (2a, a)\text{-intersecting system,} \\ |A \cap B| = a \text{ for all } A \in \mathcal{A}, B \in \mathcal{B}, \\ |\mathcal{A}| = |S|, |\mathcal{B}| = |T| - 1. \end{cases}$$

Indeed, label a point $j_0 \in T$ by $\emptyset$, the remaining points of T by the members of $\mathcal{B}$, and the points of S by the members of $\mathcal{A}$.

Lemma 24.3.4.

(i) *If* $|T| = 1$, *then* d *is hypercube embeddable.*

(ii) *If* $b \geq 2a$, *then* d *is hypercube embeddable.*

(iii) *If* $b < 2a$ *and* $2 \leq |T| \leq |S| \leq \frac{a}{2a-b} + 1$, *then* d *is hypercube embeddable.*

(iv) *If* $b < 2a$ *and* d *is hypercube embeddable, then* $\min(|T|, |S| - 1) \leq \lfloor \frac{b}{2a-b} \rfloor$.

Proof. For (i), (ii), and (iii), we construct two families $\mathcal{A}$, $\mathcal{B}$ satisfying (24.3.3). In the three cases, we take for $\mathcal{A}$ a Δ-system with parameters $(b, b-a)$ and center A_0, $|A_0| = b - a$.
In case (i), take simply $\mathcal{B} = \emptyset$. In case (ii), as $b \geq 2a$, we can find a subset B_0 of A_0 with $|B_0| = a$. Then, we take for $\mathcal{B}$ a Δ-system with parameters $(2a, a)$ and center B_0 such that $(A \setminus A_0) \cap (B \setminus B_0) = \emptyset$ for all $A \in \mathcal{A}$, $B \in \mathcal{B}$.
In case (iii), we have $a \geq (s-1)(2a-b)$ (setting $s := |S|$), i.e., for each $A \in \mathcal{A}$, we can find $s-1$ disjoint subsets $A^{(1)}, \ldots, A^{(s-1)}$, of $A \setminus A_0$, each of cardinality $2a-b$. Note that $x := b - a + s(2a-b) \leq 2a$. Let $X^{(1)}, \ldots, X^{(s-1)}$ be disjoint sets of cardinality $2a - x$, disjoint from $\bigcup_{A \in \mathcal{A}} A$. Given $A_1 \in \mathcal{A}$, we set $\mathcal{B} := \{B^{(1)}, \ldots, B^{(s-1)}\}$ where, for $1 \leq j \leq s-1$,

$$B^{(j)} = A_0 \cup A_1^{(1)} \cup \bigcup_{A \in \mathcal{A} \setminus \{A_1\}} A^{(j)} \cup X^{(j)}.$$

Then, $\mathcal{A}$, $\mathcal{B}$ satisfy (24.3.3).
(iv) Suppose that $\min(|T|, |S| - 1) \geq \lfloor \frac{b}{2a-b} \rfloor + 1$. Let k be an integer such that

$$\min(|T|, |S| - 1) \geq k \geq \left\lfloor \frac{b}{2a-b} \right\rfloor + 1.$$

Set $b_i = 1$ for $k+1$ points of S, $b_i = -1$ for k points of T, and $b_i = 0$ for the remaining points of V_n. Then,

$$\sum_{i,j \in V_n} b_i b_j d(i,j) = 2k(k(2a-b) - b) > 0.$$

Hence, d violates a $(2k+1)$-gonal inequality and, thus, is not hypercube embeddable. ∎

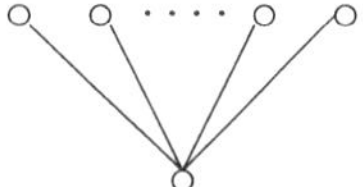
Figure 24.3.5

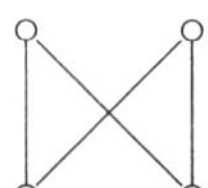
Figure 24.3.6

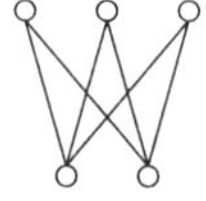
Figure 24.3.7

distance
2a
b

Proposition 24.3.8. *Let $a \leq b$ be positive integers with b odd. Let d be a distance on n points with range of values $\{2a, b\}$. If $n \geq 2a^2+2a+5$, then d is hypercube embeddable if and only if* (i) *d satisfies* (24.1.1) *and $b \geq 2a$ or* (ii) *d is the distance from Figure* 24.3.5.

Proof. Remains to show the "only if" part. Suppose that d is hypercube embeddable and $b < 2a$. Let $\mathcal{A}$ and $\mathcal{B}$ satisfying (24.3.3). By assumption, we have $|S| \geq a^2 + a + 3$. Hence, $\mathcal{A}$ is a $(b, b-a)$-intersecting system with $|\mathcal{A}| \geq a^2 + a + 3$. By Lemma 23.1.11, $\mathcal{A}$ is a Δ-system; let A_0 be its center, $|A_0| = b - a$. If $|T| \geq 2$, then $|\mathcal{B}| \geq 1$. Let $B \in \mathcal{B}$ and set $\alpha := |B \cap A_0|$. Then, $|B \cap (A \setminus A_0)| = a - \alpha$ for all $A \in \mathcal{A}$. Therefore,

$$2a = |B| \geq \alpha+|\mathcal{A}|(a-\alpha) = a|\mathcal{A}|-\alpha(|\mathcal{A}|-1) \geq a|\mathcal{A}|-(b-a)(|\mathcal{A}|-1) = (2a-b)|\mathcal{A}|+b-a,$$

which implies

$$|\mathcal{A}| \leq \frac{3a-b}{2a-b} = \frac{a}{2a-b} + 1.$$

This contradicts the fact that $|\mathcal{A}| = |S| \geq a^2 + a + 3$. Therefore, $|T| = 1$, i.e., d is the distance from Figure 24.3.5. ∎

Proposition 24.3.9. *Let $a \leq b$ be positive integers with b odd and $b \geq 2a$. Let d be a distance on n points with range of values $\{2a, b\}$. Then d is hypercube embeddable if and only if d satisfies* (24.1.1). ∎

Proposition 24.3.10. *Let $a \leq b$ be positive integers with b odd and $b < \frac{4}{3}a$. Let d be a distance with range of values $\{2a, b\}$. The following assertions are equivalent.*

(i) *d is hypercube embeddable.*

(ii) *d satisfies the parity condition* (24.1.1) *and the* 5*-gonal inequality (i.e., d does not contain as substructure the distance from Figure* 24.3.7*).*

(iii) *d is one of the distances from Figures* 24.3.5 *and* 24.3.6.

Proof. The implication (ii) $\Longrightarrow$ (iii) follows from Lemma 24.3.4 (iv), after noting that $\lfloor \frac{b}{2a-b} \rfloor = 1$ if $b < \frac{4}{3}a$. The distance from Figure 24.3.6 (i.e., the case $|S| = |T| = 2$) is indeed hypercube embeddable; label the two nodes of T by $\emptyset$ and $A \cup A'$, and the two nodes of S by $A_0 \cup A$ and $A_0 \cup A'$, where A_0, A, A' are disjoint sets of respective cardinalities $b - a, a, a$. ∎

Note that Proposition 24.1.9 is the case $a = b = 1$ of Proposition 24.3.10. So, we have a complete characterization of the hypercube embeddable distances with values in $\{2a, b\}$ (b odd) except when a, b satisfy: $\frac{4}{3}a \leq b < 2a$.

24.3.2 Distances with Values $a, b, a + b$ (a,b odd)

Let a, b be positive odd integers with $a < b$. Let d be a distance on V_n with range of values $\{a, b, a + b\}$. Suppose that d is a semimetric and satisfies the parity condition (24.1.1). Let (S, T) be the bipartition of V_n provided by Lemma 24.1.2. Hence, $d(i, j) = a + b$ for $i \neq j \in S$, $i \neq j \in T$, and $d(i, j) \in \{a, b\}$ for $i \in S, j \in T$. Moreover, the pairs ij with $d(i, j) = a$ form a matching.

Proposition 24.3.11. *If there are at least two pairs at distance a, then the following assertions are equivalent.*

(i) *d is hypercube embeddable.*

(ii) *d satisfies* (24.1.1) *and the* 5*-gonal inequality.*

(iii) *d is the distance from Figure* 24.3.12.

Proof. Let $i, i' \in S$, $j, j' \in T$ such that $d(i,j) = d(i',j') = a$. If there exists $k \in V_n \setminus \{i, i', j, j'\}$, then set $b_i = b_{i'} = b_k = 1$, $b_j = b_{j'} = -1$, and $b_h = 0$ for the remaining points. Then,

$$\sum_{i,j \in V_n} b_i b_j d(i,j) = 4a > 0,$$

i.e., d violates a 5-gonal inequality. This shows (ii) $\Longrightarrow$ (iii). If $V_n = \{i, i', j, j'\}$ then d is indeed hypercube embeddable; let A and B be disjoint sets with $|A| = a$ and $|B| = b$ and label i by A, i' by B, j by $\emptyset$, and j' by $A \cup B$. This shows (iii) $\Longrightarrow$ (i). ∎

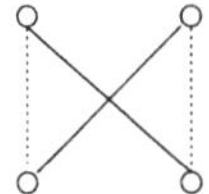

Figure 24.3.12

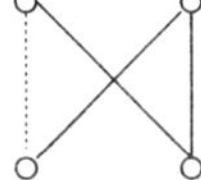

Figure 24.3.13

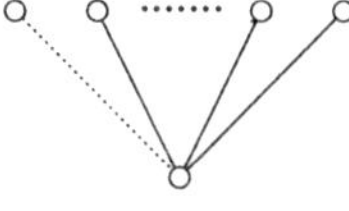

Figure 24.3.14

distance

a+b

a

b

From now on, we can suppose that there is exactly one pair (i_0, j_0) at distance a, where $i_0 \in S$, $j_0 \in T$. In an h-labeling of d, we can suppose that j_0 is labeled by $\emptyset$ and, then, i_0 should be labeled by a set A_0 of cardinality a. Therefore, an h-labeling of d exists if and only if there exist two set families $\mathcal{A}$ and $\mathcal{B}$ such that

$$(24.3.15) \qquad \begin{cases} \mathcal{A}, \mathcal{B} \text{ are } (b, \frac{b-a}{2}) - \text{intersecting systems,} \\ |A \cap B| = \frac{a+b}{2} \text{ for all } A \in \mathcal{A}, B \in \mathcal{B}, \\ A \cap A_0 = B \cap A_0 = \emptyset \text{ for all } A \in \mathcal{A}, B \in \mathcal{B}, \\ |\mathcal{A}| = |S| - 1, |\mathcal{B}| = |T| - 1. \end{cases}$$

Indeed, label the points of $S \setminus \{i_0\}$ by the members of $\mathcal{A}$ and the points of $T \setminus \{j_0\}$ by $A_0 \cup B$ where $B \in \mathcal{B}$.

Proposition 24.3.16. *Let $a < b$ be odd integers and let d be a distance on $n \geq 2(\frac{a+b}{2})^2 + a + b + 7$ points with range of values $\{a, b, a+b\}$ which is not the distance from Figure* 24.3.12. *Then, d is hypercube embeddable if and only if d is the distance from Figure* 24.3.14.

Proof. The distance from Figure 24.3.14 is hypercube embeddable; indeed choose for $\mathcal{A}$ a Δ-system. Conversely, suppose that d is hypercube embeddable. By assumption,

$$|S| \geq (\frac{a+b}{2})^2 + (\frac{a+b}{2}) + 4.$$

Hence, $\mathcal{A}$ is a $(b, \frac{b-a}{2})$-intersecting system with

$$|\mathcal{A}| \geq (\frac{a+b}{2})^2 + \frac{a+b}{2} + 3.$$

By Lemma 23.1.11, $\mathcal{A}$ is a Δ-system; let A_1 be its center, $|A_1| = \frac{b-a}{2}$. Suppose that $|T| \geq 2$ and let $B \in \mathcal{B}$. Then, $|B \cap (A \setminus A_1)| \geq a$ for all $A \in \mathcal{A}$, implying $b = |B| \geq a|\mathcal{A}|$, in contradiction with the above assumption on $|\mathcal{A}|$. Therefore, $|T| = 1$, i.e., d is the distance from Figure 24.3.14. ∎

Proposition 24.3.17. *Let a, b be odd integers such that $a < b < 2a$. Let d be a distance with range of values $\{a, b, a+b\}$. Then, d is hypercube embeddable if and only if d is one of the distances from Figures* 24.3.12, 24.3.13, *and* 24.3.14.

Proof. Suppose that d is hypercube embeddable and that d is not the distance from Figure 24.3.12. Set $k := \min(|T|, |S|-1)$. If $k \geq 2$, then $k \leq \lfloor \frac{b}{a} \rfloor$ (else, d violates a $(2k+1)$-gonal inequality). Hence, $k = 1$, which implies that d is the distance from Figures 24.3.14 or 24.3.13. ∎

24.3.3 Distances with Values $b, 2a, b+2a$ (b odd, $b < 2a$)

Proposition 24.3.18. *Let a, b be positive integers with b odd and $b < 2a$. Let d be a distance on $n \geq 2a^2 + 2a + 5$ points with range of values $\{2a, b, 2a+b\}$. The following assertions are equivalent.*

(i) *d is hypercube embeddable.*

(ii) *d is a semimetric, d satisfies* (24.1.1) *and d does not contain as substructure any of the distances from Figures* 24.3.19-24.3.26.

In particular, if $b < a$, then d is hypercube embeddable if and only if d is a semimetric and satisfies (24.1.1).

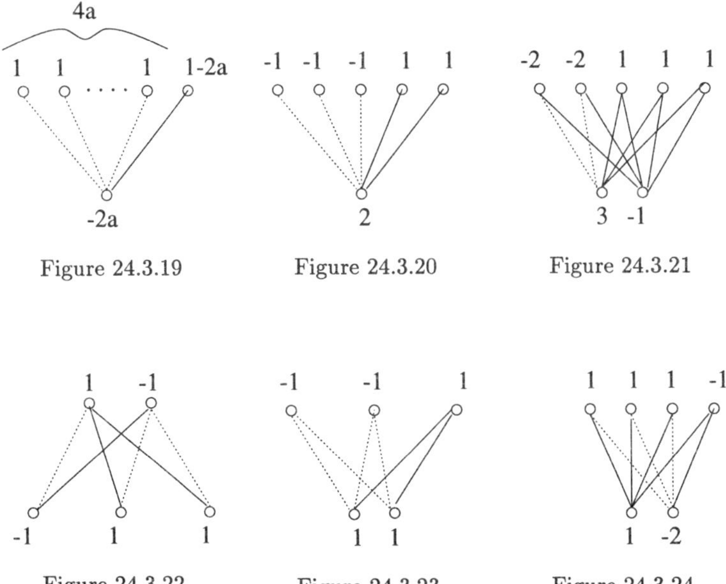

Figure 24.3.19 Figure 24.3.20 Figure 24.3.21

Figure 24.3.22 Figure 24.3.23 Figure 24.3.24

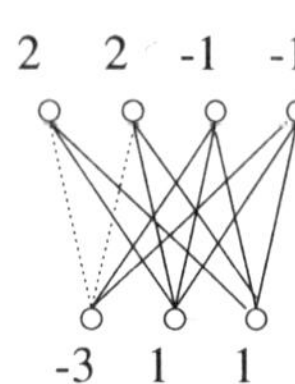

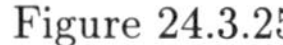

Figure 24.3.25

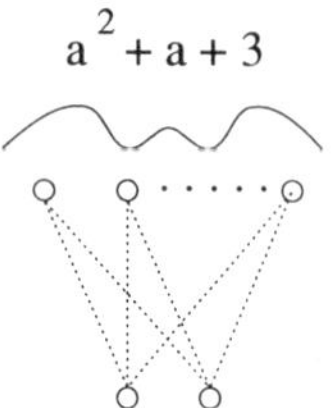

Figure 24.3.26

In Figures 24.3.19-24.3.26, a plain edge represents distance $2a+b$, a dotted edge distance b and no edge means distance $2a$.

Proof. For the implication (i) $\Longrightarrow$ (ii), we check that none of the distances from Figures 24.3.19-24.3.26 is hypercube embeddable. Indeed, the distances from Figures 24.3.19-24.3.25 violate some hypermetric inequality. The numbers assigned to the nodes in Figures 24.3.19-24.3.25 indicate a choice of integers b_i's for which the hypermetric inequality (24.1.5) is violated. For instance, for the distance from Figure 24.3.19,

$$\sum_{i,j\in V_n} b_i b_j d(i,j) = 4a(2a(2a-b)-b) \geq 4a > 0$$

since $2a - b \geq 1$. The distance from Figure 24.3.26 is not hypercube embeddable by Proposition 24.3.8 (and its proof).

We show the implication (ii) $\Longrightarrow$ (i). As d satisfies the parity condition, V_n is partitioned into $S \cup T$ with $|S| \geq |T|$, $d(i,j) = 2a$ for $(i,j) \in S^2 \cup T^2$, $d(i,j) \in \{b, b+2a\}$ for $(i,j) \in S \times T$. Set $s := |S|$. For $j \in T$, set

$$N_b(j) := \{i \in S \mid d(i,j) = b\}.$$

For $v \in \{0, 1, 2, \ldots, s-1, s\}$, set

$$T_v := \{j \in T \mid |N_b(j)| = v\}.$$

We group below several observations on the sets T_v.

(i) $T_{s-1} = \emptyset$ (since d does not contain the configuration from Figure 24.3.19).
(ii) $|T_s| \leq 1$ (since d does not contain the configuration from Figure 24.3.26).
(iii) All T_v are empty except maybe T_0, T_1, T_2, T_s (indeed, $|N_b(j)| \leq 2$ or $|N_b(j)| \geq s-1$ for all $j \in T$, since d does not contain the substructure from Figure 24.3.20).
(iv) At least one of T_0 and T_2 is empty (since d does not contain the substructure from Figure 24.3.21).
(v) If $|T_1| \geq 2$, then
(v1) either all $N_b(j)$ ($j \in T_1$) are equal,
(v2) or all $N_b(j)$ ($j \in T_1$) are distinct
(since d does not contain the substructure from Figure 24.3.22).
(vi) If $j \neq j' \in T_2$, then $|N_b(j) \cap N_b(j')| = 1$ (use Figures 24.3.22 and 24.3.23).
(vii) If $j \in T_1$ and $j' \in T_2$, then $N_b(j) \cap N_b(j') \neq \emptyset$ (by Figure 24.3.22).
(viii) If $b < a$, then $T_2 = T_s = \emptyset$ (by the triangle inequality).

We show how to construct an h-labeling of d. Let A_i ($i \in S$) be disjoint sets of cardinality a. Set $A := \cup_{i\in S} A_i$. Label the elements of S by the A_i's.

Suppose first that $b < a$. Then, by (viii), $d(i_1, j_1) = \ldots = d(i_r, j_r) = b$ for some $i_1, \ldots, i_r \in S$, $j_1, \ldots, j_r \in T$, $1 \leq r \leq |T|$. Let X, B_j ($j \in T \setminus \{j_1, \ldots, j_r\}$) be pairwise disjoints sets that are disjoint from A and satisfy $|X| = b$, $|B_j| = a$. Label $j_1, \ldots, j_r$ by $A_{i_1} \cup X, \ldots, A_{i_r} \cup X$, respectively, and $j \in T \setminus \{j_1, \ldots, j_r\}$ by $X \cup B_j$. This gives an h-labeling of d.

We now suppose that $b \geq a$. Let X be a set disjoint from A with $|X| = b - a$.
- If $T_s \neq \emptyset$ then $T_s = \{x\}$ (by (i)); label x by X.
- Label each element $j \in T_2$ by $\bigcup_{i \in N_b(j)} A_i \cup X$ (this gives already an h-labeling of the projection of d on $S \cup T_s \cup T_2$ (by (vi))).
- Suppose that all $N_b(j)$ ($j \in T_1$) are equal to, say, $\{i_0\}$, as in ($v1$). Let Y_j ($j \in T_1$) be pairwise disjoint sets that are disjoint from A and X and have cardinality a. Label $j \in T_1$ by $A_{i_0} \cup X \cup Y_j$.
If all $N_b(j)$ ($j \in T_1$) are distinct as in ($v2$), then label $j \in T_1$ by $\bigcup_{i \in N_b(j)} A_i \cup X \cup Y$, where Y is a set disjoint from A and X with $|Y| = a$.
(In both cases, we have obtained an h-labeling of the projection of d on $S \cup T_s \cup T_2 \cup T_1$ (by (vii)).)
- Suppose that $T_0 \neq \emptyset$. Then, $T_2 = \emptyset$ by (iv). Let Z_k ($k \in T_0$) be pairwise disjoint sets that are disjoint from all the sets constructed so far and have cardinality a.
If we are in case ($v1$), then $|T_1| \leq 1$ or ($|T_1| \leq 2$ and $|T_0| = 1$). (Indeed, if $|T_1|, |T_2| \geq 2$, then d contains the substructure from Figure 24.3.25 and, if $|T_1| \geq 3$, $|T_0| = 1$, then we have the substructure from Figure 24.3.24.) If $|T_1| = 1$, $T_1 = \{j\}$, label $k \in T_0$ by $X \cup Y_j \cup Z_k$. If $|T_1| = 2$, $T_1 = \{j, j'\}$, then label the unique element $k \in T_0$ by $X \cup Y_j \cup Y_{j'}$.
Else, we are in case ($v2$). Then, label $k \in T_0$ by $X \cup Y \cup Z_k$.
In both cases, we have constructed an h-labeling of d. ∎

Observe that the exclusion of the distance from Figure 24.3.26 is used only for showing that $|T_s| \leq 1$, i.e., that at most one point is at distance b from all points of S. Consider the distance d_s on $s + 2$ points which has the same configuration as in Figure 24.3.26 but with s nodes on the top level instead of $a^2 + a + 3$. Let $s(a, b)$ denote the largest integer s such that d_s is hypercube embeddable. Then, Proposition 24.3.18 remains valid if we exclude the distance $d_{s(a,b)+1}$ instead of excluding the distance d_{a^2+a+3} from Figure 24.3.26. Note that

$$2 \leq \frac{a}{2a - b} + 1 \leq s(a, b) \leq a^2 + a + 2,$$

with $s(a, b) = 2$ if $b < \frac{4}{3}a$ (use Proposition 24.3.10). This implies the following result.

Proposition 24.3.27. *Let a, b be positive integers with b odd and $b < \frac{4}{3}a$. Let d be a distance on $n \geq 2a^2 + 2a + 5$ points with range of values $\{2a, b, 2a + b\}$. The following assertions are equivalent.*

(i) *d is hypercube embeddable.*
(ii) *d is ℓ_1-embeddable and satisfies* (24.1.1).
(iii) *d is hypermetric and satisfies* (24.1.1).
(iv) *d satisfies* (24.1.1) *and the $(2k+1)$-gonal inequalities for $2k + 1 = 5, 7, 11, 8a - 1$.*
(v) *d is a semimetric, d satisfies* (24.1.1), *and d does not contain as substructure any of the distances from Figures* 24.3.7 *and* 24.3.19-24.3.25. ∎

Note that Proposition 24.3.27 is a direct extension of the result given in Avis [1990] for the subcase $a = b = 1$.

24.4 Truncated Distances of Graphs

We consider here the hypercube embeddability problem for the distances taking two values of the form a, $2a$, where $a \geq 1$ is an integer. Such distances can be interpreted as scale multiples of truncated distances of graphs. Indeed, for a distance d on V with values a, $2a$, let $G = (V, E)$ denote the graph with edges the pairs at distance a. Then, d coincides with ad_G^*, where d_G^* is the truncated distance of G (taking value 1 on an edge and value 2 on a non edge). (The truncated distance has already been considered earlier in Chapter 17.)

As we see here, the hypercube embeddability problem can be solved in polynomial time for the class of distances with values a, $2a$. The case when a is odd has already been settled in the previous section (cf. Proposition 24.3.10). Namely, such distances are hypercube embeddable if and only if they satisfy the parity condition (24.1.1) and the pentagonal inequalities. Hence, it suffices to verify that all subspaces on at most five points are hypercube embeddable in order to ensure that the whole distance space is hypercube embeddable. A similar finite point criterion has been discovered by Assouad and Delorme [1980, 1982] in the case when a is even. The proof uses again as a basic tool the fact that some equidistant submetrics are h-rigid, but the details are more involved in this case.

Theorem 24.4.1. *Let (X, d) be a finite distance space with range of values $\{2k, 4k\}$ where $k \geq 1$ is an integer. Set $N_k := (3M^2R+R+2M^2)(M+1)+8k+4$, where $M := 4k^2 + 4k + 3$ and R denotes the Ramsey number[1] $r(M, M)$. Then, (X, d) is hypercube embeddable if and only if (Y, d) is hypercube embeddable for every subset $Y \subseteq X$ with $|Y| \leq N_k$.*

The value of N_k given in this theorem could certainly be improved (at the cost of a more detailed analysis). For instance, Assouad and Delorme [1980] show that, in the case $k = 1$, N_2 can be replaced by 120. The value given here suffices, however, for the purpose of demonstrating polynomial-time solvability.

The rest of the section is devoted to the proof of Theorem 24.4.1. We introduce some notation. For convenience, we visualize a distance d on V with values $2k, 4k$ as $d = 2kd_G^*$, where $G = (V, E)$ is the graph with edges the pairs at distance $2k$. Then, isometric subspaces of (V, d) correspond to induced subgraphs of G. Then we introduce the class of graphs $\mathcal{F}_k$ which consists of the graphs G for which $2kd_G^*$ is hypercube embeddable. Our aim is to show that a graph G belongs to $\mathcal{F}_k$ whenever all its induced subgraphs on at most N_k nodes belong to $\mathcal{F}_k$. It is convenient to formalize this notion. A class $\mathcal{C}$ of graphs is said to have *order of congruence*[2] p if $\mathcal{C}$ is closed under taking induced subgraphs and

[1] We remind that, given integers $s, t \geq 1$, the *Ramsey number* $r(s, t)$ denotes the smallest integer n such that every graph on n nodes contains either a clique of size s or a stable set of size t. Some estimations for $r(s, s)$ are known; for instance, $2^{\frac{s}{2}} \leq r(s, s) \leq \binom{2s-2}{s-1}$. (Cf., e.g., Bondy and Murty [1976].)

[2] The notion of order of congruence has already been introduced earlier; it is, however, used

if a graph belongs to $\mathcal{C}$ whenever all its induced subgraphs on at most p nodes belong to $\mathcal{C}$, and p is the smallest such integer. Thus, we have to show that $\mathcal{F}_k$ has order of congruence $\leq N_k$. For this, we decompose $\mathcal{F}_k$ into two subfamilies $\mathcal{G}_k$ and $\mathcal{F}_k \setminus \mathcal{G}_k$ whose orders of congruence are easier to determine. Namely, let $\mathcal{G}_k$ denote the class of graphs G satisfying, either (i) $\nabla G \in \mathcal{F}_k$, or (ii) $G \in \mathcal{F}_k$ and G is a suspension graph (i.e., has a node adjacent to all other nodes). We will also consider a class $\mathcal{H}_k$ (to be described later) such that $\mathcal{F}_k \setminus \mathcal{G}_k \subseteq \mathcal{H}_k \subseteq \mathcal{F}_k$. Then, the main steps of the proof consist of showing that

(i) $\mathcal{G}_k$ has order of congruence $\leq 8k+4$,

(ii) $\mathcal{H}_k$ has order of congruence $\leq (3M^2R + R + 2M^2)(M+1)$.

The result from Theorem 24.4.1 follows in view of the following (easy) lemma.

Lemma 24.4.2. *Let $\mathcal{A}$ and $\mathcal{B}$ be families of graphs having respective orders of congruence a and b. Then, $\mathcal{A} \cup \mathcal{B}$ has order of congruence $\leq a+b$.* ∎

We now determine the order of congruence of $\mathcal{G}_k$ and the next subsection will study the order of congruence of $\mathcal{H}_k$.

Proposition 24.4.3. *The family $\mathcal{G}_k$ has order of congruence $\leq 8k+4$.*

Proof. Define $\mathcal{I}_k$ as the subfamily of $\mathcal{G}_k$ consisting of the graphs G whose suspension ∇G belongs to $\mathcal{F}_k$. In a first step, we show that

(a) $\qquad \mathcal{I}_k$ has order of congruence $\leq 4k+2$.

For this, let $G = (V, E)$ be a graph such that $G[X] \in \mathcal{I}_k$ for all $X \subseteq V$ with $|X| \leq 4k+2$; we show that $G \in \mathcal{I}_k$, i.e., that ∇G is hypercube embeddable with scale $2k$. We can assume without loss of generality that G is connected. By the assumption, we know that ∇G is ℓ_1-embeddable (as $4k+2 \geq 6$ and the family of graphs whose suspension is ℓ_1-embeddable has order of congruence 6 by Theorem 17.1.8). Therefore, G is a line graph or a subgraph of a cocktail-party graph (again by Theorem 17.1.8). If G is a line graph we are done, since $2d^*_{\nabla G}$ is hypercube embeddable. Hence, G is contained in a cocktail-party graph $K_{n\times 2}$. We can assume without loss of generality that G has exactly one pair of opposite nodes. In other words, $G = K_{n+1} \backslash e$ and $2kd^*_{\nabla G} = \text{ant}_{4k}(2k\mathbb{1}_{n+1})$. Let $c(k)$ denote the largest integer such that $\text{ant}_{4k}(2k\mathbb{1}_{c(k)})$ is hypercube embeddable. Then, $c(k) \leq 4k+1$ by Theorem 23.3.1 (i). Hence, $G \in \mathcal{I}_k$ if $n+1 \leq c(k)$. On the other hand, if $n \geq c(k)$ then we can choose a subset X of V of size $c(k)+1$ containing both end nodes of the deleted edge e. Now $G[X] \in \mathcal{I}_k$ since $|X| \leq 4k+2$. This implies that $\text{ant}_{4k}(2k\mathbb{1}_{c(k)+1})$ is hypercube embeddable, contradicting the definition of $c(k)$. Hence, (a) holds.

We can now proceed with the proof. Let $G = (V, E)$ be a graph such that $G[X] \in \mathcal{G}_k$ for all $|X| \leq 8k+4$; we show that $G \in \mathcal{G}_k$. We can suppose that $G \notin \mathcal{I}_k$, else we are done. By (a), there exists $Y \subseteq V$ with $|Y| \leq 4k+2$ such

here in a more restrictive sense.

that $G[Y] \not\in \mathcal{I}_k$, i.e., $\nabla G[Y] \not\in \mathcal{F}_k$. For any $Z \subseteq V$ with $|Z| \leq 4k+2$, we have that $G[Y \cup Z] \in \mathcal{G}_k$, which implies that $G[Y \cup Z]$ is a suspension graph with apex node $z \in Y$. Then, G is a suspension graph with apex $y \in Y$. (For, suppose that for every $y \in Y$ there exists $z_y \in Z$ not adjacent to y; then a contradiction is reached by considering $Z := \{z_y \mid y \in Y\}$.) One can now verify that $G[V \setminus \{y\}]$ belongs to $\mathcal{I}_k$ with the help of (a). This shows that $G \in \mathcal{F}_k$ and, thus, $G \in \mathcal{G}_k$. ∎

Next, we consider the class $\mathcal{F}_k \setminus \mathcal{G}_k$. Some further definitions are needed. We remind that $M := 4k^2 + 4k + 3$ and $R = r(M, M)$ is the Ramsey number. Let $G = (V, E)$ be a graph. We call *M-clique* (resp. *M-stable set*) any maximal clique (resp. stable set) of size $\geq M$. Then, V_0 denotes the union of all M-cliques and V_1 denotes the union of their pairwise intersections. A subset $L \subseteq V_0$ is called *transversal* if $G[L]$ is connected and L meets every M-clique in at most one point. Finally, V_2 denotes the union of the transversals that have a node adjacent to a node of $V \setminus V_0$.

We now introduce the class $\mathcal{H}_k$ which consists of the graphs $G \in \mathcal{F}_k$ satisfying the following conditions (i)-(vi):

(i) G has no M-stable set.

(ii) G has at most M distinct M-cliques.

(iii) Any two distinct M-cliques meet in at most two points.

(iv) A node adjacent to 4 nodes of an M-clique K belongs to K.

(v) $G[L]$ is a complete graph for every transversal L.

(vi) Maximal transversals are pairwise disjoint.

Note that, for a graph satisfying (i)-(vi), $|V \setminus V_0| \leq R$ (by the definition of V_0 and R), $|V_1| \leq M^2$, and $|V_2| \leq 3M^2R$.

We show in the next subsection that $\mathcal{F}_k \setminus \mathcal{G}_k$ is contained in $\mathcal{H}_k$ and we study the order of congruence of $\mathcal{H}_k$.

24.4.1 The Class of Graphs $\mathcal{H}_k$

Our aim here is to show that $\mathcal{F}_k \setminus \mathcal{G}_k \subseteq \mathcal{H}_k$ and to determine the order of congruence of $\mathcal{H}_k$. In what follows, $G = (V, E)$ is a graph in $\mathcal{F}_k \setminus \mathcal{G}_k$. Hence, $2kd_G^*$ can be isometrically embedded into some hypercube $H(\Omega)$. That is, there is a mapping:

$$x \in V \mapsto X \subseteq \Omega$$

such that $|X \triangle Y| = 2k$ if xy is an edge and $|X \triangle Y| = 4k$ otherwise. (We will use below the same letters x and X for denoting a node of G and the subset labeling it in the hypercube; we may sometimes identify both notions and speak of node X.)

We will use the following fact (cf. Theorem 22.0.6 or 23.1.9): The sets Y labeling the nodes y of an M-clique K form a Δ-system (because $M = 4k^2 +$

$4k+3$). That is, there exists a set C such that $C \subseteq Y$, $|Y \setminus C| = k$ for all $y \in K$ and the sets $Y \setminus C$ are pairwise disjoint; the set C is called the *center* of the M-clique K. A similar result holds for M-stable sets (as $M \geq k^2+k+3$).

Lemma 24.4.4. *G has no M-stable set.*

Proof. Suppose, for contradiction, that S is an M-stable set. We can assume without loss of generality that the nodes $y \in S$ are labeled by pairwise disjoint sets Y, each of cardinality $2k$. We claim:

(a) Every node $x \in V \setminus S$ is nonadjacent to some node of S.

For, suppose that $x \in V \setminus S$ is adjacent to all nodes of S. Then, x is labeled by $X = \emptyset$ (by the pentagonal inequality applied to the nodes $Y_1, Y_2, Y_3 \in S$, X and $\emptyset$). This implies that x is adjacent to all nodes of G, contradicting the fact that $G \notin \mathcal{G}_k$. Indeed, let $x' \in V \setminus S$ ($x' \neq x$) be labeled by X'; if $|X'| = 4k$ then $|S| \leq 4$ (since $|X' \triangle Y| \leq 4k$ implies that $|X' \cap Y| \geq k$ for each $y \in S$). Therefore, (a) holds.

Let $x \in V \setminus S$ be labeled by X and let $y \in S$ which is not adjacent to x; thus, $|X \triangle Y| = 4k$. Then, $2k \leq |X| \leq 6k$ (by the triangle inequality applied to $\emptyset$,X,Y). If $|X| > 2k$ then $|X \cap Y| = \frac{1}{2}(|X| - 2k) > 0$ for every node $y \in S$ nonadjacent to x; hence, $|X| \geq |S| - 2 > 6k$ since x can be adjacent to at most two nodes of S (by the pentagonal inequality). Therefore, $|X| = 2k$. So, we have found a hypercube embedding of $2kd^*_{\nabla G}$ (labeling the apex node by $\emptyset$). This contradicts the assumption that $G \notin \mathcal{G}_k$. ∎

Lemma 24.4.5. *Let K be an M-clique with center C and let $x \in V \setminus K$ be labeled by X. Then, $|X \triangle C| = 3k$.*

Proof. By the maximality of K, there exists $y \in K$ which is not adjacent to x; so $|X \triangle Y| = 4k$. Thus, $3k \leq |X \triangle C| \leq 5k$ (by the triangle inequality applied to X, Y, C). Clearly, $|X \triangle C| = 3k$ if x is adjacent to some $y \in K$. Suppose now that x is not adjacent to any $y \in K$. Then, for $y \in K$, $4k = |X \triangle Y| = |(X \triangle C) \triangle (C \triangle Y)|$ and, thus, $|X \triangle C| - 3k = 2|(X \setminus C) \cap (Y \setminus C)|$. This implies that $|X \triangle C| = 3k$ (else $|X \setminus C| \geq M$). ∎

Lemma 24.4.6. *Two distinct M-cliques meet in at most two points.*

Proof. Suppose that y_1, y_2, y_3 are distinct elements in two distinct cliques K, K' with respective centers C, C'. The pentagonal inequality $Q(1,1,1,-1,-1)^T x \leq 0$ applied to the distance on the points Y_1, Y_2, Y_3, C, C' (assigning 1 to Y_1, Y_2, Y_3 and -1 to C, C') implies that $C = C'$. This contradicts the triangle inequality applied to $C = C'$, $Y \in K \setminus K'$ and $Y' \in K' \setminus K$. ∎

Lemma 24.4.7. *Let K and K' be distinct cliques with respective centers C and C'. Then, $|C \triangle C'| = 2k$.*

Proof. Let $y' \in K' \backslash K$ be labeled by Y'. Then, by Lemma 24.4.5 and the triangle inequality applied to C, C', Y', we obtain that $2k \leq |C \triangle C'| \leq 4k$. Moreover, $3k = |C \triangle Y'| = |(C \triangle C') \triangle (C' \triangle Y')|$ implying that $2|(Y' \setminus C') \cap (C \setminus C')| = |C \triangle C'| - 2k$. Therefore, $|C \triangle C'| = 2k$ for, otherwise, $|C \setminus C'| \geq |K' \setminus K|$ which yields: $M \leq |K'| = |K \cap K'| + |K' \setminus K| \leq 4k + 2$. ∎

Lemma 24.4.8. *There are at most M distinct M-cliques.*

Proof. Let $K_1, \ldots, K_N$ be the distinct M-cliques with respective centers $C_1, \ldots, C_N$ and suppose that $N > M$. Then, there exists a set C such that $|C \triangle C_i| = k$ for all $i = 1, \ldots, N$ (since, by Lemma 24.4.7, we have a $2k$-valued equidistant metric on $C_1, \ldots, C_N$). We claim that $|C \triangle X| = 2k$ for all $x \in V$; this implies that $2kd^*_{\nabla G}$ is hypercube embeddable, contradicting our assumption that $G \notin \mathcal{G}_k$. First, suppose that x belongs to some M-clique K_i. Then, $|C \triangle X| \leq |C \triangle C_i| + |C_i \triangle X| = 2k$ and $3k = |X \triangle C_j| \leq |X \triangle C| + |C \triangle C_j| = |X \triangle C| + k$, which shows that $|C \triangle X| = 2k$. Suppose now that $x \notin \bigcup_{i=1}^N K_i$. Then, $|X \triangle C| \leq |X \triangle C_i| + |C_i \triangle C| = 4k$. On the other hand, $3k = |X \triangle C_i| = |(X \triangle C) \triangle (C \triangle C_i)|$ yielding $2|(X \setminus C) \cap (C_i \setminus C)| = |X \triangle C| - 2k$. Henceforth, $|X \triangle C| = 2k$ for, otherwise, $|X \setminus C| \geq M$. ∎

Lemma 24.4.9. *Let K be an M-clique and let $x \in V \setminus K$. Then, x is adjacent to at most three points of K.*

Proof. If x is adjacent to 4 points of K then their labeling sets together with the center C of K provide a hypercube embedding for the metric $\mathrm{ant}_{3k}(2k\mathbb{1}_5)$. This cannot be, since $3k < s_h(2k\mathbb{1}_5)$ (indeed, $s_h(2k\mathbb{1}_5)) \geq \frac{2k}{6}\binom{5}{2} = \frac{10}{3}k > 3$; recall Proposition 7.2.6). ∎

Lemma 24.4.10. *Every transversal is a clique and maximal transversals are pairwise disjoint.*

Proof. Suppose that x, y, z are distinct points of a transversal L such that y is adjacent to x and z and x, z are not adjacent. Let $K \neq K'$ be M-cliques containing respectively x and y and let C, C' be their centers. We obtain a contradiction by applying the pentagonal inequality to X, Z, C', Y, C (assigning 1 to X, Z, C' and -1 to Y, C).

Suppose now that L_1 and L_2 are nondisjoint maximal transversals. Let $x \in L_1 \cap L_2$, $y \in L_1 \backslash L_2$ and $z \in L_2 \backslash L_1$, let K, K' be M-cliques containing respectively x and y, with centers C and C'. We have a contradiction by applying the pentagonal inequality to Y, Z, C, X, C' (assigning 1 to Y, Z, C and -1 to X, C'). ∎

Proposition 24.4.11. *The order of congruence of the class $\mathcal{H}_k$ is less than or equal to $(3M^2R + R + 2M^2)(M + 1)$.*

Proof. Let $G = (V, E)$ be a graph such that $G[U] \in \mathcal{H}_k$ for all $U \subseteq V$ with $|U| \leq N(M+1)$ where $N := 3M^2R + R + 2M^2$; we show that $G \in \mathcal{H}_k$. The only thing to verify is that $G \in \mathcal{F}_k$, as G satisfies obviously (i)-(vi) (since $N(M+1)$ is large enough). In particular, we have that $|V \setminus V_0| \leq R$, $|V_1| \leq M^2$ and $|V_2| \leq 3M^2R$.

Let $L_1, \ldots, L_p$ denote the maximal transversals that are disjoint from V_2 (i.e., having no node adjacent to a node of $V \setminus V_0$); we have that $p \geq 1$ (else we are done since $|V| \leq N(M+1)$ as $V_0 \subseteq V_2$). Let V^* be a subset of V containing $(V \setminus V_0) \cup V_1 \cup V_2$ and meeting every M-clique of G in M points. Then, $|V^*| \leq N$. Finally, set $V^{**} := V^* \cup \bigcup_{i \in I} L_i$, where I consists of the indices $i \in [1, p]$ for which $L_i \cap V^* \neq \emptyset$ and $L_i \not\subseteq V^*$. Obviously, $|I| \leq |V^*|$ which implies that $|V^{**}| \leq N(M+1)$.

Therefore, $G[V^{**}] \in \mathcal{H}_k$ and, thus, we have an embedding $x \in V^{**} \mapsto X \subseteq \Omega^*$ in some hypercube $H(\Omega^*)$ for the distance $2kd^*_{G[V^{**}]}$. We now indicate how to extend this embedding to the whole graph G.

Let Ω_i ($i \in [1, p] \setminus I$) be disjoint sets, disjoint from Ω^*, and each having cardinality k. For each M-clique $K \cap V^{**}$ of $G[V^{**}]$, we let C_K denote its center. It remains to label the nodes of $V \setminus V^{**}$. A node $x \in V \setminus V^{**}$ belongs to a unique M-clique K and the (unique) maximal transversal containing x is L_i for some $i \in [1, p] \setminus I$. We then label x by the set $X := C_K \cup \Omega_i$. We have to verify that this gives a correct labeling, i.e., that, for $y \in V$ labeled by Y, $|X \triangle Y| = 2k$ if x, y are adjacent and $|X \triangle Y| = 4k$ otherwise.

Suppose first that x, y are not adjacent. If $y \in V^{**}$, then $|C_K \triangle Y| = 3k$ (by Lemma 24.4.5) which gives $|X \triangle Y| = 4k$. If $y \in V \setminus V^{**}$, then y is labeled by $C_{K'} \cup \Omega_j$ where $K' \neq K$ and $i \neq j$, which gives again $|X \triangle Y| = 4k$ since $|C_K \triangle C_{K'}| = 2k$ (by Lemma 24.4.7).

Suppose now that x, y are adjacent. If $y \in K \cap V^{**}$ then $|Y \triangle C_K| = k$ and, thus, $|X \triangle Y| = 2k$. If $y \in K \setminus V^{**}$ then $Y = C_K \cup \Omega_j$ with $i \neq j$ (else $x, y \in L_i \cap K$), which yields again that $|X \triangle Y| = 2k$. Now, if $y \notin K$ then $y \in L_i$ since $\{x, y\}$ is transversal. This implies that $y \notin V^{**}$ (as $i \notin [1, p]$). Let K' be the M-clique containing y. Then, $Y = C_{K'} \cup \Omega_i$ and, thus, $|X \triangle Y| = |C_K \triangle C_{K'}| = 2k$. ∎

24.5 Metrics with Restricted Extremal Graph

let d be a metric on V_n. Given distinct $i, j \in V_n$, the pair ij is said to be *extremal* for d if there does not exist $k \in V_n \setminus \{i, j\}$ such that

$$d(i,k) = d(i,j) + d(j,k) \quad \text{or} \quad d(j,k) = d(i,j) + d(i,k).$$

Then, the *extremal graph* of d is defined as the subgraph of K_n formed by the set of extremal edges of d. The notion of extremal graph turns out to be useful when studying the metrics that can be decomposed as a nonnegative (integer) sum of cut semimetrics. Assertion (i) in Theorem 24.5.1 was proved by Papernov [1976] and (ii) by Karzanov [1985].

Theorem 24.5.1. *Let d be a metric on V_n whose extremal graph is either K_4, or C_5, or a union of two stars*[3]. *Then,*

(i) *d is ℓ_1-embeddable, i.e., $d \in \mathrm{CUT}_n$.*

(ii) *d is hypercube embeddable if and only if d satisfies the parity condition (24.1.1).*

Note that it suffices to show Theorem 24.5.1 (ii), as it implies (i). The proof that we present below was given by Schrijver [1991]. It is shorter than Karzanov's original proof, but it is nonconstructive. Karzanov's proof yields an algorithm permitting to construct a $\mathbb{Z}_+$-realization of d in $O(n^3)$ time (if one exists). Schrijver shows the following result, from which Theorem 24.5.1 will then follow easily.

Theorem 24.5.2. *Let $G = (V, E)$ be a connected bipartite graph and, for $W \subseteq V$, let $H = (W, F)$ be a graph which is either K_4, C_5, or a union of two stars. Then, there exist pairwise edge disjoint cuts $\delta_G(S_1), \ldots, \delta_G(S_t)$ in G such that, for each $(r, s) \in F$, the number of cuts $\delta_G(S_h)$ $(1 \leq h \leq t)$ separating r and s is equal to the distance $d_G(r, s)$ from r to s in G.* (Here, the symbol $\delta_G(S)$ denotes the cut in G which consists of the edges of G having one endnode in S and the other endnode in $V \setminus S$.)

Proof. Suppose that the theorem does not hold. Let G be a counterexample with smallest value of $|E|$. Then,

(24.5.3) for each $\emptyset \neq S \subset V$, there exist $(r, s) \in F$ and a path P connecting r and s in G such that $|P \setminus \delta_G(S)| \leq d_G(r, s) - 2$

(where P denotes the edge set of the path). Suppose S is a subset of V for which (24.5.3) does not hold. Then, for each $(r, s) \in F$, $|P \cap \delta_G(S)| = 1$ (resp. 0) for each shortest rs-path P if $\delta_G(S)$ separates (resp. does not separate) r and s. Let G' denote the connected bipartite graph obtained from G by contracting the edges of $\delta_G(S)$. Hence, for $(r, s) \in F$, $d_{G'}(r, s) = d_G(r, s) - 1$ if $\delta_G(S)$ separates r, s and $d_{G'}(r, s) = d_G(r, s)$ otherwise. As G' has fewer edges than G, by Theorem 24.5.2, we can find paiwise edge disjoint cuts $\delta_{G'}(S'_1), \ldots, \delta_{G'}(S'_t)$ in G' such that $d_{G'}(r, s)$ is equal to the number of cuts $\delta_{G'}(S'_h)$ separating r and s. These t cuts yield t cuts $\delta_G(S_h)$ in G which, together with the cut $\delta_G(S)$, are pairwise disjoint and satisfy: for $(r, s) \in F$, the number of cuts separating r and s is equal to $d_G(r, s)$. This contradicts our assumption that G is a counterexample to Theorem 24.5.2.

Claim 24.5.4. *For all $i \neq j \in V$, there exists $(r, s) \in F$ such that $\{i, j\} \cap \{r, s\} = \emptyset$ and*

$$d_G(i, j) + d_G(r, s) \geq \max(d_G(i, r) + d_G(j, s), d_G(i, s) + d_G(j, r)).$$

[3] A graph is said to be a *union of two stars* if it has two nodes such that every edge contains one of them.

Proof of Claim 24.5.4. Let $i \neq j \in V$. Set $X := \{k \in V \mid d_G(i,j) = d_G(i,k) + d_G(j,k)\}$. Hence, $i, j \in X$.

Suppose first that $X = V$. By (24.5.3) applied to $\{i\}$, we find $(r,s) \in F$ and an rs-path P such that $|P \setminus \delta_G(\{i\})| \leq d_G(r,s) - 2$. Hence, P is a shortest rs-path and i is an internal node of P and, thus, $i \notin \{r,s\}$. Using the fact that $X = V$, one obtains that $j \notin \{r,s\}$ and $d_G(i,j) + d_G(r,s) = d_G(i,r) + d_G(j,r) + d_G(r,s) \geq d_G(r,i) + d_G(s,j)$; the other inequality of Claim 24.5.4 follows in the same way.

Suppose now that $X \neq V$. Let G' denote the graph obtained from G by contracting the edges of $\delta_G(X)$. By (24.5.3) applied to X, there exists $(r,s) \in F$ such that

$$d_{G'}(r,s) \leq d_G(r,s) - 2.$$

Moreover, we claim

$$\text{(24.5.5)} \qquad \begin{cases} d_{G'}(i,s) \geq d_G(i,s) - 1, \; d_{G'}(r,j) \geq d_G(r,j) - 1, \\ d_{G'}(j,s) \geq d_G(j,s) - 1, \; d_{G'}(r,i) \geq d_G(r,i) - 1. \end{cases}$$

We show that $d_{G'}(i,s) \geq d_G(i,s) - 1$; the other inequalities of (24.5.5) can be proved in the same way. Let P be a path connecting i and s in G such that $|P \setminus \delta_G(X)| = d_{G'}(i,s)$ and with smallest value of $|P \cap \delta_G(X)|$. Suppose that $|P \cap \delta_G(X)| \geq 2$. Let P' denote the smallest subpath of P starting at i and such that $|P' \cap \delta_G(X)| = 2$. Let k denote the other endnode of P', so $k \in X$, and set $P'' := P \setminus P'$. As P' is not contained in X, we have $d_G(i,k) \leq |P'| - 1$ and, as G is bipartite, $d_G(i,k) \leq |P'| - 2$. Let Q' be a shortest path from i to k in G. Then,

$$|P'| - 2 = d_{G'}(i,k) \leq |Q' \setminus \delta_G(X)| \leq |P'| - 2 - |Q' \cap \delta_G(X)|,$$

which implies

$$Q' \cap \delta_G(X) = \emptyset \quad \text{and} \quad |Q'| = d_G(i,k) = |P'| - 2.$$

Consider the path Q from i to s obtained by juxtaposing Q' and P''. Then,

$$|Q \setminus \delta_G(X)| = |P \setminus \delta_G(X)| \quad \text{and} \quad |Q \cap \delta_G(X)| = |P \cap \delta_G(X)| - 2,$$

contradicting our choice of P. Therefore, $|P \cap \delta_G(X)| \leq 1$. This shows that

$$d_{G'}(i,s) = |P \setminus \delta_G(X)| \geq |P| - 1 \geq d_G(i,s) - 1.$$

Hence, (24.5.5) holds.

From $d_{G'}(r,s) \leq d_G(r,s) - 2$ and (24.5.5), we deduce that $\{i,j\} \cap \{r,s\} = \emptyset$. Moreover, there exists a rs-path P in G such that $|P \setminus \delta_G(X)| = d_{G'}(r,s)$ and P contains a node $k \in X$. Hence,

$$\begin{aligned} d_G(r,s) + d_G(i,j) &\geq d_{G'}(r,s) + 2 + d_G(i,j) \\ &= d_{G'}(r,k) + d_{G'}(s,k) + 2 + d_G(i,k) + d_G(j,k) \\ &\geq d_{G'}(r,i) + d_{G'}(s,j) + 2 \geq d_G(r,i) + d_G(s,j) \end{aligned}$$

(using (24.5.5) for the last inequality). The other inequality from Claim 24.5.4 follows in the same way. ∎

From Claim 24.5.4, we deduce, in particular, that H is not a union of two stars. Hence, H is either K_4 or C_5.

Suppose first that $H = K_4$. From Claim 24.5.4, we obtain

(24.5.6) $\quad d_G(i,j) + d_G(h,k) = d_G(i,h) + d_G(j,k)$ for all distinct $i,j,h,k \in W$.

For $i \in W$, set

$$f(i) := \frac{1}{2}(d_G(i,h) + d_G(i,k) - d_G(h,k))$$

where $h \neq k \in W \setminus \{i\}$; the definition does not depend on the choice of h,k by (24.5.6). Then, $d_G(i,j) = f(i) + f(j)$ for $i \neq j \in W$. Suppose $f(i) \neq 0$. By (24.5.3) applied to $\{i\}$, there exists $(r,s) \in F$ and a rs-path P such that $|P \setminus \delta_G(\{i\})| \leq d_G(r,s) - 2$. Hence, P is a shortest rs-path passing through i. Thus, $|P| = d_G(r,s) = f(r) + f(s)$, and $|P| = d_G(i,r) + d_G(i,s) = f(r) + f(s) + 2f(i)$, implying $f(i) = 0$. We obtain a contradiction.

Suppose now that $H = C_5$. Say, $W := \{r_1, r_2, r_3, r_4, r_5\}$ and $F := \{(r_i, r_{i+1}) \mid 1 \leq i \leq 5\}$, where the indices are taken modulo 5. Applying Claim 24.5.4 to r_i, r_{i+2}, we obtain that

$$d_G(r_i, r_{i+2}) + d_G(r_{i+3}, r_{i+4}) \geq d_G(r_i, r_{i+3}) + d_G(r_{i+2}, r_{i+4}),$$

$$d_G(r_i, r_{i+2}) + d_G(r_{i+3}, r_{i+4}) \geq d_G(r_i, r_{i+4}) + d_G(r_{i+2}, r_{i+3})$$

for $1 \leq i \leq 5$ (as (r_{i+3}, r_{i+4}) is the only edge of C_5 disjoint from r_i and r_{i+2}). Adding up these ten inequalities, we obtain the same sum on both sides of the inequality sign. Hence, each of the above inequalities is, in fact, an equality. Hence, (24.5.6) holds again, yielding a contradiction as above. ∎

Proof of Theorem 24.5.1. Let d be an integral metric on V_n satisfying the parity condition (24.1.1) and whose extremal graph $H := (W,F)$ is either K_4, or C_5, or a union of two stars. We show that d can decomposed as a nonnegative integer sum of cut semimetrics. Consider the complete graph K_n on V_n. We construct a connected bipartite graph G by subdividing the edges of K_n in the following way: For all distinct $i,j \in V_n$, replace the edge ij by a path P_{ij} consisting of $d(i,j)$ edges. The fact that G is bipartite follows from the parity condition. By Theorem 24.5.2, there exist edge disjoint cuts $\delta_G(S_h)$ $(1 \leq h \leq t)$ in G such that, for each $(r,s) \in F$, $d_G(r,s)$ is equal to the number of cuts $\delta_G(S_h)$ separating r and s. Setting $T_h := S_h \cap V_n$, we obtain that, for each $(r,s) \in F$,

(24.5.7) $$d(r,s) = d_G(r,s) = \sum_{1 \leq h \leq k} \delta(T_h)(r,s).$$

Moreover, for all $i \neq j \in V_n$, we have

$$d(i,j) \geq \sum_{1 \leq h \leq t} \delta(T_h)(i,j). \tag{24.5.8}$$

Indeed, the number of cuts $\delta_G(S_h)$ separating r and s is less than or equal to the number of cuts $\delta_G(S_h)$ intersecting the path P_{ij} which, in turn, is less than or equal to the length $d(i,j)$ of P_{ij} since the cuts $\delta_G(S_h)$ are pairwise edge disjoint. In fact, equality holds in (24.5.8). To see it, let $i \neq j \in V_n$ and let $P := (i_0, \ldots, i_k)$ be a path in K_n which contains the edge (i,j) and is a geodesic for d (i.e., P is a shortest path (with respect to the length function d) between its extremities i_0 and i_k; that is, $d(i_0, i_k) = \sum_{0 \leq m \leq k-1} d(i_m, i_{m+1})$). Choose such a path P having maximum number of edges. Then, the pair (i_0, i_k) is extremal for d. For, if not, there exists $x \in V_n \setminus \{i_0, i_k\}$ such that, e.g., $d(i_0, x) = d(i_0, i_k) + d(x, i_k)$ and, then, $(i_0, \ldots, i_k, x)$ is a geodesic containing (i,j) and longer than P. Then, using (24.5.8), we have

$$d(i_0, i_k) = \sum_{m=0}^{k-1} d(i_m, i_{m+1}) \geq \sum_{m=0}^{k-1} \sum_{h=1}^{t} \delta(T_h)(i_m, i_{m+1}).$$

But,

$$\sum_{m=0}^{k-1} \sum_{h=1}^{t} \delta(T_h)(i_m, i_{m+1}) = \sum_{h=1}^{t} \sum_{m=0}^{k-1} \delta(T_h)(i_m, i_{m+1}) \geq \sum_{h=1}^{t} \delta(T_h)(i_0, i_k) = d(i_0, i_k),$$

where the last equality follows from (24.5.7) as the edge (i_0, i_k) belongs to F. Therefore, equality holds in (24.5.8) for each of the edges (i_m, i_{m+1}) of P and, in particular, for the edge (i,j). This shows that equality holds in (24.5.8) for all $i \neq j \in V_n$. Therefore, $d = \sum_{h=1}^{t} \delta(T_h)$, showing that d is hypercube embeddable. ∎

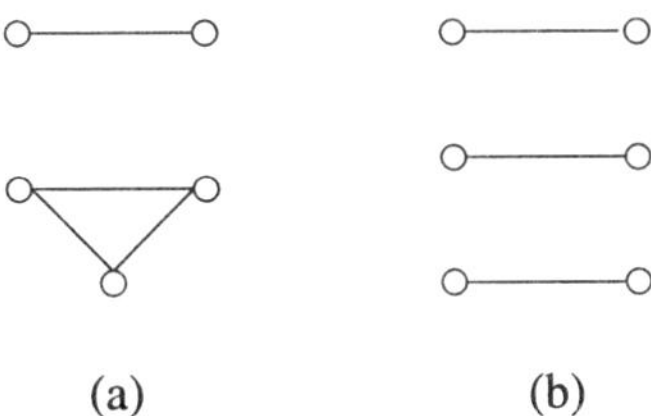

Figure 24.5.9

Remark 24.5.10. One can check that a graph H with no isolated node is K_4, C_5, or a union of two stars if and only if H does not contain as a subgraph the two graphs from Figure 24.5.9. The exclusion of these two graphs is necessary for the validity of Theorem 24.5.1. Indeed, let d_1 be the path metric of the complete bipartite graph $K_{2,3}$; then, d is not hypercube embeddable (as d does not satisfy the pentagonal inequality) and its extremal graph is the graph (a) from Figure 24.5.9. Let d_2 be the path metric of the graph $K_{3,3} \setminus e$; then its extremal graph is the graph (b) from Figure 24.5.9 and d_2 is not hypercube embeddable (as it contains d_1 as a subdistance). (In fact, both d_1 and d_2 lie on extreme rays of the semimetric cone.) ∎

Chapter 25. Cut Lattices, Quasi h-Distances and Hilbert Bases

We consider in this chapter several additional questions related to the notion of hypercube embedding. A possible way of relaxing this notion is to look for *integer* combinations rather than *nonnegative integer* combinations of cut semimetrics. In other words, one considers the lattice $\mathcal{L}_n$ generated by all cut semimetrics on V_n. We recall in Section 25.1 the characterization of $\mathcal{L}_n$. This is an easy result; namely, $\mathcal{L}_n$ consists of the integer distances satisfying the parity condition. We also present the characterization of some sublattices of $\mathcal{L}_n$, namely, of the sublattice generated by all even T-cut semimetrics and of the sublattice generated by all k-uniform cut semimetrics.

Clearly, for a distance d on V_n,

$$(25.0.1) \qquad d \text{ is hypercube embeddable } \Longrightarrow d \in \mathrm{CUT}_n \cap \mathcal{L}_n.$$

We consider in Section 25.2 quasi h-distances, which are the distances d that belong to $\mathrm{CUT}_n \cap \mathcal{L}_n$ but are not hypercube embeddable. As was mentioned in Theorem 24.1.3, the implication (25.0.1) is an equivalence for any distance d on $n \leq 5$ points. This fact can be reformulated as saying that, for $n \leq 5$, the family of cut semimetrics on V_n is a Hilbert basis. We consider in Section 25.3 the more general question of characterizing the graphs whose family of cuts is a Hilbert basis.

25.1 Cut Lattices

The set

$$\mathcal{L}_n := \{ \sum_{S \subseteq V_n} \lambda_S \delta(S) \mid \lambda_S \in \mathbb{Z} \text{ for all } S \subseteq V_n \}$$

is trivially a lattice contained in $\mathbb{Z}^{\binom{n}{2}}$, called the *cut lattice*. The next result from Assouad [1982] gives a characterization of $\mathcal{L}_n$.

Proposition 25.1.1. *Let $d \in \mathbb{Z}^{E_n}$. Then, $d \in \mathcal{L}_n$ if and only if d satisfies the parity condition* (24.1.1).

Proof. The parity condition is clearly a necessary condition for membership in $\mathcal{L}_n$. Conversely, suppose d is integral and satisfies the parity condition. Then, V_n can be partitioned into $V_n = S \cup T$ in such a way that $d(i, j)$ is odd if $i \in S, j \in T$

and $d(i,j)$ is even otherwise. Set $d' := d + \delta(S)$. Then, all components of d' are even. As

$$d' = \sum_{1 \le i < j \le n} \frac{d'(i,j)}{2} (\delta(\{i\}) + \delta(\{j\}) - \delta(\{i,j\})),$$

we deduce that $d' \in \mathcal{L}_n$ and, thus, $d = d' - \delta(S)$ belongs to $\mathcal{L}_n$ too. ∎

A basis of the cut lattice $\mathcal{L}_n$ is provided by the set

$$\{\delta(i) \mid 1 \le i \le n-1\} \cup \{\delta(\{i,j\}) \mid 1 \le i < j \le n-1\}.$$

(Indeed, every vector $d \in \mathcal{L}_n$ can be decomposed as

$$d = \sum_{1 \le i \le n-1} \alpha_i \delta(i) + \sum_{1 \le i < j \le n-1} \beta_{ij} \delta(\{i,j\}),$$

setting $\alpha_i := d_{in} - \sum_{1 \le h \le n-1, h \ne i} \beta_{ih}$ $(1 \le i \le n-1)$ and $\beta_{ij} := \frac{1}{2}(d_{in} + d_{jn} - d_{ij})$ $(1 \le i < j \le n-1)$.) Note also that the image of the cut lattice $\mathcal{L}_n$ under the covariance mapping ξ is nothing but the integer lattice $\mathbb{Z}^{\binom{n}{2}}$. For small values of n, $\mathcal{L}_n$ coincides with classical lattices; for instance, $\mathcal{L}_3 = A_3 = D_3$ (recall Example 13.2.5).

Complete characterizations are also known for several sublattices of $\mathcal{L}_n$. We consider below the sublattices generated by the k-uniform cut semimetrics, the even cut semimetrics (more generally, the even T-cut semimetrics), and the odd cut semimetrics on six points. Given $S \subseteq V_n$, the cut semimetric $\delta(S)$ is said to be *k-uniform* if $|S| \in \{k, n-k\}$. We let $\mathcal{K}_n^k$ denote the set of all k-uniform cut semimetrics on V_n. The following characterization of the k-uniform cut lattice $\mathbb{Z}(\mathcal{K}_n^k)$ is given in Deza and Laurent [1992e], based on a result of Wilson [1973].

Proposition 25.1.2. *Let k be an integer such that $2 \le k \le n$ and $k \ne \frac{n}{2}$ and let $d \in \mathbb{Z}^{E_n}$. Then, $d \in \mathbb{Z}(\mathcal{K}_n^k)$ if and only if d satisfies the following conditions:*

(i) $\sum_{1 \le i < j \le n} d(i,j) \equiv 0 \ (mod\ k(n-k))$,

(ii) $D_i := \frac{1}{n-2k} \left(\sum_{1 \le j \le n, j \ne i} d(i,j) - \frac{1}{n-k} \sum_{1 \le r < s \le n} d(r,s) \right) \in \mathbb{Z}$ *for all* $i \in V_n$,

(iii) $D_i + D_j + d(i,j) \equiv 0 \ (mod\ 2)$ *for all* $i, j \in V_n$. ∎

In the case $k = \lfloor \frac{n}{2} \rfloor$, we have the following result.

Proposition 25.1.3. *Let $d \in \mathbb{Z}^{E_n}$.*

(i) *If $n = 2k+1$, then $d \in \mathbb{Z}(\mathcal{K}_n^k)$ if and only if d satisfies the congruence relation:* $\sum_{1 \le i < j \le n} d(i,j) \equiv 0 \ (mod\ k(n-k))$.

(ii) *If $n = 2k$, then $d \in \mathbb{Z}(\mathcal{K}_n^k)$ if and only if* (iia),(iib) *hold:*

(iia) $\sum_{1\le j\le n, j\ne i} d(i,j) = \frac{1}{k} \sum_{1\le r<s\le n} d(r,s)$ *for each* $1 \le i \le n$,

(iib) $\sum_{1\le i<j\le n} d(i,j) \equiv 0$ $(mod\ k^2)$.

Proof. (i) Observe that the conditions (ii),(iii) from Proposition 25.1.2 are implied by the condition (i) of Proposition 25.1.2.
(ii) The conditions (iia),(iib) are clearly necessary for membership in $\mathbb{Z}(\mathcal{K}_n^k)$. Conversely, suppose that d satisfies (iia),(iib) and let d' denote its projection on the set $\{1, \ldots, n-1\}$. From (iia), we obtain

$$\sum_{1\le r<s\le n-1} d'(r,s) = (k-1) \sum_{1\le i\le n-1} d(i,n). \tag{25.1.4}$$

Hence, $\sum_{1\le r<s\le n-1} d'(r,s) \equiv 0 \pmod{k(k-1)}$, as $\sum_{1\le i\le n-1} d(i,n) \equiv 0 \pmod{k}$ by (iia,)(iib). Using (i), we deduce that $d' \in \mathbb{Z}(\mathcal{K}_{n-1}^k)$. Hence,

$$d' = \sum_{S\subseteq\{1,\ldots,n-1\}, |S|=k} \lambda_S \delta(S)$$

with $\lambda_S \in \mathbb{Z}$ for all S. We show that $d = \sum_S \lambda_S \delta(S)$. As

$$\sum_{1\le r<s\le n-1} d'(r,s) = k(k-1)(\sum_S \lambda_S),$$

(25.1.4) yields:

$$\sum_{1\le i\le n-1} d(i,n) = k(\sum_S \lambda_S).$$

Then, by (iia),

$$\sum_{1\le r<s\le n} d(r,s) = k^2(\sum_S \lambda_S) \text{ and } \sum_{1\le j\le n, j\ne i} d(i,j) = k(\sum_S \lambda_S)$$

for each $i = 1, \ldots, n$. We compute, for instance, $d(1,n)$. The above relations yield:

$$d(1,n) = k(\sum_S \lambda_S) - \sum_{2\le j\le n-1} d(1,j).$$

Using the value of $d(1,j) = d'(1,j)$ given by the decomposition of d', we obtain that $d(1,n) = \sum_{S|1\in S} \lambda_S$. This shows that $d = \sum_S \lambda_S \delta(S)$, i.e., that $d \in \mathbb{Z}(\mathcal{K}_n^k)$. ∎

Suppose that n is even. A cut semimetric $\delta(S)$ on V_n is said to be *even* if $|S|$ is even. Similarly, $\delta(S)$ is called *odd* if $|S|$ is odd. More generally, let $T \subseteq V_n$ such that $|T|$ is even; then, $\delta(S)$ is said to be an *even T-cut semimetric* if $|S \cap T|$ is even. We let $\mathcal{K}_n^T$ denote the set of even T-cut semimetrics on V_n. Hence, $\mathcal{K}_n^{V_n}$ is the set of even cut semimetrics. We let also $\mathcal{K}_n^{odd}$ denote the set of odd cut semimetrics. We first give a characterization of the even cut lattice $\mathbb{Z}(\mathcal{K}_n^{V_n})$, whose proof can be found in Deza, Laurent and Poljak [1992].

Proposition 25.1.5. *Let $n \geq 6$ be an even integer and let $d \in \mathbb{Z}^{E_n}$. Then, d belongs to the even cut lattice $\mathbb{Z}(\mathcal{K}_n^{V_n})$ if and only if d satisfies the parity condition* (24.1.1) *and the following conditions:*

(i) $\sum_{1 \leq i < j \leq n} d(i,j) \equiv 0 \ (mod\ 4)$.

(ii) $\sum_{i<j, i,j \in V_n \setminus \{k\}} d(i,j) - \sum_{i \in V_n \setminus \{k\}} d(i,k) \equiv 0 \ (mod\ 8)$ *for all $k \in V_n$ in the case when $n \equiv 0 \ (mod\ 4)$.*

(iii) $d(h,k) + \sum_{i<j, i,j \in V_n \setminus \{h,k\}} d(i,j) - \sum_{i \in V_n \setminus \{h,k\}} (d(i,h) + d(i,k)) \equiv 0 \ (mod\ 8)$ *for all $h \neq k \in V_n$ in the case when $n \equiv 2 \ (mod\ 4)$.* ∎

It is not difficult to extend the result for even T-cuts. Namely,

Proposition 25.1.6. *Let $T \subseteq V_n$ with $|T|$ even, $2 \leq |T| \leq n-1$. let $d \in \mathbb{Z}^{E_n}$. Suppose first that $|T| = 2$, $T := \{s,t\}$. Then, $d \in \mathbb{Z}(\mathcal{K}_n^T)$ if and only if d satisfies the parity condition* (24.1.1) *and the conditions: $d_{rs} = 0$, $d_{ri} = d_{si}$ for all $i \in V_n \setminus T$. Suppose now that $|T| \geq 4$. Then, $d \in \mathbb{Z}(\mathcal{K}_n^T)$ if and only if d satisfies* (24.1.1) *and the following conditions:*

(i) $\sum_{i<j,\ i,j \in T} d_{ij} \equiv 0 \ (mod\ 4)$.

(ii) $Q_n(\underbrace{1,\ldots,1,-1,\ldots,-1}_{T},\underbrace{2,\ldots,2,0,\ldots,0}_{V_n \setminus T})^T d \equiv 0 \ (mod\ 8)$, *where there are exactly α coefficients* 1, *and β coefficients* 2, *for the following values of α, β:*

(iia) $(\alpha = 1, \beta = 0)$ *and* $(\alpha = 2, \beta = 1)$ *if* $|T| \equiv 0 \ (mod\ 4)$,
(iib) $(\alpha = 2, \beta = 0)$ *and* $(\alpha = 1, \beta = 1)$ *if* $|T| \equiv 2 \ (mod\ 4)$. ∎

A characterization of the odd cut lattice $\mathbb{Z}(\mathcal{K}_n^{odd})$ is known only in the case $n = 6$. So, $\mathbb{Z}(\mathcal{K}_6^{odd})$ is the lattice in $\mathbb{R}^{15}$ generated by the 16 cut semimetrics $\delta(\{i\})$ $(1 \leq i \leq 6)$ and $\delta(\{1,i,j\})$ $(2 \leq i < j \leq 6)$. We need the following notation. Given distinct $a,b,c \in V_6$, let $v^{a,bc} \in \mathbb{R}^{E_6}$ be the vector defined by

$$\begin{cases} v_{ab}^{a,bc} = v_{ac}^{a,bc} = 1, \ v_{bc}^{a,bc} = 2, & \\ v_{ij}^{a,bc} = 2 & \text{for } i \neq j \in V_6 \setminus \{a,b,c\}, \\ v_{ai}^{a,bc} = -2, \ v_{bi}^{a,bc} = v_{ci}^{a,bc} = -1 & \text{for } i \in V_6 \setminus \{a,b,c\}. \end{cases}$$

Consider the conditions:

(25.1.7) $\quad (v^{a,bc})^T x \leq 0$ for all distinct $a,b,c \in V_6$,

(25.1.8) $\quad (v^{a,bc})^T x \equiv 0 \ (mod\ 4)$ for all distinct $a,b,c \in V_6$,

(25.1.9) $(v^{1,bc})^T x - (v^{1,b'c'})^T x \equiv 0 \ (mod\ 12)$ for $2 \leq b < c \leq 6$, $2 \leq b' < c' \leq 6$.

The next result from Deza and Laurent [1993a] gives the characterization of the odd cut lattice $\mathbb{Z}(\mathcal{K}_6^{odd})$, also of the cone $\mathbb{R}_+(\mathcal{K}_6^{odd})$ and of the integer cone $\mathbb{Z}_+(\mathcal{K}_6^{odd})$.

Proposition 25.1.10.

(i) *Let* $d \in \mathbb{R}_+^{E_6}$. *Then,* $d \in \mathbb{R}_+(\mathcal{K}_6^{odd})$ *if and only if* d *satisfies* (25.1.7).

(ii) *Let* $d \in \mathbb{Z}^{E_6}$. *Then,* $d \in \mathbb{Z}(\mathcal{K}_6^{odd})$ *if and only if* d *satisfies* (25.1.8) and (25.1.9).

(iii) *Let* $d \in \mathbb{Z}_+^{E_6}$. *Then,* $d \in \mathbb{Z}_+(\mathcal{K}_6^{odd})$ *if and only if* d *satisfies* (25.1.7), (25.1.8), and (25.1.9). ∎

An immediate consequence of Proposition 25.1.10 is that the family of odd cut semimetrics on V_6 forms a Hilbert basis; see Section 25.3.

Further information on sublattices of the cut lattice $\mathcal{L}_n$ can be found in Deza and Grishukhin [1996a].

25.2 Quasi h-Distances

Let d be a distance on V_n. Then, d is called a *quasi h-distance* if $d \in \mathrm{CUT}_n \cap \mathcal{L}_n$ and d is not hypercube embeddable. In other words, d can be decomposed both as a *nonnegative* combination of cut semimetrics and as an *integer* combination of cut semimetrics, but not as a *nonnegative integer* combination of cut semimetrics. We remind that the smallest integer η such that ηd is hypercube embeddable is called the minimum scale of d and is denoted by $\eta(d)$.

As stated in Theorem 24.1.3, there are no quasi h-distances on $n \leq 5$ points. We have seen already several ways of constructing quasi h-distances. Quasi h-distances can be constructed, for instance, using the antipodal extension operation (described in Section 7.2). Indeed, let d be a distance on V_n which is hypercube embeddable and let $\alpha \in \mathbb{Z}_+$ such that $s_{\ell_1}(d) \leq \alpha < s_h(d)$. Then, $\mathrm{ant}_\alpha(d)$ is a quasi h-distance. Recall that $\mathrm{ant}_\alpha(d)$ is the distance on V_{n+1} defined by $\mathrm{ant}_\alpha(d)(1, n+1) = \alpha$, $\mathrm{ant}_\alpha(d)(i, n+1) = \alpha - d(1,i)$ for $1 \leq i \leq n$, and $\mathrm{ant}_\alpha(d)(i,j) = d(i,j)$ for $1 \leq i < j \leq n$. As an example, for $n \geq 6$, the distance

$$d_n^* := 2d(K_n \backslash e)$$

(taking value 2 on all pairs except value 4 on the pair corresponding to the edge e) is a quasi h-distance (as $d_n^* = \mathrm{ant}_4(2\mathbb{1}_{n-1})$; see Example 7.2.7).

The gate extension operation (described in Section 7.1) permits also to construct quasi h-distances. If d is a distance on V_n and $\alpha \in \mathbb{R}_+$, its gate extension $\mathrm{gat}_\alpha(d)$ is the distance on V_{n+1} defined by $\mathrm{gat}_\alpha(d)(1, n+1) = \alpha$, $\mathrm{gat}_\alpha(d)(i, n+1) = \alpha + d(1,i)$ for $1 \leq i \leq n$, and $\mathrm{gat}_\alpha(d)(i,j) = d(i,j)$ for $1 \leq i < j \leq n$. Then, for $\alpha \in \mathbb{Z}_+$, $\mathrm{gat}_\alpha(d)$ is a quasi h-distance if and only if d is a quasi h-distance.

This implies, in particular, that there is an infinity of quasi h-distances on n points for all $n \geq 7$. Indeed, all gate extensions of $d_6^* = 2d(K_6\backslash e)$ are quasi h-distances.

As the following result by Laburthe [1994] indicates, other examples of quasi h-distances on 6 points can be constructed. It implies, moreover, that there is also an infinity of quasi h-distances on 6 points.

Lemma 25.2.1. *Let e be an edge of K_6 and let v be a node of K_6 which is not adjacent to e. Then, the distance $2d(K_6\backslash e) + m\delta(\{v\})$ is a quasi h-distance for each integer $m \geq 0$.*

Proof. Suppose K_6 is the complete graph on $V_6 = \{1, \ldots, 6\}$, e is the edge $(1,6)$ and v is the node 2. Set $d := 2d(K_6\backslash e) + m\delta(\{v\})$. Let $d = \sum_S \alpha_S \delta(S)$ be a $\mathbb{Z}_+$-realization of d, with $\alpha_S \in \mathbb{Z}_+$. As d satisfies the triangle equality: $d_{16} = d_{1i} + d_{i6}$ for $i = 3,4,5$, we deduce that $\alpha_S = 0$ if S is one of the sets: 3, 4, 5, 16, 23, 24, 25, 34, 35, 45, 126, 136, 146, and 156. Hence, $d = \sum_{S \in \mathcal{S}} \alpha_S \delta(S)$, where $\mathcal{S}$ may contain the sets: 1, 2, 6, 12, 13, 14, 15, 26, 36, 46, 56, 123, 124, 125, 134, 135, 145. By computing d_{12}, d_{26}, and d_{16}, we obtain, respectively,

$$m + 2 = \alpha_1 + \alpha_2 + \alpha_{13} + \alpha_{14} + \alpha_{15} + \alpha_{26} + \alpha_{134} + \alpha_{135} + \alpha_{145},$$

$$m + 2 = \alpha_2 + \alpha_6 + \alpha_{12} + \alpha_{36} + \alpha_{46} + \alpha_{56} + \alpha_{123} + \alpha_{124} + \alpha_{125},$$

$$4 = \sum_{S \in \mathcal{S}} \alpha_S - \alpha_2.$$

Adding the first two relations and subtracting the third one, we obtain that $\alpha_2 = m$. Therefore, if d is hypercube embeddable, then so is $d - m\delta(\{2\})$. This contradicts the fact that $2d(K_6\backslash e)$ is a quasi h-distance. ∎

In fact, as a consequence of Theorem 25.2.2 below, there are no other quasi h-distances on 6 points besides those described in Lemma 25.2.1. Theorem 25.2.2 and Corollary 25.2.3 were proved by Laburthe [1994, 1995]. The proof of Theorem 25.2.2 involves many technical details, so we do not give it here. (Details about the full proof can also be found in Laburthe, Deza and Laurent [1995].)

Theorem 25.2.2. *Every quasi h-distance on V_6 is a nonnegative integer sum of cuts and of the distances $2d(K_6\backslash e)$, for e edge of K_6.* ∎

Corollary 25.2.3. *The only quasi h-distances on V_6 are of those of the form $2d(K_6\backslash e) + m\delta(\{v\})$, where e is an edge of K_6, v is a node of K_6 not adjacent to e, and $m \in \mathbb{Z}_+$.*

The proof of Corollary 25.2.3 uses the identities (a)-(i) below, which show that all the perturbations of $2d(K_6\backslash e)$ (obtained by adding a cut semimetric), other than the one considered in Lemma 25.2.1, are hypercube embeddable. For $1 \leq i < j \leq n$, let e_{ij} denote the edge ij of K_6. Then,

(a) $2d(K_6\backslash e_{12}) + \delta(\{1\}) = \delta(\{2\}) + \delta(\{1,3\}) + \delta(\{1,4\}) + \delta(\{1,5\}) + \delta(\{1,6\})$,

(b) $2d(K_6\backslash e_{12}) + \delta(\{1,2\}) = 2\delta(\{1\}) + 2\delta(\{2\}) + \delta(\{3\}) + \delta(\{4\}) + \delta(\{5\}) + \delta(\{6\})$,

(c) $2d(K_6\backslash e_{12}) + \delta(\{1,3\}) = \delta(\{2\}) + \delta(\{1,3\}) + \delta(\{3,4,5\}) + \delta(\{3,4,6\}) + \delta(\{4,5,6\})$,

(d) $2d(K_6\backslash e_{12}) + \delta(\{3,4\}) = \delta(\{1\}) + \delta(\{3\}) + \delta(\{4\}) + \delta(\{2,5\}) + \delta(\{2,6\}) + \delta(\{2,3,4\})$,

(e) $2d(K_6\backslash e_{12}) + \delta(\{1,2,3\}) = \delta(\{1\}) + \delta(\{2\}) + \delta(\{4\}) + \delta(\{5\}) + \delta(\{6\}) + \delta(\{1,3\}) + \delta(\{2,3\})$,

(f) $2d(K_6\backslash e_{12}) + \delta(\{1,3,4\}) = \delta(\{1,3\}) + \delta(\{1,4\}) + \delta(\{2,5\}) + \delta(\{2,6\}) + \delta(\{1,5,6\})$,

(g) $2d(K_6\backslash e_{12}) + 2d(K_6\backslash e_{23}) = \delta(\{1\}) + \delta(\{2,3\}) + \delta(\{2,4\}) + \delta(\{2,5\}) + \delta(\{3,6\}) + \delta(\{1,2,6\}) + \delta(\{1,3,4\}) + \delta(\{1,3,5\})$,

(h) $2d(K_6\backslash e_{12}) + 2d(K_6\backslash e_{34}) = \delta(\{1\}) + \delta(\{2,3\}) + \delta(\{2,4\}) + \delta(\{3,5\}) + \delta(\{4,6\}) + \delta(\{1,3,4\}) + \delta(\{1,3,6\}) + \delta(\{1,4,5\})$,

(i) $2d(K_6\backslash e_{12}) + \delta(\{3\}) + \delta(\{4\}) = \delta(\{1,3\}) + \delta(\{2,4\}) + \delta(\{3,4\}) + \delta(\{1,4,5\}) + \delta(\{1,4,6\})$.

Proof of Corollary 25.2.3. Let d be a quasi h-distance on V_6. Then, by Theorem 25.2.2, d can be written as

$$d = \sum_S \alpha_S \delta(S) + \sum_{1 \le i < j \le 6} \beta_{ij} 2d(K_6\backslash e_{ij})$$

with $\alpha_S, \beta_{ij} \in \mathbb{Z}_+$. We can suppose that $\beta_{ij} \in \{0,1\}$ for all i,j, because $4d(K_6\backslash e_{ij})$ is hypercube embeddable. Using (g) and (h), we can rewrite d as

$$d = \sum_S \alpha'_S \delta(S) + 2d(K_6\backslash e),$$

where $\alpha'_S \in \mathbb{Z}_+$ and, for instance, e is the edge $(1,2)$. From relations (a)-(f), we deduce that $\alpha_S = 0$ if $S = \{1\}$, or $\{2\}$, or if $|S| = 2$, or 3. Therefore, using relation (i), we obtain that $d = 2d(K_6\backslash e_{12}) + m\delta(\{i\})$, where $i \in \{3,4,5,6\}$ and $m \in \mathbb{Z}_+$. ∎

As we just saw, there is an infinity of quasi h-distances on V_n, for any $n \ge 6$. However, it follows from Lemma 4.3.9 that there exists an integer η_n which is a common scale for all quasi h-distances d on V_n, i.e., such that $\eta_n d$ is hypercube embeddable for all quasi h-distances d. As an application of Corollary 25.2.3, we have:

Corollary 25.2.4. *We have:* $\eta_6 = 2$. *In other words,* $2d$ *is hypercube embeddable for every integer valued distance* d *on* 6 *points which is* ℓ_1*-embeddable and satisfies the parity condition* (24.1.1). ∎

For the class of graphic distances, the following results have been shown in Chapter 21: The minimum scale of the path metric of a connected graph on n nodes is equal to 1, or is an even integer less than or equal to $n-2$. Moreover, for an ℓ_1-rigid graph, the minimum scale is equal to 1 or 2.

Much of the treatment of Chapter 23 can be reformulated in terms of minimum scales. Indeed, consider the metric $d_n := \mathrm{ant}_2(\mathbb{1}_n)$ (this is the path metric of the graph $K_{n+1}\backslash e$)). Then,

$$2td_n = 2t\ \mathrm{ant}_2(\mathbb{1}_n) = \mathrm{ant}_{4t}(2t\mathbb{1}_n)$$

is hypercube embeddable if and only if $4t \geq s_h(2t\mathbb{1}_n)$. Therefore, the minimum scale $\eta(d_n)$ can be expressed as

$$\eta(d_n) = 2\min(t \in \mathbb{Z}_+ \mid 4t \geq s_h(2t\mathbb{1}_n)).$$

In particular, Theorem 23.3.1 (i) implies:

(i) $\eta(d_{4t}) \geq 2t$ with equality if and only if there exists a Hadamard matrix of order $4t$.

Compare (i) with the next statement (ii), which follows from Theorems 22.0.6 and 22.0.7.

(ii) $\eta^1(\mathbb{1}_{t^2+t+2}) \geq 2t$ with equality if and only if there exists a projective plane of order t (where, for a hypercube embeddable distance d, $\eta^1(d)$ denotes the smallest integer λ (if any) such that λd is not h-rigid, i.e., has at least two distinct $\mathbb{Z}_+$-realizations.)

Some quasi h-distances can also be constructed using the spherical extension operation (described in Section 7.3). The examples from Lemmas 25.2.5 and 25.2.6 below are taken from Deza and Grishukhin [1994]. Recall that, if d is a distance on V_n and $t \in \mathbb{R}_+$, its spherical t-extension is the distance $\mathrm{sph}_t(d)$ on V_{n+1} defined by $\mathrm{sph}_t(d)(i,n+1) = t$ for all $1 \leq i \leq n$, and $\mathrm{sph}_t(d)(i,j) = d(i,j)$ for all $1 \leq i < j \leq n$. If $d \in \mathrm{CUT}_n$ and $2t \geq s_{\ell_1}(d)$, then $\mathrm{sph}_t(d) \in \mathrm{CUT}_{n+1}$. As a first example, consider the distance

$$\theta_n^t := \mathrm{ant}_{2t}(\mathrm{sph}_t(2\mathbb{1}_{n-2})),$$

where n,t are positive integers, i.e., θ_n^t is the distance on V_n defined by

$$\begin{cases} \theta_n^t(n-1,n) & = 2t, & \\ \theta_n^t(i,n-1) = \theta_n^t(i,n) & = t & \text{for } 1 \leq i \leq n-2, \\ \theta_n^t(i,j) & = 2 & \text{for } 1 \leq i < j \leq n-2. \end{cases}$$

Clearly, θ_n^t admits the following decompositions:

$$\theta_n^t = \sum_{1 \leq i \leq n-2} \delta(\{i,n\}) + (t-1)\delta(\{n-1\}) + (t-n+3)\delta(\{n\}),$$

$$\theta_n^t = \frac{1}{2}\left(\sum_{1\le i\le n-2} (\delta(\{i,n-1\}) + \delta(\{i,n\})\right) + (2t-n+2)\,(\delta(\{n-1\}) + \delta(\{n\}))\,.$$

This shows that θ_n^t is hypercube embeddable if $t \ge n-3$ and that $2\theta_n^t$ is hypercube embeddable if $t \ge \frac{n-2}{2}$.

Lemma 25.2.5. *Let $t \ge 1$ be an integer.*

(i) *If $n \neq 6$, then θ_n^t is hypercube embeddable if and only if $t \ge n-3$.*

(ii) *For $n \ge 6$, if $\lceil\frac{n-2}{2}\rceil \le t \le n-4$, then θ_n^t is a quasi h-distance.*

Proof. (i) Suppose that θ_n^t is hypercube embeddable. Then, in any hypercube embedding of θ_n^t, we can suppose that each point $i \in \{1,\ldots,n-2\}$ is labeled by the singleton $\{i\}$ (as the metric $2\mathbb{1}_{n-2}$ is h-rigid if $n \neq 6$). This implies that one of the points $n-1, n$ should be labeled by a set A containing $\{1,\ldots,n-2\}$ and, thus, $|A| - 1 = t \ge n-3$.
(ii) If $t \ge \lceil\frac{n-2}{2}\rceil$, then θ_n^t is ℓ_1-embeddable. Hence, if $n \neq 6$ and $\lceil\frac{n-2}{2}\rceil \le t \le n-4$, then θ_n^t is a quasi h-distance. If $n = 6$ and $t = 2$, then θ_n coincides with the distance d_6^*, which is known to be a quasi h-distance. ∎

Given $n \ge 6$, let μ_n denote the distance on V_n defined by

$$\mu_n := \delta(\{1\}) + \delta(\{2\}) + \sum_{3\le i<j\le n-1} \delta(\{1,2,i,j\}), \text{ i.e.,}$$

$$\begin{cases} \mu_n(1,2) & = 2, & \\ \mu_n(1,n) = \mu_n(2,n) & = 1 + \binom{n-3}{2}, & \\ \mu_n(1,i) = \mu_n(2,i) & = 1 + \binom{n-4}{2} & \text{for } 3 \le i \le n-1, \\ \mu_n(i,n) & = n-4 & \text{for } 3 \le i \le n-1, \\ \mu_n(i,j) & = 2(n-5) & \text{for } 3 \le i < j \le n. \end{cases}$$

For instance, for $n = 6$, μ_6 coincides with the path metric of the graph $K_6 \backslash P$, where $P := (1,6,2)$ is a path on three nodes.

Lemma 25.2.6. *Let t, n be integers such that $n \ge 6$, $n \equiv 2$ (mod 4), and $2t \ge 2 + \binom{n-3}{2}$. Then, $\mathrm{sph}_t(\mu_n)$ is a quasi h-distance.*

Proof. It is easy to see that the condition $n \equiv 2$ (*mod* 4) ensures that all components of μ_n are even integers, which implies that $\mathrm{sph}_t(\mu_n) \in \mathcal{L}_{n+1}$. Let F denote the face of the cone CUT_n defined by the hypermetric inequality $Q(b)^T x := \sum_{1\le i<j\le n} b_i b_j x_{ij} \le 0$, where $b := (1,1,-1,\ldots,-1,n-4) \in \mathbb{R}^n$ (with $n-3$ components -1). Set

$$\mathcal{S} := \{1, 2, 1i, 2i, 12i(3 \le i \le n-1), 12ij(3 \le i < j \le n-1)\},$$

(where we denote the sets $\{1\}, \{1,i\}$ by the strings 1, $1i$, etc.). The nonzero cut semimetrics satisfying the equation $Q(b)^T x = 0$ are $\delta(S)$ for $S \in \mathcal{S}$, which are linearly independent. Hence, the face F is a simplex face of CUT_n. As the distance μ_n lies on F, we deduce that μ_n is ℓ_1-rigid and $s_{\ell_1}(\mu_n) = 2 + \binom{n-3}{2}$. Let G denote the face of the cone CUT_{n+1} defined by the hypermetric inequality $Q(b,0)^T x \leq 0$; the nonzero cut semimetrics lying on G are $\delta(S)$, $\delta(S \cup \{n+1\})$ for $S \in \mathcal{S}$ and $\delta(\{n+1\})$. As $2t \geq s_{\ell_1}(\mu_n)$, $\mathrm{sph}_t(\mu_n)$ is ℓ_1-embeddable and, in fact, $\mathrm{sph}_t(\mu_n)$ lies on the face G. Suppose that $\mathrm{sph}_t(d)$ is hypercube embeddable. Then, there exist nonnegative integers $\gamma, \alpha_S, \beta_S$ $(S \in \mathcal{S})$ such that

$$\mathrm{sph}_t(\mu_n) = \gamma\delta(\{n+1\}) + \sum_{S \in \mathcal{S}} \alpha_S \delta(S) + \beta_S \delta(S \cup \{n+1\}).$$

Then, $\sum_{S \in \mathcal{S}} (\alpha_S + \beta_S)\delta(S) = d$, which implies that $\alpha_S = \beta_S = 0$ if S is not one of the sets $\{1\}, \{2\}, \{1,2,i,j\}$, and

$$\begin{cases} \alpha_i + \beta_i & = 1 \quad \text{for } i = 1,2, \\ \alpha_{ij} + \beta_{ij} & = 1 \quad \text{for } 3 \leq i < j \leq n-1. \end{cases}$$

(setting $\alpha_{ij} = \alpha_{12ij}, \beta_{ij} = \beta_{12ij}$). Looking at the component of $\mathrm{sph}_t(\mu_n)$ indexed by the pairs $(1, n+1)$ and $(2, n+1)$, we obtain:

$$\alpha_1 + \beta_2 + \sum_{i,j} \alpha_{ij} + \gamma = t, \ \alpha_2 + \beta_1 + \sum_{i,j} \alpha_{ij} + \gamma = t,$$

which implies

$$\alpha_1 = \alpha_2, \beta_1 = \beta_2, \gamma = t - \sum_{i,j} \alpha_{ij} - 1.$$

Looking at the component indexed by $(i, n+1)$ $(3 \leq i \leq n-1)$, we obtain:

$$\sum_j \alpha_{ij} + \beta_1 + \beta_2 + \sum_{i,j} \beta_{ij} - \sum_j \beta_{ij} + \gamma = t.$$

Therefore,

$$2\sum_j \alpha_{ij} + 2\beta_1 - 2\sum_{i,j} \alpha_{ij} + \binom{n-3}{2} - n + 3 = 0.$$

Summing over $i = 3, \ldots, n-1$ yields

$$4(n-5)\sum_{i,j} \alpha_{ij} = (n-3)(4\beta_1 + (n-3)(n-6)).$$

Looking finally at the component indexed by the pair $(n, n+1)$ yields:

$$\beta_1 + \beta_2 + \sum_{i,j} \beta_{ij} + \gamma = t$$

and, thus,

$$2\sum_{i,j} \alpha_{ij} - 2\beta_1 - \binom{n-3}{2} + 1 = 0.$$

Using the fact that

$$2\sum_{i,j}\alpha_{ij} = \frac{n-3}{2(n-5)}(4\beta_1 + (n-3)(n-6)),$$

we deduce that $2\beta_1 = 1$, contradicting the fact that β_1 is integer. This shows that $\mathrm{sph}_t(\mu_n)$ is not hypercube embeddable and, therefore, is a quasi h-distance. ∎

25.3 Hilbert Bases of Cuts

Let X be a finite set of vectors in $\mathbb{Z}^k$. We remind that $\mathbb{Z}(X)$, $\mathbb{R}_+(X)$ and $\mathbb{Z}_+(X)$ denote, respectively, the lattice, the cone and the integer cone generated by X. Clearly, the following inclusion holds:

$$\mathbb{Z}_+(X) \subseteq \mathbb{R}_+(X) \cap \mathbb{Z}(X).$$

The set X is said to be a *Hilbert basis* if equality holds, i.e., if

$$\mathbb{Z}_+(X) = \mathbb{R}_+(X) \cap \mathbb{Z}(X).$$

Clearly, if X is linearly independent, then X is a Hilbert basis. We consider here the question of determining the graphs whose family of cuts is a Hilbert basis.

Given a graph G and $S \subseteq V$, the cut $\delta_G(S)$ consists of the edges $e \in E$ with one endnode in S and the other endnode in $V \setminus S$. Let $\mathcal{K}_G \subseteq \{0,1\}^E$ denote the family of the incidence vectors of the cuts of G. Then, $\mathbb{R}_+(\mathcal{K}_G)$ is called the *cut cone* of G and it is also denoted by $\mathrm{CUT}(G)$. Hence, if G is the complete graph K_n on n nodes, then $\mathbb{R}_+(\mathcal{K}_{K_n}) = \mathrm{CUT}(K_n)$ coincides with the cone CUT_n. Moreover, the integer cone $\mathbb{Z}_+(\mathcal{K}_{K_n})$ consists precisely of the distances on V_n that are hypercube embeddable (recall Proposition 4.2.4).

We are interested in the following problem:

Problem 25.3.1. *Let $\mathcal{H}$ denote the class of graphs G whose family of cuts $\mathcal{K}_G$ is a Hilbert basis. Identify the graphs G belonging to the family $\mathcal{H}$.*

We review here what is known about the class $\mathcal{H}$. This problem will be revisited in Section 27.4.3 in the more general setting of binary matroids.

By Theorem 24.1.3, the graphs K_3, K_4, K_5 belong to $\mathcal{H}$. On the other hand, the graph K_6 does not belong to $\mathcal{H}$ (as the distance $2d(K_6\backslash e)$ belongs to $\mathbb{R}_+(\mathcal{K}_{K_6}) \cap \mathbb{Z}(\mathcal{K}_{K_6})$ but not to $\mathbb{Z}_+(\mathcal{K}_{K_6})$). Moreover,

Proposition 25.3.2.

(i) *Every graph with no K_5 minor belongs to $\mathcal{H}$.*

(ii) *Every graph on at most six nodes and distinct from K_6 belongs to $\mathcal{H}$.*

(iii) *If G belongs to $\mathcal{H}$, then G does not have K_6 as a minor.* ∎

Assertion (i) in Proposition 25.3.2 is proved in Fu and Goddyn [1995] and assertions (ii), (iii) in Laurent [1996b]. The proof of the above results uses, in particular, the fact that the class $\mathcal{H}$ is closed under certain operations. Namely,

(i) $\mathcal{H}$ is closed under the clique k-sum operation for graphs ($k = 0, 1, 2, 3$).

(ii) If $G \in \mathcal{H}$ and if e is an edge of G, then the graph G/e (obtained by contracting the edge e) belongs to $\mathcal{H}$.

(iii) If $G \in \mathcal{H}$ and if e is an edge of G for which each inequality $v^T x \leq 0$ defining a facet of the cut cone $\mathrm{CUT}(G)$ satisfies:

$$v_e \in \{0, 1, -1\}, \quad \sum_{f \in \delta_G(S)} v_f \in 2\mathbb{Z} \ \text{ for all cuts } \delta_G(S),$$

then the graph $G \backslash e$ (obtained by deleting the edge e) belongs to $\mathcal{H}$.

For instance, Proposition 25.3.2 (iii) can be checked in the following way. Suppose that G is a graph that contains K_6 as a subgraph. Let $x \in \mathbb{R}^E$ be defined by $x_e = 2$ for all edges of G except $x_e = 4$ for one edge belonging to the subgraph K_6. Then, $x \in \mathbb{R}_+(\mathcal{K}_G) \cap \mathbb{Z}(\mathcal{K}_G)$ (as x can be extended to a point of $\mathrm{CUT}_n \cap \mathcal{L}_n$) and $x \notin \mathbb{Z}_+(\mathcal{K}_G)$ (because the projection of x on K_6 does not belong to $\mathbb{Z}_+(\mathcal{K}_{K_6})$). Hence, every graph $G \in \mathcal{H}$ does not contain K_6 as a subgraph. Proposition 25.3.2 (iii) follows, using the fact that $\mathcal{H}$ is closed under contracting edges.

The complete characterization of the class $\mathcal{H}$ seems a hard problem. This is partly due to the fact that the linear description of the cut cone is not known for general graphs. Many questions are yet unsolved.

For instance, is the class $\mathcal{H}$ closed under the ΔY-operation[1] ? A first example to check is whether the graph from Figure 25.3.3 belongs to $\mathcal{H}$ (this is the graph obtained by applying once the ΔY-operation to K_6, i.e., replacing a triangle by a claw $K_{1,3}$).

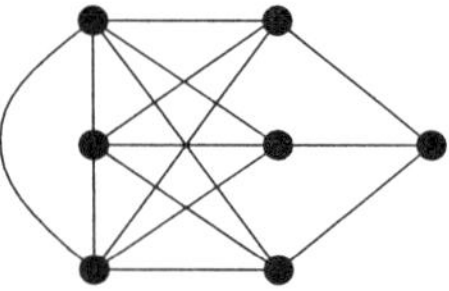

Figure 25.3.3: ΔY-transform of K_6

Is the class $\mathcal{H}$ closed under the deletion of edges ? (As mentioned above, this could be proved only if a technical assumption is made on the facets of the cut cone.)

[1] Let G be a graph having a clique on the nodes u, v, and w. The ΔY*-operation* applied to this clique consists of removing the three edges uv, uw, vw and adding a new node to G adjacent to each of the three nodes u, v, and w.

As we will see in Section 27.4.3, the question of characterizing the graphs whose family of cuts is a Hilbert basis can be posed in the more general framework of binary matroids.

Another question of interest is to determine a Hilbert basis for the cut cone on 6 points; this is the smallest case when the cuts do not form a Hilbert basis. In fact, the following result holds, which is equivalent to Theorem 25.2.2.

Theorem 25.3.4. *The* 31 *nonzero cut semimetrics on* V_6 *together with the* 15 *metrics* $2d(K_6 \backslash e)$ *(for* $e \in E(K_6)$*) form a Hilbert basis.* ∎

Finally, recall Proposition 25.1.10 which implies that the 16 odd cuts of K_6 form a Hilbert basis.

Part V

Facets of the Cut Cone and Polytope

Introduction

In this last part, we survey the results which are known about the facial structure of the cut cone and of the cut polytope. Actually, all the facets of the cut polytope can be derived from those of the cut cone, via the switching operation (see Section 26.3). Therefore, we will almost exclusively concentrate our attention to the facets of the cut cone. On the other hand, (almost) any result about the facial structure of the cut polyhedra has a direct counterpart for the correlation polyhedra, as both sets of polyhedra are in one-to-one linear correspondence (see Section 26.1).

As was explained in Part I, the members of the cut cone CUT_n are the semimetrics on $V_n = \{1, \ldots, n\}$ that can be isometrically embedded into some ℓ_1-space. Hence, the facets of CUT_n correspond to the linear inequalities characterizing ℓ_1-spaces. On the other hand, the cut polytope plays an important role in combinatorial optimization, as it permits to model the max-cut problem: $\max(c^T\delta(S) \mid S \subseteq V_n)$ (where $c \in \mathbb{R}^{E_n}$) by

$$\begin{array}{ll} \max & c^T x \\ \text{s.t.} & x \in \mathrm{CUT}_n^{\square}. \end{array}$$

(Recall Sections 4.1 and 5.1 where connections and applications are mentioned.) We will see in Section 31.2 that the facets of the cut polytope $\mathrm{CUT}_n^{\square}$ have moreover the following interesting application: They yield valid inequalities for the pairwise angles among a set of n unit vectors.

The complete description of all the facets of the cut cone CUT_n (or of the cut polytope $\mathrm{CUT}_n^{\square}$) is probably hopeless. Indeed, as the max-cut problem is NP-hard, it follows from a result of Karp and Papadimitriou [1982] that there is no polynomially concise way of describing a list of inequalities sufficient to define $\mathrm{CUT}_n^{\square}$ if NP $\neq$ co-NP.

Still a limited knowledge of some classes of facets remains interesting, from many points of view. For instance, when a separation routine is available, these classes of facets may be used in cutting plane algorithms for solving practical instances of the max-cut problem. (The reader may consult the survey by Jünger, Reinelt and Thienel [1995] for more information on solving optimization problems using cutting plane procedures.) Also, some subclasses of facets are sometimes already sufficient for handling some special classes of ℓ_1-metrics (for a list of several such cases, see Remark 6.3.5), or for the complete description of the cut polyhedra for restricted classes of graphs (see Section 27.3).

The complete description of all the facets of the cut polyhedra CUT_n and $\mathrm{CUT}_n^\square$ is known for $n \leq 8$; we list the facets of the cut polyhedra on $n \leq 7$ nodes in Section 30.6. For instance, CUT_7 has 38780 facets, while $\mathrm{CUT}_7^\square$ has 116764 facets. For $n = 8$, the number of facets is enormous: more than 217 millions, subdivided into 147 orbits !

Part V is organized as follows. We present in Chapter 26 some tools and operations for constructing facets and in Chapters 27-30 the known classes of valid inequalities and facets for the cut cone and polytope. We then group in Chapter 31 several geometric properties of the cut polytope and related objects.

Most of Part V deals with the facial structure of the cut polyhedra. However, we also visit en route some adjacent topics where cut and semimetric polyhedra are directly relevant; namely, cycle polyhedra of general binary matroids in Section 27.4, a positive semidefinite approximation of the cut polytope in Section 28.4.1, and completion problems for positive semidefinite matrices and Euclidean distance matrices in Sections 31.3 and 31.4.

Many of the inequalities that we investigate are of the form:

$$\sum_{1\leq i<j\leq n} b_i b_j x_{ij} - \sum_{ij\in E(G)} x_{ij} \leq 0,$$

where $b_1, \ldots, b_n$ are integers and G is a subgraph (possibly edge weighted) of K_n. When $\sum_{1\leq i\leq n} b_i = 1$ and G is the empty graph, we find the hypermetric inequalities. When $\sum_{1\leq i\leq n} b_i$ is odd and G is an antiweb (resp. a suspended tree), we have the clique-web inequalities (resp. the suspended-tree inequalities), considered in Chapter 29 and Section 30.1. However, three classes of facets are presented in Sections 30.2-30.4, which do not fit into this scheme.

Hypermetric inequalities constitute perhaps the most interesting known class of valid inequalities for the cut cone. They have already been considered in Part I in connection with the study of ℓ_1-metrics; we saw in Remark 6.3.5 several classes of metrics for which the hypermetric inequalities already suffice for characterizing ℓ_1-embeddability. They have also been extensively studied in Part II, in connection with Delaunay polytopes in lattices. We focus in Chapter 28 on the study of hypermetric inequalities as facets of the cut polyhedra. Note that all the facets of the cut cone on $n \leq 6$ points are, in fact, hypermetric.

Triangle inequalities, which are a very special case of hypermetric inequalities, are treated in detail in Chapter 27. In Chapter 29, we consider the clique-web inequalities, which are a generalization of hypermetric inequalities. Other classes of facets not covered by this vast class are given in Chapter 30.

We describe in Chapter 31 several properties of geometric type for the cut polytope $\mathrm{CUT}_n^\square$ and related objects. We study in Sections 31.6-31.8 questions dealing with adjacency properties and small dimensional faces of the cut polytope and of the semimetric polytope, or with the 'shape' of $\mathrm{CUT}_n^\square$ in terms of distances of its facets to the barycentrum, or with its simplex facets. We describe in Section 31.2 an interesting application of the linear description of the cut cone CUT_n for finding valid inequalities for the pairwise angles among any set of n

vectors in $\mathbb{R}^n$. Section 31.1 indicates how cuts can be used for disproving a long standing conjecture by Borsuk. Finally, Sections 31.3 and 31.4 consider the completion problems for partial positive semidefinite matrices and partial Euclidean distance matrices. It turns out that several of the polyhedra studied in this book, including the cut polytope, the semimetric polytope and the negative type cone, play an important role in these two problems.

The symbol $\delta(S)$, which was called "a cut semimetric" so far in the book, will be most often called "a *cut vector*" in Part V. This reflects the fact that we are not any more concerned here with the study of semimetrics but rather with the cut polyhedra CUT_n and $\mathrm{CUT}_n^\square$ as geometrical objects. We remind that the symbol $\delta_{K_n}(S)$ denotes the cut in K_n determined by S, that is, the set of edges in K_n having exactly one endnode in S. Hence, the incidence vector of the edge set $\delta_{K_n}(S)$ is the cut vector $\delta(S)$. We may sometimes use interchangeably the notation $\delta(S)$ and $\delta_{K_n}(S)$ and, e.g., speak of the "cut $\delta(S)$".

We will occasionally consider some other cones and polytopes, generated by restricted cut families, such as equicuts, even cuts, etc. The notions of even T-cut, k-uniform cut and odd cut semimetrics have already been introduced in Section 25.1, where we have studied the lattices they span. For convenience, we recall the definitions here. Let S be a subset of V_n. Then, $\delta(S)$ is called an *even cut vector* (or even cut semimetric) if $|S|$ and $|V_n \setminus S|$ are both even; $\delta(S)$ is called an *equicut vector* (resp. an *inequicut vector*) if $|S| = \lfloor \frac{n}{2} \rfloor, \lceil \frac{n}{2} \rceil$ (resp. $|S| \neq \lfloor \frac{n}{2} \rfloor, \lceil \frac{n}{2} \rceil$). Given an integer $k \leq n$, $\delta(S)$ is a *k-uniform cut vector* if $|S| = k, n-k$. We let ECUT_n (resp. ICUT_n, UCUT_n^k) denote the cone in $\mathbb{R}^{E_n}$ generated by all even cut (resp. inequicut, k-uniform cut) vectors; $\mathrm{EQCUT}_n^\square$ denotes the polytope in $\mathbb{R}^{E_n}$ defined as the convex hull of all equicut vectors in K_n. These polyhedra are called in the obvious manner, i.e., $\mathrm{EQCUT}_n^\square$ is the *equicut polytope*, ECUT_n is the *even cut cone*, ICUT_n is the *inequicut cone*, and UCUT_n^k is the *k-uniform cut cone.* Note that $\mathrm{EQCUT}_n^\square$ coincides with the face of $\mathrm{CUT}_n^\square$ defined by the inequality:

$$\text{(a)} \qquad \sum_{1 \leq i < j \leq n} x_{ij} \leq \left\lfloor \frac{n}{2} \right\rfloor \left\lceil \frac{n}{2} \right\rceil .$$

When n is odd, $n = 2p+1$, then $\mathrm{EQCUT}_{2p+1}^\square$ is a facet of $\mathrm{CUT}_{2p+1}^\square$ (as the inequality (a) is a switching of the pure hypermetric inequality, which is facet defining by Corollary 28.2.5 (i)). When n is even, $n = 2p$, then $\mathrm{EQCUT}_{2p}^\square$ is a face of dimension $\binom{2p}{2} - p$ of $\mathrm{CUT}_{2p}^\square$.

We also consider the *multicut polytope* $\mathrm{MC}_n^\square$, which is defined as the convex hull of all multicut vectors $\delta(S_1, \ldots, S_p)$ (for any partition of V_n into an arbitrary number of parts). (Note that the cone generated by all multicut vectors coincides with the cut cone CUT_n, as $\delta(S_1, \ldots, S_p) = \frac{1}{2} \sum_{i=1}^p \delta(S_i)$ for any partition of V_n into $S_1 \cup \ldots \cup S_p$.)

We will only give very occasional information on the above polyhedra. For more details concerning the equicut polytope, see Conforti, Rao and Sassano [1990a, 1990b], Deza, Fukuda and Laurent [1993], De Souza [1993], De Souza and Laurent [1995]; see Deza, Grötschel and Laurent [1991, 1992], Chopra and Rao [1995] and further references therein for the multicut polytope; the inequicut, even cut and k-uniform cut cones are considered, respectively, in Deza, Fukuda and Laurent [1993], and Deza and Laurent [1993b, 1992e].

We will use the definitions about polyhedra (faces, facets, valid inequalities, etc.) from Section 2.2. We also use the following additional definitions for roots, pure inequalities and edgeweight vectors. Given an inequality $v^T x \leq v_0$ which is valid for the cut polytope $\mathrm{CUT}_n^\square$, where $v \in \mathbb{R}^{E_n}$ and $v_0 \in \mathbb{R}$, a cut vector $\delta(S)$ is called a *root* of the inequality $v^T x \leq v_0$ if it satisfies it at equality, i.e., if $v^T \delta(S) = v_0$. Similarly, the cut vectors that belong to a face F of CUT_n or of $\mathrm{CUT}_n^\square$ are called its roots. We let $R(F)$ denote the set of roots of the face F.

Given $v \in \mathbb{R}^{E_n}$ and $v_0 \in \mathbb{R}$, the inequality $v^T x \leq v_0$ is said to be *pure* if $v_{ij} \in \{0, 1, -1\}$ for all $ij \in E_n$. For any cut vector $\delta(S)$, we use the notation:

$$v(\delta(S)) := v^T \delta(S) = \sum_{ij \in \delta(S)} v_{ij}.$$

Given $v \in \mathbb{R}^{E_n}$, we can define an associated weighted graph G_v with node set $\{1, \ldots, n\}$, whose edges are the pairs ij for which $v_{ij} \neq 0$; to the edge ij we assign the weight v_{ij}. The graph G_v is called the *support graph* of v. Conversely, if $G = (V, E)$ is an edge weighted graph, that is a graph G together with some weights v_{ij} ($ij \in E$) on its edges, then we define its *edgeweight vector* as the vector $v \in \mathbb{R}^{E_n}$ whose components are the weights v_{ij} for the edges $ij \in E$ and setting $v_{ij} := 0$ if ij is not an edge of G.

Let $v^T x \leq 0$ be a valid inequality for the cut cone CUT_n. As CUT_n is a full-dimensional cone in the space $\mathbb{R}^{\binom{n}{2}}$, the inequality $v^T x \leq 0$ defines a facet of CUT_n if it has $\binom{n}{2} - 1$ linearly independent roots, i.e., if there exist $\binom{n}{2} - 1$ linearly independent cut vectors $\delta(S)$ such that $v^T \delta(S) = 0$. Hence, one way to show that $v^T x \leq 0$ is facet defining for CUT_n is by exhibiting such a set of cut vectors. This direct method can be applied, for instance, for the small values of n where linear independence can be tested by hand or by computer.

Equivalently, one may show that the inequality $v^T x \leq 0$ is facet inducing for CUT_n in the following way. Let $a \in \mathbb{R}^{E_n}$ be a vector such that

$$\{x \in \mathrm{CUT}_n \mid v^T x = 0\} \subseteq \{x \in \mathrm{CUT}_n \mid a^T x = 0\}$$

or, equivalently, such that $a(\delta(S)) = 0$ whenever $v(\delta(S)) = 0$ for $S \subseteq V_n$. If for every such vector a there exists a scalar α such that $a = \alpha v$, then the inequality $v^T x \leq 0$ is facet inducing for CUT_n. (Note that we may restrict ourselves to showing this property for all $a \in \mathbb{R}^{E_n}$ for which the inequality $a^T x \leq 0$ is valid for CUT_n.) We will see in Section 26.5 some lifting techniques permitting to construct facets in an iterative manner.

Chapter 26. Operations on Valid Inequalities and Facets

In this chapter we present several operations on valid inequalities and facets of the cut polytope. One of the basic properties of the cut polytope $\mathrm{CUT}_n^\square$ is that all its facets can be deduced from the facets of the cut cone CUT_n using the so-called switching operation (cf. Section 26.3.2). In fact, switchings and permutations constitute the whole group of symmetries of the cut polytope (cf. Section 26.3.3). A general technique for constructing new facets of CUT_{n+1} from given facets of CUT_n is by applying the so-called lifting operation; we describe conditions of application of this technique in Section 26.5 and the converse operation: collapsing, in Section 26.4. Another technique for constructing facets (by looking at projections) is mentioned in Section 26.6.

26.1 Cut and Correlation Vectors

As was already explained in Section 5.2, cut vectors are in one-to-one linear correspondence with correlation vectors, via the correlation mapping. This fact will be very often used here, so we recall the details.

Let $V_n = \{1, \ldots, n\}$, $V_{n+1} = V_n \cup \{n+1\}$, and let E_n, E_{n+1} denote the set of unordered pairs of elements of V_n, V_{n+1}, respectively. The *covariance mapping*

$$\xi : \mathbb{R}^{E_{n+1}} \longrightarrow \mathbb{R}^{V_n \cup E_n}$$

is defined as follows. For $x = (x_{ij})_{1 \le i < j \le n+1} \in \mathbb{R}^{E_{n+1}}$ and $p = (p_{ij})_{1 \le i \le j \le n} \in \mathbb{R}^{V_n \cup E_n}$, let $p = \xi(x)$ be defined by

$$\begin{array}{ll} p_{ii} = x_{i,n+1} & \text{for } 1 \le i \le n, \\ p_{ij} = \frac{1}{2}(x_{i,n+1} + x_{j,n+1} - x_{ij}) & \text{for } 1 \le i < j \le n. \end{array}$$

Given a subset $S \subseteq V_n$, the vector $\pi(S) := \xi(\delta(S))$ is called the *correlation vector of S pointed at position* $n+1$; so, $\pi(S)_{ij} = 1$ if $i, j \in S$ and $\pi(S)_{ij} = 0$ otherwise for $1 \le i \le j \le n$. In the above definition, we have distinguished the position $n+1$; we also denote the covariance mapping ξ by ξ_{n+1} if we want to stress this fact. Of course, any other position $i \in V_{n+1}$ could be distinguished as well with the covariance mapping ξ_i being analogously defined.

The covariance mapping ξ is clearly linear and bijective. Therefore, given subsets $S_1, \ldots, S_k \subseteq V_n$, the set of cut vectors $\{\delta(S_1), \ldots, \delta(S_k)\}$ is linearly independent if and only if the corresponding set $\{\pi(S_1), \ldots, \pi(S_k)\}$ of correlation vectors is linearly independent. Often, when we have to check the linear independence of a set of cut vectors, we will work with the associated correlation vectors whose manipulation is generally easier.

Recall that the correlation cone COR_n and the correlation polytope $\mathrm{COR}_n^\square$ are defined, respectively, as the conic hull and the convex hull of the set of correlation vectors $\pi(S)$ for $S \subseteq V_n$. Hence,

$$\mathrm{COR}_n = \xi(\mathrm{CUT}_{n+1}) \quad \text{and} \quad \mathrm{COR}_n^\square = \xi(\mathrm{CUT}_{n+1}^\square).$$

As a consequence, any result on the facial structure of the cut polytope can be translated into a result on the facial structure of the correlation polytope and vice versa. We recall the following result, which was already formulated in Proposition 5.2.7.

Proposition 26.1.1. *Let $c \in \mathbb{R}^{E_{n+1}}$ and $a \in \mathbb{R}^{V_n \cup E_n}$ be related by*

$$\begin{array}{lll} a_{ii} = \sum_{1 \leq j \leq n+1, j \neq i} c_{ij} & \text{for } 1 \leq i \leq n, \\ a_{ij} = -2c_{ij} & \text{for } 1 \leq i < j \leq n. \end{array}$$

Then, the inequality:

$$\sum_{1 \leq i < j \leq n+1} c_{ij} x_{ij} \leq \alpha$$

defines a valid inequality (resp. a facet) of the cut polytope $\mathrm{CUT}_{n+1}^\square$ if and only if the inequality:

$$\sum_{1 \leq i \leq n} a_{ii} x_i + \sum_{1 \leq i < j \leq n} a_{ij} x_{ij} \leq \alpha$$

defines a valid inequality (resp. a facet) of the correlation polytope $\mathrm{COR}_n^\square$. ∎

We have chosen to present most of our results in the context of cuts. The corresponding results for the correlation polyhedra can be easily deduced using Proposition 26.1.1. The two forms taken by several classes of inequalities (triangle, hypermetric, negative type inequalities) in both the "cut" and "correlation" contexts have been shown in Figure 5.2.6. One reason for our choice of presenting inequalities for cut polyhedra rather than for correlation polyhedra is that they have often a much simpler form. For instance, we find it easier to handle the triangle inequality:

$$x_{12} + x_{13} + x_{23} \leq 2$$

rather than the corresponding inequality:

$$p_{11} + p_{22} + p_{33} - p_{12} - p_{13} - p_{23} \leq 1.$$

There are, however, some exceptions to this rule. There exist indeed some results whose formulation is easier in the context of correlation polyhedra than in the context of cuts; we will see one such result in Section 26.6.

26.2 The Permutation Operation

Since we are working with the complete graph K_n, all the faces of CUT_n or of $\mathrm{CUT}_n^{\square}$ are clearly preserved under any permutation of the nodes. Let $\mathrm{Sym}(n)$ denote the group of permutations of the set $\{1, 2, \ldots, n\}$, called the symmetric group of $\{1, \ldots, n\}$. Given a permutation $\sigma \in \mathrm{Sym}(n)$ and a vector $v \in \mathbb{R}^{E_n}$, define the vector $\sigma(v) \in \mathbb{R}^{E_n}$ by

$$\sigma(v)_{ij} := v_{\sigma(i)\sigma(j)} \text{ for } ij \in E_n.$$

The following result is trivial.

Lemma 26.2.1. *Given $v \in \mathbb{R}^{E_n}$, $v_0 \in \mathbb{R}$ and $\sigma \in \mathrm{Sym}(n)$, the following statements are equivalent.*

(i) *The inequality $v^T x \leq v_0$ is valid (resp. facet inducing) for $\mathrm{CUT}_n^{\square}$.*

(ii) *The inequality $\sigma(v)^T x \leq v_0$ is valid (resp. facet inducing) for $\mathrm{CUT}_n^{\square}$.* ∎

We have a similar statement about the cut cone CUT_n when applying Lemma 26.2.1 to homogeneous inequalities, i.e., to inequalities of the form $v^T x \leq 0$. Let F be the face of $\mathrm{CUT}_n^{\square}$ induced by the valid inequality $v^T x \leq v_0$, then

$$\sigma(F) := \{\sigma(x) \mid x \in F\}$$

is the face of $\mathrm{CUT}_n^{\square}$ induced by the inequality $\sigma(v)^T x \leq v_0$. We say that F and $\sigma(F)$ are *permutation equivalent*. The two faces F and $\sigma(F)$ have obviously the same dimension.

26.3 The Switching Operation

The cut polytope has the remarkable property that if, for some vertex, all the facets containing this vertex are known, then all the facets of the whole polytope can be easily derived, using the so-called switching operation. This is a consequence of the simple fact that the symmetric difference of two cuts is again a cut. This property applies more generally to the set families that are closed under taking the symmetric difference. We first present the switching operation in the general setting of set families, which will enable us to apply it to several instances of polyhedra, and then specialize it to cut polyhedra. We describe in Section 26.3.3 the full symmetry group of the cut polytope.

26.3.1 Switching: A General Definition

Let $\mathcal{A}$ be a family of subsets of a given finite set E, let $B \subseteq E$, and set

$$\mathcal{A}_B := \{A \triangle B \mid A \in \mathcal{A}\}.$$

Let $P(\mathcal{A})$ (resp. $P(\mathcal{A}_B)$) denote the polytope in $\mathbb{R}^E$, which is defined as the convex hull of the incidence vectors of the members of $\mathcal{A}$ (resp. of $\mathcal{A}_B$). A linear description of the polytope $P(\mathcal{A}_B)$ can be easily deduced if one knows a linear description of the polytope $P(\mathcal{A})$; see Corollaries 26.3.4 and 26.3.5.

For a vector $v \in \mathbb{R}^E$, let $v^B \in \mathbb{R}^E$ be defined by

$$v^B_e := \begin{cases} -v_e & \text{if } e \in B, \\ v_e & \text{if } e \in E \setminus B. \end{cases} \tag{26.3.1}$$

Consider the mapping $r_B : \mathbb{R}^E \longrightarrow \mathbb{R}^E$ defined by $r_B(x) := x^B + \chi^B$ for $x \in \mathbb{R}^E$, i.e.,

$$(r_B(x))_e = \begin{cases} 1 - x_e & \text{if } e \in B, \\ x_e & \text{if } e \in E \setminus B. \end{cases} \tag{26.3.2}$$

The mapping r_B is an affine bijection of the space $\mathbb{R}^E$, called *switching mapping.* The following can be easily checked:

Lemma 26.3.3. *Let $A, B, A_1, \ldots, A_k$ be subsets of E and let $v \in \mathbb{R}^E$. Then,*

(i) $r_B(\chi^A) = \chi^{A \triangle B}$.

(ii) $v^B(A) = v(A \triangle B) - v(B)$.

(iii) $\{\chi^{A_1}, \ldots, \chi^{A_k}\}$ *is affinely independent if and only if* $\{\chi^{A_1 \triangle B}, \ldots, \chi^{A_k \triangle B}\}$ *is affinely independent.* ∎

Corollary 26.3.4. $P(\mathcal{A}_B) = r_B(P(\mathcal{A}))$. ∎

Corollary 26.3.5. *Given $v \in \mathbb{R}^E, v_0 \in \mathbb{R}$ and $B \subseteq E$, the following assertions are equivalent.*

(i) *The inequality $v^T x \leq v_0$ is valid or facet inducing for the polytope $P(\mathcal{A})$, respectively.*

(ii) *The inequality $(v^B)^T x \leq v_0 - v(B)$ is valid or facet inducing for the polytope $P(\mathcal{A}_B)$, respectively.* ∎

We say that the inequality:

$$(v^B)^T x \leq v_0 - v(B)$$

is obtained by *switching the inequality $v^T x \leq v_0$ by the set B.* Hence, the list of inequalities defining $P(\mathcal{A}_B)$ is obtained from the list of inequalities defining $P(\mathcal{A})$ by switching each of them by the set B.

If $v_0 = 0$ and if the set B defines a root of the inequality $v^T x \leq v_0$, i.e., if $v(B) = 0$, then the switched inequality reads: $(v^B)^T x \leq 0$. In other words, the *"switching by roots"* operation preserves homogeneous inequalities.

Let F denote the face of $P(\mathcal{A})$ induced by the valid inequality $v^T x \leq v_0$, then

$$r_B(F) := \{r_B(x) \mid x \in F\}$$

is the face of $P(\mathcal{A}_B)$ induced by the switched inequality $(v^B)^T x \le v_0 - v(B)$. By Lemma 26.3.3, both faces F, $r_B(F)$ have the same dimension and the roots of $r_B(F)$ (i.e., the members $C \in \mathcal{A}_B$ for which $\chi^C \in r_B(F)$) are exactly the vectors $\chi^{A \triangle B}$, for $\chi^A \in F$. Hence, the switching mapping establishes a 1-1 correspondence between the face lattices of the polytopes $P(\mathcal{A})$ and $P(\mathcal{A}_B)$.

Let us now suppose that the family $\mathcal{A}$ is closed under taking the symmetric difference, i.e., that $A \triangle B \in \mathcal{A}$ for all $A, B \in \mathcal{A}$; hence, $\emptyset \in \mathcal{A}$. Then, the class of valid inequalities for $P(\mathcal{A})$ is closed under switching. Moreover, the full list of facets of $P(\mathcal{A})$ can be derived from the list of facets containing any given point χ^A (where $A \in \mathcal{A}$). In particular, the full facial structure of the polytope $P(\mathcal{A})$ can be deduced from that of the cone $C(\mathcal{A})$, which is defined as the cone generated by the vectors χ^A for $A \in \mathcal{A}$. The next proposition summarizes these facts.

Proposition 26.3.6. *Let $\mathcal{A}$ be a collection of subsets of E that is closed under the symmetric difference. Suppose that*

$$C(\mathcal{A}) = \{x \in \mathbb{R}^E \mid v_i^T x \le 0 \text{ for } i = 1, \ldots, m\}.$$

Then,

$$P(\mathcal{A}) = \{x \in \mathbb{R}^E \mid (v_i^B)^T x \le -v_i(B) \text{ for } i = 1, \ldots, m, \text{ and } B \in \mathcal{A}\}.$$

∎

A typical example of a set family that is closed under taking the symmetric difference is the set of cuts in a graph, or the set of cycles in a graph. In fact, the set families that are closed under taking the symmetric difference are precisely the cycle spaces of binary matroids. The switching operation was defined in this general framework by Barahona and Grötschel [1986]. We will return to cycle spaces of binary matroids in Section 27.4.

The switching operation has been discovered independently by several other authors. In particular, by McRae and Davidson [1972] and Pitowsky [1989, 1991] in the context of the correlation polytope $\mathrm{COR}_n^\square$, by Deza [1973a] in the context of the cut cone CUT_n, by Barahona and Mahjoub [1986] in the context of the cut polytope of an arbitrary graph.

26.3.2 Switching: Cut Polytope versus Cut Cone

All the features of the switching operation described above apply to the special case of the cut polytope $\mathrm{CUT}_n^\square$ and of the cut cone CUT_n, as the set of cuts is closed under taking the symmetric difference; Corollary 26.3.5 and Proposition 26.3.6 can be reformulated as follows[1].

---

[1]Given $v \in \mathbb{R}^{E_n}$ and a cut vector $\delta(A)$ in K_n, the vector $v^{\delta(A)}$ is defined as in (26.3.1) by $v_{ij}^{\delta(A)} = -v_{ij}$ if $\delta(A)_{ij} = 1$ and $v_{ij}^{\delta(A)} = v_{ij}$ if $\delta(A)_{ij} = 0$. In other words, $v^{\delta(A)}$ stands for $v^{\delta_{K_n}(A)}$.

Corollary 26.3.7. *Given $v \in \mathbb{R}^{E_n}, v_0 \in \mathbb{R}$ and a cut vector $\delta(A)$, the following assertions are equivalent.*

(i) *The inequality $v^T x \leq v_0$ is valid or facet inducing for $\mathrm{CUT}_n^\square$, respectively.*

(ii) *The inequality $(v^{\delta(A)})^T x \leq v_0 - v^T \delta(A)$ is valid or facet inducing for $\mathrm{CUT}_n^\square$, respectively.* ∎

Corollary 26.3.8. *Suppose that*

$$\mathrm{CUT}_n = \{x \in \mathbb{R}^{E_n} \mid v_i^T x \leq 0 \ \textit{ for } i = 1, \ldots, m\}.$$

Then,

$$\mathrm{CUT}_n^\square = \{x \in \mathbb{R}^{E_n} \mid (v_i^{\delta(A)})^T x \leq -v_i^T \delta(A) \ \textit{ for } i = 1, \ldots, m, \textit{ and } A \subseteq V_n\}.$$

∎

There is clearly an analogue of switching for the correlation polytope $\mathrm{COR}_n^\square$, as this polytope is in linear bijection with the cut polytope $\mathrm{CUT}_{n+1}^\square$. We indicate explicitly in the next remark how switching applies to the correlation polytope.

Remark 26.3.9. Analogue of switching for the correlation polytope. Given a subset A of V_n, consider the mapping:

$$\varrho_{\delta(A)} := \xi r_{\delta(A)} \xi^{-1}. \tag{26.3.10}$$

This mapping acts as follows:

$$\mathrm{COR}_n^\square \xrightarrow{\xi^{-1}} \mathrm{CUT}_{n+1}^\square \xrightarrow{r_{\delta(A)}} \mathrm{CUT}_{n+1}^\square \xrightarrow{\xi} \mathrm{COR}_n^\square.$$

Here, $\delta(A)$ is considered as a cut vector in K_{n+1}, $r_{\delta(A)}$ is the corresponding switching mapping of the space $\mathbb{R}^{E_{n+1}}$, and ξ is the covariance mapping. Therefore, the mapping $\varrho_{\delta(A)}$ preserves the polytope $\mathrm{COR}_n^\square$, i.e.,

$$\varrho_{\delta(A)}(\mathrm{COR}_n^\square) = \mathrm{COR}_n^\square.$$

One can easily verify that

$$p' := \varrho_{\delta(A)}(p)$$

is defined by

$$p'_{ii} = \begin{cases} 1 - p_{ii} & \text{if } i \in A, \\ p_{ii} & \text{if } i \notin A, \end{cases}$$

for $i \in V_n$ and

$$p'_{ij} = \begin{cases} 1 - p_{ii} - p_{jj} + p_{ij} & \text{if } i, j \in A, \\ p_{ii} - p_{ij} & \text{if } i \notin A, j \in A, \\ p_{jj} - p_{ij} & \text{if } i \in A, j \notin A, \\ p_{ij} & \text{if } i, j \notin A \end{cases}$$

for $i \neq j \in V_n$. Note, therefore, that switching has a much simpler form when applied in the context of the cut polyhedra rather than in the context of the correlation polyhedra. ∎

Both the switching and the permutation operations map faces of $\mathrm{CUT}_n^\square$ to faces of $\mathrm{CUT}_n^\square$; in more technical terms, they are both symmetries of $\mathrm{CUT}_n^\square$, i.e., orthogonal linear transformations of $\mathbb{R}^{E_n}$ that map $\mathrm{CUT}_n^\square$ to itself. In Section 26.3.3, we will see that the group of symmetries of $\mathrm{CUT}_n^\square$ $(n \neq 4)$ is generated by these two operations.

Clearly, if the complete description of CUT_n we start with in Corollary 26.3.8 is nonredundant, then all of the inequalities describing $\mathrm{CUT}_n^\square$ are facet defining. But many of these inequalities may appear repeatedly. However, as observed by Grötschel [1994], there is a (theoretically) easy way to compute the number of facets of $\mathrm{CUT}_n^\square$ from the number of facets of CUT_n and vice versa, as indicated in Lemma 26.3.11 below.

Let us call, for a given face F of $\mathrm{CUT}_n^\square$, the set of all the faces of $\mathrm{CUT}_n^\square$ that can be obtained from F by applying the permutation and switching operations the *orbit* $\Omega^\square(F)$ of F. We similarly define the orbit $\Omega(F)$ of a face F of CUT_n where, instead of general switching, we only allow switching by roots of F. Recall that $R(F)$ denotes the set of roots of the face F.

Lemma 26.3.11. *Let $v^T x \leq 0$ be a valid inequality for CUT_n, it defines a face F of CUT_n and a face $F^\square$ of $\mathrm{CUT}_n^\square$. Then,*

$$|\Omega^\square(F^\square)||R(F^\square)| = |\Omega(F)|2^{n-1}.$$

Proof. Let us define a $|\Omega^\square(F^\square)| \times |2^{V_{n-1}}|$ matrix $B = (b_{HS})$ by setting $b_{HS} = 1$ if the cut vector $\delta(S)$ is a root of the face H of $\Omega^\square(F^\square)$ and $b_{HS} = 0$ else. Then, counting row- and columnwise, we obtain

$$\sum_{H \in \Omega^\square(F^\square)} \sum_{S \subseteq V_{n-1}} b_{HS} = \sum_{H \in \Omega^\square(F^\square)} \left(\sum_{S \subseteq V_{n-1}} b_{HS} \right) = |\Omega^\square(F^\square)||R(F^\square)|,$$

$$\sum_{H \in \Omega^\square(F^\square)} \sum_{S \subseteq V_{n-1}} b_{HS} = \sum_{S \subseteq V_{n-1}} \left(\sum_{H \in \Omega^\square(F^\square)} b_{HS} \right) = 2^{n-1}|\Omega(F)|.$$

∎

Hence, $|\Omega^\square(F^\square)|$ can be deduced once we know $|\Omega(F)|$ and the number of roots of $F^\square$. Of course, it is neither trivial to compute the cardinality of an orbit nor to determine the number of roots of a face. We shall as often as possible explicitly describe the roots of the valid inequalities treated in this part. (See, for instance, Section 30.4 where we present the parachute inequality whose number of roots is related to the Fibonacci sequence.) We now state an upper bound on the number of roots, given in Deza and Deza [1994a].

Proposition 26.3.12. *For any facet F of the cut polytope* $\mathrm{CUT}_n^\square$ *we have*

$$|R(F)| \leq 3 \cdot 2^{n-3},$$

with equality if and only if F is defined by a triangle inequality.

Proof. Let F be a facet of $\mathrm{CUT}_n^\square$ induced by, say, the inequality $v^T x \leq \alpha$. Applying switching, we can assume that $\alpha = 0$. Suppose v_{ij} is a nonzero component of v. If S is any subset of $V_n \setminus \{i, j\}$ then

$$v^T\delta(S \cup \{i\}) + v^T\delta(S \cup \{j\}) - v^T\delta(S) - v^T\delta(S \cup \{i, j\}) = 2v_{ij}.$$

This implies that at most three of these four cut vectors are roots of F. Therefore, F has no more than $\frac{3}{4}2^{n-1} = 3 \cdot 2^{n-3}$ roots. Suppose now that F has exactly $3 \cdot 2^{n-3}$ roots. We show that F is defined by a triangle inequality. It is not difficult to check that v has at least three nonzero coordinates, say, $v_{ij}, v_{hk}, v_{st} \neq 0$. We have:

(a) $v^T\delta(A \cup \{i\}) + v^T\delta(A \cup \{j\}) - v^T\delta(A) - v^T\delta(A \cup \{i, j\}) = 2v_{ij}$,

(b) $v^T\delta(B \cup \{h\}) + v^T\delta(B \cup \{k\}) - v^T\delta(B) - v^T\delta(B \cup \{h, k\}) = 2v_{hk}$,

(c) $v^T\delta(C \cup \{s\}) + v^T\delta(C \cup \{t\}) - v^T\delta(C) - v^T\delta(C \cup \{s, t\}) = 2v_{st}$

where $A \subseteq V_n \setminus \{i, j\}$, $B \subseteq V_n \setminus \{h, k\}$, and $C \subseteq V_n \setminus \{s, t\}$. Since F contains 3/4 of the total number of cuts, exactly three terms of the left hand side of each of the equations (a),(b),(c) are equal to 0. We can suppose that v_{ij} and v_{hk} have the same sign. Suppose first that $|\{i, j, h, k, s, t\}| = 6$. We have that $v^T\delta(\{i\}) = 0$ (for, if not, considering the equation (a) with $A = \emptyset$ and the equation (b) with $B = \{i\}$ yields $v^T\delta(\{i\}) = 2v_{ij}$ and $-v^T\delta(\{i\}) = 2v_{hk}$, contradicting the fact that v_{ij} and v_{hk} have the same sign). In the same way, $v^T\delta(\{i, s\}) = 0$, $v^T\delta(\{i, t\}) = 0$, $v^T\delta(\{i, s, t\}) = 0$. Hence, equation (c) with $C = \{i\}$ yields $v_{st} = 0$, a contradiction. Similar arguments lead to a contradiction if $|\{i, j, h, k, s, t\}| = 5$ or 4. Hence, v has exactly three nonzero coordinates v_{ij}, v_{ik}, and v_{jk}. Using (a) for $A = \emptyset$ and (b) (replacing h by i) for $B = \{j\}$, we deduce that $v^T\delta(\{j\}) = 0$, i.e., $v_{ij} = -v_{jk}$. In the same way, $v^T\delta(\{k\}) = 0$, i.e., $v_{ik} = -v_{jk}$. Therefore, $v^T x \leq 0$ is a multiple of the triangle inequality $x_{jk} - x_{ik} - x_{ij} \leq 0$. ∎

Counting the number of elements in the orbit $\Omega(F)$ of a face F of CUT_n can be done in the following way (see Deza, Grishukhin and Laurent [1991] and Deza and Laurent [1992c]). Let

$$\Omega_p(F) := \{\sigma(F) \mid \sigma \in \mathrm{Sym}(n)\}$$

denote the set of faces that are permutation equivalent to F and set

$$\mathrm{Aut}(F) := \{\sigma \in \mathrm{Sym}(n) \mid \sigma(F) = F\};$$

the members of $\mathrm{Aut}(F)$ are called the *automorphisms* of F. Then, it is easy to check that

$$|\Omega_p(F)| = \frac{n!}{|\mathrm{Aut}(F)|}.$$

Let k denote the number of distinct switchings (by roots) of F that are pairwise not permutation equivalent and let $F_0 := F, F_1, \ldots, F_{k-1}$ denote these k switchings. Then,

$$|\Omega(F)| = \sum_{0 \leq i \leq k-1} \frac{n!}{|\mathrm{Aut}(F_i)|}.$$

Hence, determining the number of elements in the orbit $\Omega(F)$ of F amounts to counting the number of automorphisms of any switching of F. This has been done in Deza and Laurent [1992c] for several classes of facets and, in particular, for all the facets of the cut cone CUT_n for $n \leq 7$. As an application, one can count the exact number of facets of CUT_n and $\mathrm{CUT}_n^{\square}$ for $n \leq 7$; see Section 30.6.

Although we have introduced orbits here only as a tool for enumerating faces we would like to remark that this concept deserves more attention. For instance, if we prove that some inequality defines a facet we automatically obtain that all the faces in its orbit are, in fact, facets. The corresponding defining inequalities are obtained from the original one by switching and permuting. Proving that an inequality defines a facet is (often) laborious work that heavily uses apparent structures and symmetries of a given inequality and it is of great importance to choose, among all possible inequalities defining the facets of the orbit, one that has an "exploitable" shape or some "nice and easily understandable form". Of course, this is a matter of taste, but nevertheless, as we have seen in our own work it is helpful to find a convenient representative inequality of an orbit, not only for proof technical purposes but also for further generalizations and the investigation of other issues.

26.3.3 The Symmetry Group of the Cut Polytope

We describe here the symmetry group of the cut polytope $\mathrm{CUT}_n^{\square}$. The results are taken from Deza, Grishukhin and Laurent [1991].

A mapping $f : \mathbb{R}^{E_n} \longrightarrow \mathbb{R}^{E_n}$ is called a *symmetry* of $\mathrm{CUT}_n^{\square}$ if it is an isometry satisfying: $f(\mathrm{CUT}_n^{\square}) = \mathrm{CUT}_n^{\square}$; an isometry of $\mathbb{R}^{E_n}$ being a linear mapping preserving the Euclidean distance. Let $\mathrm{Is}(\mathrm{CUT}_n^{\square})$ denote the set of symmetries of $\mathrm{CUT}_n^{\square}$. Let

$$G_n := \{\sigma r_{\delta(A)} \mid \sigma \in \mathrm{Sym}(n),\ A \subseteq V_n\}$$

denote the group generated by all permutation and switching mappings. The commutation rules in G_n are:

$$r_{\delta(A)} r_{\delta(B)} = r_{\delta(A \triangle B)},\ r_{\delta(A)}\sigma = \sigma r_{\delta(\sigma(A))}.$$

(In fact, G_n is nothing but the quotient of the symmetry group of the n-dimensional hypercube by a subgroup of order 2.) Therefore, $|G_n| = 2^{n-1}n!$. As we have seen earlier,

$$G_n \subseteq \mathrm{Is}(\mathrm{CUT}_n^{\square}).$$

Let $H(n)$ denote the graph whose nodes are the cut vectors $\delta(A)$ for $A \subseteq V_n$, with two cut vectors $\delta(A)$, $\delta(B)$ being adjacent if $(\| \delta(A) - \delta(B) \|_2)^2 = n-1$, i.e., if $(\| \delta(A\triangle B) \|_2)^2 = n-1$ or, equivalently, if $|A\triangle B| = 1, n-1$. Obviously, every symmetry of $\mathrm{CUT}_n^\square$ induces an automorphism of the graph $H(n)$. Therefore,

$$\mathrm{Is}(\mathrm{CUT}_n^\square) \subseteq \mathrm{Aut}(H(n)).$$

The graph $H(n)$ is known as the *folded n-cube graph.* Its automorphism group is G_n for $n > 4$, $(\mathrm{Sym}(4) \times \mathrm{Sym}(4))\mathrm{Sym}(2)$ for $n = 4$, and $\mathrm{Sym}(4)$ for $n = 3$ (see Brouwer, Cohen and Neumaier [1989] or Deza, Grishukhin and Laurent [1991]). We have the following result (only (ii) needs a proof):

Theorem 26.3.13.
(i) *For* $n \neq 4$, $\mathrm{Is}(\mathrm{CUT}_n^\square) = G_n$, *with order* $2^{n-1}n!$.
(ii) *For* $n = 4$, $\mathrm{Is}(\mathrm{CUT}_4^\square) \approx (\mathrm{Sym}(4) \times \mathrm{Sym}(4))\mathrm{Sym}(2)$, *with order* $2(4!)^2$.
Hence, $\mathrm{Is}(\mathrm{CUT}_n^\square) \approx \mathrm{Aut}(H(n))$ *for* $n \geq 3$. ∎

In other words, for $n \neq 4$, switchings and permutations are the only symmetries of $\mathrm{CUT}_n^\square$. For $n = 4$, there are are some additional symmetries. Note that $H(4)$ is the complete bipartite graph $K_{4,4}$ with bipartition of its node set (X_1, X_2), where $X_1 := \{\delta(i) \mid i = 1,2,3,4\}$ and $X_2 := \{\delta(\emptyset), \delta(\{1,i\}) \mid i = 2,3,4\}$. Every automorphism of $H(4)$ acts in the following way: Permute the elements within X_1, permute the elements within X_2, and exchange elements from X_1 to X_2. In fact, one can show that every such operation yields a symmetry of $\mathrm{CUT}_4^\square$.

In the same way, one may ask what are the symmetries of the correlation polytope $\mathrm{COR}_n^\square$. In fact, the only symmetries of $\mathrm{COR}_n^\square$ are permutations, i.e.,

$$\mathrm{Is}(\mathrm{COR}_n^\square) \approx \mathrm{Sym}(n) \quad \text{for all } n.$$

Indeed, even though the mapping $\varrho_{\delta(A)}(= \xi r_{\delta(A)} \xi^{-1})$ (the analogue of switching) preserves the polytope $\mathrm{COR}_n^\square$, it is not a symmetry of it because the mapping $\varrho_{\delta(A)}$ is not an isometry. It is shown in Laurent [1996c] that the semimetric polytope $\mathrm{MET}_n^\square$ has the same group of symmetries as $\mathrm{CUT}_n^\square$. The description of the symmetry groups of other cut polyhedra (such as the equicut polytope, the multicut polytope) can be found in Deza, Grishukhin and Laurent [1991].

26.4 The Collapsing Operation

We describe in this section the collapsing operation, which permits to construct valid inequalities for $\mathrm{CUT}_m^\square$ (or CUT_m) from valid inequalities for $\mathrm{CUT}_n^\square$ (or CUT_n), where $m < n$. The collapsing operation was introduced in De Simone, Deza and Laurent [1994] and Deza and Laurent [1992b].

Let $\pi = (I_1, \ldots, I_p)$ be a partition of $V_n = \{1, \ldots, n\}$ into p parts, i.e., $I_1, \ldots, I_p$ are nonempty disjoint subsets such that $I_1 \cup \ldots \cup I_p = V_n$. For $v \in \mathbb{R}^{E_n}$, we define its π-*collapse* $v_\pi \in \mathbb{R}^{E_p}$ by

$$(v_\pi)_{hk} := \sum_{i \in I_h, j \in I_k} v_{ij} \quad \text{for } hk \in E_p.$$

When the partition π consists only of singletons except one pair $\{i,j\}$ (where $i \neq j$), i.e., when $\pi = \{\{i,j\},\{k\}$ for $1 \leq k \leq n,\ i \neq k \neq j \neq i\}$, then $v_\pi \in \mathbb{R}^{E_{n-1}}$ and we say that v_π is obtained from v by *collapsing the two nodes* i,j *into a single node.*

If G is an edge weighted graph on V_n with edgeweight vector v, then the π*-collapse* of G is the graph G_π on V_p whose edgeweight vector is v_π. In other words, we obtain G_π from G by contracting all nodes from a common partition class into a single node and adding the edgeweights correspondingly.

If S is a subset of $V_p = \{1,2,\ldots,p\}$, define the subset $S^\pi := \bigcup_{h \in S} I_h$ of V_n. The relation

$$v_\pi^T \delta(S) = v^T \delta(S^\pi) \quad \text{for } S \subseteq \{1,\ldots,p\}$$

can be easily checked, implying immediately the following lemma.

Lemma 26.4.1. *Given $v \in \mathbb{R}^{E_n}$, $v_0 \in \mathbb{R}$ and a partition $\pi = (I_1,\ldots,I_p)$ of V_n into p parts, the following statements hold.*

(i) *If the inequality $v^T x \leq v_0$ is valid for $\mathrm{CUT}_n^\square$, then the inequality $v_\pi^T x \leq v_0$ is valid for $\mathrm{CUT}_p^\square$.*

(ii) *The roots of the inequality $v_\pi^T x \leq v_0$ are the cut vectors $\delta(S)$ such that $S \subseteq \{1,\ldots,p\}$ and $\delta(S^\pi)$ is a root of the inequality $v^T x \leq v_0$.* ∎

We present in the next result a case when the collapsing and the switching operations commute.

Lemma 26.4.2. *Let $\pi = (I_1,\ldots,I_p)$ be a partition of $\{1,\ldots,n\}$, S be a subset of V_p, and $v \in \mathbb{R}^{E_n}$. Then,*

$$(v_\pi)^{\delta(S)} = \left(v^{\delta(S^\pi)}\right)_\pi .$$

∎

Observe that the collapsing operation preserves validity, but it may not always preserve facets. Also, it may be that the collapse of a nonfacet inducing valid inequality is facet inducing. For example, for $n \geq 3$, the (triangle) inequality $x_{23} - x_{12} - x_{13} \leq 0$ is facet inducing for CUT_n but the inequality $-2x_{12} \leq 0$, obtained by collapsing the two nodes 2,3 into a single node 2, is valid but not facet inducing. For $n \geq 5$, the inequality $(x_{23} - x_{12} - x_{13}) + (x_{45} - x_{14} - x_{15}) \leq 0$ is valid but not facet inducing, while the inequality $x_{23} - x_{12} - x_{13} \leq 0$, obtained by collapsing the two nodes 1,5 into a single node 1, is facet inducing for CUT_n.

Nevertheless, collapsing may be a useful tool for the construction of facets; Theorems 26.4.3 and 26.4.4 below state results of the form: "Take a valid inequality, assume that some collapsings of it are facet inducing and ..., then the inequality is facet inducing". The next result is shown in De Simone, Deza and Laurent [1994].

Theorem 26.4.3. *Let $v \in \mathbb{R}^{E_n}$ and i_1, i_2, i_3 be distinct nodes in $\{1, 2, \ldots, n\}$. Assume that the following conditions hold.*

(i) *The inequality $v^T x \le 0$ is valid for CUT_n.*

(ii) $v^T \delta(\{i_1\}) = 0$.

(iii) *The two inequalities obtained from $v^T x \le 0$ by collapsing the nodes $\{i_1, i_2\}$, and the nodes $\{i_1, i_3\}$, respectively, are facet inducing for CUT_{n-1}.*

(iv) *For some distinct $r, s \in \{1, \ldots, n\} \setminus \{i_1, i_2, i_3\}$, $v_{rs} \neq 0$.*

Then, the inequality $v^T x \le 0$ is facet inducing for CUT_n.

Proof. For ease of notation, we may assume that $i_1 = 1,\ i_2 = 2,\ i_3 = 3$. Denote by $v^{1,2}$ and $v^{1,3}$ the vector obtained from v by collapsing the nodes $\{1,2\}$ and the nodes $\{1,3\}$, respectively. Take a valid inequality $a^T x \le 0$ for CUT_n such that

$$\{x \in \mathrm{CUT}_n \mid v^T x = 0\} \subseteq \{x \in \mathrm{CUT}_n \mid a^T x = 0\}.$$

In order to prove that $v^T x \le 0$ is facet defining, we show that $v = \alpha a$ for some scalar $\alpha > 0$. Denote analogously by $a^{1,2}$ and $a^{1,3}$ the vector obtained from a by collapsing the nodes $\{1,2\}$ and $\{1,3\}$, respectively. It is easy to see that

$$\{x \in \mathrm{CUT}_n \mid (v^{1,i})^T x = 0\} \subseteq \{x \in \mathrm{CUT}_n \mid (a^{1,i})^T x = 0\}$$

for $i = 2, 3$. Since $(v^{1,i})^T x \le 0$ is facet inducing by assumption (iii), there exists a scalar $\alpha_i > 0$ such that $v^{1,i} = \alpha_i a^{1,i}$ for $i = 2, 3$. We deduce that $\alpha_1 = \alpha_2 := \alpha$ from assumption (iv). Hence, we already have that $v_{rs} = \alpha a_{rs}$ for $3 \le r < s \le n$, or $r = 2$ and $4 \le s \le n$ and, also,

(a) $v_{1i} + v_{2i} = \alpha(a_{1i} + a_{2i})$ for $i \ge 3$,

(b) $v_{1i} + v_{3i} = \alpha(a_{1i} + a_{3i})$ for $i = 2, i \ge 4$.

Hence, $v_{1i} = a_{1i}$ for $i \ge 4$. It remains to check that $v_{12} = \alpha a_{12}$, $v_{13} = \alpha a_{13}$ and $v_{23} = \alpha a_{23}$ in order to deduce that $v = \alpha a$. By assumption (ii), we have $v^T \delta(\{1\}) = 0$, implying that

$$v_{12} + v_{13} = -\sum_{4 \le i \le n} v_{1i} = -\sum_{4 \le i \le n} \alpha a_{1i} = \alpha(a_{12} + a_{13}).$$

This relation, together with (a), (b) for $i = 3$, yields that $v_{23} = \alpha a_{23}$ and, then, $v_{13} = \alpha a_{13}$, $v_{12} = \alpha a_{12}$. ∎

The following result was proved in Deza, Grötschel and Laurent [1992] in the more general context of multicuts.

Theorem 26.4.4. *Let $v \in \mathbb{R}^{E_n}$, $v_0 \in \mathbb{R}$, and i_1, i_2, i_3, i_4 be distinct elements in $\{1, \ldots, n\}$. Assume that the following conditions hold.*

(i) *The inequality $v^T x \le v_0$ is valid for $\mathrm{CUT}_n^{\square}$.*

(ii) $v_0 \neq 0$, *or there exist distinct* $r, s \in \{1, \ldots, n\} \setminus \{i_1, i_2, i_3, i_4\}$ *such that* $v_{rs} \neq 0$.

(iii) *The three inequalities obtained from the inequality* $v^T x \leq v_0$ *by collapsing the nodes* $\{i_1, i_2\}$, *the nodes* $\{i_1, i_3\}$, *and the nodes* $\{i_1, i_4\}$, *respectively, are facet inducing for* $\mathrm{CUT}_{n-1}^{\square}$.

Then, the inequality $v^T x \leq v_0$ *is facet inducing for* $\mathrm{CUT}_n^{\square}$.

Proof. Let us assume that $i_1 = 1,\ i_2 = 2,\ i_3 = 3,\ i_4 = 4$. Let $a^T x \leq a_0$ be a valid inequality for $\mathrm{CUT}_n^{\square}$ such that

$$\{x \in \mathrm{CUT}_n^{\square} \mid v^T x = v_0\} \subseteq \{x \in \mathrm{CUT}_n^{\square} \mid a^T x = a_0\}.$$

We show the existence of a scalar $\alpha > 0$ such that $v = \alpha a,\ v_0 = \alpha a_0$. For $i = 2, 3, 4$, denote by $v^{1,i}$ (resp. $a^{1,i}$) the vector obtained from v (resp. from a) by collapsing the nodes $\{1, i\}$. Clearly, every root of the inequality $(v^{1,i})^T x \leq v_0$ is a root of the inequality $(a^{1,i})^T x \leq a_0$. Hence, there exists a scalar $\alpha_i > 0$ such that $v^{1,i} = \alpha_i a^{1,i}$ and $v_0 = \alpha_i a_0$, for $i = 2, 3, 4$. By assumption (ii), we deduce that $\alpha_1 = \alpha_2 = \alpha_3 =: \alpha$. We already have that $v_{rs} = \alpha a_{rs}$ for $3 \leq r < s \leq n$, or $r = 2$ and $4 \leq s \leq n$, or $\{r, s\} = \{2, 3\}$. Also, $v_{1i} + v_{2i} = \alpha(a_{1i} + a_{2i})$ for $i \geq 3$, from which we deduce that $v_{1i} = \alpha a_{1i}$ for $3 \leq i \leq n$. Finally, $v_{1i} + v_{4i} = \alpha(a_{1i} + a_{4i})$ for $i = 2, 3, 5, \ldots, n$, implying that $v_{12} = \alpha a_{12}$. Therefore, $v = \alpha a$ holds and, thus, the inequality $v^T x \leq v_0$ is facet inducing. ∎

26.5 The Lifting Operation

The collapsing operation, which is described in the preceding section, permits to construct certain valid inequalities of CUT_{n-1} from a given valid inequality of CUT_n. Conversely, by lifting, we mean any general procedure for constructing a valid inequality of CUT_{n+1} (preferably, facet inducing) from a given valid inequality (or facet) for CUT_n.

The simplest case of lifting is that of 0-lifting. Namely, for $v \in \mathbb{R}^{E_n}$, define its *0-lifting* $v' \in \mathbb{R}^{E_{n+1}}$ by

$$\begin{aligned} v'_{ij} &= v_{ij} && \text{for } ij \in E_n, \\ v'_{i,n+1} &= 0 && \text{for } 1 \leq i \leq n. \end{aligned}$$

In the same way, we say that the inequality $(v')^T x \leq v_0$ is obtained by *0-lifting* the inequality $v^T x \leq v_0$. It is immediate to see that 0-lifting preserves validity; a nice feature of 0-lifting is that it also preserves facets.

Theorem 26.5.1. *Given* $v_0 \in \mathbb{R}$, $v \in \mathbb{R}^{E_n}$ *and its 0-lifting* $v' \in \mathbb{R}^{E_{n+1}}$, *the following assertions are equivalent.*

(i) *The inequality* $v^T x \leq v_0$ *is facet inducing for* $\mathrm{CUT}_n^{\square}$.

(ii) *The inequality* $(v')^T x \leq v_0$ *is facet inducing for* $\mathrm{CUT}_{n+1}^{\square}$.

The proof of Theorem 26.5.1 is based on the following Lemma 26.5.2. Both results were given in Deza [1973a] (see also Deza and Laurent [1992a] for the full proofs). Given a subset F of E_n, set $\overline{F} := E_n \setminus F$. If $x \in \mathbb{R}^{E_n}$, let

$$x_F := (x_e)_{e \in F}$$

denote the projection of x on the subspace $\mathbb{R}^F$ indexed by F and, if X is a subset of $\mathbb{R}^{E_n}$, set

$$X_F := \{x_F \mid x \in X\}, \ X^F := \{x \in X \mid x_F = 0\}.$$

Lemma 26.5.2. *Let $v^T x \leq 0$ be a valid inequality for* CUT_n *and let $R(v)$ denote its set of roots. Let F be a subset of E_n.*

(i) *If* $\mathrm{rank}(R(v)_F) = |F|$ *and* $\mathrm{rank}(R(v)^F) = |\overline{F}| - 1$, *then the inequality $v^T x \leq 0$ is facet inducing.*

(ii) *If the inequality $v^T x \leq 0$ is facet inducing and $v_{\overline{F}} \neq 0$ (resp. $v_{\overline{F}} = 0$), then* $\mathrm{rank}(R(v)_F) = |F|$ *(resp.* $\mathrm{rank}(R(v)_F) = |F| - 1$).

Proof. (i) By the assumption, we can find a set $A \subseteq R(v)$ of $|F|$ vectors whose projections on F are linearly independent and a set $B \subseteq R(v)$ of $|\overline{F}| - 1$ linearly independent vectors whose projections on F are zero. It is immediate to see that the set $A \cup B$ is linearly independent. This implies that $v^T x \leq 0$ is facet inducing.

(ii) Since $v^T x \leq 0$ is facet inducing, we can find a set $A \subseteq R(v)$ of $\binom{n}{2} - 1$ linearly independent roots. Let M denote the $(\binom{n}{2} - 1) \times \binom{n}{2}$ matrix whose rows are the vectors of A. Hence, all the columns of M but one are linearly independent. We distinguish two cases:

(a) either, all the columns of M that are indexed by F are linearly independent and, then, $\mathrm{rank}(A_F) = |F|$,

(b) or, $\mathrm{rank}(A_F) = |F| - 1$, implying that $\mathrm{rank}(R(v)_F) = |F| - 1$.

Suppose first that we are in the case (b). Let $T_1 \subseteq A$ be a subset of $|F| - 1$ vectors whose projections on F are linearly independent, set $T_2 := A^F$ and $T_3 := A \setminus (T_1 \cup T_2)$. Hence, $|T_2 \cup T_3| = |\overline{F}|$. For $x \in T_3$, its projection x_F on F can be written as a linear combination of the projections on F of the vectors in T_1, say

$$x_F = \sum_{a \in T_1} \lambda_a^x a_F.$$

Set

$$x' := x - \sum_{a \in T_1} \lambda_a^x a$$

and $T_3' := \{x' \mid x \in T_3\}$. It is easy to check that the set $T_2 \cup T_3'$ is linearly independent. Note that, for any $x \in T_2 \cup T_3'$, $x_F = 0$ and $v^T x = 0$; this implies that $v_{\overline{F}} = 0$. Suppose now that we are in the case (a). Then, $\mathrm{rank}(R(v)_{\overline{F}}) = |\overline{F}| - 1$ and, by the above reasoning, $v_F = 0$, which implies that $v_{\overline{F}} \neq 0$. ∎

Proof of Theorem 26.5.1. We can assume, without loss of generality, that $v_0 = 0$; else, switch the inequality by a root and apply the result for the switched inequality.
Suppose first that the inequality $(v')^T x \leq 0$ is facet inducing for CUT_{n+1}. Consider the set $F := \{(i, n+1) \mid 1 \leq i \leq n\}$ and let $\overline{F} := E_n$ denote its complement in the set E_{n+1}. By construction, $v'_F = 0$. Hence, we deduce from Lemma 26.5.2 (ii) that $\mathrm{rank}(R(v')_{\overline{F}}) = |\overline{F}| - 1 = \binom{n}{2} - 1$. Therefore, the inequality $v^T x \leq 0$ is facet inducing for CUT_n.
Suppose now that the inequality $v^T x \leq 0$ is facet inducing for CUT_n. Since $v \neq 0$, we can suppose without loss of generality that $v_{\overline{F}} \neq 0$, where $F := \{12, 13, \ldots, 1n\}$ and $\overline{F} := E_n \setminus F$. By Lemma 26.5.2 (ii), $\mathrm{rank}(R(v)_F) = |F|$. Therefore, we can find $n-1$ roots $\delta(T_k)$ $(1 \leq k \leq n-1)$ of $v^T x \leq 0$ whose projections on F are linearly independent. Since $v^T x \leq 0$ is facet inducing, we can also find $\binom{n}{2} - 1$ linearly independent roots $\delta(S_j)$ $(1 \leq j \leq \binom{n}{2} - 1)$ of $v^T x \leq 0$. Without loss of generality, we can suppose that the element 1 does not belong to any of the sets T_k and S_j. So, we have a set

$$\mathcal{C} := \left\{\delta(S_j) \mid 1 \leq j \leq \binom{n}{2} - 1\right\} \cup \{\delta(T_k \cup \{n+1\}) \mid 1 \leq k \leq n-1\} \cup \{\delta(\{n+1\})\}$$

of $\binom{n+1}{2} - 1$ cut vectors (in K_{n+1}), which are roots of the inequality $(v')^T x \leq 0$. We show that the set $\mathcal{C}$ is linearly independent. For this, we verify that the correlation vectors (pointed at position 1) associated with the cut vectors in $\mathcal{C}$ are linearly independent. Let us consider the square matrix M of order $\binom{n+1}{2} - 1$, whose rows are: first, the $\binom{n}{2} - 1$ vectors $\pi(S_j)$; then, the $n-1$ vectors $\pi(T_k \cup \{n+1\})$ and, finally, the vector $\pi(\{n+1\})$. The columns of M are indexed by the set $I \cup J \cup K$, where $I := \{(i,j) \mid 2 \leq i \leq j \leq n\}$, $J := \{(i, n+1) \mid 2 \leq i \leq n\}$, and $K := \{(n+1, n+1)\}$. The matrix M is of the form:

$$M = \begin{pmatrix} X & 0 & 0 \\ Z & Y & e^T \\ 0 & 0 & 1 \end{pmatrix} \quad (\text{columns } I, J, K)$$

(where e denotes the all-ones vector). The matrix X has full row rank, since the vectors $\pi(S_j)$ are linearly independent. The matrix Y has full rank, since its rows are the vectors $\pi(T_k \cup \{n+1\})_J = \delta(T_k)_F$, which are linearly independent. Hence, the matrix M has full row rank, which implies that the inequality $(v')^T x \leq 0$ is facet inducing. ∎

Lifting is a very general methodology for constructing facets of polyhedra, which can be described as follows. Suppose we are given a vector $v \in \mathbb{R}^{E_n}$ for which the inequality $v^T x \leq 0$ is valid for CUT_n. Then, lifting the inequality $v^T x \leq 0$ means finding a vector $v' \in \mathbb{R}^{E_{n+1}}$ (obtained by adding n new coordinates to v, after possibly altering some of its coordinates) such that the inequality $(v')^T x \leq 0$ is valid for CUT_{n+1}. Of course, a desirable objective is to produce in this way some new facet of CUT_{n+1} starting from a given facet of CUT_n. The next lemma contains a set of conditions that are sufficient for achieving this goal.

Lemma 26.5.3. (**Lifting Lemma**) *Let $v \in \mathbb{R}^{E_n}$ and $v' \in \mathbb{R}^{E_{n+1}}$. Suppose that the following assertions hold.*

(i) *The inequality $v^T x \leq 0$ is facet inducing for CUT_n and the inequality $(v')^T x \leq 0$ is valid for CUT_{n+1}.*

(ii) *There exist $\binom{n}{2} - 1$ subsets S_j of $\{2, 3, \ldots, n\}$ such that the cut vectors $\delta(S_j)$ (in K_n) are linearly independent roots of $v^T x \leq 0$ and the cut vectors $\delta(S_j)$ (in K_{n+1}) are roots of $(v')^T x \leq 0$.*

(iii) *There exist n subsets T_k of $\{2, 3, \ldots, n, n+1\}$ with $n+1 \in T_k$ such that the cut vectors $\delta(T_k)$ (in K_{n+1}) are roots of $(v')^T x \leq 0$ and the incidence vectors of the sets T_k are linearly independent.*

Then, the inequality $(v')^T x \leq 0$ is facet inducing for CUT_{n+1}.

Proof. It suffices to check that the $\binom{n+1}{2} - 1$ cut vectors $\delta(S_j)$ and $\delta(T_k)$ are linearly independent. This can be done in the same way as in the proof of Theorem 26.5.1. ∎

In the sequel, we only consider a special case of lifting, known as *node splitting*. The node splitting operation is, in fact, converse to the collapsing operation from Section 26.4; it is defined as follows. Let $v \in \mathbb{R}^{E_n}$ and $v' \in \mathbb{R}^{E_{n+1}}$ satisfy the conditions:

$$\begin{array}{ll} v_{ij} = v'_{ij} & \text{for } 2 \leq i < j \leq n, \\ v_{1i} = v'_{1i} + v'_{i\ n+1} & \text{for } 2 \leq i \leq n. \end{array}$$

So, v' is obtained from v by splitting node 1 into two nodes $1, n+1$ and correspondingly splitting the edgeweight v_{1i} into v'_{1i} and $v'_{i\ n+1}$, the other components remaining unchanged. In other words, v comes from v' by collapsing the nodes 1 and $n+1$ into a single node 1. In this case, if $v^T x \leq 0$ is facet inducing, then the condition (ii) of Lemma 26.5.3 automatically holds. In our concrete applications of the Lifting Lemma 26.5.3, the condition (i) will hold by construction of v' and, therefore, the crucial point will be to check condition (iii), i.e., to find n additional "good" roots.

We will see in the next sections many applications of the Lifting Lemma 26.5.3 and of Theorems 26.4.3 and 26.4.4 on collapsing.

26.6 Facets by Projection

We present here another tool for showing that a given valid inequality is facet defining. This method works when some projections of the given inequality have some prescribed properties; it is described by Boissin [1994]. The result turns out to have a simpler formulation in the context of correlation polyhedra. So, we first state it for correlation polyhedra and, then, we reformulate it for cut polyhedra.

Observe first that, if an inequality

$$\sum_{1 \leq i \leq j \leq n} a_{ij} p_{ij} \leq \alpha$$

is valid for $\mathrm{COR}_n^\square$, then its projection

$$\sum_{1 \leq i \leq j \leq n-1} a_{ij} p_{ij} \leq \alpha$$

on the set (of pairs including diagonal pairs from) $V_n \setminus \{n\}$ is obviously valid for $\mathrm{COR}_{n-1}^\square$. However, it may be that the projected inequality defines a facet of $\mathrm{COR}_{n-1}^\square$ without the initial inequality to be facet defining. This is the case, for instance, for the (nonfacet defining) inequality $p_{11} + p_{22} \geq 0$ whose projection $p_{11} \geq 0$ is facet defining. The next result shows, however, that if we suppose that several projections are facet defining together with some additional conditions, then we can conclude that the initial inequality is facet defining.

Proposition 26.6.1. *Let $W_1, \ldots, W_k \subseteq V_n$ such that $V_n = W_1 \cup \ldots \cup W_k$ and each pair of elements of V_n belongs to some W_r $(1 \leq r \leq k)$. Let $a \in \mathbb{R}^{V_n \cup E_n}$ and $\alpha \in \mathbb{R}$. Suppose that the following conditions hold.*

(i) *The inequality $a^T p := \sum_{1 \leq i \leq j \leq n} a_{ij} p_{ij} \leq \alpha$ is valid for the correlation polytope $\mathrm{COR}_n^\square$.*

(ii) *The graph with node set $\{1, \ldots, k\}$ and whose edges are the pairs rs for which there exist $i, j \in W_r \cap W_s$ such that $a_{ij} \neq 0$ is connected.*

(iii) *For each $r = 1, \ldots, k$, the inequality $\sum_{i \leq j | i,j \in W_r} a_{ij} p_{ij} \leq \alpha$ (obtained as the projection of $a^T p \leq \alpha$ on W_r) defines a facet of $\mathrm{COR}^\square(W_r)$.*

Then, the inequality $a^T p \leq \alpha$ defines a facet of $\mathrm{COR}_n^\square$.

Proof. Let $b \in \mathbb{R}^{V_n \cup E_n}, \beta \in \mathbb{R}$ such that

$$\{p \in \mathrm{COR}_n^\square \mid a^T p = \alpha\} \subseteq \{p \in \mathrm{COR}_n^\square \mid b^T p = \beta\}.$$

We show that $b = \lambda a$ and $\beta = \lambda \alpha$ for some $\lambda \in \mathbb{R}$. As each vector $q \in \mathrm{COR}^\square(W_r)$ can be extended to the vector $p := (q, 0, \ldots, 0)$ of $\mathrm{COR}_n^\square$, we deduce that

$$\{q \in \mathrm{COR}^\square(W_r) \mid \sum_{i \leq j | i,j \in W_r} a_{ij} q_{ij} = \alpha\} \subseteq \{q \in \mathrm{COR}^\square(W_r) \mid \sum_{i \leq j | i,j \in W_r} b_{ij} q_{ij} = \beta\}.$$

From (iii), we obtain that there exists $\lambda_r \in \mathbb{R}$ such that $b_{ij} = \lambda_r a_{ij}$ for $i, j \in W_r$ and $\beta = \lambda_r \alpha$. We deduce easily from (ii) that all λ_r's are equal. ∎

The following is a reformulation in the context of cut polyhedra.

Corollary 26.6.2. *Let $W_1, \ldots, W_k \subseteq V_n$ such that $V_n = W_1 \cup \ldots \cup W_k$ and each pair of elements of V_n belongs to some W_r $(1 \leq r \leq k)$. Let $c \in \mathbb{R}^{E_{n+1}}$ and $\alpha \in \mathbb{R}$. Suppose that the following conditions hold.*

(i) *The inequality* $c^T x := \sum_{1 \le i < j \le n+1} c_{ij} x_{ij} \le \alpha$ *is valid for the cut polytope* $\mathrm{CUT}^\square_{n+1}$.

(ii) *The graph with node set* $\{1, \ldots, k\}$ *and whose edges are the pairs* rs *for which, either there exist* $i \ne j \in W_r \cap W_s$ *such that* $c_{ij} \ne 0$*, or there exists* $i \in W_r \cap W_s$ *such that* $c^T \delta(\{i\}) \ne 0$*, is connected.*

(iii) *The inequality* $\sum_{i \in W_r} (c_{i,n+1} + \sum_{j \in V_n \setminus W_r} c_{ij}) x_{i,n+1} + \sum_{i<j | i,j \in W_r} c_{ij} x_{ij} \le \alpha$ *defines a facet of* $\mathrm{CUT}^\square_{|W_r|+1}$*, for each* $r = 1, \ldots, k$.

Then, the inequality $c^T x \le \alpha$ *defines a facet of* $\mathrm{CUT}^\square_{n+1}$. ∎

For example, Corollary 26.6.2 permits to derive that the pentagonal inequality:

$$Q(1,1,1,-1,-1)^T x \le 0$$

defines a facet of CUT_5, from the fact that the triangle inequalities define facets of CUT_4. Indeed, consider the subsets $W_1 := \{1,2,4\}$, $W_2 := \{1,3,4\}$ and $W_3 := \{2,3,4\}$ of $\{1,2,3,4\}$. Then, the inequalities from Corollary 26.6.2 (iii) are, respectively, the triangle inequalities: $x_{12} - x_{14} - x_{24} \le 0$, $x_{13} - x_{14} - x_{34} \le 0$ and $x_{23} - x_{24} - x_{34} \le 0$. More generally, Corollary 26.6.2 permits to derive that any pure hypermetric inequality: $Q(1, \ldots, 1, -1, \ldots, -1)^T x \le 0$ is facet defining from the fact that triangle inequalities are facet defining.

A slight modification of Proposition 26.6.1 yields the following construction for facets of the correlation polytope. Given $a \in \mathbb{R}^{V_n \cup E_n}$ and $\alpha \in \mathbb{R}$, suppose that the inequality $a^T p := \sum_{1 \le i \le j \le n} a_{ij} p_{ij} \le \alpha$ defines a facet of $\mathrm{COR}^\square_n$. Consider $\beta \in \mathbb{R}$ and two elements $i^* \in V_n$ and $j^* \notin V_n$; set $V_{n+1} = V_n \cup \{j^*\}$ and define the vector $b \in \mathbb{R}^{V_{n+1} \cup E_{n+1}}$ by

$$(26.6.3) \qquad \begin{cases} b_{ij} & = a_{ij} & \text{if } i,j \in V_n, \\ b_{i,j^*} & = a_{i,i^*} & \text{if } i \in V_n \setminus \{i^*\}, \\ b_{i^*,j^*} & = \beta, \\ b_{j^*,j^*} & = a_{i^*,i^*}. \end{cases}$$

We say that the inequality $b^T p \le \alpha$ is obtained from the inequality $a^T p \le \alpha$ by *duplicating the node* i^* *as the node* j^*. Note that this operation can be seen as a special case of lifting. The next result indicates what value of β should be taken in order to ensure that $b^T p \le \alpha$ defines a facet of $\mathrm{COR}^\square_{n+1}$.

Proposition 26.6.4. *Suppose that the inequality* $a^T p \le \alpha$ *defines a facet of* $\mathrm{COR}^\square_n$ *and let* $b \in \mathbb{R}^{V_{n+1} \cup E_{n+1}}$ *be defined by* (26.6.3). *Then the inequality* $b^T p \le \alpha$ *defines a facet of* $\mathrm{COR}^\square_{n+1}$ *if and only if*

$$\beta = \alpha - \max_{S \subseteq V_n \setminus \{i^*\}} \left(a^T \pi(S \cup \{i^*\}) + \sum_{i \in S \cup \{i^*\}} a_{i,i^*} \right).$$

Proof. Let β_0 denote the right hand side in the relation defining β. We first check that $b^T p \le \alpha$ is valid for $\mathrm{COR}^\square_{n+1}$ if and only if $\beta \le \beta_0$. Indeed, let T be a subset of V_{n+1}. If $j^* \notin T$, or if $j^* \in T$ and $i^* \notin T$, then $b^T\pi(T) = a^T\pi(T) \leq \alpha$, by construction of b. If $i^*, j^* \in T$ then, setting $S := T \setminus \{i^*, j^*\}$, we have

$$b^T\pi(T) = a^T\pi(S \cup \{i^*\}) + \sum_{i \in S \cup \{i^*\}} a_{i,i^*} + \beta.$$

Hence, $b^T\pi(T) \leq \alpha$ holds for all $T \subseteq V_{n+1}$ if and only if $\beta \leq \beta_0$. If $\beta < \beta_0$, then every root of $b^Tp = \alpha$ satisfies the equation $p_{i^*,j^*} = 0$; this shows that $b^Tp \leq \alpha$ does not define a facet of $\mathrm{COR}^{\square}_{n+1}$. Suppose now that $\beta = \beta_0$. We show that $b^Tp \leq \alpha$ defines a facet of $\mathrm{COR}^{\square}_{n+1}$. For this, let $(b')^Tp \leq \alpha'$ be another inequality such that

$$\{p \in \mathrm{COR}^{\square}_{n+1} \mid b^Tp = \alpha\} \subseteq \{p \in \mathrm{COR}^{\square}_{n+1} \mid (b')^Tp = \alpha'\}.$$

By the argument of Proposition 26.6.1, there exists a scalar λ such that $\alpha' = \lambda\alpha$, $b'_{ij} = \lambda b_{ij}$ for all $i, j \in V_{n+1}$ except maybe for the pair i^*, j^*. As $\beta = \beta_0$, we can find a root $\pi(T)$ of $b^Tp = \alpha$ such that $i^*, j^* \in T$. As $(b')^T\pi(T) = \alpha'$, we deduce that $(b')_{i^*,j^*} = \lambda b_{i^*,j^*}$. This shows $b' = \lambda b$. ∎

For instance, the inequality:

$$p_{i^*,i^*} + p_{ii} - p_{i,i^*} \leq 1$$

defines a facet of $\mathrm{COR}^{\square}_3$. Applying Proposition 26.6.4, we obtain that the inequality:

$$p_{i^*,i^*} + p_{ii} + p_{j,j^*} - p_{i,i^*} - p_{i,j^*} - p_{i^*,j^*} \leq 1$$

defines a facet of $\mathrm{COR}^{\square}_4$. (Note that these two inequalities correspond to triangle inequalities for the cut polytope; recall Figure 5.2.6.) We leave it to the reader to reformulate Proposition 26.6.4 for the cut polytope. See Section 30.4 for an application of Proposition 26.6.4 to the parachute facet.

Chapter 27. Triangle Inequalities

As we will see throughout Part V, the cut polytope $\mathrm{CUT}_n^{\square}$ has many different types of facets, most of them having a rather complicated structure. Among them, the most simple ones are the triangle facets, i.e., those defined by the following triangle inequalities:

$$x_{ij} - x_{ik} - x_{jk} \leq 0, \tag{27.0.1}$$

$$x_{ij} + x_{ik} + x_{jk} \leq 2, \tag{27.0.2}$$

for distinct $i, j, k \in V_n$. The inequality (27.0.1) is a homogeneous triangle inequality, while (27.0.2) is nonhomogeneous. The homogeneous triangle inequalities have already been considered in previous chapters. Note that (27.0.2) arises from (27.0.1) by switching, e.g., by the cut $\delta(\{i\})$; hence, the class of triangle inequalities is closed under switching. The cone in $\mathbb{R}^{E_n}$ defined by the homogeneous triangle inequalities is the semimetric cone MET_n, already considered earlier. The polytope in $\mathbb{R}^{E_n}$ defined by all triangle inequalities (27.0.1) and (27.0.2) is the *semimetric polytope* and is denoted by $\mathrm{MET}_n^{\square}$. Hence,

$$\mathrm{CUT}_n \subseteq \mathrm{MET}_n \subseteq \mathbb{R}_+^{E_n} \quad \text{and} \quad \mathrm{CUT}_n^{\square} \subseteq \mathrm{MET}_n^{\square} \subseteq [0,1]^{E_n}.$$

The terminology used for the polyhedra MET_n and $\mathrm{MET}_n^{\square}$ comes, of course, from the fact that the distances on V_n that satisfy the homogeneous triangle inequalities are precisely the semimetrics.

Clearly, the semimetric polytope $\mathrm{MET}_n^{\square}$ is preserved by permutation and switching (as the class of triangle inequalities is closed under switching), a property also enjoyed by the cut polytope $\mathrm{CUT}_n^{\square}$. In fact, as shown in Laurent [1996c], both polytopes $\mathrm{MET}_n^{\square}$ and $\mathrm{CUT}_n^{\square}$ have the same group of symmetries; that is,

$$\mathrm{Is}(\mathrm{MET}_n^{\square}) = \mathrm{Is}(\mathrm{CUT}_n^{\square}).$$

The group $\mathrm{Is}(\mathrm{CUT}_n^{\square})$ has been described in Section 26.3.3.

Every triangle inequality defines a facet of the cut polytope. To see it, it suffices (in view of the results of the preceding section on permutation, switching and 0-lifting) to show that the inequality:

$$x_{12} - x_{13} - x_{23} \leq 0$$

defines a facet of CUT_3. This is indeed the case as the cut vectors $\delta(\{1\})$ and $\delta(\{2\})$ are two $(=\binom{3}{2}-1)$ linearly independent roots of this inequality. Hence, there are $3\binom{n}{3}$ triangle facets for CUT_n and $4\binom{n}{3}$ triangle facets for $\mathrm{CUT}_n^{\square}$.

The triangle inequalities are sufficient for describing the cut polyhedra for $n \leq 4$, i.e.,

$$\mathrm{CUT}_n = \mathrm{MET}_n \text{ and } \mathrm{CUT}_n^{\square} = \mathrm{MET}_n^{\square} \text{ for } n = 3, 4,$$

but $\mathrm{CUT}_n \subset \mathrm{MET}_n$, $\mathrm{CUT}_n^{\square} \subset \mathrm{MET}_n^{\square}$ for $n \geq 5$. To see it, observe that the vector $(\frac{2}{3}, \ldots, \frac{2}{3}) \in \mathbb{R}^{E_n}$ belongs to $\mathrm{MET}_n^{\square}$ but not to $\mathrm{CUT}_n^{\square}$ if $n \geq 5$. Alternatively, the pentagonal inequality (6.1.9) defines a (nontriangle) facet of the cut polytope on ≥ 5 elements. In some sense, K_5 is the unique "minimal obstruction" for the following property: The triangle inequalities form a linear description of the cut polytope. Indeed, the triangle inequalities (in fact, their projections) form the whole linear description of the cut polytope of a graph G if and only if G has no K_5-minor; see Theorem 27.3.6.

Every cut vector is a vertex of the semimetric polytope; moreover, the cut vectors are the only integral vectors of $\mathrm{MET}_n^{\square}$ (see Proposition 27.2.1). However, for $n \geq 5$, $\mathrm{MET}_n^{\square}$ has lots of additional vertices, all of them having some fractional component. Hence, a linear description of $\mathrm{CUT}_n^{\square}$ arises from that of $\mathrm{MET}_n^{\square}$ by adding constraints that cut off these fractional vertices. The vertices of $\mathrm{MET}_n^{\square}$ are studied, in particular, in Laurent [1996c] and Laurent and Poljak [1992].

On the other hand, the semimetric cone MET_n contains integral points that are not cut vectors. Indeed, given any partition $(S_1, \ldots, S_k)$ of V_n, the multicut vector $\delta(S_1, \ldots, S_k)$ is a $(0,1)$-valued member of MET_n. On the other hand, it is easy to check that the only integral points of MET_n are the multicut semimetrics. Note that

$$\delta(S_1, \ldots, S_k) = \frac{1}{2} \sum_{1 \leq h \leq k} \delta(S_h).$$

Therefore, the only integral points of MET_n that lie on an extreme ray of MET_n are the usual cut vectors $\delta(S)$ for $S \subseteq V_n$. The extreme rays of MET_n have been studied, in particular, in Avis [1977, 1980a, 1980b], Lomonosov [1978, 1985], Howe, Johnson and Lawrence [1986], Grishukhin [1992a]. The order of magnitude of the number of extreme rays of MET_n is known; Avis [1980b] gives a lower bound in $2^{n^2/2-O(n^{3/2})}$ and Graham, Yao and Yao [1980] prove the upper bound $2^{2.72n^2}$. One of the important motivations for the study of the semimetric cone comes from its role in the feasibility problem for multicommodity flows (Iri [1971]; see also Lomonosov [1985], Avis and Deza [1991]).

Although the triangle facets represent, in general, only a tiny fraction of all the facets of the cut polytope, they seem to play nevertheless an important role. This is indicated by their many geometric properties. We now discuss several properties and features of the triangle inequalities. Further geometric properties will be described in Chapter 31.

27.1 Triangle Inequalities for the Correlation Polytope

Let us recall how the triangle inequalities look like, when formulated in the context of the correlation polyhedra. Let us call these reformulated inequalities the *correlation triangle inequalities.* This information has already been given in Figure 5.2.6; we reproduce part of it here, for convenience. (We remind that the inequalities on the "cut side" are defined in the space $\mathbb{R}^{E_{n+1}}$ while the inequalities on the "correlation side" live in the space $\mathbb{R}^{V_n \cup E_n}$.)

"cut side"	"correlation side"
$d \in \mathbb{R}^{E_{n+1}}$ $\delta(S)$ (for $S \subseteq V_n$) CUT_{n+1} $\mathrm{CUT}^{\square}_{n+1}$	$p \in \mathbb{R}^{V_n \cup E_n}$ $\pi(S)$ COR_n $\mathrm{COR}^{\square}_n$
(Rooted) triangle inequalities: $d(i,j) - d(i,n+1) - d(j,n+1) \leq 0$ $d(i,n+1) - d(j,n+1) - d(i,j) \leq 0$ $d(j,n+1) - d(i,n+1) - d(i,j) \leq 0$ $d(i,n+1) + d(j,n+1) + d(i,j) \leq 2$ $(i,j \in V_n)$	(Rooted) correlation triangle inequalities: $0 \leq p_{ij}$ $p_{ij} \leq p_{ii}$ $p_{ij} \leq p_{jj}$ $p_{ii} + p_{jj} - p_{ij} \leq 1$
(Unrooted) triangle inequalities: $d(i,j) - d(i,k) - d(j,k) \leq 0$ $d(i,j) + d(i,k) + d(j,k) \leq 2$ $(i,j,k \in V_n)$	(Unrooted) correlation triangle inequalities: $-p_{kk} - p_{ij} + p_{ik} + p_{jk} \leq 0$ $p_{ii} + p_{jj} + p_{kk} - p_{ij} - p_{ik} - p_{jk} \leq 1$

Figure 27.1.1: Triangle inequalities for cut and correlation polyhedra

The rooted triangle inequalities are those that use the element $n+1$; they will be considered in detail in the next subsection.

The correlation triangle inequalities have a nice interpretation in terms of probabilities. Indeed, recall from Proposition 5.3.4 that every vector $p \in \mathrm{COR}^{\square}_n$ represents the joint correlations $\mu(A_i \cap A_j)$ $(1 \leq i \leq j \leq n)$ of n events $A_1, \ldots, A_n$ in some probability space. Then, the rooted correlation triangle inequalities simply express the following basic properties:

- The joint probability $\mu(A_i \cap A_j)$ of two events is nonnegative and less than or

equal to the probability of each of the two events, and
• the probability $\mu(A_i \cup A_j)$ $(= \mu(A_i) + \mu(A_j) - \mu(A_i \cap A_j))$ of their union is less than or equal to 1.
The unrooted inequalities can be expressed as follows:
• $-\mu(A_k) - \mu(A_i \cap A_j) + \mu(A_i \cap A_k) + \mu(A_j \cap A_k) = -\mu(A_k) - \mu(A_i \cap A_j) + \mu(A_k \cap (A_i \cup A_j)) + \mu(A_k \cap A_i \cap A_j)$, which is clearly ≤ 0, and
• $\mu(A_i) + \mu(A_j) + \mu(A_k) - \mu(A_i \cap A_j) - \mu(A_i \cap A_k) - \mu(A_j \cap A_k) = \mu(A_i \cup A_j \cup A_k) - \mu(A_i \cap A_j \cap A_k)$, which is clearly ≤ 1.
This shows again the validity of the correlation triangle inequalities for $\mathrm{COR}_n^\square$.

There is another natural way of generating the rooted correlation triangle inequalities. Remember that the polytope $\mathrm{COR}_n^\square$ can be expressed as

$$\mathrm{COR}_n^\square = \mathrm{Conv}\{(x_i x_j)_{1 \leq i \leq j \leq n} \mid x \in \{0,1\}^n\}.$$

Hence, valid inequalities for $\mathrm{COR}_n^\square$ can be generated in the following way. Suppose that

$$a^T x \geq \alpha \quad \text{and} \quad b^T x \geq \beta$$

hold for each $x \in \{0,1\}^n$. Then, the inequality

$$(a^T x - \alpha)(b^T x - \beta) \geq 0$$

also holds for each $x \in \{0,1\}^n$. If we develop the quantity $(a^T x - \alpha)(b^T x - \beta)$ and linearize, i.e., if we replace each $x_i x_j$ $(i \neq j)$ by the variable p_{ij} and each $x_i x_i$ by the variable p_{ii}, then we obtain an inequality which is clearly valid for $\mathrm{COR}_n^\square$.

We illustrate the method, starting from the inequalities:

$$x_i \geq 0, \;\; 1 - x_i \geq 0, \;\; x_j \geq 0, \;\; 1 - x_j \geq 0$$

which hold trivially for $x \in \{0,1\}^n$. By multiplying them pairwise, we obtain the inequalities:

$$\begin{array}{ll}
x_i x_j \geq 0, \text{ i.e.,} & p_{ij} \geq 0, \\
x_i(1 - x_j) \geq 0, \text{ i.e.,} & p_{ii} \geq p_{ij}, \\
x_j(1 - x_i) \geq 0, \text{ i.e.,} & p_{jj} \geq p_{ij}, \\
(1 - x_i)(1 - x_j) \geq 0, \text{ i.e.,} & p_{ii} + p_{jj} - p_{ij} \leq 1.
\end{array}$$

So, we have generated the rooted correlation triangle inequalities. This method is, in fact, described in Lovász and Schrijver [1991] and Balas, Ceria and Cornuejols [1993] as a way of generating a tighter relaxation from a given linear relaxation of a 01-polytope.

27.2 Rooted Triangle Inequalities

In this section, we consider the triangle inequalities for the polytope $\mathrm{CUT}_{n+1}^\square$; hence, they are defined on the $n+1$ elements of the set $V_{n+1} = \{1, \ldots, n, n+1\}$.

Then, a triangle inequality is called a *rooted triangle inequality* if it uses the element $n+1$, i.e., if it is of the form: $x_{ij} - x_{ik} - x_{jk} \leq 0$ or $x_{ij} + x_{ik} + x_{jk} \leq 2$, for some $i, j, k \in V_{n+1}$ such that $n+1 \in \{i, j, k\}$.

The rooted triangle inequalities define a polytope, called the *rooted semimetric polytope*, and denoted by $\mathrm{RMET}_{n+1}^{\square}$. Therefore,

$$\mathrm{CUT}_{n+1}^{\square} \subseteq \mathrm{MET}_{n+1}^{\square} \subseteq \mathrm{RMET}_{n+1}^{\square}.$$

Although the rooted polytope $\mathrm{RMET}_{n+1}^{\square}$ is a weaker relaxation of $\mathrm{CUT}_{n+1}^{\square}$ than $\mathrm{MET}_{n+1}^{\square}$, it contains already a lot of information. In particular, as stated in Proposition 27.2.1 below, it constitutes an integer programming formulation for the cut polytope.

In fact, the rooted semimetric polytope has some nice properties, that the usual semimetric polytope does not have. For instance, Padberg [1989] shows that every vertex of $\mathrm{RMET}_{n+1}^{\square}$ is half-integral, i.e., has components $0, 1, \frac{1}{2}$. In contrast, $\mathrm{MET}_{n+1}^{\square}$ has very complicated vertices, with arbitrarily large denominator. In fact, the unrooted triangle inequalities are already sufficient for cutting off the fractional vertices of $\mathrm{RMET}_{n+1}^{\square}$, as no vertex of $\mathrm{MET}_{n+1}^{\square}$ is half-integral.

We present below several properties of the rooted semimetric polytope. Some of them will be given for convenience in the context of correlations; that is, for $\xi(\mathrm{RMET}_{n+1}^{\square})$, the polytope defined by the rooted correlation triangle inequalities (see Figure 27.1.1).

27.2.1 An Integer Programming Formulation for Max-Cut

Proposition 27.2.1. *The only integral vectors of* $\mathrm{RMET}_n^{\square}$ *are the cut vectors* $\delta(S)$ *for* $S \subseteq V_n$. *Moreover, every cut vector is a vertex of* $\mathrm{RMET}_n^{\square}$.

Proof. Let $x \in \mathrm{RMET}_n^{\square} \cap \{0,1\}^{E_n}$. Set $I := \{i \in V_n \mid x_{i,n+1} = 0\}$ and $J := \{i \in V_n \mid x_{i,n+1} = 1\}$. As x satisfies the rooted triangle inequalities, we obtain that $x_{ij} = 0$ for all $i \neq j$ such that $i, j \in I$ or $i, j \in J$, and $x_{ij} = 1$ for all $i \in I$, $j \in J$. This shows that x is equal to the cut vector $\delta(J)$. We now show that every cut vector is a vertex of $\mathrm{RMET}_n^{\square}$. In view of the symmetry properties of $\mathrm{RMET}_n^{\square}$, it suffices to show that the origin is a vertex of $\mathrm{RMET}_n^{\square}$. This follows from the fact that the homogeneous rooted triangle inequalities (which all contain the origin) have full rank $\binom{n}{2}$. ∎

Hence, the system which consists of the rooted triangle inequalities together with the integrality constraint:

$$x_{ij} = 0, 1 \quad \text{for all } ij$$

forms an integer programming formulation for the max-cut problem. In other words, given a weight function $c \in \mathbb{R}^{E_{n+1}}$, the max-cut problem:

$$\max(c^T \delta(S) \mid S \subseteq V_{n+1})$$

can be reformulated as the problem:

$$\begin{array}{ll} \max & c^T x \\ & x \in \mathrm{RMET}_{n+1}^{\square} \\ & x \in \{0,1\}^{E_{n+1}}. \end{array}$$

Hence, if one wishes to solve the max-cut problem using linear programming techniques, one faces the problem of finding what are the additional constraints needed to be added to the above formulation in order to eliminate the integrality condition. This is the question of finding the facet defining inequalities for the cut polytope, which forms the main topic of Part V.

27.2.2 Volume of the Rooted Semimetric Polytope

Computing the exact volume of a polytope is, in general, a difficult problem. A nice property of the rooted triangle inequalities is that one can compute exactly the volume of the rooted semimetric polytope. This was done by Ko, Lee and Steingrimsson [1997] who showed, using the switching symmetries, how the problem can be reduced to that of computing the volume of an order polytope.

Theorem 27.2.2. *Set $d := \binom{n+1}{2}$. Then, the d-dimensional volume of the rooted semimetric polytope $\mathrm{RMET}_{n+1}^{\square}$ is equal to*

$$\mathrm{vol}\ \mathrm{RMET}_{n+1}^{\square} = \frac{n!}{(2n)!}2^n.$$

The d-dimensional volume of its image under the covariance mapping is

$$\mathrm{vol}\ \xi(\mathrm{RMET}_{n+1}^{\square}) = \frac{n!}{(2n)!}2^{2n-d}.$$

Proof. To simplify the notation, set $Q_n := \xi(\mathrm{RMET}_{n+1}^{\square})$. In a first step, let us compute the d-dimensional volume of Q_n. The polytope Q_n is defined by the rooted triangle inequalities, namely,

(i) $p_{ij} \geq 0$, $p_{ij} \leq p_{ii}$, $p_{ij} \leq p_{jj}$,

(ii) $p_{ii} + p_{jj} \leq 1 + p_{ij}$

for all $1 \leq i < j \leq n$. For $a \in \{0,1\}^n$, set

$$C_a := \{p \in Q_n \mid a_i \leq p_{ii} \leq a_i + \frac{1}{2}\ (i = 1, \ldots, n)\}.$$

Then, for distinct $a, b \in \{0,1\}^n$, the d-dimensional volume of $C_a \cap C_b$ is equal to 0 (as $C_a \cap C_b$ has dimension $< d$). Clearly, Q_n is the union of the polytopes C_a (for $a \in \{0,1\}^n$). Hence,

$$\mathrm{vol}\ Q_n = \sum_{a \in \{0,1\}^n} \mathrm{vol}\ C_a.$$

On the other hand, all C_a's have the same volume. Indeed, if A denotes the set of positions in which the coordinates of a and b differ, then C_b is the image of C_a under the mapping $\varrho_{\delta(A)}$ (the analogue of switching, defined in (26.3.10)). This implies that vol C_a = vol C_b, as the mapping $\varrho_{\delta(A)}$ is unimodular (i.e., its matrix has determinant ± 1). Therefore,

$$\text{vol } Q_n = 2^n \text{vol } C_0.$$

We show below that

$$\text{vol } C_0 = \frac{n!}{(2n)!} 2^{n-d}.$$

This implies immediately the value of the volume of Q_n. Finally, as $\text{MET}^{\square}_{n+1} = \xi^{-1}(Q_n)$ and as the determinant of the matrix of the linear mapping ξ^{-1} is equal to $2^{\binom{n}{2}}$ (in absolute value), we have

$$\text{vol MET}^{\square}_{n+1} = \frac{n!}{(2n)!} 2^n.$$

We now proceed to computing the volume of C_0. Let $C'_0 := 2C_0$ denote the polytope C_0 scaled by a factor 2. Then, as the inequality (ii) becomes redundant in the description of C_0, the polytope C'_0 is defined by the inequalities (i) and $0 \le p_{ii} \le 1$ (for all i). Let $(S_n, \prec)$ denote the partially ordered set on

$$S_n := \{p_{ij} \mid 1 \le i \le j \le n\}$$

with $p_{ij} \prec p_{ii}$ and $p_{ij} \prec p_{jj}$ as partial order. Then, C'_0 is the order polytope[1] of the poset $(S_n, \prec)$. Let $e(S_n, \prec)$ denote the number of linear extensions[2] of $(S_n, \prec)$. Then,

$$\text{vol } C'_0 = \frac{e(S_n, \prec)}{d!}.$$

Define an *ordered extension* to be a linear extension of $(S_n, \prec)$ in which $p_{11}, \ldots, p_{nn}$ occur in that order. Hence, an ordered extension is a permutation of the p_{ij}'s in which $p_{11}, \ldots, p_{nn}$ appear in that order and p_{ij} appears at the right of both p_{ii} and p_{jj}. Their number can be computed as follows. Suppose that $p_{i,k+1}$ $(i \le k)$, $\ldots, p_{in}$ $(i \le n-1)$ have already been positioned. We now try to place $p_{1k}, \ldots, p_{k-1,k}$. The first element p_{1k} should be placed at the right of p_{kk}. As there are already

$$f_k := (n-k+1) + k + (k+1) + \ldots + (n-1) = n-k+1 + \binom{n}{2} - \binom{k}{2}$$

elements of S_n placed at the right of p_{kk} (including p_{kk}), there are f_k possibilities for chosing the position of p_{1k}. Then, there are $f_k + 1$ possibilities for placing

[1]The *order polytope* P of a partially ordered set $(E, \prec)$ is the polytope in the space $\mathbb{R}^E$, which is defined by the inequalities: $0 \le x_e \le 1$ $(e \in E)$ and $x_e \le x_f$ whenever $e \prec f$ $(e, f \in E)$.

[2]A *linear extension* of a poset $(E, \prec)$ is any total order on E extending the partial order $\prec$. Stanley [1986] proves that vol $P = \frac{e(E,\prec)}{|E|!}$, if $e(E, \prec)$ is the number of linear extensions of $(E, \prec)$ and P is the order polytope of $(E, \prec)$.

p_{2k}, up to $f_k + k - 2$ possibilities for placing $p_{k-1,k}$. In total, the number of ordered extensions is

$$\prod_{k=2}^{n}\prod_{i=0}^{k-2}(f_k+i) = \prod_{k=2}^{n}\frac{(f_k+k-2)!}{(f_k-1)!} = \prod_{k=2}^{n}\frac{(\binom{n+1}{2}-\binom{k}{2}-1)!}{(\binom{n+1}{2}-\binom{k+1}{2})!}$$

which can be easily checked to be equal to $\frac{d!}{(2n)!}2^n$. Therefore, $e(S_n, \prec) = \frac{n!d!}{(2n)!}2^n$, as the number of linear extensions is equal to $n!$ times the number of ordered extensions. Hence, vol $C_0' = \frac{n!}{(2n)!}2^n$ and vol $C_0 = 2^{-d}$vol $C_0' = \frac{n!}{(2n)!}2^{n-d}$. ∎

27.2.3 Additional Notes

Chvátal Cuts of Rooted Triangle Inequalities. Boros, Crama and Hammer [1992] show that the Chvátal closure of the polytope $\xi(\mathrm{RMET}_{n+1}^{\square})$ is precisely the polytope $\xi(\mathrm{MET}_{n+1}^{\square})$; we mention their result (without proof) in Theorem 27.2.3 below. We recall the definition for the notion of Chvátal closure.

Let P be a polytope in $\mathbb{R}^k$ and let P_I denote the convex hull of the integral points of P; so, $P_I \subseteq P$. Given $a \in \mathbb{Z}^k$ and $\alpha \in \mathbb{Z}$, if the inequality

$$a^T x < \alpha + 1$$

is valid for P, then the inequality

$$a^T x \leq \alpha$$

is valid for P_I. This second inequality is called a *Chvátal cut* of P. Then, P' denotes the polytope which is defined by all the possible Chvátal cuts; it is called the *Chvátal closure* of P. Setting $P^{(0)} := P$ and $P^{(k+1)} = (P^{(k)})'$ for $k \geq 0$, we obtain a decreasing sequence of polytopes:

$$P^{(0)} \supseteq P^{(1)} \supseteq \ldots \supseteq P_I.$$

Chvátal [1973] (see, e.g., Schrijver [1986]) showed that there exists a finite index k such that $P^{(k)} = P_I$. The smallest such k is called the *Chvátal rank* of P.

Theorem 27.2.3. *The Chvátal closure of the rooted correlation semimetric polytope $\xi(\mathrm{RMET}_{n+1}^{\square})$ is the correlation semimetric polytope $\xi(\mathrm{MET}_{n+1}^{\square})$; that is, $(\xi(\mathrm{RMET}_{n+1}^{\square}))' = \xi(\mathrm{MET}_{n+1}^{\square})$.* ∎

As an example, we indicate how to obtain the unrooted correlation triangle inequality:

$$p_{11} + p_{22} + p_{33} - p_{12} - p_{13} - p_{23} \leq 1$$

as a Chvátal cut. Consider the following rooted correlation triangle inequalities:

$$\begin{aligned} p_{11} + p_{22} - p_{12} &\leq 1, \\ p_{11} + p_{33} - p_{13} &\leq 1, \\ p_{22} + p_{33} - p_{23} &\leq 1, \\ -p_{12} &\leq 0, \\ -p_{13} &\leq 0, \\ -p_{23} &\leq 0. \end{aligned}$$

Summing them up and dividing by 2, we obtain:

$$p_{11} + p_{22} + p_{33} - p_{12} - p_{13} - p_{23} \leq \frac{3}{2} < 2$$

which yields the Chvátal cut:

$$p_{11} + p_{22} + p_{33} - p_{12} - p_{13} - p_{23} \leq 1.$$

Similarly, the inequality: $-p_{11} + p_{12} + p_{13} - p_{23} \leq 0$ arises as Chvátal cut from the following inequalities:

$$\begin{aligned} -p_{11} + p_{12} &\leq 0, \\ -p_{22} + p_{12} &\leq 0, \\ -p_{11} + p_{13} &\leq 0, \\ -p_{33} + p_{13} &\leq 0, \\ p_{22} + p_{33} - p_{23} &\leq 1, \\ -p_{23} &\leq 0. \end{aligned}$$

Observe that Theorem 27.2.3 does not hold on the "cut side", i.e., the Chvátal closure of the rooted semimetric polytope $\mathrm{RMET}_{n+1}^{\square}$ is not equal to the semimetric polytope $\mathrm{MET}_{n+1}^{\square}$. For instance, the inequality

$$x_{12} + x_{13} + x_{23} \leq 2$$

is not a Chvátal cut of $\mathrm{RMET}_{n+1}^{\square}$. Indeed, the inequality

$$x_{12} + x_{13} + x_{23} < 3$$

is not valid for $\mathrm{RMET}_{n+1}^{\square}$ as it is violated by the point $x \in \mathrm{RMET}_{n+1}^{\square}$ defined by $x_{i,n+1} = \frac{1}{2}$ $(1 \leq i \leq n)$ and $x_{ij} = 1$ $(1 \leq i < j \leq n)$.

The following lower bound on the Chvátal rank of the semimetric polytope was given by Chvátal, Cook and Hartman [1989].

Theorem 27.2.4. *The Chvátal rank of the semimetric polytope* $\mathrm{MET}_n^{\square}$ *is greater than or equal to* $\frac{1}{4}(n-4)$. ∎

The Roof Duality Bound. Given weights $c \in \mathbb{R}^{V_n \cup E_n}$, consider the unconstrained quadratic 0-1 programming problem:

$$(27.2.5)\qquad C_n := \begin{array}{ll} \max & \sum_{1\le i\le j\le n} c_{ij}x_ix_j \\ \text{s.t.} & x\in\{0,1\}^n \end{array} = \begin{array}{ll} \max & c^Tp \\ \text{s.t.} & p\in \text{COR}_n^{\square}. \end{array}$$

As $\text{COR}_n^{\square} \subseteq \xi(\text{RMET}_{n+1}^{\square})$, the program

$$C_2 := \max(c^Tp \mid p \in \xi(\text{RMET}_{n+1}^{\square}))$$

gives an upper bound for the optimum value C_n of (27.2.5). The bound C_2 is known as the roof duality bound. Several equivalent formulations of C_2 are given in Hammer, Hansen and Simeone [1984]. In fact, a sequence of bounds C_k $(k = 2, \ldots, n-1)$ has been formulated, verifying:

$$C_n \le C_{n-1} \le \ldots \le C_3 \le C_2$$

and having also several equivalent formulations; see Boros, Crama and Hammer [1990], also Adams and Dearing [1994]. In particular, C_3 is the optimum value obtained when optimizing over the polytope $\xi(\text{MET}_{n+1}^{\square})$. Hence, C_2 and C_3 can be computed in time polynomial in n. More generally, C_k can be computed by solving a linear programming problem whose size is polynomial in n but exponential in k. For more details we refer, e.g., to Boros and Hammer [1991], Boros, Crama and Hammer [1992] and references therein.

27.3 Projecting the Triangle Inequalities

Let $G = (V_n, E)$ be a graph on n nodes. Let MET(G) denote the projection of the semimetric cone MET_n on the subspace $\mathbb{R}^E$ indexed by the edge set of G; MET(G) is called the *semimetric cone* of G. Similarly, let $\text{MET}^{\square}(G)$ denote the projection of $\text{MET}_n^{\square}$ on $\mathbb{R}^E$; it is called the *semimetric polytope* of G. In the same way, CUT(G) (resp. $\text{CUT}^{\square}(G)$) denotes the projection of CUT_n (resp. $\text{CUT}_n^{\square}$) on $\mathbb{R}^E$. By the definitions,

$$(27.3.1)\qquad \text{CUT}(G) \subseteq \text{MET}(G) \text{ and } \text{CUT}^{\square}(G) \subseteq \text{MET}^{\square}(G).$$

We recall that, for $S \subseteq V_n$, $\delta_G(S)$ denotes the cut in G which is the subset of E consisting of the edges $e \in E$ having exactly one endnode in S. Hence, the *cut cone* CUT(G) of G coincides with the cone in $\mathbb{R}^E$ generated by the vectors $\chi^{\delta_G(S)}$ (for $S \subseteq V_n$) and the *cut polytope* $\text{CUT}^{\square}(G)$ coincides with the convex hull of the vectors $\chi^{\delta_G(S)}$ (for $S \subseteq V_n$).

As the collection of cuts in G is closed under the symmetric difference then, by the results of Section 26.3, the switching operation applies to the cut polytope $\text{CUT}^{\square}(G)$ of an arbitrary graph G; it also applies to the semimetric polytope $\text{MET}^{\square}(G)$. Namely, for any $S \subseteq V_n$,

$$r_{\delta_G(S)}(\text{CUT}^{\square}(G)) = \text{CUT}^{\square}(G) \text{ and } r_{\delta_G(S)}(\text{MET}^{\square}(G)) = \text{MET}^{\square}(G).$$

Note that $\text{MET}^{\square}(G)$ contains no other integral vectors besides the incidence vectors of the cuts $\delta_G(S)$ $(S \subseteq V_n)$ (this follows easily from Proposition 27.2.1).

The graphs G for which $\mathrm{MET}^{\square}(G)$ has only integral vertices, i.e., for which equality holds in (27.3.1), will be characterized in Theorem 27.3.6. In general, $\mathrm{MET}^{\square}(G)$ has lots of nonintegral vertices. It is easy to see that no vertex of $\mathrm{MET}^{\square}(G)$ can have denominator 2. Hence, denominator 3 is the next case after integrality. Laurent and Poljak [1995a] study[3] the graphs G for which all the vertices of $\mathrm{MET}^{\square}(G)$ have denominator ≤ 3; such graphs are completely characterized up to 7 nodes.

27.3.1 The Semimetric Polytope of a Graph

We present here a linear description for the semimetric cone $\mathrm{MET}(G)$ and polytope $\mathrm{MET}^{\square}(G)$. As we know a linear description of the polytope $\mathrm{MET}_n^{\square}$, a linear description of $\mathrm{MET}^{\square}(G)$ can be deduced from that of $\mathrm{MET}_n^{\square}$ by applying, e.g., the Fourier-Motzkin elimination method. This method consists of combining the linear inequalities defining $\mathrm{MET}_n^{\square}$ so as to eliminate the variables x_e ($e \in E_n \setminus E$) that do not occur in $\mathrm{CUT}^{\square}(G)$; see, e.g., Schrijver [1986], Ziegler [1995].

Computing explicitly the projection of a polyhedron is, in general, a difficult task as the Fourier-Motzkin elimination method becomes very often intractable in practice. For instance, the correlation polytope[4] $\mathrm{COR}_n^{\square}$ can be obtained as the projection of a simplex lying in the space of dimension 2^n (namely, of the simplex $\mathrm{COR}_n^{\square}(2^{V_n})$; recall Section 5.4). Even though finding the facial structure of a simplex is trivial, finding all the facets of $\mathrm{COR}_n^{\square}$ is a hard task ! However, in the case of the semimetric polyhedra $\mathrm{MET}(G)$ and $\mathrm{MET}^{\square}(G)$, explicit descriptions can be found fairly easily, as the results below indicate.

We recall that a cycle is any graph which can be decomposed as the edge disjoint union of circuits. Let C be a circuit and let e be an edge that does not belong to C; then, e is said to be a *chord* of C if it joins two nodes of C. The circuit C is said to be *chordless* if it has no chord.

Let C be a cycle in G and let $F \subseteq C$ be a subset of C such that $|F|$ is odd. The inequality:

$$\sum_{e \in F} x_e - \sum_{e \in C \setminus F} x_e \leq |F| - 1 \tag{27.3.2}$$

is called a *cycle inequality*. Note that the triangle inequalities are special cases of cycle inequalities (obtained for $|C| = 3$). Note also that the class of cycle inequalities is closed under the operation of switching by a cut. (This follows from the fact that a cut and a cycle intersect in an even number of edges.) Theorem 27.3.3 below shows that the cycle inequalities form a linear description of the semimetric polyhedra; (i) is proved in Barahona [1993] and (ii), (iii) in Barahona and Mahjoub [1986].

---

[3] An extension in the context of binary matroids is considered in Gerards and Laurent [1995].

[4] Therefore, up to a linear transformation, the cut polytope $\mathrm{CUT}_n^{\square}$ can also be obtained as the projection of a simplex. We will see in Example 27.4.4 another construction permitting to realize any cut polytope $\mathrm{CUT}^{\square}(G)$ as projection of a simplex.

Theorem 27.3.3. *Let $G = (V_n, E)$ be a graph.*

(i) $\mathrm{MET}(G) = \{x \in \mathbb{R}_+^E \mid x_e - x(C \setminus \{e\}) \leq 0$ *for* C *cycle of* $G, e \in C\}$,

$$\mathrm{MET}^{\square}(G) = \{x \in \mathbb{R}_+^E \mid \; x_e \leq 1 \text{ for } e \in E,\ x(F) - x(C \setminus F) \leq |F| - 1 \text{ for } C \text{ cycle of } G, F \subseteq C, |F| \text{ odd}\}.$$

(ii) *Let C be a cycle in G, $e \in C$, and $F \subseteq C$ with $|F|$ odd. The inequality $x_e - x(C \setminus \{e\}) \leq 0$ (resp. $x(F) - x(C \setminus F) \leq |F| - 1$) defines a facet of* MET(G) *(resp.* MET$^{\square}(G)$*) if and only if C is a chordless circuit.*

(iii) *Let $e \in E$. The inequality $x_e \geq 0$ (resp. $x_e \leq 1$) defines a facet of* MET(G) *(resp.* MET$^{\square}(G)$*) if and only if e does not belong to any triangle of G.*

Proof. A first observation is that, due to switching, it is sufficient to prove the statements relevant to the cone MET(G). In a first step, we show that, if C is a cycle in G which is not a chordless circuit, then the inequality: $x_e - x(C \setminus \{e\}) \leq 0$ (where $e \in C$) follows from other cycle inequalities and the nonnegativity condition $x \geq 0$. Indeed, suppose that $C = C_1 \cup \ldots \cup C_p$, where the C_i's are edge disjoint circuits. We can suppose that $e \in C_1$. Then, the inequality: $x_e - x(C \setminus \{e\}) \leq 0$ follows by summing up the inequalities: $x_e - x(C_1 \setminus \{e\}) \leq 0$ and $-x(C_i) \leq 0$ ($i = 2, \ldots, p$). Suppose now that C is a circuit having a chord f. Then, this chord determines a partition of C into two paths P_1, P_2 such that $C_i := P_i \cup \{f\}$ is a circuit for $i = 1, 2$. Then, assuming that $e \in P_1$, the inequality: $x_e - x(C \setminus \{e\}) \leq 0$ follows by summing up the inequalities: $x_e - x(C_1 \setminus \{e\}) \leq 0$ and $x_f - x(C_2 \setminus \{f\}) \leq 0$.

Proof of (i). Set $Q(G) := \{x \in \mathbb{R}_+^E \mid x_e - x(C \setminus \{e\}) \leq 0$ for C cycle of $G, e \in C\}$. We first check that MET$(G) \subseteq Q(G)$. Let $x \in$ MET(G), let C be a cycle of G and let $e \in C$. We show that $x_e - x(C \setminus \{e\}) \leq 0$. By the above observation, we can suppose that C is a circuit, say, $C = (1, 2, \ldots, p)$ and that $e := 12$. By the definition of MET(G), there exists $y \in$ MET$_n$ whose projection on $\mathbb{R}^E$ is x. Then,

$$x_e - x(C \setminus \{e\}) = x_{12} - x_{23} - \ldots - x_{p-1,p} = \sum_{2 \leq i \leq p-1} y_{1i} - y_{1,i+1} - y_{i,i+1} \leq 0.$$

We show the converse inclusion: $Q(G) \subseteq$ MET(G) by induction on $|E|$. This inclusion holds trivially if $G = K_n$. Suppose now that $G \neq K_n$. Let $e := uv$ be an edge of K_n that does not belong to E and let $x \in Q(G)$. Then, x can be extended to a vector $y \in \mathbb{R}^{E \cup \{e\}}$ which belongs to $Q(G + e)$ ($G + e$ denoting the graph obtained by adding the edge e to G). For this, it suffices to take $y_e := \alpha$, where

$$\max_{P \mid f \in P} (x_f - x(P \setminus \{f\})) \leq \alpha \leq \min_Q x(Q),$$

where P, Q run over all paths joining u and v in G. (Such an α exists as $x \in Q(G)$.) By the induction assumption, $Q(G + e) =$ MET$(G + e)$. Therefore, y is the projection on $\mathbb{R}^{E \cup \{e\}}$ of some $z \in$ MET$_n$. Hence, x is the projection of z on R^E, which shows that $x \in$ MET(G).

The proofs of (ii),(iii) can be found in Barahona and Mahjoub [1986]; in fact, it is shown there that, under the same assumptions, the inequality in question defines a facet of the cut polytope $\mathrm{CUT}^{\square}(G)$. ∎

The list of inequalities that define the semimetric polytope $\mathrm{MET}^{\square}(G)$ can be exponentially long (in terms of n). Nevertheless, Barahona and Mahjoub [1986] showed that the separation problem for the system:

$$(27.3.4) \qquad \begin{cases} 0 \le x_e \le 1 & \text{for } e \in E, \\ x(F) - x(C \setminus F) \le |F| - 1 & \text{for } C \text{ cycle of } G, F \subseteq C, |F| \text{ odd} \end{cases}$$

can be solved in polynomial time (in terms of n and the size of x). This problem can be formulated as follows:

> *Given a vector $x \in \mathbb{Q}^E$, decide whether x satisfies all the inequalities from the system* (27.3.4). *If not, find an inequality from* (27.3.4) *that is violated by x.*

This problem can be solved in the following way. First, check whether $0 \le x_e \le 1$ holds for all edges $e \in E$. If not, then we have found a violated inequality. Otherwise, we can suppose that $0 \le x_e \le 1$ for all $e \in E$. Note that the cycle inequality (27.3.2) can be rewritten as

$$\sum_{e \in C \setminus F} x_e + \sum_{e \in F} (1 - x_e) \ge 1.$$

We form a new graph G' with two nodes i' and i'' for each node i of G. For each edge $ij \in E$, we introduce in G' the edges $i'j'$, $i''j''$ with weight x_{ij}, and the edges $i'j''$, $i''j'$ with weight $1 - x_{ij}$. So, we have defined a weight function on the edges of G'. Now, for each node i of G, we compute a shortest (with respect to this weight function) path in G' from i' to i''. Then, the minimum over all nodes i of G of these shortest paths gives the minimum value of $\sum_{e \in C \setminus F} x_e + \sum_{e \in F}(1 - x_e)$ over all cycles C, $F \subseteq C$, $|F|$ odd. If this minimum is less than one, then we have found a violated inequality; else, x satisfies all the cycle inequalities. As the computation of a shortest path can be done in $O(n^2)$, the separation problem for the system (27.3.4) can be solved in $O(n^3)$.

As an application (using the ellipsoid method, as exposed in Grötschel, Lovász and Schrijver [1988]), we deduce that one can optimize a linear objective function over the system (27.3.4) in polynomial time.

Proposition 27.3.5. *Given $c \in \mathbb{Q}^E$, the optimization problem:*

$$\max(c^T x \mid x \text{ satisfies the system } (27.3.4))$$

can be solved in polynomial time (polynomial in n and in the size of c). ∎

In fact, there is a much simpler argument for proving Proposition 27.3.5, based on Theorem 27.3.3. Indeed, as $\mathrm{MET}^{\square}(G)$ is the projection on $\mathbb{R}^E$ of $\mathrm{MET}^{\square}_n$, the two problems:

$$\max(c^T x \mid x \in \mathrm{MET}^{\square}(G)) \quad \text{and} \quad \max(c^T x \mid x \in \mathrm{MET}^{\square}_n),$$

where we extend c to $\mathbb{R}^{E_n}$ by setting $c_e := 0$ if $e \in E_n \setminus E$, have the same optimum value. Now, the latter problem is a linear programming problem with $\binom{n}{2}$ variables and $4\binom{n}{3}$ constraints. Hence, it can clearly be solved in polynomial time.

27.3.2 The Cut Polytope for Graphs with no K_5-Minor

We now return to the question of characterizing the graphs G for which equality holds in (27.3.1). Due to switching,

$$\mathrm{CUT}(G) = \mathrm{MET}(G) \Longleftrightarrow \mathrm{CUT}^{\square}(G) = \mathrm{MET}^{\square}(G).$$

As was already mentioned earlier, equality does not hold for the graph $G = K_5$. The next result shows that K_5 is the unique minimal (in the sense of graph minors) exception.

Theorem 27.3.6. $\mathrm{CUT}(G) = \mathrm{MET}(G)$ *or, equivalently,* $\mathrm{CUT}^{\square}(G) = \mathrm{MET}^{\square}(G)$ *for a graph G if and only if G does not have any K_5-minor.* ∎

Seymour [1981] proved the result concerning the cones and Barahona and Mahjoub [1986] derived the result for the polytopes using switching. Another proof for Theorem 27.3.6 is given by Barahona [1983]; it is based on a decomposition result due to Wagner [1937] for the graphs with no K_5-minor, together with a result showing how to derive a linear description of the cut polytope of a graph which is a clique k-sum of two smaller graphs ($k \leq 3$). As an application of Proposition 27.3.5 and Theorem 27.3.6, we obtain:

Theorem 27.3.7. *The max-cut problem:*

$$\max(c^T x \mid x \in \mathrm{CUT}^{\square}(G))$$

(where $c \in \mathbb{Q}^E$) can be solved in polynomial time for the class of graphs with no K_5-minor. ∎

In particular, the max-cut problem can be solved in polynomial time for the class of planar graphs.

As correlation polyhedra and cut polyhedra are in one-to-one linear correspondence, the above results have immediate counterparts for the correlation polyhedra of arbitrary graphs.

Given a graph $G = (V_n, E)$, its *correlation cone* $\mathrm{COR}(G)$ and its *correlation polytope* $\mathrm{COR}^{\square}(G)$ are defined in the following way: $\mathrm{COR}(G)$ (resp. $\mathrm{COR}^{\square}(G)$)

is the projection of the correlation cone COR_n (resp. of the correlation polytope $\mathrm{COR}_n^\square$) on the subspace $\mathbb{R}^{E \cup V_n}$, where V_n is identified with the set of diagonal pairs ii for $i \in V_n$.

Let ∇G denote the suspension graph of G obtained by adding a new node, say $n+1$, to G and making it adjacent to all nodes in V_n. Hence, the edge set $E(\nabla G)$ of ∇G is $E \cup \{(i, n+1) \mid i \in V_n\}$. We can define a one-to-one mapping ξ between the space indexed by $E(\nabla G)$ and the space indexed by $E \cup V_n$ in the following manner: For $x \in \mathbb{R}^{E(\nabla G)}$, $y \in \mathbb{R}^{E \cup V_n}$, $y = \xi(x)$ if

$$\begin{array}{lll} (27.3.8) \qquad & y_{ii} = x_{i,n+1} & \text{for } i \in V_n, \\ & y_{ij} = \frac{1}{2}(x_{i,n+1} + x_{j,n+1} - x_{ij}) & \text{for } ij \in E. \end{array}$$

Hence, when G is the complete graph K_n, then ξ is the usual covariance mapping (pointed at position $n+1$), as defined in Section 26.1. Clearly, the cut polyhedra for ∇G and the correlation polyhedra for G are in one-to-one correspondence. Namely,

$$\mathrm{COR}(G) = \xi(\mathrm{CUT}(\nabla G)) \quad \text{and} \quad \mathrm{COR}^\square(G) = \xi(\mathrm{CUT}^\square(\nabla G)).$$

In particular, Theorem 27.3.6 implies the following result for the correlation polyhedra, established in Padberg [1989].

Theorem 27.3.9. *For a graph G, $\mathrm{COR}(G) = \xi(\mathrm{MET}(\nabla G))$ or, equivalently, $\mathrm{COR}^\square(G) = \xi(\mathrm{MET}^\square(\nabla G))$ if and only if G has no K_4-minor.* ∎

The inequalities defining the polytope $\xi(\mathrm{MET}^\square(\nabla G))$ arise as projections of the correlation triangle inequalities. A linear system defining the polytope $\xi(\mathrm{MET}^\square(\nabla G))$ can be easily deduced from the linear description of $\mathrm{MET}^\square(\nabla G)$ presented in Theorem 27.3.3 (i) by applying the transformation ξ (an explicit description can be found, e.g., in Padberg [1989]).

As a direct application of Theorem 27.3.7, the unconstrained quadratic 0-1 programming problem (5.1.4) can be solved in polynomial time when the graph supporting the linear objective function has no K_4-minor.

27.4 An Excursion to Cycle Polytopes of Binary Matroids

As we remarked earlier, some properties of the cut polyhedra are valid for more general set families than cuts. Indeed, we saw in Section 26.3.1 that the switching operation applies to general set families under the only assumption that they are closed under taking symmetric differences. Such set families are known in the literature as cycle spaces of binary matroids. Cut polyhedra are, thus, special instances of cycle polyhedra of binary matroids. Cycles in graphs yield other interesting instances of binary matroids. Therefore, binary matroids constitute a unified framework for a variety of combinatorial objects. We cannot go here too much in detail into matroid theory as a detailed treatment falls out of the scope of the present book. We will therefore restrict ourselves to presenting without proof some of the main known results relevant to cycle polyhedra.

We recall in the first subsection some necessary definitions[5] about binary matroids. Section 27.4.2 reviews results about the cycle cone and polytope of a binary matroid. We then group in Section 27.4.3 several additional questions and results related, in particular, to the lattice and the integer cone generated by cycles of binary matroids.

27.4.1 Preliminaries on Binary Matroids

Cycles, Cocycles and Representation Matrix. A *binary matroid* $\mathcal{M}$ consists of a pair $(E, \mathcal{C})$, where E is a finite set (the *groundset* of $\mathcal{M}$) and $\mathcal{C}$ is a collection of subsets of E that is closed under taking symmetric differences, i.e., such that

$$C \triangle C' \in \mathcal{C} \text{ for all } C, C' \in \mathcal{C}.$$

The members of $\mathcal{C}$ are called the *cycles* of $\mathcal{M}$ and $\mathcal{C}$ is the *cycle space* of $\mathcal{M}$. Note that $\emptyset$ is always a cycle.

Two examples of binary matroids can be constructed from graphs. Let $G = (V, E)$ be a graph. As the symmetric difference of two Eulerian subgraphs (cycles) of G remains a Eulerian subgraph of G, we have a first matroid on E, denoted as $\mathcal{M}(G)$ and called the *graphic matroid* of G, whose cycle space is the set of Eulerian subgraphs of G. The symmetric difference of two cuts in G is again a cut; therefore, we have a second binary matroid on E, denoted as $\mathcal{M}^*(G)$ and called the *cographic matroid* of G, with cycle space the set of cuts of G.

Let $\mathcal{M} = (E, \mathcal{C})$ be a binary matroid. Set

$$\mathcal{C}^* := \{D \subseteq E : |C \cap D| \text{ is even for all } C \in \mathcal{C}\}.$$

Then, $\mathcal{C}^*$ is obviously closed under the symmetric difference. Hence, $\mathcal{M}^* := (E, \mathcal{C}^*)$ is again a binary matroid, called the *dual matroid* of $\mathcal{M}$. The members of $\mathcal{C}^*$ (the cycles of $\mathcal{M}^*$) are also called the *cocycles* of $\mathcal{M}$. One can check that the dual of $\mathcal{M}^*$ coincides with $\mathcal{M}$, i.e., $(\mathcal{M}^*)^* = \mathcal{M}$. As an example, the dual of the graphic matroid $\mathcal{M}(G)$ of a graph G is its cographic matroid $\mathcal{M}^*(G)$ (since a cut and a cycle have an even intersection).

The minimal nonempty cycles of $\mathcal{M}$ are called the *circuits* of $\mathcal{M}$ and the minimal nonempty cocycles are called its *cocircuits*. Every nonempty cycle can be decomposed as a disjoint union of circuits. The matroid $\mathcal{M}$ is said to be *cosimple* if no cocircuit has cardinality 1 or 2.

Binary matroids can alternatively be viewed as linear spaces over the field with two elements $GF(2) := \{0, 1\}$. Indeed, for any subsets $C, C' \subseteq E$, $\chi^{C \triangle C'} = \chi^C + \chi^{C'}$ (modulo 2). Hence, the set family $\mathcal{C}$ is closed under the symmetric difference if and only if the set $\{\chi^C \mid C \in \mathcal{C}\}$ is a linear subspace of the binary

[5]The reader may consult Welsh [1976] or Oxley [1992] for general information about matroids. We do not consider here matroids in full generality but only binary matroids.

space $GF(2)^E$. Hence, identifying a set and its incidence vector, the cycle spaces of binary matroids are nothing but the linear subspaces over $GF(2)$. In this terminology, the cocycle space $\mathcal{C}^*$ of $\mathcal{M}$ is the orthogonal complement of the cycle space $\mathcal{C}$ in $GF(2)^E$.

Hence, the cycle space of a binary matroid $\mathcal{M}$ on E can be realized as the set of solutions $x \in \{0,1\}^E$ of a linear equation:

$$Mx = 0 \text{ (modulo 2)}$$

where M is a zero-one matrix whose columns are indexed by E. Such matrix M is called a *representation matrix* of $\mathcal{M}$. The maximum number of columns of M that are linearly independent over $GF(2)$ is called the *rank* of $\mathcal{M}$. If $\mathcal{M}$ has rank r, then a representation matrix can be found for $\mathcal{M}$ having the form $(I_r \mid A)$, where I_r is the $r \times r$ identity matrix. Moreover, the matrix $(A^T \mid I_{|E|-r})$ is then a representation matrix for the dual $\mathcal{M}^*$ of $\mathcal{M}$.

Minors. Let $\mathcal{M} = (E, \mathcal{C})$ be a binary matroid and let $e \in E$. Set

$$\mathcal{C}\backslash e := \{C \subseteq E \setminus \{e\} \mid C \in \mathcal{C}\}, \quad \mathcal{C}/e := \{C \setminus \{e\} \mid C \in \mathcal{C}\}.$$

In the language of binary spaces, $\mathcal{C}\backslash e$ arises from $\mathcal{C}$ by taking its intersection with the hyperplane $x_e = 0$, while $\mathcal{C}/e$ arises from $\mathcal{C}$ by taking its projection on $\mathbb{R}^{E\setminus\{e\}}$. Both $\mathcal{C}\backslash e$ and $\mathcal{C}/e$ are again binary spaces. Hence, $\mathcal{M}\backslash e := (E\setminus\{e\}, \mathcal{C}\backslash e)$ and $\mathcal{M}/e := (E \setminus \{e\}, \mathcal{C}/e)$ are both binary matroids. One says that $\mathcal{M}\backslash e$ is obtained from $\mathcal{M}$ by *deleting* the element e and that $\mathcal{M}/e$ is obtained from $\mathcal{M}$ by *contracting* the element e. The deletion and contraction operations commute with respect to taking duals, namely,

$$(\mathcal{M}\backslash e)^* = \mathcal{M}^*/e, \quad (\mathcal{M}/e)^* = \mathcal{M}^*\backslash e.$$

If $\mathcal{N}$ is a binary matroid that can be obtained from $\mathcal{M}$ by a series of deletions and contractions, one says that $\mathcal{N}$ is a *minor* of $\mathcal{M}$. Every minor of $\mathcal{M}$ is of the form: $\mathcal{M}\backslash X/Y$, where X, Y are two disjoint subsets of E (as the deletion and contraction operations commute, i.e., $\mathcal{M}\backslash e/f = \mathcal{M}/f\backslash e$). Observe that the deletion and contraction operations, when applied to the graphic matroid $\mathcal{M}(G)$ of a graph G, correspond to the usual operations of deleting and contracting an edge in G. Hence, the class of graphic matroids is closed under taking minors; the same holds for the class of cographic matroids.

We now present several concrete examples of binary matroids that we will need.

The Fano Matroid. The *Fano matroid* F_7 is the binary matroid on the set $E := \{1,2,3,4,5,6,7\}$ whose cycles are $\emptyset$, E, and the sets

$$124, 135, 167, 236, 257, 347, 456$$

together with their complements. (Here we denote a set $\{1,2,4\}$ by the string 124.) Note that the cycles of size 3 of F_7 can be viewed as the lines of the

Fano plane, shown in Figure 27.4.1. A representation matrix for F_7 is also shown there. The *dual Fano matroid* F_7^* is the dual of the Fano matroid F_7; its nonempty cycles are the complements of the lines in the Fano plane.

$$\begin{matrix} 1 & 2 & 3 & 4 & 5 & 6 & 7 \end{matrix}$$
$$\begin{pmatrix} 1 & 0 & 0 & 1 & 1 & 0 & 1 \\ 0 & 1 & 0 & 1 & 0 & 1 & 1 \\ 0 & 0 & 1 & 0 & 1 & 1 & 1 \end{pmatrix}$$

Representation matrix for F_7

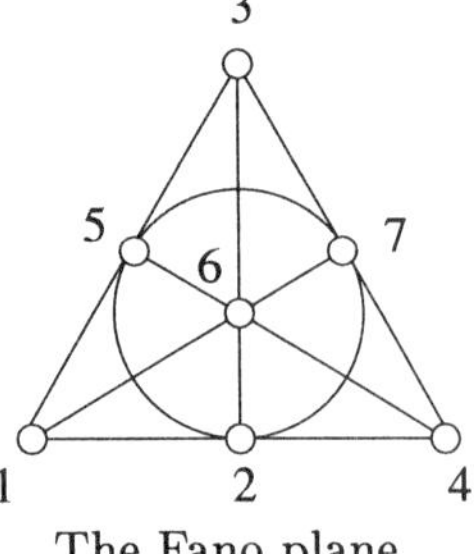

The Fano plane

Figure 27.4.1

The Matroid R_{10}. Let $E := \{e_{ij} \mid 1 \leq i < j \leq 5\}$ denote the edge set of the complete graph K_5. Then, R_{10} denotes the binary matroid on E whose cycles are the sets $C \subseteq E$ forming a cycle of even length in K_5. The cocycles are the cuts of K_5 together with their complements. Note that R_{10} is self-dual; that is, R_{10} is isomorphic to its dual R_{10}^*.

The Binary Projective Space $\mathcal{P}_r$. Let $\mathcal{P}_r$ denote the binary matroid which is represented by the $r \times (2^r - 1)$ matrix whose columns are all possible nonzero binary vectors of length r. Hence, $\mathcal{P}_3$ coincides with the Fano matroid F_7. One can verify that $\mathcal{P}_r$ has $2^r - 1$ nonempty cocycles, each having size 2^{r-1}, and that their incidence vectors are linearly independent. The dual matroid $\mathcal{P}_r^*$ is represented, for instance, by the matrix $(I_{2^r-r-1} \mid A_r)$, where I_{2^r-r-1} is the identity matrix of order $2^r - r - 1$ and A_r is the matrix whose rows are all binary vectors of length r having at least two nonzero components.

27.4.2 The Cycle Cone and the Cycle Polytope

Let $\mathcal{M} = (E, \mathcal{C})$ be a binary matroid. The *cycle polytope* of $\mathcal{M}$ is the polytope $\text{CYC}^\square(\mathcal{M})$, which is defined as the convex hull of the incidence vectors of the cycles of $\mathcal{M}$, i.e.,

$$\text{CYC}^\square(\mathcal{M}) := \text{Conv}(\{\chi^C \mid C \in \mathcal{C}\}).$$

The *cycle cone* of $\mathcal{M}$ is the cone $\text{CYC}(\mathcal{M})$, which is defined as the conic hull of the incidence vectors of the cycles of $\mathcal{M}$, i.e.,

$$\text{CYC}(\mathcal{M}) := \{\sum_{C \in \mathcal{C}} \lambda_C \chi^C \mid \lambda_C \geq 0 \text{ for all } C \in \mathcal{C}\}.$$

Therefore, if $\mathcal{M} = \mathcal{M}^*(G)$ is the cographic matroid of a graph G, then its cycle cone and polytope coincide, respectively, with the cut cone $\mathrm{CUT}(G)$ and the cut polytope $\mathrm{CUT}^\square(G)$ of G.

For the study of cycle polyhedra, we may clearly assume to deal with cosimple matroids. Indeed, if $\{e\}$ is a cocircuit of $\mathcal{M}$, then no cycle of $\mathcal{M}$ contains e and, thus, the cycle cone and polytope are contained in the hyperplane $x_e = 0$. Similarly, if $\{e, f\}$ is a cocircuit of $\mathcal{M}$, then the cycle polyhedra lie in the hyperplane $x_e - x_f = 0$. Moreover, the cycle polyhedra of a cosimple binary matroid are full-dimensional (cf. Barahona and Grötschel [1986]).

By the results of Section 26.3.1, the switching operation preserves the cycle polytope $\mathrm{CYC}^\square(\mathcal{M})$ of any binary matroid $\mathcal{M}$; that is, $r_C(\mathrm{CYC}^\square(\mathcal{M})) = \mathrm{CYC}^\square(\mathcal{M})$ for any cycle C of $\mathcal{M}$. Moreover, a linear description of the cycle polytope $\mathrm{CYC}^\square(\mathcal{M})$ can be derived from a linear description of the cycle cone $\mathrm{CYC}(\mathcal{M})$. (Recall Proposition 26.3.6.)

Some valid inequalities for the cycle polytope $\mathrm{CYC}^\square(\mathcal{M})$ can be easily defined as follows. First, the inequalities: $0 \leq x_e \leq 1$ (for $e \in E$) are trivially valid for $\mathrm{CYC}^\square(\mathcal{M})$. Let D be a cocycle of $\mathcal{M}$ and let $F \subseteq D$ with $|F|$ odd; then, the *cocycle inequality*:

$$x(F) - x(D \setminus F) \leq |F| - 1 \tag{27.4.2}$$

is valid for $\mathrm{CYC}^\square(\mathcal{M})$. (This follows from the fact that every cycle has an even intersection with the cocycle D). Let $\mathrm{MET}^\square(\mathcal{M})$ denote the polytope in $\mathbb{R}^E$, which is defined by the inequalities: $0 \leq x_e \leq 1$ ($e \in E$) together with the cocycle inequalities (27.4.2) for $D \in \mathcal{C}^*$, $F \subseteq D$ with $|F|$ odd. Similarly, define the cone:

$$\mathrm{MET}(\mathcal{M}) := \{x \in \mathbb{R}_+^E \mid x_e - x(D \setminus \{e\}) \leq 0 \text{ for all } D \in \mathcal{C}^*, e \in D\}.$$

In the case when $\mathcal{M}$ is the cographic matroid $\mathcal{M}^*(G)$ of a graph G, then $\mathrm{MET}^\square(\mathcal{M}^*(G)) = \mathrm{MET}^\square(G)$ and $\mathrm{MET}(\mathcal{M}^*(G)) = \mathrm{MET}(G)$ are the usual semimetric polyhedra of the graph G (which explains our notation).

We have the inclusions:

$$\mathrm{CYC}^\square(\mathcal{M}) \subseteq \mathrm{MET}^\square(\mathcal{M}),\ \mathrm{CYC}(\mathcal{M}) \subseteq \mathrm{MET}(\mathcal{M}).$$

Hence arises the question of characterizing the binary matroids for which equality holds. Due to switching,

$$\mathrm{CYC}^\square(\mathcal{M}) = \mathrm{MET}^\square(\mathcal{M}) \Longleftrightarrow \mathrm{CYC}(\mathcal{M}) = \mathrm{MET}(\mathcal{M}).$$

Following Seymour [1981], the binary matroids $\mathcal{M}$ for which equality: $\mathrm{CYC}(\mathcal{M}) = \mathrm{MET}(\mathcal{M})$ holds are said to have the *'sums of circuits property'*.

Observe that none of the binary matroids F_7^*, R_{10}, or $\mathcal{M}^*(K_5)$ has the sums of circuits property. Indeed, the vector $(\frac{2}{3}, \ldots, \frac{2}{3})$ is a fractional vertex of the polytope $\mathrm{MET}^\square(\mathcal{M})$ for $\mathcal{M} = \mathcal{M}^*(K_5)$ or F_7^* and the vector $(\frac{3}{4}, \ldots, \frac{3}{4})$ is a fractional vertex of the polytope $\text{MET}^{\square}(R_{10})$. In fact, the exclusion of these three matroids F_7^*, R_{10}, and $\mathcal{M}^*(K_5)$ as minors characterizes the sums of circuits property.

Theorem 27.4.3. *A binary matroid $\mathcal{M}$ has the sums of circuits property, that is,* $\text{CYC}(\mathcal{M}) = \text{MET}(\mathcal{M})$ *or, equivalently,* $\text{CYC}^{\square}(\mathcal{M}) = \text{MET}^{\square}(\mathcal{M})$*, if and only if $\mathcal{M}$ does not have F_7^*, $\mathcal{M}^*(K_5)$, or R_{10} as a minor.* ∎

In particular, the cographic matroid of a graph with no K_5 minor has the sums of circuits property, a result already mentioned in Theorem 27.3.6, and any graphic matroid has the sums of circuits property, a result established earlier by Seymour [1979].

Theorem 27.4.3 was proved by Seymour [1981], who considered the sums of circuits property along with several other properties related to multicommodity flows; the result follows there from more general considerations. A more direct proof was given by Grötschel and Truemper [1989a].

The proof of Theorem 27.4.3 relies essentially on some decomposition results for binary matroids, involving an operation on matroids which can be seen as an analogue of the clique sum operation for graphs. Roughly speaking, a matroid with no F_7^*, $\mathcal{M}^*(K_5)$, or R_{10} minor can be decomposed into pieces that are either cographic matroids with no $\mathcal{M}^*(K_5)$ minor, or graphic matroids, or copies of F_7. Thus the proof can be sketched as follows: show that the sums of circuits property is preserved under taking minors and under the 'clique sum operation'.

Further results about the facial structure of cycle polytopes can be found in Barahona and Grötschel [1986] and Grötschel and Truemper [1989b]. For instance, Barahona and Grötschel give conditions under which the cocycle inequality (27.4.2) defines a facet of the cycle polytope. Clearly, if the inequality (27.4.2) is facet defining, then D must be a cocircuit without a chord (a *chord* of D being an element $e \in E$ for which there exist two cocircuits D_1 and D_2 such that $D_1 \cap D_2 = \{e\}$ and $D = D_1 \triangle D_2$). Conversely, if $\mathcal{M}$ has no F_7^* minor and if D is chordless cocircuit, then the inequality (27.4.2) defines a facet of $\text{CYC}^{\square}(\mathcal{M})$.

Example 27.4.4. The cycle polytope of the dual projective space $\mathcal{P}_r^*$. The dual projective space $\mathcal{P}_r^*$ is defined on a groundset of cardinality $2^r - 1$; it has $2^r - 1$ nonempty cycles, each of cardinality 2^{r-1}, and whose incidence vectors are linearly independent. Therefore, its cycle polytope $\text{CYC}^{\square}(\mathcal{P}_r^*)$ is a full-dimensional simplex with 2^r vertices, whose facets are defined by the inequality:

$$\sum_{e \in E} x_e \leq 2^{r-1}$$

together with its switchings by the $2^r - 1$ nonempty cycles. These switched inequalities all have a right hand side zero and they constitute the full linear description of the cone $\text{CYC}(\mathcal{P}_r^*)$. As an example, we obtain that the cycle polytope of the dual Fano matroid $F_7^* = \mathcal{P}_3^*$ is defined by the inequality:

$$\sum_{e \in E} x_e \leq 4$$

together with its switchings by the seven circuits of F_7^* (complements of the Fano lines). We note, therefore, that the cocycle inequalities (27.4.2) do not define facets of the cycle polytope of F_7^*.

Hence, the cycle polytope of a dual projective space, being a simplex, has a very simple facial structure. In fact, as we see below, *any* cycle polytope can be realized as the projection of such a simplex. But finding the facial structure of a cycle polytope is a task which, in general, is far from being easy ! (It is already difficult in the special case when the binary matroid in question is the cographic matroid of the complete graph; then we have the problem of describing the facial structure of the cut polytope $\mathrm{CUT}_n^\square$ which forms, in fact, the main objective of this Part V.)

Following Grötschel and Truemper [1989b], we now indicate how any cycle polytope can be realized as projection of a simplex. Let $\mathcal{M}$ be a cosimple binary matroid. Consider a representation matrix of $\mathcal{M}$ of the form $(I \mid A)$ where A is a 0,1-matrix having two units at least per row. Say, A has r columns. Recall that $\mathcal{P}_r^*$ is represented by the matrix $(I_{2^r-r-1} \mid A_r)$, where the rows of A_r are all binary vectors of length r with two units at least. Now, A is a row submatrix of A_r. Let Y denote the index set for the rows of A_r not present in A. Then, $\mathcal{M}$ coincides with the contraction minor $\mathcal{P}_r^*/Y$ of $\mathcal{P}_r^*$. Therefore, its cycle polytope $\mathrm{CYC}^\square(\mathcal{M})$ can be obtained from the cycle polytope $\mathrm{CYC}^\square(\mathcal{P}_r^*)$ of $\mathcal{P}_r^*$ by projecting out the variables x_e ($e \in Y$). ∎

27.4.3 More about Cycle Spaces

We mention here some questions and results dealing with other relevant aspects of binary matroids. In particular, we mention results concerning optimization over cycle spaces. We also consider the lattice $\mathbb{Z}(\mathcal{M})$ and the integer cone $\mathbb{Z}_+(\mathcal{M})$ generated by the cycle space of a binary matroid $\mathcal{M}$. In this setting, we find again two problems raised earlier concerning the existence of nonbasic Delaunay polytopes and the study of Hilbert bases.

Indeed, as every cycle polytope $\mathrm{CYC}^\square(\mathcal{M})$ is a Delaunay polytope in the lattice $\mathbb{Z}(\mathcal{M})$, Problem 13.2.3 raises the question of existence of a basis of $\mathbb{Z}(\mathcal{M})$ consisting only of cycles. This question remains open for general binary matroids but several classes of binary matroids are known for which it has a positive answer. Goddyn [1993] raised the question of characterizing the binary matroids whose cycle space is a Hilbert basis, which contains the question posed in Section 25.3 about Hilbert bases of cuts. We review what is known about this problem.

The Maximum Weight Cycle Problem. Let $\mathcal{M} = (E, \mathcal{C})$ be a binary matroid and $w \in \mathbb{Q}^E$. The *maximum weight cycle problem* consists of finding a cycle $C \in \mathcal{C}$ whose weight $\sum_{e \in C} w_e$ is maximum. This problem is NP-hard as it contains the max-cut problem as a special instance. However, this problem becomes polynomial-time solvable for several classes of binary matroids. This is the case, for instance, for cographic matroids of graphs with no K_5-minor (recall Theorem 27.3.7). This is also the case for graphic matroids, by the results of Edmonds and Johnson [1973]. Grötschel and Truemper [1989a] show that the maximum weight cycle problem can be solved in polynomial time for the larger class of binary matroids having the sums of circuits property.

The latter result is based on showing that the separation problem for the polytope $\mathrm{MET}^{\square}(\mathcal{M})$ can be solved in polynomial time if $\mathcal{M}$ has the sums of circuits property. Using decomposition results, Grötschel and Truemper reduce this question to the special cases when $\mathcal{M}$ is graphic or cographic. A separation algorithm was given by Padberg and Rao [1982] in the graphic case and by Barahona and Mahjoub [1986] in the cographic case (the latter algorithm has been described in Section 27.3.1).

The Cycle Lattice. For a binary matroid $\mathcal{M} = (E, \mathcal{C})$, let

$$\mathbb{Z}(\mathcal{M}) := \{\sum_{C \in \mathcal{C}} \lambda_C \chi^C \mid \lambda_C \in \mathbb{Z}\ \forall C \in \mathcal{C}\}$$

denote the lattice generated by its cycle space, called the *cycle lattice* of $\mathcal{M}$. If $\mathcal{M}$ is the cographic matroid of the complete graph K_n, then $\mathbb{Z}(\mathcal{M})$ coincides with the cut lattice $\mathcal{L}_n$ introduced earlier. As we saw in Proposition 25.1.1, the cut lattice $\mathcal{L}_n$ has a very easy description; namely, an integer vector x belongs to $\mathcal{L}_n$ if and only if $x_{ij} + x_{ik} + x_{jk} \in 2\mathbb{Z}$ for all $i, j, k \in \{1, \ldots, n\}$. One may wonder whether a similar result holds for any cycle lattice.

As cycles and cocycles have an even intersection, the following *parity condition:*

$$(27.4.5) \qquad x(D) \in 2\mathbb{Z} \text{ for every cocycle } D \text{ of } \mathcal{M}$$

is obviously necessary for a vector $x \in \mathbb{Z}^E$ to belong to $\mathbb{Z}(\mathcal{M})$. However, this condition does not suffice in general for characterizing $\mathbb{Z}(\mathcal{M})$. For instance, if $\mathcal{M}$ is the dual Fano matroid F_7^*, then every vector $x \in \mathbb{Z}(F_7^*)$ should, in fact, satisfy the congruence relation: $\sum_{e \in E} x_e = 0$ (modulo 4). The next result follows from the work of Cunningham [1977].

Theorem 27.4.6. *The following assertions are equivalent for a cosimple binary matroid $\mathcal{M}$.*

(i) *$\mathbb{Z}(\mathcal{M})$ is completely characterized by the parity condition* (27.4.5) *and the same holds for every minor of $\mathcal{M}$.*

(ii) *$\mathcal{M}$ does not have F_7^* as a minor.* ∎

This leaves unanswered the question of characterizing the binary matroids $\mathcal{M}$ whose cycle space is described by the parity condition (while not requiring this property for minors of $\mathcal{M}$). Lovász and Seress [1993] give a complete answer to this question. They give in fact several equivalent characterizations for such matroids; we mention in Theorem 27.4.7 below one of them. Further results on cycle lattices can be found in Lovász and Seress [1993, 1995].

We need a definition. Given a matrix B with rows $b_1, \ldots, b_r$, let B' denote the matrix with rows $b_1, \ldots, b_r$ and $b_i \circ b_j$ $(1 \leq i < j \leq r)$ where, for two vectors $a, b \in \mathbb{R}^n$, $a \circ b$ stands for the vector with components $a_i b_i$ $(1 \leq i \leq n)$.

Theorem 27.4.7. *Let $\mathcal{M}$ be a cosimple binary matroid with representation matrix B. Then, the cycle lattice $\mathbb{Z}(\mathcal{M})$ is completely characterized by the parity condition* (27.4.5) *if and only if the matrix B' has full column rank over $GF(2)$.* ∎

Delaunay Polytopes in Cycle Lattices. We saw in Example 13.2.5 that the cut polytope $\mathrm{CUT}_n^{\square}$ is a Delaunay Polytope in the cut lattice $\mathcal{L}_n$. More generally, as noted in Hochstättler, Laurent and Loebl [1996], the following holds.

Lemma 27.4.8. *Let $\mathcal{M}$ be a cosimple binary matroid. Then, its cycle polytope $\mathrm{CYC}^{\square}(\mathcal{M})$ is a Delaunay polytope in the cycle lattice $\mathbb{Z}(\mathcal{M})$.*

Proof. We verify that the polytope $\mathrm{CYC}^{\square}(\mathcal{M})$ is inscribed on a sphere S which is empty in $\mathbb{Z}(\mathcal{M})$. Indeed, $\mathrm{CYC}^{\square}(\mathcal{M})$ is inscribed on the sphere S with center $c := (\frac{1}{2}, \ldots, \frac{1}{2})$ and radius $r := \frac{1}{2}\sqrt{|E|}$. Let $x \in \mathbb{Z}(\mathcal{M})$. Then,

$$(\| x - c \|_2)^2 - r^2 = \sum_{e \in E} (x_e - \frac{1}{2})^2 - \frac{1}{4}|E| = \sum_{e \in E} x_e(x_e - 1) \geq 0$$

with equality if and only if x is 0, 1-valued. Now the only binary vectors in $\mathbb{Z}(\mathcal{M})$ are the incidence vectors of cycles. This shows that the sphere S is empty in $\mathbb{Z}(\mathcal{M})$. ∎

We asked in Problem 13.2.3 whether every Delaunay polytope is basic. This question remains already unsettled for Delaunay polytopes arising from cycle lattices. That is, we have the following question:

Problem 27.4.9. *Given a binary matroid $\mathcal{M}$, can one find a basis $\mathcal{B}$ for the cycle lattice $\mathbb{Z}(\mathcal{M})$ consisting only of cycles ?*

The following information is known. Gallucio and Loebl [1997] show that Problem 27.4.9 has a positive answer for graphic matroids. Gallucio and Loebl [1996] show that the same holds for matroids with no F_7^* minor (using decomposition results). A short elementary proof for this result is given by Hochstättler, Laurent and Loebl [1996]; moreover, these authors extend the result to one-element extensions of binary matroids with no F_7^* minor. Problem 27.4.9 is further studied in Hochstättler and Loebl [1995].

The Integer Cycle Cone. For a binary matroid $\mathcal{M} = (E, \mathcal{C})$, let

$$\mathbb{Z}_+(\mathcal{M}) := \{\sum_{C \in \mathcal{C}} \lambda_C \chi^C \mid \lambda_C \in \mathbb{Z}_+ \ \forall C \in \mathcal{C}\}$$

denote the integer cone generated by its cycle space, called the *integer cycle cone* of $\mathcal{M}$. If $\mathcal{M}$ is the cographic matroid of the complete graph K_n, then $\mathbb{Z}_+(\mathcal{M})$ consists, in fact, of the distances on n points that are isometrically hypercube embeddable (recall Proposition 4.2.4). Clearly,

$$\mathbb{Z}_+(\mathcal{M}) \subseteq \mathbb{Z}(\mathcal{M}) \cap \mathrm{CYC}(\mathcal{M}).$$

As in Section 25.3, we say that $\mathcal{C}$ is a Hilbert basis if equality holds in the above inclusion. Characterizing the binary matroids whose cycle space $\mathcal{C}$ is a Hilbert basis is an open problem, already within cographic matroids. We summarize below what is known about this question.

Alspach, Goddyn and Zhang [1994] answer this question for graphic matroids. Namely, they show that the family of cycles of a graph G is a Hilbert basis if and only if G does not have the Petersen graph as a minor. (Indeed, let x denote the vector indexed by the edge set of the Petersen graph taking value 2 on a perfect matching and value 1 on the remaining edges. Then, $x \in \mathbb{Z}(\mathcal{M}(P_{10})) \cap \mathrm{CYC}(\mathcal{M}(P_{10}))$ but $x \notin \mathbb{Z}_+(\mathcal{M}(P_{10}))$, which shows that the cycles of P_{10} do not form a Hilbert basis.)

Note that $\mathrm{CYC}(\mathcal{M}) = \mathrm{MET}(\mathcal{M})$ for graphic matroids. More generally, this equality holds for matroids with the sums of circuits property. Fu and Goddyn [1995] have characterized the Hilbert basis property within this class. Namely, they show that

$$\mathbb{Z}_+(\mathcal{M}) = \mathbb{Z}(\mathcal{M}) \cap \mathrm{MET}(\mathcal{M})$$

if and only if $\mathcal{M}$ is a binary matroid with no F_7^*, R_{10}, $\mathcal{M}^*(K_5)$, or $\mathcal{M}(P_{10})$ minor. At this point let us recall the result of Laurent [1996b] for cographic matroids, already mentioned in Section 25.3: The cocycle space of $\mathcal{M}(G)$ is a Hilbert basis for any graph $G \neq K_6$ on at most 6 nodes and it is not a Hilbert basis if G has a K_6 minor.

Chapter 28. Hypermetric Inequalities

In this chapter we study in detail the class of hypermetric inequalities. In particular, we present several subclasses of hypermetric inequalities that define facets of the cut cone. We also address the separation problem for hypermetric inequalities. Although its exact complexity status is not known, there are several results that indicate that it is very likely to be a hard problem. In particular, the problem of finding the smallest violated hypermetric inequality is NP-hard.

Hypermetric inequalities belong, in fact, to the larger class of gap inequalities. Unfortunately, very little is known about these more general inequalities. In particular, it is not known whether they contain new facets (besides the hypermetric ones). Moreover, computing their exact right hand sides turns out to be an NP-complete problem ! One way to avoid this difficulty is by relaxing the right hand sides by a larger number (which is easy to compute). In this manner, one obtains a class of valid inequalities which defines a weaker relaxation of the cut polytope. This relaxation forms a (nonpolyhedral) convex body which has the property that one can optimize over it in polynomial time. Moreover, optimizing over this convex body yields a very tight approximation for the max-cut problem (see Section 28.4.1).

28.1 Hypermetric Inequalities: Validity

Hypermetric inequalities constitute the first nontrivial class of valid inequalities for the cut cone. As was already mentioned earlier in Section 6.1 (and Remark 5.4.11) they were discovered independently by several authors with different mathematical backgrounds and motivation. Hypermetric inequalities have already been introduced in Chapter 6 in the context of ℓ_1- and ℓ_2-metrics. We refer to Part II for a detailed study of some of their properties, in particular, in connection with geometry of numbers. We concentrate here on the question of identifying hypermetric facets. In order to make this chapter self-contained, we recall below the definitions and basic properties of hypermetric inequalities.

To define hypermetric and further types of inequalities, it is convenient to use the following notation. For $b \in \mathbb{R}^n$, we remind that $Q(b)$ denotes the vector in $\mathbb{R}^{E_n}$ whose ij-th component is equal to the product $b_i b_j$. Hence, for $x \in \mathbb{R}^{E_n}$,

$$Q(b)^T x = \sum_{ij \in E_n} b_i b_j x_{ij}.$$

This notation gives us a convenient way of defining left hand sides of inequalities with $\binom{n}{2}$ coefficients based on a given vector b of length n. In case we want to highlight the fact that the vector b is in $\mathbb{R}^n$, we will write $Q_n(b)$ or $Q_n(b_1, \ldots, b_n)$ instead of $Q(b)$.

Definition 28.1.1. *Let $b = (b_1, \ldots, b_n)$ be an integral vector satisfying $\sum_{i=1}^n b_i = 1$. Then, the inequality:*

$$(28.1.2) \qquad Q(b)^T x = \sum_{1 \le i < j \le n} b_i b_j x_{ij} \le 0.$$

is called the hypermetric inequality *defined by b. A hypermetric inequality* (28.1.2) *is said to be* $(2k+1)$-gonal *if $\sum_{i|b_i<0} |b_i| = k$ or, equivalently, if $\sum_{i=1}^n |b_i| = 2k+1$.*

Lemma 28.1.3. *Every hypermetric inequality* (28.1.2) *is valid for the cut cone CUT_n. Moreover, the roots of a hypermetric inequality are the cut vectors $\delta(S)$ $(S \subseteq V_n)$ for which $b(S) := \sum_{i \in S} b_i$ is equal to 0 or 1.*

Proof. Given $S \subseteq V_n$, $\sum_{ij \in E_n} b_i b_j \delta(S)_{ij} = b(S)(1 - b(S)) \le 0$, since $b(S)$ is an integer. ∎

We remind that hypermetric inequalities contain as a special case the triangle inequality:

$$x_{ij} - x_{ik} - x_{jk} \le 0$$

(which is obtained by taking $b_i = b_j := 1$, $b_k = -1$ and $b_h := 0$ for $h \in V_n \setminus \{i, j, k\}$).

Since the 0-lifting operation produces facets from facets, we obtain that the inequality $Q(b)^T x \le 0$ is facet inducing if and only if the inequality $Q(b')^T x \le 0$ is facet inducing, where b' is any vector obtained from b by adding zero components. If we apply the permutation operation to a hypermetric inequality or if we switch it by one of its roots, then we obtain again a hypermetric inequality. More precisely, a permutation of $Q(b)^T x \le 0$ amounts to permuting the b_i's. Switching the inequality $Q(b)^T x \le 0$ by the cut $\delta(S)$ with $b(S) = 0$ yields the inequality $Q(b')^T x \le 0$, where $b'_i = -b_i$ if $i \in S$ and $b'_i = b_i$ otherwise. In other words, switching a hypermetric inequality by a root amounts to changing the signs of some coefficients of b.

If we switch the inequality $Q(b)^T x \le 0$ by a cut vector $\delta(S)$ which is not a root of it, we obtain an inequality which is valid for the cut polytope $\mathrm{CUT}_n^\square$ (but not for the cut cone CUT_n). For example, the inequality:

$$\sum_{1 \le i < j \le 2k+1} x_{ij} \le k(k+1)$$

is a switching of the inequality $Q_{2k+1}(1, \ldots, 1, -1, \ldots, -1)^T x \le 0$.

Hypermetric inequalities are, by definition, the inequalities of the form:

$$Q(b)^T x \leq 0,$$

where $b \in \mathbb{Z}^n$ and the sum $\sigma(b) := \sum_{i=1}^n b_i$ of its components is equal to 1. One may wonder what happens if we relax the condition on the sum $\sigma(b)$ or if we allow a nonzero right-hand side. Can the hypermetric inequality be modified so as to yield another valid inequality ? This question can be answered positively in several ways.

Maybe we should start with reminding that the inequality $Q(b)^T x \leq 0$ in the case $\sigma(b) = 0$ is nothing but the negative type inequality, which is valid but not facet defining for CUT_n (recall Corollary 6.1.4).

If $b \in \mathbb{Z}^n$ has an arbitrary sum $\sigma(b)$ then one can always construct an inequality:

$$Q(b)^T x \leq v_0$$

which is valid for $\mathrm{CUT}_n^{\square}$; it suffices to define the right-hand side v_0 in a suitable manner. We have, then, the class of gap inequalities which will be discussed in Section 28.4.

On the other hand, if $\sigma(b) \geq 2$ but, yet, one wants to construct a homogeneous inequality, then one has to modify the quantity $Q(b)^T x$ in order to preserve validity. Clique-web inequalities and suspended-tree inequalities are inequalities that fall into this category.

There is yet another way of generalizing hypermetric inequalities, which consists of asking validity not for *all* cut vectors but only for a restricted subset of them. For instance, if we allow $\sigma(b) \geq 2$ then the inequality $Q(b)^T x \leq 0$ is no longer valid for all cut vectors but it remains valid for all the cut vectors $\delta(S)$ such that $|S| \notin \{1, \ldots, \sigma(b) - 1\}$. When $\sigma(b) = 2$, the inequality $Q(b)^T x \leq 0$ is valid (and, sometimes, facet defining) for the even cut cone ECUT_n. Other such examples will be given in Section 28.5.

28.2 Hypermetric Facets

There are several known classes of hypermetric inequalities that define facets of CUT_n. Here are some examples that can be derived from results presented later:

- $Q_3(1, 1, -1)^T x \leq 0$ (triangle facet),
- $Q_5(1, 1, 1, -1, -1)^T x \leq 0$ (pentagonal facet),
- $Q_{2k+1}(1, \ldots, 1, -1, \ldots, -1)^T x \leq 0$ for $k \geq 1$ (pure hypermetric facet),
- $Q_n(b_1, \ldots, b_p, -1, \ldots, -1)^T x \leq 0$ for $3 \leq p \leq n-3$, $b_1, \ldots, b_p > 0$,
- $Q_{11}(2, 2, 2, -2, -2, -2, 1, 1, 1, -1, -1)^T x \leq 0$,
- $Q_{15}(3, 3, -3, -3, -3, 1, 1, 1, 1, 1, 1, 1, -1, -1, -1)^T x \leq 0$,
- $Q_{19}(4, 4, -4, -4, 3, 3, -3, -3, -3, 1, 1, 1, 1, 1, 1, 1, -1, -1, -1)^T x \leq 0$.

Of course, by permuting, switching, and 0-lifting we can produce whole classes of hypermetric facets from these examples. For instance, for all cut polytopes $\mathrm{CUT}_n^{\square}$, all triangle facets can be obtained from $Q(1,1,-1)^T x \leq 0$ using these operations.

In this section we will provide a number of sufficient and/or necessary conditions for a hypermetric inequality to define a facet of CUT_n. At present there is no complete characterization of all hypermetric facets, i.e., of all integer vectors $b = (b_1, \ldots, b_n)$ with $\sum_{i=1}^n b_i = 1$ and for which $Q(b)^T x \leq 0$ is facet inducing. However, a complete characterization of the hypermetric facets is known for the following classes of parameters $b = (b_1, \ldots, b_n)$:

- $b_1 \geq \ldots \geq b_p > 0 > b_{p+1} \geq \ldots \geq b_n$ with $b_n = -1$ or with $b_{n-1} = -1$ (see Theorem 28.2.4)(i.e., all negative b_i's except at most one are equal to -1),
- $b_i \in \{w, -w, 1, -1\}$ for all $i \in \{1, \ldots, n\}$, for some integer $w \geq 2$ (see Theorem 28.2.9).

There are infinitely many hypermetric inequalities. All of them are supporting and, thus, define nonempty faces of CUT_n. However, since CUT_n is a polyhedral cone, it has only finitely many faces, i.e., there are some faces that are induced by infinitely many different hypermetric inequalities. Moreover, since CUT_n is full-dimensional, each of its facets is defined by an inequality that is unique up to positive scaling. Because of the condition $\sum_{1 \leq i \leq n} b_i = 1$, no hypermetric inequality is a positive multiple of another. Thus, among all hypermetric inequalities $Q_n(b)^T x \leq 0$, there are only finitely many ones that define facets of the cut cone CUT_n. We recall the following stronger result from Section 14.2: The hypermetric cone is polyhedral, i.e., among the hypermetric inequalities only a finite subset of them is not redundant.

The next result states a necessary condition for a hypermetric inequality to define a facet of the cut cone.

Proposition 28.2.1. *Let $b \in \mathbb{Z}^n$ with $\sum_{i=1}^n b_i = 1$. Suppose that b_i, b_j are positive coefficients and that b_k is a negative coefficient of b. Set $P := \{r \mid b_r > 0\} \setminus \{i,j\}$. If*

$$\sum_{r \in P} b_r + b_k < 0$$

then the face F of CUT_n defined by $Q(b)^T x \leq 0$ is not a facet and F is contained in the facet defined by the triangle inequality: $x_{ij} - x_{ik} - x_{jk} \leq 0$.

Proof. Let $\delta(S)$ be any cut vector. We may assume that $k \notin S$. If $i, j \in S$, then

$$\begin{aligned} b(S) \geq \ & b_i + b_j - b_k + \textstyle\sum_{r|b_r<0} b_r = b_i + b_j - b_k + 1 - \sum_{r|b_r>0} b_r \\ & = 1 - b_k - \textstyle\sum_{r \in P} b_r > 1. \end{aligned}$$

Hence, $\delta(S)$ is not a root of F since $b(S) \notin \{0,1\}$. Therefore, $|S \cap \{i,j\}| \leq 1$ for every root $\delta(S)$ of F. This shows that every root of F is a root of the triangle inequality: $x_{ij} - x_{ik} - x_{jk} \leq 0$. ∎

This observation yields that, if $Q(b)^T x \leq 0$ is facet defining, then the sum of the two largest coefficients of b and the absolute value of the smallest coefficient cannot be larger than the sum of all positive coefficients of b. In fact, using general polyhedral theory (see Grötschel, Lovász and Schrijver [1988]), one can prove that the coefficients b_i cannot be larger than 2^{n^4}. Using the fact that $b(S) \in \{0,1\}$ for all roots $\delta(S)$ of a hypermetric inequality, one can improve this bound to $n2^{n^2}$. Let us recall the bound:

$$\max_{1 \leq i \leq n} |b_i| \leq \frac{2^{n-1}}{\binom{2n-2}{n-1}}(n-1)!$$

if the inequality $Q(b)^T x \leq 0$ defines a facet of CUT_n; it was obtained in Part II (see Proposition 14.2.4) as a byproduct of the interpretation of hypermetrics in terms of geometry of numbers. This latter bound is better than $n2^{n^2}$. Note, however, that for all the known hypermetric facets, $\max_{1 \leq i \leq n} |b_i|$ is only quadratic in n (see Example 28.2.6).

The main tool for constructing hypermetric facets is the lifting procedure described in Section 26.5 and, more precisely, the Lifting Lemma 26.5.3. Assume that the inequality:

$$v^T x := Q_n(b_1, \ldots, b_n)^T x \leq 0$$

is facet inducing for CUT_n and, given an integer c, consider the inequality:

$$(v')^T x := Q_{n+1}(b_1 - c, b_2, \ldots, b_n, c)^T x \leq 0.$$

Hence, the inequality $(v')^T x \leq 0$ is obtained from the inequality $v^T x \leq 0$ by splitting the node 1 into nodes $1, n+1$. As mentioned in Section 26.5, in order to show that $(v')^T x \leq 0$ is facet inducing for CUT_{n+1}, it suffices to verify that the condition (iii) from Lemma 26.5.3 holds. Hence, it suffices to exhibit n subsets T_k $(1 \leq k \leq n)$ of $\{2, \ldots, n, n+1\}$ containing the element $n+1$, such that the cut vectors $\delta(T_k)$ are roots of $(v')^T x \leq 0$ and such that the incidence vectors of the sets T_k are linearly independent.

We mention below several lifting theorems for hypermetric facets which are based on this procedure (see Lemma 28.2.3 and Theorem 28.2.4). As an application, we can construct several classes of hypermetric facets. First, as a direct application of Theorem 26.4.3, we have the following result.

Proposition 28.2.2. *Let $b_1, \ldots, b_n$ be integers with $\sum_{i=1}^n b_i = 1$ and $b_2 = b_1 - 1$. If the inequality:*

$$Q_n(b_1, b_1 - 1, b_3, \ldots, b_n)^T x \leq 0$$

defines a facet of CUT_n*, then the inequality:*

$$Q_{n+1}(b_1 - 1, b_1 - 1, b_3, \ldots, b_n, 1)^T x \leq 0$$

defines a facet of CUT_{n+1}*. In particular, if $\sum_{i=2}^n b_i = 2$ and if the inequality:*

$$Q_n(-1, b_2, \ldots, b_n)^T x \leq 0$$

defines a facet of CUT_n*, then the inequality:*

$$Q_{n+2}(-1, b_2, \ldots, b_n, 1, -1)^T x \leq 0$$

defines a facet of CUT_{n+2}. ∎

We use thereafter the notation $[p, q]$ to denote the set of all integers i with $p \leq i \leq q$, where $1 \leq p \leq q$ are some integers. Lemma 28.2.3 and Theorem 28.2.4 below are due to Deza [1973a](see also Deza and Laurent [1992a] for full proofs).

Lemma 28.2.3. *Let* $b_1, \ldots, b_n$ *be integers with* $\sum_{i=1}^n b_i = 1$. *Suppose that* $b_2 \geq b_3 \geq \ldots \geq b_p > 0$ *and* $b_{p+1} = \ldots = b_n = -1$ *where* $p \geq 2$, $n \geq 4$. *Suppose furthermore that* $Q_n(b_1, \ldots, b_n)^T x \leq 0$ *is facet inducing. Then,*

(i) $Q_{n+1}(b_1 + 1, b_2, \ldots, b_n, -1)^T x \leq 0$ *is facet inducing.*

(ii) $Q_{n+1}(b_1 - c, b_2, \ldots, b_n, c)^T x \leq 0$ *is facet inducing, for any integer* $c \in [1, n - p - b_2]$.

Proof. Let us give the proof of (ii) in order to illustrate the proof technique. Consider the n cut vectors $\delta(S)$, where the sets S are as follows:

$$\begin{array}{ll} S = \{i, n+1\} \cup [p+1, p+b_i+c-1] & \text{for } 2 \leq i \leq p, \\ S = \{2, n+1\} \cup [p+1, p+b_2+c] - \{i\} & \text{for } p+1 \leq i \leq p+b_2+c-1, \\ S = \{2, n+1\} \cup [p+1, p+b_2+c-1] \cup \{i\} & \text{for } p+b_2+c \leq i \leq n, \\ S = \{2, 3, n+1\} \cup [p+1, p+b_2+b_3+c]. & \end{array}$$

These n cut vectors $\delta(S)$ are roots of $Q_{n+1}(b_1 - c, b_2, \ldots, b_n, c)^T x \leq 0$ and one can check that the incidence vectors of the sets S are linearly independent. Therefore, by the Lifting Lemma 26.5.3, $Q_{n+1}(b_1 - c, b_2, \ldots, b_n, c)^T x \leq 0$ is facet inducing. The proof of (i) is similar but with more technical details, as one must distinguish the cases when $b_2 = 1$ and when $b_2 \geq 2$. ∎

Theorem 28.2.4. *Let* $b_1, \ldots, b_n$ *be nonzero integers such that* $\sum_{i=1}^n b_i = 1$ *and* $b_1 \geq b_2 \geq \ldots \geq b_p > 0 > b_{p+1} \geq \ldots \geq b_n$*, where* p *is some integer with* $2 \leq p \leq n-1$.

(i) *If* $p = 2$*, then* $Q(b)^T x \leq 0$ *defines a facet if and only if* $n = 3$ *and* $b_1 = b_2 = 1$, $b_3 = -1$.

(ii) *If* $2 \neq p = n - 1$*, then* $Q(b)^T x \leq 0$ *does not define a facet.*

(iii) *Suppose* $p = n - 2$.

   (iiia) *If* $Q(b)^T x \leq 0$ *defines a facet, then* $b_1 = 1$.

   (iiib) *If* $b_{n-1} = -1$*, then* $Q(b)^T x \leq 0$ *defines a facet if and only if* $b_n = -n + 4$ *and* $b_1 = \ldots = b_{n-2} = 1$.

(iv) *Suppose* $3 \leq p \leq n - 3$ *and* $b_{n-1} = -1$. *Then,* $Q(b)^T x \leq 0$ *defines a facet if and only if* $b_1 + b_2 \leq n - p - 1 + \mathrm{sign}(|b_1 - b_p|)$.

(Here, $\text{sign}(|b_1 - b_p|) = 1$ if $b_1 > b_p$ and $\text{sign}(|b_1 - b_p|) = 0$ if $b_1 = b_p$.)

Proof. Let R denote the set of roots of the inequality $Q(b)^T x \leq 0$.
(i) We suppose $p = 2$. Assume that the inequality $Q(b)^T x \leq 0$ is facet defining. We show that it coincides with a triangle inequality. We may assume that $1 \notin S$ and $2 \in S$ for every root $\delta(S)$ of $Q(b)^T x \leq 0$. Set $F := \{(1,2),(1,3),(2,3)\}$. The set R_F of the projections on F of the roots in R consists of the two vectors $(1,0,1)$ and $(1,1,0)$. Hence, the set R_F has rank 2 which, by Lemma 26.5.2, implies that the projection of $Q_n(b)^T x \leq 0$ on $\overline{F}$ is zero. Therefore, $Q_n(b)^T x \leq 0$ is the triangle inequality $x_{12} - x_{13} - x_{23} \leq 0$.
(ii) If $2 \neq p = n-1$, then R consists of the cut vectors $\delta(\{i\})$ with $b_i = 1$. Hence the rank of R is less than or equal to $n-1$, implying that $Q(b)^T x \leq 0$ is not facet defining.
(iii) Suppose $p = n-2$. We can suppose that $n \notin S$ for each root $\delta(S)$ in R. If $b_1 > 1$ then, for every root $\delta(S)$, $n-1 \in S$ whenever $1 \in S$. Set $F := \{(1, n-1), (1,n), (n, n-1)\}$. The set R_F consists of the vectors $(0,0,0)$, $(0,1,1)$ and $(1,0,1)$. By Lemma 26.5.2, if $Q(b)^T x \leq 0$ is facet defining, then the projection of $Q_n(b)^T x \leq 0$ on $\overline{F}$ is zero, which yields a contradiction. This shows (iiia). If $b_{n-1} = -1$ then one shows that $Q(b)^T x \leq 0$ is facet defining by applying iteratively the lifting procedure from Lemma 28.2.3 (i), starting, e.g., from the triangle facet $Q_3(1,1,-1)^T x \leq 0$.
We leave the details of the proof of (iv) to the reader. ∎

Corollary 28.2.5.

(i) *All pure hypermetric inequalities are facet defining.*

(ii) *Let $b \in \mathbb{Z}^n$ such that $\sum_{i=1}^n b_i = 1$. If b has at least 3 and at most $n-3$ positive entries and if all negative entries are equal to -1 then the associated hypermetric inequality $Q(b)^T x \leq 0$ is facet defining.* ∎

Example 28.2.6. (Avis and Deza [1991]) Given $n \geq 7$, set $m := \lfloor \frac{n+1}{4} \rfloor$ and let $b \in \mathbb{Z}^n$ be defined by

$$\begin{array}{ll} b_i := m & \text{for } i = 1, \ldots, n-2m-1, \\ b_i := m-1 & \text{for } i = n-2m, \\ b_i := -1 & \text{for } i = n-2m+1, \ldots, n-1, \\ b_i := m(2+2m-n)+1 & \text{for } i = n. \end{array}$$

Hence, $\sum_{i=1}^n b_i = 1$ and $\sum_{i=1}^n |b_i| = 2m(n-2m) - 3 \geq \frac{n^2}{4} - 4$. The hypermetric inequality $Q(b)^T x \leq 0$ is facet defining as it satisfies the conditions of Theorem 28.2.4 (iv). Therefore, this gives an example of a hypermetric facet of CUT_n which is k-gonal with k quadratic in n. We do not know examples yielding larger values for k. ∎

We now give a lifting result permitting to identify all hypermetric facets $Q_n(b)^T x \leq 0$ where $b_i \in \{w, -w, 1, -1\}$ for $1 \leq i \leq n$, for some integer $w \geq 2$.

Let $\alpha, \alpha', \beta, \beta', w$ be nonnegative integers such that $(\alpha - \alpha')w + \beta - \beta' = 1$. We want to characterize hypermetric facets of the form

(28.2.7) $$Q_n(w, \ldots, w, -w, \ldots, -w, 1, \ldots, 1, -1, \ldots, -1)^T x \leq 0,$$

where there are α coefficients w, α' coefficients $-w$, β coefficients 1, and β' coefficients -1. For short, we denote the above hypermetric inequality by

$$Q_n(w^{(\alpha)}, -w^{(\alpha')}, 1^{(\beta)}, -1^{(\beta')})^T x \leq 0;$$

so, $n = \alpha + \alpha' + \beta + \beta'$.

We first formulate a "double lifting" lemma. Namely, we prove that under certain conditions two coefficients of b can be "doubled" in such a way that, if a hypermetric inequality $Q(b)^T x \leq 0$ of type (28.2.7) is facet defining, then the inequality associated with the new vector also is. As an application, one can characterize the hypermetric inequalities of type (28.2.7) that define facets. Lemma 28.2.8, Theorem 28.2.9 and Lemma 28.2.10 below can be found in Deza and Laurent [1992b].

Lemma 28.2.8. *Let $b = (w^{(\alpha)}, -w^{(\alpha')}, 1^{(\beta)}, -1^{(\beta')})$ be an integral vector with $\alpha, \alpha' \geq 1, \beta, \beta'$ satisfying $(\alpha - \alpha')w + \beta - \beta' = 1$. Suppose that $Q_n(b)^T x \leq 0$ is facet defining and that*
- either, $\alpha > \alpha'$,
- or, $\alpha = \alpha'$ and $\beta \geq w$
holds. Then, for $b' := (w^{(\alpha+1)}, -w^{(\alpha'+1)}, 1^{(\beta)}, -1^{(\beta')})$, the inequality $Q_{n+2}(b')^T x \leq 0$ is facet defining. ∎

Theorem 28.2.9. *Let $\alpha, \alpha', \beta, \beta'$, $w \geq 2$ be integers satisfying* $\min(\alpha, \alpha') \geq 1$ *and $(\alpha - \alpha')w + \beta - \beta' = 1$. Set $b := (w^{(\alpha)}, -w^{(\alpha')}, 1^{(\beta)}, -1^{(\beta')})$ and $n = \alpha + \alpha' + \beta + \beta'$.*

(i) *If $\alpha = \alpha'$ (i.e., $\beta' = \beta - 1$), then $Q_n(b)^T x \leq 0$ is facet inducing if and only if* $\min(\beta, \beta') \geq w$.

(ii) *If $|\alpha - \alpha'|$ is a nonzero even number, then $Q_n(b)^T x \leq 0$ is facet inducing if and only if* $\min(\beta, \beta') \geq 1$ *or* ($\min(\beta, \beta') = 0$ *and* $|\alpha - \alpha'| \geq 4$).

(iii) *If $|\alpha - \alpha'|$ is an odd number, then $Q_n(b)^T x \leq 0$ is facet inducing if and only if $|\alpha - \alpha'| \geq 3$ or ($|\alpha - \alpha'| = 1$ and* $\min(\beta, \beta') \geq w$). ∎

Lemma 28.2.10. *If the inequality:*

$$Q_n(w^{(\alpha)}, -w^{(\alpha')}, 1^{(\beta)}, -1^{(\beta')})^T x \leq 0$$

defines a facet of CUT_n, *then the inequality:*

$$Q_{n+2\gamma}((w+1)^{(\gamma)}, (-w-1)^{(\gamma)}, w^{(\alpha)}, -w^{(\alpha')}, 1^{(\beta)}, -1^{(\beta')})^T x \leq 0$$

defines a facet of $\mathrm{CUT}_{n+2\gamma}$, *for any integer* $\gamma \geq 1$. ∎

Remark 28.2.11. Some facets introduced by Padberg [1989] for the correlation polytope correspond (up to switching and via the covariance mapping) to some subclasses of the hypermetric facets obtained in Theorem 28.2.4 (iii); this is also the case for some classes of facets given by Barahona and Mahjoub [1986] for the cut polytope (see Deza and Laurent [1992a] for a precise description of the connection). ∎

For various reasons, it is interesting to identify faces that are simplexes. For instance, for $b = (1,1,-1)$, $b = (1,1,-1,0)$, and $b = (1,1,1,-1,-1)$, the inequality $Q(b)^T x \leq 0$ defines a simplex facet of CUT_3, CUT_4, and CUT_5, respectively. These three examples belong to the class of inequalities: $Q(n-4,1,1,-1,\ldots,-1)^T x \leq 0$ $(n \geq 3)$. Any such inequality defines a simplex facet of CUT_n, as the next result from Deza and Laurent [1992a] shows.

Corollary 28.2.12. *Let* $b = (b_1, b_2, 1, 1, -1, \ldots -1) \in \mathbb{Z}^n$ *where* b_1, b_2 *are integers such that* $b_1 \geq b_2$, $b_1 + b_2 = n - 5$ *and assume that* $n \geq 6$.

(i) $Q_n(b)^T x \leq 0$ *is facet inducing if and only if* $b_1 \leq n-4$.

(ii) *The face defined by* $Q_n(b)^T x \leq 0$ *is a simplex if and only if* $b_1 \geq n-4$.

Proof. (i) follows by applying Theorem 28.2.4 and (ii) by checking that all nonzero roots are linearly independent. ∎

28.3 Separation of Hypermetric Inequalities

We consider in this section the separation problem for the class of hypermetric inequalities. This problem can be formulated as follows:

(P0) Separation of hypermetric inequalities.
Instance: An integral distance d on n points.
Question: Does d satisfy all hypermetric inequalities ? If not, find a hypermetric inequality violated by d.

We do not know what is the exact complexity of the problem (P0). However, the complexity of several related problems, described below, is known. This indicates that (P0) is very likely to be a hard problem. The complexity results for the problems (P1), (P2), (P3) have been established by Avis and Grishukhin [1993].

(P1) Testing all hypermetric inequalities.
Instance: An integral distance d on n points.
Question: Does d satisfy all hypermetric inequalities ?
Complexity: co-NP. There exists an algorithm solving (P1) whose running time is in $O(h(n)\,(n^4(\log_2(n))^2 + 2n^3\log_2(n)\log_2(\| d \|_\infty)))$, where $h(n)$ denotes the

number of hypermetric inequalities that define facets of the hypermetric cone HYP_n.

We recall (from Proposition 14.2.4) that, if $Q_n(b)^T x \leq 0$ defines a facet of the hypermetric cone HYP_n, then

$$\max_{1\leq i\leq n} |b_i| \leq B_n := \frac{2^{n-1}}{\binom{2n-2}{n-1}}(n-1)!.$$

Hence, a rough estimate of $h(n)$ is

$$h(n) \leq (2B_n+1)^{n-1}.$$

Proof for (P1). Let d be an integral distance on n points. In order to check that d satisfies all hypermetric inequalities, i.e., that $d \in \mathrm{HYP}_n$, it suffices to check whether $Q_n(b)^T d \leq 0$ holds for all $b \in \mathbb{Z}^n$ with sum 1 and such that $|b_i| \leq B_n$ for all i. Note that B_n can be represented by $O(n \log_2 n)$ bits. Hence, computing $Q_n(b)^T d$ can be done in $O(h(n)\,(n^4(\log_2(n))^2 + 2n^3 \log_2(n) \log_2(\| d \|_\infty)))$ elementary operations, which is polynomial in the size of the input. This shows that (P1) is in co-NP and the announced running time for solving (P1). ∎

The separation problem[1] for hypermetric inequalities is very likely to be a hard problem, in view of the complexity status of the next problems (P2) and (P3).

(P2) Testing all $(2m+1)$-gonal inequalities.
Instance: An integral distance d on n points and an integer m.
Question: Does d satisfy all $(2m+1)$-gonal hypermetric inequalities ?
Complexity: co-NP-complete. Also co-NP-complete if one tests only the pure $(2m+1)$-gonal inequalities.

(P3) Finding the smallest violated hypermetric inequality.
Instance: An integral distance d on n points.
Question: Does d satisfy all hypermetric inequalities ? If not, find the smallest integer k such that d violates a $(2k+1)$-gonal inequality.
Complexity: NP-hard.

In what follows, we prove the complexity of the problems (P2) and (P3). For this, we use the known complexity of the next problems.

(P4) Finding an induced complete bipartite subgraph.
Instance: A graph G on n vertices and an integer m such that $2m+1 \leq n$.

[1] A heuristic for separating hypermetric inequalities is proposed in De Simone [1992], De Simone and Rinaldi [1994]. (The main idea is to reformulate the separation problem for hypermetric inequalities as a max-cut problem over a larger complete graph, to which can then be applied any good heuristic for the max-cut problem.) Hypermetric inequalities are reported there to be effective from a computational point of view for solving max-cut problems on complete graphs of about 20-25 nodes.

Question: Does G contain the complete bipartite graph $K_{m,m+1}$ as an induced subgraph ?
Complexity: NP-complete (Garey and Johnson [1979]).

(P5) Finding the largest induced complete bipartite subgraph.
Instance: A graph G on n nodes.
Question: Find the largest integer m such that G contains $K_{m,m+1}$ as an induced subgraph.
Complexity: NP-hard (Garey and Johnson [1979]).

We show that (P4) reduces to (P2) and that (P5) reduces to at most $\frac{n-1}{2}$ questions of type (P3). We introduce some notation. Let $G = (V_n, E)$ be a graph and let $t \in \mathbb{R}_+$. We construct the distance $d^t(G)$ on V_n by setting

$$\begin{aligned} d^t(G)_{ij} &= 1 && \text{if } ij \in E, \\ d^t(G)_{ij} &= 1+t && \text{if } ij \in E_n \setminus E. \end{aligned}$$

Given $b \in \mathbb{Z}^n$, set

$$V_+(b) := \{i \in V_n \mid b_i > 0\},\ V_- = \{i \in V_n \mid b_i < 0\},\ V(b) := V_+(b) \cup V_-(b).$$

Let $E(G,b)$ denote the edge set of the subgraph of G induced by $V(b)$ and set

$$E^{\triangle}(G,b) := E(G,b) \triangle E(K_{V_+(b),V_-(b)}),$$

where $E(K_{V_+(b),V_-(b)})$ denotes the edge set of the complete bipartite graph with node bipartition $(V_+(b), V_-(b))$. We state an intermediate result.

Lemma 28.3.1. *Let $b \in \mathbb{Z}^n$ with $\sum_{i=1}^n b_i = 1$ and $\sum_{i=1}^n |b_i| = 2k+1$. Then,*

$$Q(b)^T d^t(G) = k^2 t - k - (t+1) \sum_{1 \le i \le n} \frac{|b_i|(|b_i|-1)}{2} - t \sum_{ij \in E^{\triangle}(G,b)} |b_i b_j|. \tag{28.3.2}$$

Proof. Suppose first that $E^{\triangle}(G,b) = \emptyset$. Then,

$$Q(b)^T d^t(G) = \sum_{i \in V_-(b), j \in V_+(b)} b_i b_j + (1+t)\left(\sum_{i,j \in V_-(b), i<j} b_i b_j + \sum_{i,j \in V_+(b), i<j} b_i b_j \right).$$

Using the identity:

$$\sum_{i,j \in X, i<j} b_i b_j = \frac{1}{2}\left(\left(\sum_{i \in X} b_i\right)^2 - \sum_{i \in X} (b_i)^2\right),$$

and the fact that $\sum_{i \in V_+(b)} b_i = k+1$, $\sum_{i \in V_-(b)} b_i = -k$, we obtain:

$$Q(b)^T d^t(G) = -k(k+1) + \frac{t+1}{2}\left(k^2 + (k+1)^2 - \sum_{i \in V(b)} (b_i)^2\right),$$

which coincides with the sum of the first three terms of (28.3.2). If $E^{\triangle}(G,b) \neq \emptyset$, then one easily checks that one more term should be added, which is equal to $-t\sum_{ij\in E^{\triangle}(G,b)} |b_ib_j|$. ∎

Corollary 28.3.3. *Let $G = (V_n, E)$ be a graph and let m be an integer such that $n \geq 2m+1$. Set $d_m(G) := m^3 d^t(G)$, where $t := \frac{1}{m} + \frac{1}{m^3}$. Then,*

(i) *$d_m(G)$ satisfies all $(2k+1)$-gonal inequalities, for $1 \leq k \leq m-1$.*

(ii) *$d_m(G)$ satisfies all $(2m+1)$-gonal inequalities except when G contains $K_{m,m+1}$ as an induced subgraph in which case $d_m(G)$ violates a pure $(2m+1)$-gonal inequality.*

Proof. Suppose $\sum_{i=1}^n |b_i| = 2k+1$ with $k \leq m-1$. Then, by Lemma 28.3.1, $Q(b)^T d^t(G) \leq k^2 t - k \leq 0$. Suppose now that $k = m$. By Lemma 28.3.1,

$$Q(b)^T d^t(G) \leq \frac{1}{m} - (t+1) \sum_{1\leq i\leq n} \frac{|b_i|(|b_i|-1)}{2} - t \sum_{ij\in E^{\triangle}(G,b)} |b_ib_j|.$$

If $|b_i| \geq 2$ for some i or if $E^{\triangle}(G,b) \neq \emptyset$, then $Q(b)^T d^t(G) \leq \frac{1}{m} - t < 0$. Otherwise, b is pure, G contains an induced $K_{m,m+1}$ subgraph, and $Q(b)^T d^t(G) = \frac{1}{m} > 0$. ∎

Proof for (P2). Let G be a graph on n nodes and let m be an integer such that $n \geq 2m+1$. Consider the distance $d_m(G)$. By Corollary 28.3.3, G contains an induced $K_{m,m+1}$ subgraph if and only if $d_m(G)$ violates a $(2m+1)$-gonal inequality (or, equivalently, a pure $(2m+1)$-gonal inequality). This shows that (P2) is co-NP complete. ∎

Proof for (P3). Let G be a graph on n nodes. Set $k := \lfloor \frac{n-1}{2} \rfloor$. Let m denote the largest integer such that G contains an induced $K_{m,m+1}$ subgraph. For $s \leq k$, set $t := \frac{1}{s} + \frac{1}{s^3}$ and consider the distance $d_s(G)$. If $s > m$, then G contains no induced $K_{s,s+1}$ subgraph. Hence, by Corollary 28.3.3, the answer to (P3) is, either that d_s satisfies all hypermetric inequalities, or that $d_s(G)$ violates a $(2p+1)$-gonal inequality for some $p > s$. If $s = m$, then G contains $K_{m,m+1}$ and the answer to (P3) is that the smallest inequality violated by $d_s(G)$ is $(2s+1)$-gonal. Therefore, the answers of (P3) applied successively to $s = k, k-1, \ldots, 1$, give us the value of m. This shows that (P3) is NP-hard. ∎

Remark 28.3.4. In contrast with the hypermetric case, the separation problem for the class of negative type inequalities can be solved in polynomial time. An easy way to see it is by using relation (6.1.15), which states that the negative type cone NEG_{n+1} and the positive semidefinite cone PSD_n are in one-to-one correspondence via the covariance mapping ξ. In other words, let d be a distance on $n+1$ points. Define the $n \times n$ symmetric matrix (p_{ij}) by setting

$$\begin{array}{ll} p_{ii} := d_{i,n+1} & \text{for } i = 1, \ldots, n, \\ p_{ij} := \frac{1}{2}(d_{i,n+1} + d_{j,n+1} - d_{ij}) & \text{for } i \neq j = 1, \ldots, n. \end{array}$$

Then, d satisfies all the negative type inequalities if and only if the matrix (p_{ij}) is positive semidefinite. Moreover, a negative type inequality violated by d (if some exists) can be found in polynomial time (by Proposition 2.4.3). ∎

28.4 Gap Inequalities

We describe here a large class of valid inequalities for $\mathrm{CUT}_n^\square$, which generalizes the hypermetric and negative type inequalities. It was introduced by Laurent and Poljak [1996b]. For $b \in \mathbb{Z}^n$, the quantity

$$\gamma(b) := \min_{S \subseteq V_n} |b(S) - b(\overline{S})|$$

is called the *gap* of b. (Here, $\overline{S} := V_n \setminus S$.) We also set

$$\sigma(b) := \sum_{1 \leq i \leq n} b_i.$$

Then, the inequality:

$$Q(b)^T x \leq \frac{1}{4}(\sigma(b)^2 - \gamma(b)^2) \tag{28.4.1}$$

is called a *gap inequality*. The next two lemmas show validity of the gap inequalities and invariance of the gap under switching.

Lemma 28.4.2. *Every gap inequality* (28.4.1) *is valid for* $\mathrm{CUT}_n^\square$.

Proof. Let $S \subseteq V_n$ and suppose, e.g., that $b(S) \leq \frac{\sigma(b)}{2}$. Then, by the definition of the gap $\gamma(b)$, we have that $b(S) \leq \frac{\sigma(b)-\gamma(b)}{2}$, which implies that $Q(b)^T\delta(S) = b(S)(\sigma(b) - b(S)) \leq \frac{\sigma(b)^2-\gamma(b)^2}{4}$. ∎

Lemma 28.4.3. *Let $b \in \mathbb{Z}^n$, let $A \subseteq V_n$ and let b' be obtained from b by changing the signs of its components indexed by A, i.e., $b'_i := -b_i$ if $i \in A$ and $b'_i := b_i$ if $i \in V_n \setminus A$. Then, $\gamma(b') = \gamma(b)$.*

Proof. For $S \subseteq V_n$, we have $b'(S \Delta A) = b(S \setminus A) - b(A \setminus S) = b(S) - b(A)$ and $b'(\overline{S \Delta A}) = b(\overline{S} \cap \overline{A}) - b(S \cap A) = b(\overline{S}) - b(A)$. This implies that $\gamma(b') = \gamma(b)$. ∎

Clearly, the class of inequalities (28.4.1) is closed under permutation. Lemma 28.4.3 implies that it is also closed under switching, since

$$\frac{1}{4}(\sigma(b')^2 - \gamma(b')^2) = \frac{1}{4}(\sigma(b)^2 - \gamma(b)^2) - Q(b)^T\delta(A)$$

if b' is obtained from b by changing the signs of its components indexed by the set $A \subseteq V_n$. Therefore, the class of gap inequalities is closed under permutation and switching.

Observe that $\gamma(b)$ and $\sigma(b)$ have the same parity. Trivially, $\gamma(b) \leq \sigma(b)$. One can check that

$$\gamma(b) \leq \max_i |b_i|.$$

In particular, $\gamma(b) = 1$ if $\sigma(b) = 1$, and $\gamma(b) = 0$ if $\sigma(b) = 0$. Therefore, the gap inequalities (28.4.1) with $\sigma(b) = 1$ are precisely the hypermetric inequalities, while those with $\sigma(b) = 0$ are the negative type inequalities. Using Lemma 28.4.3 we can identify which gap inequalities arise as switchings of hypermetric or negative type inequalities.

Lemma 28.4.4. *Let $b \in \mathbb{Z}^n$. Then, the inequality* (28.4.1) *can be obtained by switching from a hypermetric inequality (resp. from a negative type inequality) if and only if $\gamma(b) = 1$ (resp. $\gamma(b) = 0$) or, equivalently, if there exists a subset $S \subseteq V_n$ such that $b(S) = \frac{1}{2}(\sigma(b) - 1)$ (resp. $b(S) = \frac{1}{2}\sigma(b)$).* ∎

Hence, the orbits of faces of $\mathrm{CUT}_n^\square$ defined by hypermetric inequalities consist of those faces of $\mathrm{CUT}_n^\square$ defined by inequalities of type (28.4.1) for which $\gamma(b) = 1$ holds. In other words, Chapter 28 is devoted to the class of the gap inequalities (28.4.1) with $\gamma(b) = 1$.

We remind that no inequality (28.4.1) with gap $\gamma(b) = 0$ is facet defining (by Corollary 6.1.4). In the case of a gap $\gamma(b) = 1$, large classes of inequalities (28.4.1) defining facets have been presented in Section 28.2. The gap inequalities (28.4.1) seem difficult to study, in the case of a gap $\gamma(b) \geq 2$. In particular, we do not know any example of an integer sequence $b \in \mathbb{Z}^n$ with $\gamma(b) \geq 2$ and for which the gap inequality (28.4.1) defines a facet of $\mathrm{CUT}_n^\square$. Thus, the following problem is open.

Problem 28.4.5. *Does there exist an integer sequence b with gap $\gamma(b) \geq 2$ for which the gap inequality: $Q(b)^T x \leq \frac{1}{4}(\sigma^2(b) - \gamma^2(b))$ defines a facet of the cut polytope ?*

Only the following is known. Laurent and Poljak [1996b] show[2] that, if the components of b take only two distinct values (in absolute value) and if $\gamma(b) \geq 2$, then the gap inequality (28.4.1) is not facet inducing.

In addition, the gap inequalities present the following difficulty: It is an NP-hard problem to compute the gap of a sequence $b \in \mathbb{Z}^n$ and, thus, to compute the exact right-hand side of the inequality (28.4.1). Indeed, given $b \in \mathbb{Z}^n$, deciding whether $\gamma(b) = 0$ amounts to deciding whether b can be partitioned into two subsequences of equal sums; this is the partition problem, which is known to be NP-complete (Garey and Johnson [1979]).

In order to avoid this difficulty, one may consider the following weaker inequality:

[2]Laurent and Poljak [1996b] also present a characterization of the sequences b for which (28.4.1) is facet defining that involves only conditions expressed in the n-dimensional space instead of the dimension $\binom{n}{2}$ in which the cut polytope $\mathrm{CUT}_n^\square$ lives.

$$(28.4.6) \qquad Q(b)^T x \le \frac{1}{4}\sigma(b)^2$$

instead of (28.4.1). The inequality (28.4.6) is trivially valid for $\mathrm{CUT}_n^\square$. However, no inequality (28.4.6) defines a facet of $\mathrm{CUT}_n^\square$. (Indeed, if $\gamma(b) \neq 0$ then (28.4.6) contains no root at all and, if $\gamma(b) = 0$, then (28.4.6) is a switching of a negative type inequality.)

The inequalities (28.4.6) have nevertheless a number of interesting properties. As we see below they present, in particular, the big advantage of being much more tractable than the gap inequalities (28.4.1).

28.4.1 A Positive Semidefinite Relaxation for Max-Cut

Let $\mathcal{J}_n$ denote the set in $\mathbb{R}^{E_n}$ which is defined by the inequalities (28.4.6) for all $b \in \mathbb{Z}^n$, i.e.,

$$\mathcal{J}_n := \{x \in \mathbb{R}^{E_n} \mid Q(b)^T x \le \frac{1}{4}\sigma(b)^2 \ \text{ for all } b \in \mathbb{Z}^n\}.$$

Then, $\mathcal{J}_n$ is a convex body in $\mathbb{R}^{E_n}$. A first interesting property is that the separation problem over $\mathcal{J}_n$ can be solved in polynomial time. This is the following problem:

> *Given a vector $x \in \mathbb{Q}^{E_n}$, determine whether x belongs to $\mathcal{J}_n$. If not, find $b \in \mathbb{Z}^n$ such that $Q(b)^T x > \frac{1}{4}\sigma(b)^2$.*

This problem can solved in the following way: For $x \in \mathbb{R}^{E_n}$ define the $n \times n$ symmetric matrix X with zero diagonal and with ij-th off-diagonal entry x_{ij}. Then,

$$x \in \mathcal{J}_n \iff \text{the matrix } J - 2X \text{ is positive semidefinite},$$

where J denotes the all-ones matrix. (This follows from the fact that $b^T(J - 2X)b = \sigma(b)^2 - 4Q(b)^T x$ for all $b \in \mathbb{R}^n$.) The result now follows in view of Proposition 2.4.3.

Therefore, one can optimize any linear objective function over $\mathcal{J}_n$ in polynomial time. That is, given $c \in \mathbb{Q}^{E_n}$, one can compute the quantity:

$$B(c) := \max(c^T x \mid x \in \mathcal{J}_n)$$

in polynomial time (with an arbitrary precision); see Goemans and Williamson [1994] for more details. As the inequalities (28.4.6) are valid for the cut polytope $\mathrm{CUT}_n^\square$, the convex body $\mathcal{J}_n$ provides a relaxation of $\mathrm{CUT}_n^\square$, i.e.,

$$\mathrm{CUT}_n^\square \subseteq \mathcal{J}_n.$$

Therefore, the quantity $B(c)$ is an upper bound[3] for the value of the max-cut problem:

$$\mathrm{mc}(K_n, c) := \max(c^T x \mid x \in \mathrm{CUT}_n^{\square}).$$

That is,

$$\mathrm{mc}(K_n, c) \leq B(c)$$

for all $c \in \mathbb{Q}^{E_n}$. Goemans and Williamson [1994] show that the quantity $B(c)$ provides a very good approximation for the max-cut $\mathrm{mc}(K_n, c)$, when the weight function c is nonnegative. Namely,

Theorem 28.4.7. *Given $c \in \mathbb{Q}_+^{E_n}$, we have*

$$\frac{\mathrm{mc}(K_n, c)}{B(c)} \geq \alpha$$

where $\alpha := \min_{0 \leq \theta \leq \pi} \frac{2}{\pi} \frac{\theta}{1 - \cos\theta}$. The quantity α can be estimated as follows: $.87856 < \alpha < .87857$.

Proof. As above, for a vector $x \in \mathbb{R}^{E_n}$ we let X denote the $n \times n$ symmetric matrix with zero diagonal and with off-diagonal entries x_{ij}. Set $Y := J - 2X$. Then,

$$x \in \mathcal{J}_n \Longleftrightarrow Y \succeq 0.$$

Hence, the quantity $B(c)$ can be reformulated as

$$\begin{array}{rll} B(c) = & \max & \frac{1}{2} \sum\limits_{1 \leq i < j \leq n} c_{ij}(1 - y_{ij}) \\ & \text{s.t.} & Y = (y_{ij}) \succeq 0 \\ & & y_{ii} = 1 \ (i = 1, \ldots, n). \end{array}$$

Using the representation of positive semidefinite matrices as Gram matrices, we can further rewrite $B(c)$ as

$$\begin{array}{rll} B(c) = & \max & \frac{1}{2} \sum\limits_{1 \leq i < j \leq n} c_{ij}(1 - v_i^T v_j) \\ & \text{s.t.} & v_1, \ldots, v_n \in S, \end{array} \tag{28.4.8}$$

where $S := \{x \in \mathbb{R}^n : \| x \|_2 = 1\}$. Let $v_1, \ldots, v_n \in \mathbb{R}^n$ be unit vectors realizing the optimum in the above program, i.e., $B(c) = \frac{1}{2} \sum_{1 \leq i < j \leq n} c_{ij}(1 - v_i^T v_j)$. The crucial step in the proof consists now of constructing a 'good' random cut, i.e., whose weight is not too far from the value of the max-cut. For this, one proceeds as follows:

[3]In fact, this upper bound coincides with another upper bound $\phi(c)$ introduced earlier by Delorme and Poljak [1993a] and defined in terms of a minimization problem involving the maximum eigenvalue of the associated Laplacian matrix $L(c)$. Namely, $L(c)$ is the symmetric $n \times n$ matrix with diagonal entries $c^T \delta(i)$ (for $i = 1, \ldots, n$) and with off-diagonal entries $-c_{ij}$ (for $i \neq j$). That the two bounds $\phi(c)$ and $B(c)$ coincide has been shown by Poljak and Rendl [1995] using duality of semidefinite programming.

- Select a random unit vector $r \in \mathbb{R}^n$.
- Set $S := \{i \in V_n \mid v_i^T r \geq 0\}$.

Let $E(c^T\delta(S))$ denote the expected weight of the cut vector $\delta(S)$. Then,

$$E(c^T\delta(S)) \leq \text{mc}(K_n, c).$$

We now evaluate a lower bound for $E(c^T\delta(S))$. We have

$$E(c^T\delta(S)) = \sum_{1 \leq i < j \leq n} c_{ij} p_{ij},$$

where p_{ij} is the probability that an edge ij of K_n is cut by the partition $(S, V_n \backslash S)$. This probability is equal to the probability[4] that a random hyperplane separates the two vectors v_i and v_j. Hence,

$$p_{ij} = \frac{1}{\pi} \arccos(v_i^T v_j).$$

Therefore,

$$E(c^T\delta(S)) = \sum_{1 \leq i < j \leq n} c_{ij} \frac{\arccos(v_i^T v_j)}{\pi} \geq \alpha \sum_{1 \leq i < j \leq n} c_{ij} \frac{1 - v_i^T v_j}{2} = \alpha B(c),$$

where the last inequality follows by the definition of α. This shows the desired inequality: $\alpha B(c) \leq \text{mc}(K_n, c)$. ∎

The above proof shows the existence of a random cut whose weight is at least α times the optimum $\text{mc}(K_n, c)$. Goemans and Williamson [1994] show that the above procedure can be derandomized so as to yield a polynomial deterministic $(\alpha - \epsilon)$-approximation algorithm for the max-cut problem (for any $\epsilon > 0$). Note that the best previous result in this direction was a $\frac{1}{2}$-approximation algorithm, due to Sahni and Gonzales [1976].

Goemans and Williamson's result came as a breakthrough in the area of combinatorial optimization. It shows, indeed, how semidefinite programming can be applied successfully for designing approximation algorithms for combinatorial problems. The method has been since then applied to several other problems; see, in particular, Goemans and Williamson [1994], Karger, Motwani and Sudan [1994], Frieze and Jerrum [1995].

Consider the following instance of the max-cut problem: $n = 5$ and the weight function c takes value 1 on the edges of the 5-circuit C_5 and value 0 elsewhere. Then, $\text{mc}(K_5, c) = 4$ and $B(c) = \frac{5}{2}(1 + \cos(\frac{\pi}{5})) = \frac{25+5\sqrt{5}}{8}$ (obtained by taking in (28.4.8) the vectors $v_i = (\cos(\frac{4i\pi}{5}), \sin(\frac{4i\pi}{5}))$ for $i = 1, \ldots, 5$; see Delorme and Poljak [1993b]). Hence, $\frac{\text{mc}(K_5,c)}{B(c)} = \frac{32}{25+5\sqrt{5}}$ $(= .88445)$. Delorme and Poljak [1993b] conjecture that this is the worst case ratio, i.e., that

$$\frac{\text{mc}(K_n, c)}{B(c)} \geq \frac{32}{25 + 5\sqrt{5}} \ (= .88445)$$

[4]This probability has been, in fact, computed in Section 6.4; by relation(6.4.4), it is equal to $\arccos(v_i^T v_j)/\pi$.

for all $c \geq 0$. Hence, the result from Theorem 28.4.7 shows a lower bound, which is very close to this conjectured value.

Karloff [1996] has made recently a detailed analysis of the performance of the Goemans-Williamson algorithm. We remind the inequalities:

$$\alpha \leq \frac{E(c^T\delta(S))}{B(c)} \leq \frac{\mathrm{mc}(K_n,c)}{B(c)} \leq 1, \tag{28.4.9}$$

where $E(c^T\delta(S))$ is the expected weight of a cut constructed by the random procedure described in the proof of Theorem 28.4.7. Karloff constructs a class of instances (K_n, c) for which equality is attained in the right most inequality of (28.4.9) and also (asymptotically) in the left most inequality of (28.4.9). (Namely, given an even integer m and $b \leq \frac{m}{12}$, consider the graph $J(m, m/2, b)$ whose node set is the family $\mathcal{A}$ of subsets of $[1, m]$ of cardinality $m/2$, with an edge between $A, B \in \mathcal{A}$ if $|A \cap B| = b$. Now, define a weight function c on the edges of K_n ($n := |\mathcal{A}|$) by taking value 1 on the edges of $J(m, m/2, b)$ and value 0 elsewhere.)

The convex set underlying the computation of the bound $B(c)$ is the set $\mathcal{J}_n$ or, rather, its image $\mathcal{E}_n$ under the linear bijection: $x \mapsto 1 - 2x$. That is,

$$\mathcal{E}_n := \{Y \ n \times n \text{ symmetric} \mid Y \succeq 0,\ y_{ii} = 1 \text{ for } i = 1, \ldots, n\}.$$

The set $\mathcal{E}_n$ is called an *elliptope* (standing for *ellip*soid and poly*tope*) in Laurent and Poljak [1995b]. Hence, $\mathcal{E}_n$ can be seen as the intersection of the positive semidefinite cone PSD_n by the hyperplanes $y_{ii} = 1$ $(i = 1, \ldots, n)$. The elliptope $\mathcal{E}_3$ (or rather its 3-dimensional projection consisting of the upper triangular parts of the matrices) is shown in Figure 31.3.6.

The elliptope $\mathcal{E}_n$ has been studied in detail by Laurent and Poljak [1995b, 1996a]. For instance, all the possible dimensions for the faces of $\mathcal{E}_n$ are known as well as for its polyhedral faces (see Section 31.5); it is shown in Laurent and Poljak [1995b] that $\mathcal{E}_n$ has vertices (that is, extreme points with full-dimensional normal cone) that are precisely the 'cut matrices' xx^T for $x \in \{-1, 1\}^n$. Moreover, the link with the well-known theta function introduced by Lovász [1979] for approximating the Shannon capacity and the stability number of a graph is described in Laurent, Poljak and Rendl [1997].

The elliptope $\mathcal{E}_n$ is also relevant to the following problem, known in linear algebra as the *positive semidefinite completion problem*[5]:

> *Given a partial real symmetric matrix X whose entries are only specified on a subset F of the positions (including all diagonal positions), determine whether the missing entries can be completed so as to make X positive semidefinite.*

An easy observation is that it suffices to consider the positive semidefinite completion problem for matrices whose diagonal entries are all equal to 1. (Indeed,

[5] We refer to Johnson [1990] for a comprehensive survey on completion problems.

if a partial symmetric matrix X is completable to a positive semidefinite matrix, then its diagonal entries are nonnegative. Moreover, we can suppose that all diagonal entries are positive as, otherwise, the problem reduces to considering the submatrix of X with positive diagonal entries. Finally, if D denotes the diagonal matrix whose ith-diagonal entry is $\frac{1}{\sqrt{x_{ii}}}$, then the partial matrix $X' := DXD$ (with entries $\frac{x_{ij}}{\sqrt{x_{ii}x_{jj}}}$ for $ij \in F$) is completable to a positive semidefinite matrix if and only if the same holds for X. Now X' is a partial symmetric matrix with an all-ones diagonal.)

In the case of partial symmetric matrices with an all-ones diagonal, the positive semidefinite completion problem amounts to the problem of deciding membership in the projection $\mathcal{E}(G)$ of the elliptope $\mathcal{E}_n$ on the subspace $\mathbb{R}^E$, where $G = (V_n, E)$ denotes the graph corresponding to the specified off-diagonal positions (that is, $E := F \setminus \{ii \mid i = 1, \ldots, n\}$). A description (in closed form) of the projected elliptope $\mathcal{E}(G)$ is known for several classes of graphs G, e.g., for chordal graphs and for series-parallel graphs. Details are given in Section 31.3.

28.5 Additional Notes

Improving Faces by Subtracting Inequalities. We mention here a general technique, introduced in De Simone, Deza and Laurent [1994], for constructing from a given face F another face G containing F and of higher dimension. As an example, we apply it to some hypermetric inequalities.

Suppose that we have a valid inequality $v^T x \leq 0$ for CUT_n. Suppose also that we can find another valid inequality $w^T x \leq 0$ such that the inequality $(v-w)^T x \leq 0$ remains valid for CUT_n. Let F_v (resp. F_w, F_{v-w}) denote the face of CUT_n defined by the inequality $v^T x \leq 0$ (resp. $w^T x \leq 0$, $(v-w)^T x \leq 0$). Then,

$$F_v = F_w \cap F_{v-w}.$$

Hence, the dimension of F_w (or of F_{v-w}) is greater than or equal to the dimension of F_v. We give in Theorem 28.5.2 below an example where the face F_{v-w} is, in fact, a facet of CUT_n. We start with an easy remark.

Lemma 28.5.1. *Let $v^T x \leq 0$ and $w^T x \leq 0$ be two valid inequalities for CUT_n and let F_v, F_w denote the faces of CUT_n that they define, respectively. If $F_v \subseteq F_w$, then the inequality $(M_w v - m_v w)^T x \leq 0$ is valid for CUT_n, where m_v denotes the minimum nonzero value of $|v^T \delta(S)|$ and M_w denotes the maximum value of $|w^T \delta(S)|$ for all subsets S of $\{1, \ldots n\}$.* ∎

Observe that, with the notation of Lemma 28.5.1, the face defined by the inequality $(M_w v - m_v w)^T x \leq 0$ contains the face F_v. Moreover, if $m_v \geq M_w = 2$, then the inequality $(v-w)^T x \leq 0$ is also valid for CUT_n; for example, $m_v = M_v = 2$ for any triangle facet.

Let us look at an example. Consider $Q_7(1,3,2,-1,-1,-1,-2)^T x \leq 0$ as $v^T x \leq 0$ and the inequality $Q_7(0,1,1,0,0,0,-1)^T x \leq 0$ as $w^T x \leq 0$. By Proposition 28.2.1, we know that $F_v \subseteq F_w$. Therefore, as $m_v = M_w = 2$ and setting $b := (1,3,2,-1,-1,-1,-2)$ and $d := (0,1,1,0,0,0,-1)$, the inequality:

$$(v-w)^T x = \sum_{1 \leq i < j \leq 7} (b_i b_j - d_i d_j) x_{ij} \leq 0$$

is valid for CUT_7. In fact, it defines a facet of CUT_7 (indeed, $Q_7(b)^T x \leq 0$ has 19 linearly independent roots which, together with the cut $\delta(\{1,7\})$, form a set of 20 linearly independent roots of the above inequality). (Observe that the above inequality $(v-w)^T x \leq 0$ is, in fact, switching equivalent to the clique-web inequality $\mathrm{CW}_7^1(3,2,2,-1,-1,-1,-1)^T x \leq 0$, which will be defined in the next section.) More generally, we have the following result.

Theorem 28.5.2. *For $n \geq 7$, let $b = (2n-13,3,2,-1,-1,-1,-2,\ldots,-2)$ and let $d = (n-7,1,1,0,0,0,-1,\ldots,-1)$ be two vectors in $\mathbb{Z}^n$. The inequality:*

$$\sum_{1 \leq i < j \leq 7} (b_i b_j - d_i d_j) x_{ij} \leq 0$$

defines a facet of CUT_n. ∎

Generalization to Other Cut Families. We now indicate how hypermetric inequalities can be modified so as to yield valid inequalities for other cut polyhedra.

Inequalities for Even T-Cuts. Let $T \subseteq V_n$ be a set of even cardinality. Recall that $\delta(S)$ is an even T-cut vector if $S \cap T$ is even. Let $b \in \mathbb{Z}^n$ such that b_i is odd for all $i \in T$ and b_i is even for all $i \in V_n \setminus T$ and $\sum_{i=1}^n b_i = 2$. Then, the inequality:

$$Q(b)^T x \leq 0$$

is valid for all even T-cut vectors. (Indeed, let $\delta(S)$ be a cut vector such that $|S \cap T|$ is even. Then, $Q(b)^T \delta(S) = b(S)(2 - b(S)) \leq 0$, since $b(S) = b(S \cap T) + b(S - T) \neq 1$ as $b(S)$ is an even number.)

In the special case when $T = V_n$ and when $|b_i| = 1$ for all $i \in V_n$, then the above inequality defines a facet of the even cut cone ECUT_n (Deza and Laurent [1993b]).

Inequalities for t-Ary Cuts. Let $t \geq 2$ be an integer and suppose that $n \equiv 0 \pmod t$. Then, the cut vector $\delta(S)$ is said to be t-ary if $|S| \equiv 0 \pmod t$. Hence, the notions of 2-ary and even cut vectors are the same. Let $b \in \mathbb{Z}^n$ such that $\sum_{i=1}^n b_i = t$ and $b_i \equiv \beta \pmod t$ for all $i \in V_n$, for some $\beta \in \{1,2,\ldots t-1\}$. Then, the inequality:

$$Q(b)^T x \leq 0$$

is valid for all t-ary cut vectors. (Indeed, let $\delta(S)$ be a cut vector such that $|S| \equiv 0 \pmod t$. Then, $Q(b)^T \delta(S) = b(S)(t - b(S)) \leq 0$, since $b(S) \equiv \beta|S| \equiv 0 \pmod t$.)

Inequalities for Multicuts. Let $b \in \mathbb{Z}^n$ with $\sigma := \sum_{i=1}^n b_i \geq 1$. Then, the inequality:

$$Q(b)^T x \leq \frac{\sigma(\sigma-1)}{2}$$

is valid for the multicut polytope $\mathrm{MC}_n^\square$. Moreover, Grötschel and Wakabayashi [1990] show that this inequality defines a facet of $\mathrm{MC}_n^\square$ in the special case when $|b_i| = 1$ for all $i \in V_n$. Other classes of such facets can be found in Grötschel and Wakabayashi [1990] and Deza, Grötschel and Laurent [1992]. Another generalization of hypermetric inequalities for the multicut polytope is presented in Chopra and Rao [1995].

Observe that, if we suppose that $\sigma \geq 2$, then the inequality:

$$Q(b)^T x \leq \frac{\sigma(\sigma-2)}{2}$$

becomes valid for all even multicut vectors, i.e., the vectors $\delta(S_1, \ldots, S_p)$ corresponding to a partition of V_n where all the S_i's have an even cardinality (Deza and Laurent [1993b]).

Chapter 29. Clique-Web Inequalities

We have seen in Chapter 28 that a valid inequality for CUT_n, namely the hypermetric inequality $Q(b)^T x \leq 0$, can be constructed for any integer vector $b \in \mathbb{Z}^n$ with $\sum_{i=1}^n b_i = 1$. More generally, how can we construct a valid inequality if we have an arbitrary integer vector $b \in \mathbb{Z}^n$?

When $\sum_{i=1}^n b_i = 3$, we can construct a valid inequality for CUT_n in the following way. Let $\{1, 2, \ldots, p\}$ denote the set of indices i for which b_i is positive and suppose that $3 \leq p \leq n-1$. Let C denote a circuit with node set $\{1, \ldots, p\}$. Then, the inequality:

$$\sum_{1 \leq i < j \leq n} b_i b_j x_{ij} - \sum_{ij \in E(C)} x_{ij} \leq 0 \tag{29.0.1}$$

is valid for CUT_n. (Indeed, the value of the left hand side of (29.0.1) at a cut vector $\delta(S)$ is $b(S)(3 - b(S)) - |\delta(S) \cap E(C)|$, which is nonpositive if $b(S) \leq 0$ or if $b(S) \geq 3$; if $b(S) = 1, 2$, then $b(S)(3-b(S)) = 2$ and $1 \leq |S \cap \{1, \ldots, p\}| \leq p-1$ which implies that $|\delta(S) \cap E(C)| \geq 2$.) In fact, the inequality (29.0.1) remains valid if we replace the circuit C by an arbitrary 2-edge connected graph with node set $\{1, \ldots, p\}$. However, we will consider here only the case of a circuit. It is an open problem to characterize the 2-edge connected graphs for which (29.0.1) is facet inducing for CUT_n.

We describe in this chapter how to construct, more generally, a valid inequality for CUT_n for any $b \in \mathbb{Z}^n$ with $\sum_{i=1}^n b_i = 2r+1$ ($r \geq 1$). We present the class of clique-web inequalities $(\mathrm{CW}_n^r)^T x \leq 0$. They are of the form (29.0.1), with the circuit C being replaced by a more complicated graph, namely, a weighted antiweb. We will see later in Section 30.1 the class of suspended-tree inequalities, where the circuit C is replaced by a suspended-tree.

29.1 Pure Clique-Web Inequalities

We first introduce the clique-web inequality $\mathrm{CW}_n^r(b)^T x \leq 0$ in its pure form, i.e., in the case when $|b_i| = 1$ for all i.

Definition 29.1.1. *Given integers p and r with $p \geq 2r+3$, the* antiweb AW_p^r, *with parameters p and r, is the graph with node set $V_p = \{1, 2, \ldots p\}$ whose edges are the pairs $(i, i+1), (i, i+2), \ldots (i, i+r)$ for $i \in V_p$ (the indices being taken modulo p). The* web W_p^r *is the complement in the complete graph K_p of the* *antiweb* AW_p^r.

The web W_7^2 and the antiweb AW_8^2 are shown below.

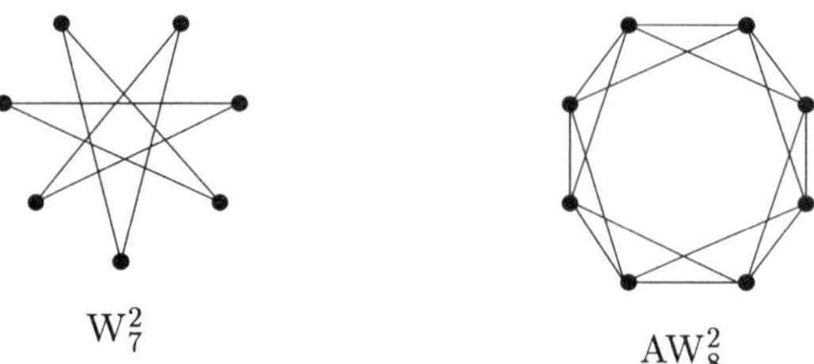

In the following we shall use the same notation AW_p^r (or W_p^r) for denoting the antiweb (or the web) graph or its edge set.

Definition 29.1.2. *Let n, p, q, r be integers satisfying*
(i) $n = p + q,\ p - q = 2r + 1,\ q \geq 2$ *or, equivalently,*
(ii) $p = \frac{n+1}{2} + r,\ q = \frac{n-1}{2} - r,\ 0 \leq r \leq \frac{n-5}{2}$.
The pure clique-web inequality $(\mathrm{CW}_n^r)^T x \leq 0$ *with parameters satisfying* (i) *is the inequality:*

$$\sum_{1 \leq i < j \leq n} b_i b_j x_{ij} - \sum_{ij \in \mathrm{AW}_p^r} x_{ij} \leq 0, \tag{29.1.3}$$

where $b := (1, \ldots 1, -1, \ldots, -1)$ *with first p coefficients equal to $+1$ and last $q = n - p$ coefficients equal to -1.*

The inequality (29.1.3) can also be written as

$$\sum_{ij \in W_p^r} x_{ij} + \sum_{p+1 \leq i < j \leq n} x_{ij} - \sum_{\substack{1 \leq i \leq p \\ p+1 \leq j \leq n}} x_{ij} \leq 0.$$

Hence, there is a web on the first p nodes (those for which $b_i = +1$) and a clique on the last q nodes (those for which $b_i = -1$), thus justifying the terminology "clique-web" inequality. We restrict our attention to the case $q \geq 2$, because inequality (29.1.3) in the case $q = 1$ takes the form:

$$-\sum_{1 \leq i \leq n-1} x_{in} + \sum_{1 \leq i \leq r+1} x_{i,r+i+1} = \sum_{1 \leq i \leq r+1} (x_{i,r+i+1} - x_{in} - x_{r+i+1,n}) \leq 0;$$

hence, it is a sum of $r + 1$ triangle inequalities, i.e., except when $r = 0$, it is not facet defining.

As an example, Figure 29.1.4 shows the support graph of $(\mathrm{CW}_{11}^2)^T x \leq 0$ (edges with weight 1 are indicated by a plain line and edges with weight -1 by a dotted line, each node of the triangle being joined to each node of the web).

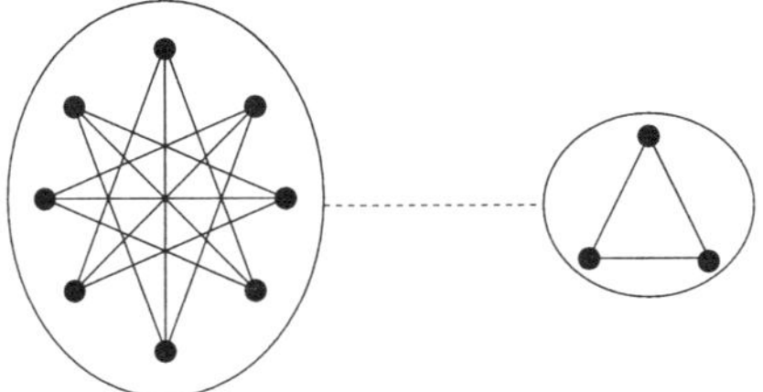

Figure 29.1.4: $(\mathrm{CW}_{11}^{2})^T x \leq 0$

Note that switching the inequality $(\mathrm{CW}_n^r)^T x \leq 0$ by the cut $\delta(\{1, 2, \ldots p\})$ yields the following inequality:

$$(29.1.5) \qquad ((\mathrm{CW}_n^r)^{\delta(\{1,\ldots,p\})})^T x := \sum_{ij \in W_p^r} x_{ij} + \sum_{p+1 \leq i < j \leq n} x_{ij} + \sum_{\substack{1 \leq i \leq p \\ p+1 \leq j \leq n}} x_{ij} \leq pq.$$

In the case when $q = 2$ (i.e., $r = \frac{n-5}{2}$, $p = 2r + 3$), the inequality (29.1.5) is also called a *bicycle odd wheel inequality*; it was introduced by Barahona, Grötschel and Mahjoub [1985] and Barahona and Mahjoub [1986]. Note that the web W_{2r+3}^r is a circuit. Figure 29.1.6 shows the graph supporting the bicycle odd wheel inequality on 7 points, i.e., the inequality (29.1.5) for $n = 2r+5$ and $r = 1$.

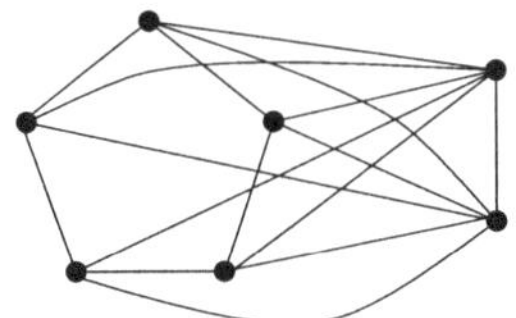

Figure 29.1.6: The bicycle odd wheel inequality on 7 points

Let us examine some special cases of the clique-web inequality $(\mathrm{CW}_n^r)^T x \leq 0$.

- For $r = 0$, the antiweb AW_p^0 is the empty graph. Therefore, the clique-web inequality $(\mathrm{CW}_n^0)^T x \leq 0$ coincides with the pure hypermetric inequality.
- For $r = 1$, the antiweb AW_p^1 is a circuit. The clique-web inequality $(\mathrm{CW}_n^1)^T x \leq 0$ is the inequality (29.0.1), which was introduced in Deza and Laurent [1992a] under the name of "cycle inequality ".
- For $r = \frac{n-5}{2}$ (the largest possible value of r), i.e., $p = 2r + 3$, or $q = 2$, the web W_{2r+3}^r is a circuit. The clique-web inequality $(\mathrm{CW}_{2r+5}^r)^T x \leq 0$ coincides (up to switching) with the bicycle odd wheel inequality.

Actually, it is the inspection of the above three cases that led us in Deza and Laurent [1992a] to the general definition of the clique-web inequalities.

29.2 General Clique-Web Inequalities

In order to define the clique-web inequality $\mathrm{CW}_n^r(b)^T x \leq 0$ for an arbitrary vector $b \in \mathbb{Z}^n$, we have to use the collapsing operation (described in Section 26.4).

Let $b = (b_1, \dots b_n)$ be integers such that $\sum_{i=1}^n b_i = 2r+1$ and suppose that

$$b_1, b_2, \dots, b_p > 0 > b_{p+1}, b_{p+2}, \dots, b_n$$

for some $2 \leq p \leq n-1$. Set $N := \sum_{i=1}^n |b_i|$ and $P := \sum_{i=1}^p b_i = \sum_{i|b_i>0} b_i$. Let $\pi(b)$ denote the partition of $V_N = \{1, 2, \dots, N\}$ into the following n classes: the p intervals $I_0 = [1, b_1]$ and $I_i = [b_1 + \dots + b_i + 1, b_1 + \dots + b_i + b_{i+1}]$ for $i = 1, \dots, p-1$ (which partition $[1, P]$) together with $n-p$ arbitrary subsets forming a partition of the set $[P+1, N] = V_N \setminus [1, P]$ and with respective sizes $|b_{p+1}|, \dots |b_n|$.

Definition 29.2.1. *Given integers $b = (b_1, \dots b_n)$, $r \geq 0$, such that $\sum_{i=1}^n b_i = 2r+1$ and setting $N := \sum_{1 \leq i \leq n} |b_i|$, the* clique-web inequality $\mathrm{CW}_n^r(b)^T x \leq 0$ *is defined as the $\pi(b)$-collapse of the (pure) clique-web inequality $(\mathrm{CW}_N^r)^T x \leq 0$.*

The clique-web inequality $\mathrm{CW}_n^r(b)^T x \leq 0$ can be described in a more explicit way using the notion of weighted antiweb. A weighted antiweb is an edge weighted graph obtained by collapsing of a (usual) antiweb. Let $b_1, \dots b_p$ be positive integers such that $\sum_{i=1}^p b_i \geq 2r+1$. Set $P := \sum_{i=1}^p b_i$ and consider the partition $\pi_0(b_1, \dots, b_p)$ of $V_P = \{1, \dots, P\}$ into the p intervals $I_0, I_1, \dots, I_p$ described above.

Definition 29.2.2. *With the above notation, the antiweb $\mathrm{AW}_p^r(b_1, \dots b_p)$ is the weighted graph obtained by $\pi_0(b_1, \dots b_p)$-collapsing the antiweb AW_P^r.*

Then the clique-web inequality $\mathrm{CW}_n^r(b)^T x \leq 0$ can be alternatively described as follows:

$$\text{(29.2.3)} \qquad \mathrm{CW}_n^r(b)^T x := \sum_{1 \leq i < j \leq n} b_i b_j x_{ij} - \sum_{ij \in \mathrm{AW}_p^r(b_1, \dots b_p)} x_{ij} \leq 0.$$

In relation (29.2.3), the quantity $\sum_{ij \in \mathrm{AW}_p^r(b_1, \dots, b_p)} x_{ij}$ should be understood as the sum $\sum_{1 \leq i < j \leq p} v_{ij} x_{ij}$, where v denotes the edgeweight vector of $\mathrm{AW}_p^r(b_1, \dots b_p)$. In the pure case, i.e., when $|b_i| = 1$ for all i, then the inequalities (29.1.3) and (29.2.3) coincide, i.e., $\mathrm{CW}_n^r(1, \dots 1, -1, \dots -1)^T x \leq 0$ and $(\mathrm{CW}_n^r)^T x \leq 0$ coincide.

Let us give some examples of weighted antiwebs.

Lemma 29.2.4. *Assume $b_i \geq r$ for $i = 1, \dots, p$. Then, the antiweb $\mathrm{AW}_p^r(b_1, \dots b_p)$ is $\frac{r(r+1)}{2} \mathrm{AW}_p^1$, i.e., it is the circuit $C(1, 2, \dots p)$ with weight $\frac{r(r+1)}{2}$ on its edges.* ∎

Lemma 29.2.5. *The antiweb* $\mathrm{AW}_p^r(2,1,\ldots,1)$ *is the weighted graph obtained from* AW_p^r *by*

(i) *deleting the edges* $(p-i, r-i)$ *for* $i = 0, 1, \ldots, r-2$, *and*

(ii) *assigning weight* 2 *to the edges* $(1, i)$ *and* $(1, p-r+i)$ *for* $i = 2, \ldots, r$. ∎

Figures 29.2.6, 29.2.7, and 29.2.8 show the antiwebs AW_8^2, $\mathrm{AW}_4^2(2,2,2,2)$, and $\mathrm{AW}_{10}^3(2,1,1,1,1,1,1,1,1,1)$. We picture in Figure 4.5 the support graph of the inequality $\mathrm{CW}_7^2(2,2,2,2,-1,-1,-1)^T x \leq 0$.

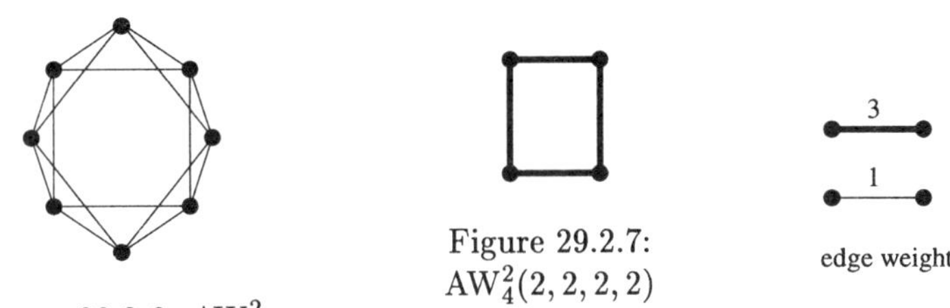

Figure 29.2.6: AW_8^2

Figure 29.2.7: $\mathrm{AW}_4^2(2,2,2,2)$

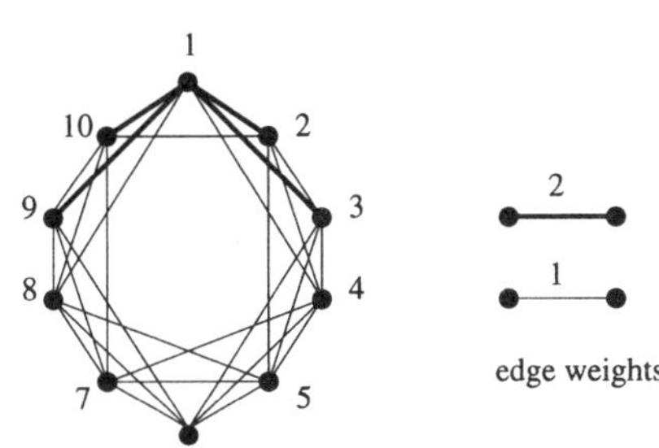

Figure 29.2.8: $\mathrm{AW}_{10}^3(2,1,1,1,1,1,1,1,1,1)$

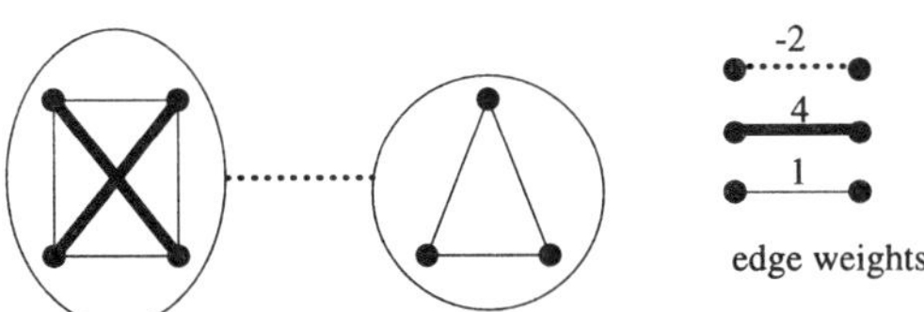

Figure 29.2.9: $\mathrm{CW}_7^2(2,2,2,2,-1,-1,-1)^T x \leq 0$

29.3 Clique-Web Inequalities: Validity and Roots

In this section, we show that the clique-web inequality $\mathrm{CW}_n^r(b)^T x \leq 0$ is valid for the cut cone CUT_n. Since the clique-web inequality $\mathrm{CW}_n^r(b)^T x \leq 0$ is defined as collapsing of a pure clique-web inequality $(\mathrm{CW}_N^r)^T x \leq 0$ ($N = \sum |b_i|$), it is sufficient to prove validity in the pure case (recall Lemma 26.4.1). We also describe the roots of the pure clique-web inequality. Validity of the pure clique-web inequality was established by Alon [1990] and the roots were described by Deza and Laurent [1992b].

Proposition 29.3.1. *Let p, r be integers such that $p \geq 2r+3$, $r \geq 1$ and let S be a subset of $\{1, \ldots, p\}$ of cardinality s.*

(i) *If $s \leq r$, then $|\delta_{K_p}(S) \cap \mathrm{AW}_p^r| \geq s(2r+1-s)$, with equality if and only if S induces a clique in AW_p^r, i.e., if any two nodes of S are adjacent in the graph AW_p^r.*

(ii) *If $r+1 \leq s \leq \frac{p}{2}$, then $|\delta_{K_p}(S) \cap \mathrm{AW}_p^r| \geq r(r+1)$, with equality if and only if S is an interval of $[1,p]$, i.e., if $S = \{i, i+1, i+2, \ldots, i+s-1\}$ for some $i \in \{1, \ldots, p\}$ (the indices being taken modulo p).*

Proof. The proof is by induction on $p \geq 2r+3$. Let us first prove the result for $p = 2r+3$. Then, $\mathrm{AW}_p^r = K_p \setminus C$, where C is the circuit with node set $[1,p]$ and with edges the pairs $(i, i+r+1)$ for $i = 1, \ldots, p$, where the indices are taken modulo p; so, $C = (1, r+2, p, r+1, p-1, r, p-2, r-1, \ldots, 2, r+3)$. Let S be a subset of $\{1, \ldots, p\}$. Then,

$$|\delta_{K_p}(S) \cap \mathrm{AW}_p^r| = s(2r+3-s) - |\delta_{K_p}(S) \cap C| \geq s(2r+1-s),$$

since $|\delta_{K_p}(S) \cap C| \leq 2s$. Moreover, the equality: $|\delta_{K_p}(S) \cap \mathrm{AW}_p^r| = s(2r+1-s)$ holds if and only if $|\delta_{K_p}(S) \cap C| = 2s$, i.e., if no two nodes of S are adjacent on C; this means that any two nodes of S are adjacent on AW_p^r, i.e., that S induces a clique in AW_p^r. When $r+1 \leq s \leq \frac{p}{2}$, i.e., $s = r+1$, it is easy to see that S induces a clique of AW_p^r if and only if S is an interval. Hence, Proposition 29.3.1 holds in the case $p = 2r+3$.
Let us now suppose that Proposition 29.3.1 holds for some given $p \geq 2r+3$. We show that it also holds for $p+1$. We first establish a connection between the graphs AW_p^r and AW_{p+1}^r. We suppose that AW_p^r and AW_{p+1}^r are defined, respectively, on the node sets $\{1, 2, \ldots, p\}$ and $\{1, 2, \ldots, p, p+1\}$ (with the nodes being arranged in that cyclic order; so, the new node $p+1$ is inserted between the two nodes p and 1 on the circuit $(1, 2, \ldots, p)$). Then,

- The edges belonging to AW_p^r, but not to AW_{p+1}^r, are the pairs $(i, p-r+i)$ for $i = 1, 2, \ldots, r$.
- The edges belonging to AW_{p+1}^r, but not to AW_p^r, are the pairs $(i, p+1)$ and $(p+1, p-r+i)$ for $i = 1, 2, \ldots, r$.

Take a subset S of $\{1, \ldots, p, p+1\}$ and set $T := S \setminus \{p+1\}$. Suppose that $s := |S| \leq \frac{p+1}{2}$. In order to make the notation easier let us denote by v the incidence vector of the cut $\delta_{K_{p+1}}(S)$. Then,

$$\begin{aligned} |\delta_{K_{p+1}}(S) \cap \mathrm{AW}^r_{p+1}| &= |\delta_{K_p}(T) \cap \mathrm{AW}^r_p| + \sum_{ij \in \mathrm{AW}^r_{p+1} \setminus \mathrm{AW}^r_p} v_{ij} - \sum_{ij \in \mathrm{AW}^r_p \setminus \mathrm{AW}^r_{p+1}} v_{ij} \\ &= |\delta_{K_p}(T) \cap \mathrm{AW}^r_p| + \sum_{1 \leq i \leq r} (v_{i,p+1} + v_{p+1,p-r+i} - v_{i,p-r+i}) \\ &= |\delta_{K_p}(T) \cap \mathrm{AW}^r_p| + \sum_{i=1}^{r} \Delta_i \geq |\delta_{K_p}(T) \cap \mathrm{AW}^r_p|, \end{aligned}$$

after setting

$$\Delta_i := v_{i,p+1} + v_{p+1,p-r+i} - v_{i,p-r+i}$$

for all $i = 1, \ldots, r$. (Note that $\Delta_i \geq 0$ for all i, by the validity of the triangle inequalities.) We distinguish two cases depending whether the node $p+1$ belongs to the set S or not.

Case 1: $p+1 \notin S$, i.e., $S = T$. Then, by the induction assumption, the inequalities from Proposition 29.3.1 hold. We now consider the equality case.

Suppose first that $s \leq r$ and $|\delta_{K_{p+1}}(S) \cap \mathrm{AW}^r_{p+1}| = s(2r+1-s)$. Then, $|\delta_{K_p}(S) \cap \mathrm{AW}^r_p| = s(2r+1-s)$ and $\Delta_i = 0$ for all $i = 1, \ldots, r$. By the induction assumption, the set S induces a clique in AW^r_p. Then, S also induces a clique in AW^r_{p+1}. Indeed, suppose that there exist $i, j \in S$ such that $ij \notin \mathrm{AW}^r_{p+1}$. Then, $ij \in \mathrm{AW}^r_p \setminus \mathrm{AW}^r_{p+1}$ and, thus, ij is of the form $(i, p-r+i)$ for some $i \in \{1, \ldots, r\}$. This implies that $\Delta_i = 2$ (since $p+1 \notin S$), yielding a contradiction.

Suppose now that $r+1 \leq s \leq \frac{p+1}{2}$ and that $|\delta_{K_{p+1}}(S) \cap \mathrm{AW}^r_{p+1}| = r(r+1)$. Then, $|\delta_{K_p}(S) \cap \mathrm{AW}^r_p| = r(r+1)$ and $\Delta_i = 0$ for all $i = 1, \ldots, r$. By the induction assumption, the set S is an interval in $[1,p]$. If the pair $\{1,p\}$ is not contained in S, then S is still an interval in $[1,p+1]$. Suppose that both nodes $1,p$ belong to S. Then, $S = [p-x+1,p] \cup [1,y]$ for some integers $x, y \geq 1$ with $x+y \geq r+1$. If $x \geq r$, then $p-r+1, 1 \in S$, implying that $\Delta_1 = 2$. If $y \geq r$, then $p, r \in S$, implying that $\Delta_r = 2$. If $x \leq y \leq r-1$, then $p-x+1, x \in S$, implying that $\Delta_x = 2$ and, finally, if $y \leq x \leq r-1$, then $p-y+1, y \in S$ implying that $\Delta_y = 2$. In all cases, we obtain a contradiction. Therefore, the pair $\{1,p\}$ is not contained in S and, thus, S is an interval in $[1,p+1]$.

Case 2: $p+1 \in S$; so, $|T| = s-1$. Suppose first that $s \leq r$. Denote by K the set of indices $i \in [1,r]$ for which $\Delta_i = 0$; set $k := |K|$. Then, $|\delta_{K_p}(T) \cap \mathrm{AW}^r_p| \geq (s-1)(2r+1-(s-1))$ by the induction assumption. Moreover, $\Delta_i = 2$ for all $i \in [1,r] \setminus K$. As $S \cap \{i, p-r+i\} \neq \emptyset$ for all $i \in K$, we obtain that $s \geq 1+k$. Therefore,

$$\begin{aligned} |\delta_{K_{p+1}}(S) \cap \mathrm{AW}^r_{p+1}| &\geq (s-1)(2r+1-(s-1)) + 2(r-k) \\ &\geq (s-1)(2r+1-(s-1)) + 2(r-s+1) = s(2r+1-s). \end{aligned}$$

This shows that the inequality from Proposition 29.3.1 (i) holds. Suppose now that equality: $|\delta_{K_{p+1}}(S) \cap \mathrm{AW}^r_{p+1}| = s(2r+1-s)$ holds. Then, $|\delta_{K_p}(T) \cap \mathrm{AW}^r_p| =$

$(s-1)(2r+1-(s-1))$ and, thus, by the induction assumption, T induces a clique in AW_p^r. Moreover, $s = 1+k$. This implies that $|S \cap \{i, p-r+i\}| = 1$ for all $i \in K$. Therefore, $S \cap \{i, p-r+i\} = \emptyset$ for $i \in [1,r] \backslash K$ and $S \cap [r+1, p-r] = \emptyset$. We show that S induces a clique in AW_{p+1}^r. Note first that the node $p+1$ is adjacent to all other nodes from S in AW_{p+1}^r. Suppose that $i \neq j \in S$ are not adjacent in AW_{p+1}^r. Then, $i, j \in T$ and, thus, ij is an edge from $\mathrm{AW}_{p+1}^r \backslash \mathrm{AW}_p^r$. Hence, $j = p-r+i$ and, thus, the set S contains both nodes i and $p-r+i$, yielding a contradiction.

Suppose now that $s \geq r+2$. Then, $|\delta_{K_p}(T) \cap \mathrm{AW}_p^r| \geq r(r+1)$ (by the induction assumption, as $|T| \geq r+1$). Hence, the inequality from Proposition 29.3.1 (ii) holds. Moreover, if $|\delta_{K_{p+1}}(S) \cap \mathrm{AW}_{p+1}^r| = r(r+1)$, then $|\delta_{K_p}(T) \cap \mathrm{AW}_p^r| = r(r+1)$ and $\Delta_i = 0$ for all $i \in [1,r]$. By the induction assumption, the set T is an interval in $[1,p]$. If $1 \in S$ or if $p \in S$, then the set S remains an interval in $[1, p+1]$. Suppose that $1, p \notin S$. Then, $T := [x,y]$ for some $2 \leq x < y \leq p-1$. For each $i \in [1,r]$, we have $|S \cap \{i, p-r+i\}| \geq 1$ as $\Delta_i = 0$. Hence, $r \in S$ since $p \notin S$, which implies that $x \leq r$. Moreover, $p-r+x-1 \in S$ since $x-1 \notin S$, which yields $p-r+x-1 \leq y$, i.e., $|T| = y-x+1 \geq p-r$. Therefore, $s \geq p-r+1$, which contradicts the fact that $s \leq \frac{p+1}{2}$.

Let us finally suppose that $s = r+1$. Then, $|\delta_{K_p}(T) \cap \mathrm{AW}_p^r| \geq r(r+1)$ (by the induction assumption), implying that $|\delta_{K_{p+1}}(S) \cap \mathrm{AW}_{p+1}^r| \geq r(r+1)$, i.e., the inequality from Proposition 29.3.1 (ii) holds. Suppose now that $|\delta_{K_{p+1}}(S) \cap \mathrm{AW}_{p+1}^r| = r(r+1)$. Then, $|\delta_{K_p}(T) \cap \mathrm{AW}_p^r| = r(r+1)$ and $\Delta_i = 0$ for all $i \in [1,r]$. Therefore, the set T induces a clique in AW_p^r. Moreover, $|S \cap \{i, p-r+i\}| = 1$ for all $i \in [1,r]$ and $S \cap [r+1, p-r] = \emptyset$. We show that S is an interval in $[1, p+1]$. If $i, j \in [1,r]$ are such that $i \in S$ and $j \notin S$, then $p-r+j \in S$, implying that $j \geq i$ (as $p-r+j$ should be adjacent to i in AW_p^r). Hence, $S \cap [1,r]$ is of the form $[1,x]$ for some $0 \leq x \leq r$. Therefore, $S = \{p+1\} \cup [1,x] \cup [p-r+x+1, p]$ is indeed an interval in $[1, p+1]$. ∎

The next results establish validity of the inequality $(\mathrm{CW}_n^r)^T x \leq 0$ and describe its roots. Given $n = p+q$ with $p-q = 2r+1$ and $q \geq 2$, the clique-web inequality $(\mathrm{CW}_n^r)^T x \leq 0$ is as defined in Definition 29.1.2.

Proposition 29.3.2. *The clique-web inequality $(\mathrm{CW}_n^r)^T x \leq 0$ is valid for* CUT_n.

Proposition 29.3.3. *Assume $r \geq 1$. The roots of $(\mathrm{CW}_n^r)^T x \leq 0$ are the cut vectors $\delta(S)$ where $S = S_+ \cup S_-$ with $S_+ \subseteq \{1, 2, \ldots, p\}$ and $S_- \subseteq \{p+1, \ldots, n\}$ and S is of one of the following two types:*

- (i) (type 1) *$S_- = \emptyset$ and S_+ induces a clique in* AW_p^r.
- (ii) (type 2) *S_+ is an interval of $[1,p]$ with $r+1 \leq |S_+| \leq p-r$ and $|S_+| - |S_-| =$* $r, r+1$.

Remark 29.3.4. Recall that, when we say that S_+ is an interval of $[1,p]$, we mean that $S_+ = \{i, i+1, \ldots, i+|S_+|-1\}$ (for some $i \in \{1, \ldots, p\}$) where the indices are taken modulo p. There are some redundancies in the presentation of the roots given in Proposition 29.3.3. It is easy to see that a nonredundant description of the roots (i.e., in which each root occurs exactly once) can be obtained by replacing in Proposition 29.3.3 the family of sets of type 2 by the family of sets of type 2′ where

(type 2′) S_+ is an interval of $[1,p]$ with $r+1 \leq |S_+| \leq p-r-2$ and $|S_+|-|S_-| = r$.

Note also that, if $S \subseteq \{1,\ldots,p\}$ induces a clique in AW_p^r, then $|S| \leq r+1$ holds. For example, the interval $[1,r+1] = \{1,2\ldots,r,r+1\}$ induces a clique in AW_p^r; therefore, any subset S of the following type 1* induces a clique in AW_p^r:

(type 1*) S is contained in an interval of size $r+1$ of $[1,p]$.

In general, there may exist other sets than those of type 1* inducing a clique in AW_p^r. For instance, for $r=3$ and $p=9$, the set $\{1,4,7\}$ induces a clique in AW_9^3. However, for $p > 3k$, the only sets inducing cliques in AW_p^r are those of type 1*. Actually, in all our proofs for clique-web facets, we shall only use roots $\delta(S)$ in which S is of type 1* or of type 2. ∎

Proof of Propositions 29.3.2 and 29.3.3. Take a subset S of $\{1,\ldots,n\}$. Then, $S = S_+ \cup S_-$, where $S_+ \subseteq \{1,\ldots,p\}$ and $S_- \subseteq \{p+1,\ldots,n\}$. Set $s := |S|$, $s_+ := |S_+|$, $s_- := |S_-|$; so, $s = s_+ + s_-$. Then,

$$\begin{aligned}(\mathrm{CW}_n^r)^T\delta(S) &= b(S)(2r+1-b(S)) - |\delta_{K_p}(S)\cap \mathrm{AW}_p^r| \\ &= (s_+ - s_-)(2r+1-(s_+ - s_-)) - |\delta_{K_p}(S)\cap \mathrm{AW}_p^r|.\end{aligned}$$

Suppose that $s_+ \leq r$. Then,

$$(\mathrm{CW}_n^r)^T\delta(S) \leq (s_+ - s_-)(2r+1-(s_+ - s_-)) - s_+(2r+1-s_+) \leq 0.$$

The first inequality follows from Proposition 29.3.1 (i) and the second one from the fact that the mapping $x \mapsto x(2r+1-x)$ is monotone nondecreasing for $x \leq r$. Therefore, if $\delta(S)$ is a root of $(\mathrm{CW}_n^r)^Tx \leq 0$, then $s_- = 0$ and, by Proposition 29.3.1 (i), S_+ induces a clique in AW_p^r. Suppose now that $r+1 \leq s_+ \leq \frac{p}{2}$. Then,

$$(\mathrm{CW}_n^r)^T\delta(S) \leq (s_+ - s_-)(2r+1-(s_+ - s_-)) - r(r+1) \leq 0.$$

The first inequality follows from Proposition 29.3.1 (ii) and the second one from the fact that $x(2r+1-x) \leq r(r+1)$ for any integer x. Therefore, if $\delta(S)$ is a root of $(\mathrm{CW}_n^r)^Tx \leq 0$, then $s_+ - s_- = r, r+1$ and, by Proposition 29.3.1 (ii), S_+ is an interval. This concludes the proof. ∎

How to find the roots of a general clique-web inequality $\mathrm{CW}_n^r(b)^Tx \leq 0$? They can be deduced from the roots in the pure case, in the manner described in Lemma 26.4.1 (ii). Let us, as an example, describe the roots of two clique-web inequalities, namely, of $\mathrm{CW}_n^r(b)^Tx \leq 0$ where $b_i \geq r$ if $b_i > 0$, and of $\mathrm{CW}_n^r(2,1,\ldots,1,-1,\ldots,-1)^Tx \leq 0$.

Proposition 29.3.5. *Suppose that $b_i \geq r$ for $i \in \{1,\ldots,p\}$ and $b_i < 0$ for $i \in \{p+1,\ldots,n\}$. Then, the roots of $\mathrm{CW}_n^r(b)^T x \leq 0$ are the cut vectors $\delta(S)$ for which $S \cap [1,p]$ is an interval of $[1,p]$ and $b(S) = r, r+1$.* ∎

Proposition 29.3.6. *The roots of $\mathrm{CW}_n^r(2,1,\ldots,1,-1,\ldots,-1)^T x \leq 0$ (where there are $p-1$ coefficients $+1$ and $n-p$ coefficients -1) are the cut vectors $\delta(S)$ where $S = S_+ \cup S_-$ with $S_+ \subseteq \{1,2,\ldots,p\}$ and $S_- \subseteq \{p+1,\ldots,n\}$ satisfying* (i) *or* (ii).

(i) *$S_- = \emptyset$, $1 \notin S$ (resp. $1 \in S$) and S (resp. $S \cup \{p+1\}$) induces a clique in AW_{p+1}^r, where AW_{p+1}^r is the antiweb defined on the $p+1$ nodes $1,2,\ldots,p,p+1$, taken in that circular order.*

(ii) *S_+ is an interval of $[1,p]$ and*
- *either, $1 \notin S_+$, $r+1 \leq |S_+| \leq p+1-r$ and $|S_+| - |S_-| = r, r+1$,*
- *or, $1 \in S_+$, $r \leq |S_+| \leq p-r$ and $|S_+| - |S_-| = r, r-1$.* ∎

In the case $r = 1$, AW_p^1 is just a circuit of length p and any interval collapsing (i.e., we collapse only consecutive nodes on the circuit) of a circuit is again a circuit. Therefore, for $r = 1$, the clique-web inequality $\mathrm{CW}_n^1(b)^T x \leq 0$ takes a particularly easy form; namely, if $b_1,\ldots,b_p > 0$, $b_{p+1},\ldots,b_n < 0$, then

$$\mathrm{CW}_n^1(b)^T x := \sum_{1 \leq i < j \leq n} b_i b_j x_{ij} - \left(\sum_{1 \leq i \leq p-1} x_{i,i+1} + x_{1p} \right) \leq 0.$$

In particular, for all roots $\delta(S)$ of $\mathrm{CW}_n^1(b)^T x \leq 0$, $S \cap [1,p]$ must be an interval. Observe also that, for any integers $r \geq 1$, $b_1,\ldots,b_p > 0 > b_{p+1},\ldots,b_n$, the inequality

$$\sum_{1 \leq i < j \leq n} b_i b_j x_{ij} - \frac{r(r+1)}{2} \left(\sum_{1 \leq i \leq p-1} x_{i,i+1} + x_{1p} \right) \leq 0 \qquad (29.3.7)$$

is valid for CUT_n. When $b_1,\ldots,b_p \geq r$, as we noted in Lemma 29.2.4, the clique-web inequality $\mathrm{CW}_n^r(b)^T x \leq 0$ is indeed of the form (29.3.7). However, if $b_i < r$ for some $i \in \{1,\ldots,p\}$, then the inequality (29.3.7) is not facet inducing, since it is dominated by $\mathrm{CW}_n^r(b)^T x \leq 0$. (Indeed, one can easily check that

$$|\delta_{K_p}(S) \cap \mathrm{AW}_p^r(b_1,\ldots,b_p)| \leq \frac{r(r+1)}{2} |\delta_{K_p}(S) \cap C|$$

holds for any subset S of $\{1,\ldots,p\}$, where C denotes the circuit $(1,2,\ldots,p)$ (using the definition of collapsing and decomposing S as a union of intervals of $[1,p]$; see Lemma 2.8 in Deza, Grötschel and Laurent [1992]).)

As an easy consequence of the description of the roots of $(\mathrm{CW}_n^r)^T x \leq 0$, we have the following result.

Proposition 29.3.8. *Assume that $r \geq 1$. Then, $b(S) \in \{1,2,\ldots,r+1\}$ for every nonzero root $\delta(S)$ of $\mathrm{CW}_n^r(b)^T x \leq 0$.* ∎

29.4 Clique-Web Facets

In this section, we describe several classes of clique-web facets, i.e., several classes of parameters b for which the inequality $\mathrm{CW}_n^r(b)^T x \leq 0$ is facet inducing. The full characterization of all clique-web facets seems a very hard problem; actually, this problem is already not solved for the special case $r = 0$ of the hypermetric inequalities. We can restrict ourselves here to the case $r \geq 1$, since the case $r = 0$ has been considered in the preceding chapter. The following classes of clique-web facets are known:

- $(\mathrm{CW}_n^r)^T x \leq 0$, i.e., $\mathrm{CW}_n^r(1,\ldots,1,-1,\ldots,-1)^T x \leq 0$ (pure case),
- $\mathrm{CW}_n^r(r,\ldots,r,-1,\ldots,-1)^T x \leq 0$ for $r \geq 1$, $p \geq 5$,
- $\mathrm{CW}_n^r(b_1,\ldots,b_p,-1,\ldots,-1)^T x \leq 0$ with $b_1,\ldots,b_p \geq r$ and some additional conditions (see Theorem 29.4.4),
- $\mathrm{CW}_n^r(2,1,\ldots,1,-1,\ldots,-1)^T x \leq 0$ for $r \geq 1$, $p \geq 2r+3$,
- $\mathrm{CW}_n^r(b_1,\ldots,b_p,-1,\ldots,-1,-2\ldots,-2)^T x \leq 0$ for $b_1,\ldots,b_p \geq r$ and some additional conditions (see Theorem 29.4.6).

Here are three examples of clique-web facets[1] $\mathrm{CW}_n^r(b)^T x \leq 0$ with $r = 1$:

- $\mathrm{CW}_7^1(3,2,2,-1,-1,-1,-1)^T x \leq 0$,
- $\mathrm{CW}_7^1(2,2,1,1,-1,-1,-1)^T x \leq 0$,
- $\mathrm{CW}_7^1(1,1,1,1,1,-1,-1)^T x \leq 0$.

For $n \leq 6$, all the facets of CUT_n are hypermetric and, for $n = 7$, the only clique-web facets of CUT_7 are either hypermetric or one of the above three facets. See Section 30.6 for a full description of the cut cone CUT_n for $n \leq 7$.

Let us start with mentioning a necessary condition for a clique-web inequality $\mathrm{CW}_n^r(b)^T x \leq 0$ to be facet defining. The next result generalizes Proposition 28.2.1, which corresponds to the case $r = 0$ of hypermetric inequalities.

Proposition 29.4.1. *Let $b \in \mathbb{Z}^n$ with $\sum_{i=1}^n b_i = 2r+1$, $r \geq 1$, and $b_1 \geq b_2 \geq \ldots \geq b_p > 0 > b_{p+1} \geq \ldots \geq b_n$. If*

$$b_1 + b_2 - b_n > \sum_{i=1}^{p} b_i - r$$

then the face F of CUT_n defined by the clique-web inequality $\mathrm{CW}_n^r(b)^T x \leq 0$ is contained in the facet defined by the triangle inequality: $x_{12} - x_{1n} - x_{2n} \leq 0$;

[1] These three clique-web facets were discovered by Assouad and Delorme [1982] (cf. Assouad [1984]). In fact, the first one had also been found earlier by Avis [1977] and (in the context of the correlation polytope) by McRae and Davidson [1972]. These facets demonstrate that the cut cone does have some facets that are not induced by hypermetric inequalities. It had, indeed, been speculated by several authors at various moments (e.g., by Davidson [1969], Deza [1973a], Pitowsky [1991]) that all facets of the cut cone are hypermetric.

hence, F is not a facet of CUT_n.

Proof. Let $\delta(S)$ be a nonzero root of $\mathrm{CW}_n^r(b)^T x \leq 0$. We can suppose without loss of generality that $n \notin S$; then, $b(S) \in \{1,2,\ldots,r+1\}$ by Proposition 29.3.8. Then, the set $\{1,2\}$ is not contained in S (else, $b(S) \geq b_1+b_2+\sum_{i=p+1}^{n-1} b_i \geq r+2$). This implies that $\delta(S)$ is also a root of the triangle inequality: $x_{12}-x_{1n}-x_{2n} \leq 0$. ∎

The inequality from Proposition 29.4.1 is best possible. For instance, the inequalities:

$$\mathrm{CW}_n^1(n-5,2,1,1,-1,\ldots,-1)^T x \leq 0,\ \mathrm{CW}_n^1(b_1,n-2-b_1,2,-1,\ldots,-1)^T x \leq 0$$

(where $2 \leq b_1 \leq n-4$) are facet defining; equality $b_1+b_2-b_n = \sum_{1\leq i\leq p} b_i - r$ holds for both of them (with $r=1$).

We now present several classes of clique-web facets. When considering a clique-web inequality $\mathrm{CW}_n^r(b)^T x \leq 0$, we shall always assume that b is an integer vector such that $\sum_{i=1}^n b_i = 2r+1$, even if this condition is not stated explicitly.

Theorem 29.4.2.

(i) $(\mathrm{CW}_n^r)^T x \leq 0$, *i.e., the pure clique-web inequality* (29.1.3), *defines a facet of* CUT_n *for any* $r \geq 0$.

(ii) $\mathrm{CW}_n^r(r,\ldots,r,-1,\ldots,-1)^T x \leq 0$ *is facet inducing for all* $r \geq 1$, $p \geq 5$.

(iii) $\mathrm{CW}_n^r(2,1,\ldots,1,-1,\ldots,-1)^T x \leq 0$ *is facet inducing for all* $r \geq 1$, $p \geq 2r+3$.

(iv) $\mathrm{CW}_n^r(r+2,r+1,r+1,-1,\ldots,-1)^T x \leq 0$ *is facet inducing for all* $r \geq 1$.

(v) $\mathrm{CW}_n^r(r+1,r+1,r,r,-1,\ldots,-1)^T x \leq 0$ *is facet inducing for all* $r \geq 1$. ∎

Theorem 29.4.2 (i)-(iii) is given in Deza and Laurent [1992b] and (iv)-(v) in Deza and Laurent [1992c]. We shall prove only the assertion (i). The proofs for (ii)-(v) are, indeed, along the same lines and with somewhat lengthy details. In order not to interrupt the text, we delay the proof of Theorem 29.4.2 (i) till Section 29.6.

Further classes of clique-web facets can be obtained by applying the lifting result from Theorem 29.4.3 below. This result follows as a direct application of Theorem 26.4.4, after noting that the two weighted antiwebs $\mathrm{AW}_p^r(b_1,b_2,\ldots,b_p)$ and $\mathrm{AW}_p^r(b'_1,b_2,\ldots,b_p)$ coincide if $b_1,b'_1 \geq r$. The class of facets from Theorem 29.4.4 is given in Deza and Laurent [1992c].

Theorem 29.4.3. *Let* $r \geq 1, b_1,\ldots,b_p > 0 > b_{p+1},\ldots,b_n$ *with* $\sum_{i=1}^n b_i = 2r+1$. *Suppose that, for some distinct* $j,k \in \{p+1,\ldots,n\}$, $b_j = b_k := d < 0$ *and that, for some* $i \in \{1,\ldots,n\} \setminus \{j,k\}$, *either* $b_i \leq d$ *or* $b_i \geq r$. *If the inequality:*

$$\mathrm{CW}_n^r(b_1,\ldots,b_n)^T x \leq 0$$

defines a facet of CUT_n*, then the inequality:*

$$\mathrm{CW}_{n+1}^r(b_1, \ldots, b_{i-1}, b_i - d, b_{i+1}, \ldots, b_n, d)^T x \leq 0$$

defines a facet of CUT_{n+1}. ∎

Theorem 29.4.4. *Given integers* $p \geq 3, r \geq 1, b_1, \ldots, b_p \geq r$*, the following assertions are equivalent.*

(i) $\mathrm{CW}_n^r(b_1, \ldots, b_p, -1, \ldots, -1)^T x \leq 0$ *is facet inducing.*

(ii) $p \geq 5$*, or* $p = 4$ *and* $b_1, b_2 \geq r+1$ *(up to a cyclic shift on* $[1,4]$*), or* $p = 3$ *and* $b_1 \geq r+2, b_2, b_3 \geq r+1$ *(up to a cyclic shift on* $[1,3]$*).*

Proof. The implication (ii) ⇒ (i) is proved by iteratively using the lifting result from Theorem 29.4.3 starting from the known facets from Theorem 29.4.2 (ii),(iv),(v). We now check the implication (i) ⇒ (ii). Suppose that the inequality $\mathrm{CW}_n^r(b_1, \ldots, b_p, -1 \ldots, -1)^T x \leq 0$ is facet inducing with $p = 3$ or 4. Consider first the case $p = 3$. We can suppose, for instance, that $b_1 \geq b_2 \geq b_3$ (up to a cyclic shift on $[1,3]$). From Proposition 29.4.1 below, we must have that $b_3 \geq r+1$. Suppose, for contradiction, that $b_1 = r+1$; so, $b_1 = b_2 = b_3 = r+1$. Then, every root of $\mathrm{CW}_n^r(r+1, r+1, r+1, -1, \ldots, -1)^T x \leq 0$ is a root of $\mathrm{CW}_n^0(1,1,1,-1,-1,0,\ldots,0)^T x \leq 0$, contradicting the fact that the former inequality is facet defining. This shows that $b_1 \geq r+2$. Consider now the case $p = 4$. One can check that the roots of the inequality $\mathrm{CW}_n^r(x, r, r, r, -1, \ldots, -1)^T x \leq 0$ (where $x \geq r$) and of the inequality $\mathrm{CW}_n^r(x, r, y, r, -1, \ldots, -1)^T x \leq 0$ (where $x, y \geq r+1$) are also roots of the inequality $\mathrm{CW}_n^0(1,0,1,0,-1,0,\ldots,0)^T x \leq 0$; hence, these two inequalities are not facet inducing. Therefore, up to a cyclic shift on $[1,4]$, we must have that $b_1, b_2 \geq r+1$. ∎

The following two further examples of lifting theorems for clique-web facets can be found in Deza and Laurent [1992c].

Theorem 29.4.5. *Given integers* $b_1, b_2, \ldots, b_p \geq r \geq 1$, $b_{p+1}, \ldots, b_n < 0$*, we assume that* (i) *and* (ii) *hold.*

(i) $\mathrm{CW}_n^r(b_1, \ldots, b_n)^T x \leq 0$ *is facet inducing for* CUT_n.

(ii) *There exist n subsets* $T_1, \ldots, T_n$ *of* $\{1, \ldots, n\}$ *such that* (iia)-(iic) *hold:*

- (iia) $T_j \cap \{1, \ldots, p\}$ *is an interval of* $\{1, \ldots, p\}$ *for all* $1 \leq j \leq n$,
- (iib) $b(T_j) = r+2$ *for all* $1 \leq j \leq n$,
- (iic) *the incidence vectors of the sets* $T_1, \ldots, T_n$ *are linearly independent.*

Then, $\mathrm{CW}_{n+1}^{r-1}(b_1, \ldots, b_n, -2)^T x \leq 0$ *defines a facet of* CUT_{n+1}. ∎

Theorem 29.4.6. *Given integers* $b_1, \ldots, b_p \geq r \geq s \geq 0$, $p \geq 5$*, the inequality:*

$$\mathrm{CW}_n^{r-s}(b_1, \ldots, b_p, -1, \ldots, -1, -2, \ldots, -2)^T x \leq 0$$

defines a facet of CUT_n. *(Here, there are* $q-s$ *components equal to* -1, s *components equal to* -2, *with* $q=n-p$ *and* $\sum_{i=1}^{p} b_i-(q-s)=2r+1$.) ∎

The proof of Theorem 29.4.6 is based on iterated applications of the lifting procedure from Theorem 29.4.5 starting from the known clique-web facet $\mathrm{CW}^r(b_1,\ldots,b_p,-1,\ldots,-1)^T x\leq 0$. The next result follows by applying Theorem 26.4.3.

Theorem 29.4.7. *Given integers* $b_1,\ldots,b_p>0>b_{p+1},\ldots,b_n$ *with* $p\geq 3$, *suppose that there exist two indices* $i,j\in\{1,\ldots,p\}$ *such that* $b_j=b_i-1$ *and* $j=i+1(\mathrm{mod}\ p)$, *e.g.,* $i=1, j=2$, *for simplicity in the notation. If the two inequalities:*

$$\mathrm{CW}_n^r(b_1,b_1-1,b_3,\ldots,b_n)^T x\leq 0 \quad \textit{and} \quad \mathrm{CW}_n^r(b_1-1,b_1,b_3,\ldots,b_n)^T x\leq 0$$

are facet inducing for CUT_n, *then the inequality:*

$$\mathrm{CW}_{n+1}^r(b_1-1,1,b_1-1,b_3,b_4,\ldots,b_n)^T x\leq 0$$

is facet inducing for CUT_{n+1}. ∎

This is about all the information we have about clique-web facets. Probably, some other classes of clique-web facets can be obtained by further applications of our various lifting theorems. However, it would be interesting to have some new proof techniques for constructing new classes of clique-web facets. We conclude this section with a few more remarks. The class of clique-web faces from Proposition 29.4.8 below is described in De Simone, Deza and Laurent [1994].

Proposition 29.4.8. *The inequality* $\mathrm{CW}_n^2(n-6,2,2,1,1,-1,\ldots,-1)^T x\leq 0$ *defines a face of* CUT_n *of dimension* $\binom{n}{2}-(n-4)$ *for all* $n\geq 8$. ∎

Consider the clique-web inequality $\mathrm{CW}_n^r(n-4,r+1,r+1,-1,\ldots,-1)^T x\leq 0$. It is facet inducing for $(r=0,n\geq 3)$ and for $(r\geq 1,n\geq r+6)$. Its number of roots is equal to $\binom{n-2}{r+2}+2n-4$. Therefore, it is a simplex facet only for $(r=0,n\geq 3)$ and $(r\geq 1,n=r+6)$; actually, for $r=0$, it is the only class of simplex hypermetric facets that we know. Note also that the number of roots is in $O(n^r)$, i.e., polynomial in n of arbitrary degree r.

Observe that the three clique-web facets defined by the inequalities:

$$\mathrm{CW}_n^1(n-4,2,2,-1,\ldots,-1)^T x\leq 0,\ \mathrm{CW}_n^1(n-5,2,1,1,-1,\ldots,-1)^T x\leq 0,$$

$$\mathrm{CW}_n^1(n-6,1,1,1,1,-1,\ldots,-1)^T x\leq 0$$

(for $n\geq 7$) all have the same number of roots, namely, $\binom{n-2}{3}+2n-4$. Therefore, these three facets are simplex facets for $n=7$ (recall that they are precisely the nonhypermetric clique-web facets of CUT_7).

Finally, let us mention a generalization of the clique-web inequalities for the multicut polytope. Let r and $b = (b_1, \ldots, b_n)$ be integers such that $\sigma := \sum_{i=1}^n b_i \geq 2r + 1$ and $b_1, \ldots, b_p > 0 > b_{p+1}, \ldots, b_n$. Then, the inequality:

$$\text{(29.4.9)} \qquad \sum_{1 \leq i < j \leq n} b_i b_j x_{ij} - \sum_{ij \in \text{AW}_p^r(b_1, \ldots, b_p)} x_{ij} \leq \frac{\sigma(\sigma - 2r - 1)}{2}$$

is valid for the multicut polytope $\text{MC}_n^\square$. It defines a facet of $\text{MC}_n^\square$ for several classes of parameters b; this is the case, for instance, if $|b_i| = 1$ for all $1 \leq i \leq n$. We refer to Deza, Grötschel and Laurent [1992] for a detailed exposition.

29.5 Separation of Clique-Web Inequalities

We address here the separation problem for clique-web inequalities. It is not known how to separate the whole class, but a polynomial algorithm is known for separating a small subclass.

Recall (see Definition 29.1.2) that the pure clique-web inequality $(\text{CW}_n^r)^T x \leq 0$ is specified by two parameters which can be n and r or, equivalently, q and r, with n being then given by $n = 2q + 2r + 1$. Gerards [1985] shows that the subclass consisting of the pure clique-web inequalities with $q = 2$ can be separated in polynomial time. Note that these inequalities can be rewritten as

$$\text{(29.5.1)} \qquad x_{uv} + \sum_{ij \in E(C)} x_{ij} - \sum_{i \in V(C)} (x_{iu} + x_{iv}) \leq 0,$$

where C is an odd circuit and u, v are two nodes that do not belong to C.

More precisely, given an integer N, let $\mathcal{S}_N^{(2)}$ denote the system consisting of the inequalities (29.5.1) (for C odd circuit with $V(C) \subseteq V_N = \{1, \ldots, N\}$ and $u, v \in V_N \setminus V(C)$) and the triangle inequalities: $x_{ij} - x_{ik} - x_{jk} \leq 0$ (for $i, j, k \in V_N$). The separation problem for the system $\mathcal{S}_N^{(2)}$ is the following:

> *Given a vector $x \in \mathbb{R}^N$, decide whether x satisfies all the inequalities from the system $\mathcal{S}_N^{(2)}$. If not, find an inequality in $\mathcal{S}_N^{(2)}$ which is violated by x.*

This problem can be solved in polynomial time (Gerards [1985]).

Proof. Pick two distinct elements $u, v \in V_N$. For $i, j \in V_N \setminus \{u, v\}$, set

$$\text{(29.5.2)} \qquad y_{ij} := -x_{ij} + \frac{1}{2}(x_{iu} + x_{iv} + x_{ju} + x_{jv}).$$

If C is a circuit with $V(C) \subseteq V_N \setminus \{u, v\}$, then

$$\sum_{ij \in E(C)} y_{ij} = - \sum_{ij \in E(C)} x_{ij} + \sum_{i \in V(C)} (x_{iu} + x_{iv}).$$

Hence, x satisfies the inequality (29.5.1) if and only if $\sum_{ij\in E(C)} y_{ij} \geq x_{uv}$ holds. So, one can solve the separation problem over $\mathcal{S}_N^{(2)}$ in the following way. First, check whether x satisfies all the triangle inequalities. If not, then a violated triangle inequality has been found. Otherwise, this shows that y is nonnegative. Then, one uses the polynomial algorithm from Grötschel and Pulleyblank [1981] for finding an odd cycle on $V_N \setminus \{u,v\}$ of minimum weight, with respect to the weights y (such a cycle will be, in fact, a circuit). Then, one verifies whether this minimum weight is greater than or equal to the constant x_{uv}. If not, then one has found a violated inequality. This computation is repeated for every choice of $u, v \in V_N$. ∎

In the same way, the class of bicycle odd wheel inequalities:

$$x_{uv} + \sum_{ij\in E(C)} x_{ij} + \sum_{i\in V(C)} (x_{iu} + x_{iv}) \leq 2|V(C)|$$

(where C is an odd circuit and $u, v \in V_N \setminus V(C)$) together with the triangle inequalities: $x_{ij} + x_{ik} + x_{jk} \leq 2$ $(i,j,k \in V_N)$ can be separated in polynomial time. (The proof is identical; it suffices to replace (29.5.2) by $y_{ij} := 2 - x_{ij} - \frac{1}{2}(x_{iu} + x_{iv} + x_{ju} + x_{jv})$.)

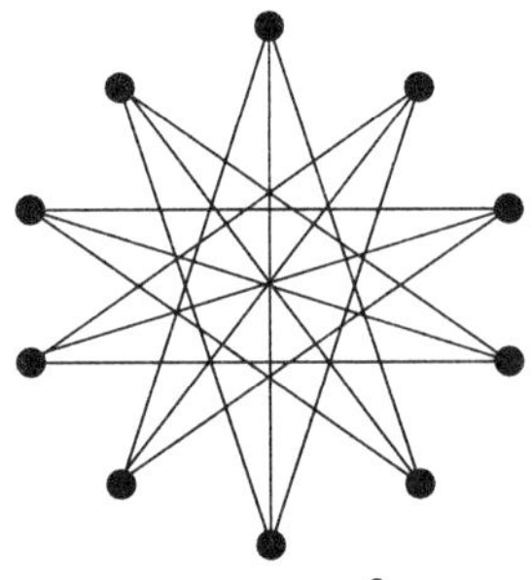

The web W_{10}^3

More generally, given an integer $q \geq 2$, let $\mathcal{S}_N^{(q)}$ denote the system consisting of the homogeneous triangle inequalities and of the inequalities:

$$\sum_{ij\in E(\mathcal{W})} x_{ij} + \sum_{ij\in E(K_Q)} x_{ij} - \sum_{i\in V(\mathcal{W}), j\in Q} x_{ij} \leq 0,$$

where $\mathcal{W}$ is a web W_{q+2r+1}^r on V_N and K_Q is a complete graph on $Q \subseteq V_N \setminus V(\mathcal{W})$ with $|Q| = q$. In order to separate the system $\mathcal{S}_N^{(q)}$, we can proceed in the same way as in the case $q = 2$ treated above. Namely, let Q be a subset of cardinality q of V_N and, for $i, j \in V_N \setminus Q$, set

$$y_{ij} := -x_{ij} + \frac{1}{q} \sum_{u\in Q} (x_{iu} + x_{ju}).$$

Then, y is nonnegative whenever x satisfies all the triangle inequalities. If $\mathcal{W}$ is a web on $q+2r+1$ nodes of $V_N \setminus Q$ with parameter r, then

$$\sum_{ij \in E(\mathcal{W})} y_{ij} = - \sum_{ij \in E(\mathcal{W})} x_{ij} + \sum_{i \in V(\mathcal{W}), u \in Q} x_{iu}.$$

Hence, if we can compute in polynomial time the minimum weight web (with respect to the weights y) on $q+2r+1$ nodes with parameter r, then we can solve the separation problem over $\mathcal{S}_N^{(q)}$ in polynomial time. Unfortunately, nothing is known about the minimum weight web problem if $q \neq 2$. For instance, for $q = 3$, the web W_{2r+4}^r is the circular graph on $2r+4$ nodes where each node i is adjacent to nodes $r+1+i$, $r+2+i$ and $r+3+i$, and we have the problem of finding such a graph structure of minimum weight. We show below the web W_{10}^3.

29.6 An Example of Proof for Clique-Web Facets

We give here the proof of Theorem 29.4.2 (i), i.e., we show that the pure clique-web inequality $(\mathrm{CW}_n^r)^T x \leq 0$ is facet inducing for CUT_n, if $r \geq 1$. (The case $r = 0$ was already treated; see, e.g., the observation after Corollary 26.6.2 for a proof.) We use the following two lemmas taken, respectively, from Barahona and Mahjoub [1986] and Barahona, Grötschel and Mahjoub [1985].

Lemma 29.6.1. *Let $a \in \mathbb{R}^{E_n}$, $i \neq j \in \{1, \ldots, n\}$, $S \subseteq \{1, \ldots, n\} \setminus \{i, j\}$ such that the cut vectors $\delta(S)$, $\delta(S \cup \{i\})$, $\delta(S \cup \{j\})$, $\delta(S \cup \{i, j\})$ satisfy the equality $a^T x = 0$. Then, $a_{ij} = 0$ holds.* ∎

Lemma 29.6.2. *Let $a \in \mathbb{R}^{E_n}$, I, J, H and S be disjoint subsets of $\{1, \ldots, n\}$ such that the cut vectors $\delta(S \cup J), \delta(S \cup H), \delta(S \cup I \cup J), \delta(S \cup I \cup H)$ satisfy the equality $a^T x = 0$. Then,*

$$\sum_{i \in I, j \in J} a_{ij} = \sum_{i \in I, h \in H} a_{ih}$$

holds. In particular, if $I = \{i\}, J = \{j\}, H = \{h\}$, then $a_{ij} = a_{ih}$ holds. ∎

In order to prove that $(\mathrm{CW}_n^r)^T x \leq 0$ is facet inducing for CUT_n, we consider a valid inequality $a^T x \leq 0$ for CUT_n such that $a(\delta(S)) = 0$ for all roots $\delta(S)$ of $(\mathrm{CW}_n^r)^T x \leq 0$. We show the existence of a scalar α such that $a^T x = \alpha (\mathrm{CW}_n^r)^T x$. For this, it suffices to prove the following statements:

(a) $a_{ij} = 0$ for all $ij \in \mathrm{AW}_p^r$.

(b) $a_{ij} = \alpha$ for all $ij \in K_p \setminus \mathrm{AW}_p^r$.

(c) $a_{i'j'} = \alpha$ for all $i'j' \in K_q$.

(d) $a_{ij'} = \alpha$ for all $i \in \{1, \ldots, p\}$, $j' \in [1', q']$.

We have chosen to denote the n nodes on which CW_n^r is defined by $[1,n] = [1,p] \cup [1',q']$ ($q = n-p$). We prove the statements (a)-(d) through the following Claims 29.6.3-29.6.8. We use the description of the roots of $(\mathrm{CW}_n^r)^T x \leq 0$ given in Proposition 29.3.3.

Claim 29.6.3. *Assertion* (a) *holds, i.e.,* $a_{ij} = 0$ *for* $ij \in \mathrm{AW}_p^r$.

Proof. Let u be an integer such that $2 \leq u \leq r+1$ and set $S := [2, u-1]$; so, $S = \emptyset$ for $u = 2$. The sets $S, S\cup\{1\}, S\cup\{u\}$ and $S \cup \{1,u\}$ define roots (of type 1) of $(\mathrm{CW}_n^r)^T x \leq 0$ and, hence, of $a^T x \leq 0$. Therefore, Lemma 29.6.1 implies that $a_{1u} = 0$ and the general result follows by symmetry. ∎

Claim 29.6.4. *For some scalar* γ, $a_{ij'} = \gamma$ *for all* $i \in \{1,\ldots,p\}$, $j' \in [1',q']$.

Proof. Take two distinct nodes i', j' in $[1',q']$ and set $S := [2, r+2]$. Then, the sets $S\cup\{i'\}, S\cup\{j'\}, S\cup\{1,i'\}, S\cup\{1,j'\}$ all define roots (of type 2). Therefore, Lemma 29.6.2 implies that $a_{1i'} = a_{1j'}$, i.e., $a_{11'} = \ldots = a_{1q'} =: \gamma_1$, for some scalar γ_1. Similarly, $a_{i1'} = \ldots = a_{iq'} =: \gamma_i$, for some scalar γ_i, for all i. Let i be an integer such that $r+2 \leq i \leq p-r$, and set $T := [2, i-1] \cup [2', (i-r-1)']$ (with $T = [2, r+1]$ if $i = r+2$). The sets $T\cup\{1\}, T\cup\{i\}, T\cup\{1,1'\}$ and $T\cup\{i,1'\}$ all define roots. Therefore, Lemma 29.6.2 implies that $a_{11'} = a_{i1'}$, i.e., $\gamma_1 = \gamma_i$. So, $\gamma_1 = \gamma_{r+2} = \gamma_{r+3} = \ldots = \gamma_{p-r}$; similarly, $\gamma_2 = \gamma_{r+3} = \ldots = \gamma_{p-r+1}$ and, therefore, $\gamma_1 = \gamma_2$. By symmetry, $\gamma_1 = \gamma_2 = \ldots = \gamma_p =: \gamma$. ∎

Claim 29.6.5. *For some scalar* α, $a_{i'j'} = \alpha$ *for all* $1' \leq i' < j' \leq q'$.

Proof. Take distinct nodes i', j', h' in $[1',q']$ and set $S := [1, r+2]$. Then, the sets $S\cup\{j'\}, S\cup\{h'\}, S\cup\{i',j'\}$ and $S\cup\{i'h'\}$ all define roots. Hence, we deduce from Lemma 29.6.2 that $a_{i'j'} = a_{i'h'}$, henceforth stating the result. ∎

From the fact that $\delta(\{i\})$ is a root for all $1 \leq i \leq p$, we have that $a^T\delta(\{i\}) = 0$. Using the above claims, we deduce that

$$(S_i) \qquad \sum_{1\leq j\leq p} a_{ij} = -q\gamma$$

where we set, by convention, $a_{ii} := 0$.

Claim 29.6.6. $\gamma = -\alpha$.

Proof. Since the set $[1, r+1] \cup \{1'\}$ defines a root, equality $a^T\delta([1,r+1]\cup\{1'\}) = 0$, together with the above claims, yields:

$$\sum_{1\leq j\leq p} (a_{1j} + a_{2j} + \ldots + a_{r+1,j}) + (r+1)(q-1)\gamma + (p-r-1)\gamma + (q-1)\alpha = 0.$$

Using relations $(S_1), \ldots, (S_{r+1})$, we deduce from the above identity that

$$-q(r+1)\gamma + (r+1)(q-1)\gamma + (p-r-1)\gamma + (q-1)\alpha = 0,$$

i.e., $\gamma(p-2r-2)+\alpha(q-1)=0$. Thus, $\gamma=-\alpha$ since $p=q+2r+1$ and $q\geq 2$. ∎

Claim 29.6.7. $a_{1,r+2}=\alpha$.

Proof. Since the set $[1,r+2]\cup\{1'\}$ defines a root, we deduce the relation:

$$\sum_{1\leq j\leq p}(a_{1j}+\ldots+a_{r+2,j})-2a_{1,r+2}-\alpha(r+2)(q-1)-\alpha(p-r-2)+\alpha(q-1)=0$$

which, using relations $(S_1),\ldots,(S_{r+2})$, yields: $a_{1,r+2}=\alpha$. ∎

In order to finish the proof, we must show that condition (b) holds. For this, it suffices to show, for instance, that $a_{ij}=\alpha$ for all $1\leq i<j\leq\frac{(p+1)}{2}$ such that $ij\not\in \mathrm{AW}_p^r$, i.e., $j\geq i+r+1$. We prove the following statement (H_u) by induction on u, $r+2\leq u\leq\frac{(p+1)}{2}$:

(H_u) $\qquad a_{ij}=\alpha$ for all $1\leq i<j\leq u$ such that $ij\not\in\mathrm{AW}_p^r$.

Assertion (H_{r+2}) holds from Claim 29.6.7. Take $u\geq r+3$ and assume that (H_i) holds for $i\leq u-1$; we show that (H_u) holds, i.e., that $a_{iu}=\alpha$ for $1\leq i\leq u-r-1$.

Claim 29.6.8. $a_{iu}=\alpha$ *for all* $1\leq i\leq u-r-1$.

Proof. Set $S:=[i+1,u]\cup[1',(u-i-r)']$. Then, both S and $S\cup\{i\}$ define roots, yielding:

$$0=a(\delta(S\cup\{i\}))-a(\delta(S))$$

and, hence, the following relation:

$$0=\sum_{j\not\in S\cup\{i\}}a_{ij}-\sum_{j\in S}a_{ij}=\sum_{1\leq j\leq p}a_{ij}-2(a_{i,i+1}+\ldots+a_{iu})-\alpha(q-u+i+r)+\alpha(u-i-r).$$

Now, $a_{i,i+1}+\ldots+a_{iu}=a_{i,i+r+1}+\ldots+a_{i,u-1}+a_{iu}=\alpha(u-i-r-1)+a_{iu}$, the latter equality following from the induction assumption (H_i) for $i\leq u-1$. Therefore, the above relation yields:

$$0=q\alpha-2\alpha(u-i-r-1)-2a_{iu}-\alpha(q-u+i+r)+\alpha(u-i-r),$$

i.e., $a_{iu}=\alpha$. ∎

Chapter 30. Other Valid Inequalities and Facets

We describe in this chapter the other main known classes of valid inequalities defining facets of the cut polytope. The complete linear description of the cut polytope $\mathrm{CUT}_n^\square$ is known only for $n \leq 7$; it is presented in Section 30.6.

30.1 Suspended-Tree Inequalities

As a generalization of the subclass of clique-web inequalities $\mathrm{CW}_n^1(b)^T x \leq 0$ (the case $r = 1$ of $\mathrm{CW}_n^r(b)^T x \leq 0$), Boros and Hammer [1993] have introduced the following class of inequalities (30.1.1).

Let $r \geq 1$ and $b = (b_1, \ldots, b_n)$ be integers such that $\sum_{i=1}^n b_i = 2r + 1$ and suppose that $b_2, \ldots, b_p > 0 > b_{p+1}, \ldots, b_n$ for some $2 \leq p \leq n - 1$. (Note that the sign of b_1 is free.) Let T be a spanning tree on the $p - 1$ nodes of the set $\{2, \ldots, p\}$ and, for each node $i \in \{2, \ldots, p\}$, let d_i denote the degree of node i in T. Consider the inequality:

$$(30.1.1) \qquad \sum_{1 \leq i < j \leq n} b_i b_j x_{ij} - \frac{r(r+1)}{2} \left(\sum_{i=2}^{p} (2 - d_i) x_{1i} + \sum_{ij \in E(T)} x_{ij} \right) \leq 0.$$

We call it a *suspended-tree inequality* and we denote it by $\mathrm{ST}_n^r(T, b)^T x \leq 0$, since the quantity

$$(30.1.2) \qquad \sum_{i=2}^{p} (2 - d_i) x_{1i} + \sum_{ij \in E(T)} x_{ij}$$

is supported by a suspended-tree (a tree plus a node joined to some nodes of the tree). See below an example of a tree T and of the corresponding quantity (30.1.2) (weight 1 is indicated by a plain edge, weight -1 by a dotted edge and no edge means weight 0).

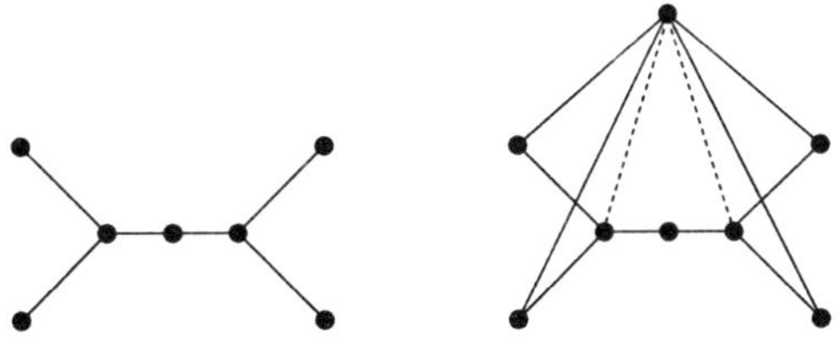

In the special case when T is the path $(2, \ldots, p)$, then $d_2 = d_p = 1$ while $d_i = 2$ for $3 \leq i \leq p-1$. Hence, the quantity (30.1.2) takes the form:

$$\sum_{1 \leq i \leq p-1} x_{i,i+1} + x_{1p},$$

i.e., its support graph is a circuit of length p. Therefore, the inequality (30.1.1) coincides with the inequality (29.3.7). This implies that, if $b_1, \ldots, b_p \geq r$, then $\mathrm{ST}_n^r(T,b)^T x \leq 0$ coincides with $\mathrm{CW}_n^r(b)^T x \leq 0$; otherwise, $\mathrm{ST}_n^r(T,b)^T x \leq 0$ is dominated by $\mathrm{CW}_n^r(b)^T x \leq 0$ and, therefore, it is not facet inducing (recall the observations after Proposition 29.3.6).

We conjecture that, for an arbitrary tree T, if $b_i < r$ for some $i \in \{1, \ldots, p\}$, then the suspended-tree inequality $\mathrm{ST}_n^r(T,b)^T x \leq 0$ is not facet inducing. We saw above that this is indeed the case if T is a path. In fact, the class of suspended-tree inequalities $\mathrm{ST}_n^r(T,b)^T x \leq 0$ with $b_i \geq r$ for all positive $b_i' s$ generalizes the class of clique-web inequalities $\mathrm{CW}_n^r(b)^T x \leq 0$ with $b_i \geq r$ for all positive $b_i' s$. On the other hand, recall that the class of clique-web inequalities has been defined as collapsing of some pure clique-web inequalities. We conjecture the existence of some large class $\mathcal{I}$ of valid inequalities, generalizing the (pure) clique-web inequalities, such that the suspended-tree inequality $\mathrm{ST}_n^r(T,b)^T x \leq 0$ with $b_i \geq r$ for all positive $b_i' s$ would occur as collapsing of some pure member of $\mathcal{I}$, while $\mathrm{ST}_n^r(T,b)^T x \leq 0$ with $b_i < r$ for some positive b_i would be dominated by some member of $\mathcal{I}$.

Proposition 30.1.3. *The inequality* (30.1.1) *is valid for* CUT_n. *Moreover, the roots of* (30.1.1) *are the cut vectors* $\delta(S)$ *for which* $S \subseteq \{2, \ldots, n\}$, $b(S) = r, r+1$ *and such that the subgraph of* T *induced by* $S \cap \{2, \ldots, p\}$ *is connected.*

Proof. Let S be a subset of $V_n \setminus \{1\}$. It is easy to check that

$$\mathrm{ST}_n^r(T,b)^T \delta(S) = b(S)(2r+1-b(S)) - r(r+1)c(S),$$

where $c(S)$ denotes the number of connected components of the subgraph of T induced by $S \cap \{2, \ldots, p\}$. Hence, $\mathrm{ST}_n^r(T,b)^T \delta(S) \leq 0$ if $c(S) \geq 1$, with equality if and only if $c(S) = 1$ and $b(S) = r, r+1$. Moreover, $\mathrm{ST}_n^r(T,b)^T \delta(S) < 0$ if $c(S) = 0$ since, then, $S \cap \{1, 2, \ldots, p\} = \emptyset$, implying that $b(S) < 0$. ∎

Theorem 30.1.4. *Let* $r \geq 1, b_1, \ldots, b_n$ *be integers such that* $\sum_{i=1}^n b_i = 2r+1$, $b_2, \ldots, b_p \geq r$, $b_{p+1} = \ldots = b_n = -1$ *for some* p *with* $3 \leq p \leq n-1$. *Let* T *be a spanning tree on* $\{2, \ldots, p\}$. *Suppose that one of the two conditions* (i), (ii) *holds.*

(i) $\sum_{i=2}^p b_i \leq n-p+r+1$ *(or, equivalently,* $b_1 \geq r$*) and there exist two disjoint sets* $S_1, S_2 \subseteq \{2, \ldots, p\}$ *such that* $b(S_l) = \sum_{i \in S_l} b_i \geq r+1$ *and the subgraph of* T *induced by* S_l *is connected, for* $l = 1, 2$.

(ii) T *is not a star and* $b(S) \leq n-p+r+1$ *for every subset* $S \subseteq \{2, \ldots, p\}$ *for which the subgraph of* T *induced by* S *is a path.*

Then, the inequality (30.1.1) *defines a facet of* CUT_n.

Proposition 30.1.3 and Theorem 30.1.4 with the condition (i) are given in Boros and Hammer [1993]; the alternative condition (ii) of Theorem 30.1.4 is proposed in Boissin [1994]. The results are presented there in the context of the correlation polyhedra. The proof below is taken from Boros and Hammer [1993].

Proof. We show that the inequality:

$$(30.1.5) \qquad \sum_{2\le i,j\le n} b_ib_jp_{ij} - (2r+1)\sum_{i=2}^{n} b_ip_{ii} + r(r+1)\left(\sum_{i=2}^{p} p_{ii} - \sum_{ij\in E(T)} p_{ij}\right) \ge 0,$$

which corresponds to the inequality (30.1.1) via the covariance mapping pointed at position 1, defines a facet of COR_{n-1}. Here, $b_2, \ldots, b_n$ are integers such that $b_2, \ldots, b_p \ge r, b_{p+1}, \ldots, b_n = -1$, T is a spanning tree on $\{2, \ldots, p\}$ and one of the conditions (i),(ii) of Theorem 30.1.4 holds. Denote the inequality (30.1.5) by $v^Tp \ge 0$. Let $w^Tp \ge 0$ be a valid inequality for COR_{n-1} such that $\{p \in \mathrm{COR}_{n-1} \mid v^Tp = 0\} \subseteq \{p \in \mathrm{COR}_{n-1} \mid w^Tp = 0\}$. We show that $w = \lambda v$ for some $\lambda > 0$.
Let $\mathcal{S}$ denote the family consisting of the sets $S \subseteq \{2, \ldots, p\}$ for which the subgraph $T[S]$ of T induced by S is connected and such that $r+1 \le b(S) \le n-p+r+1$. For $i_0 \in \{p+1, \ldots, n\}$, let $\mathcal{S}_{i_0}$ denote the family consisting of the sets $S \cup S'$, where $S \in \mathcal{S}$ and $S' \subseteq \{p+1, \ldots, n\} \setminus \{i_0\}$ with $|S'| = b(S) - r - 1$. Set $\mathcal{S}^*_{i_0} := \{A \cup \{i_0\} \mid A \in \mathcal{S}_{i_0}\}$. We prove an intermediary result.

Claim 30.1.6. *For each* $i_0 \in \{p+1, \ldots, n\}$, *the incidence vectors of the members of* $\mathcal{S}^*_{i_0}$ *have rank* $n-2$.

Proof of Claim 30.1.6. We show that the space consisting of the vectors that are orthogonal to all incidence vectors of members of $\mathcal{S}^*_{i_0}$ has dimension 1. Consider the vector $y \in \mathbb{R}^{n-1}$ defined by $y_i := b_i$ $(i = 2, \ldots, p)$, $y_{i_0} := -r-1$, and $y_i := -1$ $(i \in \{p+1, \ldots, n\} \setminus \{i_0\})$. Clearly, y is orthogonal to all incidence vectors of members of $\mathcal{S}^*_{i_0}$. Let $x \in \mathbb{R}^{n-1}$ such that $x_{i_0} = 0$ and x is orthogonal to all incidence vectors of members of $\mathcal{S}^*_{i_0}$. We show that $x = 0$. Let $S \in \mathcal{S}$. Then, $x(S) + x(S') = 0$ for every subset S' of $\{p+1, \ldots, n\} \setminus \{i_0\}$ of cardinality $b(S) - r - 1$. Hence, $x_i = \alpha$ for $i \in \{p+1, \ldots, n\} \setminus \{i_0\}$, where

$$\alpha := -\frac{x(S)}{b(S)-r-1}.$$

Therefore,

$$\text{(a)} \qquad (x + \alpha b)(S) = \alpha(r+1) \ \text{ for all } S \in \mathcal{S}.$$

Suppose, first, that the condition (i) from Theorem 30.1.4 holds. If $S \in \mathcal{S}$ and if a node $i \in \{2, \ldots, p\} \setminus S$ is adjacent to a node of S in the tree T, then $S \cup \{i\}$ still belongs to $\mathcal{S}$. Hence, $x_i + \alpha b_i = 0$ by (a) and, therefore, this relation holds for all $i \in \{2, \ldots, p\} \setminus S$. Now, taking for S the two sets S_1 and S_2 from condition (i), we deduce that $x_i + \alpha b_i = 0$ for all $i \in \{2, \ldots, p\}$. This implies that $x = 0$. Suppose now that the condition (ii) from Theorem 30.1.4 holds. Let S be the node set of a path $(i_1, \ldots, i_s)$ in T. If $|S| \geq 4$, then all the sets S, $S \setminus \{i_s\}, \ldots, S \setminus \{i_s, i_{s-1}, \ldots, i_3\}$, and $S \setminus \{i_1\}, \ldots, S \setminus \{i_1, \ldots, i_{s-2}\}$ belong to $\mathcal{S}$. Hence, we deduce from (a) that $x_i + \alpha b_i = 0$ for all $i \in S$. Therefore, $x_i = 0$ for all $i \in S$. As T is not a star, every node of T lies on a path of length at least 4. This shows that $x = 0$. ∎

Clearly, for $i_0 \in \{p+1, \ldots, n\}$ and $S \cup S' \in \mathcal{S}_{i_0}$, the correlation vectors $\pi(S \cup S')$ and $\pi(S \cup S' \cup \{i_0\})$ are both roots of the inequality (30.1.5). Hence, the relations

$$v_{i_0} + \sum_{i \in S \cup S'} v_{i,i_0} = 0, \; w_{i_0} + \sum_{i \in S \cup S'} w_{i,i_0} = 0$$

hold for all $S \cup S' \in \mathcal{S}_{i_0}$. We deduce from Claim 30.1.6 that there exists a scalar λ_{i_0} such that $w_{i_0} = \lambda_{i_0} v_{i_0}$ and $w_{i,i_0} = \lambda_{i_0} v_{i,i_0}$ for all $i \in \{2, \ldots, n\} \setminus \{i_0\}$. As $v_{ij} \neq 0$ for $i, j \in \{p+1, \ldots, n\}$, we obtain that all λ_{i_0}'s are equal to, say, λ. Hence, $w_i = \lambda v_i$ for $i = p+1, \ldots, n$ and $w_{ij} = \lambda v_{ij}$ if at least one of i, j belongs to $\{p+1, \ldots, n\}$.

Set $u := w - \lambda v$. We show that $u = 0$. For $i_0 \in \{2, \ldots, p\}$, let $\mathcal{T}_{i_0}$ denote the family consisting of the nonempty sets $S \subseteq \{2, \ldots, p\} \setminus \{i_0\}$ such that the subgraphs $T[S]$ and $T[S \cup \{i_0\}]$ are both connected and $b(S) + b_{i_0} \leq n - p + r + 1$. Clearly, if $S \in \mathcal{T}_{i_0}$ and $S', S'' \subseteq \{p+1, \ldots, n\}$ with $|S'| = b(S) - r$, $|S''| = b_{i_0} - 1$, then the correlation vectors $\pi(S \cup S')$ and $\pi(S \cup \{i_0\} \cup S' \cup S'')$ are roots of the inequality (30.1.5). This implies that

$$\text{(b)} \qquad u_{i_0} + \sum_{i \in S} u_{i,i_0} = 0 \;\text{ for all } S \in \mathcal{T}_{i_0}.$$

If i is a node of T adjacent to i_0, then $\{i\}$ belongs to $\mathcal{T}_{i_0}$ and, thus, $u_{i_0} + u_{i,i_0} = 0$. If $(i_0, i_1, \ldots, i_s)$ is a path in T, then all sets $\{i_1\}, \{i_1, i_2\}, \ldots, \{i_1, \ldots, i_s\}$ belong to $\mathcal{T}_{i_0}$. From (b), we deduce that $u_{i_0,i_2} = u_{i_0,i_3} = \ldots = u_{i_0,i_s} = 0$. This shows that $u_{i_0,i} = 0$ if i is not adjacent to i_0 and $u_{i_0,i} = -u_{i_0}$ if i is adjacent to i_0. Hence, we obtain that, for some $\alpha \in \mathbb{R}$, $u_2 = \ldots = u_p = \alpha$, $u_{ij} = -\alpha$ if ij is an edge of T and $u_{ij} = 0$ otherwise. Finally, $\alpha = 0$, which shows that $u = 0$, i.e., $w = \lambda v$. ∎

We now present some other classes of facets, which are obtained by modifying the definition of the suspended-tree inequalities. Let $b = (b_1, \ldots, b_{2p+1})$ with $b_i = 1$ if $i \in \{1, \ldots, p+2\}$ and $b_i = -1$ if $i \in \{p+3, \ldots, 2p+1\}$; so, $\sum_{i=1}^{2p+1} b_i = 3$. Let T be a spanning tree defined on the $p+1$ nodes of the set $\{2, \ldots, p+2\}$. Hence, the inequality:

$$\sum_{1 \leq i < j \leq 2p+1} b_i b_j x_{ij} - \left(\sum_{i=2}^{p+2} (2 - d_i) x_{1i} + \sum_{ij \in E(T)} x_{ij} \right) \leq 0$$

is the case $b = (1, \dots, 1, -1, \dots, -1)$ and $r = 1$ of the inequality (30.1.1). Consider the following switching of it:

$$(30.1.7) \qquad \sum_{ij \in K_{2p+1}} x_{ij} - \left(\sum_{i=2}^{p+2} (2 - d_i) x_{1i} + \sum_{ij \in E(T)} x_{ij} \right) \leq (p-1)(p+2) = p(p+1) - 2.$$

Therefore, from Theorem 30.1.4, the inequality (30.1.7) defines a facet of the cut polytope $\mathrm{CUT}^{\square}_{2p+1}$ if T is not a star. The following two generalizations of the inequality (30.1.7) were proposed by De Souza and Laurent [1995]. They allow, respectively, the tree T to be defined on p or on $p-1$ nodes but, in order to preserve validity, an additional term (whose support graph is K_3 or K_5) must be added to (30.1.7); also the right hand side has to be modified.

Consider the complete graph K_{2p+1} defined on the nodes of $\{1, \dots, 2p+1\}$. Let T be a spanning tree defined on the p nodes of $\{2, \dots, p+1\}$ and let Δ denote the complete graph K_3 defined on $\{p+2, p+3, p+4\}$. Consider the inequality:

$$(30.1.8) \qquad \sum_{ij \in K_{2p+1}} x_{ij} - \left(\sum_{i=2}^{p+1} (2 - d_i) x_{1i} + \sum_{ij \in E(T)} x_{ij} - \sum_{ij \in \Delta} x_{ij} \right) \leq p(p+1).$$

(We remind that d_i denotes the degree of node i in the tree T.) For instance, let $p = 5$ and let T be the tree from Figure 30.1.9. Then, the quantity $\sum_{i=2}^{p+1} (2 - d_i) x_{1i} + \sum_{ij \in E(T)} x_{ij} - \sum_{ij \in \Delta} x_{ij}$ is depicted in Figure 30.1.10.

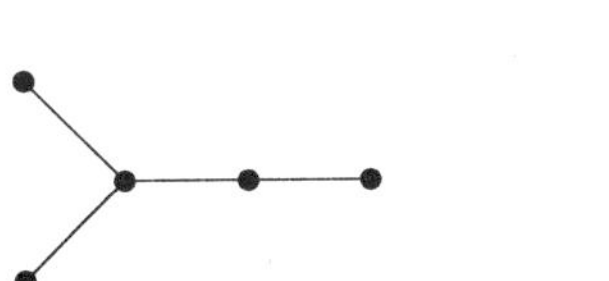

Figure 30.1.9

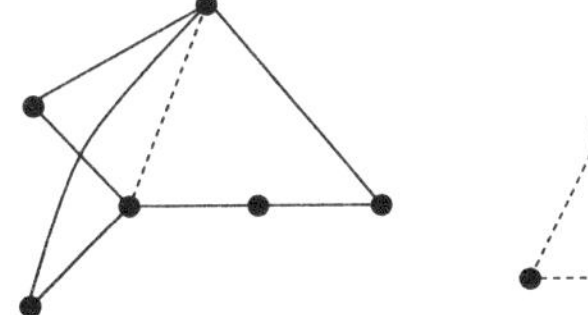

Figure 30.1.10

Suppose now that T is a spanning tree defined on the $p-1$ nodes of $\{2, \dots, p\}$. Let C denote the 5-circuit $(p+1, p+2, p+3, p+4, p+5)$ and let C' denote the 5-circuit $(p+1, p+3, p+5, p+2, p+4)$. Consider the inequality:

$$(30.1.11) \qquad \sum_{ij \in K_{2p+1}} x_{ij} - \left(\sum_{i=2}^{p} (2 - d_i) x_{1i} + \sum_{ij \in E(T)} x_{ij} + \sum_{ij \in C} x_{ij} - \sum_{ij \in C'} x_{ij} \right) \leq p(p+1).$$

If $p = 7$ and T is the tree from Figure 30.1.12, then the quantity: $\sum_{i=2}^{p}(2-d_i)x_{1i} + \sum_{ij \in E(T)} x_{ij} + \sum_{ij \in C} x_{ij} - \sum_{ij \in C'} x_{ij}$ is depicted in Figure 30.1.13.

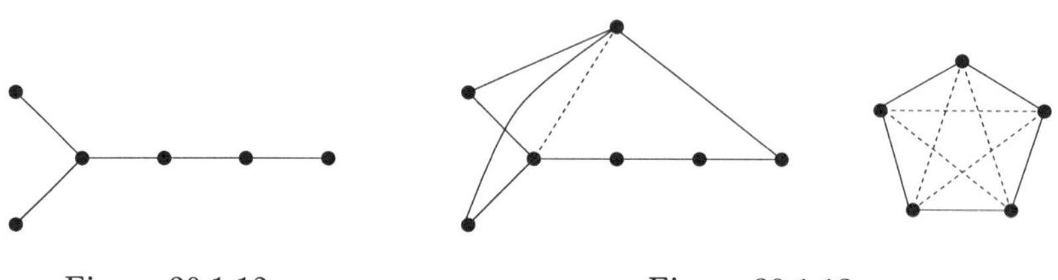

Figure 30.1.12 Figure 30.1.13

The following result can be found in De Souza and Laurent [1995].

Theorem 30.1.14.

(i) *The inequality* (30.1.8) *defines a facet of the cut polytope* $\mathrm{CUT}^{\square}_{2p+1}$ *if* $p \geq 5$ *and if* T *is not a star.*

(ii) *The inequality* (30.1.11) *defines a facet of* $\mathrm{CUT}^{\square}_{2p+1}$ *if* $p \geq 6$ *and if, for each node* u *of* T, *some connected component of* $T \setminus u$ *has at least three nodes.* ∎

The next case to consider would be when the tree T is defined on $p-2$ nodes. It is natural to conjecture the existence of a valid inequality for $\mathrm{CUT}^{\square}_{2p+1}$ which, by analogy with (30.1.8) and (30.1.11), would involve, besides the suspended tree, a graph K_7 with suitable edge weights. We refer to De Souza [1993] for a discussion on this; some examples of such inequalities are proposed there.

30.2 Path-Block-Cycle Inequalities

Given an integer $p \geq 1$, let C denote the circuit $(1, 2, \ldots, p+2)$. Then, the inequality:

$$(30.2.1) \qquad \sum_{1 \leq i < j \leq 2p+1} x_{ij} - \sum_{ij \in E(C)} x_{ij} \leq p(p+1) - 2$$

defines a facet of $\mathrm{CUT}^{\square}_{2p+1}$. Indeed, the inequality (30.2.1) coincides with a switching of the (pure) clique-web inequality $(\mathrm{CW}^{1}_{2p+1})^T x \leq 0$, namely, with its switching by the cut $\delta(\{1, 2, \ldots, p+2\})$ (expressed in (29.1.5)). In this section, we describe a class of inequalities which generalizes (30.2.1). Instead of using only one circuit, we consider a graph structure, called path-block-cycle, which is constructed from several circuits.

More precisely, a path-block-cycle is a graph defined as follows. Let $C_h = (V_h, E(C_h))$ $(1 \leq h \leq r)$ be r circuits such that $V_h \cap V_{h'} = \bigcap_{1\leq h\leq r} V_h$ for all distinct $h, h' \in \{1, \ldots, r\}$. We suppose, moreover, that the common nodes are visited in the same order along each of the circuits C_h. Then, the graph with node set $\bigcup_{1\leq h\leq r} V_h$ and with edge set $\bigcup_{1\leq h\leq r} E(C_h)$ (allowing repetition of the edges) is called a *path-block-cycle* and is abbreviated as PBC. It may contain multiple edges, if some edge is used by several of the circuits C_h. Figure 30.2.2 shows an example of a PBC graph composed of three circuits. The black nodes are those common to all three circuits; the three circuits are drawn, respectively, by a plain line, a thick line and a dotted line.

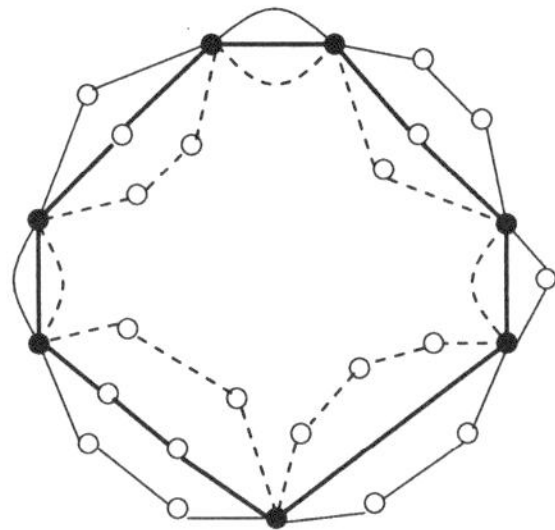

Figure 30.2.2: A path-block-cycle graph

Set $\bigcap_{1\leq h\leq r} V_h =: \{x_1, \ldots, x_t\}$, where $t \geq 1$. Hence, each circuit C_h decomposes into t subpaths P_{ih} $(1 \leq i \leq t)$, where P_{ih} starts at x_i and ends at x_{i+1} (the indices i are taken modulo t). Set

$$q_{ih} := |V(P_{ih})| - 2,$$

i.e., q_{ih} is the number of internal nodes of P_{ih}, $q_{ih} \geq 0$. Let $\pi_1, \ldots, \pi_{r-1}$ denote the $r-1$ largest values taken by q_{ih}, for $1 \leq i \leq t$, $1 \leq h \leq r$. For each integer k such that $k(k+1) \leq 2(r-1)$, set

$$Q_k := \sum_{1\leq h\leq r-1-\frac{k(k+1)}{2}} \pi_h.$$

Finally, set

$$n_0 := 2p + 1 - |\bigcup_{1\leq h\leq r} V_h|.$$

Given a PBC graph composed of r circuits $C_1, \ldots, C_r$ and which is a subgraph of K_{2p+1}, we define the following inequality:

$$\text{(30.2.3)} \qquad \sum_{1\leq i<j\leq 2p+1} x_{ij} - \sum_{1\leq h\leq r} \sum_{ij\in E(C_h)} x_{ij} \leq p(p+1) - 2r.$$

It is called a *path-block-cycle inequality*. These inequalities are introduced in their full generality in De Souza [1993]; a more restricted subclass (see below) is

considered in De Souza and Laurent [1995]. Actually, they have been introduced in the context of the equicut polytope $\mathrm{EQCUT}^{\square}_{2p+1}$.
We give first the characterization of the PBC inequalities that are valid for the cut polytope, established by De Souza [1993].

Proposition 30.2.4. *Given a PBC graph that is a subgraph of K_{2p+1}, the inequality* (30.2.3) *is valid for $\mathrm{CUT}^{\square}_{2p+1}$ if and only if $Q_k + n_0 \leq p-1-k$ for all $k \in \mathbb{Z}_+$ such that $k(k+1) \leq 2(r-1)$.*

Proof. Denote the inequality (30.2.3) by $v^T x \leq p(p+1) - 2r$.
Necessity. Suppose that k is an integer such that $Q_k + n_0 \geq p-k$ and $k(k+1) \leq 2(r-1)$. Then, we can construct a node set S of cardinality $p-k$ whose nodes are, either internal nodes of the paths P_{ih} corresponding to the $r-1-\frac{k(k+1)}{2}$ largest π_h values, or do not belong to the PBC graph. Moreover, this set S can be chosen in such a way that

$$v^T\delta(S) \geq p(p+1) - k(k+1) - 2(r-1-\frac{k(k+1)}{2}), \text{ i.e., } v^T\delta(S) \geq p(p+1) - 2r + 2.$$

Hence, the cut vector $\delta(S)$ violates the inequality (30.2.3).
Sufficiency. Suppose that $Q_k + n_0 \leq p-k-1$ for all $k \in \mathbb{Z}_+$ such that $k(k+1) \leq 2(r-1)$. Suppose also that there exists a cut vector $\delta(S)$ violating (30.2.3). Set $|S| := p-k$ for some $1 \leq k \leq p-1$. Then,

$$v^T\delta(S) = p(p+1) - k(k+1) - \sum_{1\leq h\leq r} |\delta_{K_{2p+1}}(S) \cap E(C_h)| > p(p+1) - 2r.$$

Hence, the cut $\delta_{K_{2p+1}}(S)$ cannot meet all the r circuits C_h. Therefore, the set $\bigcap_{1\leq h\leq r} V_h$ of common nodes is entirely contained in S or in its complement. This implies that $\delta_{K_{2p+1}}(S)$ can only intersect $r-\lambda$ paths P_{ih} for some integer λ, $1 \leq \lambda \leq r$. Then, $\sum_{1\leq h\leq r} |\delta_{K_{2p+1}}(S) \cap E(C_h)| = 2(r-\lambda)$, i.e., $v^T\delta(S) = p(p+1) - 2r + 2\lambda - k(k+1)$. This implies that $2\lambda > k(k+1)$, i.e., $\lambda \geq \frac{k(k+1)}{2} + 1$. Set $\alpha := |S \setminus \bigcup_{1\leq h\leq r} V_h|$ and $\beta := |S| - \alpha$. These β nodes are distributed among $r - \lambda \leq r - \frac{k(k+1)}{2} - 1$ paths P_{ih}, which implies that $\beta \leq Q_k$. On the other hand, $\alpha \leq n_0$. Therefore, $p-k = \alpha + \beta \leq n_0 + Q_k \leq p-k-1$, yielding a contradiction. This shows that (30.2.3) is valid for $\mathrm{CUT}^{\square}_{2p+1}$. ∎

Recall that the equicut polytope $\mathrm{EQCUT}^{\square}_{2p+1}$ is the facet of $\mathrm{CUT}^{\square}_{2p+1}$ defined by the inequality:

$$\sum_{1\leq i<j\leq 2p+1} x_{ij} \leq p(p+1).$$

Therefore, for any facet defining inequality for $\mathrm{EQCUT}^{\square}_{2p+1}$, there exists a suitable linear combination of it with the equation $\sum_{1\leq i<j\leq 2p+1} x_{ij} = p(p+1)$ which produces a facet defining inequality for $\mathrm{CUT}^{\square}_{2p+1}$. In fact, the inequality (30.2.3) arises in this way from the inequality:

$$\sum_{1\leq h\leq r} \sum_{ij\in E(C_h)} x_{ij} \geq 2r. \tag{30.2.5}$$

De Souza [1993] shows that (30.2.5) is valid for $\mathrm{EQCUT}^{\square}_{2p+1}$ if and only if $n_0 + Q_0 \leq p-1$; moreover, he shows that (30.2.3) defines a facet of $\mathrm{CUT}^{\square}_{2p+1}$ whenever it is valid for $\mathrm{CUT}^{\square}_{2p+1}$ and whenever (30.2.5) defines a facet of $\mathrm{EQCUT}^{\square}_{2p+1}$. A class of PBC graphs is given in De Souza and Laurent [1995] for which (30.2.5) is facet defining for $\mathrm{EQCUT}^{\square}_{2p+1}$; we describe it below in Theorem 30.2.7.

Consider a PBC graph formed by r circuits satisfying the following conditions:

(i) All the r circuits have the same length, i.e., $|V_1| = \ldots = |V_r|$.

(ii) The number of nodes common to all circuits is even, say, $t = 2s$.

(iii) For each odd $i \in \{1, 2, \ldots, 2s\}$, the path P_{ih} consists of the edge (x_i, x_{i+1}), i.e., $q_{ih} = 0$.

(iv) For each even $i \in \{1, 2, \ldots, 2s\}$, any two paths P_{ih}, $P_{ih'}$ have only their endnodes x_i, x_{i+1} in common and they have the same length $q + 1$, i.e., $q_{ih} = q$.

Hence, a PBC graph satisfying (i)-(iv) is fully determined by the parameters (s, q, r).

Set $E_r := \{x_i x_{i+1} \mid 1 \leq i \leq 2s-1, i \text{ is odd}\}$ and $E_1 := \bigcup_{1 \leq h \leq r} E(C_h) \setminus E_r$. Then, the edges of E_r belong to all r circuits while the edges of E_1 belong to exactly one circuit. Hence,

$$\sum_{1 \leq h \leq r} \sum_{ij \in E(C_h)} x_{ij} = \sum_{ij \in E_1} x_{ij} + r \sum_{ij \in E_r} x_{ij}.$$

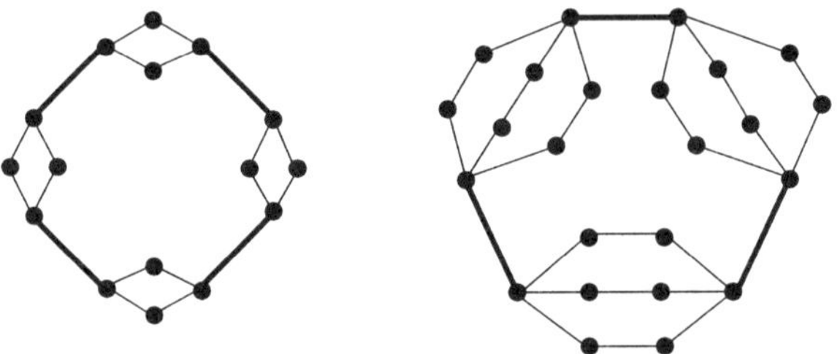

Figure 30.2.6: PBC graphs

Figure 30.2.6 shows two PBC graphs satisfying the conditions (i)-(iv) with the parameters $s = 4, q = 1, r = 2$, and $s = 3, q = 2, r = 3$, respectively; the thick edges are the edges of E_r.

Theorem 30.2.7. *Consider a PBC graph with parameters* (s, q, r) *satisfying the conditions* (i)-(iv) *above. Suppose that this PBC is a subgraph of* K_{2p+1} *and that* $n_0 = p - (r-1)q - 1$ *(or, equivalently,* $p = (s-1)(qr+2) + q$*). Then, the inequality* (30.2.3) *defines a facet of* $\mathrm{CUT}^{\square}_{2p+1}$. ∎

Remark 30.2.8. (i) De Souza [1993] proposes some inequalities that are obtained as a common generalization of suspended-tree inequalities and path-block-cycle inequalities. Observe that the graph structure used in suspended-treee inequalities is a suspended tree while the graph structure used in path-block-cycle inequalities is a collection of circuits. Note also that a circuit is a special case of suspended tree. Hence, a natural idea would be to consider inequalities involving a collection of suspended-trees. We refer to De Souza [1993] where several inequalities of this type are investigated in the context of the equicut polytope. (ii) The separation problem for suspended-tree inequalities and for path-block-cycle inequalities is probably hard. Some separation routines for path-block-cycle inequalities (or, more precisely, for the inequalities (30.2.5) which occur for the equicut polytope) are proposed by De Souza [1993]. ∎

30.3 Circulant Inequalities

In this section, we present a class of facets, whose support graphs are circulant graphs. The *circulant graph* $C(n,r)$ is the graph on the n nodes $\{1,2,\ldots,n\}$ whose edges consist of the pairs $(i,i+1),(i,i+r)$ for $i=1,\ldots,n$, where the indices are taken modulo n. Figure 30.3.1 shows the graph $C(8,3)$.

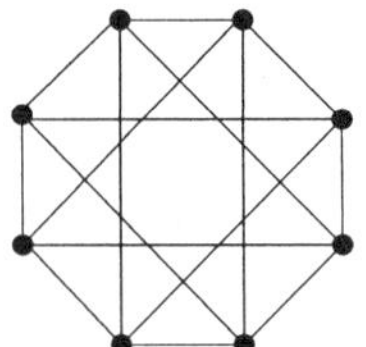

Figure 30.3.1: $C(8,3)$

Poljak and Turzik [1987, 1992] have computed the maximum cardinality of a cut in any circulant graph $C(n,r)$ and presented new classes of facets for the cut polytope and the bipartite subgraph polytope which are supported by some circulant graphs. Theorems 30.3.2 and 30.3.3 can be found in Poljak and Turzik [1992] and Theorem 30.3.4 in Poljak and Turzik [1987, 1992].

The *bipartite subgraph polytope* $\mathrm{BIP}_n^{\square}$ is the polytope in $\mathbb{R}^{E_n}$ defined as the convex hull of the incidence vectors of the edge sets $F \subseteq E_n$ for which (V_n, F) is a bipartite subgraph of the complete graph K_n.

Theorem 30.3.2. *Let n,r be integers with $n \geq 2r+1 \geq 4$. Then, the maximum cardinality of a bipartite subgraph in $C(n,r)$ is equal to $\max(2n-u_t-v_t \mid t=0,1,\ldots,r)$, where $u_t := |nt-v_tr|$ and v_t is the unique integer having the same parity as n and satisfying $nt-r \leq v_tr < nt+r$.* ∎

Theorem 30.3.3. *Let $n=kr+s$ where $0<s<r<\frac{n}{2}$, r and k are even, s is odd and g.c.d.$(n,r)=1$. Denote by E_O (resp. E_I) the set of edges $(i,i+1)$*

(resp. $(i, i+r)$) for $i = 1, \ldots, n$ of $C(n,r)$ (the indices being taken modulo n). Then, the inequality:

$$\sum_{ij \in E_I} x_{ij} + s \sum_{ij \in E_O} x_{ij} \leq (s+1)n - sk - r$$

defines a facet of the bipartite subgraph polytope $\mathrm{BIP}_n^{\square}$. ∎

Theorem 30.3.4. *Let $n = kr + 1$, where $k, r \geq 2$ are even integers. Then, the inequality:*

$$\sum_{ij \in C(n,r)} x_{ij} \leq 2n - k - r$$

defines a facet of the cut polytope $\mathrm{CUT}_n^{\square}$. ∎

Observe that, in the case $r = 2$, the circulant $C(n,2)$ coincides with the antiweb AW_n^2. Poljak and Turzik [1992] observed that the problem: "Does a graph G contain a circulant $C(n,2)$ for some n ?" is NP-complete. Hence, the separation problem for the class of inequalities :

$$\sum_{ij \in C(n,2)} x_{ij} \leq \frac{3}{2}(n-1), \quad n \text{ odd}$$

is NP-hard.

30.4 The Parachute Inequality

We describe in this section the class of parachute inequalities. They have been introduced[1] in Deza and Laurent [1992a] and further studied in Deza and Laurent [1992c, 1992d].

The parachute inequality is defined on an odd number of points, say, on $2k+1$ points. It is convenient to denote the elements of the set V_{2k+1} as $\{0, 1, 2, \ldots, k, 1', 2', \ldots, k'\}$. We define the following paths P and Q:

$$P := (k, k-1, \ldots, 2, 1, 1', 2', \ldots, (k-1)', k'),\ Q := (k-1, \ldots, 2, 1, 1', 2', \ldots, (k-1)').$$

Then, the inequality:

$$(30.4.1) \qquad (\mathrm{Par}_{2k+1})^T x := \sum_{ij \in P} x_{ij} - \sum_{1 \leq i \leq k-1} (x_{0i} + x_{0i'} + x_{ki'} + x_{k'i}) - x_{kk'} \leq 0$$

is called the *parachute inequality* and is denoted as $(\mathrm{Par}_{2k+1})^T x \leq 0$. Figure 30.4.3 shows the support graph of the parachute inequality on 7 points.

For k even, the parachute inequality is not valid for the cut cone; for example, it is violated by the cut vector $\delta(\{1, 3, \ldots, k-1\} \cup \{2', 4', \ldots, k'\})$. But, for k odd, the parachute inequality is facet inducing.

[1]The parachute inequality on seven points (in fact, a switching of it) has been introduced earlier by Assouad and Delorme [1982] (cf. Assouad [1984]).

Theorem 30.4.2. *For k odd, $k \geq 3$, the parachute inequality* (30.4.1) *defines a facet of* CUT_{2k+1}. ▮

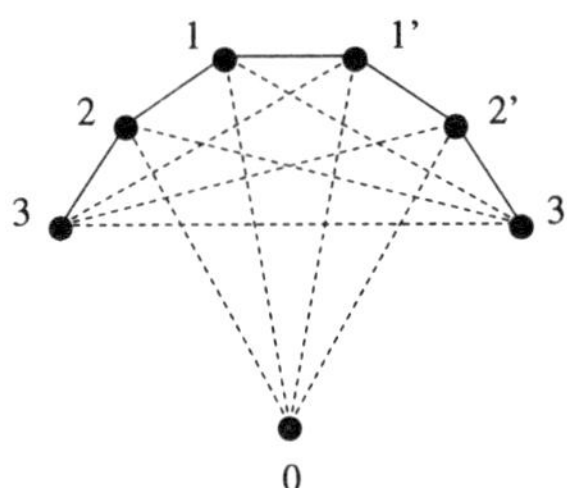

Figure 30.4.3: $(\mathrm{Par}_7)^T x \leq 0$

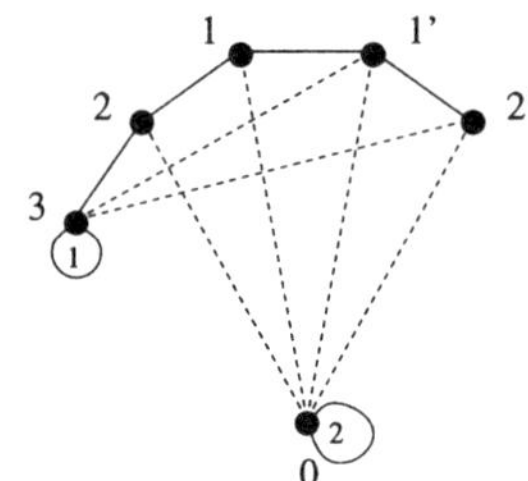

Figure 30.4.4: The analogue of $(\mathrm{Par}_7)^T x \leq 0$ for COR_6

30.4.1 Roots and Fibonacci Numbers

As we now see, the parachute inequality presents two interesting features. First, its number of roots can be expressed in terms of the Fibonacci numbers; second, there is a close connection between the parachute inequality $(\mathrm{Par}_{2k+1})^T x \leq 0$ and the following clique-web inequality: $(\mathrm{CW}_{2k+1}^{k-2}(1, \ldots, 1, -1, -1))^T x \leq 0$. Recall that the *Fibonacci sequence* is the sequence $(f_i)_{i \geq 1}$ defined recursively by

$$\begin{aligned} &f_1 = f_2 = 1, \\ &f_{i+2} = f_i + f_{i+1} \quad \text{for } i \geq 1. \end{aligned}$$

We introduce some definitions that we need for the description of the roots of the parachute inequality. Given a subset S of $V_{2k+1} = \{0, 1, \ldots, k, 1', \ldots, k'\}$, the set S is called *symmetric* if, for all $i \in \{1, \ldots, k\}$, $i \in S$ if and only if $i' \in S$. Let $A = (1, 2, \ldots, n)$ be a path. A subset S of $\{1, \ldots, n\}$ is called *alternated* along the path A if $|S \cap \{i, i+1\}| \leq 1$ for all $i = 1, 2, \ldots, n-1$, and S is called *pseudo-alternated* along the path A if $|S \cap \{i, i+1\}| = 1$ for all $i \in \{1, \ldots, n-1\} - \{j\}$ and $|S \cap \{j, j+1\}| = 0, 2$ for some $j \in \{1, \ldots, n-1\}$. One can easily check that, for n even, there are exactly $n-1$ pseudo-alternated subsets S along the path A for which $1, n \in S$. Also, an easy induction shows that the number of alternated subsets along the path $A = (1, \ldots, n)$ is equal to the Fibonacci number f_{n+2}.

Proposition 30.4.5. *If k is odd, then the parachute inequality* (30.4.1) *is valid for* CUT_{2k+1}. *For any k, the roots of the parachute inequality are the cut vectors $\delta(S)$ for which S is a subset of $\{1, 2, \ldots, k, 1', 2', \ldots, k'\}$ of one of the following four types:*

Type 1: *$k, k' \in S$ and S is pseudo-alternated along the path P.*

Type 2: *$k, k' \notin S$ and S is alternated along the path Q.*

Type 3: *For k odd, $k \in S$, $k' \notin S$ and* (a) *or* (b) *holds:*

(a) $S = \{2', 4', \ldots, (k-1)', k\} \cup T$, *where* T *is a subset of* $\{1, 2, \ldots, k-2\}$ *alternated along the path* $(1, 2, \ldots, k-2)$.

(b) $S = \{k, 1', (k-1)'\} \cup T \cup T'$, *where* T *is a subset of* $\{2, 3, \ldots, k-2\}$ *which is alternated along the path* $(2, 3, \ldots, k-2)$, *and* T' *is a subset of* $\{2', 3', \ldots, (k-2)'\}$ *for which* $T' \cup \{1', (k-1)'\}$ *is pseudo-alternated along the path* $(1', 2', \ldots, (k-1)')$.

Type 3′: *similar to* Type 3, *exchanging nodes* i, i' *for all* $i = 1, \ldots, k$.

Therefore, the total number of roots of $(\mathrm{Par}_{2k+1})^T x \leq 0$ *is equal to* $f_{2k} + 2kf_{k-1} + 2f_{k-2} + 2k - 1$ *for* k *odd and to* $f_{2k} + 2k - 1$ *for* k *even, while the number of nonzero symmetric roots of* $(\mathrm{Par}_{2k+1})^T x \leq 0$ *is equal to the Fibonacci number* f_k.

Proof. Let S be a subset of $\{1, \ldots, k, 1', \ldots, k'\}$ and set $s := |S \cap \{1, \ldots, k-1\}|$, $s' := |S \cap \{1', \ldots, (k-1)'\}|$. One checks easily that

$$\begin{aligned}(\mathrm{Par}_{2k+1})^T \delta(S) &= |\delta(S) \cap P| - 2(s + s'), \text{ if } k, k' \notin S \\ &= |\delta(S) \cap P| - 2s - k, \text{ if } k \in S, k' \notin S \\ &= |\delta(S) \cap P| - 2(k-1), \text{ if } k, k' \in S\end{aligned}$$

Then, it is easy to see that, for k odd, the parachute inequality is valid and that, for any k, the roots of $(\mathrm{Par}_{2k+1})^T x \leq 0$ are indeed of Types 1,2,3, or 3′. There are $2k-1$ roots of Type 1, f_{2k} roots of Type 2 and $f_k + (k-1)f_{k-1}$ roots of Type 3. Hence, altogether, there are $2k - 1 + f_{2k} + 2f_k + 2(k-1)f_{k-1}$ roots for k odd, and $2k - 1 + f_{2k}$ roots for k even. There is only one symmetric root of Type 1, namely, $\delta(\{k, \ldots, 3, 1, 1', 3', \ldots, k'\})$ for k odd and $\delta(\{k, \ldots, 4, 2, 2', 4', \ldots, k'\})$ for k even. The number of symmetric roots of Type 2 is equal to the number of alternated subsets along the path $(2, 3, \ldots, k-1)$, i.e., to f_k. There are no symmetric roots of Type 3 or 3′. Therefore, in total, there are f_k nonzero symmetric roots. ∎

We now show a connection between the parachute inequality $(\mathrm{Par}_{2k+1})^T x \leq 0$ and the clique-web inequality $\mathrm{CW}^{k-2}_{2k+1}(1, \ldots, 1, -1, -1)^T x \leq 0$. For this, let us first define the following inequality:

$$\sum_{ij \in Q} x_{ij} - \sum_{1 \leq i \leq k-1} (x_{0i} + x_{0i'}) - \sum_{1 \leq i \leq k-2} (x_{0'i} + x_{0'i'}) \leq 0, \tag{30.4.6}$$

which is defined on the $2k$ nodes of the set $\{0, 1, 2, \ldots, k-1, 0', 1', \ldots, (k-1)'\}$. The inequality (30.4.6) is called the *Fibonacci inequality* and is denoted as $(\mathrm{Fib}_{2k})^T x \leq 0$.

Consider now the clique-web inequality $\mathrm{CW}^{k-2}_{2k+1}(1, \ldots, 1, -1, -1)^T x \leq 0$. Then, the inequality obtained from it by collapsing a positive node (i.e., a node i for which $b_i = 1$) and a negative node (i.e., a node j for which $b_j = -1$) coincides with the Fibonacci inequality $(\mathrm{Fib}_{2k})^T x \leq 0$ (if one labels in a suitable way the points on which the clique-web inequality is defined). This shows, in particular, that the Fibonacci inequality is valid for CUT_{2k}.

Observe that the Fibonacci inequality $(\mathrm{Fib}_{2k})^T x \leq 0$ can also be obtained from the parachute inequality $(\mathrm{Par}_{2k+1})^T x \leq 0$ by collapsing the two nodes k, k'

into a single node denoted as $0'$. Therefore, the roots of $(\text{Fib}_{2k})^T x \leq 0$ are the cut vectors $\delta(S \setminus \{k, k'\} \cup \{0'\})$ for S of Type 1 and the cut vectors $\delta(S)$ for S of Type 2. Hence, $(\text{Fib}_{2k})^T x \leq 0$ has $f_{2k} + 2k - 1$ roots. It can be checked that it defines a face of CUT_{2k} of rank $\binom{2k-1}{2} + 2 = \binom{2k}{2} - (2k-3)$.

Therefore, for any $k \geq 3$, the parachute inequality $(\text{Par}_{2k+1})^T x \leq 0$ and the clique-web inequality $\text{CW}^{k-2}_{2k+1}(1, \ldots, 1, -1, -1)^T x \leq 0$ admit a common collapsing.

Figures 30.4.7 and 30.4.8 show, respectively, the support graphs of the Fibonacci inequality $(\text{Fib}_6)^T x \leq 0$ and of the clique-web inequality $(\text{CW}^1_7)^T x \leq 0$. (Collapse the nodes $0'$ and $0''$ in $(\text{CW}^1_7)^T x \leq 0$ to obtain $(\text{Fib}_6)^T x \leq 0$.)

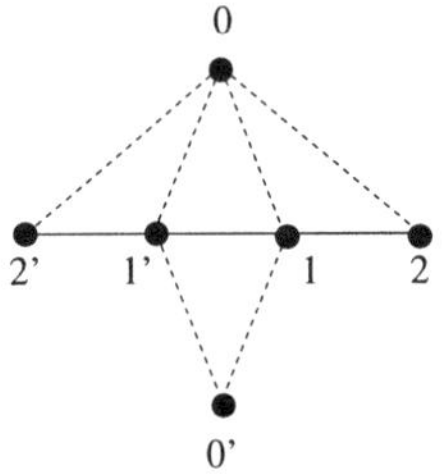

Figure 30.4.7: $(\text{Fib}_6)^T x \leq 0$

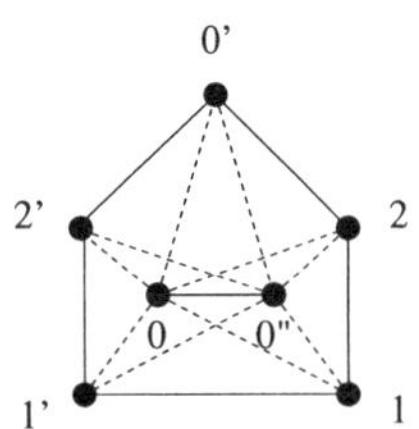

Figure 30.4.8: $(\text{CW}^1_7)^T x \leq 0$

30.4.2 Generalizing the Parachute Inequality

Finally, we present a class of inequalities from Boissin [1994] generalizing the parachute inequality. These new inequalities can be obtained by applying the operation of "duplicating a node", described in Section 26.6. As this operation is easier to apply to facets of the correlation polytope, we first reformulate the parachute inequality for the correlation polytope.

We start with the parachute inequality $(\text{Par}_{2k+1})^T x \leq 0$, which is defined on the set $V_{2k+1} = \{0, 1, 2, \ldots, k, 1', 2', \ldots, k'\}$. If we apply the covariance mapping $\xi_{k'}$ (pointed at position k'), then we obtain the following inequality $a^T p \geq 0$, which is defined on all pairs of (nonnecessarily distinct) elements of $V_{2k+1} \setminus \{k'\}$:

$$(30.4.9) \qquad \begin{aligned} a^T p := \; & (k-1)p_{00} + \tfrac{k-1}{2} p_{kk} + \sum_{ij \in E(P')} p_{ij} \\ & - \sum_{h=1}^{k-1} (p_{0h} + p_{0,h'} + p_{k,h'}) \geq 0, \end{aligned}$$

where P' denotes the path $(k, k-1, \ldots, 2, 1, 1', 2', \ldots, (k-1)')$. Figure 30.4.4 shows the quantity $a^T p$ for $k = 3$ (the loops at nodes 0 and 3 indicate the values of a_{00} and a_{33}).

Let $h \in \{1, \ldots, k-1\}$ and let $b^T p \geq 0$ denote the inequality obtained from

$a^T p \geq 0$ by duplicating node h; recall the definition of b from relation (26.6.3). Hence, if we denote the new node by h^*, then $b_{ij} = a_{ij}$ for $i,j \in V_{2k+1} \setminus \{k'\}$, $b_{h^*,h^*} = 0$, $b_{0,h^*} = -1$ and, for $i \in V_{2k+1} \setminus \{0, k'\}$, $b_{i,h^*} = 1$ if i is adjacent to h on the path P' and $b_{i,h^*} = 0$ otherwise. It is easy to check that one should set $b_{h,h^*} = 1$ in order to ensure that the inequality $b^T p \geq 0$ defines a facet of COR_{2k+1} (see Proposition 26.6.4).

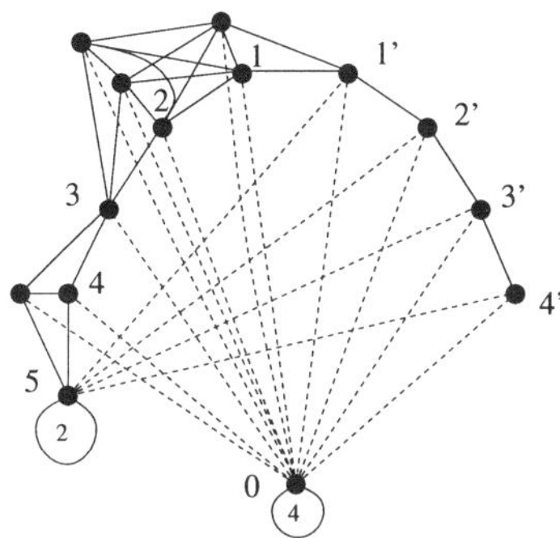

Figure 30.4.10: $B^T p \geq 0$ (facet of COR_{14})

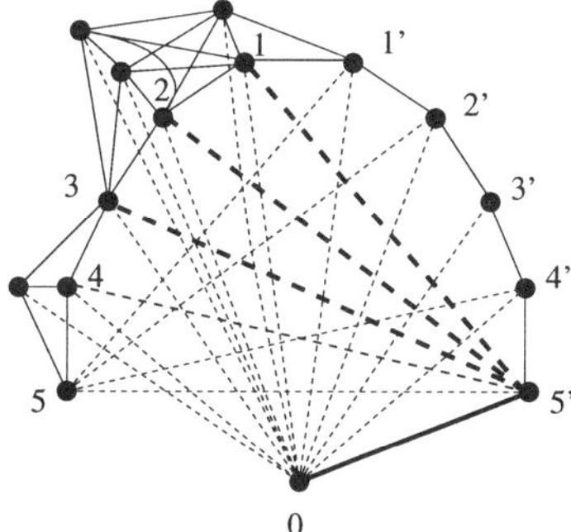

Figure 30.4.11: $C^T x \leq 0$ (facet of CUT_{15})

Of course, one can repeat this operation, i.e., introduce more nodes as duplicates of the nodes $h \in \{1, \ldots, k-1\}$. Namely, let $Q_1, \ldots, Q_{k-1}$ denote pairwise disjoint sets that are disjoint from V_{2k+1}. We build the inequality $B^T p \geq 0$, obtained by adding all nodes of Q_h successively as duplicates of node h and of the nodes of Q_h already introduced, for $h = 1, \ldots, k-1$. For instance, Figure 30.4.10 shows the quantity $B^T p$, where $k = 5$ and we have introduced one duplicate of node 1 and of node 4, and two duplicates of node 2.

As an exercise, let us formulate the inequality for the cut polytope corresponding to the inequality $B^T p \geq 0$. It is the inequality $C^T x \leq 0$, defined on the pairs of distinct elements of the set $V_{2k+1} \cup \bigcup_{1 \leq h \leq k-1} Q_h$, where

$C_{0k} = 0,$	
$C_{0,k'} = \sum_{1 \leq h \leq k-1} \lvert Q_h \rvert,$	
$C_{0i} = -1$	for $i \in Q_h \cup \{h, h'\} (1 \leq h \leq k-1),$
$C_{ij} = 1$	if ij is an edge of the path P,
$C_{i,1'} = 1$	for $i \in Q_1,$
$C_{ij} = 1$	for $i \in Q_h \cup \{h\}, j \in Q_{h+1} \cup \{h+1\}$ $(1 \leq h \leq k-1),$
$C_{ij} = 1$	for $i \neq j \in Q_h,$
$C_{k,i'} = -1$	for $i' = 1', 2', \ldots, k',$
$C_{k',i} = -(\lvert Q_{h-1} \rvert + \lvert Q_h \rvert + \lvert Q_{h+1} \rvert)$	for $i \in Q_h (1 \leq h \leq k-1),$

setting $Q_0 = Q_k = \emptyset$. For instance, the inequality $B^T p \geq 0$ from Figure 30.4.10 corresponds to the inequality $C^T x \leq 0$, which is shown in Figure 30.4.11 (with weight 4 on edge $(5', 0)$, weight -4 on edges $(5', 1), (5', 2), (5', 3)$ and weight -2 on edge $(5', 4)$). The next result is a direct application of Proposition 26.6.4.

Theorem 30.4.12. *The inequality $B^T p \geq 0$ defines a facet of the correlation polytope. Equivalently, the inequality $C^T x \leq 0$ defines a facet of the cut polytope.* ∎

30.5 Some Sporadic Examples

Grishukhin [1990] introduced the following inequality:

(30.5.1) $$\sum_{1 \leq i < j \leq 4} x_{ij} + x_{56} + x_{57} - x_{67} - x_{16} - x_{36} - x_{27} - x_{47} - 2 \sum_{1 \leq i \leq 4} x_{5i} \leq 0$$

and proved that it defines a facet of CUT_7. We also denote the inequality (30.5.1) as $(\mathrm{Gr}_7)^T x \leq 0$. Note that, if we collapse both nodes 6,7 in $(\mathrm{Gr}_7)^T x \leq 0$, then we obtain the hypermetric inequality $Q_6(1,1,1,1,-2,-1)^T x \leq 0$. In other words, $\mathrm{Gr}_7^T x \leq 0$ can be seen as a lifting of the inequality $Q_6(1,1,1,1,-2,-1)^T x \leq 0$. On the other hand, consider the following inequality (30.5.2), denoted as $(\mathrm{Gr}_8)^T x \leq 0$; it is introduced in De Simone, Deza and Laurent [1994] and shown to be a facet of CUT_8:

(30.5.2) $$\sum_{1 \leq i < j \leq 4} x_{ij} + x_{68} + x_{78} - x_{67} - x_{16} - x_{36} - x_{27} - x_{47} - \sum_{1 \leq i \leq 4} (x_{5i} + x_{8i}) \leq 0.$$

Observe that, if we collapse both nodes 5,8 in $(\mathrm{Gr}_8)^T x \leq 0$, we obtain $(\mathrm{Gr}_7)^T x \leq 0$. Hence, $(\mathrm{Gr}_7)^T x \leq 0$ is a nonpure inequality that comes as collapsing of the pure inequality $(\mathrm{Gr}_8)^T x \leq 0$. Figures 30.5.3, 30.5.4 show, respectively, the support graphs of the inequalities $(\mathrm{Gr}_7)^T x \leq 0$, $(\mathrm{Gr}_8)^T x \leq 0$. (In Figure 30.5.3, the thick dotted edge between node 5 and the circle enclosing nodes 1,2,3,4 indicates that node 5 is joined to all four nodes 1,2,3,4 by an edge with weight -2.)

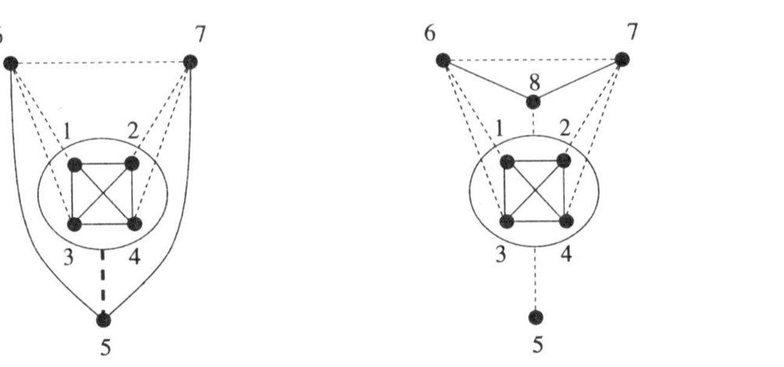

Figure 30.5.3: $(\mathrm{Gr}_7)^T x \leq 0$ Figure 30.5.4: $(\mathrm{Gr}_8)^T x \leq 0$

Kelly (unpublished manuscript) introduced the following class of valid inequalities. Consider a partition of the set $\{1, \ldots, n\}$ into $P \cup Q \cup \{n\}$, where $|P| = p, |Q| = q$ with $p,q \geq 2$ and $p + q + 1 = n$. Let K_p (resp. K_q) denote the complete graph on the set P (resp. Q). Set $t := pq - p^2 + 1$. Consider the following inequality, denoted as $(\mathrm{Kel}_n(p))^T x \leq 0$:

$$(p-1) \sum_{ij \in K_q} x_{ij} + (p+1) \sum_{ij \in K_p} x_{ij} - p \sum_{\substack{i \in Q \\ j \in P}} x_{ij} + (q-p-t) \sum_{i \in Q} x_{in} + t \sum_{i \in P} x_{in} \leq 0.$$

The following can be found in Deza and Laurent [1992a].

Proposition 30.5.5. *For $n \geq 5$, the inequality $(\mathrm{Kel}_n(p))^T x \leq 0$ is valid for* CUT_n. ∎

It is an open question to determine what are the parameters p and n for which the inequality $(\mathrm{Kel}_n(p))^T x \leq 0$ is facet inducing. Here is some partial information.

Proposition 30.5.6. *Assume $n \geq 7$. Then,*

(i) *The inequality $(\mathrm{Kel}_n(2))^T x \leq 0$ coincides (up to permutation) with the clique-web inequality $\mathrm{CW}_n^1(n-4, 2, 2, -1, \ldots, -1)^T x \leq 0$; hence, it is facet inducing for* CUT_n.

(ii) *The inequality $(\mathrm{Kel}_n(n-3))^T x \leq 0$ defines a simplex face of CUT_n of dimension $\binom{n}{2} - 3$.* ∎

30.6 Complete Description of CUT_n and $\mathrm{CUT}_n^\square$ for $n \leq 7$.

We present here the complete linear description[2] of the cut cone CUT_n and the cut polytope $\mathrm{CUT}_n^\square$ for $n \leq 7$.

For $n = 3, 4$, the only facet defining inequalities for CUT_n are the triangle inequalities, i.e.,

$$x_{ij} - x_{ik} - x_{jk} \leq 0$$

for distinct i, j, k in $\{1, \ldots, n\}$. Hence, CUT_3 (resp. $\mathrm{CUT}_3^\square$) has 3 (resp. 4) facets, while CUT_4 (resp. $\mathrm{CUT}_4^\square$) has 12 (resp. 16) facets.

For $n = 5$, the facets of CUT_5 are, up to permutation and switching, induced by one of the following inequalities:

[2] This linear description was obtained independently by several authors. In particular, the linear description of CUT_5 was obtained by Deza [1960, 1973a], Davidson [1969]; that of CUT_6 by Baranovskii [1971], McRae and Davidson [1972], Avis [1989]; and that of CUT_7 by Grishukhin [1990]. In fact, McRae and Davidson [1972] had already found the list of facets for CUT_7 and conjectured that it was complete.

1. $Q_5(1,1,-1,0,0)^T x \le 0$ (triangle inequality),
2. $Q_5(1,1,1,-1,-1)^T x \le 0$ (pentagonal inequality).

In total, CUT_5 (resp. $\mathrm{CUT}_5^{\square}$) has 30+10=40 facets (resp. 40+16=56 facets).

For $n = 6$, the facets of CUT_6 are, up to permutation and switching, induced by one of the following inequalities:

1. $Q_6(1,1,-1,0,0,0)^T x \le 0$,
2. $Q_6(1,1,1,-1,-1,0)^T x \le 0$,
3. $Q_6(2,1,1,-1,-1,-1)^T x \le 0$.

In total, CUT_6 has 60+60+90=210 facets and $\mathrm{CUT}_6^{\square}$ has 80+96+192=368 facets.

For $n = 7$, the facets of CUT_7 are, up to permutation and switching, induced by one of the following eleven inequalities:

1. $Q_7(1,1,-1,0,0,0,0)^T x \le 0$,
2. $Q_7(1,1,1,-1,-1,0,0)^T x \le 0$,
3. $Q_7(2,1,1,-1,-1,-1,0)^T x \le 0$,
4. $Q_7(1,1,1,1,-1,-1,-1)^T x \le 0$,
5. $Q_7(2,2,1,-1,-1,-1,-1)^T x \le 0$,
6. $Q_7(3,1,1,-1,-1,-1,-1)^T x \le 0$,
7. $\mathrm{CW}_7^1(1,1,1,1,1,-1,-1)^T x \le 0$,
8. $\mathrm{CW}_7^1(2,2,1,1,-1,-1,-1)^T x \le 0$,
9. $\mathrm{CW}_7^1(3,2,2,-1,-1,-1,-1)^T x \le 0$,
10. $(\mathrm{Par}_7)^T x \le 0$,
11. $(\mathrm{Gr}_7)^T x \le 0$.

Among the 11 types of facets of CUT_7, the first five are not simplices, the last five are not hypermetric, and five of them are pure (i.e., have all their coefficients equal to $0, 1, -1$) (namely, the $1^{st}, 2^{nd}, 4^{th}, 7^{th}$ and 10th ones).

Let F_i denote the facet of CUT_7 defined by the i-th inequality, for $i = 1, \ldots, 11$. It has been computed in Deza and Laurent [1992c] that the orbit $\Omega(F_i)$ (which consists of all the facets of CUT_7 that can be obtained from F_i by (root) switching and/or permutation) contains, respectively, 105, 210, 630, 35, 546, 147, 5292, 8820, 2205, 7560, and 13230 elements, for $i = 1, \ldots, 11$. Hence, among the 11 types of facets of CUT_7, the facet defined by the inequality $(\mathrm{Gr}_7)^T x \le 0$ is the one that has the largest number of distinct permutations and switchings, namely it has 13230 ones !

Therefore, CUT_7 has $\sum_{i=1}^{11} |\Omega(F_i)| = 105 + 210 + \ldots + 13230 = 38780$ facets. Using Lemma 26.3.11, one can compute that $\mathrm{CUT}_7^{\square}$ has 116764 facets.

The number of distinct (up to permutation) facets of CUT_7 has been computed in De Simone, Deza and Laurent [1994]; it is equal to 36. More precisely, let $\nu(F_i)$ denote the number of (root) switchings of F_i that are pairwise not permutation equivalent. For instance, $\nu(F_1) = 1$ as any two (root) switchings of the triangle inequality is again a triangle inequality. In fact, $\nu(F_i) = 1, 1, 2, 1, 3, 2, 4, 7, 5, 3, 7$, respectively, for $i = 1, \ldots, 11$. The distinct switchings of each F_i are described in detail in De Simone, Deza and Laurent [1994].

We summarize in Figure 30.6.1 some information on the facets of CUT_7. Namely, for each facet F_i ($i = 1, \ldots, 11$), we give:
- its number (S) of distinct (up to permutation) switchings by roots,
- its number (R) of roots,
- the size (o) $|\Omega(F_i)|$ of its orbit in CUT_7 (i.e., the number of distinct facets of CUT_7 that can be obtained from F_i by permutation and/or root switching),
- the size (O) $|\Omega^{\square}(F_i)|$ of its orbit in $\mathrm{CUT}_7^{\square}$ (i.e., the number of distinct facets of $\mathrm{CUT}_7^{\square}$ that can be obtained from F_i by permutation and/or switching).

F	F_1	F_2	F_3	F_4	F_5	F_6	F_7	F_8	F_9	F_{10}	F_{11}	total
S	1	1	2	1	3	2	4	7	5	3	7	36
R	48	40	30	35	26	21	21	21	21	21	21	
o	105	210	630	35	546	147	5292	8820	2205	7560	13230	38780
O	140	336	1344	64	1344	448	16128	26880	6720	23040	40320	116764

Figure 30.6.1: Data on the facets of CUT_7

Christof and Reinelt[3] [1996] have recently computed the facial description of the cut polytope on $n \leq 8$ points. For $n = 8$, they obtain $217,093,472$ facets for the cut polytope $\mathrm{CUT}_8^{\square}$ and $49,604,520$ facets for the cut cone CUT_8, that are subdivided into 147 orbits. The structure of these facets has not been analyzed and we cannot list them here as there are too many. So, the number of facets

[3]Christof and Reinelt [1996] provide a list of the distinct facets (up to permutation and switching) and compute for each facet its number of roots and the cardinality of its orbit in the cut cone and in the cut polytope. Therefore, the data from Figure 30.6.1 are reconfirmed. These informations are available on the following WWW site, which the reader may consult for the description of CUT_8 and $\mathrm{CUT}_8^{\square}$:
http://www.iwr.uni-heidelberg.de/iwr/comopt/soft/SMAPO/SMAPO.html.

grows dramatically fast from $n = 7$ to $n = 8$; $\mathrm{CUT}_8^\square$ has more than thousand times more facets than $\mathrm{CUT}_7^\square$! This is an indication that the structure of the facets of CUT_n is becoming more and more complicated with increasing values of n. Inequalities have no apparent symmetries and are seemingly very difficult to generalize. In fact, the facet $(\mathrm{Gr}_7)^T x \leq 0$ (from (5.9)) is the smallest (and unique for $n \leq 7$) example of a facet for which we could not find a proper generalization. This phenomenon of increasing complexity of the facial structure for large n is general for polytopes arising from hard optimization problems. For instance, the facial structure of the symmetric traveling salesman polytope is known for $n = 8$ (Boyd and Cunningham [1991], Christof, Jünger and Reinelt [1991]); it has been recently computed for $n = 9, 10$ by Padberg [1995] and by Christof and Reinelt [1996]. (There are 42,104,442 facets for $n = 9$ and more than 51,043,900,866 facets for $n = 10$.)

30.7 Additional Notes

We mention here some other interesting questions related to the study of the facets of the cut cone. First, we consider some subcones of CUT_n generated by subfamilies of cuts; we show that they inherit, in a sense, all the facets of CUT_n. Then, we consider the following three questions related to the collapsing operation in the cut cone: Does every facet collapse to some triangle facet ? Does every (nonpure) facet arise as collapsing of some pure facet ? Does every facet have the parity property ?

Transport of Facets to Other Subcones. Let $\mathcal{K}$ be a subset of the set of all cut vectors in K_n. One may also be interested in finding the facial structure of the cone $\mathbb{R}_+(\mathcal{K})$ or of the polytope $\mathrm{Conv}(\mathcal{K})$ for some specific cut families $\mathcal{K}$. A general problem is as follows: Which facets of the cut cone CUT_n do the polyhedra $\mathbb{R}_+(\mathcal{K})$ and $\mathrm{Conv}(\mathcal{K})$ inherit ? Clearly, any inequality which is facet inducing for CUT_n is valid for $\mathbb{R}_+(\mathcal{K})$ and $\mathrm{Conv}(\mathcal{K})$, but when does it induce a facet of the latter polyhedra ? Such a question has been looked at in the case when $\mathcal{K}$ consists of the even cut vectors, or of the inequicut vectors, or of the equicut vectors. A surprising feature of the even cut and inequicut cones ECUT_n, ICUT_n, and of the equicut polytope EQCUT_n is that they already "contain" all the facets of the cut cone. More precisely, every inequality defining a facet of the cut cone CUT_n can be zero-lifted to some facet of ECUT_m, ICUT_m, $\mathrm{EQCUT}_m^\square$, for any m large enough. The assertion (i) in Theorem 30.7.1 below is proved in Deza and Laurent [1993b] and (ii), (iii) in Deza, Fukuda and Laurent [1993].

Theorem 30.7.1. *Given $v \in \mathbb{R}^{E_n}$, integers $m \geq n$, define $v' \in \mathbb{R}^{\binom{m}{2}}$ by setting $v'_{ij} = v_{ij}$ for $1 \leq i < j \leq n$ and $v'_{ij} = 0$ for $1 \leq i \leq n < j \leq m$ and $n+1 \leq i < j \leq m$. Assume that the inequality $v^T x \leq 0$ defines a facet of the cut cone CUT_n. Then,*

(i) *The inequality $(v')^T x \leq 0$ defines a facet of the even cut cone ECUT_m for any m even, $m \geq n+5$.*

(ii) *The inequality* $(v')^T x \leq 0$ *defines a facet of the inequicut cone* ICUT_m *for any* m *such that* $n < \lfloor \frac{m}{2} \rfloor$.

(iii) *The inequality* $(v')^T x \leq 0$ *defines a facet of the equicut polytope* $\mathrm{EQCUT}_m^\square$ *for any* m *odd,* $m \geq 2n+1$. ∎

The valid inequalities of the cut cone CUT_n can also be transported to the k-uniform cut cone UCUT_n^k in the following way. Suppose that $1 \leq k \leq n-1$, $k \neq \frac{n}{2}$. Given $v \in \mathbb{R}^{E_{n+1}}$, define $v^* \in \mathbb{R}^{E_n}$ by setting

$$v_{ij}^* := v_{ij} + \frac{v_{i,n+1} + v_{j,n+1}}{n-2k} - \frac{v^T \delta(\{n+1\})}{(n-k)(n-2k)}$$

for $1 \leq i < j \leq n$. If the inequality $v^T x \leq 0$ is valid for the cut cone CUT_{n+1}, then the inequality $(v^*)^T x \leq 0$ is valid for the k-uniform cut cone UCUT_n^k (Deza and Laurent [1992e]). For example, for $1 \leq i < j \leq n$, if $v^T x := x_{i,n+1} - x_{j,n+1} - x_{ij} \leq 0$ is a triangle inequality, then $(v^*)^T x \leq 0$ is

$$\sum_{1 \leq h \leq n, h \neq i,j} (x_{ih} - x_{jh}) - (n-2k) x_{ij} \leq 0.$$

If $v^T x := x_{ij} - x_{i,n+1} - x_{j,n+1}$, then $(v^*)^T x \leq 0$ is

$$2 \sum_{1 \leq h < l \leq n} x_{hl} - (n-k) \sum_{1 \leq h \leq n, h \neq i,j} (x_{ih} + x_{jh}) + (n-k)(n-2k-2) x_{ij} \leq 0.$$

In fact, the 2-uniform cut cone UCUT_n^2 is a simplex cone, which is completely described by the latter $\binom{n}{2}$ inequalities (Deza, Fukuda and Laurent [1993]).

Questions Related to the Collapsing Operation. We address now the following three questions about the facets of CUT_n:
Question 1: Does every facet of the cut cone collapse to some triangle facet ?
Question 2: Does every (nonpure) facet arise as collapsing of some pure facet ?
Question 3: If the inequality $v^T x \leq 0$ defines a facet of CUT_n, is it the case that $v^T \delta(S)$ is an even number for every cut vector $\delta(S)$?

Recall that the collapsing operation preserves valid inequalities but not necessarly facets. Call a facet *tight* if none of its collapsings is facet inducing. Hence, the answer to Question 1 is "yes" precisely if the only tight facet of the cut cone CUT_n (for any n) is the triangle facet. A probably more reasonable conjecture is the following: The number of tight facets of the cut cone is finite.

We have checked that most of the known classes of facets of CUT_n do indeed collapse to some triangle facet. (As an example, let us consider the pure clique-web inequality: $(\mathrm{CW}_n^r)^T x \leq 0$; the inequality obtained from it by collapsing all the nodes from the set $\{1, \ldots, n\} \setminus \{1, r+2\}$ into a single node, say u, is precisely the triangle inequality $x_{1,r+2} - x_{1u} - x_{u,r+2} \leq 0$.) In fact, there are often several ways of collapsing a given facet to some triangle facet. Collapsings to some triangle facet are given explicitly for all the facets of CUT_7 in De Simone, Deza and Laurent [1994].

Given a facet inducing inequality $v^T x \leq 0$, a *purification* of it is any pure inequality valid for a larger cut cone and admitting a collapsing which is precisely $v^T x \leq 0$. As observed in De Simone [1992], such a purification always exists. But, Question 2 asks whether every facet admits a purification which is facet inducing. The answer is "yes", by construction, for the class of clique-web facets. Also, in Section 30.5, we mentioned explicitly the inequality (30.5.2), which is a purification of the nonpure inequality (30.5.1). However, we do not know the answer to Question 2 for the classes of suspended-tree inequalities, or of path-block-cycle inequalities. It is a challenging problem to find some large class of pure facets from which the facets (30.1.1),(30.1.7),(30.1.8), or (30.1.11) could be deduced by collapsing (recall the remark preceding Proposition 30.1.3).

If the answer to Question 3 is "yes", we say that the inequality $v^T x \leq 0$ has the *parity property*. We also say that the vector v has the parity property if $v^T \delta(S)$ is even for all cut vectors $\delta(S)$. Let e_{ij} $(1 \leq i < j \leq n)$ denote the coordinate vectors in $\mathbb{R}^{E_n}$. It can be easily checked that $v \in \mathbb{R}^{E_n}$ has the parity property if and only if v can be written as an integer combination of the triangle vectors $e_{ij} + e_{ik} + e_{jk}$ (for $1 \leq i < j < k \leq n$) and of the double edge vectors $2e_{ij}$ (for $1 \leq i < j \leq n$), i.e., if

$$v \in \mathbb{Z}(e_{ij} + e_{ik} + e_{jk} \ (1 \leq i < j < k \leq n), \ 2e_{ij} \ (1 \leq i < j \leq n)).$$

Observe that the parity property is preserved under switching and collapsing. We have checked that every known class of facets of the cut cone enjoys the parity property.

It is an interesting problem to look for a facet of $\mathrm{CUT}_n^{\square}$ that does not have the parity property; a good candidate would be some inequality of the form

$$v^T x := \sum_{ij \in E} x_{ij} \leq v_0,$$

where E is the edge set of a regular graph of odd degree and v_0 is the maximum size of a cut in the graph. (Note that both assumptions of validity and full rank are necessary for the parity property. Indeed, it is easy to construct some valid inequality which is not facet inducing and does not have the parity property, or some nonvalid inequality whose set of roots has full rank and which does not have the parity property.)

As an illustration, we describe below the explicit decomposition of some facet defining inequalities in the lattice $\mathbb{Z}(e_{ij}+e_{ik}+e_{jk} \ (1 \leq i < j < k \leq n), \ 2e_{ij} \ (1 \leq i < j \leq n))$. We set

$$T(i,j;k) := x_{ij} - x_{ik} - x_{jk}.$$

- For the facet defined by the parachute inequality: $(\mathrm{Par}_{2k+1})^T x \leq 0$ (k odd),

$$(\mathrm{Par}_{2k+1})^T x = \sum_{1 \leq i \leq k-1} (T(i,i+1;a_{i'}) + T(i',(i+1)';a_i)) + T(1,1';0) - T(k,k';0),$$

where $a_i := k$, $a_{i'} := k'$ for i odd, and $a_i = a_{i'} := 0$ for i even.

- For the facet defined by the inequality: $(\mathrm{Gr}_7)^T x \leq 0$,

$$\begin{aligned}(\mathrm{Gr}_7)^T x = \; & T(1,2;5) + T(1,3;5) + T(1,4;6) + T(2,3;7) \\ & + T(2,4;5) + T(3,4;5) - T(6,7;5).\end{aligned}$$

- For the hypermetric facet: $Q_n(-(n-4), -1, 1, \ldots, 1)^T x \leq 0$,

$$Q_n(-(n-4), -1, 1, \ldots, 1)^T x = -(\left\lfloor \frac{n}{2} \right\rfloor - 2) 2x_{12} + \sum_{3 \leq i < j \leq n} T(i,j;a_{ij}) - T_0,$$

where $a_{ij} := 2$ if $ij = (2t+1, 2t+2)$ for $1 \leq t \leq \lfloor \frac{n}{2} \rfloor - 1$, and $a_{ij} := 1$ otherwise, and $T_0 := T(2,n;1)$ if n is odd and $T_0 := 0$ if n is even.

- For the facet defined by the clique-web inequality: $(\mathrm{CW}^r_{2r+5})^T x \leq 0$,

$$\begin{aligned}(\mathrm{CW}^r_{2r+5})^T x = \; & \textstyle\sum_{i=1}^{r+1} (T(i, i+r+1; 2r+4) + T(i, i+r+2; 2r+5)) \\ & + T(r+2, 2r+3; 2r+4) - T(r+2, 2r+5; 2r+4).\end{aligned}$$

Chapter 31. Geometric Properties

This chapter contains several results of geometric type for the cut polytope $\mathrm{CUT}_n^{\square}$. One of our objectives here is to study the geometric shape of $\mathrm{CUT}_n^{\square}$, in particular, in connection with its linear relaxation by the semimetric polytope $\mathrm{MET}_n^{\square}$ and with its convex (nonpolyhedral) relaxation by the elliptope $\mathcal{E}_n$.

We have already seen (in Section 26.3.3) that the polytope $\mathrm{CUT}_n^{\square}$ has a lot of symmetries. We are interested, for instance, in the following further questions: What are the edges of the polytope $\mathrm{CUT}_n^{\square}$? More generally, what is the structure of its faces of small dimension ? We can, in some sense, give an answer to this question up to dimension $\log_2 n$. Indeed, it turns out that $\mathrm{CUT}_n^{\square}$ has a lot of faces of dimension up to $\log_2 n$ in common with its relaxations $\mathrm{MET}_n^{\square}$ and $\mathcal{E}_n$ that arise by taking sets of cuts in general position (see Theorems 31.5.9 and 31.6.4).

As we have seen in the rest of Part V, $\mathrm{CUT}_n^{\square}$ has a great variety of facets, most of them having a very complicated structure. A legitimate question to ask is which ones are the most important among them ? Giving a precise definition of the word "important" in this question is not an easy task. However, it is intuitively clear that some facets are more essential than others; some facets have indeed a "big area" while some others contribute only to rounding off some little corners of the polytope. One way of measuring the importance of a facet is by computing the Euclidean distance of the hyperplane containing the facet to the barycentrum of $\mathrm{CUT}_n^{\square}$. It seems intuitively clear that facets that are close to the barycentrum are more important than facets that are far apart. It is conjectured that the triangle facets are the closest facets to the barycentrum; see Section 31.7 for results related to this conjecture. We remind from Chapter 27 that triangle inequalities share several other interesting properties.

The cut polytope is not a simplicial polytope (if $n \geq 5$) as some of its facets are not simplices. However, it seems that the great majority of its facets are simplices. This has been verified for $n \leq 7$, where it has been computed that about 97% of the facets are simplices. We group in Section 31.8 results on the simplex facets of $\mathrm{CUT}_n^{\square}$.

Section 31.5 presents several geometric properties of the elliptope $\mathcal{E}_n$, which was defined in Section 28.4.1 as the set of $n \times n$ symmetric positive semidefinite matrices with an all-ones diagonal. Up to a simple transformation, $\mathcal{E}_n$ is a (nonpolyhedral) relaxation of the cut polytope $\mathrm{CUT}_n^{\square}$.

One more interesting interpretation of the cut polytope is mentioned in Section 31.2; namely, the fact that the valid inequalities for $\mathrm{CUT}_n^{\square}$ yield inequalities for the pairwise angles among a set of n unit vectors in $\mathbb{R}^n$. (This is essentially a reformulation of the fact, stated in Section 6.4, that spherical distance spaces are ℓ_1-embeddable.) We describe in Section 31.3 some further implications of this result in connection with the completion problem for partial positive semidefinite matrices. In fact, this problem amounts to the description of projections of the elliptope $\mathcal{E}_n$. In general, the projected elliptope $\mathcal{E}(G)$ is contained in the image of $\mathrm{CUT}^{\square}(G)$ under the mapping $x \mapsto \cos(\pi x)$. It turns out that both bodies coincide when the graph G has no K_4-minor (see Theorem 31.3.7). Further results are given for larger classes of graphs in Section 31.3.

In Section 31.4 we consider the analogue completion problem for Euclidean distance matrices. In fact, this problem is nothing but the problem of describing projections of the negative type cone NEG_n. It turns out that there are several results for this problem, which are in perfect analogy with the known results for the positive semidefinite completion problem. We mention in Section 31.4.2 how the two completion problems can be linked (using, in particular, one of the metric transforms which was exposed in Chapter 9, namely, the Schoenberg transform).

In Section 31.1 we describe how cuts have been used for disproving a long standing conjecture of Borsuk.

31.1 Disproval of a Conjecture of Borsuk Using Cuts

The following question was asked by Borsuk [1933] more than sixty years ago:

> *Given a set X of points in $\mathbb{R}^d$, is it always possible to partition X into $d+1$ subsets, each having a smaller diameter than X ?*

We recall that the *diameter*[1] of a set $X \subseteq \mathbb{R}^d$ is defined as

$$\mathrm{diam}(X) := \max_{x,y \in X} \parallel x - y \parallel_2,$$

the maximum Euclidean distance between any two points of X. Borsuk's question has been answered in the negative by Kahn and Kalai [1993], who constructed a counterexample using cut vectors. We present here a variation of their counterexample, which is due to Nilli [1994].

Let $n = 4p$ where p is an odd prime integer, and $d := \binom{n}{2}$. As set of points $X \subseteq \mathbb{R}^d$, we take the set

$$X := \{\delta(S) \mid S \subseteq V_n,\ |S| \text{ is even and } 1 \in S\}$$

of all even cut vectors in K_n; hence, $|X| = 2^{n-2}$. Then, X provides a counterexample to Borsuk's question in the case when

[1]For a polytope P, there is another notion of diameter besides the geometric notion considered here. Namely, the diameter of P is also sometimes defined as the diameter of its 1-skeleton graph; for instance, the diameter (of the 1-skeleton graph of) the cut polytope is 1 (see Section 31.6).

(31.1.1) $$\frac{2^{n-2}}{\sum_{i=0}^{p-1}\binom{n-1}{i}} > \binom{n}{2} + 1.$$

The smallest counterexample occurs in dimension $d = \binom{44}{2} = 946$ for $n = 44$, $p = 11$. The proof is based on the following result of Nilli [1994].

Lemma 31.1.2. *Let $n = 4p$ with p odd prime and let $\mathcal{E}$ denote the set of vectors $x \in \{\pm 1\}^n$ such that $x_1 = 1$ and x has an even number of positive components. If $\mathcal{F} \subseteq \mathcal{E}$ contains no two orthogonal vectors, then $|\mathcal{F}| \leq \sum_{p=0}^{p-1}\binom{n-1}{i}$.*

Proof. Observe that the scalar product of two elements $a, b \in \mathcal{E}$ is divisible by 4. Hence, by the assumption, $a^T b \not\equiv 0 \pmod{p}$ for any $a \neq b \in \mathcal{F}$. For each $a \in \mathcal{F}$, we consider the polynomial P_a in the variables $X_1, \ldots, X_n$ defined by

$$P_a(X) := \prod_{i=1}^{p-1}(\sum_{j=1}^{n} a_j X_j - i).$$

Then,

(i) $P_a(b) \equiv 0 \pmod{p}$ for all $a \neq b \in \mathcal{F}$,

(ii) $P_a(a) \not\equiv 0 \pmod{p}$ for all $a \in \mathcal{F}$.

Let Q_a denote the polynomial obtained from P_a by developing it and repeatedly replacing the product X_i^2 by 1 for each $i = 1, \ldots, n$. Hence, $Q_a(x) = P_a(x)$ for all $x \in \{\pm 1\}^n$. Therefore, Q_a also satisfies the relations (i),(ii) above. These relations permit to check that the set $\{Q_a \mid a \in \mathcal{F}\}$ is linearly independent over the field $GF(p)$. Hence, $|\mathcal{F}|$ is less than or equal to the dimension of the space of polynomials in $n-1$ variables (as $x_1 = 1$) of degree at most $p-1$ over $GF(p)$, which is precisely $\sum_{i=0}^{p-1}\binom{n-1}{i}$. ∎

We now show that the set X of all even cut vectors cannot be partitioned into $d+1$ subsets of smaller diameter. It turns out to be more convenient to work with ± 1-valued vectors rather than with the $(0,1)$-valued cut vectors. In other words, we show that the set

$$X_1 := \{xx^T \mid x \in \mathcal{E}\}$$

cannot be partitioned into $d+1$ subsets of smaller diameter if the condition (31.1.1) holds ($\mathcal{E}$ is defined as in Lemma 31.1.2). (Note that xx^T is the $n \times n$ symmetric matrix with entries $x_i x_j$ and, thus, all its diagonal entries are equal to 1. Hence, the vectors xx^T ($x \in \{\pm 1\}^n$) lie, in fact, in the space of dimension d.) Given $x, y \in \mathcal{E}$, we have

$$(\| xx^T - yy^T \|_2)^2 = 2n^2 - 2(x^Ty)^2 \leq 2n^2$$

with equality if $x^T y = 0$. Hence, the diameter of X_1 is equal to $n\sqrt{2}$. Suppose that X_1 is partitioned into s subsets $Y^1 \cup \ldots \cup Y^s$, where each Y^i has diameter $< n\sqrt{2}$. Then, no two vectors in Y^i are orthogonal. We deduce from Lemma 31.1.2

that $|Y^i| \leq \sum_{j=0}^{p-1}\binom{n-1}{j}$ for all i. This implies that $2^{n-2} \leq s\sum_{j=0}^{p-1}\binom{n-1}{j}$. Therefore, the condition (31.1.1) implies that $s > \binom{n}{2} + 1 = d + 1$. This shows that, under the condition (31.1.1), the set X_1 (or X) cannot be partitioned into $d+1$ subsets of smaller diameter.

31.2 Inequalities for Angles of Vectors

Let $v_1, \ldots, v_n$ be n unit vectors in $\mathbb{R}^m$ $(m \geq 1)$. Set

$$\theta_{ij} := \arccos(v_i^T v_j) \quad \text{for } 1 \leq i < j \leq n.$$

We consider the question of determining valid inequalities that are satisfied by the angles θ_{ij}. A classical result in 3-dimensional geometry asserts that

$$\theta_{12} \leq \theta_{13} + \theta_{23},\ \theta_{13} \leq \theta_{12} + \theta_{23},\ \theta_{23} \leq \theta_{12} + \theta_{13},\ \theta_{12} + \theta_{13} + \theta_{23} \leq 2\pi$$

for the pairwise angles among three vectors in $\mathbb{R}^3$ (see Theorem 31.2.2 below). Observe that the above inequalities are nothing but the triangle inequalities (for the variable $\frac{\theta}{\pi}$). An analogue result holds in any dimension $m \geq 3$, as was shown in Theorem 6.4.5. We repeat the result here for convenience.

Theorem 31.2.1. *Let $v_1, \ldots, v_n$ be n unit vectors in $\mathbb{R}^m$ $(n \geq 3,\ m \geq 1)$. Let $a \in \mathbb{R}^{E_n}$ and $a_0 \in \mathbb{R}$ such that the inequality $a^T x \leq a_0$ is valid for the cut polytope $\mathrm{CUT}_n^{\square}$. Then,*

$$\sum_{1 \leq i < j \leq n} a_{ij} \arccos(v_i^T v_j) \leq \pi a_0.$$

∎

Therefore, the valid inequalities for the cut polytope $\mathrm{CUT}_n^{\square}$ have the following nice interpretation: They yield valid inequalities for the pairwise angles among a set of n unit vectors. A whole wealth of such inequalities have been presented in the preceding paragraphs. As an example,

$$\sum_{1 \leq i < j \leq n} \arccos(v_i^T v_j) \leq \left\lfloor \frac{n}{2} \right\rfloor \left\lceil \frac{n}{2} \right\rceil \pi$$

for any n unit vectors $v_1, \ldots, v_n$. The question of determining the maximum value for the sum of pairwise angles among a set of vectors was first asked by Fejes Tóth [1959]; he conjectured that the above inequality holds and proved that this is the case for $n \leq 6$. The even case $n = 2p$ was settled by Sperling [1960] and the general case by Kelly [1970b].

In case $n = 3$ the statement from Theorem 31.2.1 can, in fact, be formulated as an equivalence[2].

Theorem 31.2.2. *The following assertions are equivalent for $\alpha, \beta, \gamma \in [0, \pi]$.*

[2]This fact has been known since long; see, e.g., Blumenthal [1953] (Lemma 43.1), or Berger [1987] (Corollary 18.6.10) or, more recently, Barrett, Johnson and Tarazaga [1993].

(i) *The matrix*

$$A := \begin{pmatrix} 1 & \cos\alpha & \cos\beta \\ \cos\alpha & 1 & \cos\gamma \\ \cos\beta & \cos\gamma & 1 \end{pmatrix}$$

is positive semidefinite.

(ii) *There exist three unit vectors* $v_1, v_2, v_3 \in \mathbb{R}^3$ *such that* $\alpha = \arccos(v_1^T v_2)$, $\beta = \arccos(v_1^T v_3)$ *and* $\gamma = \arccos(v_2^T v_3)$.

(iii) $\alpha \leq \beta + \gamma$, $\beta \leq \alpha + \gamma$, $\gamma \leq \alpha + \beta$ *and* $\alpha + \beta + \gamma \leq 2\pi$.

Proof. Clearly, (i) $\Longleftrightarrow$ (ii). Now, $\det A$ can be expressed as:

$$\begin{aligned}
\det A &= 1 + 2\cos\alpha \cdot \cos\beta \cdot \cos\gamma - \cos^2\alpha - \cos^2\beta - \cos^2\gamma \\
&= (1 - \cos^2\beta - \cos^2\gamma + \cos^2\beta \cdot \cos^2\gamma) \\
&\quad - (\cos^2\alpha + \cos^2\beta \cdot \cos^2\gamma - 2\cos\alpha \cdot \cos\beta \cdot \cos\gamma) \\
&= (1 - \cos^2\beta)(1 - \cos^2\gamma) - (\cos\alpha - \cos\beta \cdot \cos\gamma)^2 \\
&= \sin^2\beta \cdot \sin^2\gamma - (\cos\alpha - \cos\beta \cdot \cos\gamma)^2 \\
&= (\cos(\beta - \gamma) - \cos\alpha) \cdot (\cos\alpha - \cos(\beta + \gamma))
\end{aligned}$$

Hence, $A \succeq 0 \Longleftrightarrow \det A \geq 0 \Longleftrightarrow |\beta - \gamma| \leq \alpha \leq \beta + \gamma \leq 2\pi - \alpha$, which is equivalent to (iii). ∎

Some generalizations of this result will be presented in the next subsection; see, in particular, Theorem 31.3.7.

31.3 The Positive Semidefinite Completion Problem

We consider here the elliptope $\mathcal{E}_n$ and its projections on subsets of the entries. We recall from Section 28.4 that

$$\mathcal{E}_n = \{Y \ n \times n \text{ symmetric matrix} \mid Y \succeq 0,\ y_{ii} = 1\ \forall i = 1, \ldots, n\}.$$

Given a subset E of $E_n := \{ij \mid 1 \leq i < j \leq n\}$, consider the graph $G := (V_n, E)$ and the projection $\mathcal{E}(G)$ of $\mathcal{E}_n$ on the subspace $\mathbb{R}^E$, i.e.,

$$\mathcal{E}(G) := \{x \in \mathbb{R}^E \mid \exists Y = (y_{ij}) \in \mathcal{E}_n \text{ such that } x_{ij} = y_{ij}\ \forall ij \in E\}.$$

Hence, $\mathcal{E}_n$ and $\mathcal{E}(K_n)$ are in one-to-one correspondence as the elements of $\mathcal{E}(K_n)$ are precisely the upper triangular parts of the matrices in $\mathcal{E}_n$.

Given a graph $G = (V_n, E)$ and $x \in \mathbb{R}^E$, denote by X the partial symmetric $n \times n$ matrix whose off-diagonal entries are specified only on the positions corresponding to edges in G (and the symmetric ones); the ijth-entry of X is x_{ij} for $ij \in E$ and the diagonal entries of X are all equal to 1. Then, $x \in \mathcal{E}(G)$ if and only if the partial matrix X can be completed to a positive semidefinite matrix. Hence, the positive semidefinite completion problem, which was introduced in Section 28.4, is the problem of testing membership in the elliptope $\mathcal{E}(G)$.

This problem has received a lot of attention in the literature, especially within the community of linear algebra. This is due, in particular, to its many applications (e.g., to probability and statistics, engineering, etc.) and to its close connection with other important matrix properties such as Euclidean distance matrices. (See, e.g., the survey of Johnson [1990] for a broad survey on completion problems.) We present here some results about the positive semidefinite completion problem that are most relevant to the topic of this book, namely, to cut and semimetric polyhedra. Indeed, it turns out that, for some graphs, the elliptope $\mathcal{E}(G)$ has a closed form description involving the cut and semimetric polytopes of G. We give here a compact presentation covering results obtained by several authors. The exposition in this section as well as in the next Section 31.4 follows essentially the survey paper by Laurent [1997d].

31.3.1 Results

Let $G = (V_n, E)$ be a graph and let $x \in \mathbb{R}^E$ with corresponding partial matrix X. Clearly, if $x \in \mathcal{E}(G)$ then every principal submatrix of X whose entries are all specified is positive semidefinite. In other words, if $K \subseteq V_n$ induces a clique in G then the projection x_K of x on the edge set of $G[K]$ belongs to the elliptope $\mathcal{E}(K)$ of the clique K. (Here, we use the same letter K for denoting the clique as a node set or as a graph.) Hence,

(31.3.1) $\quad x_K \in \mathcal{E}(K)$ for each clique K in G

is a necessary condition for $x \in \mathcal{E}(G)$, called *clique condition.* Another necessary condition for membership in $\mathcal{E}(G)$ can be deduced from the result in Section 31.2. Clearly, all the components of $x \in \mathcal{E}(G)$ belong to the interval $[-1, 1]$; hence, x can be parametrized as

$$x = \cos(\pi a), \quad \text{i.e.,} \quad x_e = \cos(\pi a_e) \text{ for all } e \in E,$$

where $0 \leq a_e \leq 1$ for all $e \in E$. Then, Theorem 31.2.1 can be reformulated as

$$\mathcal{E}(K_n) \subseteq \cos(\pi \mathrm{CUT}_n^\square) := \{\cos(\pi a) \mid a \in \mathrm{CUT}_n^\square\}.$$

By taking the projections of both sides on the subspace $\mathbb{R}^E$ indexed by the edge set of G, we obtain

$$\mathcal{E}(G) \subseteq \cos(\pi \mathrm{CUT}^\square(G)) := \{\cos(\pi a) \mid a \in \mathrm{CUT}^\square(G)\}.$$

In other words,

(31.3.2) $\quad a \in \mathrm{CUT}^\square(G)$

is a necessary condition for $x = \cos(\pi a) \in \mathcal{E}(G)$, called *cut condition.* As $\mathrm{CUT}^\square(G) \subseteq \mathrm{MET}^\square(G)$ (by (27.3.1)) we deduce that

(31.3.3) $\quad a \in \mathrm{MET}^\square(G)$

is also a necessary condition for $x = \cos(\pi a) \in \mathcal{E}(G)$, called *metric condition.*

None of the conditions (31.3.1), (31.3.2), or (31.3.3) suffices for characterizing $\mathcal{E}(G)$ in general. For instance, let $C = (V_n, E)$ be a circuit on $n \geq 4$ nodes and let $x \in \mathbb{R}^E$ be defined by $x_e := 1$ for all edges except $x_e := -1$ for one edge of C. Then, x satisfies (31.3.1) but $x \notin \mathcal{E}(C)$. As another example, consider the 4×4 matrix X with diagonal entries 1 and with off-diagonal entries $-\frac{1}{2}$. Then, $X \notin \mathcal{E}_4$ (as X is not positive semidefinite because $Xe = -\frac{1}{2}e$, where e denotes the all ones vector). Hence, the vector $x := (-\frac{1}{2}, \ldots, -\frac{1}{2}) \in \mathbb{R}^{E(K_4)}$ does not belong to $\mathcal{E}(K_4)$, while $\frac{1}{\pi} \arccos x = (\frac{2}{3}, \ldots, \frac{2}{3})$ belongs to $\mathrm{MET}^{\square}(K_4) = \mathrm{CUT}^{\square}(K_4)$.

Hence arises the question of characterizing the graphs G for which the conditions (31.3.1), (31.3.2), (31.3.3) (taken together or separately) suffice for the description of $\mathcal{E}(G)$. Let $\mathcal{P}_K$ (resp. $\mathcal{P}_M$, $\mathcal{P}_C$) denote the class of graphs G for which the clique condition (31.3.1) (resp. the metric condition (31.3.3), the cut condition (31.3.2)) is sufficient for the description of $\mathcal{E}(G)$.

We start with the description of the class $\mathcal{P}_K$. Recall that a graph is said to be *chordal* if every circuit of length ≥ 4 has a chord. We will also use the following characterization from Dirac [Di61]: A graph is chordal if and only if it can be obtained from cliques by means of clique sums.

Clearly, every graph $G \in \mathcal{P}_K$ must be chordal. (For, suppose that C is a chordless circuit in G of length ≥ 4; define $x \in \mathbb{R}^E$ by setting $x_e := 1$ for all edges e in C except $x_{e_0} := -1$ for one edge e_0 in C, and $x_e := 0$ for all remaining edges in G. Then, x satisfies (31.3.1) but $x \notin \mathcal{E}(G)$.) Grone, Johnson, Sá, and Wolkowicz [1984] show that $\mathcal{P}_K$ consists precisely of the chordal graphs. Namely,

Theorem 31.3.4. *For a graph $G = (V, E)$, we have*

$$\mathcal{E}(G) = \{x \in \mathbb{R}^E \mid x_K \in \mathcal{E}(K) \quad \forall K \text{ clique in } G\}$$

if and only if G is chordal.

The proof relies upon Lemma 31.3.5 below, since cliques belong trivially to $\mathcal{P}_K$ and every chordal graph can be build from cliques by taking clique sums.

Lemma 31.3.5. *The class $\mathcal{P}_K$ is closed under taking clique sums.*

Proof. Let $G = (V, E)$ be the clique sum of two graphs $G_1 = (V_1, E_1)$ and $G_2 = (V_2, E_2)$. Suppose that $G_1, G_2 \in \mathcal{P}_K$; we show that $G \in \mathcal{P}_K$. For this, let $x \in \mathbb{R}^E$ such that $x_K \in \mathcal{E}(K)$ for every clique K in G. Then, for $i = 1, 2$, the projection of x on the subspace $\mathbb{R}^{E_i}$ belongs to $\mathcal{E}(G_i)$ and, thus, can be completed to a positive semidefinite matrix of order $|V_i|$. Hence, we can find vectors $u_j \in \mathbb{R}^k$ $(j \in V_1)$ and $v_j \in \mathbb{R}^k$ $(j \in V_2)$ such that $x_{ij} = u_i^T u_j$ for all $i, j \in V_1$ and $x_{ij} = v_i^T v_j$ for all $i, j \in V_2$. Now, by looking at the values on the common clique $V_1 \cap V_2$, we have that $u_i^T u_j = v_i^T v_j$ for all $i, j \in V_1 \cap V_2$. Hence, there exists an orthogonal $k \times k$ matrix A such that $Au_i = v_i$ for all $i \in V_1 \cap V_2$.

Now, the Gram matrix of the system of vectors: Au_i $(i \in V_1)$, v_i $(i \in V_2 \setminus V_1)$ provides a positive semidefinite completion of x, which shows that $x \in \mathcal{E}(G)$. ∎

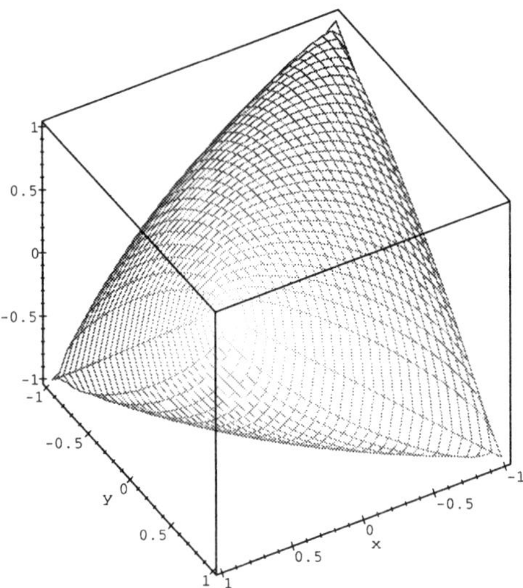

Figure 31.3.6: The elliptope $\mathcal{E}(K_3)$ of the complete graph on 3 nodes

We now turn to the description of the classes $\mathcal{P}_M$ and $\mathcal{P}_C$. Obviously,

$$\mathcal{P}_M \subseteq \mathcal{P}_C.$$

By Theorem 31.2.2 the graph K_3 belongs to $\mathcal{P}_M$. In other words, $\mathcal{E}(K_3) = \cos(\pi \mathrm{MET}_3^\square)$. Thus, $\mathcal{E}(K_3)$ is a 'deformation' via the cosine mapping of the 3-dimensional simplex $\mathrm{MET}_3^\square$; see Figure 31.3.6 for a picture of the elliptope $\mathcal{E}(K_3)$. As was observed earlier, the graph K_4 does not belong to $\mathcal{P}_C$. Laurent [1997b] shows that the classes $\mathcal{P}_M$ and $\mathcal{P}_C$ are identical and consist precisely of the graphs with no K_4-minor.

Theorem 31.3.7. *The following assertions are equivalent for a graph G:*

(i) $\mathcal{E}(G) = \{x = \cos(\pi a) \mid a \in \mathrm{CUT}^\square(G)\}$.

(ii) $\mathcal{E}(G) = \{x = \cos(\pi a) \mid a \in \mathrm{MET}^\square(G)\}$.

(iii) *G has no K_4-minor.*

The proof relies essentially upon the following decomposition result for graphs with no K_4-minor[3] [4] (see Duffin [1965]): A graph G has no K_4-minor if and

[3]A graph with no K_4-minor is also known under the name of (simple) *series-parallel graph*. We stress 'simple' as series-parallel graphs are allowed in general to contain loops and multiple edges. But, here, we consider only simple graphs.

[4]From this follows that every graph with no K_4-minor is a subgraph of a chordal graph (on the same node set) containing no clique of size 4.

only if $G = K_3$, or G is a subgraph of a clique k-sum ($k = 0, 1, 2$) of two smaller graphs (i.e., with less nodes than G), each having no K_4-minor. We state two intermediary results.

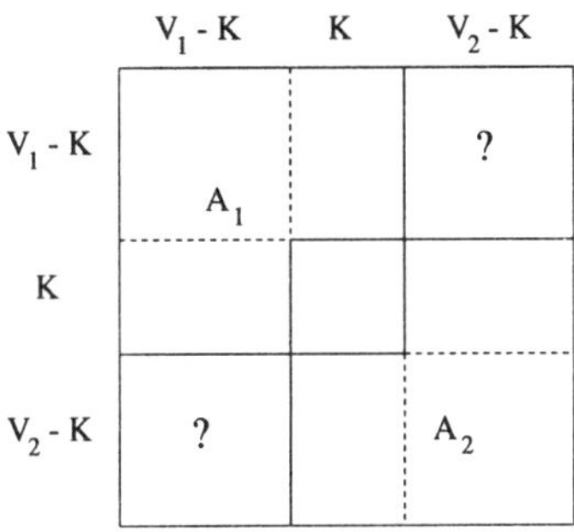

Figure 31.3.8

Lemma 31.3.9. *Each of the classes $\mathcal{P}_M$ and $\mathcal{P}_C$ is closed under taking minors.*

Proof. Let $G = (V, E)$ be a graph on $n = |V|$ nodes, let $e = uv$ be an edge in G and let G' be the graph obtained from G by deleting or contracting the edge e. We show that $G' \in \mathcal{P}_M$ (resp. $G' \in \mathcal{P}_C$) whenever $G \in \mathcal{P}_M$ (resp. $G \in \mathcal{P}_C$).

We first consider the case when $G' = G \backslash e$ is obtained by deleting e. We suppose first that $G \in \mathcal{P}_M$; we show that $G' \in \mathcal{P}_M$. For this, let $a \in \mathrm{MET}^{\square}(G')$; we show that $\cos(\pi a) \in \mathcal{E}(G')$. Let $b \in \mathrm{MET}^{\square}(G)$ whose projection on the edge set of G' is a. Then, $\cos(\pi b) \in \mathcal{E}(G)$ as $G \in \mathcal{P}_M$, which implies that its projection $\cos(\pi a)$ on the edge set of G' belongs to $\mathcal{E}(G')$.

Suppose now that $G \in \mathcal{P}_C$; we show that $G' \in \mathcal{P}_C$. The reasoning is similar. Indeed, if $a \in \mathrm{CUT}^{\square}(G')$, let $b \in \mathrm{CUT}^{\square}(G)$ whose projection on the edge set of G' is a; then, $\cos(\pi b) \in \mathcal{E}(G)$ which implies that $\cos(\pi a) \in \mathcal{E}(G')$.

We consider now the case when $G' = G/e$ is obtained by contracting edge e. Let w denote the node of G' obtained by contraction of edge $e = uv$. The proof is based on the following simple observation: Given $a \in \mathbb{R}^{E'}$ define $b \in \mathbb{R}^{E}$ by setting $b_{uv} := 0$, $b_{iu} := a_{iw}$ if i is adjacent to u in G, $b_{iv} := a_{iw}$ if i is adjacent to v in G, and $b_f := a_f$ for all remaining edges f of G. Then, $b \in \mathrm{MET}^{\square}(G)$ (resp. $b \in \mathrm{CUT}^{\square}(G)$) whenever $a \in \mathrm{MET}^{\square}(G')$ (resp. $a \in \mathrm{CUT}^{\square}(G')$). Suppose that $G \in \mathcal{P}_M$, let $a \in \mathrm{MET}^{\square}(G')$ and let $b \in \mathrm{MET}^{\square}(G')$ be defined as above. Then, $\cos(\pi b) \in \mathcal{E}(G)$. Hence, there exists a matrix $B \in \mathcal{E}_n$ extending $\cos(\pi b)$. If A denotes the matrix obtained from B by deleting the row and column indexed by u and renaming v as w, then $A \in \mathcal{E}_{n-1}$ and A extends $\cos(\pi a)$, which shows that $\cos(\pi a) \in \mathcal{E}(G')$. The proof is identical in the case of $\mathcal{P}_C$. ∎

Lemma 31.3.10. *The class $\mathcal{P}_M$ is closed under taking clique sums.*

Proof. Let $G_1 = (V_1, E_1)$ and $G_2 = (V_2, E_2)$ be two graphs in $\mathcal{P}_M$ such that $K := V_1 \cap V_2$ induces a clique in both G_1 and G_2 and there are no edges between

$V_1 \setminus V_2$ and $V_2 \setminus V_1$. Let $G = (V_1 \cup V_2, E_1 \cup E_2)$ denote their clique sum. We show that $G \in \mathcal{P}_M$. For this, let $a \in \mathrm{MET}^{\square}(G)$. The projection a_i of a on $\mathbb{R}^{E_i}$ belongs to $\mathrm{MET}^{\square}(G_i)$, which implies that $\cos(\pi a_i) \in \mathcal{E}(G_i)$ for $i = 1, 2$. Hence, there exists a matrix $A_i \in \mathcal{E}_{n_i}$ ($n_i := |V_i|$) extending $\cos(\pi a_i)$. Consider the partial symmetric matrix M shown in Figure 31.3.8, whose entries m_{uv} ($u \in V_1 \setminus V_2$, $v \in V_2 \setminus V_1$) remain to be specified. Hence, the entries of M are specified on the graph H defined as the clique sum (along K) of two complete graphs with respective node sets V_1 and V_2. As H is chordal, we deduce from Theorem 31.3.4 that M can be completed to a positive semidefinite matrix. This shows that $\cos(\pi a) \in \mathcal{E}(G)$ as M extends $\cos(\pi a)$. ∎

Proof of Theorem 31.3.7. As $\mathcal{P}_M \subseteq \mathcal{P}_C$, it suffices to verify that a graph in $\mathcal{P}_C$ has no K_4-minor and that a graph with no K_4-minor belongs to $\mathcal{P}_M$. The statement that a graph in $\mathcal{P}_C$ has no K_4-minor follows from Lemma 31.3.9 and the fact that $K_4 \notin \mathcal{P}_C$. Conversely, suppose that G has no K_4-minor. We show that $G \in \mathcal{P}_M$ by induction on the number of nodes. If $G = K_3$ then $G \in \mathcal{P}_M$ by Theorem 31.2.2. Otherwise, G is a subgraph of a clique sum of two smaller graphs G_1 and G_2 with no K_4-minors. Now, G_1 and G_2 belong to $\mathcal{P}_M$ by the induction assumption. This implies that $G \in \mathcal{P}_M$, using Lemmas 31.3.9 and 31.3.10. ∎

Let us now consider the class $\mathcal{P}_{KM}$ (resp. $\mathcal{P}_{KC}$) consisting of the graphs G for which the clique and metric conditions (31.3.1), (31.3.3) (resp. the clique and cut conditions (31.3.1), (31.3.2)) taken together suffice for the description of $\mathcal{E}(G)$. In view of the above results, it suffices here to assume that the clique condition (31.3.1) holds for all cliques of size ≥ 4. Obviously,

$$\mathcal{P}_{KM} \subseteq \mathcal{P}_{KC}.$$

In fact, the two classes $\mathcal{P}_{KM}$ and $\mathcal{P}_{KC}$ coincide. Several equivalent characterizations for the graphs in this class are known; they are presented below. First, we need some definitions.

Call *splitting* the converse operation to that of contracting an edge; hence, splitting a node u in a graph means replacing u by two adjacent nodes u' and u'' and replacing every edge uv in an arbitrary manner, either by $u'v$, or by $u''v$ (but in such a way that each of u' and u'' is adjacent to at least one node). (This operation can be seen as a special case of the splitting operation defined in Section 26.5.) See Figure 31.3.12 for an example. *Subdividing* an edge $e = uv$ means inserting a new node w and replacing edge e by the two edges uw and wv. Hence, this is a special case of splitting. A graph that can be constructed from a given graph G by subdividing its edges is called a *homeomorph* of G. Note that splitting a node of degree 2 or 3 amounts to subdividing one of the edges incident to that node. (Therefore, homeomorphs of K_4 and splittings of K_4 are the same notions; in particular, a graph has no K_4-minor if and only if it contains no homeomorph of K_4 as a subgraph.) Figure 31.3.11 shows a homeomorph of K_4; the dotted lines indicate paths.

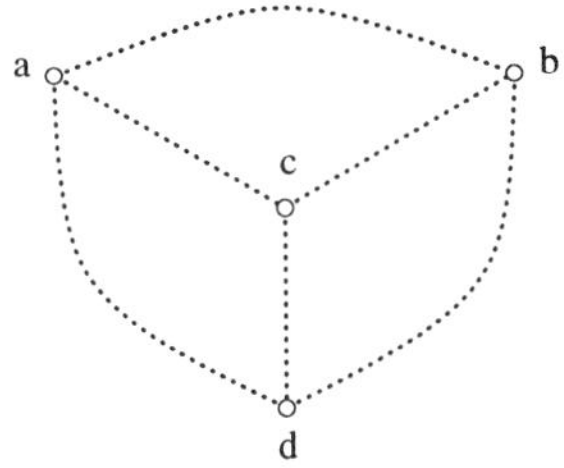

Figure 31.3.11: A homeomorph of K_4

Let $W_n := \nabla C_{n-1}$ denote the *wheel* on n nodes, obtained by adding a new node adjacent to all nodes of a circuit of length $n-1$. Hence, $W_4 = K_4$. Figure 31.3.12 (a) shows the wheel W_7 and (c) shows the graph $\widehat{W_4}$ obtained from K_4 by splitting one node. Clearly, W_n ($n \geq 5$) and any splitting of W_n ($n \geq 4$) do not belong to $\mathcal{P}_{KC}$ (by Theorem 31.3.7, since these graphs do not contain cliques of size 4 while having a K_4-minor).

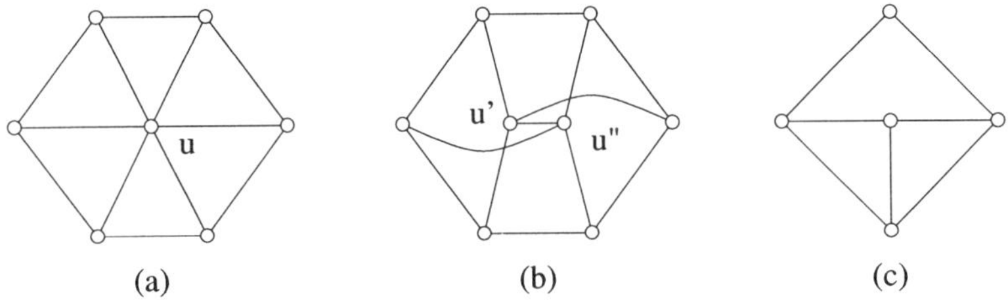

Figure 31.3.12: (a) The wheel W_7; (b) Splitting node u in W_7; (c) The graph $\widehat{W_4}$

Several equivalent characterizations for the graphs in $\mathcal{P}_{KM}$ have been discovered by Barrett, Johnson and Loewy [1996]; more precisely, they show the equivalence of assertions (i), (iii), (iv), (v) in Theorem 31.3.13 below. Building upon their result, Johnson and McKee [1996] show the equivalence of (i) and (vi); in other words, the graphs in $\mathcal{P}_{KM}$ arise from the graphs in $\mathcal{P}_K$ and $\mathcal{P}_M$ by taking clique sums. Laurent [1997c] observes moreover the equivalence of (i) and (ii); hence, the two classes $\mathcal{P}_{KM}$ and $\mathcal{P}_{KC}$ coincide even though the cut condition (31.3.2) is stronger than the metric condition (31.3.3). We delay the proof of the next result till Section 31.3.2.

Theorem 31.3.13. *The following assertions are equivalent for a graph G:*

(i) *$G \in \mathcal{P}_{KM}$, i.e., $\mathcal{E}(G)$ consists of the vectors $x = \cos(\pi a)$ such that $a \in \mathrm{MET}^{\square}(G)$ and $x_K \in \mathcal{E}(K)$ for every clique K in G.*

(ii) *$G \in \mathcal{P}_{KC}$, i.e., $\mathcal{E}(G)$ consists of the vectors $x = \cos(\pi a)$ such that $a \in \mathrm{CUT}^{\square}(G)$ and $x_K \in \mathcal{E}(K)$ for every clique K in G.*

(iii) *No induced subgraph of G is W_n ($n \geq 5$) or a splitting of W_n ($n \geq 4$).*

(iv) *Every induced subgraph of G that contains a homeomorph of K_4 contains a clique of size 4.*

(v) *There exists a chordal graph G' containing G as a subgraph and having no new clique of size 4.*

(vi) *G can be obtained by means of clique sums from chordal graphs and graphs with no K_4-minor.*

We close the section with a result concerning the graphs whose elliptope is a polytope. It turns out that this occurs only in the most trivial case, when $\mathcal{E}(G) = [-1,1]^E$. Set

$$Q_n := \text{Conv}(xx^T \mid x \in \{\pm 1\}^n)$$

and, for a graph $G = (V_n, E)$, let $Q(G)$ denote the projection of Q_n on the subspace $\mathbb{R}^E$ indexed by the edge set of G. Hence, Q_n (resp. $Q(G)$) is nothing but the image of the cut polytope $\text{CUT}_n^\square$ (resp. $\text{CUT}^\square(G)$) under the mapping $x \mapsto 1 - 2x$. Clearly,

$$Q_n \subseteq \mathcal{E}_n, \quad Q(G) \subseteq \mathcal{E}(G).$$

The following result of Laurent [1997b] characterizes the graphs for which equality $Q(G) = \mathcal{E}(G)$ holds; its proof is along the same lines as that of Theorem 31.3.7.

Theorem 31.3.14. *For a graph G, equality $Q(G) = \mathcal{E}(G)$ holds if and only if G has no K_3-minor, i.e., if G is a forest. Then, $\mathcal{E}(G) = [-1,1]^E$.* ∎

As the class of graphs G for which $\mathcal{E}(G)$ is a polytope is closed under taking minors, we deduce:

Corollary 31.3.15. *The elliptope $\mathcal{E}(G)$ of a graph G is a polytope if and only if G is a forest; then, $\mathcal{E}(G) = [-1,1]^E$.* ∎

31.3.2 Characterizing Graphs with Excluded Induced Wheels

We give here the full proof[5] of Theorem 31.3.13, which states several equivalent characterizations for the graphs containing no splittings of wheels as induced subgraphs. We show the following implications:

$$\text{(ii)} \Longrightarrow \text{(iii)} \Longrightarrow \text{(iv)} \Longrightarrow \text{(v)} \Longrightarrow \text{(i) and (i)} \Longleftrightarrow \text{(vi)},$$

the implication (i) $\Longrightarrow$ (ii) being obvious.

[5]The proof given here follows the exposition in Laurent [1997d]. It is based essentially on the original proofs of Barrett, Johnson and Loewy [1996] and Johnson and McKee [1996]. However, several parts have been simplified and shortened; in particular, the implications (iv) $\Longrightarrow$ (v) $\Longrightarrow$ (i).

The following notion of 'path avoiding a clique' will be useful in the proof. Let $G = (V, E)$ be a graph, let K be a clique in G and let $a \in K$, $x \in V \setminus K$. A path P joining the two nodes a and x is said to *avoid the clique* K if P contains no other node of K besides a.

We start with some preliminary results.

Lemma 31.3.16. *The class $\mathcal{P}_{KC}$ is closed under taking induced subgraphs.*

Proof. Suppose $G = (V, E)$ belongs to $\mathcal{P}_{KC}$ and let $H = G[U]$ be an induced subgraph of G, where $U \subseteq V$. We show that $H \in \mathcal{P}_{KC}$. Let x be a vector indexed by the edge set of H satisfying (31.3.1) and (31.3.2); we show that $x \in \mathcal{E}(H)$. For this we extend x to a vector y indexed by the edge set of G by setting $y_{uv} := 0$ for an edge $uv \in E$ with $u \in U$, $v \in V \setminus U$ and $y_{uv} := 1$ for an edge $uv \in E$ contained in $V \setminus U$. It is clear that y satisfies (31.3.1). By assumption, $a := \frac{1}{\pi} \arccos x \in \mathrm{CUT}^{\square}(H)$; we verify that $b := \frac{1}{\pi} \arccos y \in \mathrm{CUT}^{\square}(G)$. Indeed, say

$$a = \sum_{S \subseteq U} \lambda_S \delta_H(S)$$

where $\lambda_S \geq 0$, $\sum_S \lambda_S = 1$. Then,

$$b = \frac{1}{2} \sum_{S \subseteq U} \lambda_S \left(\delta_G(S) + \delta_G(U \setminus S)\right),$$

which shows that $b \in \mathrm{CUT}^{\square}(G)$. Hence, y satisfies (31.3.2). Therefore, $y \in \mathcal{E}(G)$ which implies that $x \in \mathcal{E}(H)$. ∎

Lemma 31.3.17. *The class $\mathcal{P}_{KM}$ is closed under taking clique sums.* ∎

We omit the proof which is analogue to that of Lemma 31.3.10.

Lemma 31.3.18. *Let $G = (V, E)$ be a graph in which every induced subgraph containing a homeomorph of K_4 also contains a clique of size 4. Let K be a clique in G with $|K| \geq 4$, let $a, b, c \in K$, $v \in V \setminus K$, and let P_a (resp. P_b, P_c) be a path from a (resp. from b, c) to v avoiding the clique K. Then, there exists a node $w \in V \setminus K$ lying on one of the paths P_a, P_b or P_c which is adjacent to all three nodes a, b and c.*

Proof. Let W denote the set of nodes lying on the paths P_a, P_b or P_c. Clearly, there is a path avoiding K from every node $w \in W$ to each node in $\{a, b, c\}$. For $w \in W$, define $d(w)$ as the smallest sum $|Q_a| + |Q_b| + |Q_c|$, where Q_a, Q_b, Q_c are paths avoiding K that join w to a, b, c, respectively, in the graph $G[W]$. Suppose w is a node in W for which $d(w)$ is minimum and let Q_a, Q_b, Q_c be the corresponding paths, as defined above. Let $W_0 \subseteq W$ denote the set of nodes lying on Q_a, Q_b or Q_c. Then,

$$V(Q_a) \cap V(Q_b) = V(Q_a) \cap V(Q_c) = V(Q_b) \cap V(Q_c) = \{w\}.$$

Indeed, if z is a node in $V(Q_a) \cap V(Q_b)$ distinct from w, then it is easy to see that $d(z) < d(w)$. Hence, the three paths Q_a, Q_b, Q_c together with the edges ab, ac and bc form a homeomorph of K_4 contained in $G[W_0]$. By the assumption, $G[W_0]$ must contain a clique S of size 4. We show that

$$S = \{w, a, b, c\}.$$

Suppose that $w \notin S$. Then, S contains two nodes r, s that lie on a common path, say, on Q_a; say, w, r, s, a lie in that order along Q_a. Let $t \in S \setminus \{r, s\}$. We can suppose that t lies on Q_b (as t does not lie on Q_a, by minimality of $d(w)$). Then, $t = b$ (else, we would have $d(t) < d(w)$). Hence, S is of the form $\{r, s, b, c\}$ which implies that $d(r) < d(w)$, a contra diction. Therefore, the set S contains w; so, $S = \{w, r, s, t\}$ where r, s, t lie on Q_a, Q_b, Q_c, respectively. Now, $r = a$ (else, $d(r) < d(w)$); similarly, $s = b$ and $t = c$. This shows that $S = \{w, a, b, c\}$. ∎

Proposition 31.3.19. *Let $G = (V, E)$ be a graph satisfying the following conditions:*

(i) *Every induced subgraph of G containing a homeomorph of K_4 contains a clique of size* 4.

(ii) *G contains a clique of size* 4.

(iii) *For every maximal clique K in G, $a \in K$ and $v \in V \setminus K$, there exists a path avoiding K from a to v.*

Then, G is chordal.

Proof. We show the result by induction on the number n of nodes in G. The result holds trivially if $n = 4$ (as $G = K_4$). Let $n \geq 5$ and let K be a maximal clique in G of size ≥ 4. We can assume that the subgraph $G[V \setminus K]$ induced by $V \setminus K$ is connected. (Else, letting $W_1, \ldots, W_p$ denote the connected components of $G[V \setminus K]$, then $G_i := G[K \cup W_i]$ is chordal for each $i = 1, \ldots, p$, by the induction assumption. Hence, G is chordal as it is a clique sum of chordal graphs.) We show that $K = V$, i.e., that G is a complete graph. For this, suppose $K \neq V$. For each $x \in V \setminus K$, let $N(x)$ denote the set of nodes in K that are adjacent to x. We claim:

(a) If $x, y \in V \setminus K$ are adjacent and if $N(x) \not\subseteq N(y), N(y) \not\subseteq N(x)$, then $N(x) \cap N(y) = \emptyset, |N(x)| = |N(y)| = 1$.

Indeed, let $a \in N(x) \setminus N(y)$ and $b \in N(y) \setminus N(x)$. Suppose first that there exists $c \in N(x) \cap N(y)$. Then, the subgraph of G induced by $\{a, b, c, x, y\}$ contains a homeomorph of K_4 but no clique of size 4, contradicting (i). If $|N(x)| \geq 2$, we obtain again a contradiction with (i) by choosing now c in $N(x) \setminus \{a\}$. This shows (a). Next, we have:

(b) If $x \in V \setminus K$ and $|N(x)| = 1$, then $N(x) \subset N(y)$ for some $y \in V \setminus K$.

Say, $N(x) = \{a\}$. Let $b, c \in K \setminus \{a\}$ and let P_a, P_b, P_c be paths from x to a, b, c, respectively, that avoid K. By Lemma 31.3.18, there exists a node $y \in V \setminus K$ lying on one of these paths which is adjacent to a, b and c. Hence, $N(x) \subset N(y)$.

Call a set $N(x)$ $(x \in V \setminus K)$ *maximal* if $N(x) = N(y)$ whenever $N(x) \subseteq N(y)$ for $y \in V \setminus K$. We show:

(c) Let $x \neq y \in V \setminus K$ for which $N(x)$ and $N(y)$ are both maximal. Then, $N(x) = N(y)$.

Suppose that $N(x) \neq N(y)$. Then, by (a) and (b), x and y are not adjacent. Let $(x, z_1, \ldots, z_p, y)$ be a path of shortest length joining x and y in $G[V \setminus K]$. Then, $N(z_1) \subseteq N(x)$ and $N(z_p) \subseteq N(y)$. Let us first assume that $N(z_i) \not\subseteq N(x)$ for some $i = 1, \ldots, p$. Let i be the smallest such index. Then, $N(z_1) \cup \ldots \cup N(z_{i-1}) \subseteq N(x)$ and $N(z_i) \not\subseteq N(x)$. Let $a \in N(x) \setminus N(z_i)$ and $b \in N(z_i) \setminus N(x)$. We claim that $N(z_1) \cup \ldots \cup N(z_{i-1}) \subseteq \{a\}$. For, suppose that there exists an element $a' \in N(z_1) \cup \ldots \cup N(z_{i-1})$ with $a' \neq a$. Then, applying Lemma 31.3.18, we find a node $w \in \{x, z_1, \ldots, z_{i-1}, z_i\}$ which is adjacent to all three nodes a, b and a'. This implies that $w = z_j$ $(j < i)$ and, thus, $b \in N(z_j) \subseteq N(x)$, a contradiction. Therefore, $N(z_1) \cup \ldots \cup N(z_{i-1}) \subseteq \{a\}$. Let $c \in N(x) \setminus \{a\}$; then the subgraph of G induced by $\{a, b, c, x, z_1, \ldots, z_{i-1}, z_i\}$ contains a homeomorph of K_4 but no clique of size 4, contradicting (i). When $N(z_i) \not\subseteq N(y)$ for some $i = 1, \ldots, p$, we obtain a contradiction in the same manner as above. Hence, we have that $N(z_1) \cup \ldots \cup N(z_p) \subseteq N(x) \cap N(y)$. Taking $a \in N(x) \setminus N(y)$, $b \in N(y) \setminus N(x)$, $c \in N(x) \setminus \{a\}$, the subgraph of G induced by $\{a, b, c, x, z_1, \ldots, z_p, y\}$ contains a homeomorph of K_4 but no clique of size 4, yielding again a contradiction. Hence, (c) holds.

We can now conclude the proof. Let $N(x_0)$ denote the unique maximal set of the form $N(x)$ $(x \in V \setminus K)$. Then, $N(x_0) = K$ (by (iii)). Hence, $K \cup \{x_0\}$ is a clique, which contradicts the maximality of K. ∎

Proof of Theorem 31.3.13.
The implication (ii) $\Longrightarrow$ (iii) follows from Lemma 31.3.16 since W_n $(n \geq 5)$ and a splitting of W_n $(n \geq 4)$ do not belong to $\mathcal{P}_{KC}$. The implication (vi) $\Longrightarrow$ (i) follows from Lemma 31.3.17 and the fact that chordal graphs and graphs with no K_4-minor belong to $\mathcal{P}_{KM}$.

(v) $\Longrightarrow$ (i) Suppose $G = (V, E)$ is a graph satisfying (v). Let $G' = (V, E')$ be a chordal graph such that $E \subseteq E'$ and every clique of size 4 in G' is, in fact, a clique in G. We show that $G \in \mathcal{P}_{KM}$. For this, let $x = \cos(\pi a) \in \mathbb{R}^E$ such that $a \in \text{MET}^{\square}(G)$ and $x_K \in \mathcal{E}(K)$ for all cliques K in G. Let $b \in \text{MET}^{\square}(G')$ extending a and set $y := \cos(\pi b)$. Then, y satisfies the clique condition (31.3.1) (as $y_K = x_K \in \mathcal{E}(K)$ for each clique K of size ≥ 4 in G'). As G' is chordal, we deduce that $y \in \mathcal{E}(G')$ and, thus, $x \in \mathcal{E}(G)$.

(iii) $\Longrightarrow$ (iv) Suppose $G = (V, E)$ is a graph for which there exists a subset $U \subseteq V$ such that $G[U]$ contains a homeomorph of K_4 and contains no clique of size 4. Choose such U of minimum cardinality; set $G' := G[U] := (U, E')$. Moreover, let

$H = (W, F)$ be a homeomorph of K_4 contained in G' having minimum number of edges. Then, $W = U$ (by minimality of $|U|$) and $H \neq K_4$ (by assumption). To fix ideas, suppose H is the graph shown in Figure 31.3.11; so, H consists of the six paths P_{ab}, P_{ac}, P_{bc}, P_{ad}, P_{bd} and P_{cd} (where P_{ab} denotes the path joining the nodes a and b, etc.); let us refer to the nodes a, b, c, d as the 'corners' of H.

We show that G' is a wheel or a splitting of a wheel. This is obvious if $|E' \setminus F| \leq 1$. So, we can suppose that $|E' \setminus F| \geq 2$. A first observation is:

(a) The end nodes of an edge $e \in E' \setminus F$ do not lie on a common path in H.

Indeed, suppose that the end nodes x and y of e lie, say, on the path P_{ab}. Let $P_{ab}(x, y)$ denote the subpath of P_{ab} joining x and y. Then, the graph obtained from H by deleting $P_{ab}(x, y)$ and adding the edge e is again a homeomorph of K_4 contained in G' but having less edges than H. This contradicts the minimality of H. Hence, (a) holds.

There are two possibilities for an edge $e = xy \in E' \setminus F$: Either, (I) e lies within a face of H (i.e., x and y lie on two paths in H sharing a common end node) or, (II) e connects two disjoint paths in H. We make two observations:

(b) Let $e = xy \in E' \setminus F$ where x, y are internal nodes in P_{ab}, P_{cd}, respectively. Then, $|P_{ac}| = |P_{bc}| = |P_{ad}| = |P_{bd}| = 1$.

Indeed, suppose $|P_{ac}| > 1$. Then, the graph obtained from H by adding e and deleting P_{ac} is a homeomorph of K_4 (with corners x, y, b, d) contained in G' with less edges than H. Similarly,

(c) Let $e = xy \in E' \setminus F$ lying in a face of H. Say, x, y lie on P_{ab}, P_{ac}, respectively. Then, (ci) $xa, ya \in E, |P_{bc}| = |P_{bd}| = |P_{cd}| = 1$, or (cii) $y = c, |P_{ac}| = |P_{bc}| == |P_{cd}| = 1$, or (ciii) $x = b, |P_{ab}| = |P_{bc}| = |P_{bd}| = 1$.

Suppose first that there exists an edge $e \in E' \setminus F$ of type (II). Say, $e = xy$ where x, y are internal nodes on P_{ab}, P_{cd}, respectively. Let $e' = x'y'$ be another edge in $E' \setminus F$. Then, e' is of type (I). (Indeed, if e' is of type (II) then e' connects the same paths P_{ab} and P_{cd} - this follows from (b) and the fact that $H \neq K_4$. Say, $x \neq x'$ and d, y', y, c lie in that order along P_{cd}. Then, adding e, e' to H and deleting P_{ad}, P_{bd} and the subpath $P_{cd}(d, y')$ creates a homeomorph of K_4 with less edges than H.) We can suppose without loss of generality that e' lies within the face of H containing a, b, c. By (c), e' is of the form cz where z lies on P_{ab}. Say, z lies between a and x. Then, adding e, e' to H and deleting P_{ad}, P_{ac} and $P_{ab}(a, z)$ creates a smaller homeomorph of K_4 than H.

Hence, we can now suppose that every edge in $E' \setminus F$ is of type (I), i.e., lies within a face of H. If $E' \setminus F$ contains an edge as in (ci), then it is easy to see that one can always find a smaller homeomorph of K_4 in G'. Hence, we can suppose that all edges in $E' \setminus F$ are as in (cii) or (ciii). Let $e = cx \in E' \setminus F$, where x is an internal node of P_{ab}. This implies easily that every other edge $e' \in E' \setminus F$

is of the form cz, where z lies on P_{ab}, P_{bd} or P_{ad}. Therefore, G' is a wheel or a splitting of a wheel.

(iv) $\Longrightarrow$ (v) Suppose that G satisfies the assumption (iv). We show that (v) holds by induction on the number of nodes in G. We can suppose that G contains a homeomorph of K_4; else, the result holds. By (iv), G has a clique of size 4. We can suppose, moreover, that there exist a maximal clique K in G, $a_0 \in K$, and $x_0 \in V \setminus K$ such that no path avoiding K from a to x exists; for, if not, G is chordal by Proposition 31.3.19 and we are done. Let S denote the set of nodes $b \in K$ for which there exists a path from x_0 to b avoiding K. Moreover, let T denote the set of nodes $x \in V \setminus K$ that can be joined to all nodes of S by some path avoiding K, and that cannot be joined to any other point of $K \setminus S$ by a path avoiding K. Then, $S \neq K$ (as $a_0 \notin S$) and $T \neq \emptyset$ (as $x_0 \in T$). Moreover, there is no edge between T and $(V \setminus K) \setminus T$, or $K \setminus S$. Consider the induced subgraphs $G[S \cup T]$ and $G[V \setminus T]$; both are proper subgraphs of G. By the induction assumption, there exists a chordal graph H_1 (resp. H_2) containing $G[S \cup T]$ (resp. $G[V \setminus T]$) as a subgraph and having no new clique of size 4. Let $H := H_1 \cup H_2$ denote the graph with edge set $E(H_1) \cup E(H_2)$. Then, H contains G as a subgraph. Moreover, H is chordal and H contains no new clique of size 4. This follows from the fact that H is, in fact, the clique sum of the two graphs H_1 and H_2 (along the clique S). Hence, G satisfies (v).

(i) $\Longrightarrow$ (vi). Let G be a graph in $\mathcal{P}_{KM}$. We show that G satisfies (vi) by induction on the number of nodes. We can suppose that G is connected (else, the result follows by induction) and that G contains a homeomorph of K_4. It suffices now to show that G contains a clique cutset, i.e., a clique K such that $G[V \setminus K]$ is disconnected. If G contains a simplicial[6] node v, then the set of neighbors of v forms a clique cutset. Suppose now that G contains no simplicial node. Using the implication (i) $\Longrightarrow$ (iv) (already shown above), we know that G contains a clique of size 4. Let K be a maximal clique in G of size ≥ 4 such that $G[V \setminus K]$ is connected (else, we are done). Observe that, for every $a \in K$ and $x \in V \setminus K$, there exists a path from a to x avoiding K. (Indeed, as a is not a simplicial node, a is adjacent to some node $w \in V \setminus K$. Now, v and w can be joined by some path in $G[V \setminus K]$, which yields a path from v to a avoiding K.) Hence, the graph G satisfies the conditions (i)-(iii) from Proposition 31.3.19. Therefore, G is chordal. This yields a contradiction as every chordal graph contains a simplicial vertex. This concludes the proof for (i) $\Longrightarrow$ (vi). ∎

31.4 The Euclidean Distance Matrix Completion Problem

Let (V_n, d) be a distance space with associated distance matrix D. We remind that D is said to be a Euclidean distance matrix when $(V_n, \sqrt{d})$ is isometrically ℓ_2-embeddable; that is, when d belongs to the negative type cone NEG_n.

[6] A node v in graph G is said to be *simplicial* if its set of neighbors induces a clique in G.

Given a subset E of $E_n = \{ij \mid 1 \leq i < j \leq n\}$, consider the graph $G = (V_n, E)$. Denote by NEG(G) the projection of the negative type cone NEG$_n$ on the subspace $\mathbb{R}^E$ indexed by the edge set E of G. Hence, a vector $d = (d_{ij})_{ij \in E}$ belongs to NEG(G) if and only if there exist vectors $u_1, \ldots, u_n \in \mathbb{R}^m$ (for some $m \geq 1$) such that

$$\text{(31.4.1)} \qquad \sqrt{d_{ij}} = \| u_i - u_j \|_2 \quad \text{for all } ij \in E.$$

To $d \in \mathbb{R}^E$ corresponds a partial symmetric $n \times n$ matrix $M = (m_{ij})$ whose entries are specified only on the diagonal positions and on the positions corresponding to edges in E; namely, $m_{ii} := 0$ for all $i \in V_n$ and $m_{ij} = m_{ji} := d_{ij}$ for all $ij \in E$. Then, $d \in$ NEG(G) if the unspecified entries of M can be chosen in such a way that one obtains a Euclidean distance matrix; that is, if M can be completed to a Euclidean distance matrix. Therefore, the completion problem for partial Euclidean distance matrices is that of characterizing membership in projections of the negative type cone.

Barvinok [1995] shows that, for $d \in$ NEG(G), there exists a system of vectors $u_1, \ldots, u_n \in \mathbb{R}^m$ satisfying (31.4.1) in dimension m bounded by

$$\text{(31.4.2)} \qquad m \leq \left\lfloor \frac{\sqrt{8|E|+1} - 1}{2} \right\rfloor .$$

A short proof for this fact can be given using Theorem 31.5.3 from the next section.

Proof of relation (31.4.2). For $d \in \mathbb{R}^E$ we have:

$$\exists\, u_1, \ldots, u_n \in \mathbb{R}^m \text{ such that } d_{ij} = (\| u_i - u_j \|_2)^2 \text{ for all } ij \in E$$
$$\Updownarrow$$
$$\exists \text{ symmetric } n \times n \text{ matrix } A \succeq 0 \text{ with rank} \leq m \text{ such that}$$
$$d_{ij} = a_{ii} + a_{jj} - 2a_{ij} \text{ for all } ij \in E.$$

Consider the convex set $K := \{X \mid X \succeq 0,\ x_{ii} + x_{jj} - 2x_{ij} = d_{ij} \text{ for } ij \in E\}$. If $K \neq \emptyset$ (that is, if $d \in$ NEG(G)) and if $d \neq 0$ (then, K has extreme points), then any matrix $A \in K$ which is an extreme point of K has rank r satisfying $\binom{r+1}{2} \leq |E|$ (by Theorem 31.5.3). This condition is equivalent to the inequality in (31.4.2). ∎

We present in this section a closed form description of the projected negative type cone NEG(G) for several classes of graphs. In fact, one can formulate necessary conditions for membership in NEG(G) that are similar to the conditions (31.3.1), (31.3.2) and (31.3.3) considered in Section 31.3 for the positive semidefinite completion problem. Moreover, these conditions are sufficient for precisely the same classes of graphs as those coming up in Section 31.3.

In a first step, we formulate the results concerning the Euclidean distance matrix completion problem. Then, we show how they can be derived from the corresponding results for the positive semidefinite completion problem; here are used essentially the techniques on metric transforms developed in Chapter 9.

The exposition in this section follows again essentially the survey paper of Laurent [1997d].

31.4.1 Results

We formulate here some results for the Euclidean distance matrix completion problem; proofs are delayed till Section 31.4.2.

Let $K \subseteq V_n$ be a subset of nodes that induces a clique in G. For $d \in \mathbb{R}^E$ denote by d_K its projection on the edge set of $G[K]$. Clearly, if $d \in \mathrm{NEG}(G)$ then $d_K \in \mathrm{NEG}(K)$. Therefore, the condition

(31.4.3) $\quad d_K \in \mathrm{NEG}(K)$ for every clique K in G

is a necessary condition for $d \in \mathrm{NEG}(G)$, again called *clique condition.* Bakonyi and Johnson [1995] characterize the graphs G for which the condition (31.4.3) is sufficient for the description of $\mathrm{NEG}(G)$. They show:

Theorem 31.4.4. *For a graph $G = (V_n, E)$, we have*

$$\mathrm{NEG}(G) = \{d \in \mathbb{R}^E \mid d_K \in \mathrm{NEG}(K) \ \ \forall K \text{ clique in } G\}$$

if and only if G is chordal.

The condition (31.4.3) is not sufficient for the description of $\mathrm{NEG}(G)$ when G is not chordal. Indeed, suppose that G has a chordless circuit C of length ≥ 4. Let $x \in \mathbb{R}^E$ be defined by $x_e := 0$ for all edges e in C except $x_{e_0} := 1$ for one edge e_0 in C, $x_{ij} := 1$ for all edges ij with $i \in V(C)$, $j \in V \setminus V(C)$, and $x_{ij} := 0$ for all edges ij with $i, j \in V \setminus V(C)$. Then, x satisfies (31.4.3) but $x \notin \mathrm{NEG}(G)$. Another necessary condition can be easily formulated in terms of the cut cone. Namely, the condition

(31.4.5) $\quad \sqrt{d} \in \mathrm{CUT}(G)$

is a necessary condition for $d \in \mathrm{NEG}(G)$, called *cut condition*; this follows from the fact that "$\ell_2 \Longrightarrow \ell_1$" (recall Proposition 6.4.12) and taking projections. Therefore,

(31.4.6) $\quad \sqrt{d} \in \mathrm{MET}(G)$

is also a necessary condition for $d \in \mathrm{NEG}(G)$, called *metric condition.* The condition (31.4.6) characterizes $\mathrm{NEG}(G)$ in the case when $G = K_3$. This result has, in fact, already been mentioned in Remark 6.2.12; we repeat the proof for clarity.

Lemma 31.4.7. $\mathrm{NEG}_3 = \{d \in \mathbb{R}^3_+ \mid \sqrt{d} \in \mathrm{MET}_3\}$.

Proof. Let d be a distance on V_3 and set $d_{12} := a$, $d_{13} := b$, $d_{23} := c$. Let us consider the image of d under the covariance mapping (pointed at position 3) and the corresponding symmetric matrix

$$P := \begin{pmatrix} b & \frac{b+c-a}{2} \\ \frac{b+c-a}{2} & c \end{pmatrix}.$$

We use the fact that $d \in \mathrm{NEG}_3$ if and only if $P \succeq 0$ (recall Figure 6.2.3). Now, $P \succeq 0$ if and only if $\det P \geq 0$, i.e., if $4bc - (b+c-a)^2 \geq 0$. The latter condition can be rewritten as: $a^2 - 2a(b+c) + (b-c)^2 \leq 0$, which is equivalent to $b+c-2\sqrt{bc} = (\sqrt{b}-\sqrt{c})^2 \leq a \leq b+c+2\sqrt{bc} = (\sqrt{b}+\sqrt{c})^2$. Hence, we find the condition that $\sqrt{d} \in \mathrm{MET}_3$. ∎

More generally, Bakonyi and Johnson [1995] observe that the condition (31.4.6) suffices for the description of NEG(G) if G is a circuit. In fact, the following result holds, which is an analogue of Theorem 31.3.7 (Laurent [1997c]).

Theorem 31.4.8. *The following assertions are equivalent for a graph G:*

(i) $\mathrm{NEG}(G) = \{d \in \mathbb{R}^E_+ \mid \sqrt{d} \in \mathrm{CUT}(G)\}$.

(ii) $\mathrm{NEG}(G) = \{d \in \mathbb{R}^E_+ \mid \sqrt{d} \in \mathrm{MET}(G)\}$.

(iii) *G has no K_4-minor.*

The next result identifies the graphs for which the clique and metric conditions (resp. clique and cut conditions) suffice for the description of the cone NEG(G). The equivalence of (i) and (iii) is due to Johnson, Jones and Kroschel [1995] and that of (ii) and (iii) to Laurent [1997c].

Theorem 31.4.9. *The following assertions are equivalent for a graph G:*

(i) $\mathrm{NEG}(G) = \{d \in \mathbb{R}^E_+ \mid \sqrt{d} \in \mathrm{MET}(G)$ *and* $d_K \in \mathrm{NEG}(K)\ \forall K$ *clique in* $G\}$.

(ii) $\mathrm{NEG}(G) = \{d \in \mathbb{R}^E_+ \mid \sqrt{d} \in \mathrm{CUT}(G)$ *and* $d_K \in \mathrm{NEG}(K)\ \forall K$ *clique in* $G\}$.

(iii) *No induced subgraph of G is a wheel W_n ($n \geq 5$) or a splitting of a wheel W_n ($n \geq 4$).*

We conclude this section with a result of geometric flavor given in Bakonyi and Johnson [1995]; it follows as a direct application of Theorem 31.4.4.

Proposition 31.4.10. *Let $G = (V_n, E)$ be a chordal graph, let $K_1, \ldots, K_s$ denote its maximal cliques and let $d \in \mathbb{R}^E$, $R > 0$. Suppose that there exist vectors $u_1, \ldots, u_n \in \mathbb{R}^n$ satisfying* (i) *and* (ii):

(i) $\| u_i - u_j \|_2 = d_{ij}$ *for all* $ij \in E$,

(ii) *for every* $r = 1, \ldots, s$, *the vectors* u_i $(i \in K_r)$ *lie on a sphere of radius* R.

Then there exist vectors $v_1, \ldots, v_n \in \mathbb{R}^n$ *satisfying* (i) *and all of them lying on a sphere of radius* R. ∎

31.4.2 Links Between the Two Completion Problems

There is an obvious analogy between the above results for the Euclidean distance matrix completion problem and the results from Section 31.3 for the positive semidefinite completion problem. Compare, in particular, Theorems 31.3.4 and 31.4.4, as well as Theorems 31.3.7 and 31.4.8, and Theorems 31.3.13 and 31.4.9. Following Laurent [1997c], we indicate here how to derive the results for the Euclidean distance matrix completion problem from those for the positive semidefinite completion problem.

For convenience let us introduce the following classes of graphs: $\mathcal{D}_K$ (resp. $\mathcal{D}_M$, $\mathcal{D}_C$) denotes the class of graphs for which the clique condition (31.4.3) (resp. metric condition (31.4.6), cut condition (31.4.5)) suffices for the description of NEG(G); and $\mathcal{D}_{KM}$ (resp. $\mathcal{D}_{KC}$) denotes the class of graphs for which the clique and metric (resp. clique and cut) conditions taken together suffice for the description of NEG(G).

It is also convenient to introduce a notation for the following classes of graphs, already encountered in the previous section. The class $\mathcal{G}_{ch}$ consists of all chordal graphs; the class $\mathcal{G}_{K4}$ consists of the graphs that do not contain K_4 as a minor; and the class $\mathcal{G}_{wh}$ consists of the graphs that do not contain a wheel W_n $(n \geq 5)$ or a splitting of a wheel W_n $(n \geq 4)$ as an induced subgraph.

Proving Theorems 31.4.4, 31.4.8 and 31.4.9 amounts to showing the equalities: $\mathcal{D}_K = \mathcal{G}_{ch}$, $\mathcal{D}_M = \mathcal{D}_C = \mathcal{G}_{K4}$, and $\mathcal{D}_{KM} = \mathcal{D}_{KC} = \mathcal{G}_{wh}$. For this, it suffices to verify the inclusions: $\mathcal{D}_K \subseteq \mathcal{G}_{ch}$, $\mathcal{P}_K \subseteq \mathcal{D}_K$; $\mathcal{D}_C \subseteq \mathcal{G}_{K4}$, $\mathcal{P}_M \subseteq \mathcal{D}_M$; and $\mathcal{D}_{KC} \subseteq \mathcal{G}_{wh}$, $\mathcal{P}_{KM} \subseteq \mathcal{D}_{KM}$. We do so in Lemmas 31.4.16 and 31.4.17 below.

Crucial for the proof are some links between the negative type cone and the elliptope. A first obvious link between the cone NEG(∇G) and the elliptope $\mathcal{E}(G)$ is provided by the covariance mapping (as defined in (27.3.8)). Namely, given vectors $x \in \mathbb{R}^E$ and $d \in \mathbb{R}^{E(\nabla G)}$ satisfying: $d_{i,n+1} = 1$ for all $i \in V_n$ and $d_{ij} = 2 - 2x_{ij}$ for all $ij \in E$, then

$$(31.4.11) \qquad x \in \mathcal{E}(G) \Longleftrightarrow d \in \text{NEG}(\nabla G).$$

Another essential tool is the following property of the Schoenberg transform from Theorem 9.1.1: For $d \in \mathbb{R}^{E_n}$,

$$(31.4.12) \quad d \in \text{NEG}(K_n) \Longleftrightarrow \exp(-\lambda d) \in \mathcal{E}(K_n) \ \text{ for all } \lambda > 0.$$

(We remind that the notation $\exp(-\lambda d)$ means applying the exponential function componentwise, i.e., $\exp(-\lambda d) = (\exp(-\lambda d_{ij}))_{ij}$.) This relation remains valid at the level of arbitrary graphs. Namely,

Proposition 31.4.13. *Let $G = (V_n, E)$ be a graph and $d \in \mathbb{R}^E$. The following assertions are equivalent.*

(i) $d \in \mathrm{NEG}(G)$.

(ii) $\exp(-\lambda d) \in \mathcal{E}(G)$ *for all* $\lambda > 0$.

(iii) $1 - \exp(-\lambda d) \in \mathrm{NEG}(G)$ *for all* $\lambda > 0$.

Proof. (i) $\Longrightarrow$ (ii) follows from (31.4.12) and taking projections.
(ii) $\Longrightarrow$ (iii) Given $\lambda > 0$, define the vector $D \in \mathbb{R}^{E(\nabla G)}$ by $D_{i,n+1} = 1$ for $i \in V_n$ and $D_{ij} = 2 - 2\exp(-\lambda d_{ij})$ for $ij \in E$. Then, $D \in \mathrm{NEG}(\nabla G)$ (by relation (31.4.11)) which implies that $1 - \exp(-\lambda d) \in \mathrm{NEG}(G)$.
(iii) $\Longrightarrow$ (i) Let $v^T x \leq 0$ be a valid inequality for the cone $\mathrm{NEG}(G)$. We show that $v^T d \leq 0$. By assumption, $v^T(1 - \exp(-\lambda d)) \leq 0$. Expanding in series the exponential function, we obtain:

$$\begin{aligned} v^T(1 - \exp(-\lambda d)) &= \textstyle\sum_{ij \in E} v_{ij} (\sum_{p \geq 1} \frac{(-1)^{p-1}}{p!} \lambda^p d_{ij}^p) \\ &= \textstyle\sum_{p \geq 1} \frac{(-1)^{p-1} \lambda^p}{p!} \sum_{ij \in E} v_{ij} d_{ij}^p \leq 0. \end{aligned}$$

Dividing by λ and, then, letting $\lambda \longrightarrow 0$ yields: $\sum_{ij \in E} v_{ij} d_{ij} \leq 0$. This shows that $d \in \mathrm{NEG}(G)$, as d satisfies all the valid inequalities for $\mathrm{NEG}(G)$. ∎

From this we can derive the following result[7] permitting to link the two metric conditions (31.3.3) and (31.4.6).

Lemma 31.4.14. *Let $G = (V_n, E)$ be a graph and $d \in \mathbb{R}_+^E$. Then,*

$$\sqrt{d} \in \mathrm{MET}(G) \Longrightarrow \frac{1}{\pi} \arccos(e^{-\lambda d}) \in \mathrm{MET}^{\square}(G) \quad \textit{for all } \lambda > 0.$$

Proof. Note first that it suffices to show the result in the case when $G = K_n$ (as the general result will then follow by taking projections). Next, observe that it suffices to show the result in the case $n = 3$ (as $\mathrm{MET}(K_n)$ and $\mathrm{MET}^{\square}(K_n)$ are defined by inequalities that involve only three points). Now, we have: $\sqrt{d} \in \mathrm{MET}(K_3) \Longleftrightarrow d \in \mathrm{NEG}(K_3)$ (by Lemma 31.4.7); $d \in \mathrm{NEG}(K_3) \Longleftrightarrow \exp(-\lambda d) \in \mathcal{E}(K_3)$ for all $\lambda > 0$ (by Proposition 31.4.13); finally, $\exp(-\lambda d) \in \mathcal{E}(K_3) \Longleftrightarrow \frac{1}{\pi} \arccos(e^{-\lambda d}) \in \mathrm{MET}^{\square}(K_3)$ (by Theorem 31.2.2). ∎

One more useful preliminary result is the following.

[7] The implication in Lemma 31.4.14 holds, in fact, as an equivalence. The converse implication can be shown using the mean value theorem applied to the function $f(t) = \arccos(e^{-t^2})$ and letting λ tend to zero.

Lemma 31.4.15. *Let $W_n := \nabla C$ be a wheel on n nodes, with center u_0 and circuit C. Consider the vector d indexed by the edge set of W_n and defined by $d(u_0, u) := 1$ for each node u of C, $d(u,v) := 4$ for each edge uv of C. Then, $d \in \mathrm{NEG}(W_n) \iff n$ is odd.*

Proof. Let x be the vector indexed by the edge set of C and taking value -1 on every edge. By (31.4.11), $d \in \mathrm{NEG}(W_n)$ if and only if $x \in \mathcal{E}(C)$. The latter holds if and only if $\frac{1}{\pi} \arccos x \in \mathrm{MET}^{\square}(C)$, that is, if and only if C has an even length. ∎

Lemma 31.4.16. *We have: $\mathcal{D}_K \subseteq \mathcal{G}_{ch}$, $\mathcal{D}_C \subseteq \mathcal{G}_{K4}$, and $\mathcal{D}_{KC} \subseteq \mathcal{G}_{wh}$.*

Proof. We show the inclusion: $\mathcal{D}_K \subseteq \mathcal{G}_{ch}$. For this, let $G = (V, E)$ be a non-chordal graph and let $C = (V(C), E(C))$ be a chordless circuit of length ≥ 4 in G. We define a vector $d \in \mathbb{R}^E$ satisfying (31.4.3) and such that $d \notin \mathrm{NEG}(G)$ by setting $d_e := 0$ for all edges $e \in E(C)$ except $d_{e_0} := 1$ for one edge e_0 in C; $d_e := 1$ for every edge e joining a node of C to a node of $V \setminus V(C)$; and $d_e := 0$ for every edge e joining two nodes of $V \setminus V(C)$.
The example from Lemma 31.4.15 above shows that $K_4 = W_4$ does not belong to $\mathcal{D}_C$. The inclusion: $\mathcal{D}_C \subseteq \mathcal{G}_{K4}$ now follows after noting that $\mathcal{D}_C$ is closed under taking minors.
We finally check the inclusion: $\mathcal{D}_{KC} \subseteq \mathcal{G}_{wh}$. For this, let $G = (V, E)$ be a graph in $\mathcal{D}_{KC}$ and let $H := G[U]$ be an induced subgraph of G where $U \subseteq V$. Suppose in a first step that H is a wheel $W_n := \nabla C$ ($n \geq 5$) with center u_0. Consider the vector d indexed by the edge set of G and defined in the following manner: d takes value 4 on every edge of the circuit C excepet value 0 on one edge if n is odd; d takes value 1 on every edge joining the center u_0 of the wheel to a node of C; d takes value 1 on an edge between a node of C and a node outside the wheel; d takes value 0 on every remaining edge (i.e., an edge joining u_0 to a node outside the wheel or an edge joining two nodes outside the wheel). Then d satisfies (31.4.3) and $d \notin \mathrm{NEG}(G)$ (by Lemma 31.4.15). Moreover d satisfies (31.4.5), i.e., $\sqrt{d} \in \mathrm{CUT}(G)$. Indeed, say C is the circuit $(u_1, \ldots, u_{n-1})$. Then, $\sqrt{d} = \sum_{i=1}^{n-1} \delta_G(u_i)$ if n is even and $\sqrt{d} = \delta_G(\{u_1, u_{n-1}\}) + \sum_{i=2}^{n-2} \delta_G(u_i)$ if n is odd and (u_1, u_{n-1}) is the edge of C on which d takes value 0. Finally, if H is a splitting of a wheel W_n ($n \geq 4$), extend the above vector d by assigning value 0 to every new edge created during the splitting process. ∎

Lemma 31.4.17. *$\mathcal{P}_K \subseteq \mathcal{D}_K$, $\mathcal{P}_M \subseteq \mathcal{D}_M$, and $\mathcal{P}_{PM} \subseteq \mathcal{D}_{KM}$.*

Proof. We first verify the inclusion: $\mathcal{P}_K \subseteq \mathcal{D}_K$. Let G be a graph in $\mathcal{P}_K$; we show that $G \in \mathcal{D}_K$. For this, let $d \in \mathbb{R}^E$ satisfying (31.4.3); we show that $d \in \mathrm{NEG}(G)$. By Proposition 31.4.13, $\exp(-\lambda d_K) \in \mathcal{E}(K)$ for every clique K in G and every $\lambda > 0$. As $G \in \mathcal{P}_K$, this implies that $\exp(-\lambda d) \in \mathcal{E}(G)$ for all $\lambda > 0$. Using again Proposition 31.4.13, we obtain that $d \in \mathrm{NEG}(G)$.
Suppose now that $G \in \mathcal{P}_M$; we show that $G \in \mathcal{D}_M$. Let $d \in \mathbb{R}^E$ satisfy-

ing (31.4.6), i.e., $\sqrt{d} \in \mathrm{MET}(G)$. Then, by Lemma 31.4.14, $\frac{1}{\pi}\arccos(e^{-\lambda d}) \in \mathrm{MET}^{\square}(G)$ for all $\lambda > 0$. As $G \in \mathcal{P}_M$, this implies that $\exp(-\lambda d) \in \mathcal{E}(G)$ for all $\lambda > 0$. By Proposition 31.4.13, we obtain that $d \in \mathrm{NEG}(G)$.
The inclusion $\mathcal{P}_{KM} \subseteq \mathcal{D}_{KM}$ follows by combining the above arguments. ∎

31.5 Geometry of the Elliptope

In Section 28.4.1 was introduced the convex body $\mathcal{J}_n$ as a (nonpolyhedral) relaxation of the cut polytope $\mathrm{CUT}_n^{\square}$. We remind that

$$\begin{aligned}\mathcal{J}_n &= \{x \in \mathbb{R}^{E_n} \mid \sum_{1 \le i < j \le n} b_i b_j x_{ij} \le \frac{1}{4}(\sum_{i=1}^{n} b_i)^2 \text{ for all } b \in \mathbb{Z}^n\} \\ &= \{x \in \mathbb{R}^{E_n} \mid J - 2X \succeq 0\}\end{aligned}$$

where J is the all-ones matrix and, for $x \in \mathbb{R}^{E_n}$, X is the symmetric $n \times n$ matrix with zero diagonal and off-diagonal entries x_{ij}. We also remind that the elliptope $\mathcal{E}_n$ is defined as the set of $n \times n$ symmetric positive semidefinite matrices with an all-ones diagonal. Therefore,

$$x \in \mathcal{J}_n \Longleftrightarrow J - 2X \in \mathcal{E}_n.$$

Hence, the two convex sets $\mathcal{J}_n$ and $\mathcal{E}_n$ are essentially identical (up to the transformation $x \mapsto 1 - 2x$). The convex body $\mathcal{J}_n$ is a relaxation of $\mathrm{CUT}_n^{\square}$, i.e.,

$$\mathrm{CUT}_n^{\square} \subseteq \mathcal{J}_n.$$

Moreover, $\mathcal{J}_n$ provides a good approximation for $\mathrm{CUT}_n^{\square}$ in the sense of optimization (recall Theorem 28.4.7). In fact, the convex body $\mathcal{J}_n$ presents several geometric features, which may explain and provide further insight for its good behaviour in optimization. One such property is, for instance, the fact that the only vertices of $\mathcal{J}_n$ are the cut vectors. This result is given below as well as several other geometric properties. For convenience we will work with the elliptope $\mathcal{E}_n$ rather than with $\mathcal{J}_n$ itself.

We start with recalling some definitions. Let K be a convex set in $\mathbb{R}^d$. Given a boundary point x_0 of K, its *normal cone* $N(K, x_0)$ is defined as

$$N(K, x_0) := \{c \in \mathbb{R}^d \mid c^T x \le c^T x_0 \text{ for all } x \in K\}.$$

Hence, $N(K, x_0)$ consists of the normal vectors to the supporting hyperplanes of K at x_0. Then, the *supporting cone* at x_0 is defined by

$$C(K, x_0) := \{x \in \mathbb{R}^d \mid c^T x \le 0 \text{ for all } c \in N(K, x_0)\}.$$

The dimension of the normal cone permits to classify the boundary points. In particular, a boundary point x_0 is called a *vertex* of K if its normal cone $N(K, x_0)$ is full-dimensional. A subset $F \subseteq K$ is a face of K if, for all $x \in F$, $y, z \in K$ and

$0 \leq \alpha \leq 1$, $x = \alpha y + (1-\alpha)z$ implies that $y, z \in F$. In particular, an element $x_0 \in K$ is called *an extreme point* of K if the set $\{x_0\}$ is a face of K. In what follows we consider the two convex sets $\mathcal{E}_n$ and $\mathcal{J}_n$. When dealing with $\mathcal{E}_n$ we take the space of symmetric $n \times n$ matrices as ambient space, equipped with the inner product:

$$\langle A, B \rangle := \sum_{i,j=1}^{n} a_{ij} b_{ij} \quad \text{for two symmetric } n \times n \text{ matrices } A, B$$

and, when dealing with $\mathcal{J}_n$, the ambient space is the usual Euclidean space $\mathbb{R}^{\binom{n+1}{2}}$. We remind that $\mathrm{Tr}\, A := \sum_{i=1}^{n} a_{ii}$ for an $n \times n$ matrix A.

We begin with the description of the polar of $\mathcal{E}_n$ and of its normal cones. These results are established by Laurent and Poljak [1995b, 1996a]; proofs can be found there.

Theorem 31.5.1. *The polar of $\mathcal{E}_n$ is given by*

$$(\mathcal{E}_n)^\circ = \{D - M \mid M \succeq 0, D \textit{ diagonal matrix with } \mathrm{Tr}\, D = 1\}.$$

For $A \in \mathcal{E}_n$, its normal cone is defined by

$$N(\mathcal{E}_n, A) = \{D - M \mid M \succeq 0, \langle M, A \rangle = 0, D \textit{ diagonal matrix}\}.$$

Moreover, $\dim N(\mathcal{E}_n, A) = n + \binom{n-r+1}{2}$, *where r is the rank of A.* ∎

Corollary 31.5.2. *The only vertices of $\mathcal{E}_n$ are the 'cut matrices' xx^T, for $x \in \{\pm 1\}^n$. In other words, the convex body $\mathcal{J}_n$ has 2^{n-1} vertices, namely, the cut vectors $\delta(S)$ for $S \subseteq V_n$.* ∎

We remind that, given $c \in \mathbb{R}^{E_n}$, $\max(c^T x \mid x \in \mathcal{J}_n)$ is an upper bound for the max-cut problem: $\max(c^T x \mid x \in \mathrm{CUT}_n^\square)$. Equality holds between the bound and the max-cut precisely when c belongs to the normal cone of one of the cut vectors. That the cut vectors are the only boundary points having a full dimensional normal cone supports the idea that $\mathcal{J}_n$ approximates well $\mathrm{CUT}_n^\square$. From Theorem 31.5.1 one obtains that the supporting cone $C(\mathcal{E}_n, A)$ at $A \in \mathcal{E}_n$ is the set

$$\{X \text{ symmetric } n \times n \mid x_{ii} = 0 \ \forall i = 1, \ldots, n, \ b^T X b \geq 0 \ \text{ for all } b \in \mathrm{Ker} A\}.$$

In particular, at $A = J$ (the all-ones matrix), the supporting cone is $-\mathrm{NEG}_n$. At every other vertex of $\mathcal{E}_n$, the supporting cone is an affine image of the negative type cone NEG_n (under the switching mapping). So, this makes one more connection between the elliptope and the negative type cone.

We now turn to the description of the faces of $\mathcal{E}_n$. We remind that $\mathcal{E}_n$ is obtained by taking the intersection of the cone PSD_n of positive semidefinite matrices with the linear space $W := \{X \mid x_{ii} = 1 \ \ \forall i = 1, \ldots, n\}$. The facial structure of the cone PSD_n is well understood (see Hill and Waters [1987]). It is, in some sense, rather simple. Indeed, given a matrix $A \in \mathrm{PSD}_n$ with rank r the smallest face $F_{\mathrm{PSD}}(A)$ of PSD_n that contains A is given by

$$F_{\mathrm{PSD}}(A) = \{X \in \mathrm{PSD}_n \mid \mathrm{Ker}X \supseteq \mathrm{Ker}A\}.$$

Hence, $F_{\mathrm{PSD}}(A)$ is isomorphic to the cone PSD_r and, thus, has dimension $\binom{r+1}{2}$. From this follows the description of the faces of $\mathcal{E}_n$. For $A \in \mathcal{E}_n$, the smallest face $F_{\mathcal{E}}(A)$ of $\mathcal{E}_n$ that contains A is equal to $F_{\mathrm{PSD}}(A) \cap W$ (as W is the only face of W). In other words,

$$F_{\mathcal{E}}(A) = \{X \in \mathcal{E}_n \mid \mathrm{Ker}X \supseteq \mathrm{Ker}A\}.$$

However, computing the dimension of $F_{\mathcal{E}}(A)$ requires more care[8]. This has been done by Li and Tam [1994]. For convenience, we state their result in a more general setting.

Theorem 31.5.3. *Let $A_1, \ldots, A_m$ be $n \times n$ symmetric matrices and $b_1, \ldots, b_m \in \mathbb{R}$. Consider the convex set*

$$K := \{X \in \mathrm{PSD}_n \mid \langle X, A_j \rangle = b_j \ \ \forall j = 1, \ldots, m\}.$$

Let $A \in K$ and let $F_K(A)$ be the smallest face of K that contains A. Suppose that A has rank r and that $A = QQ^T$, where Q is an $n \times r$ matrix of rank r. Then,

$$\dim\ F_K(A) = \binom{r+1}{2} - \mathrm{rank}\ \{Q^T A_j Q \mid j = 1, \ldots, m\}.$$

Proof. Call a symmetric matrix B a *perturbation* of A if $A \pm \lambda B \in K$ for some $\lambda > 0$. Then, dim $F_K(A)$ is equal to the rank of the set of perturbations of A. We claim:

(a) B is a perturbation of $A \iff B = QRQ^T$ for some $r \times r$ symmetric matrix R and $\langle B, A_j \rangle = 0$ for all $j = 1, \ldots, m$.

If $B = QRQ^T$ then $A \pm \lambda B = Q(I \pm \lambda R)Q^T$ is clearly positive semidefinite if $\lambda > 0$ is small enough. Moreover, the condition: $\langle B, A_j \rangle = 0$ for all j ensures that $A \pm \lambda B \in K$. Conversely, suppose that B is a perturbation of A. So, $A \pm \lambda B \in K$ for some $\lambda > 0$. This implies that $\langle B, A_j \rangle = 0$ for all j. Complete Q to an $n \times n$ nonsingular matrix P. Set $C := P^{-1}B(P^{-1})^T$; that is, $B = PCP^T$. Then,

$$A \pm \lambda B = P \begin{pmatrix} I_r & 0 \\ 0 & 0 \end{pmatrix} P^T \pm \lambda PCP^T = P \left(\begin{pmatrix} I_r & 0 \\ 0 & 0 \end{pmatrix} \pm \lambda \begin{pmatrix} C_{11} & C_{12} \\ C_{12} & C_{22} \end{pmatrix} \right) P^T,$$

[8]Here arises also the question of characterizing the linear subspaces V of $\mathbb{R}^n$ such that $V \subseteq \mathrm{Ker}A$ for some $A \in \mathcal{E}_n$. Delorme and Poljak [1993b] show that a vector $b \in \mathbb{R}^n$ belongs to the kernel of some matrix $A \in \mathcal{E}_n$ if and only if b satisfies: $|b_i| \leq \sum_{1 \leq j \leq n,\ j \neq i} |b_j|$ for all $i = 1, \ldots, n$. An analogue combinatorial characterization for higher dimensional spaces is not known.

setting $C := \begin{pmatrix} C_{11} & C_{12} \\ C_{12} & C_{22} \end{pmatrix}$. Hence, $\begin{pmatrix} I_r & 0 \\ 0 & 0 \end{pmatrix} \pm \lambda \begin{pmatrix} C_{11} & C_{12} \\ C_{12} & C_{22} \end{pmatrix} \succeq 0$. This implies that $C_{12} = C_{22} = 0$. Therefore, $B = QC_{11}Q^T$, where C_{11} is a symmetric $r \times r$ matrix. Hence, (a) holds.

Now, every perturbation of A is of the form $B = QRQ^T$ with $\langle B, A_j \rangle = 0$ for all j; that is, $\langle R, Q^T A_i Q \rangle = 0$ for all j. Hence, the dimension of $F_K(A)$ is equal to the dimension of the orthogonal complement of $\{Q^T A_j Q \mid j = 1, \ldots, m\}$ in the space of symmetric $r \times r$ matrices. Hence, we have the desired formula for dim $F_K(A)$. ∎

Corollary 31.5.4. *Let $A \in \mathcal{E}_n$ with rank r, let $F_{\mathcal{E}}(A)$ denote the smallest face of $\mathcal{E}_n$ containing A, and suppose that A is the Gram matrix of the vectors $u_1, \ldots, u_n \in \mathbb{R}^r$. Then,*

$$\text{(31.5.5)} \qquad \dim\ F_{\mathcal{E}}(A) = \binom{r+1}{2} - \text{rank}\ \{u_i u_i^T \mid i = 1, \ldots, n\}.$$

∎

In particular, one obtains bounds for the rank of extreme matrices[9] of $\mathcal{E}_n$.

Corollary 31.5.6. *Let $A \in \mathcal{E}_n$ with rank r. If A is an extreme point of $\mathcal{E}_n$ then $\binom{r+1}{2} \leq n$.* ∎

Moreover, as we see below, for every r such that $\binom{r+1}{2} \leq n$ there exists an extreme matrix in $\mathcal{E}_n$ having rank r. The formula (31.5.5) can be used for finding the possible dimensions for the faces of $\mathcal{E}_n$, as observed in Laurent and Poljak [1996a]. Namely,

Proposition 31.5.7. *Let $A \in \mathcal{E}_n$ with rank r and set $k :=$ dim $F_{\mathcal{E}}(A)$. Then,*

$$\max(0, \binom{r+1}{2} - n) \leq k \leq \binom{r}{2}.$$

Moreover, for every integers $r, k \geq 0$ satisfying the above inequality, there exists a matrix $A \in \mathcal{E}_n$ with rank r and with dim $F_{\mathcal{E}}(A) = k$.

Proof. The inequality from Proposition 31.5.7 follows from (31.5.5), after noting that $r \leq$ rank $\{u_1 u_1^T, \ldots, u_n u_n^T\} \leq n$. The existence part relies essentially on a construction proposed in Grone, Pierce and Watkins [1990], which goes as follows. Let $e_1, \ldots, e_r$ denote the coordinate vectors in $\mathbb{R}^r$ and set $w_{ij} := \frac{1}{\sqrt{2}}(e_i + e_j)$ for $1 \leq i < j \leq r$. Then, the $\binom{r+1}{2}$ matrices: $e_i e_i^T$ $(i = 1, \ldots, r)$ and $w_{ij} w_{ij}^T$ $(1 \leq i < j \leq r)$ are linearly independent. Suppose first that $n = \binom{r+1}{2} - k$ where

[9] Solving this question has been the subject of several papers in the linear algebra literature; for example, by Christensen and Vesterstrøm [1979], Loewy [1980], Grone, Pierce and Watkins [1990].

$k \leq \binom{r}{2}$. Hence, $r \leq n \leq \binom{r+1}{2}$. Define A as the Gram matrix of the following n vectors: $e_1, \ldots, e_r$ together with $n-r$ of the vectors w_{ij}. By construction, A has rank r and dim $F_{\mathcal{E}}(A) = \binom{r+1}{2} - n = k$. When $n > \binom{r+1}{2} - k$, we can take as matrix A the Gram matrix of the following n vectors: e_1 repeated $n - \binom{r+1}{2} + k + 1$ times, $e_2 \ldots e_r$, and $\binom{r}{2} - k$ of the w_{ij}'s. ∎

Therefore, the range $\mathcal{D}_n$ of the possible values for the dimension of the faces of $\mathcal{E}_n$ is given by:

$$\mathcal{D}_n = [0, \binom{k_n}{2}] \cup \bigcup_{r=k_n+1}^{n} [\binom{r+1}{2} - n, \binom{r}{2}],$$

where k_n is the smallest integer k such that $\binom{k+2}{2} - n > \binom{k_n}{2} + 1$, i.e., $2k_n > n$; that is, $k_n = \lfloor \frac{n}{2} \rfloor + 1$. For instance,

$$k_3 = 2, \ \mathcal{D}_3 = [0,1] \cup \{3\},$$

$$k_4 = 3, \ \mathcal{D}_4 = [0,3] \cup \{6\},$$

$$k_5 = 3, \ \mathcal{D}_5 = [0,3] \cup [5,6] \cup \{10\},$$

$$k_6 = 4, \ \mathcal{D}_6 = [0,6] \cup [9,10] \cup \{15\},$$

$$k_7 = 4, \ \mathcal{D}_7 = [0,6] \cup [8,10] \cup [14,15] \cup \{21\}.$$

One can verify on Figure 31.3.6 that the proper faces of $\mathcal{E}_3$ have dimension 0 (extreme points) or 1 (an edge between two cut vectors; there are six such faces). A detailed description of the faces of $\mathcal{E}_n$ can be found in Laurent and Poljak [1995b, 1996a] for $n = 3$ and $n = 4$, respectively.

Finally, the possible dimensions for the polyhedral faces of $\mathcal{E}_n$ are as follows; they were computed by Laurent and Poljak [1996a].

Theorem 31.5.8. *If F is a polyhedral face of $\mathcal{E}_n$ with dimension k, then $\binom{k+1}{2} \leq n-1$. Moreover, if all the vertices of F are cut matrices then F is a simplex. Conversely, for every integer $k \geq 1$ such that $\binom{k+1}{2} \leq n-1$, $\mathcal{E}_n$ has a polyhedral face of dimension k (which can be chosen to be a simplex with cut matrices as vertices).* ∎

Every polyhedral face of $\mathcal{E}_n$ with cut matrices as vertices yields clearly a face of the cut polytope. We describe below a construction for such polyhedral faces, due to Laurent and Poljak [1996a]. We need a definition in order to state the result.

Let $S_1, \ldots, S_k$ be k subsets of V_n. The cut vectors $\delta(S_1), \ldots, \delta(S_k)$ are said to be in *general position* if the set

$$\bigcap_{i \in I} S_i \cap \bigcap_{i \notin I} (V_n \setminus S_i)$$

is nonempty, for every subset $I \subseteq \{1, \ldots, k\}$. This implies that $2^k \leq n$, i.e., $k \leq \log_2 n$. Moreover, the cut vectors $\delta(S_1), \ldots, \delta(S_k)$ are linearly independent.

Theorem 31.5.9. *Let $\delta(S_1), \ldots, \delta(S_k)$ be k cuts in general position. Then, the set $F := \mathrm{Conv}(\delta(S_1), \ldots, \delta(S_k))$ is a face of the convex body $\mathcal{J}_n$. (Equivalently, the set $\mathrm{Conv}(x_1 x_1^T, \ldots, x_k x_k^T)$ is a face of $\mathcal{E}_n$, where $x_h \in \mathbb{R}^n$ is defined by $x_h(i) := 1$ if $i \in S_h$ and $x_h(i) := -1$ if $i \in V_n \setminus S_h$, for $h = 1, \ldots, k$.) Therefore, F is also a face of the cut polytope $\mathrm{CUT}_n^{\square}$.* ∎

This result shows that $\mathcal{J}_n$ and $\mathrm{CUT}_n^{\square}$ share fairly many common faces, up to dimension $\lfloor \log_2 n \rfloor$. This supports again the idea that $\mathcal{J}_n$ approximates well the cut polytope $\mathrm{CUT}_n^{\square}$. In fact, the faces considered in Theorem 31.5.9 are also faces in common with the semimetric polytope $\mathrm{MET}_n^{\square}$; see Theorem 31.6.4.

31.6 Adjacency Properties

We now return to the study of the geometry of the cut polytope itself, as well as with respect to its linear relaxation by the semimetric polytope. We mention first some results on the faces of low dimension and, then, facts and questions about the small cut and semimetric polytopes.

31.6.1 Low Dimension Faces

A striking property of the cut polytope $\mathrm{CUT}_n^{\square}$ is that any two of its vertices form an edge of $\mathrm{CUT}_n^{\square}$. In fact, much more is true. In order the formulate the results, we need some definitions.

Let P be a polytope with set of vertices V. Given an integer $k \geq 1$, the polytope P is said to be *k-neighborly* if, for any subset $W \subseteq V$ of vertices such that $|W| \leq k$, the set $\mathrm{Conv}(W)$ is a a face of P. This implies, in particular, that every k vertices of P are affinely independent. Hence, every polytope is 1-neighborly and a polytope is 2-neighborly precisely when its 1-skeleton graph is a complete graph.

Given an integer d and a polyhedron P, we let $\phi_d(P)$ denote the set of d-dimensional faces of P.

Barahona and Mahjoub [1986] show that $\mathrm{CUT}_n^{\square}$ is 2-neighborly, i.e., that any two cut vectors are adjacent on $\mathrm{CUT}_n^{\square}$. In other words, the 1-skeleton graph of $\mathrm{CUT}_n^{\square}$ is a complete graph. Padberg [1989] shows the following stronger result: Any two cut vectors are adjacent on the rooted semimetric polytope $\mathrm{RMET}_n^{\square}$ (defined by the triangle inequalities going through a given node; recall Section 27.2). More generally, Deza, Laurent and Poljak [1992] show the following result.

Theorem 31.6.1. *Let W be a set of cut vectors such that $|W| \leq 3$. Then, the set $\mathrm{Conv}(W)$ is a simplex face of the semimetric polytope $\mathrm{MET}_n^{\square}$.*

Proof. Due to switching, we can suppose that the set W contains the zero cut vector $\delta(\emptyset)$. Let us first consider the case when $|W| = 2$; say, $W = \{\delta(\emptyset), \delta(S)\}$, where $S \neq \emptyset, V_n$. In order to show that the set $\text{Conv}(W)$ is a face of $\text{MET}_n^\square$, it suffices to find a vector $w \in \mathbb{R}^{E_n}$ satisfying the following property:

(a) $w^T x \leq 0$ for all $x \in \text{MET}_n^\square$, with equality if and only if $x \in \text{Conv}(W)$.

For this, set $w_{ij} := 0$ if $\delta(S)_{ij} = 1$ and $w_{ij} := -1$ otherwise. It is immediate to verify that w satisfies the desired property.

We now consider the case when $|W| = 3$; say, $W = \{\delta(\emptyset), \delta(S), \delta(T)\}$, where $\delta(S)$ and $\delta(T)$ are distinct and nonzero. Set $A := S \cap T$, $B := \overline{S} \cap T$, $C := S \cap \overline{T}$, and $D := \overline{S} \cap \overline{T}$. Again, we should find $w \in \mathbb{R}^{E_n}$ satisfying (a). Let us first suppose that the sets A, B, C, D are nonempty. Let $a \in A$, $b \in B$, $c \in C$, and $d \in D$. Define $w \in \mathbb{R}^{E_n}$ by setting $w_{ab} = w_{ac} = w_{bd} = w_{cd} := -1$, $w_{ad} = w_{bc} := 1$, $w_{ij} := -1$ if $i \neq j$ both belong to A, or B, or C, or D (denote by E the set of these pairs ij), and $w_{ij} := 0$ otherwise. Then, $w^T\delta(S) = w^T\delta(T) = 0$. Let $x \in \text{MET}_n^\square$. Then,

$$w^T x = -\sum_{ij \in E} x_{ij} + \sigma,$$

where

$$\sigma := x_{ad} + x_{bc} - x_{ab} - x_{bd} - x_{cd} - x_{ac}.$$

We have the relations:

(i) $\sigma = (x_{ad} - x_{ac} - x_{cd}) + (x_{bc} - x_{cd} - x_{bd}) + x_{cd} - x_{ab} \leq x_{cd} - x_{ab}$,

(ii) $\sigma = (x_{ad} - x_{ab} - x_{bd}) + (x_{bc} - x_{ab} - x_{ac}) + x_{ab} - x_{cd} \leq x_{ab} - x_{cd}$,

(iii) $\sigma = (x_{ad} - x_{ac} - x_{cd}) + (x_{bc} - x_{ab} - x_{ac}) + x_{ac} - b_{bd} \leq x_{ac} - b_{bd}$,

(iv) $\sigma = (x_{ad} - x_{ab} - x_{bd}) + (x_{bc} - x_{bd} - x_{cd}) + x_{bd} - x_{ac} \leq x_{bd} - x_{ac}$.

From (i)-(iv) we deduce that $\sigma \leq 0$. Therefore, $w^T x \leq 0$. Moreover, if $w^T x = 0$, then $x_{ij} = 0$ for all $ij \in E$ and $\sigma = 0$. Hence, using (i)-(iv), $x_{ab} = x_{cd} := \alpha$, $x_{ac} = x_{bd} := \beta$ for some $\alpha, \beta \geq 0$, $\alpha + \beta \leq 1$, and $x_{ad} = x_{bc} = \alpha + \beta$. From this follows easily that $x = \alpha\delta(S) + \beta\delta(T)$, which shows that $x \in \text{Conv}(W)$.

Finally, let us suppose that one of the sets A, B, C, D is empty. Say, $D = \emptyset$. Then, $A, B, C \neq \emptyset$; let $a \in A$, $b \in B$ and $c \in C$. We now define $w \in \mathbb{R}^{E_n}$ by $w_{ab} = w_{ac} := -1$, $w_{bc} := 1$, $w_{ij} := -1$ if $i \neq j$ both belong to A, or B, or C, and $w_{ij} := 0$ otherwise. It can be verified as above that w satisfies (a). ∎

Corollary 31.6.2. *The cut polytope* $\text{CUT}_n^\square$ *is 3-neighborly.* ∎

Corollary 31.6.3.

(i) *For* $n \geq 4$, *every face of* $\text{CUT}_n^\square$ *of dimension* $d \leq 5$ *is a simplex.*

(ii) $\phi_d(\text{CUT}_n^\square) \subseteq \phi_d(\text{MET}_n^\square)$, *for* $d = 0, 1, 2$. ∎

The results from Corollaries 31.6.2 and 31.6.3 (i) are best possible; that is, $\mathrm{CUT}_n^\square$ is not 4-neighborly and there exists a 6-dimensional face of $\mathrm{CUT}_n^\square$ ($n \geq 4$) which is not a simplex. Indeed, for $n = 4$, $\mathrm{CUT}_4^\square$ itself is a nonsimplex 6-dimensional face. For $n \geq 5$, consider the face F of $\mathrm{CUT}_n^\square$ which is defined by the inequality:

$$\sum_{4 \leq i < j \leq n} x_{ij} \geq 0.$$

Then, F contains the following eight cut vectors $\delta(S)$ for $S = \emptyset$, $\{2\}$, $\{3\}$, $\{1,2\}$, $\{1,3\}$, $\{2,3\}$, and $\{1,2,3\}$. They are not affinely independent as they satisfy:

$$\delta(\{1\}) + \delta(\{2\}) + \delta(\{3\}) + \delta(\{1,2,3\}) = \delta(\{1,2\}) + \delta(\{1,3\}) + \delta(\{2,3\}).$$

Hence, F is a nonsimplex face of dimension 6 of $\mathrm{CUT}_n^\square$. (In fact, one can check that F is also a face of $\mathrm{MET}_n^\square$.) Hence, the four cut vectors $\delta(\emptyset)$, $\delta(\{1,2\})$, $\delta(\{1,3\})$, and $\delta(\{2,3\})$ do not form a face of $\mathrm{CUT}_n^\square$. This shows that $\mathrm{CUT}_n^\square$ is not 4-neighborly.

The result of Corollary 31.6.3 (ii) is also best possible, i.e., there exists a 3-dimensional face of $\mathrm{CUT}_n^\square$ which is not a face of $\mathrm{MET}_n^\square$ (for $n \geq 5$). The following example is given in Deza and Deza [1995]. Let $n \geq 5$. Consider the face F of $\mathrm{CUT}_n^\square$ which is defined by $F := \mathrm{CUT}_5^\square$ for $n = 5$ and

$$F := \{x \in \mathrm{CUT}_n^\square \mid x_{1i} + x_{2i} + x_{12} = 2 \text{ and } x_{1i} - x_{2i} - x_{12} = 0 \text{ for } i = 6, \ldots, n\}$$

for $n \geq 6$. The cut vectors lying in F are of the form $\delta(S \cup \{1\})$, where $S \subseteq \{2,3,4,5\}$. Therefore, $F \approx \mathrm{CUT}_5^\square$ is a 10-dimensional face of $\mathrm{CUT}_n^\square$ which is not a face of $\mathrm{MET}_n^\square$. Consider the set

$$G := \mathrm{Conv}(\delta(\{1,2\}), \delta(\{1,3\}), \delta(\{1,4\}), \delta(\{1,5\}))$$

and let H denote the face of $\mathrm{MET}_n^\square$ which is defined by the triangle inequalities:

$$x_{1i} + x_{2i} + x_{12} = 2,\ x_{1i} - x_{2i} - x_{12} = 0 \ (i = 6, \ldots, n)$$

$$\text{and } \ x_{1i} + x_{1j} + x_{ij} = 2 \ (2 \leq i < j \leq 5).$$

Then, G is a 3-dimensional face of $\mathrm{CUT}_n^\square$ as

$$G = \{x \in F \mid \sum_{1 \leq i < j \leq 5} x_{ij} = 6 \text{ and } x_{1i} + x_{1j} + x_{ij} = 2 \text{ for } 2 \leq i < j \leq 5\}.$$

But, G is not a face of $\mathrm{MET}_n^\square$. To see it, consider the point $x \in \mathbb{R}^{E_n}$ defined by $x_{1i} = 1$, $x_{2i} = x_{3i} = x_{4i} = x_{5i} = \frac{1}{3}$ for $i = 6, \ldots, n$, $x_{ij} = 0$ for $6 \leq i < j \leq n$, and $x_{ij} = \frac{2}{3}$ for $1 \leq i < j \leq 5$. Then, $x \in H \setminus G$. If G is a face of $\mathrm{MET}_n^\square$, then G is a face of H and, thus, there exists a triangle inequality valid for G and violated by x. Now one can easily check that no such inequality exists. This shows that G is not a face of $\mathrm{MET}_n^\square$.

Even though not every d-dimensional face of $\mathrm{CUT}_n^\square$ is a face of $\mathrm{MET}_n^\square$ when $d \geq 3$, the next result shows that a lot of them remain faces of $\mathrm{MET}_n^\square$ when $d \leq \log_2 n$.

Given $S_1, \ldots, S_k \subseteq V_n$, recall that the cut vectors $\delta(S_1), \ldots, \delta(S_k)$ are said to be in general position if the set

$$\bigcap_{i \in I} S_i \cap \bigcap_{i \notin I} (V_n \setminus S_i)$$

is nonempty, for every subset $I \subseteq \{1, \ldots, k\}$. Then, $k \leq \log_2 n$ and the cut vectors $\delta(S_1), \ldots, \delta(S_k)$ are linearly independent. Deza, Laurent and Poljak [1992] show that cuts in general position form a face; the proof of this result is along the same lines as that of Theorem 31.6.1, but with more technical details. Compare the results in Theorems 31.6.4 and 31.5.9.

Theorem 31.6.4. *Let $\delta(S_1), \ldots, \delta(S_k)$ be k cut vectors in general position. Then, the set* $\text{Conv}(\delta(S_1), \ldots, \delta(S_k))$ *is a face of* $\text{MET}_n^\square$ *and, thus, of* $\text{CUT}_n^\square$. ∎

Therefore, $\text{CUT}_n^\square$ and $\text{MET}_n^\square$ share a lot of common faces, at least up to dimension $\lfloor \log_2 n \rfloor$. This is an indication that the semimetric polytope is wrapped quite tightly around the cut polytope.

31.6.2 Small Polytopes

We group here some results and questions related to facets/vertices of the cut polytope $\text{CUT}_n^\square$ and the semimetric polytope $\text{MET}_n^\square$, especially for the small values of n, $n \leq 7$. The reader may consult Deza [1994, 1996] for a detailed survey on various combinatorial and geometric properties of these polyhedra.

n	# facets of CUT_n	# facets of $\text{CUT}_n^\square$	# orbits of facets
3	3	4	1
4	12	16	1
5	40	56	2
6	210	368	3
7	38,780	116,764	11
8	49,604,520	217,093,472	147

Figure 31.6.5: Number of facets of cut polyhedra for $n \leq 8$

All the facets of the cut cone CUT_n and the cut polytope $\text{CUT}_n^\square$ are known for $n \leq 7$; they were described in Section 30.6. The extreme rays of MET_n and the vertices of $\text{MET}_n^\square$ are also known for $n \leq 7$; the extreme rays of MET_7 were computed by Grishukhin [1992a] and the vertices of $\text{MET}_7^\square$ by Deza, Deza and Fukuda [1996]. For $n \leq 6$, they are very simple. Namely, besides the cut vectors (that are all the integral vertices), all of them arise from the vector $(2/3, \ldots, 2/3)$ after possibly applying switching[10] and gate 0-extensions[11]. Fig-

[10]The metric polytope being preserved under the switching operation, its set of vertices is partitioned into switching classes. Namely, if x is a vertex of $\text{MET}_n^\square$, then all vectors in its switching class $\{r_{\delta(A)}(x) \mid A \subseteq V_n\}$ are also vertices of $\text{MET}_n^\square$. For instance, the cut vectors form a single switching class.

[11]Given $x \in \mathbb{R}^{E_n}$, we remind that its gate 0-extension is the vector $y \in \mathbb{R}^{E_{n+1}}$ defined by $y_{ij} := x_{ij}$ for $ij \in E_n$, $y_{1,n+1} := 0$, $y_{i,n+1} := x_{1i}$ for $i = 2, \ldots, n$. It can be easily verified that

ures 31.6.5 and 31.6.6 summarize information on the number of facets/vertices of the cut and semimetric polyhedra. Data for CUT_8 and $\mathrm{CUT}_8^\square$ come from Christof and Reinelt [1996]. (We remind that orbits are obtained by action of switching and permutations.)

n	# extreme rays of MET_n	# vertices of $\mathrm{MET}_n^\square$	# orbits of vertices
3	3	4	1
4	7	8	1
5	25	32	2
6	296	544	3
7	55,226	275,840	13

Figure 31.6.6: Number of extreme rays/vertices of semimetric polyhedra for $n \le 7$

Much information is known about the 1-skeleton graph of $\mathrm{MET}_n^\square$ and about the ridge graphs[12] of $\mathrm{MET}_n^\square$ and $\mathrm{CUT}_n^\square$. We quote here some facts and questions and refer to the original papers or to the survey by Deza [1996] for more details.

The 1-Skeleton Graph of the Semimetric Polytope. As the semimetric polytope $\mathrm{MET}_n^\square$ is preserved under the switching operation, this induces a partition of its vertices into switching classes. The cut vectors form a single switching class, which is a clique in the 1-skeleton graph of $\mathrm{MET}_n^\square$ (by Theorem 31.6.1). On the other hand, it is shown in Laurent [1996c] that every other switching class of vertices is a stable set in the 1-skeleton graph of $\mathrm{MET}_n^\square$; that is, no two nonintegral switching equivalent vertices of $\mathrm{MET}_n^\square$ form an edge on $\mathrm{MET}_n^\square$. The following conjecture is posed by Laurent and Poljak [1992].

Conjecture 31.6.7. *Every fractional vertex of* $\mathrm{MET}_n^\square$ *is adjacent to some cut vector (i.e., to some integral vertex of* $\mathrm{MET}_n^\square$*). Equivalently, for every fractional vertex x of* $\mathrm{MET}_n^\square$*, some switching $r_{\delta(S)}(x)$ of it lies on an extreme ray of* MET_n.

This can be seen as an analogue of the following property, shared by the facets of the cut polytope: For every facet of the cut polytope there exists a switching of it that contains the origin. A consequence of Conjecture 31.6.7 would be that the 1-skeleton graph of $\mathrm{MET}_n^\square$ has diameter ≤ 3. Conjecture 31.6.7 has been verified for several classes of vertices (see Laurent [1996c]) and for $n \le 7$ (see Deza, Deza and Fukuda [1996]).

Adjacency has been analyzed in detail for some classes of vertices. Given a subset $S \subseteq V_n$, let $d(K_{S,V_n\setminus S})$ denote the path metric of the complete bipartite graph with node bipartition $(S, V_n \setminus S)$. Then, $x_S := \frac{1}{3}d(K_{S,V_n\setminus S})$ is a vertex of $\mathrm{MET}_n^\square$ (taking value $\frac{1}{3}$ on the edges of the bipartition and value $\frac{2}{3}$ elsewhere).

y is a vertex of $\mathrm{MET}_{n+1}^\square$ whenever x is a vertex of $\mathrm{MET}_n^\square$.

[12]Let P be a d-dimensional polyhedron. Its *ridge graph* is the graph with node set the set of facets of P and with two facets being adjacent if their intersection has dimension $d-2$.

The vertices x_S ($S \subseteq V_n$) form a switching class. The adjacency relations between the cut vectors $\delta(S)$ and the vertices x_T (for $S, T \subseteq V_n$) are described in Deza and Deza [1994b]. Namely, the two vertices $\delta(S)$ and x_T are adjacent on $\mathrm{MET}_n^\square$ if and only if the cut vectors $\delta(S)$ and $\delta(T)$ are not adjacent in the folded n-cube graph, i.e., if $|S \triangle T| \neq 1, n-1$. In the case $n = 5$, the cut vectors $\delta(S)$ and the vectors x_T (for $S, T \subseteq V_5$) form all the vertices of $\mathrm{MET}_5^\square$. Hence, the 1-skeleton graph of $\mathrm{MET}_5^\square$ is completely known; its diameter is equal to 2.

Deza and Deza [1994b] analyze adjacency among further vertices of the form: cut vectors $\delta(S)$, x_T ($S, T \subseteq V_n$) and their gate extensions. This permits, in particular, to describe the 1-skeleton graph of $\mathrm{MET}_6^\square$, whose diameter is equal to 2.

The vertices of $\mathrm{MET}_7^\square$ and their adjacencies are described in Deza, Deza and Fukuda [1996]; in particular, the 1-skeleton graph of $\mathrm{MET}_7^\square$ has diameter 3. Figure 31.6.8 shows the 13 orbits of vertices of $\mathrm{MET}_7^\square$; for each orbit O_i, a representative vertex v_i is given as well as its cardinality $|O_i|$, the number A_i of neighbors of v_i in the 1-skeleton graph and the number I_i of triangle facets containing v_i.

Orbit	Representative vertex v_i	$\lvert O_i\rvert$	I_i	A_i
O_1	$(0,0,0,0,0,0,0,0,0,0,0,0,0,0,0,0,0,0,0,0,0)$	64	105	55 226
O_2	$\frac{2}{3}(1,1,1,1,1,1,1,1,1,1,1,1,1,1,1,1,1,1,1,1,1)$	64	35	896
O_3	$\frac{2}{3}(1,1,1,1,1,0,1,1,1,1,1,1,1,1,1,1,1,1,1,1,1)$	1 344	40	763
O_4	$\frac{2}{3}(1,1,1,1,0,1,1,1,1,1,0,1,1,1,1,1,1,1,1,1,1)$	6 720	45	594
O_5	$\frac{2}{3}(1,1,1,1,0,0,1,1,1,1,1,1,1,1,1,1,1,1,1,1,0)$	2 240	49	496
O_6	$\frac{1}{4}(1,2,3,1,2,1,1,2,2,1,2,1,1,2,3,2,3,2,1,2,1)$	20 160	30	96
O_7	$\frac{1}{3}(1,1,1,1,1,1,2,2,1,1,1,2,1,1,1,1,1,1,2,2,2)$	4 480	26	76
O_8	$\frac{2}{5}(2,1,1,1,1,2,2,1,1,1,1,2,1,1,1,2,1,1,2,1,2)$	23 040	28	57
O_9	$\frac{1}{3}(2,2,1,1,1,2,2,1,1,1,1,2,1,1,1,2,1,1,2,1,2)$	40 320	22	46
O_{10}	$\frac{1}{3}(1,1,1,1,1,1,2,2,1,1,1,2,1,1,1,2,1,1,2,2,2)$	40 320	23	39
O_{11}	$\frac{2}{7}(1,2,3,2,1,2,1,2,1,2,1,1,2,1,1,1,2,2,1,1,1)$	40 320	25	30
O_{12}	$\frac{1}{5}(3,2,3,3,1,1,1,2,2,2,2,3,3,3,3,4,4,2,2,4,2)$	16 128	25	27
O_{13}	$\frac{1}{6}(1,2,4,2,2,2,1,3,3,3,3,2,2,2,4,2,2,2,4,4,4)$	80 640	23	24
Total		275 840		

Figure 31.6.8: The orbits of vertices of $\mathrm{MET}_7^\square$

The Ridge Graph of the Semimetric Polytope. The ridge graph G_n of the semimetric polytope $\mathrm{MET}_n^\square$ is studied in detail in Deza and Deza [1994b]. The graph G_n has $4\binom{n}{3}$ vertices and, for $n \geq 4$, two triangle facets are adjacent in G_n if and only if they are nonconflicting. (Two triangle inequalities are said to be *conflicting* if there exists a pair ij such that the two inequalities have nonzero coordinates of distinct signs at the position ij.) For instance, $G_3 = K_4$ and

$G_4 = K_4 \times K_4$. More generally, for $n \geq 4$ the complement of G_n is locally[13] the bouquet[14] of $n-3$ copies of $K_3 \times K_3$ along a common K_3; its valency is $k = 3(2n-5)$, two adjacent nodes have $\lambda \in \{2(n-2), 4\}$ common neighbors, while two nonadjacent nodes have μ common neighbors with $\mu \in \{4,6\}$ for $n=5$ and $\mu \in \{0,4,6\}$ for $n \geq 6$. In particular, the diameter of G_n is equal to 2 for $n \geq 4$. Note that the complement of the ridge graph G_5 of $\mathrm{MET}_5^\square$ provides an example[15] of a regular graph of diameter 2 in which the number of common neighbors to two arbitrary nodes belongs to $\{\lambda, \mu\} = \{4,6\}$.

Deza and Deza [1994b] also describe the ridge graph G'_n of the semimetric cone MET_n, which is an induced subgraph of G_n. Namely, for $n \geq 4$, the complement of G'_n is locally the bouquet of $n-3$ copies of the circuit C_6 along a common edge. The graph G'_n has diameter 2 for $n \geq 4$.

The Ridge Graph of the Cut Polytope. The ridge graph of the cut polytope $\mathrm{CUT}_n^\square$ is studied in Deza and Deza [1994a]. (It suffices to consider the case $n \geq 5$ as $\mathrm{CUT}_n^\square = \mathrm{MET}_n^\square$ for $n \leq 4$.) The ridge graph of $\mathrm{CUT}_n^\square$ is described there for $n \leq 7$. In particular, two facets of $\mathrm{CUT}_5^\square$ are adjacent in the ridge graph if and only if they are nonconflicting, but this is not true for $n \geq 6$. The ridge graph of $\mathrm{CUT}_n^\square$ has diameter 2 for $n = 4,5$, diameter 3 for $n = 6$ and its diameter belongs to $\{3,4\}$ for $n = 7$. The following conjecture is posed in Deza and Deza [1994a].

Conjecture 31.6.9. *Every facet of* $\mathrm{CUT}_n^\square$ *is adjacent to at least one triangle facet in the ridge graph of* $\mathrm{CUT}_n^\square$.

This conjecture would imply that the ridge graph of $\mathrm{CUT}_n^\square$ has diameter ≤ 4. The conjecture is shown to hold for $n \leq 7$. Further properties and questions, also concerning the ridge graph of the cut cone, can be found in Deza and Deza [1994a].

n	vol $\mathrm{MET}_n^\square$	vol $\mathrm{CUT}_n^\square$	ratio ρ_n
3	1/3	1/3	100%
4	2/45	2/45	100%
5	4/1701	32/14,175	$\sim$ 96%
6	71,936/1,477,701,225	2384/58,046,625	$\sim$ 84%

Figure 31.6.10: Volumes of cut and semimetric polytopes for $n \leq 6$

Further combinatorial properties of cut and semimetric polyhedra have been studied. For instance, Deza and Deza [1995] have completely described the face lattices of both the cut polytope $\mathrm{CUT}_n^\square$ and the semimetric polytope $\mathrm{MET}_n^\square$

[13] The local structure of a graph G is the subgraph induced by the neighbors of any given vertex, assuming that these induced subgraphs are the same at all the vertices.

[14] Let $G = (V,E)$ be a graph and, for $U \subset V$, let $H = G[U]$ be an induced subgraph of G. Let $G_i = (V_i, E_i)$ $(i = 1, \ldots, k)$ be k isomorphic copies of G such that $V_i \cap V_j = U$ for all $i \neq j$. Then, the graph $(\cup_{i=1}^k V_i, \cup_{i=1}^k E_i)$ is called the bouquet of the k copies of G along H.

[15] This generalization of the notion of strongly regular graph is studied in Erickson et al. [1996].

for $n \leq 5$. Deza, Deza and Fukuda [1996] give the edge connectivity of the adjacency and ridge graphs for cut and semimetric polytopes. To conclude we mention some facts about the volume of cut and semimetric polyhedra.

A way of measuring the tightness of the relaxation of $\mathrm{CUT}_n^{\square}$ by $\mathrm{MET}_n^{\square}$ could be by considering the ratio

$$\rho_n := \frac{\mathrm{vol}\ \mathrm{CUT}_n^{\square}}{\mathrm{vol}\ \mathrm{MET}_n^{\square}}$$

of their volumes. Unfortunately, computing the volume of a polytope is a hard task in general. These volumes have been computed in the case $n \leq 6$ in Deza, Deza and Fukuda [1996]; we report the results in Figure 31.6.10.

31.7 Distance of Facets to the Barycentrum

We are interested here in evaluating what is the minimum possible distance of a facet to the barycentrum of $\mathrm{CUT}_n^{\square}$. Most of the results here come from Deza, Laurent and Poljak [1992].

Let $b := \left(\sum_{S \subseteq V_n | 1 \notin S} \delta(S)\right) / 2^{n-1}$ denote the barycentrum of $\mathrm{CUT}_n^{\square}$. Then,

$$b = (1/2, \ldots, 1/2).$$

The Euclidean distance from b to the hyperplane defined by the equation: $v^T x = \alpha$ is given by the formula:

$$\frac{|v^T b - \alpha|}{\| v \|_2}.$$

It can be easily checked that the distance from b to a facet F remains invariant if we replace F by a switching of it. In particular, the distance from b to any triangle facet is equal to $\frac{1}{2\sqrt{3}}$. The following conjecture is posed by Deza, Laurent and Poljak [1992].

Conjecture 31.7.1. *The distance from the barycentrum b to any facet of $\mathrm{CUT}_n^{\square}$ is greater than or equal to $\frac{1}{2\sqrt{3}}$, this smallest distance being attained precisely by the triangle facets.*

They show that this conjecture holds for all pure facets, i.e., for all the facets that are defined by an inequality with $0, \pm 1$-coefficients.

Theorem 31.7.2. *Let $v^T x \leq \alpha$ be an inequality defining a facet of $\mathrm{CUT}_n^{\square}$ and such that $v \in \{-1, 0, 1\}^{E_n}$. Then, the distance from this facet to the barycentrum b is greater than or equal to $(2\sqrt{3})^{-1}$. Moreover, this smallest distance is realized precisely when $v^T x \leq \alpha$ is a triangle inequality.* ∎

The proof of Theorem 31.7.2 relies on establishing a good lower bound for the max-cut problem in the graph K_n with edge weights v. For a vector $v \in \mathbb{R}^{E_n}$, set

$$\mathrm{mc}(K_n, v) := \max(v^T \delta(S) \mid S \subseteq V_n).$$

Then, it shown in Deza, Laurent and Poljak [1992] that

$$\mathrm{mc}(K_n, v) \geq \left(\sum_{1 \leq i < j \leq n} v_{ij} \right) /2 + \parallel v \parallel_2 (2\sqrt{3})^{-1}$$

for every $v \in \{0, \pm 1\}^{E_n}$. Note that, if one can prove that this inequality remains valid for *any* $v \in \mathbb{R}^{E_n}$, then Conjecture 31.7.1 would follow.

It may be instructive to evaluate the exact distance to the barycentrum for some concrete classes of facets. For instance, let $D(r, p)$ denote the distance from the barycentrum b to the hyperplane defined by the clique-web inequality:

$$\mathrm{CW}^r_{2p-2r-1}(1, \ldots, 1, -1, \ldots, -1)^T x \leq 0$$

(with p coefficients $+1$ and $p - 2r - 1$ coefficients -1). Then,

$$D(r,p) = \frac{r+1}{2} \sqrt{\frac{p - 2r - 1}{2p - r - 1}}.$$

Hence, for $r = 0$ (hypermetric case), $D(0,p) = \frac{1}{2}\sqrt{\frac{p-1}{2p-1}}$, which is asymptotically $\frac{1}{2\sqrt{2}} (> \frac{1}{2\sqrt{3}})$ when $p \longrightarrow \infty$. In the case $p = 2r + 3$ (the case of the bicycle odd wheel inequality), $D(r, 2r + 3) = \frac{r+1}{\sqrt{6r+10}}$, which tends to $\frac{1}{\sqrt{6}} (> \frac{1}{2\sqrt{3}})$ as $r \longrightarrow \infty$.

One can also check that the distance from b to the hyperplane defined by the (nonpure) clique-web inequality:

$$\mathrm{CW}^r_{p(r+1)-2r-1}(r, \ldots, r, -1, \ldots, -1)^T x \leq 0$$

(with p coefficients r and $pr - 2r - 1$ coefficients -1) is asymptotically $\frac{1}{\sqrt{2}}$ as $r, p \longrightarrow \infty$.

We show in Figure 31.7.3 what is the exact distance to the barycentrum for each of the eleven types of facets of $\mathrm{CUT}^{\square}_7$. These eleven types of facets are listed as in Section 30.6 as F_i for $i = 1, \ldots, 11$. The second row in Figure 31.7.3 gives the exact value for the distance $D(F)$ from the barycentrum b to the hyperplane containing the facet F. The third row gives an approximate value for $D(F) \cdot 2\sqrt{3}$, that is, the ratio $\frac{D(F)}{D(F_1)}$, where F_1 is the triangle facet. Hence, the pentagonal facet is the next closest facet, while the facet F_8 is the farthest one.

F	F_1	F_2	F_3	F_4	F_5	F_6	F_7	F_8	F_9	F_{10}	F_{11}
dist.	$\frac{1}{2\sqrt{3}}$	$\frac{1}{\sqrt{10}}$	$\frac{2}{\sqrt{31}}$	$\frac{1}{2}\sqrt{\frac{3}{7}}$	$\frac{1}{2}\sqrt{\frac{6}{11}}$	$\frac{7}{2\sqrt{69}}$	$\frac{1}{2}$	$\frac{5}{2\sqrt{11}}$	$\frac{9}{2\sqrt{133}}$	$\sqrt{\frac{2}{7}}$	$\frac{5}{2\sqrt{29}}$
ratio	1	1.09	1.24	1.13	1.28	1.46	1.73	2.61	1.35	1.85	1.61

Figure 31.7.3: Distance to the barycentrum of the facets of CUT_7

We conclude with two related results, concerning the width and the diameter of the cut polytope. Given a polytope P, its *width* is defined as

$$\mathrm{width}(P) := \min_{\|c\|_2=1} (\max_{x\in P} c^T x - \min_{x\in P} c^T x).$$

The diameter of P has already been defined in Section 31.1 as $\max_{x,y\in P} \| x-y \|_2$. It is easy to see that it can be alternatively defined as

$$\mathrm{diam}(P) = \max_{\|c\|_2=1} (\max_{x\in P} c^T x - \min_{x\in P} c^T x).$$

In other words, the width and the diameter are, respectively, the smallest and the largest distance between two supporting hyperplanes for P. G. Rote (personal communication) has computed the width of $\mathrm{CUT}_n^\square$.

Proposition 31.7.4. *The width of the cut polytope* $\mathrm{CUT}_n^\square$ *is equal to* 1.

Proof. The proof is based on the following inequality: Let $a_1, \ldots, a_N \in \mathbb{R}$ be N scalars such that $\sum_{i=1}^N a_i = 0$ and $\sum_{i=1}^N a_i^2 = N$. Then,

$$\max_i a_i - \min_i a_i \geq 2.$$

(Indeed, say, $a_1 \leq \ldots \leq a_N$ and set $s := \frac{a_1+a_N}{2}$. If $a_N - a_1 < 2$ then $|a_i - s| < 1$ for all i. Hence, $a_i^2 + s^2 - 2a_i s < 1$ for all i. By summing over i, we obtain that $N + Ns^2 < N$, a contradiction.)
Let $c \in \mathbb{R}^{E_n}$ with $\| c \|_2 = 1$. For $S \subseteq V_n$, set $x_S := e - 2\delta(S)$ (where e denotes the all-ones vector) and set $a_S := c^T x_S$. Then, it is easy to check that $\sum_S a_S = 0$ and $\sum_S (a_S)^2 = N (:= 2^{n-1})$. Applying the above inequality, we obtain that $2 \leq \max_S c^T x_S - \min_S c^T x_S$. This shows that $1 \leq \max_S c^T \delta(S) - \min_S c^T \delta(S)$. Hence, the width of $\mathrm{CUT}_n^\square$ is greater than or equal to 1. The value 1 is attained, for instance, by taking for c a coordinate vector. Hence, $\mathrm{CUT}_n^\square$ has width 1. ∎

Hence, the cut polytope has the same width as the unit hypercube. Poljak and Tuza [1995] have computed the diameter of the cut polytope.

Proposition 31.7.5. *The diameter of* $\mathrm{CUT}_n^\square$ *is equal to* $\sqrt{\lfloor \frac{n}{2} \rfloor \lceil \frac{n}{2} \rceil}$, *that is, to* $\frac{n}{2}$ *if n is even and to* $\frac{\sqrt{n^2-1}}{2}$ *if n is odd.*

Proof. Set $\alpha := \sqrt{\lfloor \frac{n}{2} \rfloor \lceil \frac{n}{2} \rceil}$. Let $c \in \mathbb{R}^{E_n}$ with Euclidean norm 1. Let $\delta(S)$ and $\delta(T)$ be two cut vectors realizing, respectively, the maximum and the minimum of $c^T x$ over $x \in \mathrm{CUT}_n^\square$. Define $c' := c^{\delta(T)}$, i.e., $c'_{ij} := -c_{ij}$ if $|T \cap \{i,j\}| = 1$ and $c'_{ij} = c_{ij}$ otherwise. Then,

$$c^T \delta(S) - c^T \delta(T) = (c')^T \delta(S \Delta T) \leq \sqrt{|\delta(S \Delta T)|} \leq \alpha$$

(the last but one inequality follows from the fact that $\sum_{1 \leq i \leq n} u_i \leq \sqrt{n}$ for any vector $u \in \mathbb{R}^n$ of Euclidean norm 1). On the other hand, the following vector

c realizes equality. Let $S \subseteq V_n$ with $|S| = \lfloor \frac{n}{2} \rfloor$. Set $c_{ij} := \frac{1}{\alpha}$ if $\delta(S)_{ij} = 1$ and $c_{ij} := 0$ if $\delta(S)_{ij} = 0$. Then, $\max c^T x = \alpha$ is attained at $\delta(S)$ and $\min c^T x = 0$ is attained at $\delta(\emptyset)$. ∎

31.8 Simplex Facets

We give here some more information on the simplex faces of $\mathrm{CUT}_n^\square$. We have seen in Section 31.6 that $\mathrm{CUT}_n^\square$ has lots of simplex faces of dimension up to $\lfloor \log_2 n \rfloor$. In fact, $\mathrm{CUT}_n^\square$ has also fairly many simplex facets.

Let us summarize the known classes of simplex facets of $\mathrm{CUT}_n^\square$; for more details, we refer to Deza and Laurent [1993a].

For $n \geq 3$, the hypermetric inequality:

(31.8.1) $$Q_n(n-4, 1, 1, -1, \ldots, -1)^T x \leq 0$$

defines a simplex facet of $\mathrm{CUT}_n^\square$ (Deza and Rosenberg [1984]; recall Corollary 28.2.12).

For $n \geq 6$, the clique-web inequality:

(31.8.2) $$\mathrm{CW}_n^{n-6}(n-4, n-5, n-5, -1, \ldots, -1)^T x \leq 0$$

defines a simplex facet of $\mathrm{CUT}_n^\square$ (Deza and Laurent [1992c]). For $n = 6$, the two inequalities (31.8.1) and (31.8.2) coincide. Actually, for $n \leq 6$, all the simplex facets of $\mathrm{CUT}_n^\square$ arise from (31.8.2) (up to permutation and switching).

For $n = 7$, in addition to the simplex facets that can be derived from (31.8.1) and (31.8.2) by permutation and switching, there are four more groups of simplex facets; namely, the clique-web facets defined by the two inequalities:

$$\mathrm{CW}_7^1(2, 2, 1, 1, -1, -, 1, -1)^T x \leq 0,$$

$$\mathrm{CW}_7^1(1, 1, 1, 1, 1, -1, -1)^T x \leq 0,$$

the facet defined by the parachute inequality $(\mathrm{Par}_7)^T x \leq 0$ (recall (30.4.1)), and the facet defined by Grishukhin's inequality $(\mathrm{Gr}_7)^T x \leq 0$ (recall (30.5.1)).

Hence, among the eleven types of facets of $\mathrm{CUT}_7^\square$, six of them are simplices, namely, the ones numbered 6 to 11 in Section 30.6. Therefore, using the data from Figure 30.6.1, one can count the exact number of simplex facets of $\mathrm{CUT}_7^\square$. Among its 116764 facets, $\mathrm{CUT}_7^\square$ has 113536 simplex facets. Hence, about 97.2% of the total number of facets are simplices ! Deza and Deza [1994a] conjecture that this phenomenon is general, i.e., that the great majority of facets of $\mathrm{CUT}_n^\square$ are simplices. They state the following as an attempt to understand the global shape of the cut polytope:

> "We think that the shape of the cut polytope is essentially given by the nonsimplex facets, in particular, by its triangle facets, and that the huge majority of the facets of $\mathrm{CUT}_n^{\square}$ are simplices which only 'polish' it."

Interestingly, each of the simplex facets described above has the following property (31.8.3) (see Deza and Laurent [1993a] for a proof). Let F denote such a simplex facet and let $\delta(S_k)$ $(1 \leq k \leq \binom{n}{2})$ denote its roots. Let $d \in F$ with decomposition $d = \sum_{1 \leq k \leq \binom{n}{2}} \lambda_k \delta(S_k)$ where $\lambda_k \geq 0$ for all k. Then,

$$\text{(31.8.3)} \qquad \begin{array}{c} d \text{ belongs to the cut lattice } \mathcal{L}_n \\ \text{(i.e., if } d \in \mathbb{Z}^{E_n} \text{ and satisfies the parity condition (24.1.1)),} \\ \Downarrow \\ \text{all } \lambda_k^{\prime} s \text{ are integers.} \end{array}$$

In other words, in the terminology of Part IV, the parity condition suffices for ensuring hypercube embeddability for the class of distances $d \in F$.

Bibliography

[1994] W.P. Adams and P.M. Dearing. On the equivalence between roof duality and Lagrangian duality for unconstrained 01-quadratic programming problems. *Discrete Applied Mathematics*, 48:1–20, 1994. [430]

[1985] F. Afrati, C.H. Papadimitriou, and G. Papageorgiou. The complexity of cubical graphs. *Information and Control*, 66:53–60, 1985. [294]

[1989] F. Afrati, C.H. Papadimitriou, and G. Papageorgiou. The complexity of cubical graphs - Corrigendum. *Information and Computation*, 82:350–353, 1989. [294]

[1977] R. Alexander. The width and diameter of a simplex. *Geometriae Dedicata*, 6:87–94, 1977. [82]

[1978] R. Alexander. Planes for which the lines are the shortest paths between points. *Illinois Journal of Mathematics*, 22:177–190, 1978. [111]

[1988] R. Alexander. Zonoid theory and Hilbert's fourth problem. *Geometriae Dedicata*, 28:199–211, 1988. [111]

[1990] N. Alon. The CW-inequalities for vectors in ℓ_1. *European Journal of Combinatorics*, 11:1–5, 1990. [472]

[1994] B. Alspach, L. Goddyn, and C.-Q. Zhang. Graphs with the circuit cover property. *Transactions of the American Mathematical Society*, 344:131–154, 1994. [444]

[1988] I. Althöfer. On optimal realizations of finite metric spaces by graphs. *Discrete and Computational Geometry*, 3:103–122, 1988. [310]

[1982] R.V. Ambartzumian. *Combinatorial Integral Geometry - with Applications to Mathematical Stereology.* John Wiley & Sons, Chichester, 1982. [111,162]

[1987] M. Aschbacher, P. Baldi, E.B. Baum, and R.M. Wilson. Embeddings of ultrametric spaces in finite dimensional structures. *SIAM Journal on Algebraic and Discrete Methods*, 8:564–577, 1987. [311]

[1977] P. Assouad. Un espace hypermétrique non plongeable dans un espace L_1. *Comptes Rendus de l'Académie des Sciences de Paris*, 285(A):361–363, 1977. [84]

[1979] P. Assouad. Produit tensoriel, distances extrémales et réalisation de covariances. *Comptes Rendus de l'Académie des Sciences de Paris*, 288(A):649–652 and 675–677, 1979. [54,58,107,108,114]

[1980a] P. Assouad. Caractérisations de sous-espaces normés de L_1 de dimension finie. Number 19 in *Séminaire d'Analyse Fonctionnelle.* Ecole Polytechnique Palaiseau, 1979-1980. [110]

[1980b] P. Assouad. Plongements isométriques dans L^1: aspect analytique. Number 14 in *Séminaire d'Initiation à l'Analyse* (G. Choquet, H. Fakhoury, J. Saint-Raymond). Université Paris VI, 1979-1980. [39,54,58,108,114,119]

[1982] P. Assouad. Sous espaces de L^1 et inégalités hypermétriques. *Comptes Rendus de l'Académie des Sciences de Paris*, 294(A):439–442, 1982. [193,198,381]

[1984] P. Assouad. Sur les inégalités valides dans L^1. *European Journal of Combinatorics*, 5:99–112, 1984. [72,110,111,172,193,477,497]

[1980] P. Assouad and C. Delorme. Graphes plongeables dans L^1. *Comptes Rendus de l'Académie des Sciences de Paris*, 291(A):369–372, 1980. [253,255,262,370]

[1982] P. Assouad and C. Delorme. Distances sur les graphes et plongements dans L^1, I et II. Technical Report, Université d'Orsay, 1982. [255,257,261,262,370,477,497]

[1980] P. Assouad and M Deza. Espaces métriques plongeables dans un hypercube: aspects combinatoires. In M. Deza and I.G. Rosenberg, editors, *Combinatorics 79 - Part I*, volume 8 of *Annals of Discrete Mathematics*, pages 197–210. North-Holland, Amsterdam, 1980. [356]

[1982] P. Assouad and M. Deza. Metric subspaces of L^1. Number 82-03 in *Publications Mathématiques d'Orsay*. Université de Paris-Sud, Orsay, 1982. [39]

[1997] Y. Aumann and Y. Rabani. An $O(\log k)$ approximate min-cut max-flow theorem and approximation algorithm. *SIAM Journal on Computing*, 1997 (to appear). [134]

[1991] F. Aurenhammer and J. Hagauer. Recognizing binary Hamming graphs in $O(n^2 \log n)$ time. In R.H. Möhring, editor, *Graph-Theoretic Concepts in Computer Science*, volume 484 of *Lecture Notes in Computer Science*, pages 90–98. Springer-Verlag, Berlin, 1991. [286]

[1990] F. Aurenhammer, J. Hagauer, and W. Imrich. Factoring cartesian-product graphs at logarithmic cost per edge. In R. Kannan and W.R. Pulleyblank, editors, *Integer Programming and Combinatorial Optimization*, pages 29–44. University of Waterloo Press, 1990. (Updated version in: *Computational Complexity*, 2:331–349, 1992.) [306]

[1977] D. Avis. *Some Polyhedral Cones Related to Metric Spaces*. PhD thesis, Stanford University, 1977. [37,39,422,477]

[1980a] D. Avis. On the extreme rays of the metric cone. *Canadian Journal of Mathematics*, 32:126–144, 1980. [422]

[1980b] D. Avis. Extremal metrics induced by graphs. In M. Deza and I.G. Rosenberg, editors, *Combinatorics 79 - Part I*, volume 8 of *Annals of Discrete Mathematics*, pages 217–220. North-Holland, Amsterdam, 1980. [422]

[1981] D. Avis. Hypermetric spaces and the Hamming cone. *Canadian Journal of Mathematics*, 33:795–802, 1981. [84,286]

[1989] D. Avis (and Mutt). All facets of the six point Hamming cone. *European Journal of Combinatorics*, 10:309–312, 1989. [229,235,503]

[1990] D. Avis. On the complexity of isometric embedding in the hypercube. In T. Asano, T. Ibaraki, H. Imai, and T. Nishizeki, editors, *Algorithms*, volume 450 of *Lecture Notes in Computer Science*, pages 348–357. Springer-Verlag, Berlin, 1990. [362,369]

[1991] D. Avis and M. Deza. The cut cone, L^1 embeddability, complexity and multicommodity flows. *Networks*, 21:595–617, 1991. [49,422,451]

[1993] D. Avis and V.P. Grishukhin. A bound on the k-gonality of facets of the hypermetric cone and related complexity problems. *Computational Geometry: Theory and Applications*, 2:241–254, 1993. [204,205,226,453]

[1994] D. Avis and H. Maehara. Metric extensions and the L_1 hierarchy. *Discrete Mathematics*, 131:17–28, 1994. [214,289]

[1995] M. Bakonyi and C.R. Johnson. The Euclidian distance matrix completion problem. *SIAM Journal on Matrix Analysis and Applications*, 16:646–654, 1995. [529,530]

[1993] E. Balas, S. Ceria, and G. Cornuejols. A lift-and-project cutting plane algorithm for mixed 0-1 programs. *Mathematical Programming*, 58:295–324, 1993. [424]

[1987] K. Ball. Inequalities and sphere-packings in ℓ_p. *Israel Journal of Mathematics*, 58:243–256, 1987. [34,159]

[1990] K. Ball. Isometric embedding in ℓ_p-spaces. *European Journal of Combinatorics*, 11:305–311, 1990. [72,157]

[1990] H-J. Bandelt. Recognition of tree metrics. *SIAM Journal on Discrete Mathematics*, 3:1–6, 1990. [311]

[1996a] H-J. Bandelt and V.D. Chepoi. Embedding metric spaces in the rectilinear plane: a six-point criterion. *Discrete and Computational Geometry*, 15:107–117, 1996. [141,148,152]

[1996b] H-J. Bandelt and V.D. Chepoi. Embedding into the rectilinear grid. Preprint, 1996. [155,156]

[1996c] H-J. Bandelt and V.D. Chepoi. Weakly median graphs: decomposition, ℓ_1-embedding and algebraic characterization. Preprint, 1996. [330]

[1997] H-J. Bandelt, V.D. Chepoi, and M. Laurent. Embedding into rectilinear spaces. *Discrete and Computational Geometry*, 1997 (to appear). [141,144]

[1992] H-J. Bandelt and A.W.M. Dress. A canonical decomposition theory for metrics on a finite set. *Advances in Mathematics*, 92:47–105, 1992. [28,56,92,145,146,147]

[1991] H-J. Bandelt and H.M. Mulder. Cartesian factorization of interval-regular graphs having no long isometric odd cycles. In Y. Alavi et al., editor, *Graph Theory, Combinatorics and Applications*, volume 1, pages 55–75. John Wiley & Sons, New York, 1991. [309]

[1981] F. Barahona. *Balancing signed graphs of fixed genus in polynomial time.* Technical Report, Departamento de Matematicas, Universidad de Chile, 1981. [48]

[1982] F. Barahona. On the computational complexity of Ising spin glass models. *Journal of Physics A, Mathematical and General*, 15:3241–3253, 1982. [48,52]

[1983] F. Barahona. The max-cut problem on graphs not contractible to K_5. *Operations Research Letters*, 2:107–111, 1983. [48,434]

[1993] F. Barahona. On cuts and matchings in planar graphs. *Mathematical Programming*, 60:53–68, 1993. [431]

[1994] F. Barahona. Ground-state magnetization of Ising spin glasses. *Physical Review B*, 49(18):12864–12867, 1994. [52]

[1986] F. Barahona and M. Grötschel. On the cycle polytope of a binary matroid. *Journal of Combinatorial Theory B*, 40:40–62, 1986. [405,439,440]

[1988] F. Barahona, M. Grötschel, M. Jünger, and G. Reinelt. An application of combinatorial optimization to statistical physics and circuit layout design. *Operations Research*, 36:493–513, 1988. [38,51,52]

[1985] F. Barahona, M. Grötschel, and A.R. Mahjoub. Facets of the bipartite subgraph polytope. *Mathematics of Operations Research*, 10:340–358, 1985. [469,483]

[1989] F. Barahona, M. Jünger, and G. Reinelt. Experiments in quadratic 0-1 programming. *Mathematical Programming*, 44:127–137, 1989. [52,56]

[1986] F. Barahona and A.R. Mahjoub. On the cut polytope. *Mathematical Programming*, 36:157–173, 1986. [405,431,433,434,453,469,483,539]

[1982] F. Barahona, R. Maynard, R. Rammal, and J.-P. Uhry. Morphology of ground states of a two dimensional frustration model. *Journal of Physics A*, 15:L673–L699, 1982. [52]

[1971] E.P. Baranovskii. Simplexes of L-subdivisions of Euclidean spaces. *Mathematical Notes*, 10:827–834, 1971. (Translated from: *Matematicheskie Zametki* 10:659–671, 1971.) [69,229,235,503]

[1973] E.P. Baranovskii. Theorem on L-partitions of point lattices. *Mathematical Notes*, 13:364–370, 1973. (Translated from: *Matematicheskie Zametki* 13:605–616, 1973.) [69,235]

[1992] E.P. Baranovskii. Partitioning of Euclidean space into L-polytopes of some perfect lattices. In S.P. Novikov, S.S. Ryshkov, N.P. Dolbilin, and M.A. Shtan'ko, editors, *Discrete Geometry and Topology*, volume 196(4) of *Proceedings of the Steklov Institute of Mathematics*, pages 29–51. American Mathematical Society, Providence, Rhode Island, 1992. [181]

[1995] E.P. Baranovskii. In the Second International Conference 'Algebraic, Probabilistic, Geometrical, Combinatorial, and Functional Methods in the Theory of Numbers', September 1995, Voronezh, Russia. [229]

[1996] W.W. Barrett, C.R. Johnson, and R. Loewy. The real positive definite completion problem: cycle completability. *Memoirs of the American Mathematical Society*, Number 584, 69 pages, 1996. [521,522]

[1993] W. Barrett, C.R. Johnson, and P. Tarazaga. The real positive definite completion problem for a simple cycle. *Linear Algebra and its Applications*, 192:3–31, 1993. [514]

[1995] A.I. Barvinok. Problems of distance geometry and convex properties of quadratic maps. *Discrete and Computational Geometry*, 13:189–202, 1995. [528]

[1970] L.W. Beineke. Characterization of derived graphs. *Journal of Combinatorial Theory*, 9:129–135, 1970. [254]

[1976] C. Berg, J.P.R. Christensen, and P. Ressel. Positive definite functions on abelian semigroups. *Mathematische Annalen*, 223:253–272, 1976. [107]

[1987] M. Berger. *Geometry II*. Springer-Verlag, Berlin, 1987. [514]

[1980] I. Bieche, R. Maynard, R. Rammal, and J.-P. Uhry. On the ground states of of the frustration model of a spin glass by a matching method of graph theory. *Journal of Physics A, Mathematical and General*, 13:2553–2571, 1980. [52]

[1981] N.L. Biggs and T. Ito. Covering graphs and symmetric designs. In P.J. Cameron, J.W.P. Hirschfeld, and D.R. Hughes, editors, *Finite Geometries and Designs*, volume 49 of *London Mathematical Society Lecture Note Series*, pages 40–51. Cambridge University Press, Cambridge, 1981. [344]

[1967] G. Birkhoff. *Lattice Theory*. American Mathematical Society, Providence, Rhode Island, 1967. [105,106]

[1973] I.F. Blake and J.H. Gilchrist. Addresses for graphs. *IEEE Transactions on Information Theory*, IT-19:683–688, 1973. [100,278,288,314]

[1953] L.M. Blumenthal. *Theory and Applications of Distance Geometry*. Oxford University Press, Oxford, 1953. (Second edition: Chelsea, New York, 1970.) [25,29,74,76,78,80,86,87,88,113,120,514]

[1994] H.L. Bodlaender and K. Jansen. On the complexity of the maximum cut problem. In P. Enjalbert, E.W.Mayr, and K.W. Wagner, editors, *Proceedings of the 11th Annual Symposium on Theoretical Aspects of Computer Science*, volume 775 of *Lecture Notes in Computer Science*, pages 769–780. Springer-Verlag, 1994. [48]

[1994] N. Boissin. *Optimisation des Fonctions Quadratiques en Variables Bivalentes.* PhD thesis, Centre National d'Etudes des Télécommunications, Issy Les Moulineaux, France, 1994. [54,416,489,500]

[1969] E.D. Bolker. A class of convex bodies. *Transactions of the American Mathematical Society*, 145:323–345, 1969. [110]

[1976] J.A. Bondy and U.S.R. Murty. *Graph Theory with Applications.* American Elsevier, New York, and Macmillan, London, 1976. [11]

[1936] C.E. Bonferroni. Il calcolo delle assicurazioni su grouppi di teste. *Studi in Onore del Professor S.O. Carboni (Roma)*, 1936. [65]

[1854] G. Boole. *An Investigation of the Laws of Thought on Which Are Founded the Mathematical Theories of Logic and Probabilities.* Dover Publications, New York, original edition 1854. [61]

[1990] E. Boros, Y. Crama, and P.L. Hammer. Upper bounds for quadratic 0-1 optimization problems. *Operations Research Letters*, 9:73–79, 1990. [430]

[1992] E. Boros, Y. Crama, and P.L. Hammer. Chvátal cuts and odd cycle inequalities in quadratic 0-1 optimization. *SIAM Journal on Discrete Mathematics*, 5:163–177, 1992. [428,430]

[1991] E. Boros and P.L. Hammer. The max-cut problem and quadratic 0-1 optimization: polyhedral aspects, relaxations and bounds. *Annals of Operations Research*, 33:151–180, 1991. [54,430]

[1993] E. Boros and P.L. Hammer. Cut polytopes, boolean quadratic polytopes and nonnegative quadratic pseudo-boolean functions. *Mathematics of Operations Research*, 18:245–253, 1993. [54,487,489]

[1989] E. Boros and A. Prékopa. Closed form two-sided bounds for probabilities that at least r and exactly r out of n events occur. *Mathematics of Operations Research*, 14:317–342, 1989. [65]

[1933] K. Borsuk. Drei Sätze über die n-dimensionale Euklidische Sphäre. *Fundamenta Mathematicae*, 20:177–190, 1933. [10,512]

[1985] J. Bourgain. On Lipschitz embedding of finite metric spaces in Hilbert space. *Israel Journal of Mathematics*, 52:46–52, 1985. [125]

[1991] S. Boyd and W.H. Cunningham. Small travelling salesman polytopes. *Mathematics of Operations Research*, 16:259–271, 1991. [506]

[1966] J. Bretagnolle, D. Dacunha Castelle, and J.-L. Krivine. Lois stables et espaces L^p. *Annales de l'Institut Henri Poincaré*, 2:231–259, 1966. [33,34,110]

[1989] A.E. Brouwer, A.M. Cohen, and A. Neumaier. *Distance-Regular Graphs.* Springer-Verlag, Berlin, 1989. [96,206,239,257,268,308,410]

[1971] P. Buneman. The recovery of trees from measures of dissimilarity. In F.R. Hodson et al., editor, *Mathematics in the Archeological and Historical Sciences*, pages 387–395, Edinburgh University Press, Edinburgh, 1971. [147]

[1974] P. Buneman. A note on metric properties of trees. *Journal of Combinatorial Theory B*, 17:48–50, 1974. [310]

[1995] G. Burosch and P.V. Ceccherini. Isometric embeddings into cube-hypergraphs. *Discrete Mathematics*, 137:77–85, 1995. [294]

[1976] F.C. Bussemaker, D.M. Cvetković, and J.J. Seidel. Graphs related to exceptional root systems. TH-report 76-WSK-05, Technical University Eindhoven, The Netherlands, 1976. [267]

[1976] P.J. Cameron, J.M. Goethals, J.J. Seidel, and E.E. Shult. Line graphs, root systems, and elliptic geometry. *Journal of Algebra*, 43:305–327, 1976. [93,267]

[1959] J.W.S. Cassels. *An Introduction to the Geometry of Numbers*, volume 99 of *Die Grundlehren der mathematischen Wissenschaften in Einzeldarstellungen*. Springer-Verlag, Berlin, 1959. [190]

[1841] A. Cayley. On a theorem in the geometry of position. *Cambridge Mathematical Journal*, vol. II:267–271, 1841. (Also in *The Collected Mathematical Papers of Arthur Cayley*, Cambridge University Press, pages 1–4, 1889.) [25,73]

[1986] P.V. Ceccherini and A. Sappa. A new characterization of hypercubes. In A. Barlotti et al., editor, *Combinatorics 84*, Annals of Discrete Mathematics 30, pages 137–142. North-Holland, Amsterdam, 1986. [309]

[1947] P.L. Chebyshev. *Complete Collected Works - Vol.2*. Moscow-Leningrad (in Russian), 1947. [213]

[1996] V. Chepoi, M. Deza, and V.P. Grishukhin. Clin d'oeil on L_1-embeddable planar graphs. Rapport LIENS-96-11, Ecole Normale Supérieure, Paris, 1996. [325,326,330]

[1996] V. Chepoi and B. Fichet. A note on circular decomposable metrics. Preprint, 1996. [91]

[1995] S. Chopra and M.R. Rao. Facets of the k-partition polytope. *Discrete Applied Mathematics*, 61:27–48, 1995. [400,465]

[1979] J.P.R. Christensen and J. Vesterstrøm. A note on extreme positive definite matrices. *Mathematische Annalen*, 244:65–68, 1979. [537]

[1991] T. Christof, M. Jünger, and G. Reinelt. A complete description of the traveling salesman polytope on 8 nodes. *Operations Research Letters*, 10:497–500, 1991. [506]

[1996] T. Christof and G. Reinelt. Personal communication, 1996. [505,543]

[1996] G. Christopher, M. Farach, and M.A. Trick. The structure of circular decomposable metrics. In J. Diaz and M. Serna, editors, *Algorithms - ESA '96*, volume 1136 of *Lecture Notes in Computer Science*, pages 486–500. Springer-Verlag, Berlin, 1996. [91]

[1941] K.L. Chung. On the probability of the occurrence of at least m events among n arbitrary events. *Annals of Mathematical Statistics*, 12:328–338, 1941. [63]

[1973] V. Chvátal. Edmonds polytopes and a hierarchy of combinatorial problems. *Discrete Mathematics*, 4:305–337, 1973. [428]

[1980] V. Chvátal. Recognizing intersection patterns. In M. Deza and I.G. Rosenberg, editors, *Combinatorics 79 - Part I*, volume 8 of *Annals of Discrete Mathematics*, pages 249–251. North-Holland, Amsterdam, 1980. [356]

[1989] V. Chvátal, W. Cook, and M. Hartmann. On cutting-plane proofs in combinatorial optimization. *Linear Algebra and its Applications*, 114/115:455–499, 1989. [429]

[1990a] M. Conforti, M.R. Rao, and A. Sassano. The equipartition polytope: part I. *Mathematical Programming*, 49:49–70, 1990. [399]

[1990b] M. Conforti, M.R. Rao, and A. Sassano. The equipartition polytope: part II. *Mathematical Programming*, 49:71–91, 1990. [399]

[1988] J.H. Conway and N.J.A. Sloane. *Sphere Packings, Lattices and Groups*, volume 290 of *Grundlehren der mathematischen Wissenschaften*. Springer-Verlag, New York, 1988. [172,206,239]

[1991] J.H. Conway and N.J.A. Sloane. The cell structures of certain lattices. In P. Hilton, F. Hirzebruch and R. Remmert, editor, *Miscellanea Mathematica*, pages 71–108. Springer-Verlag, Berlin, 1991. [206,239]

[1983] R.J. Cook and D.G. Pryce. Uniformly geodetic graphs. *Ars Combinatoria*, 16-A:55–59, 1983. [309]

[1993] M. Coornaert and A. Papadopoulos. *Symbolic Dynamics and Hyperbolic Groups*, volume 1539 of *Lecture Notes in Mathematics*. Springer-Verlag, Berlin, 1993. [56]

[1973] H.S.M. Coxeter. *Regular Polytopes*. Dover Publications, New York, 1973. [186]

[1988] F. Critchley. On certain linear mappings between inner-product and squared-distance matrices. *Linear Algebra and its Applications*, 105:91–107, 1988. [56]

[1994] F. Crichtley and B. Fichet. The partial order by inclusion of the principal classes of dissimilarity on a finite set, and some of their basic properties. In B. van Cutsem, editor, volume 93 of *Classification and Dissimilarity Analysis, Lecture Notes in Statistics*, pages 5–65. Springer-Verlag, New York, 1994. [163]

[1988] G.M. Crippen and T.F. Havel. *Distance Geometry and Molecular Conformation*. Research Studies Press, Taunton, Somerset, England, 1988. [74]

[1978] J.P. Cunningham. Free trees and bidirectional trees as representations of psychological distance. *Journal of Mathematical Psychology*, 17:165–188, 1978. [310]

[1977] W.H. Cunningham. Chords and disjoint paths in matroids. *Discrete Mathematics*, 19:7–15, 1977. [442]

[1994] B. van Cutsem, editor. *Classification and Dissimilarity Analysis*, volume 93 of *Lecture Notes in Statistics*, pages 5–65. Springer-Verlag, New York, 1994. [163]

[1962] L. Danzer and B. Grünbaum. Über zwei Probleme bezüglich konvexer Körper von P. Erdös and von V.L. Klee. *Mathematische Zeitschrift*, 79:95–99, 1962. [183]

[1969] E. Davidson. Linear inequalities for density matrices. *Journal of Mathematical Physics*, 10(4):725–734, 1969. [64,69,477,503]

[1967] D.A. Dawson and D. Sankoff. An inequality for probabilities. *Proceedings of the American Mathematical Society*, 18:504–507, 1967. [63]

[1991] M. De Berg. On rectilinear link distances. *Computational Geometry: Theory and Applications*, 1:13–34, 1991. [165]

[1995] V. Deineko, R. Rudolf, and G. Woeginger. Sometimes travelling is easy: the master tour problem. In P. Spirakis, editor, *Algorithms - ESA'95*, volume 979 of *Lecture Notes in Computer Science*, pages 128–141. Springer-Verlag, Berlin, 1995. [92]

[1982] J. de Leeuw and W. Heiser. Theory of multidimensional scaling. In P.R. Krishnaiah and L.N. Kanal, eds, *Handbook of Statistics*, Vol. 2, pages 285–316. North-Holland, Amsterdam, 1982. [74]

[1993a] C. Delorme and S. Poljak. Laplacian eigenvalues and the maximum cut problem. *Mathematical Programming*, 62:557–574, 1993. [460]

[1993b] C. Delorme and S. Poljak. Combinatorial properties and the complexity of a max-cut approximation. *European Journal of Combinatorics*, 14:313–333, 1993. [461,536]

[1989] C. De Simone. The cut polytope and the boolean quadric polytope. *Discrete Mathematics*, 79:71–75, 1989/90. [54,56]

[1992] C. De Simone. *The Max Cut Problem*. PhD thesis, Rutgers University, New Brunswick, USA, 1992. [454,508]

[1994] C. De Simone, M. Deza, and M. Laurent. Collapsing and lifting for the cut cone. *Discrete Mathematics*, 127:105–130, 1994. [410,411,463,480,502,505,508]

[1995] C. De Simone, M. Diehl, M. Jünger, P. Mutzel, G. Reinelt, and G. Rinaldi. Exact ground states of Ising spin glasses: new experimental results with a branch-and-cut algorithm. *Journal of Statistical Physics*, 80(1-2):487–496, 1995. [52]

[1996] C. De Simone, M. Diehl, M. Jünger, P. Mutzel, G. Reinelt, and G. Rinaldi. Exact ground states of two-dimensional $\pm J$ Ising spin glasses. *Journal of Statistical Physics*, 84(5-6):1363–1371, 1996. [52]

[1994] C. De Simone and G. Rinaldi. A cutting plane algorithm for the max-cut problem. *Optimization Methods and Software*, 3:195–214, 1994. [454]

[1993] C.C. De Souza. *The Graph Equipartition Problem: Optimal Solutions, Extensions and Applications.* PhD thesis, Université Catholique de Louvain, Belgique, 1993. [399,492,493,495,496]

[1995] C.C. De Souza and M. Laurent. Some new classes of facets for the equicut polytope. *Discrete Applied Mathematics*, 62:167–191, 1995. [400,491,492,494,495]

[1980] A.K. Dewdney. The embedding dimension of a graph. *Ars Combinatoria*, 9:77–90, 1980. [159]

[1994] A. Deza. *Graphes et Faces de Polyèdres Combinatoires.* PhD thesis, Université de Paris-Sud, Orsay, France, 1994. [542]

[1996] A. Deza. *Metric Polyhedra: Combinatorial Structure and Optimization.* PhD thesis, Tokyo Institute of Technology, Tokyo, Japan, 1996. [542,543]

[1994a] A. Deza and M. Deza. On the skeleton of the dual cut polytope. In H. Barcelo and G. Kalai, editors, *Jerusalem Combinatorics '93*, volume 178 of *Contemporary Mathematics*, pages 101–111, 1994. [407,545,549]

[1994b] A. Deza and M. Deza. The ridge graph of the metric polytope and some relatives. In *Polytopes: Abstract, Convex and Computational*, volume 440 of *NATO ASI Series C: Mathematical and Physical Sciences*, pages 359–372. Kluwer Academic Publishers, Dordrecht, 1994. [544,545]

[1995] A. Deza and M. Deza. The combinatorial structure of small cut and metric polytopes. In T.H. Ku, editor, *Combinatorics and Graph Theory 95*, World Scientific Publisher, Singapour, pages 70–88, 1995. [541,545]

[1996] A. Deza, M. Deza and K. Fukuda. On skeletons, diameters and volumes of metric polyhedra. In M. Deza, R. Euler, and Y. Manoussakis, editors, *Combinatorics and Computer Science*, volume 1120 of *Lecture Notes in Computer Science*, pages 112–128. Springer-Verlag, Berlin, 1996. [542,543,544,546]

[1996] A. Deza, M. Deza and V.P. Grishukhin. Embeddings of fullerenes, virus capsids, geodesic domes and coordinations polyhedra into half-cubes. Technical Report DMA RO 960701, Ecole Polytechnique Fédérale de Lausanne, Département de Mathématiques, 1996. [325,329]

[1960] M.E. Tylkin (=M. Deza). On Hamming geometry of unitary cubes (in Russian). *Doklady Akademii Nauk SSSR* 134 (1960) 1037-1040. (English translation in: *Cybernetics and Control Theory*, 134(5): 940–943, 1961.) [69,235,355,503]

[1962] M.E. Tylkin (=M. Deza). Realizability of distance matrices in unit cubes (in Russian). *Problemy Kybernetiki*, 7:31–42, 1962. [69,69]

[1973a] M. Deza. Matrices des formes quadratiques non négatives pour des arguments binaires. *Comptes rendus de l'Académie des Sciences de Paris*, 277(A):873–875, 1973. [54,56,405,414,450,477,503]

[1973b] M. Deza. Une propriété extrémale des plans projectifs finis dans une classe de codes equidistants. *Discrete Mathematics*, 6:343–352, 1973. [337,345]

[1974] M. Deza. Solution d'un problème de Erdös-Lovász. *Journal of Combinatorial Theory B*, 16:166–167, 1974. [345]

[1982] M. Deza. Small pentagonal spaces. *Rendiconti del Seminario Matematico di Brescia*, 7:269–282, 1982. [355]

[1978] M. Deza, P. Erdös, and P. Frankl. Intersection properties of systems of finite sets. *Proceedings of the London Mathematical Society*, 3:369–384, 1978. [340]

[1993] M. Deza, K. Fukuda, and M. Laurent. The inequicut cone. *Discrete Mathematics*, 119:21–48, 1993. [399,506,507]

[1993] M. Deza and V.P. Grishukhin. Hypermetric graphs. *The Quarterly Journal of Mathematics Oxford*, 2:399–433, 1993. [172,198,210,211,212, 262,264,269]

[1994] M. Deza and V.P. Grishukhin. Lattice points of cut cones. *Combinatorics, Probability and Computing*, 3:191–214, 1994. [97,100,388]

[1995a] M. Deza and V.P. Grishukhin. L-polytopes and equiangular lines. *Discrete Applied Mathematics*, 56:181–214, 1995. [238]

[1995b] M. Deza and V.P. Grishukhin. Delaunay polytopes of cut lattices. *Linear Algebra and its Applications*, 226:667–686, 1995. [181]

[1996a] M. Deza and V.P. Grishukhin. Cut lattices and equiangular lines. *European Journal of Combinatorics*, 17:143–156, 1996. [237,385]

[1996b] M. Deza and V.P. Grishukhin. Bounds on the covering radius of a lattice. *Mathematika*, 43:159–164, 1996. [214]

[1996c] M. Deza and V.P. Grishukhin. A zoo of ℓ_1-embeddable polytopal graphs. Preprint 96-006, Universität Bielefeld, 1996. [325,329]

[1991] M. Deza, V.P. Grishukhin, and M. Laurent. The symmetries of the cut polytope and of some relatives. In P. Gritzmann and B. Sturmfels, editors, *Applied Geometry and Discrete Mathematics - The Victor Klee Festschrift*, volume 4 of *DIMACS Series in Discrete Mathematics and Theoretical Computer Science*, pages 205–220, 1991. [408,409,410]

[1992] M. Deza, V.P. Grishukhin, and M. Laurent. Extreme hypermetrics and L-polytopes. In G. Halász et al., editor, *Sets, Graphs and Numbers, Budapest (Hungary), 1991*, volume 60 of *Colloquia Mathematica Societatis János Bolyai*, pages 157–209, 1992. [172,184,217,222,230,236,240,243,244, 245]

[1993] M. Deza, V.P. Grishukhin, and M. Laurent. The hypermetric cone is polyhedral. *Combinatorica*, 13:397–411, 1993. [172,190,199]

[1995] M. Deza, V.P. Grishukhin, and M. Laurent. Hypermetrics in geometry of numbers. In W. Cook, L. Lovász, and P.D. Seymour, editors, *Combinatorial Optimization*, volume 20 of *DIMACS Series in Discrete Mathematics and Theoretical Computer Science*, pages 1–109. American Mathematical Society, 1995. [vi,172]

[1991] M. Deza, M. Grötschel, and M. Laurent. Complete descriptions of small multicut polytopes. In P. Gritzmann and B. Sturmfels, editors, *Applied Geometry and Discrete Mathematics - The Victor Klee Festschrift*, volume 4 of *DIMACS Series in Discrete Mathematics and Theoretical Computer Science*, pages 221–252, 1991. [400]

[1992] M. Deza, M. Grötschel, and M. Laurent. Clique-web facets for multicut polytopes. *Mathematics of Operations Research*, 17:981–1000, 1992. [400,412,465,476,481]

[1996a] M. Deza and T. Huang. Complementary ℓ_1-graphs and related combinatorial structures. In M. Deza, R. Euler, and Y. Manoussakis, editors, *Combinatorics and Computer Science*, volume 1120 of *Lecture Notes in Computer Science*, pages 74–90. Springer-Verlag, Berlin, 1996. [328]

[1996b] M. Deza and T. Huang. ℓ_1-embeddability of some block graphs and cycloids. *Bulletin of the Institute of Mathematics, Academia Sinica*, 24:87–102, 1996. [328]

[1992a] M. Deza and M. Laurent. Facets for the cut cone I. *Mathematical Programming*, 56:121–160, 1992. [255,414,450,453,469,497,503]

[1992b] M. Deza and M. Laurent. Facets for the cut cone II: clique-web facets. *Mathematical Programming*, 56:161–188, 1992. [410,452,472,478]

[1992c] M. Deza and M. Laurent. New results on facets of the cut cone. *Journal of Combinatorics, Information and System Sciences*, 17:19–38, 1992. [408,409,478,479, 497,504,549]

[1992d] M. Deza and M. Laurent. The Fibonacci and parachute inequalities for ℓ_1-metrics. *The Fibonacci Quarterly*, 30:54–61, 1992. [497]

[1992e] M. Deza and M. Laurent. Extension operations for cuts. *Discrete Mathematics*, 106/107:163–179, 1992. [94,382,400,507]

[1993a] M. Deza and M. Laurent. The cut cone: simplicial faces and linear dependencies. *Bulletin of the Institute of Mathematics, Academia Sinica*, 21:143–182, 1993. [385,549,550]

[1993b] M. Deza and M. Laurent. The even and odd cut polytopes. *Discrete Mathematics*, 119:49–66, 1993. [400,464,465,506]

[1993c] M. Deza and M. Laurent. Variety of hypercube embeddings of the equidistant metric and designs. *Journal of Combinatorics, Information and System Sciences*, 18:293–320, 1993. [341,349,350]

[1994a] M. Deza and M. Laurent. ℓ_1-rigid graphs. *Journal of Algebraic Combinatorics*, 3:153–175, 1994. [96,284,325]

[1994b] M. Deza and M. Laurent. Applications of cut polyhedra - I. *Journal of Computational and Applied Mathematics*, 55:191-216, 1994. [55]

[1994c] M. Deza and M. Laurent. Applications of cut polyhedra - II. *Journal of Computational and Applied Mathematics*, 55:217–247, 1994. [55]

[1995a] M. Deza and M. Laurent. Hypercube embedding of generalized bipartite metrics. *Discrete Applied Mathematics*, 56:215–230, 1995. [356,358,362]

[1992] M. Deza, M. Laurent, and S. Poljak. The cut cone III: on the role of triangle facets. *Graphs and Combinatorics*, 8:125–142, 1992. (Updated version in: *Graphs and Combinatorics*, 9:135–152, 1993.) [383,539,542,546]

[1990] M. Deza and H. Maehara. Metric transforms and Euclidean embeddings. *Transactions of the American Mathematical Society*, 317:661–671, 1990. [120,124]

[1994] M. Deza and H. Maehara. A few applications of negative-type inequalities. *Graphs and Combinatorics*, 10:255–262, 1994. [85,86]

[1990] M. Deza, D.K. Ray-Chaudhuri, and N.M. Singhi. Positive independence and enumeration of codes with a given distance pattern. In A. Friedman and W. Miller, series editors, *Coding Theory and Design Theory, part I, Coding Theory*, volume 20 of *The IMA volumes in Mathematics and its Applications*, pages 93–101. Springer-Verlag, New York, 1990. [357]

[1984] M. Deza and I.G. Rosenberg. Intersection and distance patterns. *Utilitas Mathematica*, 25:191–214, 1984. [549]

[1996] M. Deza and S. Shpectorov. Recognition of the ℓ_1-graphs with complexity $O(nm)$, or football in a hypercube. *European Journal of Combinatorics*, 17:279–289, 1996. [315,323]

[1996] M. Deza and M. Stogrin. Scale-isometric embeddings of Archimedean tilings, polytopes and their duals into cubic lattices and hypercubes. Preprint, 1996. [329]

[1996] M. Deza and J. Tuma. A note on ℓ_1-rigid planar graphs. *European Journal of Combinatorics*, 17:157–160, 1996. [329]

[Di61] G.A. Dirac. On rigid circuit graphs. *Abhandlungen aus dem Mathematischen Seminar der Universität Hamburg*, 25:71–76, 1961. [517]

[1973] D.Z. Djokovic. Distance preserving subgraphs of hypercubes. *Journal of Combinatorial Theory B*, 14:263–267, 1973. [279,283]

[1987a] Y. Dodge. An introduction to statistical data analysis L_1-norm based. In Y. Dodge, editor, *Statistical Data Analysis Based on the L_1-Norm and Related Methods*, pages 1–22. Elsevier Science Publishers, North-Holland, Amsterdam, 1987. [163]

[1987b] Y. Dodge, editor. *Statistical Data Analysis Based on the L_1-Norm and Related Methods.* Elsevier Science Publishers, North-Holland, Amsterdam, 1987. [163]

[1992] Y. Dodge, editor. *L_1-Statistical Analysis and Related Methods.* Elsevier Science Publishers, North-Holland, Amsterdam, 1992. [163]

[1976] L.E. Dor. Potentials and isometric embeddings in L_1. *Israel Journal of Mathematics*, 24:260–268, 1976. [34]

[1965] R. Duffin. Topology of series-parallel networks. *Journal of Mathematical Analysis and Applications*, 10:303–318, 1965. [518]

[1987] H. Edelsbrunner. *Algorithms in Combinatorial Geometry.* Springer-Verlag, Berlin, 1987. [171]

[1977] J. Edmonds and R. Giles. A min-max relation for submodular functions on graphs. In P.L. Hammer et al., editors, *Studies in Integer Programming*, volume 1 of *Annals of Discrete Mathematics*, pages 185–204, 1977. [147]

[1973] J. Edmonds and E.L. Johnson. Matching, Euler tours and the Chinese postman. *Mathematical Programming*, 5:88–124, 1973. [442]

[1974] R.M. Erdahl. A convex set of second order inhomogeneous polynomials with applications to reduced density matrices. In R.M. Erdahl, editor, *Reduced Density Operators with Applications to Physical and Chimical Systems II*, volume 40 of *Queen's papers in pure and applied mathematics*, pages 28–35. Queen's University, Kingston, 1974. [188]

[1987] R.M. Erdahl. Representability conditions. In R.M. Erdahl and V.H. Smith, editors, *Density Matrices and Density Functionals*, pages 51–75. D. Reidel, 1987. [55,64,69,188]

[1992] R.M. Erdahl. A cone of inhomogeneous second-order polynomials. *Discrete and Computational Geometry*, 8:387–416, 1992. [182,188,189,235,240]

[1987] R.M. Erdahl and S.S. Ryshkov. The empty sphere. *Canadian Journal of Mathematics*, 39:794–824, 1987. [172,186,188]

[1948] P. Erdös. Problem 4306. *American Mathematical Monthly*, 55:431, 1948. [183]

[1957] P. Erdös. Some unsolved problems. *Michigan Journal of Mathematics*, 4:291–300, 1957. [183]

[1996] M. Erickson, S. Fernando, W.H. Haemers, D. Hardy, and J. Hemmeter. Deza graphs. Preprint, 1996. [545]

[1992] T. Feder. Product graph representations. *Journal of Graph Theory*, 16:467–488, 1992. [286,301,302,306]

[1985] J. Feigenbaum, J. Hershberger, and A.A. Schäffer. A polynomial time algorithm for finding the prime factors of cartesian product graphs. *Discrete Applied Mathematics*, 12:123–138, 1985. [305]

[1959] L. Fejes Tóth. Über eine Punktverteilung auf der Kugel. *Acta Mathematica Academiae Scientiarum Hungaricae*, 10:13–19, 1959. [10,514]

[1987a] B. Fichet. The role played by L_1 in data analysis. In Y. Dodge, editor, *Statistical Data analysis Based on the L_1-Norm and Related Methods*, pages 185–193. Elsevier Science Publishers, North-Holland, Amsterdam, 1987. [28,56,163]

[1987b] B. Fichet. Data analysis: geometric and algebraic structures. In Yu. Prohorov and V.V. Sazonov, editors, *Proceedings of the 1st World Congress of the Bernouilli Society, volume 2, Mathematical Statistics, Theory and Applications,* pages 123–132. VNU Science Press, Utrecht, 1987. [163]

[1988] B. Fichet. L_p-spaces in data analysis. In H.H. Bock, editor, *Classification and Related Methods of Data Analysis*, pages 439–444. North-Holland, Amsterdam, 1988. [157]

[1992] B. Fichet. The notion of sphericity for finite L_1-figures of data analysis. In Y. Dodge, editor, *L_1-Statistical Analaysis and Related Methods*, pages 129–144. Elsevier Science Publishers, North-Holland, Amsterdam, 1992. [98,163]

[1994] B. Fichet. Dimensionality problems in L_1-norm representations. In B. van Cutsem, editor, *Classification and Dissimilarity Analysis*, volume 93 of *Lecture Notes in Statistics*, pages 201–224. Springer-Verlag, New York, 1994. [33,35,157,163]

[1977] S. Foldes. A characterization of hypercubes. *Discrete Mathematics*, 17:155–159, 1977. [309]

[1962] L.R. Ford Jr. and D.R. Fulkerson. *Flows in Networks.* Princeton University Press, Princeton, 1962. [134]

[1995] S. Fortune. Voronoi diagrams and Delaunay triangulations. In D.-Z. Du and F. Hwang, editors, *Computing in Euclidean Geometry*, pages 225–265. World Scientific, Singapore, 1995 (2nd edition). [172]

[1990] A. Frank. Packing paths, circuits and cuts - A survey. In B. Korte, L. Lovász, H.J. Prömel, and A. Schrijver, editors, *Paths, Flows, and VLSI-Layout*, pages 47–100. Springer-Verlag, Berlin, 1990. [v]

[1995] A. Frieze and M. Jerrum. Improved approximation algorithms for MAX k-CUT and MAX BISECTION. In E. Balas and J. Clausen, editors, *Integer Programming and Combinatorial Optimization*, volume 920 of *Lecture Notes in Computer Science*, pages 1–13. Springer-Verlag, Berlin, 1995. [461]

[1995] X. Fu and L. Goddyn. Matroids with the circuit cover property. Preprint, 1995. [392,444]

[1977] J. Galambos. Bonferroni inequalities. *Annals of Probability*, 5:577–581, 1977. [63]

[1997] A. Gallucio and M. Loebl. (p,q)-odd digraphs. *Journal of Graph Theory*, 1997 (to appear). [443]

[1996] A. Gallucio and M. Loebl. Cycles of binary matroids without an F_7^*-minor. Preprint, 1996. [443]

[1975] M.R. Garey and R.L. Graham. On cubical graphs. *Journal of Combinatorial Theory B*, 18:84–95, 1975. [294]

[1979] M.R. Garey and D.S. Johnson. *Computers and Intractability: A Guide to the Theory of NP-Completeness.* Freeman, San Francisco, 1979. [11,18,455,458]

[1995a] N. Garg. A deterministic $O(\log k)$-approximation algorithm for the sparsest cut. Preprint, 1995. [134]

[1995b] N. Garg. On the distortion of Bourgain's embedding. Preprint, 1995. [125]

[1993] N. Garg, V.V. Vazirani, and M. Yannakakis. Approximate max-flow min-(multi)cut theorems and their applications. *Proceedings of the 25th Annual ACM Symposium on the Theory of Computing*, pages 698–707, 1993. (Updated version in: *SIAM Journal on Computing*, 25:235–251, 1996.) [134]

[1979] A.V. Geramita and J. Seberry. *Orthogonal Designs.* Marcel Dekker, New York, 1979. [343]

[1985] A.M.H. Gerards. Testing the bicycle odd wheel inequalities for the bipartite subgraph polytope. *Mathematics of Operations Research*, 10:359–360, 1985. [481]

[1995] A.M.H. Gerards and M. Laurent. A characterization of box $1/d$-integral binary clutters. *Journal of Combinatorial Theory B*, 65:186–207, 1995. [431]

[1993] L. Goddyn. Cones, lattices and Hilbert bases of circuits and perfect matchings. In N. Robertson and P.D. Seymour, editors, *Graph Structure Theory*, volume 147 of *Contemporary Mathematics*, pages 419–440, 1993. [441]

[1995] C.D. Godsil, M. Grötschel, and D.J.A. Welsh. Combinatorics in statistical physics. In R.L. Graham, M. Grötschel, and L. Lovász, editors, *Handbook of Combinatorics*, Chapter 37 in volume II, pages 1925–1954. North-Holland, Elsevier, Amsterdam, 1995. [51]

[1994] M.X. Goemans and D.P. Williamson. 0.878-approximation algorithms for MAX CUT and MAX 2SAT. *Proceedings of the 26th Annual ACM Symposium on the Theory of Computing*, pages 422–431, 1994. (Updated version as: Improved approximation algorithms for maximum cut and satisfiability problems using semidefinite programming. *Journal of the Association for Computing Machinery*, 42:1115–1145, 1995.) [459,460,461]

[1987] A.D. Gordon. A review of hierarchical classification. *Journal of the Royal Statistical Society A*, 150 - Part 2:119–137, 1987. [311]

[1982] J.C. Gower. Euclidean distance geometry. *The Mathematical Scientist*, 7:1–14, 1982. [80]

[1985] J.C. Gower. Properties of Euclidean and non-Euclidean distance matrices. *Linear Algebra and its Applications*, 67:81–97, 1985. [75]

[1993] D.A. Grable. Sharpened Bonferroni inequalities. *Journal of Combinatorial Theory B*, 57:131–137, 1993. [65]

[1988] R.L. Graham. Isometric embeddings of graphs. In L.W. Beineke and R.J. Wilson, editors, *Selected Topics in Graph Theory 3*, pages 133–150. Academic Press, London, 1988. [297]

[1978] R.L. Graham and L. Lovász. Distance matrix polynomials of trees. *Advances in Mathematics*, 29:60–88, 1978. [286]

[1971] R.L. Graham and H.O. Pollack. On the addressing problem for loop switching. *The Bell System Technical Journal*, 50:2495–2519, 1971. [278,285,293]

[1972] R.L. Graham and H.O. Pollack. On embedding graphs in squashed cubes. In Y. Alavi, D.R. Lick, and A.T. White, editors, *Graph Theory and Applications*, volume 303 of *Lecture Notes in Mathematics*, pages 90–110. Springer-Verlag, Berlin, 1972. [294]

[1985] R.L. Graham and P.M. Winkler. On isometric embeddings of graphs. *Transactions of the American Mathematical Society*, 288:527–536, 1985. [80,291,297]

[1980] R. Graham, A. Yao, and F. Yao. Information bounds are weak for the shortest distance problem. *Journal of the Association for Computing Machinery*, 27:428–444, 1980. [422]

[1990] V.P. Grishukhin. All facets of the cut cone C_n for $n = 7$ are known. *European Journal of Combinatorics*, 11:115–117, 1990. [502,503]

[1992a] V.P. Grishukhin. Computing extreme rays of the metric cone for seven points. *European Journal of Combinatorics*, 13:153–165, 1992. [422,542]

[1992b] V.P. Grishukhin. On a t-extension of a distance space. Technical report SOCS-92.5, School of Computer Science, Mc Gill University, Montreal, 1992. [213,214]

[1993] V.P. Grishukhin. L-polytopes, even unimodular lattices and perfect lattices. Rapport LIENS-93-1, Ecole Normale Supérieure, Paris, 1993. [248]

[1984] R. Grone, C.R. Johnson, E.M. Sá, and H. Wolkowicz. Positive definite completions of partial Hermitian matrices. *Linear Algebra and its Applications*, 58:109–124, 1984. [517]

[1990] R. Grone, S. Pierce, and W. Watkins. Extremal correlation matrices. *Linear Algebra and its Applications*, 134:63–70, 1990. [537]

[1994] M. Grötschel. Personal communication. 1994. [407]

[1987] M. Grötschel, M. Jünger, and G. Reinelt. Calculating exact ground states of spin glasses, a polyhedral approach. In J.L. van Hemmen and I. Morgenstern, editors, *Heidelberg Colloquium on Spin Glasses*, pages 325–353. Springer-Verlag, Berlin, 1987. [52]

[1988] M. Grötschel, L. Lovász, and A. Schrijver. *Geometric Algorithms and Combinatorial Optimization.* Springer-Verlag, Berlin, 1988. [20,21,49,130,433,449]

[1981] M. Grötschel and W.R. Pulleyblank. Weakly bipartite graphs and the max-cut problem. *Operations Research Letters*, 1:23–27, 1981. [482]

[1989a] M. Grötschel and K. Truemper. Decomposition and optimization over cycles in binary matroids. *Journal of Combinatorial Theory B*, 46:306–337, 1989. [440,442]

[1989b] M. Grötschel and K. Truemper. Master polytopes for cycles of binary matroids. *Linear Algebra and its Applications*, 114/115:523–540, 1989. [440,441]

[1990] M. Grötschel and Y. Wakabayashi. Facets of the clique-partitioning polytope. *Mathematical Programming*, 47:367–387, 1990. [465]

[1967] B. Grünbaum. *Convex Polytopes.* John Wiley & Sons, New York, 1967. [11]

[1975] F. Hadlock. Finding a maximum cut of a planar graph in polynomial time. *SIAM Journal on Computing*, 4:221–225, 1975. [48,52]

[1978] F. Hadlock and F. Hoffman. Manhattan trees. *Utilitas Mathematica*, 13:55–67, 1978. [142]

[1964] S.L. Hakimi and S.S. Yau. Distance matrix of a graph and its realizability. *Quarterly Journal of Applied Mathematics*, 22:305–317, 1964. [311]

[1977] J.I. Hall. Bounds for equidistant codes and partial projective planes. *Discrete Mathematics*, 17:85–94, 1977. [336,337]

[1977] J.I. Hall, A.J.E.M. Jansen, A.W.J. Kolen, and J.H. van Lint. Equidistant codes with distance 12. *Discrete Mathematics*, 17:71–83, 1977. [340]

[1967] M. Hall Jr. *Combinatorial Theory.* John Wiley & Sons, New York, 1967. [336]

[1965] P.L. Hammer. Some network flow problems solved with pseudo-boolean programming. *Operations Research*, 13:388–399, 1965. [56]

[1984] P.L. Hammer, P. Hansen, and B. Simeone. Roof duality, complementation and persistency in quadratic 01-optimization. *Mathematical Programming*, 28:121–155, 1984. [430]

[1975] H. Hanani. Balanced incomplete block designs and related designs. *Discrete Mathematics*, 11:255–369, 1975. [343]

[1934] G.H. Hardy, J.E. Littlewood, and G. Pólya. *Inequalities.* Cambridge University Press, Cambridge, 1934. [31]

[1972] I. Havel and P. Liebl. Embedding the dichotomic tree into the cube. *Časopis pro Pestovani Matematiky*, 97:201–205, 1972. [294]

[1973] I. Havel and P. Liebl. Embedding the polytomic tree into the n-cube. *Časopis pro Pestovani Matematiky*, 98:307–314, 1973. [294]

[1993] T.L. Hayden and P. Tarazaga. Distance matrices and regular figures. *Linear Algebra and its Applications*, 195:9–16, 1993. [82]

[1988] T.L. Hayden and J. Wells. Approximations by matrices positive semidefinite on a subspace. *Linear Algebra and its Applications*, 109:115–130, 1988. [80]

[1991] T.L. Hayden, J. Wells, Wei-Min Liu, and P. Tarazaga. The cone of distance matrices. *Linear Algebra and its Applications*, 144:153–169, 1991. [75]

[1987] R.D. Hill and S.R. Waters. On the cone of positive semidefinite matrices. *Linear Algebra and its Applications*, 90:81–88, 1987. [536]

[1996] W. Hochstättler, M. Laurent and M. Loebl. Cycle Bases for lattices of matroids without Fano dual minor and their one-element extensions. Preprint, 1996. [443]

[1995] W. Hochstättler and M. Loebl. On bases of circuit lattices. Report No. 95.202, Angewandte Mathematik und Informatik, Universität zu Köln, 1995. [443]

[1978] W. Holsztysnki. $\mathbb{R}^n$ as universal metric space. *Notices of the American Mathematical Society*, 25:A–367, 1978. [156]

[1986] B.K.P. Horn. *Robot Vision.* M.I.T. Press, Cambridge, Massachussets, 1986. [163]

[1991] R.A. Horn and C.R. Johnson. *Topics in Matrix Analysis.* Cambridge University Press, Cambridge, 1991. [113,116]

[1986] E. Howe, C.R. Johnson, and J. Lawrence. The structure of distances in networks. *Networks*, 16:87–106, 1986. [422]

[1963] T.C. Hu. Multi-commodity network flows. *Operations Research*, 11:344–360, 1963. [134]

[1984] W. Imrich, J.M.S. Simões-Pereira, and C.M. Zamfirescu. On optimal embeddings of metrics in graphs. *Journal of Combinatorial Theory B*, 36:1–15, 1984. [310]

[1989] A.N. Isachenko. On the structure of the quadratic boolean problem polytope. In *Combinatorics and Graph Theory*, volume 25 of *Banach Center Publications*, pages 87–91. P.W.N., Warszawa, 1989. [54]

[1971] M. Iri. On an extension of the maximum-flow minimum-cut theorem to multicommodity flows. *Journal of the Operations Research Society of Japan*, 13:129–135, 1970-71. [422]

[1990] C.R. Johnson. Matrix completion problems: a survey. In C.R. Johnson, editor, *Matrix Theory and Applications*, volume 40 of *Proceedings of Symposia in Applied Mathematics*, pages 171–198. American Mathematical Society, Providence, Rhode Island, 1990. [462,516]

[1995] C.R. Johnson, C. Jones, and B. Kroschel. The distance matrix completion problem: cycle completability. *Linear and Multilinear Algebra*, 39:195–207, 1995. [530]

[1996] C.R. Johnson and T.A. McKee. Structural conditions for cycle completable graphs. *Discrete Mathematics*, 159:155–160, 1996. [521,522]

[1967] S.C. Johnson. Hierarchical clustering schemes. *Psychometrika*, 32:241–254, 1967. [163]

[1984] W.B. Johnson and J. Lindenstrauss. Extensions of Lipschitz mappings into a Hilbert space. In R. Beals et al., editors, *Conference in Modern Analysis and Probability*, volume 26 of *Contemporary Mathematics*, pages 189–206. American Mathematical Society, Providence, Rhode Island, 1984. [129]

[1995] M. Jünger, G. Reinelt, and S. Thienel. Practical problem solving with cutting plane algorithms in combinatorial optimization. In W. Cook, L. Lovász, and P.D. Seymour, editors, *Combinatorial Optimization*, volume 20 of *DIMACS Series in Discrete Mathematics and Theoretical Computer Science*, pages 111–152. American Mathematical Society, 1995. [397]

[1993] J. Kahn and G. Kalai. A counterexample to Borsuk's conjecture. *Bulletin of the American Mathematical Society*, 29:60–62, July 1993. [10,512]

[1997] J. Kahn, N. Linial, and A. Samorodnitsky. Inclusion-exclusion: exact and approximate. Combinatorica, 1997 (to appear). [65]

[1975] K. Kalmanson. Edgeconvex circuits and the traveling salesman problem. *Canadian Journal of Mathematics*, 27:1000–1010, 1975. [92]

[1994] D. Karger, R. Motwani, and M. Sudan. Approximate graph colouring by semidefinite programming. *Proceedings of the 35th IEEE Symposium on Foundations of Computer Science*, pages 2–13. Computer Society Press, 1994. [461]

[1996] H. Karloff. How good is the Goemans-Williamson MAX CUT algorithm ? *Proceedings of the 28th Annual ACM Symposium on the Theory of Computing*, pages 427–434, 1996. [462]

[1972] R.M. Karp. Reducibility among combinatorial problems. In R.E. Miller and J.W. Thatcher, editors, *Complexity of Computer Computations*, pages 85–103. Plenum Press, New York, 1972. [48]

[1982] R.M. Karp and C.H. Papadimitriou. On linear characterization of combinatorial optimization problem. *SIAM Journal on Computing*, 11:620–632, 1982. [50,397]

[1985] A.V. Karzanov. Metrics and undirected cuts. *Mathematical Programming*, 32:183–198, 1985. [49,353,375]

[1968] J.B. Kelly. Products of zero-one matrices. *Canadian Journal of Mathematics*, 20:298–329, 1968. [64,69,356]

[1970a] J.B. Kelly. Metric inequalities and symmetric differences. In O.Shisha, editor, *Inequalities II*, pages 193–212. Academic Press, New York, 1970. [69,72,111,106]

[1970b] J.B. Kelly. Combinatorial inequalities. In R. Guy, H. Hanani, N. Sauer and J. Schonheim, editors, *Combinatorial Structures and their Applications*, pages 201–208. Gordon and Breach, New York, 1970. [87,514]

[1972] J.B. Kelly. Hypermetric spaces and metric transforms. In O. Shisha, editor, *Inequalities III*, pages 149–158. Academic Press, New York, 1972. [118]

[1975] J.B. Kelly. Hypermetric spaces. In L.M. Kelly, editor, *The Geometry of Metric and Linear Spaces*, volume 490 of *Lecture Notes in Mathematics*, pages 17–31. Springer-Verlag, Berlin, 1975. [90]

[1990] P. Klein, A. Agrawal, R. Ravi, and S. Rao. Approximation through multicommodity flow. In *31st Annual Symposium on Foundations of Computer Science*, volume 2, pages 726–737, 1990. (Updated version in: *Combinatorica*, 15:187–202, 1995.) [134]

[1997] Chun-Wa Ko, J. Lee, and E. Steingrimsson. The volume of relaxed boolean-quadric and cut polytopes. *Discrete Mathematics*, 163:293–298, 1997. [426]

[1990] J. Koolen. On metric properties of regular graphs. Master's thesis, Technische Universiteit Eindhoven, 1990. [269,308]

[1993] J. Koolen. On uniformely geodetic graphs. *Graphs and Combinatorics*, 9:325–333, 1993. [308,309]

[1994] J. Koolen. *Euclidean Representations and Substructures of Distance-Regular Graphs.* PhD thesis, Technische Universiteit Eindhoven, 1994. [269,308]

[1994] J. Koolen and S.V. Shpectorov. Distance-regular graphs the distance matrix of which has only one positive eigenvalue. *European Journal of Combinatorics*, 15:269–275, 1994. [268,308]

[1976] S. Kounias and J. Marin. Best linear Bonferroni bounds. *SIAM Journal on Applied Mathematics*, 30:307–323, 1976. [61,64,69]

[1986] E.F. Krause. *Taxicab Geometry - An Adventure in Non-Euclidean Geometry.* Dover Publications, New York, 1986. [164]

[1986] D.W. Krumme, K.N. Venkataraman, and G. Cybenko. Hypercube embedding is NP-complete. In M.T. Heath, editor, *Hypercube Multiprocessors 1986*, pages 148–157. SIAM, Philadelphia, Pennsylvania, 1986. [294]

[1994] F. Laburthe. Etude de la base de Hilbert du cone des coupes à six sommets. Mémoire de DEA, Ecole Normale Supérieure, Paris, 1994. [386]

[1995] F. Laburthe. The Hilbert basis of the cut cone over the complete graph K_6. In E. Balas and J. Clausen, editors, *Integer Programming and Combinatorial Optimization*, volume 920 of *Lecture Notes in Computer Science*, pages 252–266. Springer-Verlag, Berlin, 1995. [386]

[1995] F. Laburthe, M. Deza, and M. Laurent. The Hilbert basis of the cut cone over the complete graph K_6. Rapport LIENS-95-7, Ecole Normale Supérieure, Paris, 1995. [386]

[1995] J.C. Lagarias. Point lattices. In R.L. Graham, M. Grötschel, and L. Lovász, editors, *Handbook of Combinatorics*, Chapter 19 in volume I, pages 919–966. North-Holland, Elsevier, Amsterdam, 1995. [177]

[1989] C.W.H. Lam, L.H. Thiel, and S. Swierzc. The nonexistence of finite projective planes of order 10. *Canadian Journal of Mathematics*, 41:1117–1123, 1989. [336]

[1985] P. Lancaster and M. Tismenetsky. *The Theory of Matrices. Second Edition, with Applications.* Computer Science and Applied Mathematics. Academic Press, Orlando, 1985. [11]

[1994] M. Laurent. Hypercube embedding of distances with few values. In H. Barcelo and G. Kalai, editors, *Jerusalem Combinatorics '93*, volume 178 of *Contemporary Mathematics*, pages 179–207, 1994. [362,363]

[1996a] M. Laurent. Delaunay transformations of a Delaunay polytope. *Journal of Algebraic Combinatorics*, 5:37–46, 1996. [225]

[1996b] M. Laurent. Hilbert bases of cuts. *Discrete Mathematics*, 150:257–279, 1996. [392,444]

[1996c] M. Laurent. Graphic vertices of the metric polytope. *Discrete Mathematics*, 151:131–153, 1996. [410,421,422,543]

[1997a] M. Laurent. Max-cut problem. In M. Dell'Amico, F. Maffioli, and S. Martello, editors, chapter in *Annotated Bibliographies in Combinatorial Optimization*. John Wiley & Sons, 1997 (to appear). [38]

[1997b] M. Laurent. The real positive semidefinite completion problem for series-parallel graphs. *Linear Algebra and its Applications*, 252:347–366, 1997. [518,522]

[1997c] M. Laurent. A connection between positive semidefinite and Euclidean distance matrix completion problems. *Linear Algebra and its Applications*, 1997 (to appear). [521,530,531]

[1997d] M. Laurent. A tour d'horizon on positive semidefinite and Euclidean distance matrix completion problems. In P. Pardalos and H. Wolkowicz, editors, *Topics in Semidefinite and Interior-Point Methods, The Fields Institute for Research in Mathematical Science, Communications Series*. Providence, Rhode Island, 1997 (to appear). [516,522,529]

[1992] M. Laurent and S. Poljak. The metric polytope. In E. Balas, G. Cornuejols, and R. Kannan, editors, *Integer Programming and Combinatorial Optimization*, pages 274–286. Carnegie Mellon University, GSIA, Pittsburgh, USA, 1992. [422,543]

[1995a] M. Laurent and S. Poljak. One-third-integrality in the metric polytope. *Mathematical Programming*, 71:29–50, 1995. [431]

[1995b] M. Laurent and S. Poljak. On a positive semidefinite relaxation of the cut polytope. *Linear Algebra and its Applications*, 223/224:439–461, 1995. [462,535,538]

[1996a] M. Laurent and S. Poljak. On the facial structure of the set of correlation matrices. *SIAM Journal on Matrix Analysis and Applications*, 17:530–547, 1996. [462,535,537,538]

[1996b] M. Laurent and S. Poljak. Gap inequalities for the cut polytope. *European Journal of Combinatorics*, 17:233–254, 1996. [457,458]

[1997] M. Laurent, S. Poljak, and F. Rendl. Connections between semidefinite relaxations of the max-cut and stable set problems. *Mathematical Programming*, 77:101–121, 1997. [462]

[1985] E.L. Lawler, J.K. Lenstra, A.H.G. Rinnoy Kan, and D.B. Shmoys. *The Traveling Salesman Problem - A Guided Tour of Combinatorial Optimization*, John Wiley & Sons, Chichester, 1985. [92]

[1987] G. Le Calve. L_1-embeddings of a data structure (I, D). In Y. Dodge, editor, *Statistical Data Analysis Based on the L_1-Norm and Related Methods*, pages 195–202. Elsevier Science Publishers, North-Holland, Amsterdam, 1987. [28,163]

[1975] P.G.H. Lehot. An optimal algorithm to detect a line graph and output its root graph. *Journal of the Association for Computing Machinery*, 21:569–575, 1975. [324]

[1988] T. Leighton and S. Rao. An approximate max-flow min-cut theorem for uniform multicommodity flow problems with applications to approximation algorithms. In *29th Annual Symposium on Foundations of Computer Science*, pages 422–431, 1988. [134]

[1973] P.W.H. Lemmens and J.J. Seidel. Equiangular lines. *Journal of Algebra*, 24:494–512, 1973. [236,237]

[1994] Chi-Kwong Li and Bit-Shun Tam. A note on extreme correlation matrices. *SIAM Journal on Matrix Analysis and its Appplications*, 15:903–908, 1994. [536]

[1993] V.Y. Lin and A. Pinkus. Fundamentality of ridge functions. *Journal of Approximation Theory*, 75:295–311, 1993. [84]

[1994] N. Linial, E. London, and Y. Rabinovich. The geometry of graphs and some of its algorithmic applications. In *35th Annual Symposium on Foundations of Computer Science*, pages 577–591, 1994. (Updated version in: *Combinatorica*, 15:215–245, 1995.) [125,128,129,134,159]

[1990] N. Linial and N. Nisan. Approximate inclusion-exclusion. *Combinatorica*, 10:349–365, 1990. [65]

[1973] J.H. van Lint. A theorem on equidistant codes. *Discrete Mathematics*, 6:353–358, 1973. [337]

[1980] R. Loewy. Extreme points of a convex subset of the cone of positive semidefinite matrices. *Mathematische Annalen*, 253:227–232, 1980. [537]

[1978] M.V. Lomonosov. On a system of flows in a network (in Russian). *Problemy Peredatchi Informatzii*, 14:60–73, 1978. [422]

[1985] M.V. Lomonosov. Combinatorial approaches to multiflow problems. *Discrete Applied Mathematics*, 11:1–93, 1985. [422]

[1993] M. Lomonosov and A. Sebö. On the geodesic-structure of graphs: a polyhedral approach to metric decomposition. In G. Rinaldi and L.A. Wolsey, editors, *Integer Programming and Combinatorial Optimization*, pages 221–234, 1993. [298,301,306]

[1979] L. Lovász. On the Shannon capacity of a graph. *IEEE Transactions on Information Theory*, 25:1–7, 1979. [462]

[1994] L. Lovász. Personal communication, 1994. [184,199,204]

[1991] L. Lovász and A. Schrijver. Cones of matrices and set-functions and $0-1$ optimization. *SIAM Journal on Optimization*, 1:166–190, 1991. [424]

[1993] L. Lovász and A. Seress. The cocycle lattice of binary matroids. *European Journal of Combinatorics*, 14:241–250, 1993. [443]

[1995] L. Lovász and A. Seress. The cocycle lattice of binary matroids, II. *Linear Algebra and its Applications*, 226/228:553–565, 1995. [443]

[1955] P.O. Löwdin. Quantum theory of many-particles systems. I. Physical interpretation by means of density matrices, natural spin-orbitals, and convergence problems in the method of configurational interaction. *Physical Review*, 97(6):1474–1489, 1955. [54]

[1991] M. Luby and B. Veličković. On deterministic approximation of DNF. *Proceedings of the 23rd Annual ACM Symposium on the Theory of Computing*, pages 430–438, 1991. [65]

[1976] E. Luczak and A. Rosenfeld. Distance on a hexagonal grid. *IEEE Transactions on Computers*, C-25(5): 532–533, 1976. [165]

[1993] Y.I. Lyubich and L.N. Vaserstein. Isometric embeddings between classical Banach spaces, cubature formulas, and spherical designs. *Geometriae Dedicata*, 47:327–362, 1993. [34]

[1986] H. Maehara. Metric transforms of finite spaces and connected graphs. *Discrete Mathematics*, 61:235–246, 1986. [124]

[1992] S.M. Malitz and J.I. Malitz. A bounded compactness theorem for L^1-embeddability of metric spaces in the plane. *Discrete Computational Geometry*, 8:373–385, 1992. [33,141,156]

[1958] F. Marczewski and H. Steinhaus. On a certain distance of sets and the corresponding distance of functions. *Colloquium Mathematicum*, 6:319–327, 1958. [119]

[1976] D. McCarthy, R.C. Mullin, P.J. Schellenberg, R.G. Stanton, and S.A. Vanstone. An approximation to a projective plane of order 6. *Ars Combinatoria*, 2:169–189, 1976. [350]

[1977] D. McCarthy, R.C. Mullin, P.J. Schellenberg, R.G. Stanton, and S.A. Vanstone. Towards the nonexistence of $(7,1)$-designs with 31 varieties. In B.L. Hartnell and H.C. Williams, editors, *Proceedings of the Sixth Manitoba Conference on Numerical Mathematics*, volume 18 of *Congressus Numerantium*, pages 265–285. Utilitas Mathematica, Winnipeg, 1977. [350]

[1977] D. McCarthy and S.A. Vanstone. Embedding $(r,1)$-designs in finite projective planes. *Discrete Mathematics*, 19:67–76, 1977. [347]

[1972] W.B. McRae and E.R. Davidson. Linear inequalities for density matrices II. *Journal of Mathematical Physics*, 13:1527–1538, 1972. [54,64,69,405,477,503]

[1928] K. Menger. Untersuchungen über allgemeine Metrik. *Mathematische Annalen*, 100:75–163, 1928. [25,73,78,141]

[1931] K. Menger. New foundation of Euclidean geometry. *American Journal of Mathematics*, 53:721–745, 1931. [25,78,80]

[1954] K. Menger. *Géométrie Générale.* Mémorial des Sciences Mathématiques CXXIV, Académie des Sciences de Paris, Gauthier-Villars, Paris, 1954. [25,77,78]

[1987] M.M. Mestechkin. On the diagonal N-representability problem. In R.M. Erdahl and V.H. Smith, editors, *Density Matrices and Density Functionals*, pages 77–87. D. Reidel, 1987. [64,69]

[1987] M. Mezard, G. Parisi, and M.A. Virasoro. *Spin Glass Theory and Beyond.* World Scientific, Singapore, 1987. [51]

[1990] W.H. Mills. Balanced incomplete block designs with $\lambda = 1$. *Congressus Numerantium*, 73:175–180, 1990. [343]

[1980] H.M. Mulder. *The Interval Function of a Graph.* Mathematical Centre Tracts 132, Mathematisch Centrum, Amsterdam, 1980. [309]

[1982] H.M. Mulder. Interval-regular graphs. *Discrete Mathematics*, 41:253–269, 1982. [309]

[1987] A. Neumaier. Some relations between roots, holes and pillars. *European Journal of Combinatorics*, 8:29–33, 1987. [237]

[1941] J. von Neumann and I.J. Schoenberg. Fourier integrals and metric geometry. *Transactions of the American Mathematical Society*, 50:226–251, 1941. [113,117]

[1994] A. Nilli. On Borsuk's problem. In H. Barcelo and G. Kalai, editors, *Jerusalem Combinatorics '93*, volume 178 of *Contemporary Mathematics*, pages 209–210, 1994. [512]

[1987] A. Noda. Generalized Radon transform and Lévy's Brownian motion I and II. *Nagoya Mathematical Journal*, 105:71–87 and 89–107, 1987. [162]

[1989] A. Noda. White noise representations of some Gaussian random fields. *The Bulletin of Aichi University of Education*, 38:35–45, 1989. [162]

[1992] S.P. Novikov, S.S. Ryshkov, N.P. Dolbilin, and M.A. Shtan'ko, editors. *Discrete Geometry and Topology. Proceedings of the Steklov Institute of Mathematics*, 196 (4), 1992. [171]

[1972] G.I. Orlova and Y.G. Dorfman. Finding the maximum cut in a graph. *Engineering Cybernetics*, 10:502–506, 1972. [48,52]

[1992] J.G. Oxley. *Matroid Theory.* Oxford University Press, London, 1992. [436]

[1989] M.W. Padberg. The boolean quadric polytope: some characteristics, facets and relatives. *Mathematical Programming*, 45:139–172, 1989. [54,425,435,453,539]

[1995] M.W. Padberg. Personal communication, 1995. [506]

[1982] M.W. Padberg and M.R. Rao. Odd minimum cut-sets and *b*-matchings. *Mathematics of Operations Research*, 7:67–80, 1982. [442]

[1982] C.H. Papadimitriou and K. Steiglitz. *Combinatorial Optimization: Algorithms and Complexity.* Prentice-Hall, Englewood Cliffs, New Jersey, 1982. [18]

[1976] B.A. Papernov. On existence of multicommodity flows (in Russian). *Studies in Discrete Optimization*, Nauka, Moscow, pages 230–261, 1976. [375]

[1982] D. Penny, L.R. Foulds, and M.D. Hendy. Testing the theory of evolution by comparing phylogenetic trees constructed from five different protein sequences. *Nature*, 297:197–200, 1982. [310]

[1972] J.R. Pierce. Network for block switching of data. *The Bell System Technical Journal*, 51:1133–1145, 1972. [278]

[1986] I. Pitowsky. The range of quantum probability. *Journal of Mathematical Physics*, 27:1556–1565, 1986. [54,59]

[1989] I. Pitowsky. *Quantum Probability - Quantum Logic*, volume 321 of *Lecture Notes in Physics.* Springer-Verlag, Berlin, 1989. [54]

[1991] I. Pitowsky. Correlation polytopes: their geometry and complexity. *Mathematical Programming*, 50:395–414, 1991. [54,61,64,69,405,477]

[1993] S.A. Plotkin and É. Tardos. Improved bounds on the max-flow min-cut ratio for multicommodity flows. *Proceedings of the 25th Annual ACM Symposium on the Theory of Computing*, pages 691–697, 1993. [134]

[1995] S. Poljak and F. Rendl. Nonpolyhedral relaxations of graph-bisection problems. *SIAM Journal on Optimization*, 5:467–487, 1995. [460]

[1987] S. Poljak and D. Turzik. On a facet of the balanced subgraph polytope. *Časopis Pro Pestovani Matematiky*, 112:373–380, 1987. [496]

[1992] S. Poljak and D. Turzik. Max-cut in circulant graphs. *Discrete Mathematics*, 108:379–392, 1992. [496,497]

[1995] S. Poljak and Z. Tuza. Maximum cuts and largest bipartite subgraphs. In W. Cook, L. Lovász, and P.D. Seymour, editors, *Combinatorial Optimization*, volume 20 of *DIMACS Series in Discrete Mathematics and Theoretical Computer Science*, pages 181–244. American Mathematical Society, 1995. [v,38,548]

[1990] K.F. Prisăcaru, P.S. Soltan, and V. Chepoi. On embeddings of planar graphs into hypercubes (in Russian). *Buletinal Academiei de Stiinte a R.S.S. Moldovenesti, Matematica*, 1:43–50, 1990. [329]

[1991] S.T. Rachev. *Probability Metrics and the Stability of Stochastic Models.* John Wiley & Sons, Chichester, 1991. [161,162]

[1955] R.A. Rankin. On packing spheres in Hilbert space. *Proceedings of the Glasgow Mathematical Association*, 2:145–146, 1955. [85]

[1993] L. Reid and X. Sun. Distance matrices and ridge function interpolation. *Canadian Journal of Mathematics*, 45:1313–1323, 1993. [84]

[1964] C.A. Rogers. *Packing and Covering*, volume 54 of *Cambridge tracts in Mathematics and Mathematical Physics.* Cambridge University Press, Cambridge, 1964. [172]

[1957] C.A. Rogers and G.C. Shephard. The difference body of a convex body. *Archiv der Mathematik*, 8:220–233, 1957. [205]

[1976] A. Rosenfeld and A.C. Kak. *Digital Picture Processing.* Academic Press, New York, 1976. [163]

[1986] R.L. Roth and P.M. Winkler. Collapse of the metric hierarchy for bipartite graphs. *European Journal of Combinatorics*, 7:371–375, 1986. [286,288]

[1966] W. Rudin. *Real and Complex Analysis.* McGraw-Hill, New York, 1966. [32]

[1963] H.J. Ryser. *Combinatorial Mathematics*, volume 14 of *The Carus Mathematical Monographs.* The Mathematical Association of America, 1963. [342]

[1979] S.S. Ryshkov and E.P. Baranovskii. Classical methods in the theory of lattice packings. *Russian Mathematical Surveys*, 34(4):1–68, 1979. [172,248]

[1988] S.S. Ryshkov and R.M. Erdahl. The empty sphere - Part II. *Canadian Journal of Mathematics*, 40:1058–1073, 1988. [172,188]

[1989] S.S. Ryshkov and R.M. Erdahl. Layerwise construction of L-bodies of lattices (in Russian). *Uspekhi Matematicheskikh Nauk*, 44 (2):241–242, 1989. (English translation in: *Russian Mathematical Surveys*, 44(2):293–294, 1989.) [185]

[1960] G. Sabidussi. Graph multiplication. *Mathematische Zeitschrift*, 72:446–457, 1960. [305]

[1976] S. Sahni and T. Gonzalez. P-complete approximations problems. *Journal of the Association for Computing Machinery*, 23:555–565, 1976. [461]

[1990] R. Scapellato. On F-geodetic graphs. *Discrete Mathematics*, 80:313–325, 1990. [309]

[1983] R. Schneider and W. Weil. Zonoids and related topics. In P.M. Gruber and J.M. Wills, editors, *Convexity and its Applications*, pages 296–317. Birkhäuser-Verlag, Basel, 1983. [110]

[1935] I.J. Schoenberg. Remarks to Maurice Fréchet's article "Sur la définition axiomatique d'une classe d'espace distanciés vectoriellement applicable sur l'espace de Hilbert". *Annals of Mathematics*, 36:724–732, 1935. [25,73,74,89]

[1937] I.J. Schoenberg. On certain metric spaces arising from Euclidean spaces by a change of metric and their imbedding in Hilbert space. *Annals of Mathematics*, 38:787–793, 1937. [25,69,83,113,117]

[1938a] I.J. Schoenberg. Metric spaces and completely monotone functions. *Annals of Mathematics*, 39:811–841, 1938. [25,69,83,113,114,117]

[1938b] I.J. Schoenberg. Metric spaces and positive definite functions. *Transactions of the American Mathematical Society*, 44:522–536, 1938. [25,69,74,83,115,117]

[1986] A. Schrijver. *Theory of Linear and Integer Programming.* John Wiley & Sons, Chichester, 1986. [11,49,428,431]

[1991] A. Schrijver. Short proofs on multicommodity flows and cuts. *Journal of Combinatorial Theory B*, 53:32–39, 1991. [376]

[1995] A. Schrijver. Personal communication, 1995. [150]

[1975] J.J. Seidel. Metric problems in elliptic geometry. In L.M. Kelly, editor, *The Geometry of Metric and Linear Spaces*, volume 490 of *Lecture Notes in Mathematics*, pages 32–43. Springer-Verlag, Berlin, 1975. [87]

[1979] P.D. Seymour. Sums of circuits. J.A. Bondy and U.S.R. Murty, editors, *Graph Theory and Related Topics*, pages 341–355. Academic Press, New York, 1979. [440]

[1981] P.D. Seymour. Matroids and multicommodity flows. *European Journal of Combinatorics*, 2:257–290, 1981. [434,439,440]

[1993] S.V. Shpectorov. On scale embeddings of graphs into hypercubes. *European Journal of Combinatorics*, 14:117–130, 1993. [100,211,252,313,314,315]

[1996] S.V. Shpectorov. Distance-regular isometric subgraphs of the halved cubes. Preprint, 1996. [269]

[1989] C.L. Siegel. *Lectures on the Geometry of Numbers.* Springer-Verlag, Berlin, 1989. [205]

[1960] G. Sperling. Lösung einer elementargeometrischen Frage von Fejes Tóth. *Archiv der Mathematik*, 11:69–71, 1960. [514]

[1986] R. Stanley. Two order polytopes. *Discrete and Computational Geometry*, 1:9–23, 1986. [427]

[1986] Z.I. Szabó. Hilbert's fourth problem, I. *Advances in Mathematics*, 59:185–301, 1986. [111]

[1987] P. Terwiliger and M. Deza. Classification of finite connected hypermetric spaces. *Graphs and Combinatorics*, 3:293–298, 1987. [211]

[1952] W.S. Torgerson. Multidimensional scaling: I. theory and method. *Psychometrika*, 17:401–419, 1952. [81]

[1977] G. Toulouse. Theory of the frustration effect in spin glasses: I. *Communications in Physics*, 2:115–119, 1977. [52]

[1987] I. Vajda. L_1-distances in statistical inference: comparison of topological functional and statistical properties. In Y. Dodge, editor, *Statistical Data Analysis Based on the L_1-Norm and Related Methods*, pages 177–184. Elsevier Science Publishers, North-Holland, Amsterdam, 1987. [163]

[1908] G.F. Voronoi. Nouvelles applications des paramètres continus à la théorie des formes quadratiques - Deuxième mémoire. *Journal für die Reine und Angewandte Mathematik*, 134:198–287, 1908. [172,188,189,226]

[1909] G.F. Voronoi. Nouvelles applications des paramètres continus à la théorie des formes quadratiques - Deuxième mémoire - Seconde Partie. *Journal für die Reine und Angewandte Mathematik*, 136:67–181, 1909. [172,188,189,226]

[1937] K. Wagner. Über eine Eigenschaft der ebene Komplexe. *Mathematische Annalen*, 114:570–590, 1937. [434]

[1990] A. Wagner and D.G. Corneil. Embedding trees in a hypercube is NP-complete. *SIAM Journal on Computing*, 19:570–590, 1990. [294]

[1993] A. Wagner and D.G. Corneil. On the complexity of the embedding problem for hypercube related graphs. *Discrete Applied Mathematics*, 43:75–95, 1993. [294]

[1992] M. Waldschmidt, P. Moussa, J.-M. Luck, and C. Itzykson, editors. *From Number Theory to Physics.* Springer-Verlag, Berlin, 1992. [187]

[1988] W.D. Wallis. *Combinatorial Designs.* Marcel Dekker, New York, 1988. [343]

[1992] P.M. Weichsel. Distance regular subgraphs of a cube. *Discrete Mathematics*, 109:297–306, 1992. [269,308]

[1975] J.H. Wells and L.R. Williams. *Embeddings and Extensions in Analysis.* Springer-Verlag, Berlin, 1975. [34,113,117]

[1976] D.J.A. Welsh. *Matroid Theory.* Academic Press, London, 1976. [436]

[1990] E. Wilkeit. Isometric embeddings in Hamming graphs. *Journal of Combinatorial Theory B*, 50:179–197, 1990. [304]

[1973] R.M. Wilson. The necessary conditions for t-designs are sufficient for something. *Utilitas Mathematica*, 4:207–217, 1973. [382]

[1975] R.M. Wilson. An existence theory for pairwise balanced designs III: proof of the existence conjectures. *Journal of Combinatorial Theory A*, 18:71–79, 1975. [343]

[1983] P.M. Winkler. Proof of the squashed cube conjecture. *Combinatorica*, 3:135–139, 1983. [278,294]

[1984] P.M. Winkler. Isometric embeddings in products of complete graphs. *Discrete Applied Mathematics*, 7:221–225, 1984. [304]

[1985] P.M. Winkler. On graphs which are metric spaces of negative type. In Y. Alavi, G. Chartrand, and L. Lesniak, editors, *Graph Theory with Applications to Algorithms and Computer Science*, pages 801–810. Wiley, New York, 1985. [290]

[1987a] P.M. Winkler. Factoring a graph in polynomial time. *European Journal of Combinatorics*, 8:209–212, 1987. [305]

[1987b] P.M. Winkler. The metric structure of graphs: theory and applications. In C. Whitehead, editor, *Surveys in Combinatorics 1987*, volume 123 of *London Mathematical Society Lecture Note Series*, pages 197–221. Cambridge University Press, Cambridge, 1987. [297]

[1988] P.M. Winkler. The complexity of metric realization. *SIAM Journal on Discrete Mathematics*, 1:552–559, 1988. [310]

[1978] H.S. Witsenhausen. A support characterization of zonotopes. *Mathematika*, 25:13–16, 1978. [110]

[1967] D. Wolfe. Imbedding of a finite metric set in an n-dimensional Minkowski space. *Proceedings of the Koninklijke Nederlandse Akademie van Wetenschappen, Series A, Mathematical Sciences*, 70:136–140, 1967. [156]

[1978] M. Yannakakis. Node-and edge-deletion NP-complete problems. *Proceedings of the 10th Annual ACM Symposium on the Theory of Computing*, pages 253–264, 1978. [48]

[1938] Gale Young and A.S. Householder. Discussion of a set of points in terms of their mutual distances. *Psychometrika*, 3:19–22, 1938. [74]

[1970] M.L. Yoseloff. *A Combinatorial Approach to the Diagonal N-Representability Problem.* PhD thesis, Princeton University, 1970. [64,69]

[1969] M.L. Yoseloff and H.W. Kuhn. Combinatorial approach to the N-representability of p-density matrices. *Journal of Mathematical Physics*, 10:703–706, 1969. [54]

[1965] K.A. Zaretskii. Factorization of an unconnected graph into a cartesian product. English translation in *Cybernetics*, 1(2):92, 1965. [305]

[1995] G.M. Ziegler. *Lectures on Polytopes*, Graduate Texts in Mathematics 152. Springer-Verlag, New York, 1995. [11,431]

Notation Index[1]

$G = (V, E)$	11	graph
$e = (u, v) = uv$	11	edge in a graph joining nodes u and v
V_n	11	set of elements $1, \ldots, n$
E_n	11	set of pairs ij for $1 \leq i < j \leq n$
$K_n = (V_n, E_n)$	11	complete graph on n nodes
K_{n_1,n_2}	11	complete bipartite graph
$\delta_G(S)$	11	cut in graph G
$G[W]$	12	subgraph of G induced by W
$G \backslash e$	12	subgraph of G obtained by deleting edge e
G/e	12	subgraph of G obtained by contracting edge e
P_n	12	path on n nodes
C_n	12	circuit on n nodes
$H(n, 2)$	12	hypercube graph
$\frac{1}{2}H(n, 2)$	12	half-cube graph
$K_{n \times 2}$	12	cocktail-party graph
P_{10}	13	Petersen graph
$G_1 \times G_2$	13	Cartesian product of two graphs G_1 and G_2
∇G	14	suspension graph of G
$L(G)$	14	line graph of G
$\mathbb{R}$	14	set of real numbers
$\mathbb{Q}$	14	set of rational numbers
$\mathbb{Z}$	14	set of integer numbers
$\mathbb{N}$	14	set of natural numbers
χ^S	14	incidence vector of a set S
$A \triangle B$	14	symmetric difference of two sets A and B
x^T, M^T	14	transpose of vector x, matrix M
$x^T y$	14	scalar product of vectors x and y
$\dim(X)$	14	dimension of a set X
$\text{rank}(X)$	14	rank of a set X
$\mathbb{K}(X)$	14	linear hull of X with coefficients in $\mathbb{K}$
$\text{Conv}(X)$	14	convex hull of X
K°	14	polar of convex set K
CUT_n	15	cut cone
$\text{CUT}_n^\square$	15	cut polytope
α_n	16	n-dimensional simplex
β_n	16	n-dimensional cross-polytope
γ_n	16	n-dimensional hypercube
P, NP, co-NP	18	complexity classes

[1]Symbols are listed in the order of first occurrence in the text.

Subject Index[1]

[1]Boldface numbers refer to the pages where items are introduced.

Coláiste Oideachais Mhuire Gan Smal
Luimneach